Lecture Notes in Control and Information Sciences

Edited by M. Thoma and A. Wyner

160

P. V. Kokotović (Ed.)

Foundations of Adaptive Control

Springer-Verlag
Berlin Heidelberg New York
London Paris Tokyo
Hong Kong Barcelona Budapest

Editor
Prof. Petar V. Kokotović
ECE
University of California
Santa Barbara
CA 93106
USA

ISBN 3-540-54020-2 Springer-Verlag Berlin Heidelberg New York
ISBN 0-387-54020-2 Springer-Verlag New York Berlin Heidelberg

Typesetting: Camera ready by author
Printing: Mercedes-Druck, Berlin
Binding: B. Helm, Berlin
61/3020-543210 Printed on acid-free paper.

Acknowledgements

While many people contributed to the organization of the 1990 Grainger Lectures and the publication of this volume, these tasks would not have been accomplished without the tireless work of Peter Sauer and Ioannis Kanellakopoulos.

Peter Sauer, Grainger Associate, was in charge of administration, finances, publicity and social programs for the Lectures.

Ioannis Kanellakopoulos, Grainger Fellow, performed all the editorial work for the volume, including proofreading, correcting and formatting the texts.

Financial support for the Lectures was provided by the Grainger Foundation. Personal attention of **David Grainger**, President, was extremely helpful.

Petar V. Kokotović
Grainger Professor

Preface

In the fourth decade of its development, adaptive control has reached a level at which a critical reexamination of its foundations is useful for both theoretical and practical purposes. A series of sixteen lectures, The 1990 Grainger Lectures, was dedicated to this task. The lectures were delivered at the University of Illinois, Urbana-Champaign, September 28–October 1, 1990. In the creative atmosphere of intense discussions among the lecturers, the scope and content of the lectures were expanded by the addition of new topics and most recent developments.

The result of this activity are the revised and enlarged texts in this volume. As it becomes clear even after a cursory review of its contents, the volume has gone far beyond its original scope of reexamining the foundations of the field. New solutions are presented for some long-standing open problems, earlier successful approaches are unified, new problems are formulated and some of them solved. Indeed, the title of the volume could have been changed into "Recent Advances in Adaptive Control". However, as all of the new results address fundamental issues, the title **Foundations of Adaptive Control** is still appropriate.

The two parts of the volume deal with adaptive control of linear and nonlinear systems, respectively. Part I contains unifications, reappraisals and new results in adaptive linear control. Most nonlinear problem formulations in Part II are new, and some of their solutions are likely to be further extended and generalized.

Part I opens with a view of the history of the field by one of the pioneers of Model Reference Adaptive Control, Bob Narendra. He starts with a wise caveat that even for a professional historian it is hard or impossible to be objective, let alone for a deeply involved participant of many exciting events. In **The Maturing of Adaptive Control**, the reader will find the wealth of information and freshness of views that only an eyewitness report can convey.

Adaptive control has always been characterized by a great diversity of concepts and algorithms and a continuous search for their unification. As such, it has been a fertile ground for both scientists and inventors who at times speak different languages. What both groups have been lacking is a common set of fundamental concepts in which to communicate the properties of their inventions and theories. **A Conceptual Framework for Parameter Adaptive Control** by Steve Morse is a major step in this direction. In the near future, his concept of tunability may become as common as the concept of controllability.

Just as the proofs of stability and convergence dominated the 1970s, the issues of robustness were central throughout the 1980s. As the next three papers in this volume show, the robustness issues will continue to be among the important research topics in the 1990s. Petros Ioannou, one of the leading authors on this subject, and Aniruddha Datta have given us a self-contained detailed presentation of the state-of-the-art in **Robust Adaptive Control: Design, Analysis and Robustness Bounds**. What in the 1980s appeared as a collection of unrelated modifications of adaptive algorithms is now a unified methodology for preventing destabilizing effects of disturbances and unmodeled dynamics. The robustness proofs are at the same time simpler and more powerful. The next two papers present new robustness results for continuous and discrete time systems. The main message of both **Robust Continuous-Time Adaptive Control by Parameter Projection** by Naik, Kumar, and Ydstie, and **Stability of the Direct Self-Tuning Regulator** by Ydstie, is that to achieve boundedness of all the signals in the presense of disturbances and unmodeled dynamics it is sufficient to introduce a parameter projection modification into the usual gradient update law. Although the continuous-time result by Naik, Kumar and Ydstie was inspired by the discrete-time result of Ydstie, their order is reversed in order to preserve the connection with the preceding continuous-time papers.

The contribution of the Soviet Academician Tsypkin, a pioneer of both control theory and stochastic adaptive control is at the same time historical and innovative. In **Adaptive-Invariant Discrete Control Systems**, he first reminds us that what today we call *disturbance-rejection* conditions appeared in the 1940s as *selective-invariance* conditions. This type of invariance is achievable for disturbances generated from known models and with complete knowledge of the plant. Tsypkin then introduces the notion of adaptive-invariant systems, which have an analogous invariance property when the parameters of the plant and disturbance generator are unknown. He restricts his presentation to the so-called interval plants, whose parameters are known to belong to given intervals.

One of the leading authors in the field of stochastic adaptive control, P. R. Kumar, was invited to summarize the field and reexamine its foundations. In **Stochastic Adaptive System Theory: Recent Advances and a Reappraisal**, he and Wei Ren have not only accomplished this task, but have also completed the solution of several long-standing open problems. They first provide a unified treatment of identification, prediction and control based on the ARMAX model and prove self-tuning and convergence properties for several adaptive schemes. They then give new general results for parallel model adaptation problems, including output-error identification and adaptive noise and echo cancelling. It is expected that these recent results will give a new impetus to both simpler and more fundamental developments in stochastic adaptive control.

Part II of this volume consists of five papers dealing with adaptive control of nonlinear systems. All the authors responded to the invitation to address truly new nonlinear problems without any linear constraints imposed on the nonlinearities. (Since the unknown parameters appear linearly, imposing linear growth constraints or global Lipschitz conditions would make the problems tractable by the methods of adaptive linear control.) The first three papers assume that the full state vector is available for feedback. More realistic situations, where only an output is available for measurement, are considered in the last two papers.

In **Adaptive Feedback Linearization of Nonlinear Systems,** Kokotović, Kanellakopoulos, and Morse first survey some recent results and then design a new adaptive scheme, applicable to a much larger class of systems than before. The difficulties of the global adaptive stabilization and tracking problems are shown to increase due to two factors: level of uncertainty and nonlinear complexity. The new systematic design procedure is applicable to the highest level of uncertainty, but it limits the nonlinear complexity by the so-called pure-feedback condition. Whenever this assumption is globally satisfied, the results on adaptive stabilization and tracking are global, that is, there is no loss of globality caused by adaptation.

A conceptually broader approach to adaptive control of nonlinear systems is presented in **Adaptive Stabilization of Nonlinear Systems** by Praly, Bastin, Pomet, and Jiang. They unify and generalize most earlier results and classify them according to additional assumptions such as matching conditions (i.e., uncertainty levels) and linear growth constraints. The unification is achieved by a novel Lyapunov approach to the design of direct schemes and by generalizations of equation error filtering and regressor filtering for indirect schemes. The key assumption in this approach is that a Lyapunov-like function exists and depends on the unknown parameters in a particular way. Depending on the properties of this function, various designs are possible, including feedback linearization designs when this function is quadratic in the transformed coordinates. Possibilities of either global or only regional adaptive stabilization are also examined.

Processes involving transfer of electromagnetic into mechanical energy require essentially nonlinear models. One of the most practical representatives of these processes is considered in **Nonlinear Adaptive Control of Induction Motors via Extended Matching** by Marino, Peresada, and Valigi. Their adaptive design encompasses, as special cases, some earlier designs and patents. However, the assumption of full state measurement is a practical disadvantage which motivates the development of adaptive schemes using only output measurements.

Adaptive output-feedback control of nonlinear systems is a new problem of major theoretical and practical interest. In **Global Adaptive Observers and Output-Feedback Stabilization for a Class of Nonlinear Systems** by Marino and Tomei, this problem is addressed in two stages. First, sufficient conditions are given for the construction of global adaptive observers for single-output systems. The construction makes use of novel filtered transformations, that is, nonlinear changes of coordinates driven by auxiliary filters. At the second stage, an observer-based adaptive output-feedback control is designed for a somewhat narrower class of systems.

A different approach to adaptive output-feedback control of nonlinear systems is presented in **Adaptive Output-Feedback Control of Systems with Output Nonlinearities** by Kanellakopoulos, Kokotović and Morse. The main result of this paper is a nonlinear extension of a 1978 paper by Feuer and Morse. Because of its complexity, this early paper is less well known than the papers mentioned in the historical survey at the beginning of this volume. However, in contrast to some other results of adaptive linear control, its nonlinear generalization does not impose linear constraints on system nonlinearities.

This brief preview suffices to show that the papers in this volume herald a decade of new breakthroughs in adaptive control research in the decade of the 1990s. With the already achieved unification of robust adaptive control, more research will undoubtedly be focused on performance. A particularly challenging task will be to develop methods to improve transients and eliminate bursting phenomena. In stochastic adaptive control, the most recent proofs of stability and convergence make it possible to address the robustness issue in a new way. In nonlinear systems, adaptive control has just made its first encouraging steps. They are likely to stimulate research leading to results applicable to larger classes of systems. Rigorous analytical methods of adaptive control may also become a theoretical basis for some developments in neural networks.

The intellectual depth and scientific content of adaptive control are growing at an impressive pace. Its mature vitality is as exciting as was its youthful impetuousness.

Petar V. Kokotović

Urbana, Illinois
February 1991

Contents

Part I

Adaptive Linear Control

The Maturing of Adaptive Control

K. S. Narendra

Department of Electrical Engineering
Center for Systems Science
Yale University
New Haven, CT 06520, USA.

Prologue

History is neither purely subjective nor purely objective. According to Edward Hallett Carr [1], the historian is engaged in a continuous process of molding his facts to his interpretation and his interpretation to his facts. The following paper is a brief history of adaptive control theory during the period 1955–1990 as interpreted by the author, who was also a participant.

1 Historical Background

Control theory is concerned with modifying the behavior of dynamical systems so as to achieve desired goals. These goals include maintaining relevant outputs of a system around desired constant values, assuring that the outputs follow specified trajectories, or more generally ensuring that the overall system optimizes a specified performance criterion. The goal is achieved by computing a suitable control input based on the observed outputs of the system. The fundamental processes involved in controlling a dynamical system consequently include the mathematical modeling of the system, identification of the system based on experimental data, processing the outputs of the system into mathematically convenient forms, using them in turn to synthesize control inputs, and applying the latter to the system to achieve the desired behavior.

In the early stages of its development, control theory was primarily concerned with linear time-invariant systems with known parameters. Around the 1940s, electrical, mechanical and chemical engineers were designing automatic control devices in their own fields, using very similar methods but disguised under different terminologies – a situation which is prevalent even today in many of the newer areas of systems theory. Gradually it became apparent that the concepts had a common basis and towards the end of World War II a frequency-domain theory of control that was mathematically elegant and universal in its scope came into existence. During the following decade, even as this theory was being used successfully for the systematic design of innumerable industrial feedback control systems, it became clear that a new methodology would be needed when substantial amounts of uncertainty are present in such systems. It was at this stage in the history of control theory, in the 1950s, when frequency-domain methods were well established and time-domain methods based on state representation of dynamical systems had not yet been introduced, that the field of adaptive control was born.

In the early 1950s there was considerable interest in the design of autopilots for high-performance aircraft. Such aircraft operated over a wide range of speeds and altitudes (according to the standards prevailing at that time). The existing feedback theory was adequate to design an efficient controller at any one operating condition, but could not cope with problems that arose when there were rapid changes in operating conditions. One obvious solution was to store the controller parameter values for the different operating regions and switch between them as the conditions changed. But this process, referred to as gain scheduling, became infeasible when the number of possible operating regimes became very large, and it was generally acknowledged that a more sophisticated approach would be needed to cope with the problem. It also became apparent in the following years that the difficulties encountered in aircraft systems were generic in nature and that very similar problems were encountered over the entire spectrum of systems applications. As in aircraft systems, parameters also tend to vary in most practical systems – ships, cement and paper mills, distillation columns, engines, chemical processes and power systems, to name but a few. The frequency-domain methods available at this time [2,3], even when used efficiently, were inadequate to achieve satisfactory performance in the entire range over which the characteristics of the system varied. The demands of a rapidly growing technology for faster and more accurate controllers have always had a strong influence on the progress of automatic control theory and when it became apparent that the existing theory was inadequate to successfully tackle the problems arising in industry, interest gradually shifted to adaptive systems which would adjust their performance with changes in the environments.

The term "adaptation" is defined in the dictionary as the modification of an organism or its parts that fits it better for the conditions of its environment. Inspired by this, in 1957, Drenick and Shahbender [4] introduced the term "adaptive system" into control theory to represent control systems that monitor their own performance and adjust their control mechanisms in the direction of improved performance. The following years witnessed considerable debate on the differences between conventional linear control systems and adaptive systems. Further, numerous feedback systems were designed which their proponents considered to possess properties peculiar to adaptive systems. Many of these were collected and presented in survey papers by Aseltine et al. [5] and Stromer [6] indicating that research on adaptive systems was flourishing even in the 1950s. An attempt was made in [5] to categorize adaptive systems into the following five classes depending upon the manner in which adaptive behavior was achieved: (i) Passive Adaptation, (ii) Input-Signal Adaptation, (iii) System Variable Adaptation, (iv) System Characteristic Adaptation, and (v) Extremum Adaptation. The first class contains those systems in which the clever design of time-invariant controllers results in satisfactory performance over wide variations in the environment. High-gain controllers designed to cope with plant parameter variations belong to this class which, according to the present terminology, would be described as robust rather than adaptive. Class (ii) includes systems which adjust their parameters in accordance with input signal characteristics while classes (iii) and (iv) contain systems in which the control input is based on system outputs, errors or their derivatives or alternately the observed impulse response or frequency response. It is worth remembering that system representation at this time was mainly in terms of transfer functions, and distinctions were made only between inputs, outputs and parameters. Even when the parameters were adjusted on-line, the overall system was analyzed as a linear time-varying system. Adaptive systems belonging to class (v) received more attention than the others during the 1960s and attempted to seek the extremum of a performance criterion (cf. Section 2). Even though the classification of adaptive systems given above is no longer in vogue, the concepts that led to such a classification have endured for over three decades and have been rediscovered by successive generations of systems theorists in their attempts to deal with uncertainty.

In the early 1960s, many sessions in the major conferences were devoted to defining adaptive control and numerous definitions were proposed [7,8,9,10]. Yet, thirty years later a universal definition of an adaptive system is still elusive. This is because adaptation is multifaceted in nature and cannot be compressed into a simple statement without vital loss of content. Hence, the many definitions proposed merely reflect the variety of personal visions encountered in the field. The seeming divergence between the different viewpoints may be partly attributed to the fact that, in the ultimate analysis, an adaptive system is merely a complex feedback system. A designer who sets out to control a system in the presence of a certain class of uncertainty and produces a controller which achieves its objective may conclude that he has designed an adaptive system. Yet, an observer unfamiliar with the process prior to the design of the controller merely observes a nonlinear control system. Hence, among the various definitions proposed, the one due to Truxal [10], that an adaptive system is one which is designed from an adaptive viewpoint, captures much of the difficulty encountered in defining adaptive systems.

The fact that most adaptive systems are nonlinear systems came to be realized in the 1960s, when attempts were made to describe them using state equations. For example, if a plant is described by the vector differential equation

$$\begin{aligned} \dot{x} &= A(p,\theta)x + B(p,\theta)r \\ y &= C(p,\theta)x\,, \end{aligned} \tag{1}$$

with $r(t) \in \mathbb{R}$, $x(t) \in \mathbb{R}^n$, $y(t) \in \mathbb{R}^m$, $p \in \mathbb{R}^{m_1}$ and $\theta(t) \in \mathbb{R}^{m_2}$, where r, x, y are respectively the input, state and output, p is an unknown constant vector, and $\theta(t)$ a time-varying control parameter vector which is adjusted using the measured signals of the system as

$$\dot{\theta} = g(y, \theta, t)\,, \tag{2}$$

it follows that the components $\theta_i(t)$ of $\theta(t)$ $(i = 1, 2, \ldots, m_2)$ are no longer "parameters", but are state variables of the overall system described by the equations (1) and (2). Hence, any adaptive system, no matter how complex, is merely a nonlinear feedback system involving estimation and control, and the design of adaptive systems consequently corresponds to the design of special classes of nonlinear systems [11].

During the past three decades great advances have been made in adaptive control theory and adaptive controllers have been used in numerous industrial problems. While we are still very far from designing autonomous systems which will take care of themselves, a considerable amount of insight has been gained regarding the nature of the new problems and the concepts and theoretical questions associated with them. During the early years some of the research efforts led to dead ends and new starts had to be made. There were also periods when the field appeared to be in the doldrums, and these were invariably followed by breakthroughs which led to a resurgence of interest in the field. The main objective of this paper is to trace some of these major developments, particularly those related to continuous-time adaptive systems, and to provide the reader with a sense of the excitement and frustration that have alternately marked the growth of the field.

2 Developments in the early 1960s

The early days of adaptive control coincided with one of the most active periods in the history of control theory. The state representation of dynamical systems was introduced at this time and the relationship between frequency- and time-domain methods was a subject of frequent discussion. The maximum principle of Pontryagin [12] which had been introduced in the 1950s and which had opened the way for a systematic study of optimal trajectories in the presence of state and control constraints was gradually being absorbed into the mainstream of control theory. The contributions of Lurie, Popov, Kalman, Yakubovich and others stimulated a great deal of interest in the absolute stability problem and more generally in Lyapunov's methods for establishing the stability or instability of nonlinear systems. This was also the period when advances in stochastic estimation were accelerated by the introduction of the Kalman-Bucy filter, when the linear optimal regulator problem with a quadratic performance index was resolved and when Bellman's dynamic programming method was being increasingly used for sequential planning and optimal resource allocation. All these had a profound impact on researchers in the area of adaptive control and on their understanding of adaptive control processes.

Even at this early stage, following the work of Feldbaum [13], it was generally realized that two philosophically different approaches exist for the control of plants with unknown parameters. In the first approach [Fig. 1a] referred to as indirect control (or explicit identification) the unknown plant parameters are estimated on-line and the controller parameters are adjusted based on these estimates. In the second approach, referred to as direct control [Fig. 1b] (or

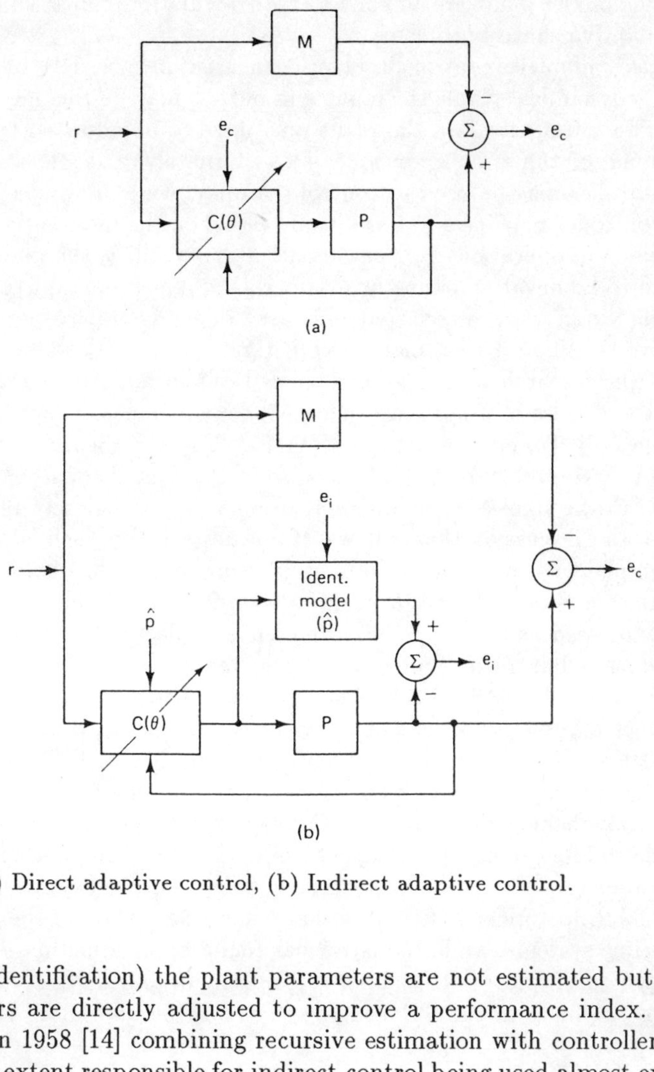

Fig. 1. (a) Direct adaptive control, (b) Indirect adaptive control.

implicit identification) the plant parameters are not estimated but the control parameters are directly adjusted to improve a performance index. A paper by Kalman in 1958 [14] combining recursive estimation with controller design was to a large extent responsible for indirect control being used almost exclusively in the late 1950s. This is reflected in the first book on adaptive control by Mishkin and Braun [15] published in the United States in 1961. Soon after this, however, direct methods also became popular for adaptive control.

2.1 MRAS and STR

As mentioned in Section 1, the aim of control is to keep the relevant outputs of a given dynamical plant within prescribed limits. If $y_{\rm p}$ denotes the output of the plant and $y_{\rm m}$ the desired output, the objective is to keep the error $e_1 = y_{\rm p} - y_{\rm m}$ between desired values. If $y_{\rm m}$ is a constant, the problem is one of regulation, and when $y_{\rm m}$ is a function of time the problem is referred to as tracking. When the characteristics of the plant are unknown, both regulation and tracking can be viewed as adaptive control problems.

In practice, adaptive control problems can arise in a variety of ways. For instance, in a dynamical plant, the input and output may be the only accessible variables and a parameter θ in the plant may have to be adjusted to minimize some functional of the error $e_1 = y_p - y_m$. Alternately, a feedforward and/or feedback controller may be used to control the unknown plant and the parameters of the controller may have to be adjusted based on the information obtained as the system is in operation. For mathematical tractability, the plant is generally assumed to be linear and time-invariant with unknown parameters. In such a case whether indirect or direct control is used depends on whether or not the parameters of the plant are estimated explicitly.

Much of the research that has been carried out in adaptive control is concerned with two classes of systems called Model Reference Adaptive Systems (MRAS) and Self Tuning Regulators (STR). While the former evolved from deterministic servo problems, the latter arose in the context of stochastic regulation problems. In spite of their different origins they are closely related as is evident from the discussion that follows. If an indirect approach is to be used, the unknown plant parameters are estimated, using a model of the plant, before a control input is chosen. If the problem is one of tracking, the desired output y_m can be expressed as the output of a reference model (ref. sections 3 and 5). Adaptive systems that make explicit use of such models for identification or control purposes are called MRAS [16]. The Self Tuning Regulator, which is based on the indirect approach, consists of a parameter estimator, a linear controller and a mechanism for computing the controller parameters from those of the plant parameter estimates. Since any estimation scheme can be combined with a suitable control scheme, various possibilities exist for designing such a controller. It is clear that STRs can also be designed based on a direct method in which the regulator parameters are directly updated. In spite of the seeming similarity of MRAS and STR, historically the former has been used mainly in the analysis of continuous-time systems, while the latter has found application in discrete-time systems. Since our interest in this paper is primarily in continuous-time adaptive systems, the emphasis in the following sections will be on MRAS rather than STR.

For the sake of historical continuity, we first deal in this section with parameter perturbation methods as well as methods based on sensitivity models which represented the two principal approaches to adaptive control in the 1960s. All subsequent sections deal with problems which arise in MRAS.

2.2 Parameter Perturbation Method

Extremum adaptation, mentioned in Section 1 was perhaps the most popular method for adaptive control in the 1960s. It had considerable appeal to researchers because of its simplicity and applicability to nonlinear plants and since it did not require explicit identification of plant parameters. In 1951, Draper and Li [17] had demonstrated that the performance of an internal combustion engine could be optimized using a parameter perturbation method involving perturbation, correlation and adjustment. Assuming that a single parameter value has to be chosen, the parameter is perturbed around a nominal value, the parameter variations are correlated with changes in the performance index and used in turn to adjust the parameter in the direction of optimum performance. Numerous papers [18,19,20,21] appeared on this subject and detailed analyses were carried out regarding the amplitude and frequency of the perturbations, the time over which the correlation is to be carried out and the amount by which the parameter is to be adjusted. The analysis in [21], for example, revealed that sophisticated perturbation methods involving differential equations with multiple time-scales would be needed for a precise analysis of the behavior of even a second-order system with a single adjustable parameter. The problem became significantly more complex when several parameters had to be adjusted simultaneously. To determine the gradient of a performance index in parameter space, the frequencies of perturbation of the different parameters had to be separated and this in turn resulted in slow convergence. The inability to use the method successfully in practical situations where multiple parameters had to be adjusted, along with the difficulty encountered in determining conditions for their stability, resulted in the rapid demise of parameter perturbation methods in the mid 1960s.

2.3 Sensitivity Models

The parameter perturbation method is a direct method for estimating the gradient of a performance index. An alternate gradient approach, based on significantly more prior information concerning the plant, used sensitivity models for the generation of the desired partial derivatives. Given a system with an input-output pair $\{u, y\}$ and a parameter θ, the partial derivative $\frac{\partial y}{\partial \theta}$ of the output y with respect to θ can be computed if the structure of the system is known. For example, if a system is described by the second-order differential equation $\ddot{y} + \theta \dot{y} + y = u$, the partial derivative $\frac{\partial y}{\partial \theta} = z$ satisfies the differential equation

$$\ddot{z} + \theta \dot{z} + z = -\dot{y} . \tag{3}$$

By constructing a model described by equation (3) with $-\dot{y}$ as its input the desired partial derivative $\frac{\partial y}{\partial \theta}$ can be generated. Similarly, using a model corresponding to each adjustable parameter, the gradient of y with respect to θ can be determined and in turn used to adjust the parameter vector to improve the performance [22,23,24].

The method using sensitivity models gained great popularity in the 1960s and was extensively used in industrial problems for the optimal adjustment of parameters. Since substantially more information is assumed in this method as compared to the parameter perturbation method, the convergence of the parameters was significantly faster than in the latter case. In many adaptive situations the structure of plant may not be known (i.e., a differential equation of the form (3) may not be available). In such cases an identification model of the plant has to be constructed and used in the sensitivity model. The output of the sensitivity model then corresponds to an estimated rather than a true gradient of the system.

2.4 M.I.T. Rule

The first MRAS was proposed in 1958 by Whitaker et al. [25] and contained a reference model whose output represented the desired output of the plant. The parameters θ of the controller were adjusted based on the error e between the output of the reference model and that of the plant. The adaptive law for the adjustment of θ was based on the fact that if the parameters θ changed much slower than the system variables, then a reasonable approach was to change θ along the negative gradient of e^2:

$$\frac{d\theta}{dt} = -\gamma e \frac{\partial e}{\partial \theta} . \tag{4}$$

However, since the plant parameters were unknown, the components of the sensitivity vector $\frac{\partial e}{\partial \theta}$ could not be obtained using the sensitivity models described earlier. Instead, they were replaced by some available signals according to a "rule" which came to be known as the M.I.T. rule.

3 Adaptive Control of Simple Systems (late 1960s)

3.1 Stable Adaptive Systems

In the gradient methods described thus far for the study of adaptive systems, the emphasis was on their performance. The control parameters were adjusted along the gradient of a performance index. Once the design of the controller was completed, the stability of the overall system was analyzed and conditions were established for local stability. Since adaptive systems are nonlinear nonautonomous systems, the derivation of sufficient conditions for even local stability was not possible in many cases.

Since stability is a fundamental requirement for the satisfactory operation of any system, it was suggested in 1963 by Grayson [26] that a reversal of the procedure adopted earlier would be more efficient. He argued that adaptive systems should be designed to be globally stable for all values of a parameter γ belonging to a set S_γ. The optimization of the system could then be carried out by choosing the optimum value $\gamma_{\mathrm{opt}} \in S_\gamma$.

The above idea can be mathematically stated as follows. Let the differential equation governing an adaptive system be

$$\begin{aligned} \dot{x} &= f[x, p, \theta, t] \\ y &= h[x, p, \theta, t] , \end{aligned} \tag{5}$$

where $x(t)$ and $y(t)$ are the state and output of the system respectively, p is a constant but unknown parameter vector, and $\theta(t)$ is a control parameter vector.

Let $y_{\mathrm{m}}(t)$ be a desired trajectory. The choice of $\theta(t)$ is at the discretion of the designer. Let an adaptive law of the form

$$\dot{\theta} = g[y, y_{\mathrm{m}}, \gamma, t] \tag{6}$$

exist, such that the nonlinear system described by (5)–(6) is globally stable for all values of the parameter γ in some set S_γ. Then, as stated earlier, the performance of the system can be improved according to some criterion by the choice of γ.

The above procedure is very similar to that adopted in the design of optimal linear time-invariant control systems, where the control parameters are chosen subject to the constraint that the poles of the overall system lie in the open left-half of the complex plane.

The suggestion that adaptive control systems should be designed from a stability viewpoint was enthusiastically received by the adaptive control community, particularly since by this time the difficulties associated with establishing the stability of gradient-based schemes was well recognized. Numerous contributions were made to the design of stable adaptive systems, the most notable of which are due to Shackcloth and Butchart [27], Parks [28] Monopoli [29], Philipson [30], and Winsor and Roy [31]. The strict positive real condition, which subsequently became an important design tool, made its first adaptive appearance in [27,28]. It should be borne in mind that in the 1960s many of the subtle questions of stability were not yet well understood. Hence, while the contributions of the early authors were undoubtedly seminal in nature, some of them contained technical flaws which were eventually eliminated in more rigorous analyses [28,32,33].

The 1966 paper by Parks [28], which is of great historical significance in adaptive control, clearly demonstrated using a specific example that gradient methods of adjusting the parameters in an adaptive system can result in instability. At the same time it was also shown that the system could be made globally stable by using a design procedure based on Lyapunov's method. This clear demonstration that gradient-based adaptive systems could be unstable, eventually tolled the death knell of such systems and witnessed a gradual shift on the part of researchers to the design of adaptive systems based on stability methods.

3.2 Lyapunov Design of Simple Adaptive Systems

The concepts used in the design of stable adaptive systems are best illustrated by considering plants which are described by first-order differential equations with unknown parameters. Following this, it can be readily shown that the same concepts carry over to higher-order systems when the state vectors of the plants are accessible. More general cases, where plants of order greater than one have to be controlled using only input and output information, are treated later in Section 5. For a detailed treatment of these topics the reader is referred to [11].

Let a plant with input-output pair $\{u(\cdot), x_{\mathrm{p}}(\cdot)\}$ be described by the scalar differential equation

$$\dot{x}_{\mathrm{p}} = a_{\mathrm{p}} x_{\mathrm{p}}(t) + k_{\mathrm{p}} u(t), \tag{7}$$

where a_p and k_p are constants and $x_p, u_p : \mathbb{R}^+ \to \mathbb{R}$. Let the control input $u(t)$ be of the form

$$u(t) = \theta x_p(t) + kr(t), \tag{8}$$

where θ and k are adjustable parameters. It is assumed that the plant parameters a_p and k_p are unknown, but that the sign of k_p is known. The aim of adaptive control is to adjust the control parameters $\theta(t)$ and $k(t)$ so that the output $x_p(t)$ follows asymptotically the output $x_m(t)$ of a reference model described by

$$\dot{x}_m(t) = a_m x_m(t) + k_m r(t), \quad a_m < 0, \tag{9}$$

where $r(t)$ is a known bounded piecewise-continuous function of time. Defining $e(t) \triangleq x_p(t) - x_m(t)$, $\phi(t) \triangleq \theta(t) - \theta^*$, and $\psi(t) \triangleq k(t) - k^*$, where $\theta^* = (a_m - a_p)/k_p$ and $k^* = k_m/k_p$, the differential equation describing $e(t)$ can be expressed as

$$\dot{e}(t) = a_m e(t) + k_p \phi x_p(t) + k_p \psi(t) r(t). \tag{10}$$

The realization that the analysis is best carried out by considering the error equations, in which only the parametric errors ϕ and ψ and the output error $e(t)$ feature, was slow in coming. The importance of the error equation is that the adaptive control problem can be stated as the stability problem of the equilibrium state of a set of error differential equations. For instance, if $\theta(t)$ and $k(t)$ are to be adjusted adaptively, the problem can be posed as follows: Determine the adaptive laws

$$\begin{aligned} \dot{\phi} = \dot{\theta} &= f_1(e, x_p, r) \\ \dot{\psi} = \dot{k} &= f_2(e, x_p, r), \end{aligned} \tag{11}$$

so that the equilibrium state of the third-order system described by (10)–(11) is uniformly stable.

Choosing as a Lyapunov candidate $V(e, \phi, \psi) = \frac{1}{2}[e^2 + |k_p|(\phi^2 + \psi^2)]$ and evaluating the time derivative of V along the solutions of equations (10) and (11) leads to

$$\dot{V}(e, \phi, \psi) = a_m e^2 + k_p \phi e x_p + k_p \psi e r - |k_p|[\phi\dot{\phi} + \psi\dot{\psi}]. \tag{12}$$

Since ϕ and ψ are parameter errors which can never be known, it follows that $\dot{\phi}$ and $\dot{\psi}$ should be chosen to cancel the terms in ϕ and ψ if $\dot{V}$ is to be non-positive for all e, ϕ and ψ. This is accomplished by choosing

$$\begin{aligned} \dot{\phi} &= -\mathrm{sgn}(k_p) e x_p \\ \dot{\psi} &= -\mathrm{sgn}(k_p) e r, \end{aligned} \tag{13}$$

where $\mathrm{sgn}(k_p) = 1$ if $k_p \geq 0$ and $\mathrm{sgn}(k_p) = -1$ if $k_p < 0$. Since $V(e, \phi, \psi)$ is indeed a Lyapunov function of the system, it follows that the system is stable. Since $\dot{V} = a_m e^2 \leq 0$, $e \in \mathcal{L}^2$. Furthermore, since by equation (10) $\dot{e}$ is bounded, we have $\lim_{t\to\infty} e(t) = 0$.

The proof given above for a first-order system forms the essence of the proofs given for more general cases. For example, let an nth-order plant be described by the vector differential equation

$$\dot{x}_p = A_p x_p + b_m u \,, \tag{14}$$

where $A_p \in \mathbb{R}^{n\times n}$, $b_m \in \mathbb{R}^n$, $x_p : \mathbb{R}^+ \to \mathbb{R}^n$, and $u : \mathbb{R}^+ \to \mathbb{R}$. Let b_m be a known vector and A_p an unknown matrix. Let a reference model be described by the differential equation

$$\dot{x}_m = A_m x_m + b_m r \,, \tag{15}$$

where A_m is an asymptotically stable matrix and r a specified scalar input which is bounded and piecewise-continuous. The objective, as in the scalar case, is to determine the input to the plant so that $\lim_{t\to\infty} e(t) = 0$ where $e(t) = x_p(t) - x_m(t)$.

To assure the existence of a solution, we assume that a vector k^* exists so that if $u = k^{*T} x_p(t) + r$, both $x_p(t)$ and $x_m(t)$ satisfy identical differential equations. Since A_m is asymptotically stable it follows that a symmetric positive definite matrix P exists such that

$$A_m^T P + P A_m = -Q \,, \quad Q = Q^T > 0 \,. \tag{16}$$

Using $V(e,\phi) = e^T P e + \phi^T \phi$ as a candidate for a Lyapunov function, (where $k(t) - k^* = \phi$), it can be shown, using the same approach as in the scalar case, that the adaptive law

$$\dot{k}(t) = -e^T(t) P b_m x_p \tag{17}$$

yields $\dot{V}(e,\phi) = -e^T Q e \leq 0$, thereby assuring global stability. The procedure outlined above was also used for the identification of dynamical systems [11] and later for both indirect control as well as in combined direct-indirect methods, for the adaptive control of systems whose state variables are accessible.

4 Adaptive Observers (early 1970s)

In linear systems theory it is well known that a controllable linear time-invariant system can be stabilized by state feedback. This formed the theoretical basis for the adaptive control methods discussed in the previous section. In more realistic cases the entire state vector cannot be measured and adaptation has to be carried out using only input-output measurements. For linear systems with known parameters, the use of a Luenberger observer to estimate the state variables of the system from input-output measurements had been established by this time. This in turn motivated researchers in the adaptive control field to attempt to estimate the state variables of linear systems from similar data even when the system parameters were unknown. Since the estimation of the state variables using an observer requires knowledge of the system parameters, and estimation of the latter requires knowledge of the state variables, it soon became evident that the problem was substantially more complex in the adaptive context. An adaptive observer was defined as one which simultaneously estimates the parameters and state variables of a linear time-invariant system from input-output measurements.

Around the late 1960s, the field of adaptive control was not particularly popular in the United States. Most of the work on continuous time adaptive observers in the country was being carried out at two universities – the University of Connecticut under the direction of Lindorff and Yale University under the direction of Narendra. There was considerable interaction between these two groups and the University of Massachusetts, where Monopoli headed the control activity. During this period it became evident that the structure of the adaptive observer, as well as the adaptive laws, would have to be chosen judiciously if the adaptive observer was to be stable. At the same time, no one was confident that even the existence of such a stable adaptive observer could be established. In the author's opinion, the period 1972-74 constituted one of the important stages in the development of continuous-time adaptive control.

The mid 1970s witnessed a renaissance in adaptive control. In Europe, Tsypkin [34] made fundamental contributions to the field and showed that many of the methods proposed for learning and adaptive control could be described in a common framework as special types of recursive equations. Identification methods were extensively investigated and different estimation methods were made the basis of design procedures. Åström and Wittenmark introduced the self-tuning idea in a seminal paper [35], and studied the asymptotic properties of the self-tuning regulator. On the practical side, minicomputers were used to implement adaptive concepts. In the United States major strides were made in the development of adaptive observers and a comprehensive survey on the subject by Narendra and Kudva [32,33] appeared in 1974. In view of the impact of adaptive observers on later developments in adaptive systems, a brief description of the nature of the problems as well as the solutions proposed are included in this section.

The first solution to the problem was given by Carroll and Lindorff [36]. Using the Kalman-Yakubovich Lemma, a similar adaptive observer was later proposed by Kudva and Narendra [37] (Fig. 2). Various canonical representations of unknown plants were suggested by Luders and Narendra [38,39], which eventually led to a substantially simplified non-minimal representation of adaptive observers.

4.1 Minimal Realization of Adaptive Observers

Any single-input single-output linear time-invariant plant which is observable can be parametrized as

$$\dot{x}_{\mathrm{p}} = [a|\bar{A}]x_{\mathrm{p}} + bu$$

$$y = x_{\mathrm{p}1}$$

where $a^{\mathrm{T}} = [a_1, a_2, \ldots, a_n]$ and $b^{\mathrm{T}} = [b_1, b_2, \ldots, b_n]$ are unknown, and $\bar{A}$ is a known matrix [11]. The objective of the adaptive observer is to estimate the vectors a and b as well as the state vector $x_{\mathrm{p}}(t)$ from the input-output data. A more convenient parametrization for the design of the adaptive observer has the form

$$\begin{aligned} \dot{x}_p &= [-\alpha|\bar{A}]x_p + gy_p + bu \\ y_p &= h^T x_p = x_1 , \end{aligned} \tag{18}$$

where α is a known constant vector which makes $[-\alpha|\bar{A}] = K$ a stable matrix, and g and b are unknown vectors in $\mathbb{R}^n$ which have to be estimated.

From the earlier work on simple adaptive systems (Section 3) it was clear that $2n$-independent signals would be needed to estimate the $2n$ unknown parameters. However, since only two signals u and y are accessible, the required signals had to be generated as part of the adaptive process. The observer was chosen to have the form

$$\dot{\hat{x}}_p = K\hat{x}_p + \hat{g}(t)y_p + \hat{b}(t)u + v_1(t) + v_2(t) , \tag{19}$$

where $v_1(t)$ and $v_2(t)$ are vectors in $\mathbb{R}^n$ which are at the discretion of the designer and $\hat{x}_p(t)$, $\hat{g}(t)$ and $\hat{b}(t)$ represent the estimates of the state vector and unknown parameter vectors respectively. If $e(t) = \hat{x}_p(t) - x_p(t)$ and $e_1(t) = \hat{y}_p(t) - y_p(t)$, the error equation (subtracting (18) from (19)) has the form

$$\dot{e} = Ke + \phi y_p + \psi u + v_1 + v_2 . \tag{20}$$

In [37] the vectors $v_1(t)$ and $v_2(t)$ are generated as the state variables of two identical controllable linear time-invariant systems with $u(t)$ and $y(t)$ as inputs respectively. The adaptive laws for adjusting $\hat{g}(t)$ and $\hat{b}(t)$ are then derived using the signals $v_1(t)$ and $v_2(t)$ and it is demonstrated that a Lyapunov function exists which assures the stability of the system. The structure of the adaptive observer is shown in Fig. 2.

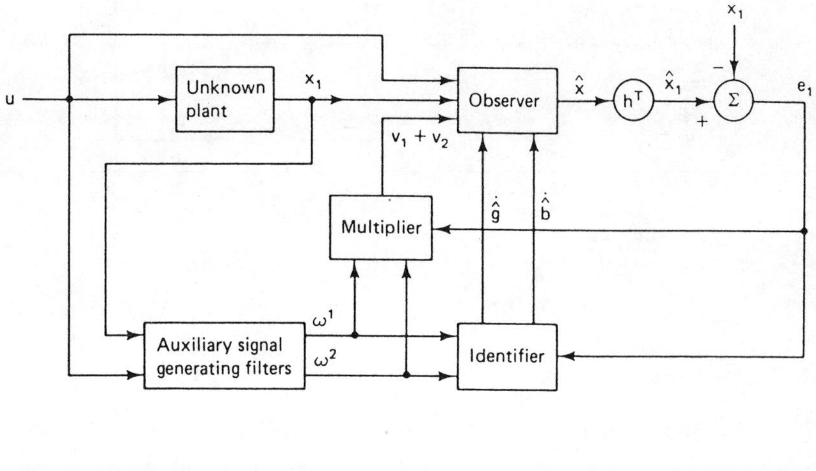

Fig. 2. The minimal adaptive observer.

4.2 Nonminimal Realization of the Adaptive Observer

The adaptive observer in Fig. 2 is unwieldy because of the signals v_1 and v_2. The question was raised, after the papers [36] and [37] appeared, as to whether it would be possible to eliminate these signals entirely by choosing a different parametrization of the adaptive observer. Several different parametrizations [38, 39] were partially successful and eventually led to a final structure based on a nonminimal representation of the model. This structure, suggested by Luders and Narendra [39] is shown in Fig. 3. Using the input u and output y and two identical controllable systems of dimension n, the signals $\bar{\omega}_1$ and $\bar{\omega}_2$ are first generated, and a linear combination of these signals (using $2n$ parameters) is chosen as the input to a stable first-order system. The output of the latter was used as the estimate of the output of the plant. The output error e_1 together with $\bar{\omega}_1$ and $\bar{\omega}_2$ determined the adaptive laws for the estimation of the unknown parameters.

In the late 1970s, Kreisselmeier [40] made numerous contributions to adaptive observers. He proposed different structures for such observers and derived conditions under which the parameters would converge arbitrarily fast. One of the significant improvements suggested by him led to an integral adaptive law which improved convergence rates substantially. From a theoretical standpoint, the integral laws enabled the procedures developed for continuous-time systems to be related to those generated using discrete-time systems.

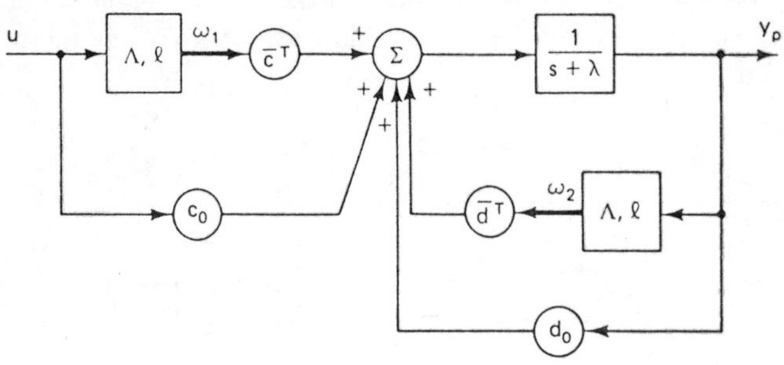

Fig. 3. The non-minimal adaptive observer.

4.3 Persistent Excitation

It was realized even in the early 1970s that the output error in an adaptive system could tend to zero without the parameters converging to their true values. The convergence of the parameter estimates was known to depend upon the properties of certain signals in the system which in turn depend on the nature of the reference input. The property of the signals which results in the parameter errors tending to zero is referred to as persistent excitation and is consequently a central concept in adaptive control.

The concept arose in the 1960s in the context of system identification of discrete systems, and was coined to express the qualitative notion that all the modes of the plant are excited. In 1966, Åström and Bohlin [41] gave a formal statement relating persistent excitation to the convergence of the identification model. Several equivalent definitions, related in one form or another to the existence of a solution to a set of linear equations, appeared soon after. As will be shown in the following sections, persistent excitation also plays a central role in the convergence of adaptive control parameters, as well as in the robustness of such systems in the presence of bounded disturbances.

In the late 1970s, numerous workers, including Sondhi and Mitra [42], Morgan and Narendra [43,44], Yuan and Wonham [45] and Anderson [46], addressed the question of persistent excitation in continuous-time systems. While all these papers contain essentially similar results, we describe here those contained in [43] and [44], which pertain to two classes of differential equations and which give rise to different definitions of persistent excitation. These are discussed briefly here since they are relevant for some of the discussions in later sections.

The error equations in the adaptive observers discussed in Section 3 and many identification problems have the form

$$\phi^{\mathrm{T}}(t)u(t) = e_1(t)\,, \tag{21}$$

where $\phi(t), u(t) \in \mathbb{R}^n$, $u(t)$ is a known input and $\phi(t)$ is the parameter error vector. If the adaptive law

$$\dot{\phi}(t) = -e_1(t)u(t) \tag{22}$$

is used, we have

$$\dot{\phi} = -u(t)u^{\mathrm{T}}(t)\phi(t)\,. \tag{23}$$

In [43], necessary and sufficient conditions were derived for the uniform asymptotic stability of the equilibrium state of the differential equation (23) (i.e., the exponential convergence of the parameter error vector to zero). These conditions are also termed the conditions for the persistent excitation of the input vector u. One of several equivalent conditions for the persistent excitation of u is that it satisfy the inequality

$$\int_t^{t+T_0} u(\tau)u^{\mathrm{T}}(\tau)d\tau \geq \alpha I\,, \quad \forall t \geq t_0 \tag{24}$$

for some positive constants t_0, T_0 and α.

The second differential equation, analyzed in [44], has the form

$$\dot{x} = \begin{bmatrix} A & bu^{\mathrm{T}}(t) \\ -u(t)b^{\mathrm{T}} & 0 \end{bmatrix} x\,, \tag{25}$$

where $A + A^{\mathrm{T}} < 0$, (A, b) is controllable and $u : \mathbb{R}^+ \to \mathbb{R}^n$ is a bounded piecewise-continuous vector. The necessary and sufficient conditions for the asymptotic stability of the equilibrium state of (25) is that positive constants T_0, δ_0, and ϵ_1 exist such that for some $t_2 \in [t, t + T_0]$

$$\left| \frac{1}{T_0} \int_{t_2}^{t_2+\delta_0} u^{\mathrm{T}}(\tau) W d\tau \right| \geq \epsilon_1 \quad \forall\, t \geq t_0 \tag{26}$$

for every vector $W \in \mathbb{R}^n$. In the context of the differential equation (25), persistent excitation of u would be defined in terms of the inequality (26). The constant ϵ_1 in (26) is referred to as the degree of persistent excitation.

In continuous-time adaptive systems both equations (23) and (25) find application, and persistent excitation in both cases implies the exponential convergence of the parameter errors to zero.

For a long time it had been realized that if the reference input contains n distinct frequencies, the vector $\omega(t) \in \mathbb{R}^{2n}$ used in the identification of an LTI plant would be persistently exciting, and hence the parameters would converge to their true values. A general statement of this condition was derived in 1983 by Boyd and Sastry [47] using the concept of spectral lines.

5 Breakthroughs in the late 1970s

The results described in Section 3 revealed that linear time-invariant systems whose state vectors are accessible can be controlled in a stable adaptive fashion even when the system parameters are unknown. When it was realized that stable adaptive observers could be designed, which estimate the parameters as well as the state variables of a linear time-invariant system, there was renewed interest in adaptive control in the mid 1970s. Based on well-known results in linear systems theory when the parameters of the system are known, it was felt that the estimates obtained by adaptive observers could be used to control systems with unknown parameters in a stable fashion. In fact, for a short period in 1974, it was even believed that the solution to the adaptive observer problem had also provided the solution to the control problem. It was only during the following months that it was realized that the control problem was more involved. In the adaptive observer, the assumption had been made that the plant is stable and that the input and output to the observed plant are uniformly bounded. But this is no longer valid in the control problem where it is precisely the boundedness of these signals that has to be established. In fact, it was during one of the sessions at the IFAC World Congress held at Boston in 1975 that it became evident to the leading research workers in the field of adaptive control that the stability problem of adaptive control systems had not been resolved. The next five years witnessed tremendous research activity and culminated in the resolution of the problem in the ideal case in 1980. This result represents the most important development in adaptive systems theory in three decades.

The Adaptive Control Problem. A plant P is represented by a linear time-invariant differential equation

$$\begin{aligned} \dot{x}_{\mathrm{p}} &= A_{\mathrm{p}} x_{\mathrm{p}} + b_{\mathrm{p}} u \\ y_{\mathrm{p}} &= h_{\mathrm{p}}^{\mathrm{T}} x_{\mathrm{p}} , \end{aligned} \tag{27}$$

where $u : \mathbb{R}^+ \to \mathbb{R}$ is the input $y_{\mathrm{p}} : \mathbb{R}^+ \to \mathbb{R}$ is the output and $x_{\mathrm{p}} : \mathbb{R}^+ \to \mathbb{R}^n$ is the state of the plant. The triple $\{h_{\mathrm{p}}^{\mathrm{T}}, A_{\mathrm{p}}, b_{\mathrm{p}}\}$ is assumed to be controllable and observable. The elements of $h_{\mathrm{p}}, A_{\mathrm{p}}$ and b_{p} are assumed to be unknown. Using only input and output information, the objective is to determine a control function u, using a differentiator-free controller, which stabilizes the overall system. This is referred to as the adaptive regulation problem. If $y_{\mathrm{m}}(t)$ is a uniformly bounded desired output signal that is specified, the problem of tracking is defined as the determination of a control input u, using a differentiator-free controller, such that $\lim_{t\to\infty} |y_{\mathrm{p}}(t) - y_{\mathrm{m}}(t)| = 0$.

As described earlier, in model reference adaptive control (MRAC) the desired output $y_{\mathrm{m}}(t)$ is specified as the output of a reference model whose input $r(t)$ is a uniformly bounded piecewise-continuous function of time.

It is worth pointing out that in the initial stages, around 1976, even precise statements of the regulation and tracking problems were not available. In fact, as stated above, it was not clear that a solution existed for either of the two problems. It was only four years later, with perfect hindsight, that a precise problem statement along with a solution was available.

The adaptive control problem can also be stated in the frequency domain as follows: A linear time-invariant plant is described by a transfer function $W_{\mathrm{p}}(s)$ where

$$W_{\mathrm{p}}(s) = \frac{K_{\mathrm{p}} Z_{\mathrm{p}}(s)}{R_{\mathrm{p}}(s)} , \tag{28}$$

where $Z_{\mathrm{p}}(s)$ and $R_{\mathrm{p}}(s)$ are monic polynomials in s of degrees $m(\leq n-1)$ and n respectively and K_{p} is a constant. The coefficients of the polynomials as well as K_{p} are assumed to be unknown. A reference model with an input-output pair $\{r, y_{\mathrm{m}}\}$ which is linear and time-invariant with an asymptotically stable transfer function $W_{\mathrm{m}}(s)$, where

$$W_{\mathrm{m}}(s) = \frac{K_{\mathrm{m}} Z_{\mathrm{m}}(s)}{R_{\mathrm{m}}(s)} , \tag{29}$$

is specified. The objective is then to determine, as mentioned earlier, a control input u to the unknown plant such that

$$\lim_{t\to\infty} |y_{\mathrm{p}}(t) - y_{\mathrm{m}}(t)| = 0 . \tag{30}$$

Once again, the prior information regarding the plant as well as the transfer function $W_{\mathrm{m}}(s)$ that would be sufficient to solve the problem was by no means clear in the early stages. As numerous workers attempted the solution of the problem, the requirements became more evident. It is also worth pointing out that, over a decade after the adaptive control problem for the ideal case was solved in 1980, numerous workers in the field are concerned at present with questions of existence of solutions for other adaptive control problems.

During the period 1975–1978 it was generally realized by those working on the adaptive control of continuous-time systems, that the control problem for the case when the relative degree n^* (number of poles – number of zeros) of the plant is unity, is substantially simpler than that for the case $n^* \geq 2$. For such systems, the reference model could be chosen to be strictly positive real and the powerful results of stability theory developed in the 1960s could be used to prove global stability.

The Adaptive Control Problem for $n^* = 1$. All the idealized adaptive control problems attempted in the 1970s can be conveniently divided into two parts. These are generally referred to as (i) the algebraic part and (ii) the analytic part. The algebraic part is concerned with the existence of a solution to the adaptive control problem. It consists in demonstrating that a control structure with constant parameters $\theta_i^*(i = 1, 2, \ldots)$ exists such that the plant together with the controller has a transfer function identical to that of the reference model. Once such a controller is shown to exist, the analytic part consists in determining stable adaptive laws for the adjustment of the control parameters $\theta_i, i = 1, 2, \ldots$, so that they evolve to the desired (but unknown) constant values θ_i^*. Even though work on adaptive systems had been in progress for several years, it was only in 1975 that it was realized that the algebraic and analytic parts of the problem could be treated separately.

The structure of a controller for the adaptive control problem defined in this section is shown in Fig. 4. The input to the plant u as well as the output y are processed through identical $(n-1)$-dimensional stable systems described by the controllable pair $[\Lambda, \ell]$. The output of the former is the $(n-1)$-vector $\omega_1(t)$, while the output of the latter is the $(n-1)$-vector $\omega_2(t)$. Together with the reference input $r(t)$ and the output $y_\mathrm{p}(t)$, these constitute the $2n$ signals fed back to the plant. The control input to the plant is then expressed as a linear combination of the above signals as follows:

$$\begin{aligned} u(t) &= \theta^\mathrm{T}(t)\omega(t) \\ &= k(t)r(t) + \theta_1^\mathrm{T}(t)\omega_1(t) + \theta_2^\mathrm{T}(t)\omega_2(t) + \theta_0(t)y_\mathrm{p}(t) \\ &= [k(t), \theta_1^\mathrm{T}(t), \theta_2^\mathrm{T}(t), \theta_0(t)]\omega(t) \\ \omega^\mathrm{T}(t) &= [r(t), \omega_1^\mathrm{T}(t), \omega_2^\mathrm{T}(t), y_\mathrm{p}(t)]\,. \end{aligned} \tag{31}$$

The algebraic part consisted in showing that constant values of the parameters exist such that the transfer function of the plant together with a controller having the structure given by (31) and these parameters would have a transfer function identical to that of the reference model. Such a controller parameter vector is denoted by θ^*. Expressing the parameter error vector as $\phi(t) = \theta(t) - \theta^*$, the output error $e_1(t) = y_\mathrm{p}(t) - y_\mathrm{m}(t)$ can be expressed in terms of the parameter errors as

$$W_\mathrm{m}(s)\phi^\mathrm{T}(t)\omega(t) = e_1(t)\,. \tag{32}$$

From equation (32) it is clear that when the parameter error is zero, the output error will also tend to zero. The analytic part of the problem was to determine adaptive laws for adjusting $\theta(t)$ (or equivalently $\phi(t)$) so that $\lim_{t\to\infty} e_1(t) = 0$.

Several research workers working independently showed that the following prior information concerning the plant was sufficient to assure the existence of θ^* as well as to develop stable adaptive control laws. The prior information is stated as a set of four assumptions:

Assumptions (I)

(i) The sign of the high-frequency gain k_p of the plant,
(ii) an upper bound on the order of the plant, and
(iii) the relative degree of $W_p(s)$ are known, and
(iv) $Z_p(s)$ is a Hurwitz polynomial in s.

Assumptions (I), under which this idealized problem could be solved, became important in adaptive control and formed the starting point for many investigations in the 1980s.

Returning to equation (31), it was shown by Narendra and Valavani [48] that if the transfer function $W_m(s)$ is strictly positive real, the adaptive laws

$$\dot{\theta}(t) = -\text{sgn}(k_p)e_1(t)\omega(t) \tag{33}$$

would result in all the signals in the adaptive system being bounded and the output error $e_1(t)$ tending to zero asymptotically.

The Adaptive Control Problem for $n^* \geq 2$. Following the solution of the adaptive control problem for the case $n^* = 1$, there was world-wide interest in solving the problem for the general case when $n^* \geq 2$. A relative degree of the plant greater than or equal to two implies that the reference model transfer function $W_m(s)$ cannot be positive real, since the latter must have a relative degree greater than or equal to that of the plant. This in turn implies that in the error equation (32) $W_m(s)$ is not strictly positive real so that the same procedure used in the case when $n^* = 1$ can no longer be adopted. During this period it almost seemed impossible to solve the problem. The major breakthrough came from a suggestion made by Monopoli [49] that an augmented error (rather than the true error between plant and model outputs) should be used to generate the adaptive laws. The process involved the addition of auxiliary signals to the error equation (32) to reduce it to a mathematically tractable form. As acknowledged by Monopoli to the author, this idea was strongly influenced by the earlier work on adaptive observers in the early 1970s.

Defining

$$\phi^T(t)W_m(s)\omega(t) - W_m(s)\phi^T(t)\omega(t) = e_2(t)\,, \tag{34}$$

it follows that

$$e_1(t) + e_2(t) = \varepsilon_1(t) = \phi^T(t)W_m(s)\omega(t) = \phi^T(t)\zeta(t)\,, \tag{35}$$

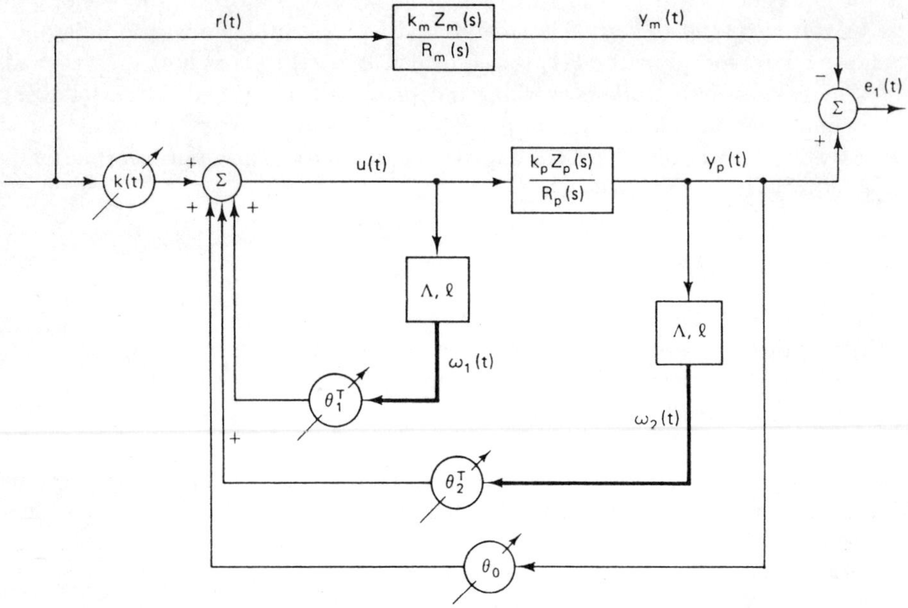

Fig. 4. The adaptive controller for $n^* = 1$.

where $e_2(t)$ is referred to as the auxiliary error and $\varepsilon_1(t)$ is the augmented error. From equation (35) it follows that if $e_2(t)$ is added to the true error $e_1(t)$, then the resulting augmented error $\varepsilon_1(t)$ is related to the parameter error vector $\phi(t)$ by the equation

$$\phi^T(t)\zeta(t) = \varepsilon_1(t), \tag{36}$$

which has been extensively studied in the context of adaptive systems. It was therefore concluded that adaptation should be carried out using the augmented error $\varepsilon_1(t)$ and the vector signal $\zeta(t)$ rather than the true error signal $e_1(t)$ and the signal $\omega(t)$ as in the case when $n^* = 1$. This represented a significant step since the generation of the adaptive law had finally been separated from the control process itself and the signals associated with it.

Following the methods used in adaptive observers the natural tendency was to choose the adaptive law as

$$\dot{\theta}(t) = \dot{\phi}(t) = -\varepsilon_1(t)\zeta(t) = -\zeta(t)\zeta^T(t)\phi(t), \tag{37}$$

which assures the boundedness of the parameter vector. However, it was soon realized that this would not assure the boundedness of all the signals in the system. The period 1978–1979 was a period of intense activity in the adaptive control area, with different groups working in different parts of the world trying frantically to determine the modifications needed in the adaptive law (33) to assure the global stability of the adaptive system. As it has happened so often in the history of science, the breakthrough was made independently by several groups simultaneously. The first wave of solutions, by Morse [50] and Narendra et al. [51] for continuous-time systems, and by Goodwin, Ramadge and Caines [52] and Narendra and Lin [53] for discrete-time systems, appeared in the June 1980 issue of the *IEEE Transactions on Automatic Control*, followed by solutions by other authors in later issues. An early unified treatment of these results was given by Egardt [54]. The problem of adaptively controlling a linear time-invariant plant which satisfied the four assumptions given in (I) had been solved. The key to the solutions was the use of a normalization factor in the adaptive control law which assumed the form

$$\dot{\phi}(t) = \frac{-\varepsilon_1(t)\zeta(t)}{1+\zeta^{\mathrm{T}}(t)\zeta(t)} . \tag{38}$$

This ensured that $\dot{\phi} \in \mathcal{L}^2$. The boundedness of the parameter vector together with this condition on the rate of change of the control parameter vector was sufficient to demonstrate that the signals of the adaptive system would be bounded and that the output error $e_1(t)$ would tend to zero asymptotically. Since many of the results derived in the 1980s used arguments similar to those used in the ideal case, the various steps used in the proof of stability are outlined here.

(i) Using the adaptive law (38) it follows that the control parameters are bounded. This in turn implies that the state variables of the adaptive system can grow at most exponentially.
(ii) From the fact that $\dot{\phi} \in \mathcal{L}^2$ it is shown that the augmented error $\varepsilon_1(t)$ is related to $\zeta(t)$ as $\varepsilon_1(t) = \beta(t)\sqrt{1+\zeta^{\mathrm{T}}(t)\zeta(t)}$ where $\beta \in \mathcal{L}^2$.
(iii) Assuming that the signals grow in an unbounded fashion, the fact that $\dot{\phi} \in \mathcal{L}^2$ implies that $\|\omega_2\|, \|\zeta\|, \|\omega\|$ and $|y_{\mathrm{p}}|$ grow at the same rate, and that $\|\omega_2\|$ grows at a different rate than $\|\omega\|$. This leads to a contradiction, so that it is concluded that all signals in the feedback loop are bounded and that $\lim_{t\to\infty} \varepsilon_1(t) = \lim_{t\to\infty} e_1(t) = 0$.
(iv) Finally, it follows that if $\omega(t)$ is persistently exciting, the parameter vector $\theta(t)$ converges to the desired value θ^*.

For detailed stability arguments based on growth rate of signals the reader is referred to the paper by Narendra, Annaswamy and Singh [55].

6 Diversity in the 1980s

The solution of the adaptive control problem in the idealized case described in the previous section provided a tremendous impetus to the field and attracted many researchers from different areas of systems theory. During the first half of the last decade, four major directions of research emerged. The first dealt with situations where the ideal assumptions (I) are satisfied but adaptation has to be carried out in the presence of different types of perturbations. The second direction of research was towards relaxing the assumptions (I) and determining the least restrictive conditions under which regulation and tracking are possible. Extension of results derived for single-input single-output systems to multivariable problems was a third direction in which research evolved, and a fourth direction was aimed at determining the conditions for assuring stochastic stability of adaptive systems when the disturbances are random in nature. Since most of the results in stochastic adaptive control apply to discrete-time systems and this paper deals mainly with continuous-time adaptive systems, only the developments in the first three directions are discussed here. Further, during the second half of the decade, the field continued to expand rapidly and many new areas of research came into existence, so that it becomes difficult to track precisely the different advances in the field. In this section we consequently attempt to provide merely a comprehensive review of the principal developments during the 1980s.

6.1 Robust Adaptive Control

In the statement of the idealized adaptive control problem treated in Section 5, the plant parameters were assumed to be constant and the system was assumed to be disturbance-free. However, in practice, these assumptions are rarely met. No plant is truly time invariant, finite-dimensional, linear, and noise-free. The question therefore naturally arose as to whether the idealized adaptive laws would perform satisfactorily even when external disturbances and parameter variations are present and imperfect models of the plant are used. The realization that small external and internal perturbations can result in unbounded signals in the system eventually led many researchers to what is termed robust adaptive control or the search for techniques for achieving bounded response even in the presence of different disturbances.

To understand some of the theoretical difficulties encountered in robust adaptive control problems, consider the Lyapunov function $V(e,\phi)$ used to prove uniform stability in the ideal case. The time-derivative $\dot{V}(e,\phi)$ of V along the trajectories of the system is invariably negative semi-definite. When disturbances and/or unmodeled dynamics are present, $\dot{V}(e,\phi)$ is indefinite and V is no longer a Lyapunov function guaranteeing stability. The objective in such cases is to modify the adaptive laws so that $\dot{V}$ is negative semidefinite outside a compact region containing the origin in the state space, so that the response of the system is bounded.

Bounded Disturbances. For the case of arbitrary bounded disturbances, Egardt [54], Peterson and Narendra [56] and Samson [57] suggested the use of a dead-zone in the adaptive law to assure bounded solutions. This modification implies that adaptation is stopped when the error is smaller than a value determined by the prior information concerning the disturbance. Assuming that θ^*, the desired control parameter vector, lies inside a sphere S, Kreisselmeier and Narendra [58] demonstrated that the ideal adaptive law, modified only at the boundary of S would result in all the signals in the adaptive system being bounded. This was the precursor, in the continuous case, of the projection algorithms that are commonly used at present in both discrete- and continuous-time systems.

The two modifications to the adaptive law suggested above require additional information concerning the disturbance or the plant. In 1983, Ioannou and Kokotovic [59] proposed a modified law which does not require such additional information to assure robustness. In place of the adaptive law $\dot{\theta} = -e_1\omega$ used in the ideal case, the modified law has the form

$$\dot{\theta} = -e_1\omega - \sigma\theta . \tag{39}$$

As mentioned earlier, one of the major problems in the analysis of adaptive systems using the stability approach is that the time derivative $\dot{V}$ of the Lyapunov function is semidefinite even in the ideal case. The introduction of the term $-\sigma\theta$ in (39) makes $\dot{V}$ negative outside a compact region containing the origin in the state space, thereby assuring the boundedness of all the signals. In 1987, Narendra and Annaswamy [60] proposed a somewhat different adaptive law based on the same principles, while avoiding some of the shortcomings of (39). This adaptive law has the form

$$\dot{\theta} = -e_1\omega - |e_1|\theta . \tag{40}$$

The σ-modification law and the $|e_1|$-modification law given in [59] and [60] have been studied extensively and are often used in applications to assure robustness in the presence of bounded external perturbations.

In contrast to the above methods, which attempt to assure robustness by modifying the adaptive laws, efforts were also made to use the concept of persistent excitation to ensure boundedness of all the signals in the system. A significant result in this direction given by Narendra and Annaswamy [61] states that, if the reference input has a sufficiently large degree of persistent excitation relative to the magnitude of the disturbance, then all the signals in the adaptive system will be bounded.

Hybrid Adaptive Control. All the adaptive laws described so far are continuous in time. Using the same controller structure, it can be shown that the parameter vector $\theta(t)$ can also be adjusted using a hybrid algorithm while assuring global stability in the ideal case and robustness in the presence of disturbances. In such a case, $\theta(t)$ is held constant over intervals of length T and updated at the end of the interval using the information collected. Several such hybrid modifications were given by Narendra, Khalifa and Annaswamy in [62]. A typical adaptive law has the form

$$\theta(t+T) = \theta(t) - \int_t^{t+T} \gamma e_1(\tau)\omega(\tau)d\tau , \tag{41}$$

so that θ is updated along the average gradient over the interval. In extensive simulation studies, hybrid adaptation was found to be preferable to continuous adaptation, as far as robustness is concerned.

Time-Varying Systems. One of the compelling reasons for considering adaptive control is to compensate for time variations in plant parameters. While much of the research reported thus far has dealt with plants with constant parameters, interest in adaptive control in the presence of plant parameter variations has existed from the very beginning. In the early stages, most of the researchers made rather restrictive assumptions on the nature of the parameter variations. Anderson and Johnstone [63] demonstrated that if the reference input is persistently exciting and the parameter variations are sufficiently small, the signals in the system will be bounded provided the initial errors are small. Later papers by Chen and Caines [64], Hill et al. [65], and Martin-Sanchez [66], considered exponential decay of parameters to constant values, finite number of jump variations, small perturbations around constant values or a combination of these. Since time variations of plant parameters may violate the assumptions (I) instantaneously, further assumptions were made by many authors to achieve robustness. In [67,68,69,70], globally bounded solutions are established assuming that the plants vary sufficiently slowly.

Towards the end of the 1980s, several workers became interested in adaptive systems in which the parameters vary rapidly [72,71]. Tsakalis and Ioannou [72] suggested a modification in the controller structure, by which, at least in theory, the control objective can be met exactly. This important development has enabled the authors to study the behavior of time-varying systems with arbitrarily varying parameters.

Unmodeled Plant Dynamics. From the results presented in Section 4 it is clear that $2n$ controller parameters are needed to control an nth-order plant. This poses a major practical limitation since, even for plants of moderate complexity, this requires a large number of adjustable parameters. Further, from a theoretical point of view, the assumption that the plant order is known is rarely met. Hence, in the opinion of many researchers, adaptively controlling an unknown plant using a reduced-order controller was the most important problem in adaptive control in the mid 1980s.

The output η of the unmodeled part of the plant can be considered as a state-dependent disturbance on the plant. Hence, it cannot be assumed to be bounded a priori and a modified adaptive law is needed, which does not require this a priori information. Ioannou and Kokotovic [59] showed that such an update law is the σ-modification (39).

The restrictive case when the true plant together with a constant controller is SPR was analyzed by Narendra and Annaswamy [73]. They showed that the degree of persistent excitation of the modeled part should be larger than the magnitude of the output of the unmodeled part to assure the boundedness of all the signals in the system. Important contributions to the problem were made by Praly [74], who employed a normalization technique, and Kreisselmeier and Anderson [75], who used a variable dead-zone to implement the adaptive law.

Along with the development of modified adaptive laws, instability mechanisms of ideal adaptive laws were analyzed. Åström [76] used averaging, previously employed in the ODE method of Ljung [77], to give an explanation of the instability examples of Rohrs et al. [78].

Applying the method of averaging, Kokotovic et al. [79,80], Fu et al. [81], and Kosut et al. [82], analyzed adaptive control when unmodeled dynamics are present. Kokotovic and co-workers showed that the frequency content of the excitation of certain signals determines a sharp stability-instability boundary for slow adaptation. Sufficient conditions for the boundedness of signals were also derived in [82] based on the frequency range of the dominant excitation.

6.2 Relaxation of Assumptions

Many practical adaptive controllers were found to perform satisfactorily even when the ideal Assumptions (I) given in Section 5 were not valid. This eventually led to a search for the least restrictive assumptions under which a solution can be obtained for the adaptive control problem. A major step in this direction was the result of Nussbaum [83] that, even when the sign of λ is unknown, the differential equations

$$\begin{aligned} \dot{x}_1 &= ax_1 + \lambda f(x_1, x_2), \quad a > 0 \\ \dot{x}_2 &= g(x_1, x_2) \end{aligned} \tag{42}$$

can be made to have solutions for which $\lim_{t\to\infty} x_1(t) = 0$ with $x_2(t)$ bounded, by the proper choice of the functions f and g. This had an immediate impact. Mudgett and Morse [84] demonstrated soon after that the sign of the high-frequency gain is not needed to stabilize the adaptive system. When Willems and Byrnes [85] considered an nth-order plant with relative degree one and zeros in the left half plane, and demonstrated that it could be stabilized by an adaptive high-gain feedback, it became clear that at least in special cases the bound on the order of the system may not be needed. The search for weaker assumptions was on. Numerous contributions were made during the following years and many partial results were obtained. For example in [86], Morse showed that a 3-dimensional high-gain adaptive controller can stabilize any minimum phase process of any finite dimension provided that the relative degree is less than three and later showed [87] that, if in addition the high-frequency gain is known a priori, a 3-dimensional controller could stabilize a process of relative degree ≤ 3 without using an augmented error.

Indirect control methods rely on the estimation of the plant parameters to adjust the control parameters. Researchers in this area are interested in an approach that would eliminate the requirement that the zeros of the plant should lie in the open left half of the complex plane. An important contribution to relaxing the assumptions for stable adaptive control was made by Kreisselmeier and Smith [88], who showed that a plant with unknown parameters can be regulated in a stable fashion, assuming that only the order of the plant is known.

A major advance in the search for weaker assumptions on the plant to ensure stable adaptation was made in 1985 by Martensson, who showed in [89] that the knowledge of the order ℓ of an LTI stabilizing compensator is sufficient to design a stabilizing controller. While the practicability of such a controller is dubious, its importance lies in the demonstration that stable controllers can be designed in theory with very little information concerning the plant. Martensson's contribution has encouraged a number of researchers to investigate adaptive controllers which, while based on weak assumptions, are also practically feasible. Active research based on switching functions is currently in progress in this area. While a complete answer to the question of least restrictive assumptions is not available at present, it appears that stable adaptive control may be possible even when all the assumptions in (I) are violated.

6.3 Multivariable Adaptive Control

Attempts to extend stable adaptive control laws derived for SISO systems to multi-input multi-output systems (MIMO) were also initiated in the early 1980s. Since large uncertainty is generally associated with complex multivariable systems, adaptive control methodologies are even more relevant in such cases. While the adaptive control of MIMO systems leads to questions which are similar to the single variable case, the solutions to these questions are substantially more involved. In particular the issue of parametrization of the adaptive controller becomes an important problem.

The four assumptions corresponding to those in (I), which are sufficient to ensure the existence of a stable adaptive controller in the MIMO case are:

Assumptions (II)

(i) The high-frequency gain matrix K_{p} of the plant must satisfy the matrix equation

$$\Gamma K_{\mathrm{p}} + K_{\mathrm{p}}^{\mathrm{T}} \Gamma = Q_0 \, , \quad \Gamma = \Gamma^{\mathrm{T}} > 0 \, , \tag{43}$$

where Q_0 is a symmetric sign definite matrix for some positive definite symmetric matrix Γ,

(ii) the right Hermite normal form $H_{\mathrm{p}}(s)$ of the plant transfer matrix $W_{\mathrm{p}}(s)$ is known,

(iii) an upper bound ν on the observability index of $W_{\mathrm{p}}(s)$ is known, and

(iv) the zeros of $W_{\mathrm{p}}(s)$ lie in C^{-}.

As in the scalar case, assumptions (i) and (iv) are used in the analytic part of the adaptive control problem while (ii) and (iii) are needed for the algebraic part. Several adaptive control algorithms have been suggested in the literature which assure the stability of the overall system when Assumptions (II) are satisfied [90,91,92]. In [92] it is shown that the assumption (ii) concerning the Hermite normal form of $W_{\mathrm{p}}(s)$ has the practical implication that only those multivariable systems $W_{\mathrm{p}}(s)$ whose Hermite normal forms (or whose Hermite normal forms after compensation, i.e., $W_{\mathrm{p}}(s)W_c(s)$) are diagonal can be controlled in an adaptive fashion.

The same questions concerning robustness in the presence of different types of perturbations of single variable questions also arise in the multivariable case.

All the methods discussed in the earlier sections can be extended to multivariable systems with minor modifications.

6.4 Recent Developments

While the main thrust of the research in adaptive systems during the greater part of the 1980s was in the three directions described earlier in this section, many new areas of research emerged during the latter part of the decade. For the sake of completeness we merely list some of these below.

Prompted by developments in nonlinear control theory dealing with feedback linearizable systems [93], the adaptive control of special classes of nonlinear systems with unknown parameters has been attempted recently [94,95]. The techniques used in these cases are naturally strongly motivated by the results for linear systems discussed in this paper.

Another recent trend in adaptive control circles is towards the use of concepts based on variable structure switching controllers [96,97,98]. The proofs of stability using these methods follow closely along the lines indicated in Section 4 for continuous adaptive laws. While extensive simulation results have indicated that these methods result in substantially improved performance, the variable structure approach raises theoretical questions which need further investigation.

Direct and indirect methods were briefly mentioned in Section 2 as the two principal methods used in adaptive control. Both of them have been used widely in robust adaptive systems. Recently it has been demonstrated that the two methods can be combined [99] to benefit from their individual advantages.

Even as the research effort in the new areas gains momentum, work continues with increased vigor in the more established areas. Attempts to determine the least stringent conditions under which stable adaptive control is possible, the search for improved methods for controlling time-varying systems, the study of phenomena such as bursting which are peculiar to robust adaptive systems, as well as the application of adaptive control principles to practical systems, are all as active as ever. However, no single area dominates the field. The author is confident that the confluence of the different ideas that are being currently explored will lead to significant developments in the area of adaptive control in the 1990s.

7 Conclusion

Having spent its fledgling years in the 1960s in relative peace and quiet among gradient methods, and its youth in the 1970s exploring the mysteries of stable nonlinear systems, a mature adaptive control theory entered the turbulent 1980s with unbridled confidence. Ten years later, on the threshold of a new decade, a more experienced and wiser field views the prospects for the future with cautious optimism.

The introduction of new concepts and tools has significantly widened the field and radically changed its structure and boundaries. Towards the end of the last decade the field has also witnessed the proliferation of research areas, with the addition of adaptive nonlinear control, variable structure adaptive control, adaptive control using integrated methods, and switching adaptive systems. This, in turn, has made communication between experts in different areas no longer as easy as it was in the past. With this ever-increasing diversity in the field, the need for conceptual frameworks for viewing adaptive control problems and formulating the right questions is assuming greater importance.

The burgeoning of new areas, while making the field more attractive to the researcher, has not been an unmixed blessing. Many important questions in the older areas have been left behind, only partially answered. While robustness has been studied extensively, we do not currently have a convenient method of quantifying it. The importance of persistent excitation for the convergence of parameters as well as for the robustness of the adaptive system is well realized, but few prescriptive methods for choosing the reference input or the filters in the controller currently exist. This implies that the choice of parameters in the adaptive system is a matter of trial and error without theoretical support. Since all the stability proofs are asymptotic in character, very little is known at present about the transient behavior of adaptive systems. In the analysis of systems with time-varying parameters, few elegant results exist which are comparable to those derived in the 1960s for linear time-invariant systems. Finally, it is good to remember that while the trend away from gradient methods to stability-based methods was strongly motivated by the desire for faster response, all the proofs of stability that have been derived in the last two decades are based on the time derivative $\dot{\theta}(t)$ of the control parameters being sufficiently small. In light of this, the methods discarded in the 1960s no longer appear unattractive and may even be worthy of further examination.

The revolutionary changes in computer technology in the last decade have made the practical implementation of adaptive algorithms economically feasible, and there is a growing demand for industrial adaptive controllers which meet increased speed, accuracy and robustness requirements. In fact, in many areas, the scope of the application is to a large extent limited by the existing theory. During the 1970s and the 1980s, the emphasis of theoretical research was on ensuring the boundedness of all the signals in adaptive systems rather than on performance. Hence, many of the results obtained are not directly applicable to problems in technology. The trend that started in the 1970s, which shifted the emphasis from performance to stability in adaptive systems, appears to have reached the zenith. With our current knowledge of the stability of adaptive systems, it is perhaps time that the trend be reversed to some extent and greater emphasis placed in the future on performance.

Thirty-five years ago, the biological implications of the term "adaptive" lent the field of adaptive control an aura which was responsible for the great interest evinced in it. The same reasons are also responsible for the enormous attraction the field enjoys even today. During the intervening years a variety of terms borrowed from biology and psychology, such as learning, pattern recognition, self-organization, artificial intelligence, and intelligent control have found their way into the systems literature. These terms were motivated by the desire to incorporate, in artificial systems, the ability of living systems to cope with large uncertainties in their environments. The various terms have developed into major independent disciplines with wide followings. Although the sources of the problems and the terminologies in these areas are different from those of adaptive control, many of the concepts used and the theoretical difficulties encountered are quite similar. It is therefore safe to say that advances in adaptive control theory in the 1990s will have a significant impact on all the methodologies dealing with systems containing large uncertainties.

Acknowledgments

The author would like to thank Prof. A. M. Annaswamy and Dr. R. M. Wheeler, Jr., for interesting discussions. Some of the sections in this paper are summaries of chapters in the book "Stable Adaptive Systems" by K. S. Narendra and A. M. Annaswamy, published by Prentice-Hall in 1989.

References

1. E. H. Carr, *What is History?*, Penguin Books, 1981.
2. J. G. Truxal, *Automatic Feedback System Synthesis*, McGraw-Hill, New York, NY, 1955.
3. B. C. Kuo, *Automatic Control Systems*, Prentice-Hall, Englewood Cliffs, NJ, 1987.
4. R. F. Drenick and R. A. Shahbender, "Adaptive servomechanisms," *AIEE Transactions*, vol. 76, pp. 286–292, Nov. 1957.
5. J. A. Aseltine, A. R. Mancini, and C. W. Sarture, "A survey of adaptive control systems," *IRE Trans. Aut. Control*, vol. 3, pp. 102–108, Dec. 1958.
6. P. R. Stromer, "Adaptive or self-optimizing control systems – a bibliography," *IRE Trans. Aut. Control*, vol. 4, pp. 65–68, May 1959.
7. M. Margolis and C. T. Leondes, "A parameter tracking servo for adaptive control systems," *IRE Trans. Aut. Control*, vol. 4, pp. 100–111, Nov. 1959.
8. L. A. Zadeh, "On the definition of adaptivity," *Proceedings of the IEEE*, vol. 51, pp. 569–570, March 1963.
9. R. Bellman, *Adaptive Control Processes – A Guided Tour*, Princeton University Press, Princeton, NJ, 1961.
10. J. G. Truxal, "Adaptive Control," *Proc. 2nd IFAC World Congress*, Basle, Switzerland, 1963.
11. K. S. Narendra and A. M. Annaswamy, *Stable Adaptive Systems*, Prentice-Hall, Englewood Cliffs, NJ, 1989.
12. L. S. Pontryagin et al., *The Mathematical Theory of Optimal Processes*, Interscience, New York, NY, 1962.
13. A. A. Feldbaum, *Optimal Control Systems*, Academic Press, New York, NY, 1965.
14. R. E. Kalman, "Design of a self-optimizing control system," *Transactions of ASME*, vol. 80, pp. 468–478, 1958.
15. E. Mishkin and L. Braun, *Adaptive Control Systems*, McGraw-Hill, New York, NY, 1961.

16. I. D. Landau, *Adaptive Control: The Model Reference Approach*, Marcel Dekker, New York, NY, 1979.
17. C. S. Draper and Y. T. Li, "Principles of optimalizing control systems and an application to the internal combustion engine," *ASME Publication*, Sep. 1951.
18. V. W. Eveleigh, *Adaptive Control and Optimization Techniques*, McGraw-Hill, New York, NY, 1967.
19. R. J McGrath, V. Rajaraman, and V. C. Rideout, "A parameter perturbation adaptive control system," *IRE Trans. Aut. Control*, vol. 6, pp. 154–161, May 1961.
20. J. L. Douce, K. C. Ng, and M. M. Gupta, "Dynamics of the parameter perturbation processes," *Proceedings of the IEE*, vol. 113, no. 6, pp. 1077–1083, June 1966.
21. R. E. Kronauer and P. G. Drew, "Design of the feedback loop in parameter-perturbation adaptive controls," in *Theory of Self-Adaptive Control Systems, Proc. 2nd IFAC Symp.* (Teddington, UK, 1965), P. H. Hammond, ed., Plenum Press, New York, NY, 1966.
22. K. S. Narendra and L. E. McBride, Jr., "Multiparameter self-optimizing systems using correlation techniques," *IRE Trans. Aut. Control*, vol. 9, pp. 31–38, Jan. 1964.
23. P. V. Kokotovic, "Method of sensitivity points in the investigation and optimization of linear control systems," *Automation and Remote Control*, vol. 25, pp. 1512–1518, 1964.
24. J. B. Cruz, Jr., ed., *Systems Sensitivity Analysis*, Dowden, Hutchinson and Ross, Stroudsburg, PA, 1973.
25. H. P. Whitaker, J. Yamron, and A. Kezer, "Design of model-reference adaptive control systems for aircraft," Report R-164, Instrumentation Laboratory, MIT, Cambridge, MA, 1958.
26. L. P. Grayson, "Design via Lyapunov's second method," *Proc. 4th Joint Amer. Control Conf.*, 1963.
27. B. Shackcloth and R. L. Butchart, "Synthesis of model reference adaptive systems by Lyapunov's second method," in *Theory of Self-Adaptive Control Systems, Proc. 2nd IFAC Symp.* (Teddington, UK, 1965), P. H. Hammond, ed., Plenum Press, New York, NY, 1966.
28. P. C. Parks, "Lyapunov redesign of model reference adaptive control systems," *IEEE Trans. Aut. Control*, vol. AC-11, pp. 362-367, 1966.
29. R. V. Monopoli, "Lyapunov's method for adaptive control system design," *IEEE Trans. Aut. Control*, vol. AC-12, pp. 334–335, 1967.
30. P. H. Phillipson, "Design methods for model reference adaptive systems," *Proc. Inst. Mech. Engineers*, vol. 183, pp. 695–700, 1969.
31. C. A. Winsor and R. J. Roy, "Design of model reference adaptive control systems by Lyapunov's second method," *IEEE Trans. Aut. Control*, vol. AC-13, p. 204, April 1968.
32. K. S. Narendra and P. Kudva, "Stable adaptive schemes for system identification and control - part I" *IEEE Trans. Systems, Man and Cybernetics*, vol. SMC-4, pp. 541–551, Nov. 1974.
33. K. S. Narendra and P. Kudva, "Stable adaptive schemes for system identification and control - part II" *IEEE Trans. Systems, Man and Cybernetics*, vol. SMC-4, pp. 552–560, Nov. 1974.
34. Ya. Z. Tsypkin, *Adaptation and Learning in Automatic Systems*, Academic Press, New York, NY, 1971.

35. K. J. Åström and B. Wittenmark, "On self-tuning regulators," *Automatica*, vol. 9, pp. 185–199, 1973.
36. R. Carroll and D. P. Lindorff, "An adaptive observer for single-input single-output linear systems," *IEEE Trans. Aut. Control*, vol. AC-18,pp. 428–435, Oct. 1973.
37. P. Kudva and K. S. Narendra, "Synthesis of an adaptive observer using Lyapunov's direct method," *Int. J. Control*, vol. 18, pp. 1201–1210, 1973.
38. G. Luders and K. S. Narendra, "An adaptive observer and identifier for a linear system," *IEEE Trans. Aut. Control*, vol. AC-18, pp. 496–499, Oct. 1973.
39. G. Luders and K. S. Narendra, "A new canonical form for an adaptive observer," *IEEE Trans. Aut. Control*, vol. AC-19, pp. 117–119, April 1974.
40. G. Kreisselmeier, "Adaptive observers with exponential rate of convergence," *IEEE Trans. Aut. Control*, vol. AC-22, pp. 2–8, Jan. 1977.
41. K. J. Åström and T. Bohlin, "Numerical identification of linear dynamic systems from normal operating records," in *Theory of Self-Adaptive Control Systems, Proc. 2nd IFAC Symp.* (Teddington, UK, 1965), P. H. Hammond, ed., Plenum Press, New York, NY, 1966.
42. M. M. Sondhi and D. Mitra, "New results on the performance of a well-known class of adaptive filters," *Proceedings of the IEEE*, vol. 64, pp. 1583–1597, Nov. 1976.
43. A. P. Morgan and K. S. Narendra, "On the uniform asymptotic stability of certain linear non-autonomous differential equations," *SIAM J. Control Optimiz.*, vol. 15, pp. 5–24, Jan. 1977.
44. A. P. Morgan and K. S. Narendra, "On the stability of nonautonomous differential equations $\dot{x} = [A + B(t)]x$ with skew-symmetric matrix $B(t)$" *SIAM J. Control Optimiz.*, vol. 15, pp. 163–176, Jan. 1977.
45. J. S. C. Yuan and W. M. Wonham, "Probing signals for model reference identification," *IEEE Trans. Aut. Control*, vol. AC-22, pp. 530–538, Aug. 1977.
46. B. D. O. Anderson, "Exponential stability of linear equations arising in adaptive identification," *IEEE Trans. Aut. Control*, vol. AC-22, pp. 83–88, Jan. 1977.
47. S. P. Boyd and S. S. Sastry, "On parameter convergence in adaptive control," *Syst. Control Lett.*, vol. 2, pp. 311–319, 1983.
48. K. S. Narendra and L. S. Valavani, "Stable adaptive controller design–direct control," *IEEE Trans. Aut. Control*, vol. AC-23, pp. 570–583, Aug. 1978.
49. R. V. Monopoli, "Model reference adaptive control with an augmented error signal," *IEEE Trans. Aut. Control*, vol. AC-19, pp. 474–484, 1974.
50. A. S. Morse, "Global stability of parameter adaptive control systems," *IEEE Trans. Aut. Control*, vol. AC-25, pp. 433-439, June 1980.
51. K. S. Narendra, Y. H. Lin, and L. S. Valavani, "Stable adaptive controller design –part II: proof of stability," *IEEE Trans. Aut. Control*, vol. AC-25, pp. 440–448, June 1980.
52. G. C. Goodwin, P. J. Ramadge, and P. E. Caines, "Discrete time multivariable adaptive control," *IEEE Trans. Aut. Control*, vol. AC-25, pp. 449–456, June 1980.
53. K. S. Narendra and Y. H. Lin, "Stable discrete adaptive control," *IEEE Trans. Aut. Control*, vol. AC-25, pp. 456–461, June 1980.
54. B. Egardt, *Stability of Adaptive Controllers*, Springer-Verlag, Berlin, 1979.
55. K. S. Narendra, A. M. Annaswamy, and R. P. Singh, "A general approach to the stability analysis of adaptive systems," *Int. J. Control*, vol. 41, pp. 193–216, 1985.
56. B. B. Peterson and K. S. Narendra, "Bounded error adaptive control," *IEEE Trans. Aut. Control*, vol. AC-27, pp. 1161–1168, Dec. 1982.

57. C. Samson, "Stability analysis of adaptively controlled systems subject to bounded disturbances," *Automatica*, vol. 19, pp. 81–86, 1983.
58. G. Kreisselmeier and K. S. Narendra, "Stable model reference adaptive control in the presence of bounded disturbances," *IEEE Trans. Aut. Control*, vol. AC-27, pp. 1169–1175, Dec. 1982.
59. P. A. Ioannou and P. V. Kokotovic, *Adaptive Systems with Reduced Models*, Springer-Verlag, New York, NY, 1983.
60. K. S. Narendra and A. M. Annaswamy, "A new adaptive law for robust adaptive control without persistent excitation," *IEEE Trans. Aut. Control*, vol. AC-32, pp. 134–145, Feb. 1987.
61. K. S. Narendra and A. M. Annaswamy, "Robust adaptive control in the presence of bounded disturbances," *IEEE Trans. Aut. Control*, vol. AC-31, pp. 306–315, April 1986.
62. K. S. Narendra, I. H. Khalifa, and A. M. Annaswamy, "Error models for stable hybrid adaptive systems–part II," *IEEE Trans. Aut. Control*, vol. AC-30, pp. 339–347, April 1985.
63. B. D. O. Anderson and R. M. Johnstone, "Adaptive systems and time-varying plants," *Int. J. Control*, vol. 37, pp. 367–377, Feb. 1983.
64. H. F. Chen and P. E. Caines, "On the adaptive control of stochastic systems with random parameters," *Proc. 23rd IEEE Conf. Dec. Control*, 1984.
65. D. J. Hill, G. C. Goodwin, and X. Xianya, "Stochastic adaptive control for exponentially convergent time-varying systems," *Proc. 23rd IEEE Conf. Dec. Control*, 1984.
66. J. M. Martin-Sanchez, "Adaptive control of time-variant processes," *Proc. 1985 Amer. Control Conf.*
67. G. Kreisselmeier, "Adaptive control of a class of slowly time-varying plants," *Syst. Control Lett.*, vol. 8, Dec. 1986.
68. T. H. Lee and K. S. Narendra, "Stable direct adaptive control of time-varying discrete-time systems," Technical Report 8720, Center for Systems Science, Yale University, New Haven, CT, 1987.
69. K. S. Tsakalis and P. A. Ioannou, "Adaptive control of linear time-varying plants," *Automatica*, vol. 23, pp. 459–468, July 1987.
70. R. H. Middleton and G. C. Goodwin, "Adaptive control of time-varying linear systems," *IEEE Trans. Aut. Control*, vol. AC-33, pp. 150–155, Feb. 1988.
71. A. M. Annaswamy and K. S. Narendra, "Adaptive control of a first-order plant with a time-varying parameter," *Proc. 1990 Amer. Control Conf.*
72. K. S. Tsakalis and P. A. Ioannou, "Adaptive control of linear time-varying plants: a new model reference controller structure," *IEEE Trans. Aut. Control*, vol. 34, pp. 1038–1046, Oct. 1989.
73. K. S. Narendra and A. M. Annaswamy, in *Adaptive and Learning Systems*, K. S. Narendra, ed., Plenum Press, New York, NY, 1985.
74. L. Praly, "Robustness of model reference adaptive control," *Proc. 3rd Yale Workshop on Applications of Adaptive Systems Theory*, Yale University, New Haven, CT, 1983.
75. G. Kreisselmeier and B. D. O. Anderson, "Robust model reference adaptive control," *IEEE Trans. Aut. Control*, vol. AC-31, pp. 127-133, 1986.
76. K. J. Åström, "Analysis of Rohrs counterexamples to adaptive control," *Proc. 22nd IEEE Conf. Dec. Control*, 1983.
77. L. Ljung, "Analysis of recursive stochastic algorithms," *IEEE Trans. Aut. Control*, vol. AC-22, pp. 551–575, Aug. 1977.

78. C. E. Rohrs, L. Valavani, M. Athans, and G. Stein, "Robustness of adaptive control algorithms in the presence of unmodeled dynamics," *Proc. 21st IEEE Conf. Dec. Control*, 1982.

79. B. Riedle and P. V. Kokotovic, "A stability-instability boundary for disturbance-free slow adaptation and unmodeled dynamics," *Proc. 23rd IEEE Conf. Dec. Control*, 1984.

80. P. V. Kokotovic, B. Riedle, and L. Praly, "On a stability criterion for continuous slow adaptation," *Syst. Control Lett.*, vol. 6, pp. 7–14, June 1985.

81. L. C. Fu, M. Bodson, and S. S. Sastry, "New stability theorems for averaging and their applications to the convergence analysis of adaptive identification and control schemes," *Proc. 24th IEEE Conf. Dec. Control*, 1985.

82. R. L. Kosut, B. D. O. Anderson, and I. M. Y. Mareels, "Stability theory of adaptive systems: methods of averaging and persistent excitation," *Proc. 24th IEEE Conf. Dec. Control*, 1985.

83. R. D. Nussbaum, "Some remarks on a conjecture in parameter adaptive control," *Syst. Control Lett.*, vol. 3, pp. 243–246, 1983.

84. D. R. Mudgett and A. S. Morse, "Adaptive stabilization of linear systems with unknown high frequency gains," *IEEE Trans. Aut. Control*, vol. AC-30, pp. 549–554, June 1985.

85. J. C. Willems and C. J. Byrnes, "Global adaptive stabilization in the absence of information on the sign of the high frequency gain," *Proc. INRIA Conf. on Analysis and Optimization of Systems, Lecture Notes in Control and Information Sciences*, vol. 62, Springer-Verlag, Berlin, 1984, pp. 49–57.

86. A. S. Morse, "A three dimensional universal controller for the adaptive stabilization of any strictly proper minimum-phase system with relative degree not exceeding two," *IEEE Trans. Aut. Control*, vol. AC-30, pp. 1188–1191, 1985.

87. A. S. Morse, "High gain feedback algorithms for adaptive stabilization," *Proc. 5th Yale Workshop on Applications of Adaptive Systems Theory*, Yale University, New Haven, CT, 1987.

88. G. Kreisselmeier and M. C. Smith, "Stable adaptive regulation of arbitrary nth-order plants," *IEEE Trans. Aut. Control*, vol. AC-31, pp. 299–305, 1986.

89. B. Martensson, "The order of any stabilizing regulator is sufficient for adaptive stabilization," *Syst. Control Lett.*, vol. 6, pp. 87–91, July 1985.

90. G. C. Goodwin and R. S. Long, "Generalization of results on multivariable adaptive control," *IEEE Trans. Aut. Control*, vol. AC-25, pp. 449–456, 1980.

91. H. Elliott and W. A. Wolovich, "Parameter adaptive control of linear multivariable systems," *IEEE Trans. Aut. Control*, vol. AC-27, pp. 340–352, 1982.

92. R. P. Singh and K. S. Narendra, "Prior information in the design of multivariable adaptive controllers," *IEEE Trans. Aut. Control*, vol. AC-29, pp. 1108–1111, Dec. 1984.

93. A. Isidori, *Nonlinear Control Systems*, 2nd ed., Springer-Verlag, Berlin, 1989.

94. J.-B. Pomet and L. Praly, "Adaptive nonlinear regulation: equation error from the Lyapunov equation," *Proc. 28th IEEE Conf. Dec. Control*, 1989.

95. I. Kanellakopoulos, P. V. Kokotovic, and R. H. Middleton, "Observer-based adaptive control of nonlinear systems under matching conditions," *Proc. 1990 Amer. Control Conf.*

96. L. Hsu, "Variable structure model-reference adaptive control (VS-MRAC) using only input and output measurements: part II," *Proc. 27th IEEE Conf. Dec. Control*, 1988.

97. L. C. Fu, "A new robust model reference adaptive control using variable structure adaptation for plants with relative degree two," *Proc. 1990 Amer. Control Conf.*
98. K. S. Narendra and J. D. Boskovic, "A combined direct, indirect and variable structure method for robust control," *IEEE Trans. Aut. Control*, to appear.
99. M. A. Duarte and K. S. Narendra, "A new approach to model reference adaptive control," *Int. J. Adapt. Control Sig. Process.*, vol. 3, pp. 53–73, 1989.

A Conceptual Framework for Parameter Adaptive Control*

A. S. Morse

Department of Electrical Engineering
Yale University
P. O. Box 1968
New Haven, CT, 06520-1968,USA.

Abstract. A conceptual framework is described in which a parameter adaptive control system is taken to be the feedback interconnection of a process Σ_P and a parameterized controller $\Sigma_C(k)$ whose parameter vector k is adjusted by a tuner Σ_T. The framework is general enough to encompass almost all parameterized controllers proposed in the literature for stabilizing linear process models. Emphasis is placed on the importance to adaptation of one of Σ_C's outputs called a *tuning error* e_T, which is the main signal driving Σ_T. For the closed-loop parameterized system $\Sigma(k)$ consisting of Σ_P and $\Sigma_C(k)$, definitions and characterizations are given of the concepts of *weak tunability* and *tunability* of $\Sigma(k)$ on a subset $\mathcal{E}$ of the parameter space $\mathcal{P}$ in which k takes values. It proves to be necessary to know a subset $\mathcal{E}$ on which $\Sigma(k)$ is weakly tunable in order to be able to construct a tuner Σ_T which adaptively stabilizes $\Sigma(k)$. For a large class of linear multivariable process models, a properly designed certainty equivalence controller results in a tunable closed-loop parameterized system. The importance of this result to both the analysis and synthesis of parameter adaptive control systems is discussed. It is demonstrated by means of examples how the connection between certainty equivalence and tunability, together with the concept of tunability itself, can be used to markedly simplify the stability analysis of adaptive control systems. A new family of indirect parameterized controllers is described which have capabilities comparable to those of the well-known direct parameterized controllers which for a long time have served as basic building blocks for adaptive control systems of all types. The concept of implicit tuning is formalized and its potential importance to adaptive control is briefly discussed.

Introduction

In a series of recent papers [1,2], a conceptual framework is introduced for defining, discussing and analyzing parameter adaptive control systems of all types. The idea is to think of an adaptive control system as the feedback interconnection of a process Σ_P and a "parameterized controller" $\Sigma_C(k)$ whose parameter vector k is adjusted by a "tuner" Σ_T. A parameterized controller is a parameter-dependent dynamical system whose outputs include not only a control signal u_C

* This research was supported by the National Science Foundation under grant ECS-9012551.

which in closed loop serves as the feedback control to the process, but also a suitably defined "tuning error" e_T which during adaptation drives the tuner Σ_T. Depending on the application, e_T might be an identification error, an augmented error, a process output, or something similar. In this context, a tuner is any algorithm driven by e_T which adjusts k. No matter how e_T and Σ_T are defined, the purpose of Σ_T is always the same: to tune k to make e_T "small". The aim of this paper, which is an abbreviated version of [1] and [2], is to discuss some of the implications of thinking of an adaptive control system from this point-of-view.

Section 1 briefly summarizes the salient features of a parameter adaptive control system viewed in the aforementioned way. In Section 2 attention is focused on properties of the subsystem $\Sigma(k)$ consisting of Σ_P in feedback with $\Sigma_C(k)$. The concept of "tunability" is defined and discussed. Under mild conditions, tunability of Σ on a known subset $\mathcal{E}$ of the space in which k takes values, proves to be necessary for adaptive stability. The problem of choosing Σ_C so that Σ has this property is addressed in Section 3 for the case when Σ_C is "identifier-based". By an identifier-based parameterized controller Σ_C is meant a system consisting of an "identifier" Σ_I and an "internal regulator" or "certainty equivalence controller" Σ_R. Σ_R is chosen off-line to endow a suitable defined "design model" Σ_D with desirable features (e.g., internal stability) for each fixed value of the design model's parameter vector p_I. The design model is presumed to have a transfer matrix which is "close" to that of the process model, for some unknown value of p_I. The coefficient matrices defining Σ_D are related to those of Σ_I by an appropriately defined, parameter-dependent, linear transformation. By making use of this transformation, it is possible to prove that if the internal regulator stabilizes Σ_D for some set of possible design model parameter values $\mathcal{E}$, then the closed-loop system Σ consisting of Σ_P, Σ_I and Σ_R is tunable on $\mathcal{E}$–this is the Certainty-Equivalence Stabilization Theorem. For the case when Σ_P is minimum phase, to achieve tunability on $\mathcal{E}$, it turns out to be enough to require Σ_R only to "output-stabilize" Σ_D on $\mathcal{E}$. This result, called the Certainty Equivalence Output Stabilization Theorem, is the key technical link needed to explain how to achieve with indirect control those results achieved previously with direct adaptive control.

The preceding theorems, which can each be proved using very elementary arguments, have many useful consequences. Since in their proofs there is no presumption of process model matching, the theorems can be applied without having to talk about "true process model parameter values", an idea which loses much of its meaning as soon as the process under consideration is assumed to have unmodeled dynamics. Since the theorems are parameterization independent, it is possible to use them to analyze directly and indirectly parameterized adaptive controllers in essentially the same way. Since the theorems enable one to sidestep much of the extraneous, detailed structure of error models in studying system behavior, it is possible to dramatically simplify the stability analyses of many adaptive control systems. An example which demonstrates this is given at the end of Section 2.

To make things explicit, examples are given in Section 3, which show how to apply the preceding ideas to SISO parameterized design models of both the direct and indirect control types. To output stabilize the latter it is necessary to restrict its parametric structure; what results is a parameterized design model which, together with its output stabilizing internal regulator, provides a new family of indirect, parameterized controllers, with capabilities comparable to those of the well-known direct parameterized controllers which for a long time have served as basic building blocks for adaptive control systems of all types [3–8].

Section 4 formalizes the concept of implicit tuning and explains how to analyze implicitly tuned systems using the ideas developed in Section 3. By implicit tuning is meant the adjustment of an identification parameter vector k_{I} and an implicitly defined internal regulator parameter vector k_{R} to make both a tuning error (e.g., an identification error) and a suitably defined "design" error simultaneously small. Implicit tuning algorithms can be either of the "one level" or "two level" type. We present examples of each and briefly discuss some of their possible generalizations.

In the sequel, prime denotes transpose, $\mathbb{R}^n$ is the real linear space of n-component column vectors x, and $||x||$ is the norm $\sqrt{x'x}$.

1 Framework

Classically [3–8], parameter adaptive control systems have been defined and discussed in terms of error models in which parameters typically enter linearly. It turns out that for many adaptive algorithms (e.g., [9–13]) error models play no essential role and parameters enter nonlinearly. Hence, a new conceptual framework in needed to describe both the classical and more recent adaptive structures. To construct such a framework it is useful to view a parameter adaptive control system as the feedback interconnection of a process Σ_{P} and a parameterized controller $\Sigma_{\mathrm{C}}(k)$ whose parameter vector k is adjusted by a tuner Σ_{T}. The *process* is a dynamical system with control input $u \in \mathbb{R}^{n_u}$, disturbance input $w \in \mathbb{R}^{n_w}$ and measured output $y \in \mathbb{R}^{n_y}$. The *parameterized controller* is a dynamical system $\Sigma_{\mathrm{C}}(k)$ depending on a control parameter k which takes values in some parameter space $\mathcal{P} \subset \mathbb{R}^{n_p}$. The inputs to Σ_{C} are the process open-loop control u, the process output y, and possibly a reference input r. The outputs generated by Σ_{C} are a *tuning error* e_{T} which drives Σ_{T}, a control signal u_{C} which becomes the feedback control to the process when u is set equal to u_{C}, and a vector of supplementary tuning data d consisting of known functions of r, y and the parameterized controller's state. The *tuner* is an algorithm $\Sigma_{\mathrm{T}}(k_0)$ *initialized* by k_0 (i.e., $k(0) = k_0$) with inputs d and e_{T} and output k, $k(t)$ being the "tuned-value" of k at time t; $\Sigma_{\mathrm{T}}(k_0)$ is presumed to be well-defined for each $k_0 \in \mathcal{P}$.

The function of Σ_{T} is to adjust k to make e_{T} "small" in some sense. Although the specifics of Σ_{T} may vary greatly from algorithm to algorithm, in most instances tuning is carried out in one of two fundamentally different ways, depending on whether $\mathcal{P}$ is countable or not. For the countable case (e.g., see

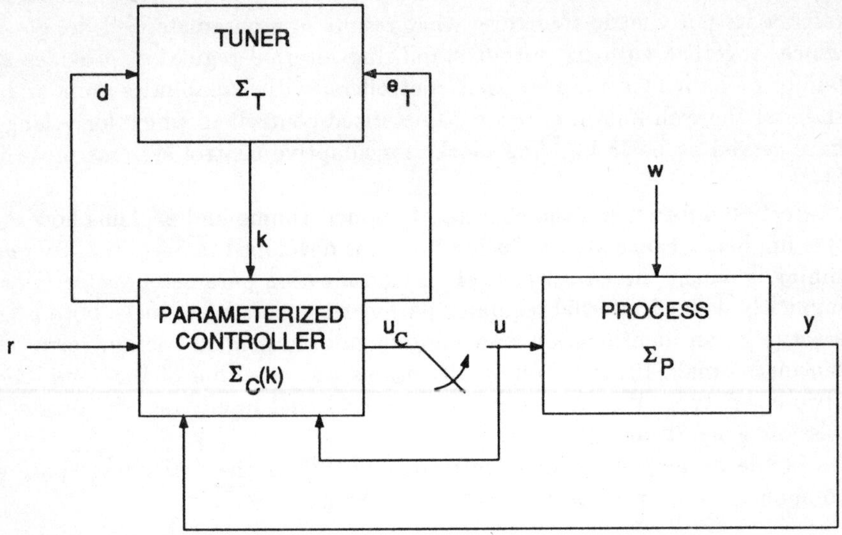

Fig. 1. Parameter adaptive control system.

[10,13]), tuning is achieved by sequentially stepping k through $\mathcal{P}$ along a predetermined path, using on-line (i.e., real-time) data to decide only when to switch k from one value along the path to the next. In contrast, for the uncountable case (e.g., see [3–8]) the path in $\mathcal{P}$ along which k is adjusted is not determined off-line but instead is computed in real time from "gradient-like" data. The main advantage of countable search algorithms over gradient-like procedures appears to be their broader applicability. On the other hand, when applicable, gradient-like algorithms are likely to exhibit far superior performance, but so far this has not been clearly demonstrated.

Since a tuner Σ_T is an algorithm driven by e_T, Σ_T will typically possess certain "rest" or "equlibrium" values of k at which tuning ceases if $e_T = 0$. To be more precise, let us agree to say that in open-loop, a tuner Σ_T is at *equilibrium value* $p_0 \in \mathcal{P}$ at time $t_0 \geq 0$, if $k(t_0) = p_0$ and if for the input $e_T(t) = 0$, $t \geq t_0$, $k(t)$ remains fixed at p_0 for $t \geq t_0$. In the sequel we assume that $\Sigma_T(k_0)$ is stationary to the extent that its possible equilibrium values at t_0 are independent of t_0 and $k_0 \in \mathcal{P}$; and we define the tuner's *equilibrium set* $\mathcal{E}_T$ to be the set of all such values in $\mathcal{P}$.

One fairly general algorithm for a tuner might be a dynamical system of the form

$$\dot{k} = \pi_1(k) + \pi_2(k, d)e_T$$
$$k(0) = k_0 \,, \tag{1}$$

where the $\pi_i(\cdot)$ are nonlinear functions and $k_0 \in \mathrm{IR}^{n_p}$. In this case $\mathcal{P}$ can be taken as $\mathbb{R}^{n_p}$ and $\mathcal{E}_T = \{p : \pi_1(p) = 0\}$. Alternatively, Σ_T might be a switching algorithm of the form [13]

$$k(t) = \begin{cases} k_0, & t \in [t_0, t_1) \\ h(i), & t \in [t_i, t_{i+1}), \quad i \geq 1, \end{cases}$$

where h is a function from the positive integers to $\mathbb{R}^{n_p}$, $k_0 \in$ image $\{h\}$, $t_0 = 0$, and for $i > 0$

$$t_i = \begin{cases} \min_{t > t_{i-1}} \{t : \int_0^t \|e_T\| d\tau = i!\}, & \text{if this set is nonempty,} \\ \infty, & \text{otherwise.} \end{cases}$$

In this case $\mathcal{P}$ is the image of h (which is countable), and $\mathcal{E}_T = \mathcal{P}$. The ideas which follow apply to both types of algorithms. Thus, unless otherwise stated, no special assumptions will be made about Σ_T, other than that it be stationary and possess a nonempty equlibrium set.

In this paper we assume w to be zero, since the presence or absence of disturbances is not important for the points we want to make. We take the process model $\Sigma_P = (C_P, A_P, B_P)$ to be a member of some known class $\mathcal{C}_P$ of stabilizable, detectable, time invariant, continuous-time linear systems; i.e.,

$$\begin{aligned} \dot{x}_P &= A_P x_P + B_P u \\ y &= C_P x_P \,. \end{aligned} \tag{2}$$

As a parameterized controller $\Sigma_C(k)$, we shall consider a system of the form

$$\begin{aligned} \dot{x}_C &= A_C(k) x_C + B_C(k) y + B_r(k) r + B_u(k) u \\ u_C &= F_C(k) x_C + G_C(k) y + G_r(k) r \\ e_T &= C_C(k) x_C + D_C(k) y \,, \end{aligned} \tag{3}$$

where $A_C(\cdot), B_C(\cdot), \cdots, D_C(\cdot)$ are matrix-valued functions on $\mathcal{P}$. An equation for d is not needed for the discussion which follows.

2 Tunability

The *closed-loop parameterized system* $\Sigma(k)$ determined by (2), (3) and the feedback connection

$$u = u_C \,, \tag{4}$$

can be concisely denoted by the equations

$$\begin{aligned} \dot{x} &= A(k) x + B(k) r \\ e_T &= C(k) x \,, \end{aligned} \tag{5}$$

where $x = [x_P', x_C']'$. Here $A(\cdot), B(\cdot)$, and $C(\cdot)$ are defied in the obvious way using (2)–(4). In this setting, the following question arises: What must be true of Σ in order for there to exist a tuning algorithm Σ_T for which the closed-loop adaptive system, consisting of Σ_T and Σ, is "stable"? In the sequel we provide some answers to this question.

With $\mathcal{E}$ any fixed, nonempty subset of Σ's parameter space $\mathcal{P}$, let us agree to call (5) *weakly tunable on* $\mathcal{E}$, if for each fixed $p \in \mathcal{E}$ and each bounded, piecewise-continuous exogenous input $r : [0,\infty) \rightarrow \mathbb{R}^{n_r}$, every possible system trajectory for which $k(t) = p$ and $e_T(t) = 0$, $t \in [0,\infty)$, is bounded on $[0,\infty)$. Call Σ *tunable on* $\mathcal{E}$ if for each $p \in \mathcal{E}$, x goes to zero as $t \rightarrow \infty$ along each trajectory on which $k(t) = p$ and both e_T and r equal zero, $t \in [0,\infty)$.

Remark 1: It is easy to verify that Σ is weakly tunable on $\mathcal{E}$ just in case, for each $p \in \mathcal{E}$, the matrix pair $(C(p), A(p))$ is weakly detectable[2] and the matrix pair obtained by restricting $C(p)$ and $A(p)$ to the controllable space of $(A(p), B(p))$ is detectable. Similarly, $(C(p), A(p))$ is tunable on $\mathcal{E}$ if and only if $(C(p), A(p))$ is detectable for each $p \in \mathcal{E}$. Thus, tunability of Σ on $\mathcal{E}$ implies weak tunability of Σ on $\mathcal{E}$ and is equivalent to weak tunability of Σ on $\mathcal{E}$ whenever Σ is controllable on $\mathcal{E}$.◇

Our aim here is to briefly explain why weak tunability is necessary for adaptive stabilization. To be specific, call $\Sigma_T(\cdot)$ an *unbiased stabilizer* of Σ if for each initialization $k_0 \in \mathcal{P}$, each bounded, piecewise-continuous input $r : [0,\infty) \rightarrow \mathbb{R}^{n_r}$ and each initial state $x(0)$, the state response x of $\Sigma(k)$, with k tuned by $\Sigma_T(k_0)$, is bounded on $[0,\infty)$.

Suppose Σ_T is a candidate tuner for $\Sigma(k)$. The definition of weak tunability implies that if $\Sigma(k)$ is not weakly tunable on Σ_T's equilibrium set $\mathcal{E}_T$, then for some exogenous input r, initial state $x(0)$, and parameter value $p_0 \in \mathcal{E}_T$, the untuned system $\Sigma(p_0)$ will admit an unbounded state response $x(t)$ along which $e_T(t) = 0$. If the same input r is applied to $\Sigma(k)$, with k tuned by $\Sigma_T(p_0)$, then clearly $k(t) = p_0$, $t \geq 0$, and $e_T(t) = 0$, $t \geq 0$. Therefore $\Sigma(k)$ will have exactly the same unbounded response to r that $\Sigma(p_0)$ has. We are led to the following

Theorem 1: *A necessary condition for the tuner $\Sigma_T(\cdot)$ to be an unbiased stabilizer of the tuned system $\Sigma(k)$, is that $\Sigma(k)$ be weakly tunable on the equilibrium set $\mathcal{E}_T$ of $\Sigma_T(\cdot)$.*

Clearly, weak tunability on $\mathcal{E}_T$ is a fundamental property that any parameter adaptive control system of the aforementioned general type must have if stability is to be assured.[3] An interesting problem then, is to determine what's required of

[2] A matrix pair (C, A) is *weakly detectable* if for each vector x for which $Ce^{At}x$ is bounded on $[0,\infty)$, it follows that $e^{At}x$ is bounded on $[0,\infty)$ as well. (C, A) is *detectable* if for each eigenvalue-eigenvector pair (λ, x), λ has a negative real part whenever $Cx = 0$. Detectability implies weak detectability but the converse is not necessarily true.

[3] For algorithms utilizing persistently exciting probing signals, weak tunability on $\mathcal{E}_T$ may well be more than is required for stability.

a process model Σ_P and its parameterized controller Σ_C for the resulting closed-loop system Σ to be weakly tunable or tunable on some given subset $\mathcal{E} \subset \mathcal{P}$. This problem is discussed further in Sect. 3. In the sequel we give some examples of tunable and untunable systems.

Example 1: Suppose for Σ_P, we take the one-dimensional system

$$\dot{y} = ay + gu , \tag{6}$$

with a and g unknown constants satisfying $a > 0$ and $g \neq 0$. To stabilize this system, consider using a control law of the form

$$u = \widehat{f}y , \tag{7}$$

where, if we had our d'ruthers, we'd choose $\widehat{f}$ so that $a + g\widehat{f} = -1$ since this would stabilize (6); but since a and g are unknown, we might instead try to choose $\widehat{f}$ in accordance with the "certainty-equivalence principle" (cf. Sect. 3} so that

$$\widehat{a} + \widehat{g}\widehat{f} = -1 , \tag{8}$$

where $\widehat{a}$ and $\widehat{g}$ are estimates of a and g respectively. However, since standard identification algorithms may cause $\widehat{g}$ to pass through zero, to avoid the possibility of 'division by zero', in place of (8) consider tuning $\widehat{f}$ with the "gradient" adjustment law

$$\dot{\widehat{f}} = -\frac{1}{2}\frac{\partial e_D^2}{\partial \widehat{f}} = -\widehat{g}e_D , \tag{9}$$

where e_D is the "design error"

$$e_D = \widehat{a} + \widehat{g}\widehat{f} + 1 \tag{10}$$

(cf. Sect. 4). Finally, to construct estimates $\widehat{a}$ and $\widehat{g}$, observe from (6) that

$$y = (a+1)\bar{y} + g\bar{u} + \epsilon ,$$

where $\epsilon = e^{-t}(y(0) - (a+1)\bar{y}(0))$, and

$$\begin{aligned} \dot{\bar{y}} + \bar{y} &= y \\ \dot{\bar{u}} + \bar{u} &= u . \end{aligned} \tag{11}$$

Thus, to generate $\widehat{g}$ and $\widehat{a}$, it makes sense to use an algorithm driven by the "identification error"

$$e_I = (\widehat{a} + 1)\bar{y} + \widehat{g}\bar{u} - y , \tag{12}$$

since this results in the familiar error equation

$$e_I = \bar{y}(\widehat{a} - a) + \bar{u}(\widehat{g} - g) - \epsilon . \tag{13}$$

If a standard identification algorithm is used, identification ceases when $e_I = 0$, in which case $\widehat{a}$ and $\widehat{g}$ become constant. Viewing this algorithm together with

(9) and (10) as a tuner Σ_T with tuning error input $e_T = e_I$ and tuned parameter $k = [\widehat{g}, \widehat{a}, \widehat{f}]'$, Σ_T's equilibrium set will be

$$\mathcal{E}_T = \left\{[p_1, p_2, p_3]' : p_1(p_2 + p_1 p_3 + 1) = 0, [p_1, p_2, p_3]' \in \mathbb{R}^3\right\} .$$

The overall parameterized controller Σ_C is thus

$$\begin{aligned} \dot{\bar{y}} &= -\bar{y} + y \\ \dot{\bar{u}} &= -\bar{u} + u \\ e_T &= (\widehat{a} + 1)\bar{y} + \widehat{g}\bar{u} - y \\ u_C &= \widehat{f} y , \end{aligned} \tag{14}$$

and $\Sigma(k)$ is the closed-loop parameterized system, described by (6), (14) and the feedback connection

$$u = u_C . \tag{15}$$

Observe that the point $[0, a, 0]' \in \mathcal{E}_T$. It is easy to verify that for k fixed at this value, Σ admits the unbounded soultion $y = e^{at}$, $\bar{y} = e^{at}/(1+a)$, $\bar{u} = 0$, $e_T = 0$, so Σ is not weakly tunable on $\mathcal{E}_T$.◇

Example 2: In Example 1, Σ is not tunable on $\mathcal{E}_T$ because $\mathcal{E}_T$ contains points $[\widehat{g}, \widehat{a}, \widehat{f}]'$ at which $\widehat{a} + \widehat{g}\widehat{f} > 0$. It is possible to eliminate this problem and to achieve stability, if sign(g) is assumed known, by using in place of (9) the adjustment law

$$\dot{\widehat{f}} = -(\text{sign}(g))e_D ,$$

together with the tuning equations[4]

$$\begin{aligned} \dot{\widehat{a}} &= -e_D - \bar{y}e_T \\ \dot{\widehat{g}} &= -\widehat{f}e_D - \bar{u}e_T . \end{aligned}$$

In this case Σ_T's equilibrium set is precisely those points $[\widehat{g}, \widehat{a}, \widehat{f}]' \in \mathbb{R}^3$ for which (8) holds. It is straightforward to check that for k fixed at any point $[p_1, p_2, p_3]' \in \mathcal{E}_T$, $y(t) = p_1(\bar{u}(0) - p_3\bar{y}(0))e^{-t}$ along any solution $[y(t), \bar{y}(t), \bar{u}(t)]'$ to (6,14,15) for which $e_T(t) \equiv 0$. Since this, (14), and (15) imply that any such solution is bounded, Σ is now tunable on $\mathcal{E}_T$.◇

Example 3: Take $\mathcal{P} = \mathbb{R}$ and let Σ be any closed-loop parameterized system with $C(k) = [1, 0]$ and

$$A(k) = \begin{bmatrix} 1 & 1 \\ 0 & k \end{bmatrix} .$$

Since $(C(\cdot), A(\cdot))$ is detectable on $\mathcal{P}$, by Remark 1 Σ must be tunable on each nonempty subset $\mathcal{E} \subset \mathcal{P}$. In spite of this, observe that no matter how k is tuned, $\Sigma(k)$ can have an unbounded state response (e.g., with $r = 0$ and $x(0) = [1, 0]'$, $x(t) = e^t[1, 0]'$), so adaptive stabilization is impossible.◇

---

[4] Motivation for these tuning equations stems from (13) and the observation that e_D can be written as $e_D = (\widehat{a} - a) + \widehat{f}(\widehat{g} - g) + g(\widehat{f} - f)$ [14]; this is an example of "single level implicit tuning" (cf. Sect. 4).

The preceding example shows that tunability of Σ on a known subset $\mathcal{E} \subset \mathbb{R}^{n_p}$, is not sufficient to insure that there is a tuner Σ_T which can adaptively stabilize $\Sigma(k)$. However, using the ideas from [10], it can be shown that if Σ is tunable on $\mathcal{E}$ and, in addition $\mathcal{E}$ contains a countable subset $\mathcal{E}^*$ containing a parameter value p_0 for which $\Sigma(p_0)$ is internally stable, then without knowing p_0 it is possible to construct a switching algorithm Σ_T, with $\mathcal{E}_T = \mathcal{E}^*$, which is an unbiased stabilizer of $\Sigma(k)$. Thus for the case when $\mathcal{P} = \mathbb{R}^{n_P}$, to achieve adaptive stability with some tuner Σ_T, it's enough to design Σ_C so that $(C(\cdot), A(\cdot))$ is detectable on a known countable subset $\mathcal{E}^* \subset \mathbb{R}^{n_p}$ containing a point p_0 which stabilizes $A(p_0)$. The following example shows that this is very easy to do without assuming very much about Σ_P.

Example 4: For a fixed integer $n > 0$, define the parameter vector $p = [p_1, p_2, \ldots, p_{2n+1}]'$, and parameterized polynomials $\beta(p,s) = s^n + p_n s^{n-1} + \ldots p_2 s + p_1$, $\alpha(p,s) = p_{2n+1}s^n + p_{2n}s^{n-1} + \ldots p_{n+2}s + p_{n+1}$. Choose $\gamma(s)$ to be any monic stable polynomial of degree n and let $\Sigma_1(p) = (A_1(p), b_1(p), c_1(p), d_1(p))$ and $\Sigma_2(p) = (A_2(p), b_2(p), c_2(p), d_2(p))$ be n-dimensional realizations of $\alpha(p,s)/\gamma(s)$ and $\gamma(s)/\beta(p,s)$ respectively. Define $\Sigma_C(k)$ to be the cascade interconnection of $\Sigma_2(k)$ with $\Sigma_1(k)$ as shown in Fig. 2.

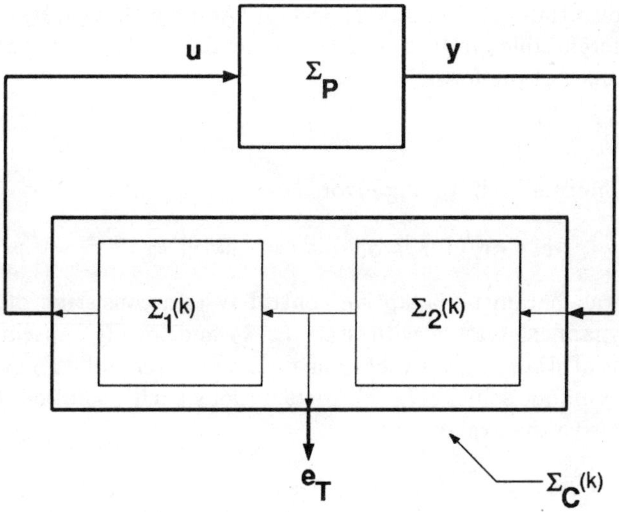

Fig. 2. Parameterized system $\Sigma(k)$.

Fix k at any value $p \in \mathbb{R}^{2n+1}$. Observe that along any loop-response for which e_T is identically zero, both u and y must go to zero since $\Sigma_1(p)$ is stable and $\Sigma_2(p)$ has a stable proper inverse. From this and the standing assumption that Σ_P is stabilizable and detectable it follows that for each $p \in \mathbb{R}^{2n+1}$, $\Sigma(p)$ must be detectable through e_T;[5] in other words, Σ is tunable on $\mathcal{P} = \mathbb{R}^{2n+1}$. Let $\mathcal{E}^*$ be a countable, dense subset of $\mathcal{P}$; clearly, Σ is also tunable on $\mathcal{E}^*$. Moreover, if in the linear space $\mathbb{R}^{1\times n_P} \oplus \mathbb{R}^{n_P \times n_P} \oplus \mathbb{R}^{n_P \times 1}$, Σ_P is sufficiently close to a system which is stabilizable, detectable and of McMillan degree not exceeding n, then there must exist a vector $p_0 \in \mathcal{E}^*$ for which $A(p_0)$ is stable. Thus $\mathcal{E}^*$ and Σ will have what's required for adaptive stabilization as long as Σ_P is close enough to a stabilizable, detectable system with McMillan degree no greater than n.◇

Before concluding this section, we illustrate by means of an example how the concept of tunability can be used in the analysis of an overall adaptive control system. Although in the example which follows, the tuner is of a very simple form, the reasoning can be extended, without major modification, to handle the most general gradient-like tuning algorithms which have been considered so far in the literature [1].

Example 5: Consider the widely studied situation in which the tuning error e_T in (5), is a scalar-valued identification error which can be reexpressed in the form

$$e_T = x_C'(k - p_0) + \epsilon\,, \tag{16}$$

where p_0 is a fixed but unknown vector in $\mathcal{P} = \mathbb{R}^{n_p}$ and ϵ is a linear combination of decaying exponentials (cf. Remark 2, Sect. 3). Assume that A, B, and C are continuously differentiable on $\mathcal{P}$, that $\Sigma(k)$ is tunable on $\mathcal{P}$, and that Σ_T is a "gradient-like" tuner of the form

$$\dot{k} = -x_C \bar{e}_T\,, \tag{17}$$

where $\bar{e}_T$ is the "normalized" tuning error

$$\bar{e}_T = e_T/(1 + ||x_C||^2) = e_T - ||x_C||^2 \bar{e}_T\,. \tag{18}$$

Because the overall parameter adaptive control system consisting of (5), (17) and (18) is a dynamical system with state (x, k) and locally Lipschitz vector field, for each initial state (x_0, k_0) there must exist an interval $[0, T)$ of maximal length on which a unique solution (x, k) exists.[6] Along such a solution, (16)–(18) can be used to derive the expressions

$$\begin{aligned} d\left(||k - p_0||^2\right)/dt &= -2((1 + ||x_C||^2)\bar{e}_T^2 - \epsilon \bar{e}_T) \\ &\le -(1 + 2||x_C||^2)\bar{e}_T^2 + \epsilon^2\,. \end{aligned}$$

[5] It is interesting to note that this will no longer be true if Σ_1 and Σ_2 are interchanged, unless Σ_P is restricted to be minimum phase.

[6] Since there is no *global* existence theory for differential equations whose vector fields are only *locally* Lipschitz, it cannot logically be concluded that $T = \infty$, unless (for example) it can *first* be established that (x, k) is bounded wherever it exists (cf. [15]).

Rearranging terms and integrating one obtains

$$||k - p_0||^2 + \int_0^T (1 + 2||x_C||^2)\bar{e}_T^2 dt \leq c_0 ,$$

where $c_0 = ||k(0) - p_0||^2 + \int_0^\infty \epsilon^2 dt$. Since $c_0 < \infty$, it follows that on $[0, T)$ k is bounded and $\bar{e}_T$ and $\dot{k}$ are square integrable; moreover, the boundedness of k together with (16) and (18) imply that $\bar{e}_T$ is bounded as well.

Up to here the analysis is straightforward and completely standard. To get beyond this point and show that x is bounded has in the past been a challenging matter requiring intricate, lengthy arguments. It is exactly here where the concept of tunability proves its worth. To show that x is bounded, we need two technical facts:

1. Because of tunability, there exists a matrix $H(p)$ which stabilizes $A(p) + H(p)C(p)$ for each fixed $p \in \mathcal{P}$; moreover, because C and A are continuously differentiable, H can be constructed (using a Riccati equation) to be continuously differentiable as well (cf. [1]).
2. The matrix $A_*(t) = A(k(t)) + H(k(t))(C(t) + [0, \dot{k}(t)'])$ is exponentially stable on [0,T); i.e., the state transmission matrix $\Phi(t, \tau)$ of $A_*(t)$ satisfies $||\Phi(t, \tau)|| \leq c_1 e^{-\lambda(t-\tau)}, 0 \leq \tau \leq t < T$, where λ and c_1 are positive constants depending on H and c_0. This result, which depends only on $A + HC$ being a continuously differentiable stability matrix, k being bounded and $\dot{k}$ being square integrable, is well-known and can easily be proved using a standard argument along the lines of [16,17,1].

Armed with these technical facts, a proof that x is bounded is at hand: All one has to do is to use the expression $\bar{e}_T = (C(k) + [0, \dot{k}'])x$, which is a consequence of (5), (17) and (18), to rewrite (5) as $\dot{x} = A_*(t)x - H(k(t))\bar{e}_T$. Since A_* is exponentially stable and H and $\bar{e}_T$ are bounded, x can be viewed as the output of an exponentially stable linear system with bounded input $-H\bar{e}_T$. Therefore x must be bounded wherever it exists. From this, the boundedness of k and the hypothesis that $[0, T)$ is the existence interval of *maximal* length it follows that $T = \infty$ and thus that both x and k must exist and be bounded on $[0, \infty)$. The reader may wish to compare the simplicity of this reasoning, with that used previously in [4,6,7,8] and elsewhere.◇

3 Identifier-Based Parameterized Controllers

For the purposes of this paper, a parameterized controller $\Sigma_C(k)$ of the "identifier-based" type consists of two subsystems, one an "identifier" $\Sigma_I(k_I)$ depending on an identification parameter vector k_I, the other an "internal regulator" $\Sigma_R(k)$ depending on a control parameter vector k. Typically $k = k_I$; in some cases however, $k = [k_I', k_R']'$ where k_R is a vector of additional internal regulator parameters (cf. Sect. 4). Depending on how Σ_I and Σ_R are defined, Σ_C may be called either a "direct" or "indirect" controller.

Although not always explicitly stated, the starting point for the development of an identifier-based parameterized controller is typically the selection of an identifier parameter space $\mathcal{P}_I \subset \mathbb{R}^{n_{p_I}}$ and a *design model* $\Sigma_D(p_I)$ upon which the design of Σ_C is to be based. $\Sigma_D(p_I)$ is chosen to have the same control input and measured output value spaces as the unknown process model Σ_P to be controlled, and also so that the process model transfer matrix from u to y can be presumed to be "close" in some sense to the transfer matrix $T_D(s, p_I)$ of $\Sigma_D(p_I)$ for some unknown $p_I \in \mathcal{P}_I$.[7] The reduced transfer matrices of $\Sigma_D(p_I)$ as p_I ranges over $\mathcal{P}_I$, are the transfer matrices which $\Sigma_D(p_I)$ *matches*. It is of considerable practical importance to develop adaptive stabilizers for design models whose classes of matchable transfer matrices are as "large" as possible.

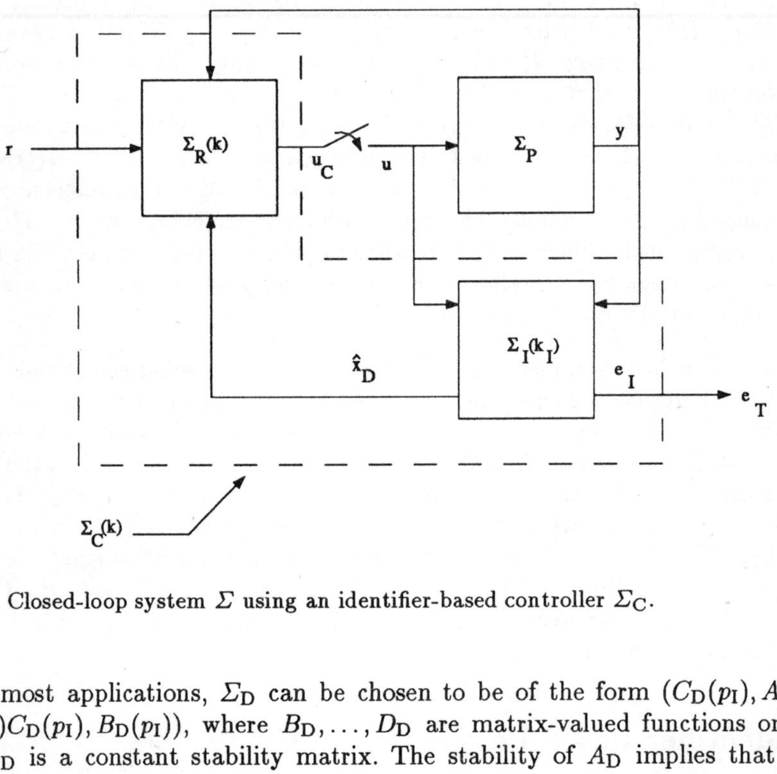

Fig. 3. Closed-loop system Σ using an identifier-based controller Σ_C.

In most applications, Σ_D can be chosen to be of the form $(C_D(p_I), A_D + D_D(p_I)C_D(p_I), B_D(p_I))$, where $B_D, \ldots, D_D$ are matrix-valued functions on $\mathcal{P}_I$ and A_D is a constant stability matrix. The stability of A_D implies that Σ_D is detectable on $\mathcal{P}_I$. On the other hand, *stabilizability* of Σ_D on $\mathcal{P}_I$ generally *cannot* be guaranteed unless special constraints are placed on $\mathcal{P}_I$ [16,17] or Σ_D is nonlinearly parameterized [18]. This has important implications which are well known [3,5,6,7,8] and which will be discussed briefly in the sequel.

[7] By exploiting the concept of a design model, it is possible to discuss adaptive control algorithms without being too specific about Σ_P ; this is of course consistent with the idea that a process model can never be known exactly.

Because of the particular structure of $\Sigma_{\mathrm{D}}(p_{\mathrm{I}})$, it is a routine matter to construct a matrix-valued function $E_{\mathrm{I}}(p_{\mathrm{I}})$ and constant matrices $A_{\mathrm{I}}, B_{\mathrm{I}}$, and D_{I}, with A_{I} stable, so that the *estimation equations*

$$\begin{aligned} E_{\mathrm{I}}(p_{\mathrm{I}})A_{\mathrm{I}} &= A_{\mathrm{D}}E_{\mathrm{I}}(p_{\mathrm{I}}) \\ E_{\mathrm{I}}(p_{\mathrm{I}})B_{\mathrm{I}} &= B_{\mathrm{D}}(p_{\mathrm{I}}) \qquad p_{\mathrm{I}} \in \mathcal{P}_{\mathrm{I}} \\ E_{\mathrm{I}}(p_{\mathrm{I}})D_{\mathrm{I}} &= D_{\mathrm{D}}(p_{\mathrm{I}}) \end{aligned} \tag{19}$$

hold.[8] These matrices together with the matrix

$$C_{\mathrm{I}}(p_{\mathrm{I}}) = C_{\mathrm{D}}(p_{\mathrm{I}})E(p_{\mathrm{I}}) \qquad p_{\mathrm{I}} \in \mathcal{P}_{\mathrm{I}} \tag{20}$$

determine an *identifier* $\Sigma_{\mathrm{I}}(k_{\mathrm{I}})$ of the form

$$\dot{x}_{\mathrm{I}} = A_{\mathrm{I}}x_{\mathrm{I}} + D_{\mathrm{I}}y + B_{\mathrm{I}}u \tag{21}$$

$$e_{\mathrm{I}} = C_{\mathrm{I}}(k_{\mathrm{I}})x_{\mathrm{I}} - y \tag{22}$$

$$\widehat{x}_{\mathrm{D}} = E_{\mathrm{I}}(k_{\mathrm{I}})x_{\mathrm{I}}\,, \tag{23}$$

where x_{I} is the identifier's state, e_{I} is an *identification error*, and $\widehat{x}_{\mathrm{D}}$ is a "surrogate" design model state which is often used in constructing the feedback control to the process.[9] The estimation equations imply that if k_{I} were held fixed at p_{I} and u and y were replaced by design system input u_{D} and output y_{D} respectively, then Σ_{I} would be a state estimator for the design system

$$\begin{aligned} \dot{x}_{\mathrm{D}} &= (A_{\mathrm{D}} + D_{\mathrm{D}}(p_{\mathrm{I}})C_{\mathrm{D}}(p_{\mathrm{I}}))x_{\mathrm{D}} + B_{\mathrm{D}}(p_{\mathrm{I}})u_{\mathrm{D}} \\ y_{\mathrm{D}} &= C_{\mathrm{D}}(p_{\mathrm{I}})x_{\mathrm{D}}\,, \end{aligned} \tag{24}$$

in that x_{D} would be given by $\widehat{x}_{\mathrm{D}}(t) + e^{A_{\mathrm{D}}t}(x_{\mathrm{D}}(0) - \widehat{x}_{\mathrm{D}}(0))$.

Example 6: Indirect Control SISO Design Model - Pick $n > 0$ and choose (c, A, b) to be any n-dimensional, controllable, observable, SISO system with A stable. For each pair $p_1, p_2 \in \mathbb{R}^n$, define $p_{\mathrm{I}} = [p_1', p_2']'$, $A_{\mathrm{D}} = A$, $B_{\mathrm{D}}(p_{\mathrm{I}}) = p_2$, $C_{\mathrm{D}}(p_{\mathrm{I}}) = c$ and $D_{\mathrm{D}}(p_{\mathrm{I}}) = p_1$. Then with $\mathcal{P}_{\mathrm{I}} = \mathbb{R}^{2n}$, Σ_{D} is the n-dimensional design model described by the equations

$$\begin{aligned} \dot{x}_{\mathrm{D}} &= (A + p_1 c)x_{\mathrm{D}} + p_2 u_{\mathrm{D}} \\ y_{\mathrm{D}} &= c x_{\mathrm{D}}\,. \end{aligned} \tag{25}$$

[8] These matrices can always be chosen so that E_{I} depends linearly on the columns of B_{D} and D_{D} and also so that each eigenvalue of A_{I} is an eigenvalue of A_{D}. One method for doing this can easily be obtained by generalizing the construction in Example 6.

[9] Note that each identifier (21)-(23) uniquely determines a parameterized design model $(C_{\mathrm{I}}(p_{\mathrm{I}}), A_{\mathrm{I}} + D_{\mathrm{I}}C_{\mathrm{I}}(p_{\mathrm{I}}), B_{\mathrm{I}}(p_{\mathrm{I}}))$, called Σ_{I}'s *natural* design model, for which (19) and (20) hold with $E_{\mathrm{I}} = I$. Although this design model's class of matchable transfer matrices is exactly the same as that of the design model which originally determined Σ_{I}, the former can often have a state space dimension which is significantly larger than that of the latter.

This is one form of the parameterized SISO design model commonly used for indirect control. As defined, Σ_{D} is observable on $\mathcal{P}_{\mathrm{I}}$ and stabilizable at those values of $p_{\mathrm{I}} \in \mathcal{P}_{\mathrm{I}}$ for which the transfer function of Σ_{D} has no unstable pole-zero cancellations. In addition, Σ_{D} can match any transfer function with McMillan degree not exceeding n.

Define identifier matrices $A_{\mathrm{I}} =$ block diag $\{A, A\}$, $B_{\mathrm{I}} = [0', b']'$, $D_{\mathrm{I}} = [b', 0']'$. It is easy to check that estimation equations (19) hold with

$$E_{\mathrm{I}}(p_{\mathrm{I}}) = [S(p_1), S(p_2)]\,, \tag{26}$$

where $S(p) = [M_1Rp, M_2Rp, \ldots, M_nRp]'$, R is the inverse of the controllability matrix of (A, b), M_i is the transpose of the observability matrix of (e_i', A) and e_i is the i-th unit vector in $\mathbb{R}^n$. A simple calculation shows that

$$cS(p) = p'Q'\,, \tag{27}$$

where Q is the parameter transformation matrix $Q = N'R$ and N is the observability matrix of (c, A). Thus, for this design model, $C_{\mathrm{I}}(p_{\mathrm{I}}) = [p_1'Q', p_2'Q']$, which is linear in p_{I}.◇

Example 7: Direct Control SISO Design Model - Let n and (c, A, b) be as in Example 6. Pick $m > 0$ and choose (A_*, b_*) to be any single-input, m-dimensional, controllable pair with A_* stable. For each $p_1, p_2 \in \mathbb{R}^n$ and $p_3 \in \mathbb{R}^m$, define $p_{\mathrm{I}} = [p_1', p_2', p_3']'$, $B_{\mathrm{D}}(p_{\mathrm{I}}) = [p_2', b_*']'$, $C_{\mathrm{D}}(p_{\mathrm{I}}) = [0_{1\times n}, p_3']$, $D_{\mathrm{D}}(p_{\mathrm{I}}) = [p_1', 0_{1\times m}]'$, and

$$A_{\mathrm{D}} = \begin{bmatrix} A & 0 \\ b_*c & A_* \end{bmatrix}.$$

With $\mathcal{P}_{\mathrm{I}} = \mathbb{R}^{2n+m}$, $\Sigma_{\mathrm{D}}(p_{\mathrm{I}})$ is the $(n+m)$-dimensional parameterized design model

$$\begin{aligned} \dot{x}_{\mathrm{D}1} &= Ax_{\mathrm{D}1} + p_1p_3'x_{\mathrm{D}2} + p_2u_{\mathrm{D}} \\ \dot{x}_{\mathrm{D}2} &= b_*cx_{\mathrm{D}1} + A_*x_{\mathrm{D}2} + b_*u_{\mathrm{D}} \\ y_{\mathrm{D}} &= p_3'x_{\mathrm{D}2}\,, \end{aligned} \tag{28}$$

which, as will be shown in the sequel, is appropriate for direct control. As defined, Σ_{D} is detectable on $\mathcal{P}_{\mathrm{I}}$ and stabilizable at those values of $p_{\mathrm{I}} \in \mathcal{P}_{\mathrm{I}}$ for which the transfer function of Σ_{D} has no unstable pole-zero cancellations. The class $\mathcal{T}$ of transfer functions which Σ_{D} can match can easily be shown to contain all transfer functions of relative degree n^* and McMillan degree not exceeding n, provided $n^* \leq n$ and the characteristic polynomial of A_* has a real factor of degree n^*. Thus, by picking $m \leq n$ and choosing A_* to have only real eigenvalues, $\mathcal{T}$ will contain all strictly proper transfer functions with relative degree not exceeding m and McMillan degree not exceeding n. If $m \leq n+1$, $\mathcal{T}$ can also be shown to contain all strictly proper transfer functions of McMillan degree equal to m.

It is not difficult to verify that as an identifier $\Sigma_{\mathrm{I}}(k_{\mathrm{I}})$ for this design model, one can use a system described by the equations

$$\begin{aligned} \dot{z}_1 &= Az_1 + by \\ \dot{z}_2 &= Az_2 + bu \\ \dot{x}_* &= A_* x_* + b_* u \\ \dot{H}_1 &= A_* H_1 + b_* z_1' \\ \dot{H}_2 &= A_* H_2 + b_* z_2' \\ \widehat{x}_{\mathrm{D}1} &= S(k_1) z_1 + S(k_2) z_2 \\ \widehat{x}_{\mathrm{D}2} &= x_* + H_1 Q k_1 + H_2 Q k_2 \\ e_{\mathrm{I}} &= k_3'(x_* + H_1 Q k_1 + H_2 Q k_2) - y\,, \end{aligned} \tag{29}$$

where $S(\cdot)$ and Q are as in Example 6, and $k_{\mathrm{I}} = [k_1', k_2', k_3']'$. For a more motivated development of this type of identifier see [3]. Identifier equations (29) can of course be written in the form of (21)-(23), with x_{I} a column vector, using the definition $x_{\mathrm{I}} = [z_1', z_2', x_*', h_1', h_2', \ldots, h_{2n}']'$ where h_j is the j-th column of $H \equiv [H_1, H_2]$; calculation of matrices $A_{\mathrm{I}}, B_{\mathrm{I}}, D_{\mathrm{I}}$ and $E_{\mathrm{I}}(p_{\mathrm{I}})$ for which (19) holds is a straightforward, but tedious task. For example, $E_{\mathrm{I}}(p_{\mathrm{I}})$ is of the form $E_{\mathrm{I}}(p_{\mathrm{I}})$ =block diag $\{[S(p_1), S(p_2)], T(p_{\mathrm{I}})\}$ where $T(p_{\mathrm{I}}) = [I, -e_1' Q p_1 I, \ldots, -e_n' Q p_1 I, -e_1' Q p_2 I, \ldots, -e_n' Q p_2 I]_{m \times (2n+1)m}$ and e_i is the i-th unit vector in $\mathbb{R}^n$. The linear dependence of $E_{\mathrm{I}}(p_{\mathrm{I}})$ on $[p_1', p_2']'$ and $C_{\mathrm{D}}(p_{\mathrm{I}})$ on p_3 imply that $C_{\mathrm{I}}(p_{\mathrm{I}})$, as defined in (20), is a bilinear function of $[p_1', p_2']'$ and p_3.◇

Remark 2: Although not central to the purpose of this paper, at this point it nevertheless seems logical to digress briefly and give an alternative formula for e_{I}, which is of particular importance to the analysis of all identifier-based parameter adaptive control systems. For this, fix $p_0 \in \mathcal{P}_{\mathrm{I}}$ and let $\Sigma_0 = (C_0, A_0, B_0)$ be an n_0-dimensional, observable, stabilizable realization of the reduced transfer matrix of $\Sigma_{\mathrm{D}}(p_0)$. Let Ω_0 be an open connected set of observable systems in $\mathbb{R}^{n_y \times n_0} \oplus \mathbb{R}^{n_0 \times n_0} \oplus \mathbb{R}^{n_0 \times n_u}$ which contains Σ_0. The expression for e_{I} given by the following theorem is the key relationship upon which the analyses of most tuning algorithms are based.◇

Identification Error Theorem: *There exists an integer $n > 0$, and analytic functions $\mathbf{A} : \Omega_0 \rightarrow \mathbb{R}^{n \times n}, \mathbf{B} : \Omega_0 \rightarrow \mathbb{R}^{n \times n_u}, \mathbf{C} : \Omega_0 \rightarrow \mathbb{R}^{n_y \times n}$, and $\mathbf{D} : \Omega_0 \rightarrow \mathbb{R}^{n_y \times n_u}$ with the following properties:*

(i) spec $(\mathbf{A}(\Sigma)) \subset$ spec $(A_{\mathrm{I}}) \cup$ {uncontrollable spectrum of Σ_0}, $\forall \Sigma \in \Omega_0$.

(ii) $\mathbf{C}(\Sigma_0) e^{\mathbf{A}(\Sigma_0)t} [\mathbf{B}(\Sigma_0), \mathbf{D}(\Sigma_0)] = 0\,, \quad t \geq 0$

(iii) If $\Sigma_{\mathrm{P}} \in \Omega_0$, then for each $x_{\mathrm{P}}(0) \in \mathbb{R}^n$, $x_{\mathrm{I}}(0) \in \mathbb{R}^{n_{\mathrm{I}}}$ and each piecewise-continuous input $u : [0, \infty) \rightarrow \mathbb{R}^{n_u}$, e_{I} can be written as

$$e_{\mathrm{I}} = e_{\mathrm{MM}} + \epsilon + \delta\,, \tag{30}$$

where

$$e_{\mathrm{MM}} = (C_{\mathrm{I}}(k_{\mathrm{I}}) - C_{\mathrm{I}}(p_{\mathrm{O}}))x_{\mathrm{I}}$$
$$\epsilon = \mathbf{C}(\Sigma_{\mathrm{P}})(e^{\mathbf{A}(\Sigma_{\mathrm{P}})t}(M_{\mathrm{P}}x_{\mathrm{P}}(0) + M_{\mathrm{I}}x_{\mathrm{I}}(0))$$
$$\delta = \int_0^t \mathbf{C}(\Sigma_{\mathrm{P}})e^{\mathbf{A}(\Sigma_{\mathrm{P}})(t-\tau)}(\mathbf{B}(\Sigma_{\mathrm{P}})u(\tau) + \mathbf{D}(\Sigma_{\mathrm{P}})y(\tau))d\tau\,,$$

and M_{P} and M_{I} are constant matrices not depending on p_{O}.

A number of expressions for e_{I} similar to (30) have been derived previously [19,20,21] and are by now reasonably well known. Equation (30) shows that if $\Sigma_{\mathrm{P}} \in \Omega_0$,[10] e_{I} can be written as the sum of three terms, the first (e_{MM}), called the "mismatch error", is due to parameter discrepancy, the second (ϵ) accounts for nonzero initial process model and identifier states and the third (δ) appears because of unmodeled dynamics. In view of property (i), ϵ must decay to zero exponentially fast so long as $\Sigma_{\mathrm{P}} \in \Omega_0$. Assuming a known upper bound on the real part of the uncontrollable spectrum of Σ_0, property (i) and the expression for δ clearly make it possible to generate a "bounding" signal β (cf. [19]), using u and y, for which $\|\delta(t)\| \leq c_1(\Sigma_{\mathrm{P}}) + c_2(\Sigma_{\mathrm{P}})\|\beta(t)\|$, $\forall\, \Sigma_{\mathrm{P}} \in \Omega_0$; here $c_1(\cdot)$ and $c_2(\cdot)$ are positive-valued analytic functions on Ω_0 which, because of property (iii), satisfy $c_i(\Sigma_0) = 0$, $i = 1,2$. Such bounding signals have been used to construct "dynamically normalized" tuning errors in an effort to develop tuning algorithms capable of handling unmodeled dynamics [20,21]. The implications of (30) have not yet been fully exploited.◇

The other subsystem of $\Sigma_{\mathrm{C}}(k)$, called an *internal regulator* $\Sigma_{\mathrm{R}}(k)$, is often simply a state feedback law of the form

$$u_{\mathrm{C}} = F_{\mathrm{R}}(k_{\mathrm{I}})\hat{x}_{\mathrm{D}} + G_{\mathrm{R}}(k_{\mathrm{I}})r$$

where $F_{\mathrm{R}}(p_{\mathrm{I}})$ and $G_{\mathrm{R}}(p_{\mathrm{I}})$ are matrix-valued functions designed to endow the closed-loop design model $(C_{\mathrm{D}}(p_{\mathrm{I}}), A_{\mathrm{D}} + D_{\mathrm{D}}(p_{\mathrm{I}})C_{\mathrm{D}}(p_{\mathrm{I}}) + B_{\mathrm{D}}(p_{\mathrm{I}})F_{\mathrm{R}}(p_{\mathrm{I}}), B_{\mathrm{R}}(p_{\mathrm{I}})G_{\mathrm{R}}(p_{\mathrm{I}}))$ with some desired properties (e.g., internal stability) for each $p_{\mathrm{I}} \in \mathcal{P}_{\mathrm{I}}$. More generally, $\Sigma_{\mathrm{R}}(k)$ might be an n_{R}-dimensional dynamic compensator of the form

$$\begin{aligned} u_{\mathrm{C}} &= C_{\mathrm{R}}(k)\hat{x}_{\mathrm{R}} + G_{\mathrm{R}}(k)r + F_{\mathrm{R}}(k)\hat{x}_{\mathrm{D}} + M_{\mathrm{R}}(k)y \\ \dot{\hat{x}}_{\mathrm{R}} &= A_{\mathrm{R}}(k)\hat{x}_{\mathrm{R}} + B_{\mathrm{R}}(k)r + D_{\mathrm{R}}(k)\hat{x}_{\mathrm{D}} + N_{\mathrm{R}}(k)y\,, \end{aligned} \tag{31}$$

where $k = [k_{\mathrm{I}}', k_{\mathrm{R}}']'$ and k_{R} is a vector of internal regulator parameters which takes values in some parameter space $\mathcal{P}_{\mathrm{R}} \subset \mathrm{IR}^{n_{p_{\mathrm{R}}}}$. In this case, internal regulator

[10] Since Σ_{P} will typically lie in many such subsets, each with a different value of p_{O}, there is nothing especially unique about the particular value of p_{O} appearing in (30), in relationship to Σ_{P}.

matrices $C_{\mathrm{R}}, G_{\mathrm{R}}, \ldots, N_{\mathrm{R}}$ are chosen to give the *closed-loop design model* $\Sigma_{\mathrm{D_{cl}}}(p)$, consisting of the design model (24) in feedback with

$$\begin{aligned} u_{\mathrm{D}} &= C_{\mathrm{R}}(p)x_{\mathrm{R}} + G_{\mathrm{R}}(p)r + F_{\mathrm{R}}(p)x_{\mathrm{D}} + M_{\mathrm{R}}(p)y_{\mathrm{D}} \\ \dot{x}_{\mathrm{R}} &= A_{\mathrm{R}}(p)x_{\mathrm{R}} + B_{\mathrm{R}}(p)r + D_{\mathrm{R}}(p)x_{\mathrm{D}} + N_{\mathrm{R}}(p)y_{\mathrm{D}}\,, \end{aligned} \tag{32}$$

desired properties for each $p \in \mathcal{P} = \mathcal{P}_{\mathrm{I}} \times \mathcal{P}_{\mathrm{R}}$. In the sequel it is assumed, without loss of generality, that $(C_{\mathrm{R}}(p), A_{\mathrm{R}}(p))$ is a detectable pair for each $p \in \mathcal{P}$.

In many situations, the tuning error e_{T} is chosen to be the same as the identification error e_{I}. There are nevertheless cases (to be discussed briefly later) in which it is useful to define e_{T} in an alternative way. For the present, however, we assume

$$e_{\mathrm{T}} = e_{\mathrm{I}} \tag{33}$$

and view an identifier-based parameterized controller $\Sigma_{\mathrm{C}}(k)$ to be as depicted in Fig. 3.

The heuristic idea of designing an internal regulator to control a design model, and then using the regulator in conjunction with an identifier to control the process is sometimes called the "certainty equivalence principle." The resulting internal regulator Σ_{R} is accordingly called a *certainty-equivalence control.* Thus, an internal regulator is the same as a certainty equivalence controller; in the sequel the two terms are used interchangeably.[11]

Our aim is to make clear, *in algorithmically independent terms,* one of the fundamental implications of certainty equivalence control. For this, set $x_{\mathrm{D_{cl}}} = [x'_{\mathrm{R}}, x'_{\mathrm{D}}]'$ and define the closed-loop design model $\Sigma_{\mathrm{D_{cl}}}(p) = (C_{\mathrm{D_{cl}}}(p), A_{\mathrm{D_{cl}}}(p), B_{\mathrm{D_{cl}}}(p))$ so that the equations

$$\begin{aligned} \dot{x}_{\mathrm{D_{cl}}} &= A_{\mathrm{D_{cl}}}(p)x_{\mathrm{D_{cl}}} + B_{\mathrm{D_{cl}}}(p)r \\ y_{\mathrm{D}} &= C_{\mathrm{D_{cl}}}(p)x_{\mathrm{D_{cl}}} \end{aligned} \tag{34}$$

model the feedback interconnection of design model (24) with (32).

Σ_{R} is said to *stabilize* the design model Σ_{D} at p if $\Sigma_{\mathrm{D_{cl}}}$ is internally stable at p; i.e., if $A_{\mathrm{D_{cl}}}(p)$ is a stability matrix at p. Let $\mathcal{E}$ be a nonempty subset of $\mathcal{P}$. A certainty equivalence controller $\Sigma_{\mathrm{R}}(k)$ is called a *stabilizing controller on* $\mathcal{E}$ if for each $p = [p'_{\mathrm{I}}, p'_{\mathrm{R}}]' \in \mathcal{E}$, $\Sigma_{\mathrm{R}}(p)$ stabilizes design model Σ_{D} at p. We can now state the

Certainty Equivalence Stabilization Theorem: *If $\Sigma_{\mathrm{R}}(k)$ is a certainty equivalence, stabilizing controller on $\mathcal{E}$, then $\Sigma(k)$ is tunable on $\mathcal{E}$.*

[11] The definition of Σ_{R} in (31) might appear more in accordance with the intuitive idea of a certainty equivalence controller, if the terms involving $\widehat{x}_{\mathrm{D}}$ were not present. We've elected to include these terms since they actually do appear in the defining equations for many certainty equivalence controllers.

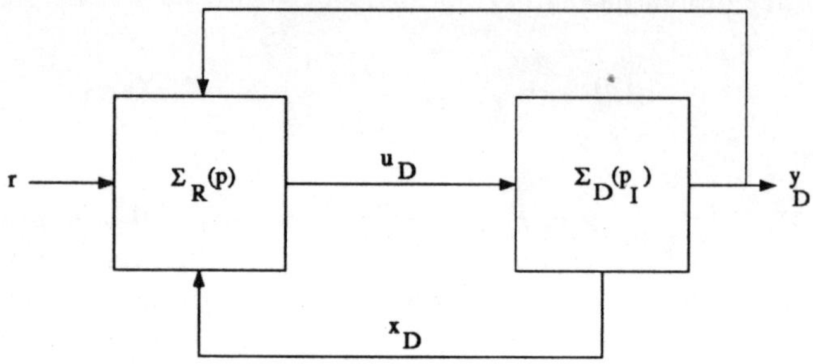

Fig. 4. Closed-loop design model $\Sigma_{D_{cl}}$.

The theorem makes clear one of the central implications of certainty equivalence control: If Σ_R is chosen to stabilize design model Σ_D on $\mathcal{E}$ (cf. Fig. 4), and is then used as a component subsystem of Σ_C, the overall parameterized system Σ (cf. Fig. 3), consisting of Σ_C in feedback with Σ_P, is tunable on $\mathcal{E}$. For this to be true it is *not* necessary for the process model Σ_P to have a transfer matrix in the class of transfer matrices matchable by Σ_D. In this context the notion of "true process model parameters" is neither necessary nor relevant.

It is usually not possible to make use of the preceding theorem to achieve tunability on all of $\mathcal{P}_I$ since $\mathcal{P}_I$ typically contains points at which Σ_D is not stabilizable. Various ideas proposed in the literature might be used to deal with this problem. One is to use a tuning algorithm which restricts k_I to a subset $\bar{\mathcal{P}}_I$ of $\mathcal{P}_I$ on which loss of stabilizability does not occur [16,17]; since this in effect reduces Σ_D's parameter space to $\bar{\mathcal{P}}_I$, Σ_D's class of matchable transfer matrices is correspondingly reduced, typically by a large amount. Another idea [8,22] is to only stabilize Σ_D at points outside of some region $\mathcal{R} \subset \mathcal{P}_I$ which contains the points at which Σ_D loses stabilizability; since tunability cannot usually be guaranteed at points inside of $\mathcal{R}$, in order to avoid instability, some form of persistent excitation must be used to make certain that k_I does not get stuck in $\mathcal{R}$. A third approach might be to use a specially parameterized design model which does not lose stabilizability at any point in its parameter space [18]; this unfortunately leads to difficult tuner design problems for which there are at present no satisfactory solutions. A fourth approach, for the most part unexplored in the literature, is to make use of implicit tuning. This will be discussed briefly in Sect. 4.

For the case when Σ_P is "minimum phase", the problem we've been discussing can often be avoided all together. To understand why, we need a few more ideas.

Recall that a constant, linear multivariable system (C, A, B) is *minimum phase* if its transfer matrix is left invertible and its transmission zeros[12] [23] are

[12] For a system (C, A, B) with left-invertible transfer matrix $T(s) = C(sI - A)^{-1}B$, a complex number λ is a *transmission zero* if there exists a nonzero vector $[x', u']'$ such that $Cx = 0$ and $(\lambda I - A)x + Bu = 0$.

all stable; as defined, minimum phase systems are necessarily detectable. Call a matrix pair (C, A) (or a system (C, A, B)) ***output stabilized*** if $Ce^{At} \to 0$ as $t \to \infty$.[13] Σ_{R} is said to ***output stabilize*** the design model Σ_{D} at p if $\Sigma_{\mathrm{D_{cl}}}$ is output stabilized at p. Let $\mathcal{E}$ be a nonempty subset of $\mathcal{P}$. A certainty equivalence controller $\Sigma_{\mathrm{R}}(k)$ is called an ***output stabilizing controller on*** $\mathcal{E}$ if for each $p = [p'_{\mathrm{I}}, p'_{\mathrm{R}}]' \in \mathcal{E}$, $\Sigma_{\mathrm{R}}(p)$ output stabilizes the design model Σ_{D} at p. The following theorem provides the key technical link needed to explain how to achieve with indirect control that previously achieved in [4,6,7,8] and elsewhere with direct adaptive control.

Certainty Equivalence Output Stabilization Theorem: *If Σ_{P} is minimum phase and $\Sigma_{\mathrm{R}}(k)$ is a certainty equivalence, output stabilizing controller on $\mathcal{E}$, then $\Sigma(k)$ is tunable on $\mathcal{E}$.*

Remark 3: The certainty equivalence output stabilization theorem does not require Σ_{P} to have a transfer matrix which can be matched by $\Sigma_{\mathrm{D}}(p_{\mathrm{I}})$ for some value of $p_{\mathrm{I}} \in \mathcal{P}_{\mathrm{I}}$. It is interesting to note, however, that in the case when exact matching is possible, to achieve tunability it is not necessary for the design model to be stabilizable, even at the "true design model parameter value" at which matching takes place! For example, the process model $\dot{y} = -4y + u$ is matched by the direct control design model $\dot{x}_{\mathrm{D1}} = -2x_{\mathrm{D1}} + p_1 p_3 x_{\mathrm{D2}} + p_2 u_{\mathrm{D}}$, $\dot{x}_{\mathrm{D2}} = -x_{\mathrm{D1}} - x_{\mathrm{D2}} + u_{\mathrm{D}}$, $y_{\mathrm{D}} = p_3 x_{\mathrm{D2}}$, at $p = [p_1, p_2, p_3]' = [-6, 3, 1]'$ and for this value of p, the design model has an uncontrollable eigenvalue at $s = 1$. Since for obvious reasons a process model Σ_{P} must always be assumed to be at least stabilizable and detectable, this example shows that a design model cannot necessarily be viewed merely as a process model with variable parameters.◇

Remark 4: Both of the preceding theorems apply to the case when $e_{\mathrm{T}} = e_{\mathrm{I}}$. In order to facilitate the tuning of k it is sometimes helpful to use an alternative definition of e_{T}. One possibility is $e_{\mathrm{T}} = [e'_{\mathrm{I}}, e'_{\mathrm{R}}]'$ where $e_{\mathrm{R}} = C_{\mathrm{IR}}(k) x_{\mathrm{R}}$ and $C_{\mathrm{IR}}(k)$ is some matrix-valued function of k; since detectability through e_{I} clearly implies detectability through e_{T}, the preceding theorems apply to this case as well. Yet another possible definition of e_{T}, useful in the case when A_{R} is a constant stability matrix, $D_{\mathrm{R}} = 0$, and e_{T} and e_{R} are of the same size, is $e_{\mathrm{T}} = e_{\mathrm{I}} + e_{\mathrm{R}}$; the preceding theorems are also applicable in this case since the stability of A_{R} and the assumption $D_{\mathrm{R}} = 0$ insure that detectability through e_{I} implies detectability through e_{T}. This definition of e_{T} finds application in model reference adaptive control as the following example illustrates.◇

[13] Note that (C, A) is output stabilized if and only if each of its unstable modes is unobservable. On the other hand, (C, A) is detectable if and only if each of its unobservable modes is stable. It follows that (C, A) is *both* output stabilized and detectable if and only if A has no unstable modes; i.e., if and only if A is a stability matrix.

Example 7 (cont'd): Observe that the control $u_D = -cx_{D1}$ causes the output y_D of design model (28) to satisfy $y_D = p_3 e^{A_* t} x_{D1}(0)$. Thus, the internal regulator matrix $F_R = [-c, 0]$ output stabilizes Σ_D on $\mathbb{R}^{2n+m}$. Therefore, by the Certainty Equivalence Output Stabilization Theorem, if Σ_P is minimum phase, the closed-loop parameterized system consisting of Σ_P , identifier (29), internal regulator $u = -c\widehat{x}_{D1} = -k_1' Q' z_1 - k_2' Q' z_2$ (cf. (27)) and tuning error $e_T = e_I$ is tunable on $\mathbb{R}^{2n+m}$. The simple dependence of this control law on k_I is the principal virtue of direct control.

The challenge in coming up with an overall adaptive control system based on this direct control design model, lies in the synthesis of an appropriate tuner since $C_I(k_I)$ (i.e., the equation for e_T) is nonlinear in k_I. The tuning problem simplifies significantly if, for fixed $n^* \leq m$, the characteristic polynomial of A_* is chosen to have a real monic factor α_* of degree n^*, and p_3 is constrained to be of the form $p_3 = p_0 c_*$, $p_0 \in \mathbb{R}$, where c_* is defined by the formula $c_*(sI - A_*)^{-1} b_* = 1/\alpha_*$. In this case $\Sigma_D([p_1', p_2', p_0 c_*]')$ has $2n + 1$ parameters and a class of matchable transfer functions containing only those of relative degree n^* with McMillan degree not exceeding n. Moreover, $e_T = k_0 c_*(x_* + H_1 Q k_1 + H_2 Q k_2) - y$ which is a form for which by now appropriate tuning algorithms are well known (e.g., see [17,19]).[14]

To convert the preceding into a model reference algorithm, we need a reference model Σ_r which we define to be

$$\begin{aligned} y_r &= c_* x_R \\ \dot{x}_R &= A_* x_R + b_* r \,, \end{aligned} \tag{35}$$

and a modified control law which we take to be $u_R = -cx_{D1} + p_R r$. This control law output stabilizes Σ_D on $\mathbb{R}^{2n+2}$ and results in the closed-loop transfer function $p_0 p_R/\alpha_*$ from r to y_D. Thus as before, the closed-loop parameterized system consisting of the minimum phase process model (2), the internal regulator $u = -c\widehat{x}_{D1} + k_R r = -k_1' Q' z_1 - k_2' Q' z_2 + k_R r$ (cf. (27)), the identifier (29) with modified identification error

$$e_I = k_0 c_*(x_* + H_1 k_1 + H_2 k_2) - y \tag{36}$$

and tuning error $e_T = e_I$, is tunable on $\mathbb{R}^{2n+2}$. To achieve model following it is necessary to change the definition of e_T. To motivate the change, note from (35) that the *tracking error* $y - y_r$ can be written as

$$y - y_r = -(e_I + e_R) + k_0 c^* \xi \,, \tag{37}$$

where e_R is the *reference parameter error* $e_R = (1 - k_0 k_R) y_r$ and $\xi = x_* + H_1 k_1 + H_2 k_2 - x_R k_R$; moreover if k is differentiable then by a simple calculation,

$$\dot{\xi} = A_* \xi + H_1 \dot{k}_1 + H_2 \dot{k}_2 + x_R \dot{k}_R \,. \tag{38}$$

[14] Things can be simplified somewhat by redefining k_1 and k_2 to be Qk_1 and Qk_2 respectively; for then, $u = -k_1' z_1 - k_2' z_2$ and $e_T = k_0 c_*(x_* + H_1 k_1 + H_2 k_2) - y$.

Equations (37) and (38) clearly show that if e_T is defined to be $e_T = [e_I, e_R]'$ then a zero tracking error will be attained asymptotically, along any system trajectory on which k becomes constant and e_T goes to zero. Equations (37) and (38) also imply that this will still be true if e_T is alternatively defined to be simply $e_I + e_R$. In this case, e_T proves to be precisely the same as the "augmented error" originally introduced by Monopoli [24]. As noted in Remark 4, neither of these definitions of e_T effect the tunability of the overall parameterized system $\Sigma(k)$.◇

Example 6 (cont'd): Simple examples show that it is impossible to output stabilize on $\mathbb{R}^{2n}$ the indirect control design model Σ_D of Example 6. One way to circumvent this problem is to modify the parameterization of Σ_D. The price paid for this, however, is a significant decrease in the size of the class of transfer functions which the modified design model can match. The modified parameterization is constructed as follows: with c and A as before, choose a nonnegative integer $m < n$ and a monic, stable polynomial $\alpha(s)$ of degree m; define vectors b_1 and b_2 so that $c(sI - A)^{-1}b_1 = 1/\beta(s)$ and $c(sI - A)^{-1}b_2 = \alpha(s)/\beta(s)$ respectively, where $\beta(s)$ is the characteristic polynomial of A. In place of (25), define the "multiplicatively parameterized" indirect control design model $\Sigma_D(p_I)$ by the equations

$$\begin{aligned} \dot{x}_D &= (A + p_0 p_1 c)x_D + (b_2 + T p_2)u_D \\ y_D &= p_0 c x_D , \end{aligned} \tag{39}$$

where

$$T = [b_1, Ab_1, \ldots, A^{m-1}b_1], \tag{40}$$

$p_1 \in \mathbb{R}^n$, $p_2 \in \mathbb{R}^m$ and $p_1 = [p_0, p_1', p_2']'$. As defined, Σ_D is detectable on $\mathcal{P}_I = \mathbb{R}^{n+m+1}$. However, unlike the previous indirect control design model which can match any transfer function of McMillan degree not exceeding n, the modified design model (39) can only match those transfer functions of McMillan degree with relative degrees exactly equal to $n - m$. This can be deduced using the identities

$$\begin{aligned} cA^{j-1}b_i &= 0, \quad j < n - m, \quad i \in \{1, 2\} \\ cA^{n-m-1}b_i &= 1, \quad i \in \{1, 2\} \end{aligned} \tag{41}$$

which hold because the transfer functions $1/\beta(s)$ and $\alpha(s)/\beta(s)$ used to define b_1 and b_2 respectively, both have relative degrees no smaller than $n - m$. What's especially important about the design model (39) is that it can be output stabilized on $\mathcal{P}_I$ by a continuously differentiable feedback function.

Proposition 1: *Let $\gamma(s)$ be any monic, stable polynomial of degree $n - m$. With*

$$f(p_I) = -c\gamma(A + p_0 p_1 c), \tag{42}$$

the control law $u_D = f(p_I)x_D$ output stabilizes the indirect control design model (39) on $\mathcal{P}_I$.

Proposition 1 gives an explicit formula for a feedback function f which output stabilizes the multiplicatively parameterized design model (39) for each $p_{\mathrm{I}} \in \mathcal{P}_{\mathrm{I}}$. The n components of f are each polynomials in the $n+1$ components of $[p_0, p_1']'$.

It can easily be shown that if f is defined by (42), then the closed-loop design model characteristic polynomial satisfies

$$\det\left(sI - A - p_0 p_1 c - (b_2 + Tp_2) f(p_{\mathrm{I}})\right) = \alpha_{\mathrm{D}}(s, p_{\mathrm{I}})\gamma(s)\,, \qquad p \in \mathcal{P}_{\mathrm{I}}\,, \tag{43}$$

where $\alpha_{\mathrm{D}}(s, p_{\mathrm{I}})$ is the m-th degree, monic numerator polynomial of the transfer function $c(sI-A)^{-1}(b_2+Tp_2)$. It should be noted, however, that merely decreeing that f should satisfy (43) does not guarantee that f will satisfy (42) or that the closed-loop design model $\Sigma_{\mathrm{D_{cl}}}$ will be output stabilized. In other words, the specification that $f(p_{\mathrm{I}})$ should assign to $A + p_o p_1 + (b_2 + Tp_2) f(p_{\mathrm{I}})$ the closed-loop characteristic polynomial $\alpha_{\mathrm{D}}(s, p_{\mathrm{I}})\gamma(s)$ does *not* pin down f enough for the purposes of indirect adaptive control.

Proposition 1 implies that to achieve tunability on $\mathcal{P}_{\mathrm{I}}$, it's enough to choose f on $\mathcal{P}_{\mathrm{I}}$ in accordance with (42) so that (43) holds, even though for some $p_{\mathrm{I}} \in \mathcal{P}_{\mathrm{I}}, \alpha_{\mathrm{D}}(s, p_{\mathrm{I}})$ may be an unstable polynomial. By recognizing that $p_0\alpha_{\mathrm{D}}(s, p_{\mathrm{I}})$ is the numerator of the transfer function of open-loop, indirect control design model (39), it can be seen that what's being done here is a kind of "adaptive pole-zero cancellation" similar to what's often implicitly done with algorithms of the direct control type.

To construct an indirect parameterized controller for adaptive model following, one can proceed from this point in essentially the same way as in the direct control case of Example 7. First, define an identifier for the multiplicatively parameterized design model (39) by modifying the identifier for the design model (25) in the obvious way; that is, by replacing c and k_2 in the identifier for (25) by $k_0 c$ and $b_2 + Tk_2$ respectively, where k_2 is an m-vector. What result are the equations

$$\begin{aligned}
\dot{z}_1 &= Az_1 + by \\
\dot{z}_2 &= Az_2 + bu \\
e_{\mathrm{I}} &= k_0(k_1'Q'z_1 + (k_2'T' + b_2')Q'z_2) - y \\
\widehat{x}_{\mathrm{D}} &= S(k_1)z_1 + S(b_2 + Tk_2)z_2\,,
\end{aligned} \tag{44}$$

where b, Q, and $S(\cdot)$ are as defined previously. Next, set $\gamma = \alpha_*$ and

$$u = f(k_{\mathrm{I}})\widehat{x}_{\mathrm{D}} + k_{\mathrm{R}} r\,, \tag{45}$$

where f is as defined in (42), $k_{\mathrm{I}} = [k_0, k_1', k_2']' \in \mathbb{R}^{n+m+1}$, k_{R} is an additional internal regulator parameter to be tuned, and r is the input to the reference model (35).

In view of the Certainty Equivalence Output Stabilization Theorem, the closed-loop parameterized system consisting of the SISO minimum phase process model Σ_{P}, the parameterized controller (44),(45) and the tuning error $e_{\mathrm{T}} = e_{\mathrm{I}}$, is tunable on $\mathbb{R}^{n+m+1}$. To develop tracking error equations analogous to (37) and (38), first use (17),(35), and (44) to write

$$y - y_r = -(e_I + e_R) + k_0(c\hat{x}_D - k_R c_* x_R), \tag{46}$$

where, as before, $e_R = (1 - k_0 k_R)y_r$. Assume that $k = [k_I', k_R]'$ is differentiable. Using the estimation equations (19) together with the definition of u in (45), one gets

$$\begin{aligned}\dot{\hat{x}}_D &= (A + k_0 k_1 c + (b_2 + Tk_2)f(k_I))\,\hat{x}_D \\ &+ (b_2 + Tk_2)k_R r + S(\dot{k}_1)z_1 + S(\dot{k}_2)z_2 - k_0 k_1 e_I\,. \end{aligned}\tag{47}$$

To proceed, we need the following result.

Lemma 1: *Let $\gamma(s)$ be as in Proposition 1 and write $(\bar{c}, \bar{A}, \bar{b})$ for the unique controllable, observable realization of $1/\gamma(s)$ with $(\bar{A}, \bar{b})$ in control canonical form. If $f(p_I)$ satisfies (42) and*

$$M(p_I) = \begin{bmatrix} c \\ c(A + p_0 p_1 c) \\ \cdot \\ \cdot \\ \cdot \\ c(A + p_0 p_1 c)^{n-m-1} \end{bmatrix},$$

then for all $p_I \in \mathbb{R}^{n+m+1}$,

$$c = \bar{c}M(p_I) \tag{48}$$

$$M(p_I)(b_2 + Tp_2) = \bar{b} \tag{49}$$

$$M(p_I)(A + p_0 p_1 c + (b_2 + Tp_2)f(p_I)) = \bar{A}M(p_I)\,. \tag{50}$$

Now set $\gamma(s) = \alpha_*(s)$, $\alpha_*(s)$ being the characteristic polynomial of A_*, and suppose without loss of generality, that (c_*, A_*, b_*) is in control canonical form; thus $(\bar{c}, \bar{A}, \bar{b}) = (c_*, A_*, b_*)$. Hence from (46) and (48),

$$y - y_r = -(e_I + e_R) + k_0 c_* \xi\,, \tag{51}$$

where $\xi = M(k_I)\hat{x}_D - k_R x_R$. Moreover from (35),(47),(49) and (50)

$$\dot{\xi} = A_*\xi + M(\dot{k}_I)\hat{x}_D + M(k_I)(S(\dot{k}_1)z_1 + S(\dot{k}_2)z_2 - k_0 k_1 e_I)\,. \tag{52}$$

The implication of (51) and (52) is clear: If k is tuned in such a way that $\dot{k}$, e_I and e_R each tend to zero as $t \to \infty$, then the tracking $y - y_r$ will tend to zero as well. Thus, an appropriate definition for e_T would be $e_T = [e_I, e_R]'$, just as in Example 7. It is interesting to note however, that unlike Example 7, zeroing the scalar tuning signal $e_T = e_I + e_R$ will not necessarily produce a zero tracking error because of the presence of e_I in (52).◇

4 Implicit Tuning

As noted in Sect. 3, one of the things to be done in constructing a certainty equivalence controller for a given design model Σ_D , is to select an internal regulator Σ_R which causes the closed-loop design model $\Sigma_{D_{cl}}$ to have certain desirable properties. This task, which might be accomplished using established synthesis techniques from linear system theory (e.g., pole placement, model following, decoupling theory, linear quadratic optimal control, etc.) can often be reduced to finding a solution p_R to some design equation of the form

$$\Delta(p_I, p_R) = 0\,, \tag{53}$$

where $p_I \in \mathcal{P}_I \subset \mathrm{IR}^{n_{p_I}}$ is a vector of design model parameters, $p_R \in \mathcal{P}_R \subset \mathrm{IR}^{n_{p_R}}$ is a vector whose elements are elements of the coefficient matrices defining Σ_R and Δ is a "design" function mapping $\mathrm{IR}^{n_{p_I}} \oplus \mathrm{IR}^{n_{p_R}}$ into $\mathrm{IR}^{n_{p_R}}$; (53) might arise from a Riccati equation, a pole placement equation or something similar. If (53) is solved explicitly for p_R as a function of p_I, then the resulting identifier-based controller's parameter space $\mathcal{P}$ is simply $\mathcal{P}_I$ and the only parameters to be tuned are the elements of k_I. If, on the other hand, p_R is taken to be a vector of additional regulator parameters, then the corresponding controller's parameter space $\mathcal{P}$ is $\mathcal{P}_I \times \mathcal{P}_R$ and the parameters to be tuned are both the elements of k_I and those of k_R; in this case the role of Σ_T is to adjust the parameter vector $k = [k_I', k_R']'$ to make *both* the tuning error e_T and the *design error*

$$e_D = \Delta(k_I, k_R) \tag{54}$$

simultaneously small. An identifier-based controller whose parameters are defined and adjusted in this manner is said to be *implicitly tuned.* [15]

Implicit tuning algorithms can be classified as being of either the "one level" or "two level" type depending on which errors k_I is being adjusted to make small. In the two level case, k_I is adjusted to make *just* the identification error e_I small whereas in the one level case k_I is tuned to make both the identification error e_I and the design error e_D small. Two level tuning is thus more in line with the traditional idea of identifier-based adaptive control than is one level tuning and thus is perhaps more intuitively appealing. In addition, two level tuning allows one to use conventional identification algorithms to adjust k_I whereas single level tuning does not.[16] In spite of this there are compelling reasons for also considering the concept of single level tuning. At the end of this section we provide an example in support of this view.

The general idea of implicit tuning has been around in one form or another for a long time. The need for dynamically tuned regulator parameters is alluded to in [3]. Some of the reparameterization techniques developed in [26] can be

[15] A more general concept of implicit tuning would allow the design function $\Delta(\cdot)$ in (54) to also be a function of controller state x_C. [21]

[16] The design of a single level implicit tuner is a nonstandard identification problem for which there is apparently almost no literature even for the interesting, highly structured case in which C is linear in k and Δ is bilinear in k_I and k_R.

construed as a form of implicit tuning. Two level implicit tuning is proposed in [27] in order to avoid some of the parametric nonlinearities which arise in connection with the multiplicative indirect control design model of Example 6. Single level implicit tuning is studied in [14] in a model reference context in an effort to achieve improved system performance. While there are no doubt many other examples of implicit tuning in the literature, the potential value of the concept does not appear to be widely appreciated. The aim of this section is to explain and illustrate by means of informal discussion and examples how the idea of implicit tuning fits into the framework of this paper and why it is likely to prove of importance to parameter adaptive control.

In the sequel we briefly sketch how the concept of tunability can be used to analyze an implicitly tuned adaptive system. For this, assume that one has chosen a design model Σ_{D} , a design function Δ and an internal regulator Σ_{R} so that for each value of the parameter $p = [p_{\mathrm{I}}', p_{\mathrm{R}}']'$ in the set

$$\mathcal{E} = \{[p_{\mathrm{I}}', p_{\mathrm{R}}']' : \Delta(p_{\mathrm{I}}, p_{\mathrm{R}}) = 0\}, \tag{55}$$

the closed-loop design model $\Sigma_{\mathrm{D_{cl}}}$ depicted in Fig. 4, is at least internally stable. Then, because of the Certainty Equivalence Stabilization Theorem, the closed-loop parameterized system Σ shown in Fig. 3 and described by

$$\dot{x} = A(k)x + B(k)r \tag{56}$$

$$e_{\mathrm{T}} = C(k)x \tag{57}$$

must be tunable on $\mathcal{E}$. For purposes of analysis define the "extended tuning error"

$$e_{\mathrm{TE}} = \begin{bmatrix} C(k) \\ C_{\mathrm{E}}(k) \end{bmatrix} x \,, \tag{58}$$

where $C_{\mathrm{E}}(k)x$ is a column representation of the elements of the outer product $\Delta(k_{\mathrm{I}}, k_{\mathrm{R}})x'$; in view of (54)

$$||C_{\mathrm{E}}(k)|| = n_{p_{\mathrm{R}}}||e_{\mathrm{D}}||\,, \tag{59}$$

where $n_{p_{\mathrm{R}}}$ is the size of p_{R}. It is known (cf. the Tunability Lemma of [1]) that tunability of Σ on $\mathcal{E}$ implies tunability on $\mathcal{P} = \mathcal{P}_{\mathrm{I}} \times \mathcal{P}_{\mathrm{R}}$ of the system described by (56)–(57) and the extended tuning error (58). Because of this and Proposition 1 of [1], we can state the following:

Observation 1: *There exists a matrix $[H(p), H_{\mathrm{E}}(p)]$, depending on p as smoothly as C, A and Δ do, for which $A(p) + H(p)C(p) + H_{\mathrm{E}}(p)C_{\mathrm{E}}(p)$ is a stability matrix for each $p \in \mathcal{P}$.*

In the light of the preceding it is not difficult to see how to modify the analysis at the end of Sect. 1 in order to establish boundedness of x for an implicitly tuned system in which both e_{D} and a normalized tuning error $\bar{e}_{\mathrm{T}}$ are simultaneously small. The following examples, which are of interest in their own right, serve to illustrate how this can be done.

Example 8: Two level implicit tuning - Assume that $\mathcal{P} = \mathbb{R}^{n_{p_I}} \oplus \mathbb{R}^{n_{p_R}}$, that the closed-loop parameterized system (56)–(57) is tunable on $\mathcal{E}$, that C and A are continuously differentiable and that e_T can be reexpressed in the form

$$e_T = w'(k_I - q) + \epsilon \,, \tag{60}$$

where $w = Lx$, L is a constant matrix, q is a constant vector, and ϵ is a linear combination of decaying exponentials (cf. Remark 2, Sect. 3). Assume, in addition, that $\Delta(p_I, p_R)$ has continuously differentiable first partials $\Delta_{p_I} = \partial\Delta/\partial p_I$ and $\Delta_{p_R} = \partial\Delta/\partial p_R$ and that for each $p = [p_I', p_R']' \in \mathcal{P}$, the smallest eigenvalue of the matrix $\Delta_{p_R}'\Delta_{p_R}$ is no smaller than some positive number μ; i.e.,

$$\Delta_{p_R}'\Delta_{p_R} \geq \mu I, \quad \forall p \in \mathcal{P} \,. \tag{61}$$

This last assumption is of course restrictive.

Consider the simple two level implicit tuning algorithm defined by the equations

$$\begin{aligned} \dot{k}_I &= -w\bar{e}_T \\ \dot{k}_R &= -\Delta_{p_R}'(k_I, k_R)e_D \,, \end{aligned} \tag{62}$$

where $\bar{e}_T$ is the normalized tuning error

$$\bar{e}_T = e_T/(1 + ||w||^2 + ||\Delta_{p_I}(k_I, k_R)w||^2) \,. \tag{63}$$

Let $[0, T)$ denote the maximal interval of existence of the dynamical system defined by (54), (55), (56), (62), and (63). Using these equations it is easy to see that

$$\begin{aligned} \frac{d}{dt}\left\{\frac{1}{\mu}||k_I - q||^2 + ||e_D||^2\right\} &= -\frac{2}{\mu}\left((1 + ||w||^2 + ||\Delta_{p_I}w||^2)\,\bar{e}_T^2 - \epsilon\bar{e}_T\right) \\ &\quad -2\left(e_D'\Delta_{p_I}w\bar{e}_T + ||\Delta_{p_R}e_D||^2\right) \\ &\leq -\frac{1}{\mu}\left((1 + 2||w||^2 + 2||\Delta_{p_I}w||^2)\,\bar{e}_T^2 - \epsilon^2\right) \\ &\quad -2\left(e_D'\Delta_{p_I}w\bar{e}_T + \mu||e_D||^2\right) \\ &\leq -\frac{1}{\mu}\left((1 + 2||w||^2 + ||\Delta_{p_I}w||^2)\,\bar{e}_T^2 - \epsilon^2\right) - \mu||e_D||^2 \,. \end{aligned}$$

Thus, by rearranging terms and integrating,

$$\frac{1}{\mu}||k_I - q||^2 + ||e_D||^2 + \int_0^t \left(\frac{1}{\mu}\left(1 + 2||w||^2 + ||\Delta_{p_I}w||^2\right)\bar{e}_T^2 + \mu||e_D||^2\right)d\tau \leq c_0 \,,$$

where $c_0 = (\frac{1}{\mu}||k_I(0) - q||^2 + ||e_D(0)||^2 + \int_0^\infty \epsilon^2 d\tau)$. This shows that k_I and e_D are bounded on $[0, T)$ and that $\bar{e}_T, w\bar{e}_T, \dot{k}_I, \Delta_{p_I}\bar{e}_T$ and e_D are in $\mathcal{L}^2(T)$, the space of square integrable functions on $[0, T)$. Moreover, since the determinant of Δ_{p_R} is a Jacobian for (53), assumption (61) and the implicit function theorem make it possible to express k_R as a continuous function of k_I and e_D; since the latter are bounded on $[0,T)$ the former must also be. Note in addition that $\bar{e}_{\mathrm{T}}$ must be bounded on $[0,T)$ because of (60) and (63).

The utility of Observation 1 becomes apparent as we turn to the problem of establishing the boundedness of x on $[0,T)$. As in Sect. 2, what we need to do first is to show that the matrix $A_*(t) = A(k(t)) + H(k(t))(C(k(t)) + \xi'(t)L)$ is exponentially stable on $[0,T)$ where $[H, H_{\mathrm{E}}]$ is as described in Observation 1 and ξ is the $\mathcal{L}^2(T)$ function $\xi = -(I + \Delta'_{p_{\mathrm{I}}}\Delta_{p_{\mathrm{I}}})w\bar{e}_{\mathrm{T}}$. This can be done by first noting that A_* can be rewritten as

$$A_* = A_{**} - H_{\mathrm{E}}C_{\mathrm{E}}\,, \tag{64}$$

where $A_{**} = A + HC + H_{\mathrm{E}}C_{\mathrm{E}} + H\xi'L$. Because of the pointwise stability of $A + HC + H_{\mathrm{E}}C_{\mathrm{E}}$ noted in Observation 1, and the fact that ξ and $\dot{k}$ are in $\mathcal{L}^2(T)$, it is easy to see that A_{**} is exponentially stable on $[0,T)$ (cf. [1,16,17]). Since (59) and $e_{\mathrm{D}} \in \mathcal{L}^2(T)$ imply that $H_{\mathrm{E}}C_{\mathrm{E}} \in \mathcal{L}^2(T)$, the same reasoning can be used again to conclude that A_* as expressed in (64) is exponentially stable on $[0,T)$. From this point on, the logic is exactly as in Sect. 2: That is, using the expression $\bar{e}_{\mathrm{T}} = (C(k) + \xi'L)x$, which is a consequence of (62) and (63), the differential equation for x in (56) can be rewritten as $\dot{x} = A_*x + Br - H\bar{e}_{\mathrm{T}}$. From this and the boundedness of r and $\bar{e}_{\mathrm{T}}$ it follows that x is bounded on $[0,T)$; and this together with the boundedness of k and the definition of T imply that $T = \infty$ and thus that (x,k) is bounded on $[0,\infty)$.◇

Example 9: Single level implicit tuning - Let Σ_{D} denote the two parameter indirect control design model

$$\begin{aligned} \dot{x}_{\mathrm{D}1} &= -x_{\mathrm{D}2} + p_1 u_{\mathrm{D}} \\ \dot{x}_{\mathrm{D}2} &= x_{\mathrm{D}1} - (p_1 + p_2)x_{\mathrm{D}2} + u_{\mathrm{D}} \\ y_{\mathrm{D}} &= x_{\mathrm{D}2}\,, \end{aligned} \tag{65}$$

defined on the parameter space $\mathcal{P}_{\mathrm{I}} = \mathbb{R}^2$. Note that Σ_{D}'s transfer function is $(s+p_1)/(s^2 + (p_1+p_2)s + 1)$ and that a pole-zero cancellation occurs whenever $p_1p_2 - 1 = 0$. Define $p_{\mathrm{I}} = [p_1, p_2]'$ and $f(p_{\mathrm{I}}) = (p_1 + p_2 - 1)[-p_1, p_1(p_1+p_2) - 1]$. A straightforward calculation shows that the state feedback law $u_{\mathrm{D}} = (1/(p_1p_2 - 1))f(p_{\mathrm{I}})x_{\mathrm{D}}$ assigns to (65) the closed-loop characteristic polynomial $s^2 + s + 1$ for every value of $p_{\mathrm{I}} \in \mathbb{R}^2$ for which $p_1p_2 - 1$ does not vanish. This implies that with u_{D} redefined to be

$$u_{\mathrm{D}} = p_3 f(p_{\mathrm{I}})x_{\mathrm{D}}\,, \tag{66}$$

the closed-loop characteristic polynomial of (65) will be $s^2 + s + 1$ along the algebraic surface in $\mathbb{R}^3$ on which $p_3(p_1p_2 - 1) - 1 = 0$. While it is tempting to define Δ to be $p_3(p_1p_2 - 1) - 1$, this leads to a difficult tuner design problem. In order to avoid this difficulty, we shall "lift" the problem to $\mathbb{R}^4$ by defining

$$\Delta(p_{\mathrm{I}}, p_{\mathrm{R}}) = \begin{bmatrix} -p_1 & 1 \\ 1 & -p_2 \end{bmatrix} p_{\mathrm{R}} + \begin{bmatrix} 0 \\ 1 \end{bmatrix}, \tag{67}$$

where $p_R = [p_3, p_4]'$. Since $p_3 = 1/(p_1 p_2 - 1)$ for every value of $p = [p_I', p_R']'$ in the set

$$\mathcal{E} = \{[p_I', p_R']' : \Delta(p_I, p_R) = 0\}\ , \tag{68}$$

the feedback law (66) stabilizes design model (65) at each point $p \in \mathcal{E}$.

Let $\Sigma_I(k_I)$ be an identifier for the design model (25), as described in Example 6. Then e_I can be written as

$$e_I = w_2(1 - k_1 - k_2) + w_3 k_1 - y\ , \tag{69}$$

where $[w_1, w_2, w_3, w_4]' = (\text{block diag}\ [Q, Q])'\ x_I$, x_I is the identifier's state, and Q is as in Example 6. In accordance with certainty equivalence, define the internal regulator state feedback law

$$u = k_3 f(k_I)\widehat{x}_D\ , \tag{70}$$

where $\widehat{x}_D$ is the identifier's surrogate state defined by (23). Let

$$\begin{aligned} e_T &= C(k)x \\ \dot{x} &= A(k)x \end{aligned} \tag{71}$$

denote the closed-loop parameterized system $\Sigma(k)$ consisting of any fixed SISO model Σ_P , Σ_I, the tuning error $e_T = e_I$ and the certainty equivalence controller (70). Since the closed-loop design model $\Sigma_{D_{cl}}(p)$ described by (65)–(66) is stable for each $p \in \mathcal{E}$, by the Certainty Equivalence Stabilization Theorem, $\Sigma(k)$ is tunable on $\mathcal{E}$.

Now assume that Σ_P has a transfer function of the form $(s + q_1)/(s^2 + (q_1 + q_2)s + 1)$ where $q = [q_1, q_2]'$ is in the set $\mathcal{R}^- = \{p_I : p_1 p_2 - 1 > 0, p_1 < 0, p_2 < 0\}$. Note that Σ_P is *nonminimum phase* and that $\mathcal{R}^-$ is an *open* set bounded by the algebraic surface on which the design model (65) loses controllability. In view of (69), the definition of e_T and Remark 2 in Sect. 3, this assumption enables us to write e_T in the familiar form

$$e_T = (w_3 - w_2)(k_1 - q_1) - w_2(k_2 - q_2) + \epsilon\ , \tag{72}$$

where ϵ is a linear combination of decaying exponentials. The assumed structure of Σ_P's transfer function also allows us to rewrite the design error

$$e_D = \Delta(k_I, k_R) \tag{73}$$

as

$$e_D = \begin{bmatrix} -k_3 & 0 \\ 0 & -k_4 \end{bmatrix} \begin{bmatrix} k_1 - q_1 \\ k_2 - q_2 \end{bmatrix} + Q_P \begin{bmatrix} k_3 - q_3 \\ k_4 - q_4 \end{bmatrix}\ , \tag{74}$$

where

$$Q_P = \begin{bmatrix} -q_1 & 1 \\ 1 & -q_2 \end{bmatrix}$$

and $[q_3, q_4] = [0, -1](Q_P^{-1})'$. Note that Q_P is positive definite for each $q \in \mathcal{R}^-$. This proves to be a consequence of $\mathcal{R}^-$ being convex, a requirement which at present appears to preclude the generalization of these particular ideas.

The structure of $e_{\rm T}$ in (72) and $e_{\rm D}$ in (74) provides plenty of motivation for using a single level implicit tuner of the form

$$\begin{aligned}\dot{k}_{\rm I} &= -\bar{w}\bar{e}_{\rm T} - k_{\rm R} \odot e_{\rm D}\\ \dot{k}_{\rm R} &= -e_{\rm D}\,,\end{aligned}$$

where $\bar{w} = [w_3 - w_2, -w_2]'$, $k_{\rm R} \odot e_{\rm D} = [k_3 e_{{\rm D}_1}, k_4 e_{{\rm D}_2}]'$ and $\bar{e}_{\rm T}$ is the normalized tuning error

$$\bar{e}_{\rm T} = \frac{e_{\rm T}}{1 + \|\bar{w}\|^2}\,.$$

Note that, unlike Example 8, $k_{\rm I}$ is being adjusted here to make both $\bar{e}_{\rm T}$ and $e_{\rm D}$ small. We leave it to the interested reader to verify that on the overall system's maximal interval of existence $[0, T)$, $k, e_{\rm D}$ and $\bar{e}_{\rm T}$ are bounded and $e_{\rm D}, \bar{e}_{\rm T}$ and $\dot{k}$ are in $\mathcal{L}^2(T)$; moreover, just as in Example 8, the equation for x in (71) can be rewritten as $\dot{x} = A_*(t)x - H\bar{e}_{\rm T}$ where H is as in Observation 1, A_* is the exponentially stable matrix $A_*(t) = A(k(t)) + H(k(t))(C(k(t)) + \xi' L)$, $\xi = k_{\rm I} + k_{\rm R} \odot e_{\rm D}$ and L is the constant matrix for which $Lx = \bar{w}$. All this of course implies, just as before, that the entire system state (x, k) exists and is bounded on $[0, \infty)$.

As noted previously, the successful development of a stabilizing tuner for this example relies on q being in a convex subset, namely $\mathcal{R}^-$, whose boundary is determined by the algebraic surface on which the underlying design model loses controllability. Examination of this surface reveals that there is another subset, $\mathcal{R}^+ = \{p_{\rm I} : p_1 p_2 - 1 > 0, p_1 > 0, p_2 > 0\}$, with the same properties. It is easy to show that, by simply changing the signs of the diagonal terms in the square matrix in (67) which defines Δ, one can obtain a stabilizing adaptive control for the case when $q \in \mathcal{R}^+$. And to handle the case when q might be in *either* $\mathcal{R}^+$ or $\mathcal{R}^-$ one can use hysteresis switching, as is demonstrated in [28].

From the preceding it is clear that it is the *detailed geometry* of the algebraic surface on which an indirect control design model loses controllability, which is of importance to adaptive control. This example thus shows that if we can learn how to implicitly tune on each of the *open* subsets of the parameter space bounded by this surface, whether or not they are convex, then we will finally have a reasonably complete solution, not relying on persistent excitation, to the long standing adaptive pole placement problem for SISO linear systems.◇

Remark 5: In the preceding examples, the problem of making both $e_{\rm T}$ and $e_{\rm D}$ simultaneously small is approached by exploiting the specific algebraic forms of both $e_{\rm T}$ and $e_{\rm D}$ viewed as functions of $k_{\rm I}$ and $k_{\rm R}$. An alternative and highly innovative approach to this problem, which does not make use of the algebraic structure of $e_{\rm D}$, has been proposed in [29] for discrete-time systems. The idea can be explained as follows. Assume that $\Sigma_{\rm D}$ is the discrete-time version of the indirect control SISO design model of Example 6 and that $\Sigma_{\rm I}$ is its associated discrete-time identifier. Using an interesting idea due to Elliot [30], it is possible to construct a discrete-time internal regulator $\Sigma_{\rm R}$ of the same form as (31), a design function $\Delta(p_{\rm I}, p_{\rm R})$ and a row vector $c_{\rm C}(p_{\rm I}, p_{\rm R})$ depending linearly on $p_{\rm R}$, with the following properties:

1. For any SISO process model Σ_P, the closed-loop parameterized system consisting of Σ_P, Σ_I, Σ_R and the tuning error $e_T = e_I$, is tunable on $\mathcal{E} = \{[p_I', p_R']' : \Delta(p_I, p_R) = 0\}$.
2. With u_D defined by the discrete time analog of (32) and $T(s, p_I, p_R)$ the closed-loop design model transfer function from r to $c_C(p_I, p_R)[x_D', x_R']'$,

$$T(s, p_I, p_R) = 0 \iff \begin{bmatrix} p_I \\ p_R \end{bmatrix} \in \mathcal{E}.$$

Assume that standard gradient-like tuning algorithms, driven by normalized versions of tuning error e_T and "control error" $e_C = c_C(k_I, k_R)[\hat{x}_D', \hat{x}_R']'$, are used to adjust k_I and k_R respectively. Since e_T and e_C depend linearly on k_I and k_R respectively, in the absence of unmodeled dynamics these algorithms will cause e_T and e_C to tend to zero. Property 2 above insures that as this occurs, $[k_I, .k_R']'$ will approach $\mathcal{E}$ (or equivalently e_D will approach zero), provided $[\hat{x}_D', \hat{x}_R']'$ is "sufficiently rich". The central contribution of [29] has been to show that $[\hat{x}_D', \hat{x}_R']'$ can be made to have this property by setting r equal to a suitably defined persistently exciting probing signal which is internally generated by an algorithm driven by the norm of e_D!

With the algorithm of [29] viewed from this perspective, it is not difficult to think of alternative algorithms, of the same type, for solving the problem under consideration. For example, since with the Elliot parameterization Δ also turns out to be linear in p_R, one might consider adjusting k_R with a gradient-like tuning algorithm driven by e_D rather than e_C. To carry out a stability analysis in this case, it would have to be shown that r is persistingly exciting enough to insure that k_I is eventually driven into a region of the parameter space for which the Jacobian matrix $\Delta_{p_R}(k_I, k_R)$ of Δ remains nonsingular. For if this can be accomplished, and it is likely that it can be, then it would be easy to show that such an overall tuner would cause both e_T and e_D to go to zero, which because of property 1 above, is all that is required for internal stability.◇

Concluding Remarks

The purpose of this paper has been to discuss some of the reasons why it is especially useful to think of a parameter adaptive control system as the feedback interconnection of a process Σ_P and a parameterized controller $\Sigma(k)$ whose parameter vector k is adjusted by a tuner Σ_T. The configuration is general enough to describe many different kinds of adaptive systems including those of the model reference, self-tuning and high-gain feedback types. While error models are not used in this setting, special emphasis is placed on the importance of a tuning error in the characterization of an adaptive system. This leads naturally to the concept of tunability - an idea of particular utility to adaptive control, as explained in Section 2.

The central conclusion to be drawn from the discussion in Section 3 is that, subject to mild conditions, certainty equivalence implies tunability. Although the section's stated results hold for a large class of continuous and discrete-time multivariable adaptive control systems, as presented they are not as general as they might be. For example, there are a large number of alternative identifier structures which could be substituted for (21)–(23) without changing the validity of the two theorems which follow. Among these are the identifier implicitly used in [3], identifiers in which e_{I} is the output of a stationary linear system with a positive real transfer matrix whose input is of the form $C_{\mathrm{I}}(k_{\mathrm{I}})x_{\mathrm{I}} - y$, and identifiers of the "equation error" [31] type in which x_{I} is generated by an equation of the form $\dot{x}_{\mathrm{I}} = (A_{\mathrm{I}} + D_{\mathrm{I}}C_{\mathrm{I}}(k_{\mathrm{I}})x_{\mathrm{I}} + B_{\mathrm{I}}u$. In addition, there is good reason to believe that these results can be extended to slowly time-varying linear systems and even to some restricted classes of nonlinear systems, using, for example, ideas along the lines of [32]. In other words, the observation that certainty equivalence implies tunability, is probably as universal as the validity of the idea of certainty equivalence itself.

The aims of Section 4 are to call attention to the concept of implicit tuning and to explain how the ideas of Section 3 might be used in the analysis of an implicitly tuned system. The section's main message is that many meaningful adaptive control problems can be reduced to nonstandard, but highly structured identification/tuner design problems in which the concepts of implicit tuning, certainty equivalence and tunability play prominent roles. Examples 8 and 9 describe two different implicit tuning methodologies. Each appears to have potential for dealing with the loss of controllability/stabilizability problem mentioned earlier, without having to resort to ad hoc methods. Clearly what's needed are new tuning or identification algorithms capable of handling the somewhat unusual combination of error equations which arise from implicit parameterizations. In most cases the tuning error equation will be linear in k_{I}; to achieve this may sometimes require a reparameterization along the lines of [26]. Usually Δ will be a continuous algebraic or polynomial function with the property that, for almost every value of p_{I}, the design equation $\Delta(p_{\mathrm{I}}, p_{\mathrm{R}}) = 0$ has a unique solution p_{R}; moreover, Δ will typically be of a form for which there is at least one well-known recursive algorithm for computing p_{R} whenever it exists. With this much known, it would seem that significant advances in this direction in adaptive control are inevitable.

References

1. A. S. Morse, "Towards a unified theory of parameter adaptive control–tunability," *IEEE Trans. Aut. Control*, vol. AC-35, pp. 1002–1012, Sept. 1990.
2. A. S. Morse, "Towards a unified theory of parameter adaptive control–Part 2: certainty equivalence and implicit tuning," *IEEE Trans. Aut. Control*, to appear.
3. A. Feuer and A. S. Morse, "Adaptive control of single-input single-output linear systems," *IEEE Trans. Aut. Control*, vol. AC-23, pp. 557–569, Aug. 1978.
4. A. S. Morse, "Global stability of parameter-adaptive control systems," *IEEE Trans. Aut. Control*, vol. AC-25, pp. 433–439, June 1980.
5. G. C. Goodwin and K. S. Sin, *Adaptive Filtering Prediction and Control*, Prentice-Hall, Inc., Englewood Cliffs, NJ, 1984.
6. K. J. Astrom and B. Wittenmark, *Adaptive Control*, Addison-Wesley Publishing Co., 1989.

7. K. S. Narendra and A. M. Annaswamy, *Stable Adaptive Systems,* Prentice-Hall, Inc., Englewood Cliffs, NJ, 1989.
8. S. S. Sastry and M. Bodson, *Adaptive Control: Stability, Convergence and Robustness*, Prentice-Hall, Inc., Englewood Cliffs, NJ, 1989.
9. R. D. Nussbaum, "Some remarks on a conjecture in parameter adaptive control," *Syst. Control Lett.*,vol. 3, pp. 243–246, 1983.
10. B. Martensson, "The order of any stabilizing regulator is sufficient a priori information for adaptive stabilizing," *Syst. Control Lett.*, vol. 6, pp. 87–91, 1985.
11. A. S. Morse, "A three dimensional universal controller for the adaptive stabilization of any strictly proper minimum-phase system with relative degree not exceeding two," *IEEE Trans. Aut. Control,* vol. AC-30, pp. 1188–1191, Dec. 1985.
12. J. C. Willems and C. I. Byrnes, "Global adaptive stabilization in the absence of information on the sign of the high-frequency gain," Lecture Notes in Control and Information Sciences, *Proc. 6th Int. Conf. on Analysis and Optimization of Systems,* Nice, June 1984, pp. 49–57.
13. D. E. Miller, *Adaptive Control of Uncertain Systems*, Doctoral Thesis, University of Toronto, Oct. 1989.
14. M. A. Duarte and K. S. Narendra, "Combined direct and indirect approach to adaptive control," Center for Systems Science, Yale University, Report No. 8711, Sept. 1987.
15. J. K. Hale, *Ordinary Differential Equations,* Wiley-Interscience, 1969.
16. G. Kreisselmeier, "An approach to stable indirect adaptive control," *Automatica,* vol. 21, pp. 425–431, July 1985.
17. R. H. Middleton, G. C. Goodwin, D. J. Hill and D. Q. Mayne, "Design Issues in Adaptive Control," *IEEE Trans. Aut. Control,* vol. 33, pp. 50–58, Jan. 1988.
18. F. Pait and A. S. Morse, "A smoothly parameterized family of stabilizable, observable linear systems containing realizations of all transfer functions of McMillan degree not exceeding n," *IEEE Trans. Aut. Control,* to appear.
19. P. Ioannou and J. Sun, "Theory and design of robust direct and indirect adaptive-control schemes," *Int. J. Control,* vol. 47, pp. 775–813, 1988.
20. B. Egardt, *Stability of Adaptive Controllers,* Springer-Verlag, Lecture Notes in Control and Information Sciences, vol. 20, 1979.
21. L. Praly, "Global stability of direct adaptive control schemes with respect to a group topology," *Adaptive and Learning Systems,* Plenum Press, 1986, pp. 1009–1014.
22. B. D. O. Anderson and R. M. Johnstone,"Global adaptive pole positioning," *IEEE Trans. Aut. Control,* vol. AC-30, pp. 11–22, Jan. 1985.
23. A. S. Morse, "Structural invariants of linear multivariable systems," *SIAM J. Control,* vol. 11, pp. 446–465, Aug. 1973.
24. R. V. Monopoli, "Model reference adaptive control with an augmented error signal," *IEEE Trans. Aut. Control,* vol. AC-19, pp. 474–484, Oct. 1974.
25. L. Praly, "Towards a globally stable direct adaptive control scheme for not necessarily minimum phase systems," *IEEE Trans. Aut. Control,* vol. AC-29, pp. 946–949, Oct. 1984.
26. S. Dasgupta and B. D. O. Anderson, "Physically based parameterizations for the design of adaptive controllers," Dept. of Systems Eng. Report, Australian National University, Canberra, 1986.
27. A. S. Morse, "Indirect adaptive control of processes satisfying the classical assumptions of direct adaptive control," *Proc. 1988 Amer. Control Conf.*, Atlanta, GA.

28. A. S. Morse, D. Q. Mayne and G. C. Goodwin, "Identifier-based, switched-parameter algorithms for the adaptive stabilization of linear systems," *Proc. Sixth Yale Workshop on Adaptive and Learning Systems*, New Haven, Aug. 1990.
29. G. Kreisselmeier and M. C. Smith, "Stable adaptive regulation of arbitrary nth-order plants," *IEEE Trans. Aut. Control*, vol. AC-31, pp. 299–305, April 1986.
30. H. Elliot, "Direct adaptive pole placement with application to nonminimum phase systems," *IEEE Trans. Aut. Control*, vol. AC-27, pp. 720–722, June 1982.
31. C. R. Johnson, Jr., *Lectures on Adaptive Parameter Estimation*, Prentice-Hall, Inc., 1987.
32. G. Bastin and M. R. Gevers, "Stable adaptive observers for nonlinear time-varying systems," *IEEE Trans. Aut. Control*, vol. AC-33, pp. 650–658, July 1988.
33. A. S. Morse and W. M. Wonham, "Status of non-interacting control," *IEEE Trans. Aut. Control*, vol. AC-16, pp. 568–581, Dec. 1971.

Robust Adaptive Control: Design, Analysis and Robustness Bounds*

Petros Ioannou and Aniruddha Datta

Department of Electrical Engineering-Systems
University of Southern California
Los Angeles, CA 90089-0781, USA.

Abstract. Despite the publication of numerous papers and several books dealing with the design, analysis and robustness properties of adaptive control, the theory of adaptive control may appear too technical and sometimes confusing to some people due to the so many different approaches, modifications and stability tools employed by various researchers in the field. The effort of this paper is to alleviate some of this confusion by presenting a procedure for designing and analyzing adaptive schemes. This procedure not only unifies most of the modifications used for stability and robustness but also clarifies why different approaches lead to the same result. The general framework that we developed can be used not only to analyze and unify most of the adaptive control schemes proposed in the literature but also to generate new ones. In addition, the analytical tools used allow the calculation of robustness margins and error bounds in a straightforward manner.

1 Introduction

The problem of designing stable adaptive controllers for a linear time-invariant system with no modeling errors was resolved as far back as 1980 [1–5]. Subsequently it was demonstrated [5–7] that in the presence of modeling errors and disturbances, an adaptive controller designed for the ideal situation, i.e., no modeling error or disturbances, could exhibit instability. Since then, for almost a decade, a considerable amount of effort has been directed towards the development of so-called "robust adaptive control schemes", i.e., adaptive control schemes which can retain certain stability properties in the presence of a wide class of modeling errors [8–19].

Due to these efforts, the 1980s have witnessed the development of a large number of robustness results in adaptive control. These results employ different robustness modifications and different methods of analysis and appear to be substantially different. However, in [13] it was shown that there does exist an underlying theory which could be used to unify within a single framework some of the robust adaptive control schemes presented in the literature. Furthermore, the results in [13] revealed that the key to a unified analysis of robust adaptive

* This work is supported in part by the National Science Foundation under Grant DMC-8452002 and in part by General Motors Foundation.

control schemes lies in first analyzing the properties of the robust adaptive law independent of the controller structure to be used and then combining the two of them together using the Certainty Equivalence Approach [20].

As far as the analysis of the adaptive laws is concerned, the ones derived in [13] are all based on gradient error minimization techniques. Indeed, such adaptive laws have the useful property that the stability analysis in the ideal case, i.e., in the absence of unmodeled dynamics and disturbances, extends, after some modifications, to the case where unmodeled dynamics and disturbances are present. Thus, it is not surprising that most of the robustness results to be found in the adaptive control literature use gradient or Newton's techniques for the derivation of the adaptive law. Nevertheless, it is well known that some of the first stability results in continuous-time adaptive control [1,2,3] were established using a positivity and Lyapunov-type synthesis approach which led to the emergence of the celebrated Strictly Positive Real (SPR) condition in adaptive control. The SPR condition was further exploited in [15] for the local analysis of adaptive systems using averaging techniques. In this paper we show that robust adaptive laws can be generated using not only gradient and Newton's techniques, but also by using the positivity and Lyapunov-type synthesis approach of [2] in a unified manner. One principal ingredient of this unification is the definition of the estimation error (known as prediction error in discrete-time algorithms) and the normalized estimation error, which clarifies the role of the artificial "augmented error" and "auxiliary" signal used in [2].

The robust adaptive laws developed in this paper can be combined, using the Certainty Equivalence Approach, with robust controller structures to yield robust adaptive control schemes. In particular, we consider the design and analysis of adaptive control schemes obtained from model reference, pole placement and linear quadratic (LQ) control structures, which are some of the most popular control structures in the adaptive control literature.

The paper is organized as follows: Section 2 contains some mathematical preliminaries. In Section 3 we develop the theory for the design and analysis of adaptive laws. Section 4 contains a discussion of commonly used adaptive controller structures and their associated robustness properties. In particular, we consider and analyze a model reference, a pole placement and a linear quadratic controller structure. In Section 5 we design and analyze robust model reference, pole placement and linear quadratic adaptive controllers by combining the appropriate controller structures of the previous section with any of the robust adaptive laws of Section 3. We illustrate the design procedure and analysis using simple examples. In Section 6, we summarize our conclusions and outline the directions for future research.

2 Mathematical Preliminaries

In this section, we give some definitions and lemmas which will be used in the subsequent sections.

Definition 2.1 *For any signal* $x : [0, \infty) \to \mathbb{R}^n$, x_t *denotes the truncation of* x *to the interval* $[0,t]$ *and is defined as*

$$x_t(\tau) = \begin{cases} x(\tau) & \text{if } \tau \leq t \\ 0 & \text{otherwise.} \end{cases} \tag{2.1}$$

Definition 2.2 *For any signal* $x : [0, \infty) \to \mathbb{R}^n$, *and for any* $\delta \geq 0$, $t \geq 0$, $\|x_t\|_2^\delta$ *is defined as*

$$\|x_t\|_2^\delta \triangleq \left(\int_0^t e^{-\delta(t-\tau)} [x^{\mathrm{T}}(\tau) x(\tau)] d\tau \right)^{\frac{1}{2}}, \tag{2.2}$$

provided that the integral in (2.2) exists.

Remark 2.1. Clearly, $\|(\cdot)_t\|_2^\delta$ represents the exponentially weighted L_2 norm of the signal truncated to $[0,t]$. When $\delta = 0$ and $t = \infty$, $\|(\cdot)_t\|_2^\delta$ becomes the usual L_2 norm and will be denoted by $\|\cdot\|_2$. It can be shown that $\|\cdot\|_2^\delta$ satisfies the usual properties of the vector norm. □

Definition 2.3 *Let* $H(s)$ *be a stable* $m \times n$ *proper transfer function matrix. Then,*

$$\|H(s)\|_\infty \triangleq \operatorname*{ess\,sup}_\omega \bar{\sigma}[H(j\omega)], \tag{2.3}$$

where $\bar{\sigma}$ *denotes the largest singular value.*

The $\|\cdot\|_\infty$ is the so-called H_∞-norm widely used in robust control [22].

Definition 2.4 *Let* $H(s)$ *be a proper transfer function matrix analytic in* $\mathrm{Re}[s] \geq -\delta/2$. *Then,*

$$\|H(s)\|_\infty^\delta \triangleq \|H(s - \delta/2)\|_\infty . \tag{2.4}$$

Definition 2.5 *Consider the signals* $x : [0,\infty) \to \mathbb{R}^n$, $y : [0,\infty) \to \mathbb{R}^+$ *and the set*

$$\mathcal{S}(y) = \left\{ x : [0,\infty) \to \mathbb{R}^n \,\middle|\, \int_t^{t+T} x^{\mathrm{T}}(\tau) x(\tau) d\tau \leq c \int_t^{t+T} y(\tau) d\tau + c_1 \right\},$$

for some $c, c_1 \geq 0$ *and* $\forall t, T \geq 0$. *We say that* x *is* y*-small in the mean if* $x \in S(y)$.

The following lemmas are key lemmas in the stability and robustness analysis of the adaptive control schemes to be presented in Sect. 5.

Lemma 2.1 *Consider the following homogeneous linear time-varying system:*

$$\dot{x} = A(t)x . \tag{2.5}$$

Let the elements of $A(t)$ *in (2.5) be differentiable bounded functions of time and assume that*

[A1] $\qquad \mathrm{Re}\{\lambda_i(A(t))\} \leq -\sigma_s, \quad \forall t \geq 0, \quad i = 1, 2, \ldots, n,$

where $\sigma_s > 0$ *is some constant.*
If any one of the following conditions

(i) $\int_t^{t+T} \left\|\dot{A}(\tau)\right\| d\tau \leq \mu T + \alpha_0$, *or*

(ii) $\int_t^{t+T} \left\|\dot{A}(\tau)\right\|^2 d\tau \leq \mu^2 T + \alpha_0$, *or*

(iii) $\left\|\dot{A}(t)\right\| \in L_2$,

where $\alpha_0 \in \mathbb{R}^+$ *and* $\mu \geq 0$, *is satisfied* $\forall t \geq 0$ *and some* $T \geq 0$, *then there exists a* $\mu^* > 0$ *such that* $\forall \mu \in [0, \mu^*)$ *the equilibrium state* $x_e = 0$ *of (2.5) is u.a.s in the large.*

Proof. From [A1] it follows that the Lyapunov equation

$$A^{\mathrm{T}}(t)P(t) + P(t)A(t) = -I \tag{2.6}$$

has a unique bounded solution $P(t)$ for each t. We consider the following Lyapunov function:

$$V(t,x) = x^{\mathrm{T}} P(t) x\,.$$

Then, along the solution of (2.5), we have:

$$\dot{V} = -\|x(t)\|^2 + x^{\mathrm{T}}(t)\dot{P}(t)x(t)\,. \tag{2.7}$$

From (2.6), $\dot{P}$ satisfies:

$$A^{\mathrm{T}}(t)\dot{P}(t) + \dot{P}(t)A(t) = -Q(t)\,, \quad \forall t \geq 0\,, \tag{2.8}$$

where $Q(t) = \dot{A}^{\mathrm{T}}(t)P(t) + P(t)\dot{A}(t)$. Due to [A1], it can be verified [23] that

$$\dot{P}(t) = \int_0^\infty e^{A^{\mathrm{T}}(t)\tau} Q(t) e^{A(t)\tau} d\tau$$

satisfies (2.8) for each $t \geq 0$. Therefore,

$$\left\|\dot{P}(t)\right\| \leq \|Q(t)\| \int_0^\infty \left\|e^{A^{\mathrm{T}}(t)\tau}\right\| \left\|e^{A(t)\tau}\right\| d\tau\,.$$

Since [A1] implies that $\left\|e^{A(t)\tau}\right\| \leq \alpha_1 e^{-\alpha_0 \tau}$ for some α_1, $\alpha_0 > 0$, it follows that

$$\left\|\dot{P}(t)\right\| \leq c\|Q(t)\|\,,$$

for some $c \geq 0$. Then,

$$\|Q(t)\| \leq 2\|P(t)\| \left\|\dot{A}(t)\right\|$$

together with $P \in L_\infty$ imply that

$$\left\|\dot{P}(t)\right\| \leq \beta \left\|\dot{A}(t)\right\|\,, \quad \forall t \geq 0 \tag{2.9}$$

for some constant $\beta \geq 0$. Using (2.9) in (2.7) and noting that P satisfies $0 < \beta_1 \leq \lambda_{\min}(P) \leq \lambda_{\max}(P) \leq \beta_2$ for some β_1, $\beta_2 > 0$, we have that:

$$\dot{V}(t) \leq -\|x(t)\|^2 + \beta \left\|\dot{A}(t)\right\| \|x(t)\|^2 \leq -\beta_2^{-1} V(t) + \beta\beta_1^{-1} \left\|\dot{A}(t)\right\| V(t)\,.$$

Therefore,

$$V(t) \leq V(t_0) \exp\left\{-\int_{t_0}^{t} \left(\beta_2^{-1} - \beta\beta_1^{-1}\left\|\dot{A}(\tau)\right\|\right) d\tau\right\}. \tag{2.10}$$

Using condition (i) in (2.10), we have

$$V(t) \leq V(t_0) \exp\left\{-\left(\beta_2^{-1} - \beta\beta_1^{-1}\mu\right)(t-t_0)\right\} \exp\left\{\beta\beta_1^{-1}\alpha_0\right\}.$$

Therefore, for $\mu^* = \frac{\beta_1}{\beta_2\beta}$ and $\forall\, \mu \in [0, \mu^*)$ $V(t) \to 0$ exponentially fast as $t \to \infty$, which implies that $x_e = 0$ is u.a.s. in the large.

In order to use condition (ii) we rewrite (2.10) as

$$V(t) \leq V(t_0) \exp\left\{-\beta_2^{-1}(t-t_0)\right\} \exp\left\{\beta\beta_1^{-1}\int_{t_0}^{t}\left\|\dot{A}(\tau)\right\| d\tau\right\}.$$

Using the Schwartz inequality and condition (ii), we have

$$\begin{aligned}\int_{t_0}^{t}\left\|\dot{A}(\tau)\right\| d\tau &\leq \left(\int_{t_0}^{t}\left\|\dot{A}(\tau)\right\|^2 d\tau\right)^{\frac{1}{2}} \sqrt{t-t_0} \\ &\leq \left[\mu^2(t-t_0)^2 + \alpha_0(t-t_0)\right]^{\frac{1}{2}} \\ &\leq \mu(t-t_0) + \sqrt{\alpha_0}\sqrt{t-t_0}.\end{aligned}$$

Therefore,

$$V(t) \leq V(t_0)\exp\left\{-\alpha(t-t_0)\right\} y(t),$$

where $\alpha = \frac{1}{2}\beta_2^{-1} - \beta\beta_1^{-1}\mu$, and

$$y(t) = \exp\left[-\frac{1}{2}\beta_2^{-1}(t-t_0) + \beta\beta_1^{-1}\sqrt{\alpha_0}\sqrt{t-t_0}\right].$$

It can be shown that

$$y(t) \leq \exp\left[\frac{1}{2}\alpha_0\frac{\beta^2\beta_2}{\beta_1^2}\right] = c, \quad \forall t \geq t_0.$$

Hence,

$$V(t) \leq cV(t_0)\exp\left\{-\alpha(t-t_0)\right\}.$$

Choosing $\mu^* = \frac{\beta_1}{2\beta_2\beta}$, we have $\forall\, \mu \in [0, \mu^*)$ that $\alpha > 0$ and, therefore, $V(t) \to 0$ exponentially fast as $t \to \infty$, which implies that $x_e = 0$ is u.a.s. in the large.

Since condition (iii) implies condition (i), the proof of part (iii) follows directly from that of part (i). □

The proof of Lemma 2.1 can also be found in [13] and it is a byproduct of results on slowly time-varying systems given in [24].

Lemma 2.2 *Let*

$$z = H(s)[u] = h * u\,,$$

where $H(s)$ is a proper transfer matrix and $h(t)$ is the corresponding impulse response.
If $H(s)$ is analytic in $\mathrm{Re}[s] \geq -\delta/2$ *for some $\delta \geq 0$ and $u \in L_{2e}$, then*

$$\|z_t\|_2^\delta \leq \|H(s)\|_\infty^\delta \|u_t\|_2^\delta\,. \tag{2.11}$$

If, in addition, $H(s)$ is strictly proper, then

$$\|z(t)\| \leq \frac{1}{\sqrt{p}} \|(s+p)H(s)\|_\infty^\delta \|u_t\|_2^\delta\,, \tag{2.12}$$

where $p \geq \delta$ is an arbitrary constant.

The proof of Lemma 2.2 follows from the results in [17,18] and is given below.

Proof. The transfer function $H(s)$ can be expressed as $H(s) = d + H_a(s)$ with

$$h(t) = \begin{cases} 0\,, & t < 0 \\ d\delta(t) + h_a(t)\,, & t \geq 0\,. \end{cases}$$

If we define

$$h_\delta(t) = \begin{cases} 0\,, & t < 0 \\ d\delta(t) + e^{\frac{\delta}{2}t} h_a(t)\,, & t \geq 0 \end{cases}$$

and $u_\delta(t) = e^{\frac{\delta}{2}t} u(t)$, then

$$z_\delta(t) = \int_0^t e^{\frac{\delta}{2}(t-\tau)} h(t-\tau) e^{\frac{\delta}{2}\tau} u(\tau) d\tau = h_\delta * u_\delta\,,$$

where $z_\delta(t) = e^{\frac{\delta}{2}t} z(t)$. Now $u \in L_{2e} \Rightarrow u_\delta \in L_{2e}$. Therefore, for the truncated signals $z_{\delta t}$, $u_{\delta t}$ at time t, we have:

$$\|z_{\delta t}\|_2 \leq \|H(s-\delta/2)\|_\infty \|u_{\delta t}\|_2\,. \tag{2.13}$$

Since $e^{-\frac{\delta}{2}t}\|z_{\delta t}\|_2 = \|z_t\|_2^\delta$ and $e^{-\frac{\delta}{2}t}\|u_{\delta t}\|_2 = \|u_t\|_2^\delta$, (2.11) follows directly from (2.13).

If $H(s)$ is strictly proper, it can be written as

$$H(s) = \frac{1}{s+p} H_p(s)$$

for some $p > 0$, where $H_p(s) = (s+p)H(s)$ is a proper transfer function and

$$z(t) = \frac{1}{s+p}[\bar{z}(t)]\,, \quad \bar{z}(t) = H_p(s)[u(t)]\,.$$

Then,

$$\|z(t)\| leq \int_0^t e^{-p(t-\tau)} \|\bar{z}(\tau)\| d\tau \leq \left(\int_0^t e^{-p(t-\tau)} d\tau\right)^{\frac{1}{2}} \left(\int_0^t e^{-p(t-\tau)} \|\bar{z}(\tau)\|^2 d\tau\right)^{\frac{1}{2}}\,,$$

where the last inequality is obtained by using the Schwartz inequality. Since $(1-e^{-pt})/p \leq 1/p$ and $p \geq \delta$, we have:

$$\|z(t)\| \leq \frac{1}{\sqrt{p}} \left(\int_0^t e^{-\delta(t-\tau)} \|\bar{z}(\tau)\|^2 d\tau \right)^{\frac{1}{2}} = \frac{1}{\sqrt{p}} \|\bar{z}_t\|_2^\delta ,$$

which, together with

$$\|\bar{z}_t\|_2^\delta \leq \|H_p(s-\delta/2)\|_\infty \|u_t\|_2^\delta = \|(s+p-\delta/2)H(s-\delta/2)\|_\infty \|u_t\|_2^\delta ,$$

imply (2.12). □

Lemma 2.3 *Consider the linear time-varying system given by*

$$\dot{x} = A(t)x + B(t)u , \quad x(0) = x_0 , \tag{2.14}$$

where $x \in \mathbb{R}^n$, $u \in \mathbb{R}^m$, and the matrices A, $B \in L_\infty$ and their elements are continuous functions of time. If the state transition matrix $\phi(t,\tau)$ of (2.14) satisfies

$$\|\phi(t,\tau)\| \leq \lambda_0 e^{-\delta_0(t-\tau)} \tag{2.15}$$

for some λ_0, $\delta_0 > 0$, and $u(t)$ is continuous, then for any $\delta \in [0, \delta_0)$, we have

(i) $\|x(t)\| \leq \frac{\lambda}{\sqrt{\delta_0}} \|u_t\|_2^\delta + \epsilon_t$, *and*

(ii) $\|x_t\|_2^\delta \leq \frac{\lambda}{\sqrt{\delta_0(\delta_0-\delta)}} \|u_t\|_2^\delta + \epsilon_t$,

where ϵ_t is an exponentially decaying to zero term due to x_0, and $\lambda = c\lambda_0$, where c is the L_∞ bound for $B(t)$.

Proof. The solution $x(t)$ of (2.14) can be expressed as

$$x(t) = \phi(t,0)x_0 + \int_0^t \phi(t,\tau)B(\tau)u(\tau)d\tau .$$

Therefore,

$$\|x(t)\| \leq \|\phi(t,0)\| \, |x_0| + \int_0^t \|\phi(t,\tau)\| \, \|B(\tau)u(\tau)\| \, d\tau .$$

Using (2.15), we have

$$\|x(t)\| \leq \epsilon_t + \lambda \int_0^t e^{-\delta_0(t-\tau)} \|u(\tau)\| d\tau , \tag{2.16}$$

where c and λ are as defined in the statement of the lemma. Applying the Schwartz inequality and using the fact that $0 \leq \delta < \delta_0$, we have

$$\|x(t)\| \leq \epsilon_t + \lambda \left(\int_0^t e^{-\delta_0(t-\tau)} d\tau \right)^{\frac{1}{2}} \left(\int_0^t e^{-\delta(t-\tau)} \|u(\tau)\|^2 d\tau \right)^{\frac{1}{2}} \leq \epsilon_t + \frac{\lambda}{\sqrt{\delta_0}} \|u_t\|_2^\delta ,$$

which completes the proof of (i). Using the triangle inequality for the $\|\cdot\|_2^\delta$-norm it follows from (2.16) that

$$\|x_t\|_2^\delta \leq \|\epsilon_t\|_2^\delta + \lambda\|Q_t\|_2^\delta\,, \tag{2.17}$$

where

$$\|Q_t\|_2^\delta \triangleq \left\| \left(\int_0^t e^{-\delta_0(t-\tau)}\|u(\tau)\|d\tau \right)_t \right\|_2^\delta$$
$$= \left[\int_0^t e^{-\delta(t-\tau)} \left(\int_0^\tau e^{-\delta_0(\tau-s)}\|u(s)\|ds \right)^2 d\tau \right]^{\frac{1}{2}} .$$

Using the Schwartz inequality, we have:

$$\|Q_t\|_2^\delta \leq \left[\int_0^t e^{-\delta(t-\tau)} \left(\int_0^\tau e^{-\delta_0(\tau-s)}ds \int_0^\tau e^{-\delta_0(\tau-s)}\|u(s)\|^2 ds \right) d\tau \right]^{\frac{1}{2}}$$
$$\leq \frac{1}{\sqrt{\delta_0}} \left[\int_0^t e^{-\delta(t-\tau)} \int_0^\tau e^{-\delta_0(\tau-s)}\|u(s)\|^2 ds d\tau \right]^{\frac{1}{2}} ,$$

where the last inequality is obtained from the use of $\int_0^\tau e^{-\delta_0(\tau-s)}ds \leq 1/\delta_0$. We now change the sequence of integration and obtain

$$\|Q_t\|_2^\delta \leq \frac{1}{\sqrt{\delta_0}} \left[e^{-\delta t} \int_0^t e^{\delta_0 s}\|u(s)\|^2 \left(\int_s^t e^{-(\delta_0-\delta)\tau} \right) ds \right]^{\frac{1}{2}}$$
$$\leq \frac{1}{\sqrt{\delta_0}} \left[e^{-\delta t} \int_0^t e^{\delta_0 s}\|u(s)\|^2 \frac{e^{-(\delta_0-\delta)s}}{\delta_0-\delta} ds \right]^{\frac{1}{2}}$$
$$\leq \frac{1}{\sqrt{\delta_0(\delta_0-\delta)}}\|u_t\|_2^\delta\,,$$

which, together with (2.17) and the fact that $\|\epsilon_t\|_2^\delta$ is also exponentially decaying to zero, complete the proof. □

Lemma 2.4 (Gronwall Lemma). *Let $\lambda(t)$, $g(t)$, $k(t)$ be nonnegative piecewise continuous functions of time. If the function $y(t)$ satisfies the inequality:*

$$y(t) \leq \lambda(t) + g(t) \int_{t_0}^t k(s)y(s)ds\,, \quad \forall t \geq t_0 \geq 0\,,$$

then

$$y(t) \leq \lambda(t) + g(t) \int_{t_0}^t \lambda(s)k(s) \left(\exp\left(\int_s^t k(\tau)g(\tau)d\tau \right) \right) ds\,, \quad \forall t \geq t_0 \geq 0\,.$$

In particular, if $\lambda(t) \equiv \lambda$ is a constant and $g(t) \equiv 1$, then

$$y(t) \leq \lambda \exp\left(\int_{t_0}^t k(s)ds \right)\,, \quad \forall t \geq t_0 \geq 0\,.$$

The proof can be found in [24].

Lemma 2.5 (Swapping Lemma). *Let $\phi, \omega : \mathbb{R}^+ \to \mathbb{R}^n$ and let ϕ be differentiable. Let $W(s)$ be a proper stable rational transfer function of order m with a minimal realization (A, b, c, d), i.e.,*

$$W(s) = c^{\mathrm{T}}(sI - A)^{-1}b + d\,.$$

Then,

$$W(s)\left[\phi^{\mathrm{T}}\omega\right] = \phi^{\mathrm{T}}W(s)[\omega] + W_c(s)\left[\left(W_b(s)\left[\omega^{\mathrm{T}}\right]\right)\dot{\phi}\right],$$

where

$$W_c(s) = -c^{\mathrm{T}}(sI - A)^{-1}, \quad W_b(s) = (sI - A)^{-1}b\,.$$

Proof. We have

$$\begin{aligned}
W(s)\left[\phi^{\mathrm{T}}\omega\right] &= W(s)\left[\omega^{\mathrm{T}}\phi\right] = d\phi^{\mathrm{T}}\omega + c^{\mathrm{T}}\int_0^t e^{A(t-\tau)}b\omega^{\mathrm{T}}\phi d\tau \\
&= d\phi^{\mathrm{T}}\omega + c^{\mathrm{T}}e^{At} \times \\
&\quad \left(\int_0^{t'} e^{-A\tau}b\omega^{\mathrm{T}}(\tau)d\tau\phi(t')\Bigg|_{t'=0}^{t'=t} - \int_0^t\int_0^{\tau} e^{-A\sigma}b\omega^{\mathrm{T}}(\sigma)d\sigma\dot{\phi}(\tau)d\tau\right) \\
&= \phi^{\mathrm{T}}\left[d\omega + c^{\mathrm{T}}\int_0^t e^{A(t-\tau)}b\omega(\tau)d\tau\right] \\
&\quad -c^{\mathrm{T}}\int_0^t e^{A(t-\tau)}\int_0^{\tau} e^{A(\tau-\sigma)}b\omega^{\mathrm{T}}(\sigma)d\sigma\dot{\phi}(\tau)d\tau \\
&= \phi^{\mathrm{T}}(t)W(s)[\omega] - c^{\mathrm{T}}(sI - A)^{-1}\left[(sI - A)^{-1}b\left[\omega^{\mathrm{T}}\right]\dot{\phi}\right].
\end{aligned}$$

□

Various other proofs for Lemma 2.5 can be found in [16,17].

Lemma 2.6 [18] *Let ϕ, $\omega : \mathbb{R}^+ \to \mathbb{R}^n$ and ϕ, ω be differentiable. Let $W(s)$ be a rational transfer function with relative degree n^* and with stable poles and zeros. Then,*

$$\begin{aligned}
\phi^{\mathrm{T}}\omega &= \Lambda_1(s,\alpha)\left[\dot{\phi}^{\mathrm{T}}\omega + \phi^{\mathrm{T}}\dot{\omega}\right] \\
&\quad +\Lambda(s,\alpha)W^{-1}(s)\left[\phi^{\mathrm{T}}W(s)[\omega] + W_c(s)\left[\left(W_b(s)\left[\omega^{\mathrm{T}}\right]\right)\dot{\phi}\right]\right], \quad (2.18)
\end{aligned}$$

where $s\Lambda_1(s) = 1 - \Lambda(s,\alpha)$, $\Lambda(s,\alpha) = \alpha^k/(s+\alpha)^k$ and $k \geq n^*$ for any arbitrary constant $\alpha > 0$. Furthermore, for large α

$$\|\Lambda_1(s,\alpha)\|_{\infty}^{\delta} \leq \frac{c}{\alpha}$$

for any $\delta \in [0, \delta_0]$, where $0 < \delta_0 << \alpha$ and $c \in \mathbb{R}^+$.

Proof. We have

$$\phi^{\mathrm{T}}\omega = \left(1 - \frac{\alpha^k}{(s+\alpha)^k}\right)\left[\phi^{\mathrm{T}}\omega\right] + \frac{\alpha^k}{(s+\alpha)^k} W^{-1}(s)W(s)\left[\phi^{\mathrm{T}}\omega\right].$$

Since

$$1 - \frac{\alpha^k}{(s+\alpha)^k} = \frac{(s+\alpha)^k - \alpha^k}{(s+\alpha)^k} = s\Lambda_1(s),$$

we have

$$\phi^{\mathrm{T}}\omega = \Lambda_1(s)\left[\dot\phi^{\mathrm{T}}\omega + \phi^{\mathrm{T}}\dot\omega\right] + \Lambda(s)W^{-1}(s)W(s)\left[\phi^{\mathrm{T}}\omega\right]. \tag{2.19}$$

Using Lemma 2.5 to substitute for $W(s)\left[\phi^{\mathrm{T}}\omega\right]$, the equality (2.18) follows directly from (2.19). Now

$$\begin{aligned}\Lambda_1(s) &= \frac{1}{s}\frac{[(s+\alpha)^k - \alpha^k]}{(s+\alpha)^k} \\ &= \frac{1}{(s+\alpha)^k}\left[s^{k-1} + \binom{k}{1}\alpha s^{k-2} + \binom{k}{2}\alpha^2 s^{k-3} + \cdots + \binom{k}{k-1}\alpha^{k-1}\right],\end{aligned}$$

where

$$\binom{k}{n} = \frac{k!}{n!(k-n)!}.$$

Hence, for large α, $\|\Lambda_1(s)\|_\infty^\delta \le \frac{c}{\alpha}$ for some $c \in \mathbb{R}^+$. □

Lemma 2.7 *Consider the system*

$$y = H(s)[u], \tag{2.20}$$

where $H(s)$ is a strictly proper rational transfer function analytic in $\mathrm{Re}[s] \ge 0$. *If $u \in \mathcal{S}(\mu^2)$ for some constant $\mu \ge 0$, then $y \in L_\infty$.*

Proof. It follows from (2.20) and the properties of $H(s)$ that

$$|y(t)|^2 \le \left(\alpha_1 \int_0^t e^{-\alpha_0(t-\tau)}\|u(\tau)\|d\tau\right)^2 \le \frac{\alpha_1^2}{\alpha_0}\int_0^t e^{-\alpha_0(t-\tau)}\|u(\tau)\|^2 d\tau$$

for some α_1, $\alpha_0 > 0$. Now

$$\begin{aligned}\Delta(t,0) &\triangleq \int_0^t e^{-\alpha_0(t-\tau)}\|u(\tau)\|^2 d\tau \le e^{-\alpha_0 t}\sum_{i=0}^{n}\int_i^{i+1} e^{\alpha_0\tau}\|u(\tau)\|^2 d\tau \\ &\le e^{-\alpha_0 t}\sum_{i=0}^{n} e^{\alpha_0(i+1)}\int_i^{i+1}\|u(\tau)\|^2 d\tau,\end{aligned}$$

where n is an integer satisfying $n \le t < n+1$. Since $u \in \mathcal{S}(\mu^2)$, it follows that

$$\Delta(t,0) \le e^{-\alpha_0 t}(\mu^2 + c)\sum_{i=0}^{n} e^{\alpha_0(i+1)} \le \frac{\mu^2 + c}{1 - e^{-\alpha_0}}e^{\alpha_0} < \infty.$$

□

3 On-line Parameter Estimation

In this section we will consider the problem of estimating on line the constant parameter vector θ^* of a certain class of dynamic systems described by

$$z(t) = f(\theta^*, t, \tau), \quad t \geq \tau \geq 0, \tag{3.1}$$

where at each time t, the response $z(\tau)$ with $\tau \leq t$ can be observed and f is some function or functional whose form may be known. If we let $\theta(t)$ to be the estimate of θ^* at time t, then the estimate $\hat{z}(t) \triangleq \hat{z}(t, \theta)$ of $z(t)$ can be constructed as

$$\hat{z}(t, \theta) = \hat{f}(\theta, t, \tau), \tag{3.2}$$

for some function or functional $\hat{f}$ which may or may not have the same form as f. The estimation problem is to design the adjustment law for θ so that $\hat{z}(t, \theta)$ is as close as possible to $z(t)$ in some sense. The form and the properties of the adjustment law which we refer to as the *adaptive law* will be very much dependent on the form of the functions f and $\hat{f}$ as well as on the criterion chosen to compare $\hat{z}(t, \theta)$ with $z(t)$. Typical choices of criteria for checking the quality of the estimation might be:

(i) $||\hat{z}(t, \theta) - z(t)|| \in L_2 \cap L_\infty$
(ii) $||\hat{z}(t, \theta) - z(t)|| \to 0$ as $t \to \infty$
(iii) $||\theta(t) - \theta^*|| \to 0$ as $t \to \infty$

or any combination of (i)–(iii). For example, in parameter identification for stable plants criterion (iii) is very crucial, whereas in adaptive control (i) and (ii) may be important but not (iii).

An important class of parametric models for θ^* of the form (3.1) is the one where θ^* appears in a linear form, i.e.,

$$z = W(s)\left[\theta^{*\mathrm{T}}\omega + \eta_0\right], \tag{3.3}$$

where

- $W(s)$ is a known proper transfer function with stable poles,
- $z \in \mathbb{R}$, $\omega \in \mathbb{R}^n$ are piecewise continuous signals which can be measured at each time t,
- θ^* is the unknown constant vector to be estimated, and
- $\eta_0 \in \mathbb{R}$ is an unknown signal due to modeling errors such as unmodeled dynamics, disturbances, noise etc.

We will refer to (3.3) as the *linear parametric model* for θ^*. Another interesting class of (3.1) is the one where the unknown parameters appear in the special bilinear form

$$z = W(s)\left[\rho^*\left(\theta^{*\mathrm{T}}\omega + z_0\right) + \eta_0\right], \tag{3.4}$$

where $z_0 \in \mathbb{R}$ is a piecewise continuous signal which can be measured at each time t and $\rho^* \in \mathbb{R}$ is an unknown constant whose sign is known. We will refer to (3.4) as the *bilinear parametric model* for θ^*, ρ^*. The general case where θ^* appears nonlinearly is a difficult one for estimation. Approximation techniques using Taylor series expansions around some known nominal value of θ^* may be used to approximate (3.1) with (3.3) or (3.4).

The following examples illustrate that (3.3) and (3.4) apply to a large class of dynamical systems.

Example 3.1. Let us consider the LTI system

$$y = G_0(s)(1 + \Delta_m(s))[u], \tag{3.5}$$

where $G_0(s) = Z_0(s)/R_0(s)$, $\Delta_m(s)$ represent the nominal plant and multiplicative uncertainty, respectively, and

$$\begin{aligned} Z_0(s) &= b_m s^m + b_{m-1}s^{m-1} + \cdots + b_0 \\ R_0(s) &= s^n + a_{n-1}s^{n-1} + \cdots + a_0 \,. \end{aligned}$$

Let a_i, b_j, $i = 0, 1, \ldots, n-1$, $j = 0, 1, \ldots, m$ be the unknown parameters to be estimated. Then, (3.5) can be expressed as

$$\begin{aligned} &(s^n + a_{n-1}s^{n-1} + \cdots + a_0)[y] \\ &\quad = (b_m s^m + b_{m-1}s^{m-1} + \cdots + b_0)[u] + Z_0(s)\Delta_m(s)[u] \,. \end{aligned} \tag{3.6}$$

If the derivatives of y, u are available, then (3.6) can be easily expressed in the form of (3.3) by defining

$$\begin{aligned} \theta^* &= [a_{n-1}, \ldots, a_0\, b_m, \ldots, b_0]^{\mathrm{T}} \\ W(s) &= 1 \\ \omega &= \left[-y^{(n-1)} - y^{(n-2)} \cdots - y\, u^{(m)} \cdots u\right]^{\mathrm{T}} \\ \eta_0 &= Z_0(s)\Delta_m(s)[u] \\ z &= y^{(n)} \,. \end{aligned}$$

In the case where the derivatives of y and u are not available, the filtered values of y and u can be used as follows: We filter each side of (3.6) with the filter $1/\Lambda(s)$, where $\Lambda(s)$ is an nth-order Hurwitz polynomial to obtain

$$\begin{aligned} &\left(s^n + a_{n-1}s^{n-1} + \cdots + a_0\right)\frac{1}{\Lambda(s)}[y] \\ &\quad = \left(b_m s^m + b_{m-1}s^{m-1} + \cdots + b_0\right)\frac{1}{\Lambda(s)}[u] + \frac{Z_0(s)\Delta_m(s)}{\Lambda(s)}[u] \,, \end{aligned} \tag{3.7}$$

or

$$\frac{s^n}{\Lambda(s)}[y] = \theta^{*\mathrm{T}}\zeta + \frac{Z_0(s)\Delta_m(s)}{\Lambda(s)}[u] \,, \tag{3.8}$$

where

$$\zeta = \left[-\frac{s^{n-1}}{\Lambda(s)}[y], \ldots, -\frac{1}{\Lambda(s)}[y], \frac{s^m}{\Lambda(s)}[u], \ldots, \frac{1}{\Lambda(s)}[u]\right]^{\mathrm{T}} ,$$

i.e., (3.8) is in the form of (3.3) with $W(s) = 1$, $z = \frac{s^n}{\Lambda(s)}[y]$, $\eta_0 = \frac{Z_0(s)\Delta_m(s)}{\Lambda(s)}[u]$ and $\omega = \zeta$.

An alternative way is to write (3.7) as

$$y = \frac{\Lambda(s) - R_0(s)}{\Lambda(s)}[y] + \frac{Z_0(s)}{\Lambda(s)}[u] + \eta_0\,, \tag{3.9}$$

by adding y on both sides of (3.7). Then, for $\Lambda(s) = s^n + \lambda_{n-1}s^{n-1} + \cdots + \lambda_0$, (3.9) may be expressed in the form (3.3) by letting

$$\begin{aligned}\theta^* &= [a_{n-1} - \lambda_{n-1}, \ldots, a_0 - \lambda_0, b_m, \ldots, b_0]^{\mathrm{T}} \\ z &= y \\ W(s) &= 1\end{aligned}$$

and $\omega = \zeta$ as defined in (3.8). □

Example 3.2. Let us consider the nonlinear system

$$c^*\dot{x} = a^*g(x) + b^*f(x)\,, \tag{3.10}$$

where a^*, b^*, c^* are unknown scalars to be estimated, $\mathrm{sgn}(c^*) = 1$, and $g(x)$, $f(x)$ are nonlinear functions which can be measured at each time t. We can write (3.10) as

$$c^*(s+1)[x] = a^*g(x) + b^*f(x) + c^*x\,.$$

Filtering each side by $\frac{1}{c^*(s+1)}$ we obtain

$$x = \frac{1}{s+1}\left[\rho^*\theta^{*\mathrm{T}}\omega\right]\,,$$

where $\rho^* = 1/c^*$, which is in the form of (3.4) with $z_0 = 0$, $\eta_0 = 0$, $\theta^* = [a^*, b^*, c^*]^{\mathrm{T}}$, $\omega = [g(x), f(x), x]^{\mathrm{T}}$ and $W(s) = 1/s + 1$. □

We should note that (3.4) can be converted to the form of the linear parametric model (3.3) by rewriting (3.4) as

$$z = W(s)\left[\bar{\theta}^{*\mathrm{T}}\bar{\omega} + \eta_0\right]\,,$$

where $\bar{\theta}^* = \left[\rho^*\theta^{*\mathrm{T}},\ \rho^*\right]$, $\bar{\omega} = [\omega^{\mathrm{T}},\ z_0]^{\mathrm{T}}$. In this case the estimate of θ^* is calculated by dividing the estimate of $\rho^*\theta^*$ by that of ρ^*. Such calculations may involve division by small numbers and can be avoided by modifying the adaptive law for ρ^* so that the estimate $\rho(t)$ of ρ^* satisfies $|\rho(t)| \geq \rho_0 > 0\ \forall t \geq 0$ and some constant $\rho_0 \leq |\rho^*|$. This modification requires the knowledge of ρ_0 and sign of ρ^* [5,10,16,26].

In the next subsections we show how θ^* in the parametric models (3.3) and (3.4) can be estimated on-line using various techniques based on Lyapunov, gradient and least-squares methods.

3.1 Lyapunov Method

With this method, a dynamic equation is first obtained in terms of error signals which include the estimation and parameter error. A certain Lyapunov-like function V is then considered whose time derivative $\dot{V}$ along the trajectories of the dynamic equation is made nonpositive for $V \geq V_0$ and some constant $V_0 \geq 0$, by properly designing the adaptive law. The properties of V and $\dot{V}$ are then used to establish the stability properties of the on-line estimation scheme. The procedure for developing the dynamic error equation and choosing V is given below.

Since θ^* is a constant vector, the linear parametric model (3.3) may be rewritten in the form

$$z = W(s)L(s)\left[\theta^{*\mathrm{T}}\zeta + \eta\right], \tag{3.11}$$

where $L(s)$, $L^{-1}(s)$ are proper transfer functions with stable poles, $\zeta = L^{-1}(s)[\omega]$, $\eta = L^{-1}(s)[\eta_0]$ and $L(s)$ is chosen so that $W(s)L(s)$ is a strictly proper SPR transfer function. For some $W(s)$ it is possible that no $L(s)$ can be found such that $W(s)L(s)$ is strictly proper and SPR. In such cases (3.3) can still be manipulated and put in the form (3.11), e.g., for $W(s) = \frac{s-1}{s+2}$ we can write (3.3) as $\bar{z} = \frac{s+1}{(s+2)(s+3)}\left[\theta^{*\mathrm{T}}\zeta + \eta\right]$, where $\zeta = \frac{s-1}{s+1}[\omega]$, $\eta = \frac{s-1}{s+1}[\eta_0]$ and $\bar{z} = \frac{1}{s+3}[z]$.

Let $\theta(t)$ be the estimate of θ^* at time t. Then the estimate $\hat{z}$ of z at time t can be constructed as

$$\hat{z} = W(s)L(s)\left[\theta^{\mathrm{T}}\zeta\right].$$

The estimation error ϵ_1 is then given by

$$\epsilon_1 = \hat{z} - z = W(s)L(s)\left[\phi^{\mathrm{T}}\zeta - \eta\right], \tag{3.12}$$

where $\phi \triangleq \theta - \theta^*$ is the parameter error, i.e., ϵ_1 is a measure of the uncertainty between $\hat{z}$ and z due to $\theta(t) \neq \theta^*$ and the modeling error term $\eta \neq 0$. We can also construct the normalized estimation error ϵ as

$$\epsilon = \epsilon_1 - W(s)L(s)\left[\epsilon n_s^2\right] = W(s)L(s)\left[\phi^{\mathrm{T}}\zeta - \epsilon n_s^2 - \eta\right], \tag{3.13}$$

where n_s is what is called the normalizing signal [11,9,13] and is designed so that

$$\text{(A1)} \qquad \frac{\zeta}{m}, \frac{\eta}{m} \in L_\infty,$$

where $m^2 = \alpha + \beta n_s^2$ for some α, $\beta > 0$. The normalizing effect of n_s is more transparent when we express (3.13) in a state-space form and solve for the quasi-steady state of ϵ, i.e.,

$$\begin{aligned} \dot{e} &= A_c e + b_c(\phi^{\mathrm{T}}\zeta - \epsilon n_s^2 - \eta) \\ \epsilon &= h_c e, \end{aligned} \tag{3.14}$$

where (h_c, A_c, b_c) is a minimal state representation of $W(s)L(s) = h_c(sI - A_c)^{-1}b_c$. Setting $\dot{e} = 0$ and solving for the quasi-steady-state response ϵ_{ss} of ϵ we obtain

$$\epsilon_{ss} = \frac{-h_c A_c^{-1} b_c(\phi^T\zeta - \eta)}{1 - h_c A_c^{-1} b_c n_s^2} = \frac{\epsilon_{1ss}}{1 - h_c A_c^{-1} b_c n_s^2} .$$

Since $W(s)L(s)$ is SPR, $h_c A_c^{-1} b_c < 0$ and therefore ϵ_{ss} is equal to the normalized quasi-steady-state response ϵ_{1ss} of the estimation error ϵ_1. Due to n_s, ϵ_{ss} can only be driven unbounded by the parameter error ϕ. In [2] ϵ_1 is called the "augmented error" and ϵn_s^2 the "auxiliary signal" and they are developed using different considerations. Treating ϵ_1, ϵ as the estimation and normalized estimation errors helps us to understand the separation of the identification and control parts and allows us to unify the various approaches based on other techniques such as gradient and least-squares methods.

It should be noted that $\phi = \theta - \theta^*$ in (3.12)–(3.14) is an unknown signal vector and therefore the second equalities in (3.12) and (3.13) cannot be used to generate ϵ_1, ϵ. The signals ϵ_1, ϵ are generated from the first equalities in (3.12) and (3.13), whereas the second equalities are used for analysis only.

The adaptive law for generating the parameter estimates $\theta(t)$ is developed by considering the Lyapunov-like function

$$V(\phi, e) = \frac{e^T P_c e}{2} + \frac{\phi^T \Gamma^{-1} \phi}{2} ,$$

where $\Gamma = \Gamma^T > 0$ and $P_c = P_c^T > 0$. The matrix P_c is chosen as the one which satisfies the algebraic equations

$$\begin{aligned} P_c A_c + A_c^T P_c &= -qq^T - \nu L \\ P_c b_c &= h_c \end{aligned} \tag{3.15}$$

for some vector q, $L = L^T > 0$ and a small constant $\nu > 0$ and whose existence is guaranteed by the SPR property of $W(s)L(s) = h_c(sI - A_c)^{-1}b_c$ [2,10]. Using (3.15), the time derivative $\dot{V}$ of V along the solution of (3.14) is given by

$$\dot{V}(\phi, e) = -\frac{1}{2}e^T qq^T e - \frac{\nu}{2}e^T Le + \epsilon\phi^T\zeta - \epsilon^2 n_s^2 - \epsilon\eta + \phi^T\Gamma^{-1}\dot{\phi} .$$

Choosing

$$\dot{\phi} = \dot{\theta} = -\Gamma\epsilon\zeta - \Gamma w(t)\theta , \tag{3.16}$$

where $w(t) \geq 0$ is a signal to be designed, referred to as the *leakage*, we have

$$\dot{V}(\phi, e) \leq -\frac{\nu}{2}e^T Le - \epsilon^2 n_s^2 - \epsilon\eta - w(t)\phi^T\theta .$$

Using $-\frac{\nu}{4}e^T Le \leq -\frac{\beta}{2}|\epsilon|^2$, where $\beta = 2\lambda_0/||h_c||^2$ and $\lambda_0 = \nu\lambda_{\min}(L)/4 > 0$, we have

$$\dot{V}(\phi, e) \leq -\lambda_0||e||^2 - \frac{|\epsilon|^2}{2}(\beta + n_s^2) + |\epsilon|\,|\eta| - \frac{|\epsilon|^2 n_s^2}{2} - w(t)\phi^T\theta .$$

Furthermore, using

$$-\frac{|\epsilon|^2}{2}(\beta+n_s^2)+|\epsilon||\eta| \le -\frac{(\beta+n_s^2)}{2}\left[\epsilon-\frac{|\eta|}{(\beta+n_s^2)}\right]^2+\frac{|\eta|^2}{m^2},$$

where

$$m^2 = 2(\beta+n_s^2), \tag{3.17}$$

we obtain

$$\dot{V}(\phi,e) \le -\lambda_0\|e\|^2 - \frac{\epsilon^2 n_s^2}{2} - w(t)\phi^{\mathrm{T}}\theta + \frac{|\eta|^2}{m^2},$$

where due to the property of n_s, $\frac{|\eta|^2}{m^2} \le C_0^2$ for some $C_0 \ge 0$. The leakage term $w(t)$ is to be chosen so that for $V \ge V_0$, and some constant $V_0 > 0$, $\dot{V} < 0$. This property of V will imply that $V \in L_\infty$ and therefore e, ϵ, $\phi \in L_\infty$.

In a similar manner, an adaptive law for estimating ρ^*, θ^* in (3.4) can be developed as follows: We rewrite (3.4) as

$$z = W(s)L(s)\left[\rho^*\left(\theta^{*\mathrm{T}}\zeta+z_1\right)+\eta\right],$$

where $L(s)$, ζ, η are as defined in (3.11) and $z_1 = L^{-1}(s)[z_0]$. Then the estimate $\hat{z}$ of z at time t can be constructed as

$$\hat{z} = W(s)L(s)\left[\rho\left(\theta^{\mathrm{T}}\zeta+z_1\right)\right],$$

where ρ, θ is the estimate of ρ^*, θ^* respectively at time t, and the estimation error can be defined as

$$\epsilon_1 = \hat{z} - z.$$

The normalized estimation error ϵ is generated as

$$\epsilon = \hat{z} - z - W(s)L(s)\left[\epsilon n_s^2\right], \tag{3.18}$$

where n_s is chosen so that

$$\text{(A1}'\text{)} \qquad \frac{\zeta}{m}, \frac{\eta}{m}, \frac{\xi}{m} \in L_\infty,$$

where $\xi = \theta^{\mathrm{T}}\zeta + z_1$ and m is given by (3.17). For analysis and design purposes (3.18) can be expressed in terms of the parameter errors ψ, ϕ defined as

$$\psi \triangleq \rho - \rho^*, \quad \phi \triangleq \theta - \theta^*.$$

From (3.18), we have

$$\epsilon = W(s)L(s)\left[\rho\left(\theta^{\mathrm{T}}\zeta+z_1\right) - \rho^*\left(\theta^{*\mathrm{T}}\zeta+z_1\right) - \eta - \epsilon n_s^2\right].$$

Since $\rho\theta^{\mathrm{T}}\zeta - \rho^*\theta^{*\mathrm{T}}\zeta + \rho^*\theta^{\mathrm{T}}\zeta - \rho^*\theta^{\mathrm{T}}\zeta = \psi\theta^{\mathrm{T}}\zeta + \rho^*\phi^{\mathrm{T}}\zeta$, we have

$$\epsilon = W(s)L(s)\left[\psi\xi + \rho^*\phi^{\mathrm{T}}\zeta - \epsilon n_s^2 - \eta\right], \tag{3.19}$$

or

$$\begin{aligned}\dot{e} &= A_c e + b_c \left(\psi\xi + \rho^* \phi^T \zeta - \epsilon n_s^2 - \eta\right)\\ \epsilon &= h_c e\,.\end{aligned}$$

Choosing

$$V(\phi, \psi, e) = \frac{e^T P_c e}{2} + \frac{\psi^2}{2\gamma} + |\rho^*| \frac{\phi^T \Gamma^{-1} \phi}{2}\,, \tag{3.20}$$

where $\gamma > 0$, it can be shown that

$$\dot{V}(\phi, \psi, e) \le -\lambda_0 \|e\|^2 - \frac{\epsilon^2 n_s^2}{2} + \epsilon\psi\xi + \epsilon\rho^* \phi^T \zeta + \frac{|\eta|^2}{m^2} + \frac{\psi\dot{\psi}}{\gamma} + |\rho^*| \phi^T \Gamma^{-1} \dot{\phi}\,,$$

where m^2 is as defined in (3.17). For

$$\begin{aligned}\dot{\rho} = \dot{\psi} &= -\gamma\epsilon\xi - \gamma w_1(t)\rho\\ \dot{\theta} = \dot{\phi} &= -\Gamma\epsilon\zeta\,\mathrm{sgn}(\rho^*) - \Gamma w(t)\theta\,,\end{aligned} \tag{3.21}$$

we have

$$\dot{V} \le -\lambda_0 \|e\|^2 - \frac{\epsilon^2 n_s^2}{2} - w_1 \psi\rho - w\phi^T \theta |\rho^*| + \frac{|\eta|^2}{m^2}\,. \tag{3.22}$$

The leakage terms w_1, $w \ge 0$ are to be chosen so that for $V \ge V_0$ and some constant $V_0 > 0$, $\dot{V} < 0$, which guarantees that e, ψ, $\phi \in L_\infty$.

A straightforward choice for $w(t)$ is $w(t) = \sigma$, where $\sigma > 0$ is a constant referred to as the σ-modification first used in [8] to improve robustness of adaptive schemes with respect to bounded disturbances and unmodeled dynamics. As indicated in [8], the σ-modification achieves robustness at the expense of destroying the ideal properties of the schemes, i.e., in the absence of plant uncertainties $\sigma > 0$ acts as a disturbance in the adaptive law which leads to non-zero estimation errors at steady state. This drawback of the σ-modification motivated two new choices for $w(t)$, the switching-σ [9] and ϵ-modification [10]. With the switching-σ, $w(t) = \sigma_s$, where $\sigma_s > 0$, when $\|\theta\|$ is larger than some constant $M_0 > \|\theta^*\|$, and $\sigma_s = 0$ when $\|\theta\| \le M_0$. As shown in [9], with this choice of w the ideal properties of the adaptive laws are maintained at the expense of knowing an upper bound for the unknown $\|\theta^*\|$. The ϵ-modification is given by $w(t) = |\epsilon m|\nu_0$ and was first introduced in [10] with the rationale that in the absence of modeling errors $w(t)$ will go to zero with the estimation error and therefore the ideal properties of the adaptive law will be guaranteed. Apparently this was not the case unless some persistently exciting conditions are used as shown in [10]. Simulations, however, indicate that with the ϵ-modification the convergence properties of the adaptive law are improved [10].

The following theorem describes the stability properties of the adaptive law (3.21) for different choices of the leakage terms $w(t)$ and $w_1(t)$.

Theorem 3.1 *The adaptive law (3.21) guarantees the following properties:*

(a) Fixed-σ [8]: *For* $w(t)=\sigma$, $w_1(t)=\sigma_1$, *where* $\sigma,\sigma_1>0$ *are constants, we have*
(i) $\epsilon,\theta,\rho\in L_\infty$, *(ii)* $\epsilon,\epsilon n_s,\dot\theta,\dot\rho\in\mathcal{S}(\sigma+\sigma_1+\frac{\eta^2}{m^2})$.

(b) Switching-σ [9]: *For* $w(t)=\sigma_s$, $w_1(t)=\sigma_{1s}$, *where*

$$\sigma_s=\begin{cases}0 & \text{if } \|\theta\|\le M_0\\ \left(\frac{\|\theta\|}{M_0}-1\right)^m\sigma_0 & \text{if } M_0<\|\theta\|\le 2M_0\\ \sigma_0 & \text{if } \|\theta\|>2M_0\end{cases}$$

$$\sigma_{1s}=\begin{cases}0 & \text{if } |\rho|\le m_1\\ \left(\frac{|\rho|}{m_1}-1\right)^m\sigma_0 & \text{if } m_1<|\rho|\le 2m_1\\ \sigma_0 & \text{if } |\rho|>2m_1\,,\end{cases}$$

with the design constants σ_0, $\sigma_{0s}>0$, $M_0>\|\theta^*\|$, $m_1>|\rho^*|$, *and* m *any finite positive integer, we have*
(i) $\epsilon,\theta,\rho\in L_\infty$ *(ii)* $\epsilon,\epsilon n_s,\dot\theta,\dot\rho\in\mathcal{S}(\frac{\eta^2}{m^2})$.

(c) ϵ-modification [10]: *For* $w(t)=|\epsilon m|\nu_0$, $w_1(t)=|\epsilon m|\nu_1$, *where* $\nu_0,\nu_1>0$ *are design constants, we have*
(i) $\epsilon,\theta,\rho\in L_\infty$ *(ii)* $\epsilon,\epsilon n_s,\dot\theta,\dot\rho\in\mathcal{S}(\nu_0+\nu_1+\frac{\eta^2}{m^2})$.

(d) Ideal Case: *For* $\eta=0$ *and* $w=w_1=0$ *or* $w=\sigma_s$, $w_1=\sigma_{1s}$, *we have*
(i) $\epsilon,\theta,\rho\in L_\infty$ *(ii)* $\epsilon,\epsilon n_s,\dot\theta,\dot\rho\in L_2$.

The properties of the adaptive law (3.16) follow directly from those of Theorem 3.1 by setting $\rho=\rho^*=1$ and $w_1=0$.

Proof. (a) For $w=\sigma$, $w_1=\sigma_1$, we have

$$\begin{aligned}-\sigma_1\psi\rho-\sigma\phi^{\mathrm T}\theta&\le-\sigma_1\left(\psi^2-|\psi|\,|\rho^*|\right)-\sigma\left(\phi^{\mathrm T}\phi-\|\phi\|\,\|\theta^*\|\right)\\&\le-\frac{\sigma_1}{2}\psi^2-\frac{\sigma}{2}\|\phi\|^2+\frac{\sigma_1}{2}|\rho^*|^2+\frac{\sigma}{2}\|\theta^*\|^2\,.\end{aligned}$$

Hence, (3.22) becomes

$$\dot V\le-\lambda_0\|e\|^2-\frac{\epsilon^2n_s^2}{2}-\frac{\sigma_1}{2}\psi^2-\frac{\sigma}{2}\|\phi\|^2+\frac{\sigma_1}{2}|\rho^*|^2+\frac{\sigma}{2}\|\theta^*\|^2+\frac{\eta^2}{m^2}\,, \tag{3.23}$$

or

$$\dot V\le-\alpha V+\frac{\sigma_1}{2}|\rho^*|^2+\frac{\sigma}{2}\|\theta^*\|^2+\frac{\eta^2}{m^2}\,, \tag{3.24}$$

where

$$\alpha=\min\left[\frac{\lambda_0}{\lambda_{\max}(P_c)},\frac{\sigma_1\gamma}{2},\frac{\sigma}{2|\rho^*|\lambda_{\max}(\Gamma^{-1})}\right].$$

Therefore, for $V>V_0=\frac{1}{\alpha}(\frac{\sigma_1}{2}|\rho^*|^2+\frac{\sigma}{2}\|\theta^*\|^2+c_0^2)$, where c_0^2 is the upper bound for $\frac{\eta^2}{m^2}$, we have $\dot V<0$, which implies that $V\in L_\infty$. Hence, $e,\phi,\psi\in L_\infty$, which implies that $\epsilon,\theta,\rho\in L_\infty$. Let us now express (3.22) in the form

$$\dot V\le-\frac{\epsilon^2m^2}{4}-\frac{\sigma_1}{2}\rho^2-\frac{\sigma}{2}\|\theta\|^2+\frac{\sigma_1}{2}|\rho^*|^2+\frac{\sigma}{2}\|\theta^*\|^2+\frac{\eta^2}{m^2}\,, \tag{3.25}$$

by using $-\lambda_0\|e\|^2 \leq -\frac{\beta}{2}\epsilon^2$ and (3.17). Integrating on both sides of (3.25), we have

$$\int_t^{t+T}\left(\frac{\epsilon^2 m^2}{4}+\frac{\sigma_1}{2}\rho^2+\frac{\sigma}{2}\|\theta\|^2\right)d\tau \leq V(t)-V(t+T) + \int_t^{t+T}\left(\frac{\sigma_1}{2}|\rho^*|^2+\frac{\sigma}{2}\|\theta^*\|^2+\frac{\eta^2}{m^2}\right)d\tau$$

$\forall t \geq 0$ and any $T \geq 0$, which together with the boundedness of V implies that ϵ, ϵn_s, $\sigma_1\rho$, $\sigma\|\theta\| \in \mathcal{S}(\sigma_1+\sigma+\frac{\eta^2}{m^2})$. Using the property of m, which guarantees that $\frac{\zeta}{m}$, $\frac{\xi}{m} \in L_\infty$, it follows from (3.21) that

$$|\dot{\rho}|^2 \leq c\left(|\epsilon m|^2+\sigma_1|\rho|^2\right)$$

$$\left\|\dot{\theta}\right\|^2 \leq c\left(|\epsilon m|^2+\sigma\|\theta\|^2\right)$$

for some $c \in \mathrm{I\!R}^+$. Therefore, $\dot{\rho}$, $\dot{\theta} \in \mathcal{S}(\sigma_1+\sigma+\frac{\eta^2}{m^2})$.

(b) For $w(t)=\sigma_s$, $w_1(t)=\sigma_{1s}$, we have

$$\sigma_s\phi^{\mathrm{T}}\theta=\sigma_s\left(\|\theta\|^2-\theta^{*\mathrm{T}}\theta\right) \geq \sigma_s\|\theta\|\left(\|\theta\|-M_0+M_0-\|\theta^*\|\right).$$

Since $\sigma_s(\|\theta\|-M_0) \geq 0$ and $M_0 > \|\theta^*\|$, it follows that

$$\sigma_s\phi^{\mathrm{T}}\theta \geq \sigma_s\|\theta\|\left(M_0-\|\theta^*\|\right) \geq 0\,,$$

i.e.,

$$\sigma_s\|\theta\| \leq \sigma_s\frac{\phi^{\mathrm{T}}\theta}{M_0-\|\theta^*\|}\,.$$

Similarly, $\sigma_{1s}\psi\rho \geq 0$ and $\sigma_{1s}|\rho| \leq \sigma_{1s}\frac{\psi\rho}{m_1-|\rho^*|}$. The inequality (3.22) for $\dot{V}$ can be written as

$$\dot{V} \leq -\lambda_0\|e\|^2-\frac{\epsilon^2 n_s^2}{2}-\sigma_{1s}\psi\rho-\sigma_s\phi^{\mathrm{T}}\theta|\rho^*|+\frac{|\eta|^2}{m^2}\,. \tag{3.26}$$

For $\|\theta\|=\|\phi+\theta^*\| > 2M_0$, we have $-\sigma_s\phi^{\mathrm{T}}\theta \leq -\frac{\sigma_0}{2}\|\phi\|^2+\frac{\sigma_0}{2}\|\theta^*\|^2$, and for $|\rho|=|\psi+\rho^*| > 2m_1$, we have $-\sigma_{1s}\psi\rho \leq -\frac{\sigma_0}{2}|\psi|^2+\frac{\sigma_0}{2}|\rho^*|^2$. Hence, following the same approach as in the proof of (a) we can show that for $V > V_0$ and for some $V_0 > 0$, $\dot{V} < 0$, i.e., $V \in L_\infty$, which implies that ϵ, θ, $\rho \in L_\infty$.

Furthermore, (3.26) also implies that ϵ, ϵn_s, $\sqrt{\sigma_{1s}\psi\rho}$, $\sqrt{\sigma_s\phi^{\mathrm{T}}\theta|\rho^*|} \in \mathcal{S}(\frac{|\eta|^2}{m^2})$. Since

$$\sigma_{1s}^2|\rho|^2 \leq c_1\sigma_{1s}\psi\rho\,,\quad \sigma_{1s}^2\|\theta\|^2 \leq c_2\sigma_s\phi^{\mathrm{T}}\theta\,,$$

where c_1, $c_2 \in \mathrm{I\!R}^+$ depend on the bounds for $\sigma_0|\rho|$ and $\sigma_0\|\theta\|$, respectively, we have

$$|\dot{\rho}|^2 \leq c(|\epsilon m|^2+\sigma_{1s}\psi\rho)\,,\quad \|\dot{\theta}\|^2 \leq c(|\epsilon m|^2+\sigma_s\phi^{\mathrm{T}}\theta)\,,$$

which imply that $\dot{\rho}$, $\dot{\theta} \in \mathcal{S}(\frac{\eta^2}{m^2})$.

(c) With $w(t)=|\epsilon m|\nu_0$, $w_1(t)=|\epsilon m|\nu_1$ from (3.20) we obtain

$$\begin{aligned}\dot V &= -\frac{e^{\mathrm T}qq^{\mathrm T}e}{2}-\frac{\nu e^{\mathrm T}Le}{2}-\epsilon^2n_s^2-\epsilon\eta-w_1\rho\psi-|\rho^*|w\phi^{\mathrm T}\theta\\ &\le -\frac{e^{\mathrm T}qq^{\mathrm T}e}{2}-\frac{\nu e^{\mathrm T}Le}{2}-\epsilon^2n_s^2+|\epsilon m|\left[\frac{|\eta|}{m}-\nu_1\rho\psi-|\rho^*|\nu_0\phi^{\mathrm T}\theta\right].\end{aligned}$$

Since

$$-\phi^{\mathrm T}\theta\le-\frac{|\phi|^2}{2}+\frac{|\theta^*|^2}{2},\quad -\psi\rho\le-\frac{\psi^2}{2}+\frac{\rho^{*2}}{2},$$

we have

$$\begin{aligned}\dot V \le& -\frac{e^{\mathrm T}qq^{\mathrm T}e}{2}-\frac{\nu e^{\mathrm T}Le}{2}-\epsilon^2n_s^2\\ &+|\epsilon m|\left[\frac{|\eta|}{m}-\nu_1\left(\frac{\psi^2}{2}-\frac{\rho^{*2}}{2}\right)-|\rho^*|\nu_0\left(\frac{\|\phi\|^2}{2}-\frac{\|\theta^*\|^2}{2}\right)\right].\end{aligned}\tag{3.27}$$

Using similar arguments as before, we can establish that for $V>V_0$ and some $V_0>0$, $\dot V\le 0$ and therefore $\epsilon,\ \theta,\ \rho\in L_\infty$. From (3.27) we can show that $\epsilon,\ \epsilon n_s\in\mathcal S(\nu_0+\nu_1+\frac{\eta^2}{m^2})$, which together with

$$|\dot\rho|^2\le c\epsilon^2m^2,\quad \left\|\dot\theta\right\|^2\le c\epsilon^2m^2,\quad m^2=2(\beta+n_s^2),$$

imply that $\dot\rho,\ \dot\theta\in\mathcal S(\nu_0+\nu_1+\frac{\eta^2}{m^2})$.

(d) For $\eta=0$, (3.22) becomes

$$\dot V\le-\lambda_0\|e\|^2-\frac{\epsilon^2n_s^2}{2}-w\phi^{\mathrm T}\theta|\rho^*|-w_1\psi\rho.\tag{3.28}$$

For $w=\sigma_s$, $w_1=\sigma_{1s}$, or for $w=w_1=0$, we have $w\phi^{\mathrm T}\theta\ge0$ and $w_1\psi\rho\ge0$. Hence, $\dot V\le0$ and $V,\ \epsilon,\ \theta,\ \rho\in L_\infty$. Since V is a nonincreasing function of time which is bounded from below, $\lim_{t\to\infty}V(t)=V_\infty$ exists. Therefore, from (3.28) we have

$$\int_0^\infty\left(\lambda_0\|e\|^2+\frac{\epsilon^2n_s^2}{2}+w\phi^{\mathrm T}\theta|\rho^*|+w_1\psi\rho\right)d\tau\le V(0)-V_\infty<\infty,$$

i.e., $e,\ \epsilon n_s,\ \epsilon,\ \sqrt{w\phi^{\mathrm T}\theta},\ \sqrt{w_1\psi\rho}\in L_2$. Since for $w=w_1=0$

$$|\dot\rho|^2\le c\epsilon^2m^2,\quad \left\|\dot\theta\right\|^2\le c\epsilon^2m^2,\quad m^2=2\left(\beta+n_s^2\right),$$

it follows that $\dot\rho,\ \dot\theta\in L_2$. For $w=\sigma_s$, $w_1=\sigma_{1s}$, we have from (b) that

$$|\dot\rho|^2\le c\left(|\epsilon m|^2+\sigma_{1s}\psi\rho\right),\quad \left\|\dot\theta\right\|^2\le c\left(|\epsilon m|^2+\sigma_s\phi^{\mathrm T}\theta\right),$$

which implies that $\dot\rho,\ \dot\theta\in L_2$. □

The use of the parametric model (3.11) and the SPR condition to derive robust adaptive laws whose properties are given by Theorem 3.1 has not been done before, and that constitutes one of the new results given in the paper. In Sect. 5, we show how such adaptive laws can be combined with control laws to form robust adaptive control schemes.

3.2 Gradient Method

This method is based on the development of an algebraic error equation and the minimization of a certain cost function $J(\theta, t)$ w.r.t. the estimated parameter θ for each time t by using the steepest descent method as shown below: Since θ^* is constant, we can rewrite (3.3) in the form

$$z = \theta^{*\mathrm{T}}\zeta + \eta\,,$$

where $\zeta = W(s)[\omega]$, $\eta = W(s)[\eta_0]$. The estimate $\hat{z}$ of z at time t is given by

$$\hat{z} = \theta^{\mathrm{T}}\zeta\,,$$

and the estimation error by

$$\epsilon_1 = \hat{z} - z = \theta^{\mathrm{T}}\zeta - z\,.$$

The normalized estimation error ϵ is constructed as

$$\epsilon = \frac{\epsilon_1}{m^2} = \frac{\theta^{\mathrm{T}}\zeta - z}{m^2}\,,$$

where $m^2 = 1 + n_s^2$ and n_s is the normalizing signal designed so that

(A1) $$\frac{\zeta}{m}, \frac{\eta}{m} \in L_\infty\,.$$

The adaptive law for updating θ is derived by minimizing various cost functions of θ w.r.t. θ. We start by considering the *instantaneous cost*

$$J(\theta, t) = \frac{\epsilon^2 m^2}{2} + \frac{w}{2}\theta^{\mathrm{T}}\theta = \frac{\left(\theta^{\mathrm{T}}\zeta - z\right)^2}{2m^2} + \frac{w}{2}\theta^{\mathrm{T}}\theta\,, \tag{3.29}$$

where $w(t) \geq 0$ is a design function which acts as a weighting. Using the gradient method we obtain

$$\dot{\theta} = -\Gamma\nabla J(\theta)\,,$$

where $\Gamma = \Gamma^{\mathrm{T}} > 0$ is a scaling matrix, and $\nabla J(\theta)$ is the gradient vector of $J(\theta, t)$ w.r.t. θ, i.e., $\nabla J(\theta) = \epsilon\zeta + w\theta$ and

$$\dot{\theta} = -\Gamma\epsilon\zeta - \Gamma w\theta\,, \quad \theta(0) = \theta_0\,. \tag{3.30}$$

The adaptive law (3.30) has the same form as (3.16) except that ϵ and ζ are defined in a different way. Similarly, we can derive an adaptive law for the bilinear model (3.4) by rewriting it in the form

$$z = \rho^*\left(\theta^{*\mathrm{T}}\zeta + z_1\right) + \eta\,,$$

where $z_1 = W(s)[z_0]$.

Then, the estimate $\hat{z}$ of z, the estimation error ϵ_1 and the normalized estimation error ϵ can be constructed as follows:

$$\hat{z} = \rho\theta^{\mathrm{T}}\zeta + \rho z_1$$
$$\epsilon_1 = \hat{z} - z = \rho^*\theta^{\mathrm{T}}\zeta + (\rho - \rho^*)\xi + \rho^* z_1 - z\,, \quad \xi = \theta^{\mathrm{T}}\zeta + z_1$$
$$\epsilon = \frac{\hat{z} - z}{m^2} = \frac{\rho^*\theta^{\mathrm{T}}\zeta + (\rho - \rho^*)\xi + \rho^* z_1 - z}{m^2}\,,$$

where $m^2 = 1 + n_s^2$ and n_s is chosen so that

$$\text{(A1}'\text{)} \qquad \frac{\zeta}{m}, \frac{\xi}{m}, \frac{\eta}{m} \in L_\infty\,,$$

and ρ, θ are as defined before. The cost function to be minimized w.r.t. ρ, θ is given by

$$\begin{aligned} J(\rho,\theta,t) &= \frac{\epsilon^2 m^2}{2} + \frac{w_1}{2}\rho^2 + \frac{w}{2}|\rho^*|\theta^{\mathrm{T}}\theta \\ &= \frac{(\rho^*\theta^{\mathrm{T}}\zeta + (\rho - \rho^*)\,\xi + \rho^* z_1 - z)^2}{2m^2} + \frac{w_1}{2}\rho^2 + \frac{w}{2}\theta^{\mathrm{T}}\theta|\rho^*|\,, \end{aligned}$$

where w, $w_1 \geq 0$ are design functions which act as weightings. Using the gradient method, we have

$$\begin{aligned} \dot{\rho} &= -\gamma\nabla J(\rho) \ = -\gamma\epsilon\xi - \gamma w_1\rho\,, && \gamma > 0 \\ \dot{\theta} &= -\Gamma_1\nabla J(\theta) = -\Gamma_1\rho^*\epsilon\zeta - |\rho^*|\Gamma_1 w\theta\,, && \Gamma_1 = \Gamma_1^{\mathrm{T}} > 0\,. \end{aligned}$$

Since Γ_1 is arbitrary, we can assume that $\Gamma_1 = \frac{\Gamma}{|\rho^*|}$ for some other arbitrary matrix $\Gamma = \Gamma^{\mathrm{T}} > 0$ and use it to get rid of the unknown parameter ρ^* in the adaptive law for θ, i.e.,

$$\begin{aligned} \dot{\theta} &= -\Gamma\epsilon\zeta\,\mathrm{sgn}(\rho^*) - \Gamma w\theta \\ \dot{\rho} &= -\gamma\epsilon\xi - \gamma w_1\rho\,. \end{aligned} \tag{3.31}$$

The stability properties of (3.31) for various choices of w, w_1 are given by the following theorem.

Theorem 3.2 *The adaptive law (3.31) guarantees the following properties:*

(a) Fixed-σ: *For $w(t) = \sigma$, $w_1(t) = \sigma_1$, where σ_1, $\sigma > 0$ are constants, we have (i) ϵ, ϵn_s, θ, ρ, $\dot{\theta}$, $\dot{\rho} \in L_\infty$ (ii) ϵ, ϵn_s, $\dot{\theta}$, $\dot{\rho} \in S(\sigma + \sigma_1 + \frac{\eta^2}{m^2})$.*

(b) Switching-σ: *For $w(t) = \sigma_s$, $w_1(t) = \sigma_{1s}$, where σ_s, σ_{1s} are as defined in Theorem 3.1(b), we have (i) ϵ, ϵn_s, θ, ρ, $\dot{\theta}$, $\dot{\rho} \in L_\infty$ (ii) ϵ, ϵn_s, $\dot{\theta}$, $\dot{\rho} \in S(\frac{\eta^2}{m^2})$.*

(c) ϵ-modification: *For $w(t) = |\epsilon m|\nu_0$, $w_1(t) = |\epsilon m|\nu_1$, and for some constants ν_0, $\nu_1 > 0$, we have (i) ϵ, ϵn_s, θ, ρ, $\dot{\theta}$, $\dot{\rho} \in L_\infty$ (ii) ϵ, ϵn_s, $\dot{\theta}$, $\dot{\rho} \in S(\nu_0 + \nu_1 + \frac{\eta^2}{m^2})$.*

(d) Ideal Case: *For* $\eta = 0$ *and* $w = w_1 = 0$, *or* $w(t) = \sigma_s$ *and* $w_1 = \sigma_{1s}$, *we have*
(i) ϵ, ϵn_s, θ, ρ, $\dot{\theta}$, $\dot{\rho} \in L_\infty$ *(ii)* ϵ, ϵn_s, $\dot{\theta}$, $\dot{\rho} \in L_2$.
If, in addition, $\frac{\dot{\zeta}}{m}, \frac{\dot{z}_1}{m}, \frac{\dot{z}}{m}, \frac{\dot{m}}{m} \in L_\infty$, *then* ϵ, ϵn_s, $\dot{\theta}$, $\dot{\rho} \to 0$ *as* $t \to \infty$.

Proof. We consider the positive definite function

$$V(\psi, \phi) = \frac{\psi^2}{2\gamma} + \frac{1}{2}|\rho^*|\phi^{\mathrm{T}}\Gamma^{-1}\phi\,, \tag{3.32}$$

which has the same form as that given by (3.20) with $P_c = 0$. The time derivative $\dot{V}$ of V along the solution of (3.31) is given by

$$\dot{V} \leq -\phi^{\mathrm{T}}\zeta\epsilon|\rho^*|\,\mathrm{sgn}(\rho^*) - \psi\xi\epsilon - |\rho^*|w\phi^{\mathrm{T}}\theta - w_1\psi\rho\,.$$

But

$$\epsilon = \frac{\rho^*\theta^{\mathrm{T}}\zeta + \psi\xi + \rho^* z_1 - z}{m^2} = \frac{\rho^*\theta^{\mathrm{T}}\zeta + \psi\xi - \rho^*\theta^{*\mathrm{T}}\zeta - \eta}{m^2} = \frac{\rho^*\phi^{\mathrm{T}}\zeta + \psi\xi - \eta}{m^2}\,. \tag{3.33}$$

Since $|\rho^*|\,\mathrm{sgn}(\rho^*) = \rho^*$, it follows that $(\rho^*\phi^{\mathrm{T}}\zeta + \psi\xi) = \epsilon m^2 + \eta$, and therefore

$$\begin{aligned} \dot{V} &\leq -\epsilon^2 m^2 - \epsilon\eta - |\rho^*|w\phi^{\mathrm{T}}\theta - w_1\psi\rho \\ &\leq -\frac{\epsilon^2}{2} - \frac{\epsilon^2 n_s^2}{2} - |\rho^*|w\phi^{\mathrm{T}}\theta - w_1\psi\rho + \frac{\eta^2}{m^2}\,. \end{aligned} \tag{3.34}$$

Equation (3.34) has exactly the same form as (3.22) except for the first terms $-\lambda_0\|e\|^2$ in (3.22) and $-\frac{\epsilon^2}{2}$ in (3.34). Since the first term does not affect the stability arguments and results, we can use Theorem 3.1 to establish that $\epsilon, \theta, \rho \in L_\infty$ and property (ii) in (a), (b), (c), (d) holds for the various choices of w, w_1. In order to complete the proof we need to establish that in (a), (b), (c), (d) we also have ϵn_s, $\dot{\theta}$, $\dot{\rho} \in L_\infty$ and in (d) that for $\frac{\dot{\zeta}}{m}, \frac{\dot{z}_1}{m}, \frac{\dot{z}}{m}, \frac{\dot{m}}{m} \in L_\infty$ we have $\epsilon, \epsilon n_s, \dot{\theta}, \dot{\rho} \to 0$ as $t \to \infty$. From (3.33) we have that $\theta, \psi, \frac{\zeta}{m}, \frac{\xi}{m}, \frac{\eta}{m} \in L_\infty$ imply that $\epsilon m \in L_\infty$, which in turn implies that $\epsilon, \epsilon n_s \in L_\infty$ and $\epsilon\zeta, \epsilon\xi \in L_\infty$. Since for all choices of w, w_1 in (a), (b), (c) and (d) $w\theta$, $w_1\rho \in L_\infty$, it follows from (3.31) that $\dot{\theta}, \dot{\rho} \in L_\infty$. Using (3.33) with $\eta = 0$ and $\dot{\theta}, \dot{\psi}, \frac{\dot{\zeta}}{m}, \frac{\dot{z}_1}{m}, \frac{\dot{z}}{m}, \frac{\dot{m}}{m} \in L_\infty$ it follows that $\dot{\epsilon}$, $\frac{d}{dt}(\epsilon m) \in L_\infty$, which together with ϵm, $\epsilon \in L_2$ imply that ϵm, ϵ, $\epsilon n_s \to 0$, and therefore $\dot{\theta}$, $\dot{\rho} \to 0$ as $t \to \infty$. □

The stability properties of (3.30) for the linear parametric model follow directly from Theorem 3.2 by setting $\rho = \rho^* = 1$, $w_1 = 0$ and $\psi = 0$.

Instead of the instantaneous cost $J(\theta, t)$, $J(\theta, \rho, t)$ one can choose various different types of cost functions which will lead to different adaptive laws. An interesting cost function to consider is the *integral cost* [21]

$$J(\theta, t) = \frac{1}{2}\int_0^t e^{-\beta(t-\tau)}\epsilon^2(t, \tau)m^2(\tau)d\tau + \frac{1}{2}w(t)\theta^{\mathrm{T}}\theta\,, \tag{3.35}$$

where $m^2 = 1 + n_s^2$ satisfies (A1), $\beta > 0$ is a design constant, $w(t) \geq 0$, $w \in L_\infty$ is a weighting coefficient,

$$\epsilon(t,\tau) = \frac{\hat{z}(t,\tau) - z(\tau)}{m^2(\tau)}, \quad \tau \leq t, \tag{3.36}$$

and $\hat{z}(t,\tau) = \theta^{\mathrm{T}}(t)\zeta(\tau)$ is the estimate of $z(\tau)$ at time τ formed by using the estimate $\theta(t)$ of θ^* at time $t \geq \tau$, i.e., $\epsilon(t,\tau)$ is the normalized estimation error at time τ based on the estimate of θ^* at time $t \geq \tau$. The design constant $\beta > 0$ acts as a forgetting factor, i.e., as t increases, the effect of the old data at time $\tau < t$ is discarded exponentially.

Using (3.36) we can express (3.35) in the form

$$J(\theta,t) = \frac{1}{2}\int_0^t e^{-\beta(t-\tau)} \frac{\left(\theta^{\mathrm{T}}(t)\zeta(\tau) - z(\tau)\right)^2}{m^2(\tau)} d\tau + \frac{1}{2} w(t)\theta^{\mathrm{T}}\theta \,. \tag{3.37}$$

Applying the gradient method to (3.37) we obtain

$$\dot{\theta} = -\Gamma \nabla J(\theta) = -\Gamma \int_0^t e^{-\beta(t-\tau)} \frac{\left(\theta^{\mathrm{T}}(t)\zeta(\tau) - z(\tau)\right)}{m^2(\tau)} \zeta(\tau) d\tau - \Gamma w \theta \,, \tag{3.38}$$

where $\Gamma = \Gamma^{\mathrm{T}} > 0$ by assuming that $\zeta(\tau)$, $z(\tau)$, $m(\tau)$, $w(t)$ are independent of $\theta(t)$ at time t. The adaptive law (3.38) can be implemented as [21,19]

$$\begin{aligned} \dot{\theta} &= -\Gamma(R(t)\theta + r(t)) - \Gamma w \theta \\ \dot{R} &= -\beta R + \frac{\zeta\zeta^{\mathrm{T}}}{m^2}, \quad R(0) = 0 \\ \dot{r} &= -\beta r - \frac{z\zeta}{m^2}, \quad r(0) = 0 \,, \end{aligned} \tag{3.39}$$

where $R \in \mathrm{IR}^{n\times n}$, $r \in \mathrm{IR}^{n\times 1}$. It is clear from (3.39) that the signal $\epsilon(t,\tau)$ does not have to be constructed. It is only used for design and analysis purposes.

Similarly, for the bilinear model we can develop the adaptive law

$$\begin{aligned} \dot{\theta} &= -\mathrm{sgn}(\rho^*)\Gamma(R\rho\theta + r) - \Gamma w \theta \\ \dot{\rho} &= -\gamma(R_1\rho\theta + r_1) - \gamma w_1 \rho \\ \dot{R} &= -\beta R + \frac{\zeta\zeta^{\mathrm{T}}}{m^2}, \quad R(0) = 0 \\ \dot{r} &= -\beta r - \frac{z\zeta}{m^2}, \quad r(0) = 0 \\ \dot{R}_1 &= -\beta R_1 + \frac{\zeta^{\mathrm{T}}\xi}{m^2}, \quad R_1(0) = 0 \\ \dot{r}_1 &= -\beta r_1 - \frac{z\xi}{m^2}, \quad r_1(0) = 0 \,, \end{aligned} \tag{3.40}$$

where $R_1 \in \mathbb{R}^{n\times 1}$, $r_1 \in \mathbb{R}$, by minimizing the cost function

$$J(\theta,\rho) = \frac{1}{2}\int_0^t e^{-\beta(t-\tau)}\frac{\left[\rho^*\theta^{\mathrm{T}}(t)\zeta(\tau) + (\rho(t)-\rho^*)\xi(\tau) + \rho^* z_1(\tau) - z(\tau)\right]^2}{m^2(\tau)}d\tau$$
$$+\frac{1}{2}w(t)\theta^{\mathrm{T}}\theta + \frac{1}{2}w_1(t)\rho^2$$

w.r.t. θ, ρ under the assumption that $\zeta(\tau)$, $\xi(\tau)$, $z(\tau)$, $w(t)$, $w_1(t)$ are independent of $\theta(t)$, $\rho(t)$ at time t.

The stability properties of (3.39) and (3.40) for different choices of the leakage w, w_1 can be developed in a slightly different way than those of (3.30) and (3.31). For clarity of presentation we confine ourselves only to the analysis of (3.39).

We consider the positive definite function $V = \frac{\phi^{\mathrm{T}}\Gamma^{-1}\phi}{2}$, whose time derivative along the solution of (3.39) is given by

$$\dot{V} = -\phi^{\mathrm{T}}[R\theta + r] - w\phi^{\mathrm{T}}\theta \tag{3.41}$$

$$= -\phi^{\mathrm{T}}R\phi + \phi^{\mathrm{T}}(t)\int_0^t e^{-\beta(t-\tau)}\frac{\zeta(\tau)\eta(\tau)}{m^2(\tau)}d\tau - w\phi^{\mathrm{T}}\theta \tag{3.42}$$

$$\le -\phi^{\mathrm{T}}R\phi + \left[\int_0^t e^{-\beta(t-\tau)}\frac{\left(\phi^{\mathrm{T}}(t)\zeta(\tau)\right)^2}{m^2(\tau)}d\tau\right]^{\frac{1}{2}}\left[\int_0^t e^{-\beta(t-\tau)}\frac{\eta^2}{m^2}d\tau\right]^{\frac{1}{2}}$$

$-w\phi^{\mathrm{T}}\theta$ (where we made use of the Schwarz Inequality)

$$= -\phi^{\mathrm{T}}R\phi + (\phi^{\mathrm{T}}R\phi)^{\frac{1}{2}}\left[\int_0^t e^{-\beta(t-\tau)}\frac{\eta^2(\tau)}{m^2(\tau)}d\tau\right]^{\frac{1}{2}} - w\phi^{\mathrm{T}}\theta\,, \tag{3.43}$$

or

$$\dot{V} \le -\frac{\phi^{\mathrm{T}}R\phi}{2} + \frac{1}{2}\left(\left\|\left(\frac{\eta}{m}\right)_t\right\|_2^{\beta}\right)^2 - w\phi^{\mathrm{T}}\theta\,, \tag{3.44}$$

where we carried out a simple completion of squares.

The leakage term $w \ge 0$ is to be chosen so that for $V \ge V_0$ and some constant $V_0 > 0$, $\dot{V} < 0$, which guarantees that $\phi \in L_\infty$. The following theorem describes the stability properties of the adaptive law (3.39) for different choices of the leakage term $w(t)$.

Theorem 3.3 *The adaptive law (3.39) guarantees the following properties:*

(a) Fixed-σ: *For $w(t) = \sigma$, where $\sigma > 0$ is a constant, we have*
(i) ϵ, ϵn_s, θ, $\dot{\theta} \in L_\infty$ (ii) ϵ, ϵn_s, $\dot{\theta} \in \mathcal{S}(\sigma + \frac{\eta^2}{m^2})$.

(b) Switching-σ: *For $w(t) = \sigma_s$, where σ_s is as defined in Theorem 3.1(b), we have*
(i) ϵ, ϵn_s, θ, $\dot{\theta} \in L_\infty$ (ii) ϵ, ϵn_s, $\dot{\theta} \in \mathcal{S}(\frac{\eta^2}{m^2})$.

(c) ϵ-modification: *For $w(t) = \|(\epsilon(t,\tau)m)_t\|_2^{\beta}\nu_0$, where $\nu_0 > 0$ is a design constant, we have*
(i) ϵ, ϵn_s, θ, $\dot{\theta} \in L_\infty$ (ii) ϵ, ϵn_s, $\dot{\theta} \in \mathcal{S}(\nu_0 + \frac{\eta^2}{m^2})$.

(d) Ideal Case: *For* $\eta = 0$ *and* $w = 0$ *or* $w = \sigma_s$*, we have*
(i) $\epsilon,\ \epsilon n_s,\ \theta,\ \dot{\theta} \in L_\infty$ *(ii)* $\epsilon,\ \epsilon n_s,\ \dot{\theta} \in L_2$.
If, in addition, $\frac{\dot{\zeta}}{m},\ \frac{\dot{m}}{m},\ \frac{\dot{z}}{m} \in L_\infty$*, then* $\epsilon,\ \epsilon n_s,\ \dot{\theta} \to 0$ *as* $t \to \infty$.

Proof. (a) For $w = \sigma$, we have

$$\begin{aligned} -\sigma\phi^{\mathrm{T}}\theta &\le -\sigma\left(\phi^{\mathrm{T}}\phi - \|\phi\|\,\|\theta^*\|\right) \\ &\le -\frac{\sigma}{2}\|\phi\|^2 + \frac{\sigma}{2}\|\theta^*\|^2 . \end{aligned}$$

Hence, (3.44) becomes

$$\dot{V} \le -\frac{\phi^{\mathrm{T}}R\phi}{2} + \frac{1}{2}\left(\left\|\left(\frac{\eta}{m}\right)_t\right\|_2^\beta\right)^2 - \frac{\sigma}{2}\|\phi\|^2 + \frac{\sigma}{2}\|\theta^*\|^2 . \tag{3.45}$$

Thus, as in the proof of Theorem 3.1(a), we can show that $\exists\, V_0 > 0$ such that for $V > V_0$, $\dot{V} < 0$, which implies that V and hence $\phi,\ \theta,\ \epsilon,\ \epsilon n_s \in L_\infty$. Since $R \in L_\infty$, from (3.39) it follows that $\dot{\theta} \in L_\infty$.

Now integrating on both sides of (3.45), we have $\forall t,\ T \ge 0$:

$$\begin{aligned} \int_t^{t+T}\left[\frac{\phi^{\mathrm{T}}R\phi}{2} + \frac{\sigma}{2}\|\phi\|^2\right]d\tau &\le V(t) - V(t+T) + \int_t^{t+T}\frac{\sigma}{2}\|\theta^*\|^2 ds \\ &\quad + \frac{1}{2}\int_t^{t+T}\left[\int_0^s e^{-\beta(s-\tau)}\frac{\eta^2}{m^2}d\tau\right]ds \\ &= V(t) - V(t+T) + \int_t^{t+T}\frac{\sigma}{2}\|\theta^*\|^2 ds \\ &\quad + \frac{1}{2}\int_t^{t+T}\left[\int_t^s e^{-\beta(s-\tau)}\frac{\eta^2}{m^2}d\tau\right]ds + c , \end{aligned}$$

where we used the fact that $\frac{|\eta|}{m} \in L_\infty$. Interchanging the order of integration in the second integral on the right hand side and using the boundedness of V, we obtain that $\sqrt{\phi^{\mathrm{T}}R\phi},\ \sqrt{\sigma}\|\phi\| \in \mathcal{S}(\sigma + \frac{\eta^2}{m^2})$. Now from (3.39) we have

$$\begin{aligned} \frac{d}{dt}\left[\phi^{\mathrm{T}}R\phi\right] &= 2\phi^{\mathrm{T}}R\dot{\phi} + \phi^{\mathrm{T}}\dot{R}\phi \\ &= -2\phi^{\mathrm{T}}R\left\{\Gamma[R\theta + r] + \Gamma\sigma\theta\right\} + \phi^{\mathrm{T}}\left[-\beta R + \frac{\zeta\zeta^{\mathrm{T}}}{m^2}\right]\phi \\ &= -2\phi^{\mathrm{T}}R\Gamma[R\theta + r + \sigma\theta] + \phi^{\mathrm{T}}\left[-\beta R + \frac{\zeta\zeta^{\mathrm{T}}}{m^2}\right]\phi \\ &= -2\phi^{\mathrm{T}}R\Gamma\left[R\phi - \int_0^t e^{-\beta(t-\tau)}\frac{\eta(\tau)\zeta(\tau)}{m^2(\tau)}d\tau + \sigma\theta\right] \\ &\quad + \phi^{\mathrm{T}}\left[-\beta R + \frac{\zeta\zeta^{\mathrm{T}}}{m^2}\right]\phi . \end{aligned} \tag{3.46}$$

Also,

$$\epsilon = \frac{\phi^{\mathrm{T}}\zeta - \eta}{m^2}$$
$$\Rightarrow \epsilon m = \frac{\phi^{\mathrm{T}}\zeta}{m} - \frac{\eta}{m}$$
$$\Rightarrow \epsilon^2 m^2 \leq 2\frac{(\phi^{\mathrm{T}}\zeta)^2}{m^2} + 2\frac{\eta^2}{m^2}\,, \tag{3.47}$$

where we used the identity $(a+b)^2 \leq 2a^2+2b^2$. Using (3.47) in (3.46), we obtain

$$\frac{d}{dt}\left[\phi^{\mathrm{T}}R\phi\right] \geq -2\phi^{\mathrm{T}}R\Gamma R\phi + 2\phi^{\mathrm{T}}R\Gamma\int_0^t e^{-\beta(t-\tau)}\frac{\eta(\tau)\zeta(\tau)}{m^2(\tau)}d\tau$$
$$-2\sigma\phi^{\mathrm{T}}R\Gamma\theta - \beta\phi^{\mathrm{T}}R\phi + \frac{\epsilon^2 m^2}{2} - \frac{\eta^2}{m^2}\,.$$

Integrating both sides of the above inequality from t to $t+T$ and making use of the fact that $\phi^{\mathrm{T}}R\phi|_t \geq 0$, we obtain an inequality involving $\phi^{\mathrm{T}}R\phi|_{t+T}$. Using this inequality in (3.42) we obtain

$$\begin{aligned}\dot V(t+T) \leq\ & 2\int_t^{t+T}\phi^{\mathrm{T}}R\Gamma R\phi d\tau - 2\int_t^{t+T}\phi^{\mathrm{T}}R\Gamma\left[\int_0^s e^{-\beta(s-\tau)}\frac{\eta(\tau)\zeta(\tau)}{m^2(\tau)}d\tau\right]ds\\ &+2\int_t^{t+T}\sigma\phi^{\mathrm{T}}R\Gamma\theta d\tau + \beta\int_t^{t+T}\phi^{\mathrm{T}}R\phi d\tau - \int_t^{t+T}\frac{\epsilon^2 m^2}{2}d\tau\\ &+\int_t^{t+T}\frac{\eta^2}{m^2}d\tau + \int_0^{t+T}e^{-\beta(t+T-\tau)}\frac{|\eta(\tau)|\ \|\phi^{\mathrm{T}}(t+T)\|\ \|\zeta(\tau)\|}{m^2(\tau)}d\tau\\ &-\sigma\phi^{\mathrm{T}}\theta|_{t+T}\,.\end{aligned} \tag{3.48}$$

Now

$$\begin{aligned}&\left|2\phi^{\mathrm{T}}R\Gamma\int_0^s e^{-\beta(s-\tau)}\frac{\eta(\tau)\zeta(\tau)}{m^2(\tau)}d\tau\right|\\ &\quad\leq 2\left\|\phi^{\mathrm{T}}R\Gamma\right\|\int_0^s e^{-\beta(s-\tau)}\frac{|\eta(\tau)||\zeta(\tau)|}{m^2(\tau)}d\tau\\ &\quad\leq \left\|\phi^{\mathrm{T}}R\Gamma\right\|^2 + \left[\int_0^s e^{-\beta(s-\tau)}\frac{\eta^2}{m^2}d\tau\right]\left[\int_0^s e^{-\beta(s-\tau)}\frac{|\zeta|^2}{m^2}d\tau\right],\end{aligned}$$

where we used the identity $2ab \leq a^2+b^2$ and the Schwarz Inequality. Integrating both sides of the above inequality and making use of R, $\frac{\zeta}{m} \in L_\infty$ and $(\phi^{\mathrm{T}}R\phi)^{\frac{1}{2}} \in \mathcal{S}(\frac{\eta^2}{m^2}+\sigma)$, we can show that

$$\sqrt{\left|2\phi^{\mathrm{T}}R\Gamma\int_0^t e^{-\beta(t-\tau)}\frac{\eta(\tau)\zeta(\tau)}{m^2(\tau)}d\tau\right|} \in \mathcal{S}\left(\frac{\eta^2}{m^2}+\sigma\right). \tag{3.49}$$

Also, $\sqrt{\sigma}\|\phi\| \in \mathcal{S}(\frac{\eta^2}{m^2}+\sigma)$ together with R, $\theta \in L_\infty$ imply that

$$\left|\sigma\phi^{\mathrm{T}}R\Gamma\theta\right|^{\frac{1}{2}} \in \mathcal{S}\left(\frac{\eta^2}{m^2}+\sigma\right). \tag{3.50}$$

Now using (3.49) and (3.50) in (3.48), and noting that $\dot{V} \in L_\infty$, $\phi^{\mathrm{T}} R\phi \in \mathcal{S}(\sigma + \frac{\eta^2}{m^2})$, it follows that ϵm and hence ϵ, $\epsilon n_s \in \mathcal{S}(\sigma + \frac{\eta^2}{m^2})$. From (3.39) we can show that

$$\left\|\dot{\theta}\right\|^2 \leq c_1 \phi^{\mathrm{T}} R^{\mathrm{T}} \Gamma^{\mathrm{T}} \Gamma R\phi + c_2 \left(\left\|\left(\frac{\zeta}{m}\right)_t\right\|_2^\beta\right)^2 \left(\left\|\left(\frac{\eta}{m}\right)_t\right\|_2^\beta\right)^2 + c_3 \sigma^2 \|\theta\|^2$$

for some c_1, c_2, $c_3 > 0$. Since R, $\frac{\zeta}{m} \in L_\infty$, $\sqrt{\phi^{\mathrm{T}} R\phi} \in \mathcal{S}(\sigma + \frac{\eta^2}{m^2})$, $\sqrt{\sigma}\|\phi\| \in \mathcal{S}(\sigma + \frac{\eta^2}{m^2})$, it follows that $\dot{\theta} \in \mathcal{S}(\sigma + \frac{\eta^2}{m^2})$. This completes the proof of (a).

(b) For $w(t) = \sigma_s$, we have

$$\sigma_s \phi^{\mathrm{T}} \theta = \sigma_s(\|\theta\|^2 - \theta^{*\mathrm{T}}\theta) \geq \sigma_s \|\theta\|(\|\theta\| - M_0 + M_0 - \|\theta^*\|)\,.$$

Since $\sigma_s(\|\theta\| - M_0) \geq 0$ and $M_0 > \|\theta^*\|$, it follows that

$$\sigma_s \phi^{\mathrm{T}} \theta \geq \sigma_s \|\theta\|(M_0 - \|\theta^*\|) \geq 0\,,$$

i.e.,

$$\sigma_s \|\theta\| \leq \sigma_s \frac{\phi^{\mathrm{T}}\theta}{M_0 - \|\theta^*\|}\,. \tag{3.51}$$

The inequality (3.44) can be written as

$$\dot{V} \leq -\frac{\phi^{\mathrm{T}} R\phi}{2} + \frac{1}{2}\left(\left\|\left(\frac{\eta}{m}\right)_t\right\|_2^\beta\right)^2 - \sigma_s \phi^{\mathrm{T}}\theta\,. \tag{3.52}$$

For $\|\theta\| = \|\phi + \theta^*\| > 2M_0$, we have

$$-\sigma_s \phi^{\mathrm{T}}\theta \leq -\frac{\sigma_0}{2}\|\phi\|^2 + \frac{\sigma_0}{2}\|\theta^*\|^2\,.$$

Hence, following the same approach as in the proof of (a) we can show that for $V > V_0$ and for some $V_0 > 0$, $\dot{V} < 0$, i.e., $V \in L_\infty$, which implies that ϵ, θ, $\epsilon n_s \in L_\infty$. Integrating on both sides of (3.52) we obtain $\sqrt{\phi^{\mathrm{T}} R\phi}$, $\sqrt{\sigma_s \phi^{\mathrm{T}}\theta} \in \mathcal{S}(\frac{\eta^2}{m^2})$. Proceeding as in the proof of (a) and making use of (3.51) we can show that ϵ, $\epsilon n_s \in \mathcal{S}(\frac{\eta^2}{m^2})$. Since

$$\sigma_s^2 \|\theta\|^2 \leq c_1 \sigma_s \phi^{\mathrm{T}}\theta\,,$$

where $c_1 \in \mathbb{R}^+$ depends on the bound for $\sigma_0\|\theta\|$, we have

$$\left\|\dot{\theta}\right\|^2 \leq c\left[\phi^{\mathrm{T}} R^{\mathrm{T}} \Gamma^{\mathrm{T}} \Gamma R\phi + \left(\left\|\left(\frac{\zeta}{m}\right)_t\right\|_2^\beta\right)^2 \left(\left\|\left(\frac{\eta}{m}\right)_t\right\|_2^\beta\right)^2 + \sigma_s \phi^{\mathrm{T}}\theta\right],$$

which implies that $\dot{\theta} \in \mathcal{S}(\frac{\eta^2}{m^2})$.

(c) With $w(t) = \|(\epsilon(t,\tau)m)_t\|_2^\beta \nu_0$, we obtain from (3.41)

$$\begin{aligned}
\dot{V} &= -\phi^{\mathrm{T}}[R\theta + r] - w\phi^{\mathrm{T}}\theta \\
&= -\phi^{\mathrm{T}}\left[\int_0^t e^{-\beta(t-\tau)}\frac{\zeta\zeta^{\mathrm{T}}\theta}{m^2}d\tau - \int_0^t e^{-\beta(t-\tau)}\frac{z\zeta}{m^2}d\tau\right] - w\phi^{\mathrm{T}}\theta \\
&= -\phi^{\mathrm{T}}\left[\int_0^t e^{-\beta(t-\tau)}\zeta(\tau)\epsilon(t,\tau)d\tau\right] - \phi^{\mathrm{T}}\theta\|(\epsilon(t,\tau)m)_t\|_2^\beta\nu_0 \\
&= -\int_0^t e^{-\beta(t-\tau)}\epsilon(t,\tau)\left[\epsilon(t,\tau)m^2(\tau) + \eta(\tau)\right]d\tau - \phi^{\mathrm{T}}\theta\|(\epsilon(t,\tau)m)_t\|_2^\beta\nu_0 \\
&\le -\left(\|(\epsilon(t,\tau)m)_t\|_2^\beta\right)^2 + \|(\epsilon(t,\tau)m)_t\|_2^\beta\left\|\left(\frac{\eta}{m}\right)_t\right\|_2^\beta - \phi^{\mathrm{T}}\theta\|(\epsilon(t,\tau)m)_t\|_2^\beta\nu_0 \\
&= -\|(\epsilon(t,\tau)m)_t\|_2^\beta\left[\|(\epsilon(t,\tau)m)_t\|_2^\beta - \left\|\left(\frac{\eta}{m}\right)_t\right\|_2^\beta + \phi^{\mathrm{T}}\theta\nu_0\right].
\end{aligned}$$

But

$$-\phi^{\mathrm{T}}\theta \le -\frac{\|\phi\|^2}{2} + \frac{\|\theta^*\|^2}{2}.$$

Thus,

$$\dot{V} \le -\|(\epsilon(t,\tau)m)_t\|_2^\beta\left[\|(\epsilon(t,\tau)m)_t\|_2^\beta - \left\|\left(\frac{\eta}{m}\right)_t\right\|_2^\beta + \nu_0\frac{\|\phi\|^2}{2} - \nu_0\frac{\|\theta^*\|^2}{2}\right].$$

Using the same arguments as before, we can establish that for $V > V_0$ and some $V_0 > 0$, $\dot{V} \le 0$ and therefore ϵ, θ, $\epsilon n_s \in L_\infty$. From (3.44) we have

$$\begin{aligned}
\dot{V} &\le -\frac{\phi^{\mathrm{T}}R\phi}{2} + \frac{1}{2}\left(\left\|\left(\frac{\eta}{m}\right)_t\right\|_2^\beta\right)^2 - \|(\epsilon(t,\tau)m)_t\|_2^\beta\nu_0\phi^{\mathrm{T}}\theta \\
&\le -\frac{\phi^{\mathrm{T}}R\phi}{2} + \frac{1}{2}\left(\left\|\left(\frac{\eta}{m}\right)_t\right\|_2^\beta\right)^2 - \|(\epsilon(t,\tau)m)_t\|_2^\beta\nu_0\left[\frac{\|\phi\|^2}{2} - \frac{\|\theta^*\|^2}{2}\right].
\end{aligned}$$

Integrating on both sides of the above equation and using the boundedness of V and ϵm, we see that $\sqrt{\phi^{\mathrm{T}}R\phi}$, $\sqrt{\nu_0}\left[\|(\epsilon(t,\tau)m)_t\|_2^\beta\right]^{\frac{1}{2}}\|\phi\| \in \mathcal{S}(\nu_0 + \frac{\eta^2}{m^2})$. Following the same steps as in (a) we can conclude that ϵ, ϵn_s, $\dot{\theta} \in \mathcal{S}(\nu_0 + \frac{\eta^2}{m^2})$ and the proof is complete.

(d) For $\eta = 0$, (3.44) becomes

$$\dot{V} \le -\frac{\phi^{\mathrm{T}}R\phi}{2} - w\phi^{\mathrm{T}}\theta. \tag{3.53}$$

For $w = \sigma_s$ or for $w = 0$, $w\phi^{\mathrm{T}}\theta \ge 0$. Hence, $\dot{V} \le 0$ and V, ϵ, $\theta \in L_\infty$. Since V is a nonincreasing function of time which is bounded from below, $\lim_{t\to\infty} V(t) = V_\infty$ exists. Therefore, from (3.53) we have $\sqrt{\phi^{\mathrm{T}}R\phi}$, $\sqrt{w\phi^{\mathrm{T}}\theta} \in L_2$. Now using the same arguments as in (a) above, we can show that ϵ, ϵn_s, $\dot{\theta} \in L_2$. The rest of the proof can be completed using exactly the same arguments as in the proof of Theorem 3.2(d). □

3.3 Least Squares

A type of integral cost function of the form (3.35) whose discrete-time version attracted considerable interest in parameter identification of discrete-time systems is the exponentially weighted function

$$\begin{aligned} J(\theta,t) &= \frac{1}{2}e^{-\beta t}(\theta-\theta_0)^{\mathrm{T}}Q_0(\theta-\theta_0) \\ &+\frac{1}{2}\int_0^t e^{-\beta(t-\tau)}\left[\epsilon^2(t,\tau)m^2(\tau)+w(\tau)\theta^{\mathrm{T}}(t)\theta(\tau)\right]d\tau\,, \end{aligned} \tag{3.54}$$

where $Q_0 = Q_0^{\mathrm{T}} > 0$, $\beta \geq 0$, $\theta_0 = \theta(0)$ and all other variables are as defined in Sect. 3.2. Using (3.36) we can rewrite (3.54) in the form

$$\begin{aligned} J(\theta,t) &= \frac{1}{2}e^{-\beta t}(\theta-\theta_0)^{\mathrm{T}}Q_0(\theta-\theta_0) \\ &+\frac{1}{2}\int_0^t e^{-\beta(t-\tau)}\left[\frac{(\theta^{\mathrm{T}}(t)\zeta(\tau)-z(\tau))^2}{m^2(\tau)}+w(\tau)\theta^{\mathrm{T}}(t)\theta(\tau)\right]d\tau\,. \end{aligned} \tag{3.55}$$

The cost (3.54) or (3.55) penalizes the square of the parameter change from $t=0$ to t weighted by $e^{-\beta t}Q_0$, an exponentially weighted average of the square of past estimation errors based on the current parameter estimate and an exponentially weighted average of the past and present parameter estimates. The exponential weighting coefficient β acts as a forgetting factor, i.e., as time t increases the effect of old data at time $\tau < t$ and of the initial parameter θ_0 are discarded exponentially.

Since $\frac{\zeta}{m}, \frac{z}{m} \in L_\infty$, $J(\theta)$ is a convex function of $\theta(t)$ over $\mathbb{R}^n$ at each time t. Hence, any local minimum is also global and satisfies

$$\nabla J(\theta) = 0\,, \quad \forall t \geq 0\,,$$

i.e.,

$$\begin{aligned} \nabla J(\theta(t)) &= e^{-\beta t}Q_0(\theta(t)-\theta_0) \\ &+\int_0^t e^{-\beta(t-\tau)}\left\{\frac{(\theta^{\mathrm{T}}(t)\zeta(\tau)-z(\tau))}{m^2(\tau)}\zeta(\tau)+w(\tau)\theta(\tau)\right\}d\tau \\ &=0\,. \end{aligned} \tag{3.56}$$

The form of $J(\theta)$ is such that it allows us to solve for the solution $\theta(t)$ of (3.56) explicitly, i.e.,

$$\theta(t) = P(t)\left[e^{-\beta t}Q_0\theta_0+\int_0^t e^{-\beta(t-\tau)}\left(\frac{z(\tau)\zeta(\tau)}{m^2(\tau)}-w(\tau)\theta(\tau)\right)d\tau\right], \tag{3.57}$$

where

$$P(t) = \left(e^{-\beta t}Q_0+\int_0^t e^{-\beta(t-\tau)}\frac{\zeta\zeta^{\mathrm{T}}}{m^2}d\tau\right)^{-1}, \quad P(0) = Q_0^{-1} = P_0\,.$$

Since $Q_0 = Q_0^{\mathrm{T}} > 0$ and $\zeta\zeta^{\mathrm{T}}$ is positive semidefinite, $P(t)$ exists at each time t. Equation (3.57) represents the so-called *non-recursive least-squares* algorithm for continuous-time systems [25,26]. A recursive version of (3.57) is developed by differentiating (3.57) w.r.t. time and using the identity

$$\frac{d}{dt}\left[PP^{-1}\right] = 0 = \dot{P}P^{-1} + P\frac{d}{dt}P^{-1}$$

to obtain

$$\dot{\theta} = -P(\epsilon\zeta + w\theta) \tag{3.58}$$

$$\dot{P} = \beta P - P\frac{\zeta\zeta^{\mathrm{T}}}{m^2}P\,, \quad P(0) = P_0 = Q_0^{-1}\,, \tag{3.59}$$

where

$$\epsilon = \epsilon(t,t) = \frac{\theta^{\mathrm{T}}(t)\zeta(t) - z(t)}{m^2(t)}\,.$$

We refer to (3.58)–(3.59) as the *continuous-time recursive least-squares* algorithm. We should note that $P(t) = [\nabla J^2(\theta)]^{-1}$, where $\nabla J^2(\theta)$ is the Hessian of $J(\theta,t)$ at θ, i.e., the least-squares solution can be obtained by applying Newton's method to the cost (3.54).

For the bilinear model we cannot develop an equivalent adaptive law using the least-squares method which can be implemented. This is due to the fact that such an adaptive law will depend on ρ^* in the same way as in the gradient method. In the gradient method, however, we were able to get rid of the unknown ρ^* by scaling the adaptive gain appropriately. In the least-squares case such a scaling is not possible because the adaptive gain $P(t)$ is no longer arbitrary but is generated from the differential equation (3.59), whose state $P(t)$ cannot be scaled by an unknown scalar.

In the identification literature, (3.58)–(3.59) with $\beta = 0$ and $w = 0$ is referred to as the *"pure" least-squares algorithm* and has a very similar form as the Kalman filter. The matrix P is usually called the *covariance matrix*.

Setting $\beta = 0$, (3.59) becomes

$$\dot{P} = -\frac{P\zeta\zeta^{\mathrm{T}}P}{m^2}\,, \quad \text{or} \quad \frac{d}{dt}P^{-1} = \frac{\zeta\zeta^{\mathrm{T}}}{m^2}\,. \tag{3.60}$$

Since $\frac{d}{dt}P^{-1} \geq 0$, P^{-1} may grow without bound; therefore, P may become arbitrarily small and slow down adaptation in some directions. This so-called *covariance wind-up* problem can be prevented by using various modifications which prevent $P(t)$ from going to zero. One such modification is the so-called *covariance resetting* defined as follows:

$$\dot{P} = -\frac{P\zeta\zeta^{\mathrm{T}}P}{m^2}\,, \quad P(t_r^+) = P_0 = \rho_0 I\,, \tag{3.61}$$

where t_r is the time for which $\lambda_{\min}(P(t)) \leq \rho_1$ and $\rho_0 \geq \rho_1 > 0$ are some design scalars. Due to (3.61), $P(t) \geq \rho_1 I \;\forall t \geq 0$.

When $\beta > 0$ in (3.59) the problem of $P(t)$ becoming arbitrarily small in some directions does no longer exist, and (3.59) is usually referred to as the *least squares with forgetting factor*. In this case, however, $P(t)$ may grow without bound due to $\beta P > 0$ and to the fact that $(P\zeta\zeta^{\mathrm{T}}P)/m^2$ is only positive semidefinite.

One way to avoid this complication is to modify (3.59) as

$$\dot{P} = \begin{cases} \beta P - \dfrac{P\zeta\zeta^{\mathrm{T}}P}{m^2} & \text{if } |P(t)| \leq R_0 \\ 0 & \text{otherwise,} \end{cases} \tag{3.62}$$

where $P(0) = P_0 = P_0^{\mathrm{T}} > 0$, $|P_0| \leq R_0$ and R_0 is a constant that serves as an upper bound for P. This modification guarantees that $P \in L_\infty$ and the adaptive law (3.58) with P given by (3.62) is referred to as the *modified least-squares with forgetting factor*. With either modification, the adaptive law (3.59) with P given by (3.61) or (3.62) behaves as a gradient algorithm with a time-varying gain, and is no longer the least-squares algorithm that we developed by setting $\nabla J(\theta) = 0$.

Despite its deficiency, the pure least-squares algorithm (3.58)–(3.59) has the unique property of guaranteeing parameter convergence to constant values in the case of $\eta = 0$ and $w = 0$ as described by the following theorem.

Theorem 3.4 *The adaptive law (3.58) with $\eta = 0$, $w = 0$, and P given by (3.59), guarantees that*

(i) θ, ϵ, ϵn_s, $P \in L_\infty$,
(ii) ϵ, ϵn_s, $\dot{\theta} \in L_2$, and
(iii) $\lim_{t\to\infty} \theta(t) = \bar{\theta}$, where $\bar{\theta}$ is a constant vector.

Proof. From (3.59) we have that $\dot{P} \leq 0$, i.e., $P(t) \leq P_0$. Since $P(t)$ is nonincreasing and bounded from below (i.e., $P = P^{\mathrm{T}} \geq 0\ \forall t \geq 0$), it has a limit, i.e.,

$$\lim_{t\to\infty} P(t) = \bar{P}\,,$$

where $\bar{P} = \bar{P}^{\mathrm{T}} \geq 0$ is a constant matrix. Let us now consider

$$\frac{d}{dt}(P^{-1}\phi) = -P^{-1}\dot{P}P^{-1}\phi + P^{-1}\dot{\phi} = \frac{\zeta\zeta^{\mathrm{T}}\phi}{m^2} - \epsilon\zeta\,.$$

Since for $\eta = 0$, $\epsilon = \frac{\phi^{\mathrm{T}}\zeta}{m^2} = \frac{\zeta^{\mathrm{T}}\phi}{m^2}$, we have

$$\frac{d}{dt}\,(P^{-1}\phi) = 0\,,$$

i.e., $P^{-1}(t)\phi(t) = P_0^{-1}\phi(0)$ and $\phi(t) = P(t)P_0^{-1}\phi(0)$. Therefore, $\lim_{t\to\infty}\phi(t) = \bar{P}P_0^{-1}\phi(0)$ and $\lim_{t\to\infty}\theta(t) = \bar{P}P_0^{-1}\phi(0) + \theta^* = \bar{\theta}$. Since $P \in L_\infty$, it follows that $\phi, \theta \in L_\infty$, which together with $\frac{\zeta}{m} \in L_\infty$ imply that ϵ, ϵm, $\epsilon n_s \in L_\infty$. Let us now choose the function

$$V(\phi) = \frac{\phi^{\mathrm{T}}P^{-1}\phi}{2} = \frac{\phi^{\mathrm{T}}(t)P_0^{-1}\phi(0)}{2}\,.$$

Then, the time derivative $\dot{V}$ of V along the solution of (3.58) is given by

$$\dot{V} = -\epsilon\phi^{\mathrm{T}}\zeta = -\epsilon^2 m^2 \leq 0\,, \tag{3.63}$$

which implies that $V \in L_\infty$. Since

$$\lim_{t\to\infty} V(\phi(t)) = \phi^{\mathrm{T}}(0)P_0^{-1}\frac{\bar{P}}{2}P_0^{-1}\phi(0) = V_\infty$$

exists, it follows from (3.63) that $\epsilon m \in L_2$, which implies that $\epsilon,\ \epsilon n_s \in L_2$. From the adaptive law

$$\dot{\theta} = -P\frac{\zeta}{m}\epsilon m$$

and the fact that $P,\ \frac{\zeta}{m} \in L_\infty$ and $\epsilon m \in L_2$, it follows that $\dot{\theta} \in L_2$ and the proof is complete. □

When $w(t) \neq 0$ the properties of Theorem 3.4 cannot be guaranteed even when $\eta = 0$. Let us now consider the adaptive law (3.58) with P given by (3.61) or (3.62). The following theorem establishes the stability properties of (3.58) for different choices of the weighting coefficient or leakage $w(t)$.

Theorem 3.5 *The adaptive law (3.58) with P given by (3.61) or (3.62) has the following properties:*

(a) Fixed-σ: *For $w = \sigma$, where $\sigma > 0$ is a fixed constant, we have* *(i)* $\epsilon,\ \epsilon n_s,\ \theta,\ \dot{\theta} \in L_\infty$ *(ii)* $\epsilon,\ \epsilon n_s,\ \dot{\theta} \in S(\sigma + \frac{\eta^2}{m^2})$.

(b) Switching-σ: *For $w(t) = \sigma_s$, where σ_s is as defined in Theorem 3.1(b), we have* *(i)* $\epsilon,\ \epsilon n_s,\ \theta,\ \dot{\theta} \in L_\infty$ *(ii)* $\epsilon,\ \epsilon n_s,\ \dot{\theta}, \in S(\frac{\eta^2}{m^2})$.

(c) ϵ-modification: *For $w(t) = |\epsilon m|\nu_0$, we have* *(i)* $\epsilon,\ \epsilon n_s,\ \theta,\ \dot{\theta} \in L_\infty$ *(ii)* $\epsilon,\ \epsilon n_s,\ \dot{\theta} \in S(\nu_0 + \frac{\eta^2}{m^2})$.

(d) Ideal Case: *For $\eta = 0$ and $w = 0$ or $w = \sigma_s$, we have* *(i)* $\epsilon,\ \epsilon n_s,\ \theta,\ \dot{\theta} \in L_\infty$ *(ii)* $\epsilon,\ \epsilon n_s, \dot{\theta} \in L_2$.

If, in addition, $\dfrac{\dot{\zeta}}{m}, \dfrac{\dot{m}}{m}, \dfrac{\dot{z}}{m} \in L_\infty$, *then* $\epsilon,\ \epsilon n_s,\ \dot{\theta} \to 0$ *as* $t \to \infty$.

Proof. Let us first examine the properties of the covariance matrix $P(t)$ given as the solution of (3.61) or (3.62). In (3.61) $P(t)$ is a discontinuous function of time whose value between discontinuities is given by the differential equation (3.61). At the discontinuity or resetting point t_r, $P(t_r^+) = P_0 = \rho_0 I$ and therefore $P^{-1}(t_r^+) = \rho_0^{-1}I$. From (3.61) we have that between discontinuities $\frac{d}{dt}P^{-1}(t) \geq 0$, i.e., $P^{-1}(t_2) - P^{-1}(t_1) \geq 0$ for $t_2 \geq t_1 \geq 0$ and for $t_r \notin [t_1,\ t_2]$, which implies that $P^{-1}(t) \geq \rho_0^{-1}I\ \forall t \geq 0$. Due to the resetting $P(t) \geq \rho_1 I\ \forall t \geq 0$. Therefore, (3.61) guarantees that

$$\rho_0 I \geq P(t) \geq \rho_1 I\,,\quad \rho_1^{-1}I \geq P^{-1}(t) \geq \rho_0^{-1}I\,. \tag{3.64}$$

In (3.62), $|P(t)| \le R_0\ \forall t \ge 0$. Using the identity $\frac{d}{dt}P^{-1} = -P^{-1}\dot{P}P^{-1}$, we have

$$\frac{d}{dt}P^{-1} = \begin{cases} -\beta P^{-1} + \dfrac{\zeta\zeta^{\mathrm{T}}}{m^2} & \text{if } |P(t)| \le R_0 \\ 0 & \text{otherwise,} \end{cases} \tag{3.65}$$

where $P^{-1}(0) = P_0^{-1}$. Since $\frac{\zeta\zeta^{\mathrm{T}}}{m^2} \in L_\infty$, (3.62) with $\beta > 0$ implies that $P^{-1} \in L_\infty$.

Therefore, (3.61)–(3.62) guarantee that P and P^{-1} are bounded positive definite symmetric matrices.

Let us now consider the function

$$V(\phi,t) = \frac{\phi^{\mathrm{T}}P^{-1}\phi}{2}\,,$$

where P is given by either (3.61) or (3.62). Since P^{-1} is a bounded positive definite symmetric matrix, it follows that V is decrescent and radially unbounded in the space of ϕ. Furthermore,

$$\dot{V} = \frac{1}{2}\phi^{\mathrm{T}}\frac{d}{dt}P^{-1}\phi + \phi^{\mathrm{T}}P^{-1}\dot{\phi} = \frac{1}{2}\phi^{\mathrm{T}}\frac{d}{dt}P^{-1}\phi - \phi^{\mathrm{T}}\epsilon\zeta - w\phi^{\mathrm{T}}\theta\,,$$

i.e.,

$$\dot{V} = -\epsilon^2 m^2 - \epsilon\eta - w\phi^{\mathrm{T}}\theta + \frac{1}{2}\phi^{\mathrm{T}}\frac{d}{dt}P^{-1}\phi\,, \tag{3.66}$$

or

$$\dot{V} \le -\frac{\epsilon^2 m^2}{2} - w\phi^{\mathrm{T}}\theta + \frac{\eta^2}{2m^2} + \frac{1}{2}\phi^{\mathrm{T}}\frac{d}{dt}P^{-1}\phi\,. \tag{3.67}$$

If P is given by (3.61), then $\frac{d}{dt}P^{-1} = \frac{\zeta\zeta^{\mathrm{T}}}{m^2}$ and

$$\frac{1}{2}\phi^{\mathrm{T}}\frac{d}{dt}P^{-1}\phi = \frac{(\phi^{\mathrm{T}}\zeta)^2}{2m^2} = \frac{\epsilon^2 m^2}{2} + \frac{\eta^2}{2m^2} + \epsilon\eta\,.$$

Therefore, using (3.66) we obtain

$$\dot{V} = -\frac{\epsilon^2 m^2}{2} - w\phi^{\mathrm{T}}\theta + \frac{\eta^2}{2m^2}\,. \tag{3.68}$$

If P is given by (3.62), then $\frac{d}{dt}P^{-1} = -\beta P^{-1} + \frac{\zeta\zeta^{\mathrm{T}}}{m^2}$ if $|P| \le R_0$ and $\frac{d}{dt}P^{-1} = 0$ otherwise. Hence, it follows from (3.65) and V that

$$\dot{V} = \begin{cases} -\dfrac{\epsilon^2 m^2}{2} - w\phi^{\mathrm{T}}\theta + \dfrac{\eta^2}{2m^2} - \dfrac{\beta}{2}\phi^{\mathrm{T}}P^{-1}\phi & \text{if } |P| \le R_0 \\ -\dfrac{\epsilon^2 m^2}{2} - w\phi^{\mathrm{T}}\theta + \dfrac{\eta^2}{2m^2} & \text{otherwise.} \end{cases}$$

Therefore, for P given by either (3.61) or (3.62), we have

$$\dot{V} \le -\frac{\epsilon^2 m^2}{2} - w\phi^{\mathrm{T}}\theta + \frac{\eta^2}{2m^2}\,. \tag{3.69}$$

The rest of the proof follows directly from the proof of Theorem 3.2 with $\rho = \rho^* = 1$, $\psi = 0$, $w_1 = 0$. □

3.4 Dead-Zones and Projections

In Sects. 3.1–3.3 we developed adaptive laws which guarantee parameter boundedness, in addition to other properties, in the presence of the modeling error η. This is achieved by using a special normalizing signal and a leakage term in the adaptive law. In this section we introduce some additional modifications, which can be either used alone or combined with leakage to form adaptive laws which can guarantee similar properties as in Sects. 3.1–3.3 in the presence of modeling errors.

Dead-Zone. Let us consider the normalized estimation error

$$\epsilon = \frac{\theta^{\mathrm{T}}\zeta - z}{m^2} = \frac{\phi^{\mathrm{T}}\zeta - \eta}{m^2},$$

as defined in the gradient and least-squares methods for the linear parametric model of θ^* which is used to "drive" the adaptive law. The signal ϵ is a measure of the parameter error ϕ, which is present in the signal $\phi^{\mathrm{T}}\zeta$, and of the modeling error η. When $\eta = 0$ and $\phi = 0$ we have $\epsilon = 0$ and no adaptation takes place. Due to $\frac{\eta}{m}, \frac{\zeta}{m} \in L_\infty$ large ϵm implies that $\frac{\phi^{\mathrm{T}}\zeta}{m}$ is large, which in turn implies that ϕ is large. In this case the effect of the modeling error η is small and the parameter estimates driven by ϵ move in a direction which reduces ϕ. When ϵm is small, however, the effect of η may be more dominant than that of the signal $\phi^{\mathrm{T}}\zeta$ and the parameter estimates may be driven in a direction that increases the parameter error ϕ or $\phi^{\mathrm{T}}\zeta$. The principal idea behind the dead-zone is to monitor the size of the estimation error and adapt only when the estimation error is large relative to the modeling error η, as shown below: We first consider the gradient algorithm for the linear parametric model

$$z = \theta^{*\mathrm{T}}\zeta + \eta .$$

Instead of the cost function (3.29), we consider

$$J(\theta, t) = \frac{\epsilon^2 m^2}{2},$$

and write

$$\dot{\theta} = \begin{cases} -\Gamma \nabla J(\theta) & \text{if } |\epsilon m| > g_0 > \frac{|\eta|}{m} \\ 0 & \text{otherwise.} \end{cases} \tag{3.70}$$

In other words, we move in the direction of steepest descent only when the estimation error is large relative to the modeling error, i.e., $|\epsilon m| > g_0$ and switch adaptation off when ϵm is small, i.e., $|\epsilon m| \le g_0$. In view of (3.70), we have

$$\dot{\theta} = -\Gamma\zeta(\epsilon + g), \quad g = \begin{cases} 0 & \text{if } |\epsilon m| > g_0 \\ -\epsilon & \text{if } |\epsilon m| \le g_0 . \end{cases} \tag{3.71}$$

This discontinuous dead-zone function $f(\epsilon, g) = \epsilon + g$ is shown in Fig. 1.

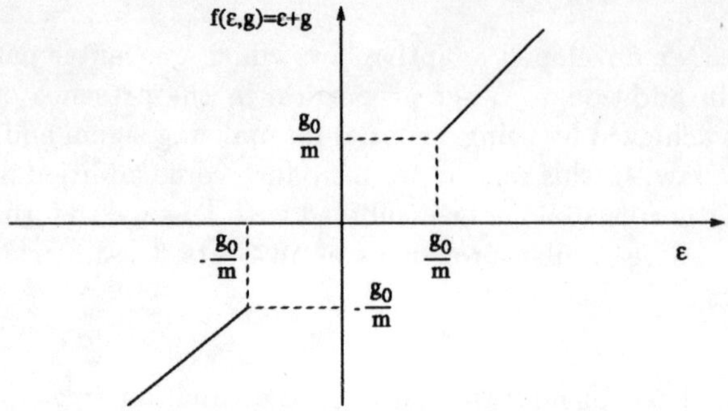

Fig. 1. Discontinuous dead-zone.

In order to avoid any implementation problems [27] which may arise due to the discontinuity in (3.71) the dead-zone function is made continuous as follows:

$$\dot{\theta} = -\Gamma\zeta(\epsilon + g), \quad g = \begin{cases} \dfrac{g_0}{m} & \text{if } \epsilon m < -g_0 \\ -\dfrac{g_0}{m} & \text{if } \epsilon m > g_0 \\ -\epsilon & \text{if } |\epsilon m| \leq g_0 \,. \end{cases} \tag{3.72}$$

This continuous dead-zone function $f(\epsilon,\, g) = \epsilon + g$ is shown in Fig. 2.

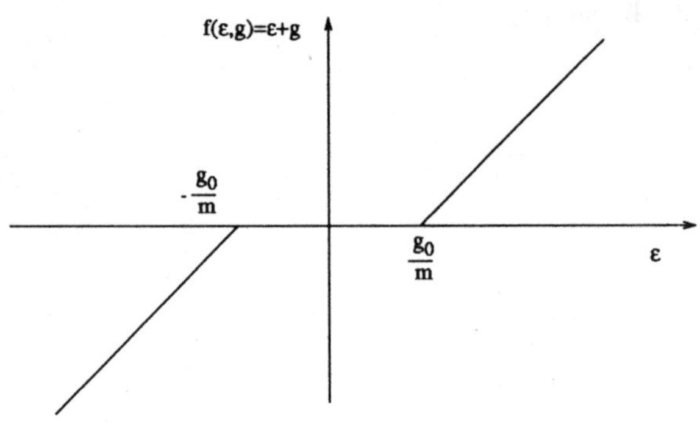

Fig. 2. Discontinuous dead-zone.

For the bilinear model (3.4) the adaptive law (3.31) with the dead-zone becomes

$$\begin{aligned} \dot{\theta} &= -\Gamma\zeta(\epsilon + g)\,\mathrm{sgn}(\rho^*) \\ \dot{\rho} &= -\gamma\xi(\epsilon + g)\,, \end{aligned} \tag{3.73}$$

where g is as defined in Fig. 2 and (3.72) but with

$$\epsilon = \frac{\rho(\theta^{\mathrm{T}}\zeta + z_1) - z}{m^2} = \frac{\rho^*\phi^{\mathrm{T}}\zeta + \psi\xi - \eta}{m^2} .$$

The least-squares adaptive law (3.58) with the dead-zone becomes

$$\dot{\theta} = -P\zeta(\epsilon + g) , \tag{3.74}$$

where ϵ, g are as defined in (3.72) and P is given by either (3.61) or (3.62).

The dead-zone modification can also be incorporated in the adaptive laws (3.39), (3.40) derived from the integral cost (3.35). For clarity of presentation, we consider the adaptive law (3.39). The principal idea behind the deadzone remains the same as before, i.e., to shut off adaptation when the normalized estimation error is small relative to the modeling error. However, this time the "shut-off" is no longer based on a pointwise-in-time comparison of the normalized estimation error and the a priori bound on the normalized modeling error. Instead, the decision to shut off adaptation or otherwise is based on the comparison of exponentially weighted L_2 norms of certain signals as shown below:

Instead of the cost function (3.37), we consider

$$J(\theta, t) = \frac{1}{2}\int_0^t e^{-\beta(t-\tau)}(\epsilon(t,\tau))^2 m^2(\tau)d\tau , \tag{3.75}$$

and write

$$\dot{\theta} = \begin{cases} -\Gamma\nabla J(\theta) & \text{if } \|(\epsilon(t,\tau)m)_t\|_2^\beta > g_0(t) \geq \left\|\left(\frac{\eta}{m}\right)_t\right\|_2^\beta + \nu , \\ & \quad \nu > 0 \text{ being arbitrary,} \\ 0 & \text{otherwise.} \end{cases} \tag{3.76}$$

In view of (3.76), we have

$$\dot{\theta} = -\Gamma\left[\int_0^t e^{-\beta(t-\tau)}\epsilon(t,\tau)\zeta(\tau)d\tau + g\right] , \tag{3.77}$$

where

$$g = \begin{cases} 0 & \text{if } \|(\epsilon(t,\tau)m)_t\|_2^\beta > g_0(t) \\ -\int_0^t e^{-\beta(t-\tau)}\epsilon(t,\tau)\zeta(\tau)d\tau & \text{otherwise.} \end{cases} \tag{3.78}$$

In order to avoid any implementation problems [27] which may arise due to the discontinuity in (3.77)–(3.78), the dead-zone function is made continuous as follows:

$$\dot{\theta} = -\Gamma\left[\int_0^t e^{-\beta(t-\tau)}\epsilon(t,\tau)\zeta(\tau)d\tau + g\right] \tag{3.79}$$

$$
g = \begin{cases} 0 & \text{if } \|(\epsilon(t,\tau)m)_t\|_2^\beta > 2g_0(t) \\ -\left(\int_0^t e^{-\beta(t-\tau)}\epsilon(t,\tau)\zeta(\tau)d\tau\right)\left[2 - \frac{\|(\epsilon(t,\tau)m)_t\|_2^\beta}{g_0}\right] & \\ \qquad \text{if } g_0(t) < \|(\epsilon(t,\tau)m)_t\|_2^\beta \leq 2g_0(t) & \\ -\int_0^t e^{-\beta(t-\tau)}\epsilon(t,\tau)\zeta(\tau)d\tau & \text{if } \|(\epsilon(t,\tau)m)_t\|_2^\beta \leq g_0(t)\,. \end{cases} \tag{3.80}
$$

The following theorem gives the stability properties of the adaptive laws (3.72), (3.73), (3.79), (3.80).

Theorem 3.6 *(a) The adaptive laws (3.72), (3.79), (3.80) for the linear parametric model guarantee:*

(i) ϵ, ϵn_s, $\dot\theta \in L_\infty$,
(ii) ϵ, $\epsilon n_s \in \mathcal{S}(g_0 + \frac{\eta^2}{m^2})$,
(iii) $\dot\theta \in L_2 \cap L_1$, *and*
(iv) $\lim_{t\to\infty} \theta = \bar\theta$.

(b) The adaptive law (3.73) for the bilinear parametric model guarantees that:

(i) ϵ, ϵn_s, $\dot\theta$, $\dot\rho \in L_\infty$,
(ii) ϵ, $\epsilon n_s \in \mathcal{S}(g_0^2 + \frac{\eta^2}{m^2})$,
(iii) $\dot\rho$, $\dot\theta \in L_2 \cap L_1$, *and*
(iv) $\lim_{t\to\infty} \theta = \bar\theta$, $\lim_{t\to\infty} \rho = \bar\rho$.

Here, $\bar\theta$ *and* $\bar\rho$ *denote a constant vector and scalar respectively.*

Proof. We start with (3.72). We consider the function

$$
V = \frac{\phi^{\mathrm{T}}\Gamma^{-1}\phi}{2}\,,
$$

whose time derivative $\dot V$ along the solution of (3.72) is given by

$$
\dot V = -\phi^{\mathrm{T}}\zeta(\epsilon+g) = -(\epsilon m^2+\eta)(\epsilon+g)\,. \tag{3.81}
$$

Now

$$
(\epsilon m^2+\eta)(\epsilon+g) = \begin{cases} (\epsilon m+g_0)^2 - \left(g_0 - \frac{\eta}{m}\right)(\epsilon m+g_0) > 0 & \text{if } \epsilon m < -g_0 \\ (\epsilon m-g_0)^2 + \left(g_0 + \frac{\eta}{m}\right)(\epsilon m-g_0) > 0 & \text{if } \epsilon m > g_0 \\ 0 & \text{if } |\epsilon m| \leq g_0\,. \end{cases} \tag{3.82}
$$

Hence, $(\epsilon m^2+\eta)(\epsilon+g) \geq 0\ \forall t$ and $\dot V \leq 0$, which implies that V, $\theta \in L_\infty$ and $\sqrt{(\epsilon m^2+\eta)(\epsilon+g)} \in L_2$. Furthermore, $\theta \in L_\infty$ implies that ϕ, ϵ, $\epsilon n_s \in L_\infty$. From the adaptive law (3.72) we have

$$
\dot\theta^{\mathrm{T}}\dot\theta = \frac{\zeta^{\mathrm{T}}\Gamma\zeta}{m^2}(\epsilon+g)^2m^2\,. \tag{3.83}
$$

But

$$(\epsilon+g)^2m^2=(\epsilon m+gm)^2=\begin{cases}(\epsilon m+g_0)^2 & \text{if } \epsilon m<-g_0\\ (\epsilon m-g_0)^2 & \text{if } \epsilon m>-g_0\\ 0 & \text{if } |\epsilon m|\le g_0\,.\end{cases}$$

which, together with (3.82), implies that

$$0\le(\epsilon+g)^2m^2\le(\epsilon m^2+\eta)(\epsilon+g)\ \Rightarrow\ (\epsilon+g)m\in L_2\,.$$

Hence, from (3.83) and $\frac{\zeta}{m}\in L_\infty$ we have that $\dot\theta\in L_2$. Equation (3.81) can also be written as

$$\dot V\le-\epsilon^2m^2+|\epsilon m|\frac{|\eta|}{m}+|\epsilon m|g_0+\frac{|\eta|}{m}g_0\,,$$

by using $|g|\le\frac{g_0}{m}$. Then, by completing the squares, we have

$$\begin{aligned}\dot V&\le-\frac{\epsilon^2m^2}{2}+\frac{|\eta|^2}{m^2}+g_0^2+\frac{|\eta|}{m}g_0\\ &\le-\frac{\epsilon^2m^2}{2}+\frac{3}{2}\frac{|\eta|^2}{m^2}+\frac{3}{2}g_0^2\,,\end{aligned}$$

i.e., $\epsilon m\in\mathcal{S}(g_0+\frac{\eta^2}{m^2})$. Since $m^2=1+n_s^2$, we have $\epsilon,\ \epsilon n_s\in\mathcal{S}(g_0+\frac{\eta^2}{m^2})$.

To show that $\lim_{t\to\infty}\theta=\bar\theta$, we use (3.81) and the fact that $(\epsilon m^2+\eta)(\epsilon+g)\ge 0$ to obtain

$$\begin{aligned}\dot V&=-(\epsilon m^2+\eta)(\epsilon+g)\\ &=-|\epsilon m+\frac{\eta}{m}||\epsilon+g|m\\ &\le-||\epsilon m|-\frac{|\eta|}{m}||\epsilon+g|m\\ &\le-|g_0-\frac{|\eta|}{m}||\epsilon+g|m\\ &\qquad(\text{since } |\epsilon+g|=0 \text{ if } |\epsilon m|\le g_0)\,.\end{aligned}$$

Since $\frac{\eta}{m}\in L_\infty$ and $g_0>\frac{|\eta|}{m}$, we integrate both sides of the above inequality and use the fact that $V\in L_\infty$ to obtain that $(\epsilon+g)m\in L_1$. Then, from the adaptive law (3.72), using the fact that $\frac{\zeta}{m}\in L_\infty$, it follows that $\|\dot\theta\|\in L_1$, which in turn implies that $\int_0^\infty\dot\theta dt$ exists and therefore θ converges to $\bar\theta$.

We next analyze the adaptive law (3.79), (3.80). This adaptive law may be rewritten as

$$\dot\theta=-\sigma_e\Gamma\int_0^t e^{-\beta(t-\tau)}\epsilon(t,\tau)\zeta(\tau)d\tau\,, \tag{3.84}$$

where

$$\sigma_e=\begin{cases}1 & \text{if } \|(\epsilon(t,\tau)m)_t\|_2^\beta>2g_0(t)\\ \dfrac{\|(\epsilon(t,\tau)m)_t\|_2^\beta}{g_0}-1 & \text{if } g_0(t)<\|(\epsilon(t,\tau)m)_t\|_2^\beta\le 2g_0(t)\\ 0 & \text{if } \|(\epsilon(t,\tau)m)_t\|_2^\beta\le g_0(t)\,.\end{cases} \tag{3.85}$$

Once again, we consider the positive definite function $V = \frac{\phi^{\mathrm{T}}\Gamma^{-1}\phi}{2}$ whose time derivative along the solution of (3.84) is given by

$$\dot{V} = -\sigma_{\mathrm{e}}\phi^{\mathrm{T}}(t)\int_0^t e^{-\beta(t-\tau)}\epsilon(t,\tau)\zeta(\tau)d\tau \tag{3.86}$$

$$= -\sigma_{\mathrm{e}}\int_0^t e^{-\beta(t-\tau)}\epsilon(t,\tau)[\epsilon(t,\tau)m^2(\tau)+\eta(\tau)]d\tau$$

$$\leq -\sigma_{\mathrm{e}}\|(\epsilon(t,\tau)m)_t\|_2^\beta\left[\|(\epsilon(t,\tau)m)_t\|_2^\beta - \left\|\left(\frac{\eta}{m}\right)_t\right\|_2^\beta\right] \tag{3.87}$$

(using the Schwarz Inequality).

From the definition of σ_{e} it follows that $\dot{V} \leq 0$, which implies that V, θ, $\epsilon \in L_\infty$ and $\lim_{t\to\infty} V(t) = V_\infty$ exists. Also, $\theta \in L_\infty$ implies that $\epsilon n_{\mathrm{s}} \in L_\infty$. From (3.87) we obtain

$$\dot{V} \leq -\sigma_{\mathrm{e}}\|(\epsilon(t,\tau)m)_t\|_2^\beta\,\nu\,.$$

Now integrating on both sides of the above inequality and using the fact that $V_\infty < \infty$, it follows that

$$\sigma_{\mathrm{e}}\|(\epsilon(t,\tau)m)_t\|_2^\beta \in L_1\,. \tag{3.88}$$

From (3.84), using the fact that $\frac{\zeta}{m} \in L_\infty$, it follows that $\|\dot{\theta}\| \in L_1$. Moreover, from (3.84) we also have $\dot{\theta} \in L_\infty$, which together with $\dot{\theta} \in L_1$ imply that $\dot{\theta} \in L_2$. Furthermore, using the same arguments as in the proof for the adaptive law (3.72), we can use the fact that $\left\|\dot{\theta}\right\| \in L_1$ to conclude that $\lim_{t\to\infty}\theta = \bar{\theta}$ for some constant vector $\bar{\theta}$.

We also need to show that ϵ, $\epsilon n_{\mathrm{s}} \in \mathcal{S}(\frac{\eta^2}{m^2}+g_0)$. This can be done as follows. First we can rewrite (3.86) as

$$\dot{V} = -\phi^{\mathrm{T}}(t)\left[\int_0^t e^{-\beta(t-\tau)}\epsilon(t,\tau)\zeta(\tau)d\tau\right] - \phi^{\mathrm{T}}(t)g$$

$$= -\phi^{\mathrm{T}}R\phi + \phi^{\mathrm{T}}(t)\int_0^t e^{-\beta(t-\tau)}\frac{\zeta(\tau)\eta(\tau)}{m^2(\tau)}d\tau - \phi^{\mathrm{T}}(t)g\,.$$

This equation being of the same form as (3.42), we can duplicate the steps in the proof of Theorem 3.3(a) to conclude that ϵ, $\epsilon n_{\mathrm{s}} \in \mathcal{S}(\frac{\eta^2}{m^2}+g_0)$ and the proof is complete. □

Parameter Projection Method. In some parameter estimation problems it is known a priori that the unknown parameter vector θ^* lies in some convex region

$$\mathcal{C} = \{\theta \in \mathbb{R}^n | g(\theta) \leq 0\}\,,$$

where $g : \mathbb{R}^n \to \mathbb{R}$ and $\mathcal{C}^\circ$, $\delta(\mathcal{C})$ denote the interior and boundary of $\mathcal{C}$. The adaptive laws developed in Sects. 3.1–3.3 can be modified so that the parameter estimate $\theta(t) \in \mathcal{C}\ \forall t \geq 0$. The principal idea behind such modifications is to choose $\theta(0) = \theta_0 \in \mathcal{C}$ and project the direction of adaptation (i.e., $\dot{\theta}$), whenever $\theta(t) \in \delta(\mathcal{C})$ has the tendency to move away from $\mathcal{C}$, so that $\theta(t)$ does not leave $\mathcal{C}$. Even though there are various ways to implement such projections and guarantee that $\theta(t) \in \mathcal{C}\, \forall t \geq 0$, we are only interested in those projection algorithms which, in addition to $\theta(t) \in \mathcal{C}\, \forall t \geq 0$, also guarantee that the stability properties of the adaptive laws without projection established in Sects. 3.1–3.3 will not change.

In this section we present some projection algorithms based on the gradient projection method which have such properties. We start by considering the following constrained minimization problem

$$\begin{aligned} &\text{minimize } J(\theta) \\ &\text{subject to } g(\theta) = 0\,. \end{aligned} \tag{3.89}$$

We start with a point θ_0 satisfying the constraint, i.e., $g(\theta_0) = 0$. To obtain an improved vector θ_1, we project the negative gradient of J at θ_0, i.e., $-\nabla J(\theta_0)$ onto the tangent plane $M(\theta_0) = \left\{\theta \in \mathbb{R}^n | \nabla_g^{\mathrm{T}}(\theta_0)\theta = 0\right\}$ obtaining the direction vector $P_{\mathrm{r}}(\theta_0)$. Then $\theta_1 = \theta_0 + \lambda P_{\mathrm{r}}(\theta_0)$, where λ is a fixed steplength or is chosen to minimize $J(\theta_1)$. The general form of this iteration is given by

$$\theta_{k+1} = \theta_k + \lambda P_{\mathrm{r}}(\theta_k)\,. \tag{3.90}$$

The explicit expression for $P_{\mathrm{r}}(\theta_k)$ can be obtained as follows: The vector $-\nabla J(\theta_k)$ can be expressed as a linear combination of the vector $P_{\mathrm{r}}(\theta_k)$ and the normal vector $\nabla g(\theta_k)$ to the tangent plane $M(\theta_k)$, i.e.,

$$-\nabla J(\theta_k) = \alpha \nabla g(\theta_k) + P_{\mathrm{r}}(\theta_k) \tag{3.91}$$

for some constant α. Since $P_{\mathrm{r}}(\theta_k)$ lies on the tangent plane $M(\theta_k)$, we have $\nabla g^{\mathrm{T}}(\theta_k) P_{\mathrm{r}}(\theta_k) = 0$, which together with (3.91) imply that

$$\alpha = -(\nabla g^{\mathrm{T}} \nabla g)^{-1} \nabla g^{\mathrm{T}} \nabla J\,.$$

Hence,

$$P_{\mathrm{r}}(\theta_k) = -\left(I - \frac{\nabla g \nabla g^{\mathrm{T}}}{\nabla g^{\mathrm{T}} \nabla g}\right) \nabla J\,. \tag{3.92}$$

We refer to $P_{\mathrm{r}}(\theta_k)$ as the projected direction onto the tangent plane $M(\theta_k)$. The gradient projection method is illustrated in Fig. 3.

It is clear from Fig. 3 that the new vector θ_{k+1} given by (3.90) may not satisfy the constraint. There are a number of techniques that can be employed to move θ_{k+1} from $M(\theta_k)$ to $g(\theta) = 0$ [28]. In our case we let $\lambda \to 0$, which gives us

$$\dot{\theta} = P_{\mathrm{r}}(\theta) = -\left[I - \frac{\nabla g \nabla g^{\mathrm{T}}}{\nabla g^{\mathrm{T}} \nabla g}\right] \nabla J\,, \quad \theta(0) = \theta_0\,, \tag{3.93}$$

where $g(\theta_0) = 0$. Since $\nabla g^{\mathrm{T}} P_{\mathrm{r}}(\theta) = 0$, i.e., $\dot{\theta}(t)$ lies on the tangent plane $M(\theta(t))$ $\forall\, t \geq 0$, the trajectory $\theta(t)$, if it exists, will satisfy $g(\theta(t)) = 0\; \forall t \geq 0$. A scaled version of (3.93) can be obtained by using the change of coordinates $\Gamma^{\frac{1}{2}}\bar{\theta} = \theta$,

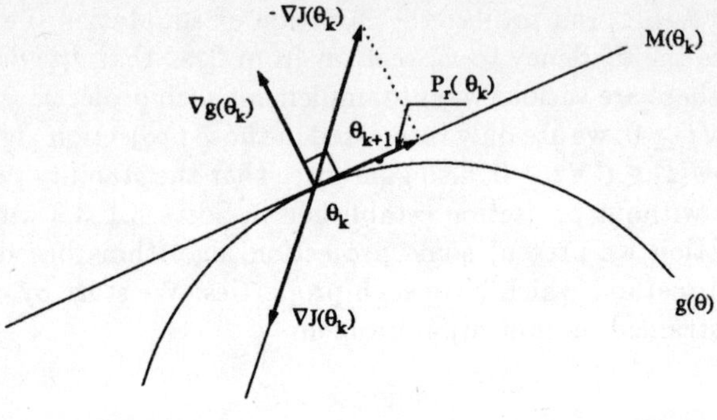

Fig. 3. Gradient projection.

where $\Gamma^{\frac{1}{2}}$ is a positive definite matrix that satisfies $\Gamma = \Gamma^{\frac{1}{2}}\Gamma^{\frac{T}{2}}$ and $\Gamma = \Gamma^{\mathrm{T}} > 0$ is a constant scaling matrix, i.e.,

$$\dot{\theta} = \bar{P}_{\mathrm{r}}(\theta) = -\Gamma^{\frac{1}{2}}\left[I - \Gamma^{\frac{T}{2}}\frac{\nabla g \nabla g^{\mathrm{T}}}{\nabla g^{\mathrm{T}}\Gamma\nabla g}\Gamma^{\frac{1}{2}}\right]\Gamma^{\frac{T}{2}}\nabla J\,. \tag{3.94}$$

The minimization problem (3.89) can now be extended to

$$\begin{aligned}&\text{minimize } J(\theta)\\ &\text{subject to } g(\theta) \le 0\,.\end{aligned} \tag{3.95}$$

The solution to (3.95) is given as follows:

$$\dot{\theta} = \begin{cases} -\nabla J & \text{if } \theta \in \mathcal{C}^0\,,\text{ or if } \theta \in \delta(\mathcal{C}) \\ & \quad\text{and } -\nabla^{\mathrm{T}} J \nabla g \le 0 \\ P_{\mathrm{r}}(\theta) = -\nabla J + \dfrac{\nabla g \nabla g^{\mathrm{T}}}{\nabla g^{\mathrm{T}}\nabla g}\nabla J & \text{otherwise,} \end{cases} \tag{3.96}$$

where $\theta(0) = \theta_0 \in \mathcal{C}^0$, or with the scaling matrix

$$\dot{\theta} = \begin{cases} -\Gamma\nabla J & \text{if } \theta \in \mathcal{C}^0\,,\text{ or if } \theta \in \delta(\mathcal{C}) \\ & \quad\text{and } -(\Gamma\nabla J)^{\mathrm{T}}\nabla g \le 0 \\ \bar{P}_{\mathrm{r}}(\theta) = -\Gamma\nabla J + \Gamma\dfrac{\nabla g \nabla g^{\mathrm{T}}}{\nabla g^{\mathrm{T}}\Gamma\nabla g}\Gamma\nabla J & \text{otherwise.} \end{cases} \tag{3.97}$$

Theorem 3.7 *The adaptive laws of Sects. 3.1–3.3, and the dead-zone adaptive law with the projection modification given in (3.96) or (3.97) retain all their stability properties that are established in the absence of projection, and in addition* $\theta \in \mathcal{C}\ \forall t \ge 0$, *provided* $\theta(0) = \theta_0 \in \mathcal{C}$.

Proof. It follows from (3.96) or (3.97) that whenever $\theta \in \delta(\mathcal{C})$, $\dot{\theta}^{\mathrm{T}}\nabla g \leq 0$, which implies that $\theta(t) \in \mathcal{C}\ \forall t \geq 0$ provided $\theta(0) = \theta_0 \in \mathcal{C}$. The adaptive laws of Sects. 3.1–3.3 and the dead-zone adaptive law are of the form

$$\dot{\theta} = -\Gamma\nabla J\,, \quad \forall t \geq 0\,. \tag{3.98}$$

Thus, projection introduces the additional term $\Gamma\frac{\nabla g\nabla g^{\mathrm{T}}}{\nabla g^{\mathrm{T}}\Gamma\nabla g}\Gamma\nabla J$ when $\theta \in \delta(\mathcal{C})$ and $-(\Gamma\nabla J)^{\mathrm{T}}\nabla g > 0$. If we use the same positive definite function to analyze the adaptive law with projection the time derivative $\dot{V}$ of V will have the additional additive term

$$q(\phi) = \frac{\phi^{\mathrm{T}}\nabla g\nabla g^{\mathrm{T}}\Gamma\nabla J}{\nabla g^{\mathrm{T}}\Gamma\nabla g} \tag{3.99}$$

when $\theta \in \delta(\mathcal{C}$ and $-(\Gamma\nabla J)^{\mathrm{T}}\nabla g > 0$. We rewrite $q(\phi)$ as

$$q(\phi) = -\phi^{\mathrm{T}}\nabla g\frac{(-(\Gamma\nabla J)^{\mathrm{T}}\nabla g)}{\nabla g^{\mathrm{T}}\Gamma\nabla g}\,.$$

Since $-(\Gamma\nabla J)^{\mathrm{T}}\nabla g > 0$ and $\theta^* \in \mathcal{C}$, $\theta \in \delta(\mathcal{C})$ implies that $\phi^{\mathrm{T}}\nabla g = (\theta - \theta^*)^{\mathrm{T}}\nabla g \geq 0$, we have $q(\phi) \leq 0$. Therefore, the term $q(\phi)$ due to projection can only make $\dot{V}$ more negative and will not affect the results developed from the properties of V and $\dot{V}$. In order to avoid repetition, we complete the proof of Theorem 3.7 only for the adaptive law (3.30) with $w(t) = \sigma_{\mathrm{s}}$, which with the projection becomes

$$\dot{\theta} = \begin{cases} -\Gamma(\epsilon\zeta + \sigma_{\mathrm{s}}\theta) & \text{if } \theta \in \mathcal{C}^0\,, \text{ or if } \theta \in \delta(\mathcal{C}) \text{ and} \\ & -[\Gamma(\epsilon\zeta + \sigma_{\mathrm{s}}\theta)]^{\mathrm{T}}\nabla g \leq 0 \\ -\Gamma(\epsilon\zeta + \sigma_{\mathrm{s}}\theta) + \Gamma\dfrac{\nabla g\nabla g^{\mathrm{T}}}{\nabla g^{\mathrm{T}}\Gamma\nabla g}\Gamma(\epsilon\zeta + \sigma_{\mathrm{s}}\theta) & \text{otherwise.} \end{cases} \tag{3.100}$$

We choose

$$V = \frac{\phi^{\mathrm{T}}\Gamma^{-1}\phi}{2}\,.$$

Then,

$$\dot{V} = \begin{cases} -\epsilon^2 m^2 - \epsilon\eta - \sigma_{\mathrm{s}}\phi^{\mathrm{T}}\theta & \text{if } \theta \in \mathcal{C}^0\,, \text{ or if } \theta \in \delta(\mathcal{C})\text{and} \\ & -[\Gamma(\epsilon\zeta + \sigma_{\mathrm{s}}\theta)]^{\mathrm{T}}\nabla g \leq 0 \\ -\epsilon^2 m^2 - \epsilon\eta - \sigma_{\mathrm{s}}\phi^{\mathrm{T}}\theta & \\ +\phi^{\mathrm{T}}\frac{\nabla g\nabla g^{\mathrm{T}}}{\nabla g^{\mathrm{T}}\Gamma\nabla g}\Gamma(\epsilon\zeta + \sigma_{\mathrm{s}}\theta) & \text{otherwise.} \end{cases}$$

Since $\phi^{\mathrm{T}}\nabla g \geq 0$ and $\nabla g^{\mathrm{T}}\Gamma(\epsilon\zeta + \sigma_{\mathrm{s}}\theta) \leq 0$ when $\theta \in \delta(\mathcal{C})$ and $-[\Gamma(\epsilon\zeta + \sigma_{\mathrm{s}}\theta)]^{\mathrm{T}}\nabla g > 0$, it follows that

$$\dot{V} \leq -\epsilon^2 m^2 - \epsilon\eta - \sigma_{\mathrm{s}}\phi^{\mathrm{T}}\theta\,.$$

Proceeding as in the proof of Theorem 3.2, it can be shown that ϵ, ϵn_s, θ, $\dot{\theta} \in L_\infty$ and ϵ, ϵn_s, $\sqrt{\sigma_s \phi^T \theta} \in \mathcal{S}(\frac{\eta^2}{m^2})$. It remains to establish that $\dot{\theta} \in \mathcal{S}(\frac{\eta^2}{m^2})$. From (3.100) we have that for $\theta \in \delta(\mathcal{C})$ and $-[\Gamma(\epsilon\zeta + \sigma_s\theta)]^T \nabla g > 0$

$$\begin{aligned} \dot{\theta}^T\dot{\theta} &= \nabla J^T \Gamma\Gamma \nabla J + 2\nabla J^T \Gamma\Gamma \nabla g \frac{(-\nabla g^T \Gamma \nabla J)}{\nabla g^T \Gamma \nabla g} \\ &+ \frac{\nabla g^T \Gamma\Gamma \nabla g}{\nabla g^T \Gamma \nabla g} \frac{(\nabla J^T \Gamma \nabla g)^2}{\nabla g^T \Gamma \nabla g} \\ &\le c\|\nabla J\|^2 \end{aligned}$$

for some $c \in \mathbb{R}^+$, where $\nabla J = \epsilon\zeta + \sigma_s\theta$. Hence,

$$\|\dot{\theta}\|^2 \le c\|\nabla J\|^2 , \quad \forall t \ge 0 ,$$

and the proof proceeds as in Theorem 3.2 to show that $\dot{\theta} \in \mathcal{S}(\frac{\eta^2}{m^2})$. □

The projection algorithm can also be used with the modified least-squares algorithms (3.61) and (3.62) as follows:

$$\dot{\theta} = \begin{cases} -P\nabla J & \text{if } \theta \in \mathcal{C}^0 \text{, or if } \theta \in \delta\mathcal{C} \\ & \text{and } -(P\nabla J)^T \nabla g \le 0 \\ \bar{P}_r(\theta) = -P\nabla J + P\dfrac{\nabla g \nabla g^T}{\nabla g^T \nabla g} P\nabla J & \text{otherwise} \end{cases}$$

$$\dot{P} = \begin{cases} 0 & \text{if } \theta \in \delta(\mathcal{C}) \\ \text{as in (3.61) or (3.62)} & \text{otherwise,} \end{cases}$$

i.e., on the boundary $\delta(\mathcal{C})$ the least-squares algorithm becomes a gradient one with constant gain since $\dot{P} = 0$ for $\theta \in \delta(\mathcal{C})$.

Further details on projection algorithms can be found in [14,16,20,26,36].

3.5 Normalizing Signals

The adaptive laws in Sects. 3.1–3.4 were developed under the assumption that a normalizing signal n_s can be constructed which can bound from above the unknown modeling error term $\eta(t)$ and the known information vector signal $\zeta(t)$. In the ideal case when $\eta = 0$ the choice of n_s is straightforward, i.e., it can be chosen as $n_s = \gamma\zeta^T P(t)\zeta$ for any constant $\gamma > 0$ where $P(t) = P^T(t) > 0$ with $P(t) \in \mathcal{L}_\infty$ etc. When $\eta \ne 0$, the design of n_s to bound the modeling error is possible provided some a priori information about η is available. For LTI plants η may represent the effect of bounded disturbances and/or unmodeled dynamics. In this case, η is the output of a transfer matrix $\Delta(s)$ whose input is the plant input u and output y, i.e.,

$$\eta = \Delta(s) \begin{bmatrix} u \\ y \end{bmatrix} + d ,$$

where d is a bounded disturbance. For such a class of modeling errors, Lemma 2.2 can be used to design the normalizing signal: Consider the normalizing signal given by

$$\dot{m}_s = -\delta_0 m_s + u^2 + y^2 \,, \quad m_s(0) = 0 \,, \tag{3.101}$$

i.e.,

$$m_s = (\|w_t\|_2^{\delta_0})^2 \quad w^2 = u^2 + y^2 \quad n_s^2 = m_s \,.$$

Using Lemma 2.2, we have

$$\frac{\|\eta\|}{m} \leq \sqrt{\frac{1}{\delta_0}} \|(s+\delta_0)\Delta(s)\|_\infty^{\delta_0} + \frac{\|d\|}{m} \,,$$

and

$$\frac{\|\eta_t\|_2^{\delta_0}}{m} \leq \|\Delta(s)\|_\infty^{\delta_0} + \frac{\|d_t\|_2^{\delta_0}}{m} \,,$$

where $m = \sqrt{1+n_s^2}$. Various other forms of normalizing signals can be found in [5,9,11,12,14].

The idea of normalization was inherent in most discrete-time identification algorithms [5,20] even before the robustness issues in adaptive control were raised. The normalizing signal used in the ideal case was proportional to the Euclidean norm of the information vector ζ and it was essential in establishing global stability. The normalizing signal used in the continuous-time case was referred to as an "auxiliary signal" [2] and had a similar form as in the discrete-time case. The use of normalizing signals of the form (3.101) was introduced in [5] in order to improve robustness in the presence of disturbances. This idea was later on used in [11] to achieve global boundedness results in the presence of unmodeled dynamics. After [11], a series of robustness results appeared, making use of normalizing signals of the form (3.101) together with leakage [6,9,10], dead-zones [5,12,13,14] and projections [11,14].

The normalizing signal slows down adaptation relative to the speed of the information vector ζ and modeling error η. A finite speed of adaptation which decreases with time in some fashion is essential in establishing stability of a closed-loop adaptive control scheme.

3.6 Persistence of Excitation and Dominant Richness

The adaptive laws developed in the previous subsections guarantee parameter boundedness and some smallness condition for the estimation error and speed of adaptation. In the absence of modeling errors these adaptive laws without modifications can also guarantee parameter convergence, i.e., $\theta(t) \to \theta^*$ as $t \to \infty$ provided $\zeta(t)$ is persistently exciting (PE), i.e., provided ζ satisfies

$$\frac{1}{T}\int_t^{t+T} \zeta(\tau)\zeta^{\mathrm{T}}(\tau)d\tau \geq \alpha_0 I$$

$\forall t \geq 0$, some $T > 0$ and $\alpha_0 > 0$ referred to as the level of excitation. The PE condition above and related ones are discussed extensively in [10,15,16,20].

Such conditions attracted considerable interest in adaptive control and parameter identification since they guarantee exponential convergence of all error signals including the parameter error to zero. In the presence of modeling errors, it is shown in [15] that the PE property of ζ guarantees local stability of the adaptive control scheme which employs an adaptive law without modifications. When plant uncertainties are present, however, the level of excitation α_0 has to be larger than the modeling error in order to establish stability. This led to the definition of ***dominantly rich*** (DR) signals in [6] and [37]. In [37] DR signals are used together with normalization and a switching-σ modification in order to establish global stability for an adaptive control scheme in the presence of unmodeled dynamics as well as exponential convergence for the tracking and parameter errors to residual sets whose sizes are of the order of the modeling error. DR signals achieve their level of excitation in the lower frequency range thus avoiding excess excitation of the unmodeled dynamics. DR signals are also used in [10] to establish boundedness in the presence of bounded disturbances for an adaptive control scheme without any modifications. In this case, the level of excitation is required to be higher than the level of the disturbance.

4 Controller Structures

An adaptive control scheme consists of two essential parts: (i) the control law which can be calculated and used to meet the control objective when the plant parameters are known and (ii) the parameter estimator or adaptive law for estimating the unknown parameters on line. The stability properties of the adaptive control scheme are very much dependent on the properties of these two parts, i.e., the control part and the estimation part. In Sect. 3, we have already designed and analyzed a wide class of adaptive laws for estimating unknown parameters on line. In this section, we study the design, analysis and robustness of several control laws which can be later on combined with appropriate adaptive laws to form adaptive control schemes. The analysis of the stability and robustness properties of these control laws in the case of ***known*** plant parameters is particularly important since it gives us a feel for what is achievable with a particular control law, given perfect parameter information. If, even in the case of known plant parameters, a given control law cannot guarantee robustness or meet the desired control objective, then it would be unreasonable to expect that such a control law would perform any better when, in the face of parametric uncertainty, it is used along with an adaptive law.

4.1 A General Feedback System

Consider the feedback system shown in Fig. 4, where $P_0(s)$ represents the plant transfer function matrix and $C_1(s)$, $C_2(s)$ the cascade and feedback compensators respectively; u denotes the control input and e_0, e_1, e_2 the inputs to the compensators C_1, C_2 and plant respectively; r, d_1, d_2 denote the reference signal, the output disturbance and input disturbance or noise respectively; y, y_0

is the plant output and the output of the compensator C_2 respectively. Due to the presence of sensor noise and output disturbances (d_1 in Fig. 4) the measured plant output e_1 is different from y.

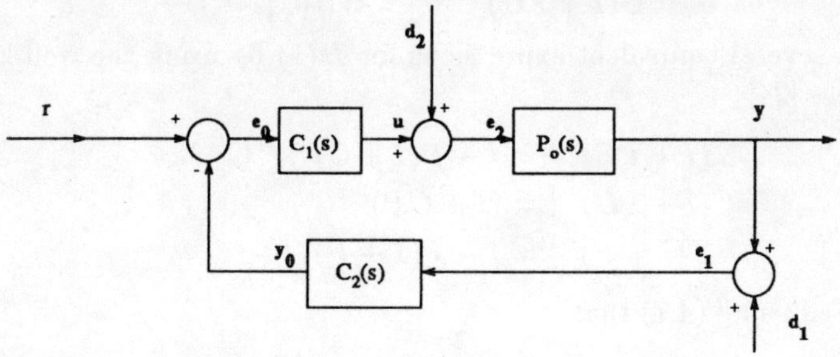

Fig. 4. Feedback system.

The matrices $P_0(s)$, $C_1(s)$, $C_2(s)$ are assumed to be proper, rational and of compatible dimensions so that the interconnections in Fig. 4 make sense.

The feedback system is described by the equations

$$\begin{bmatrix} e_0 \\ e_1 \\ e_2 \end{bmatrix} = \begin{bmatrix} r \\ d_1 \\ d_2 \end{bmatrix} - \begin{bmatrix} 0 & I & 0 \\ 0 & 0 & -I \\ -I & 0 & 0 \end{bmatrix} \begin{bmatrix} u \\ y_0 \\ y \end{bmatrix} \tag{4.1}$$

$$\begin{bmatrix} u \\ y_0 \\ y \end{bmatrix} = \begin{bmatrix} C_1 & 0 & 0 \\ 0 & C_2 & 0 \\ 0 & 0 & P_0 \end{bmatrix} \begin{bmatrix} e_0 \\ e_1 \\ e_2 \end{bmatrix}, \tag{4.2}$$

which can be written in the compact form

$$e = U - FY\,, \quad Y = G(s)[e]\,, \tag{4.3}$$

where

$$e = [e_0, e_1, e_2]^{\mathrm{T}}, \quad U = [r, d_1, d_2]^{\mathrm{T}}, \quad Y = [u, y_0, y]^{\mathrm{T}}$$

$$F = \begin{bmatrix} 0 & I & 0 \\ 0 & 0 & -I \\ -I & 0 & 0 \end{bmatrix}, \quad G = \begin{bmatrix} C_1 & 0 & 0 \\ 0 & C_2 & 0 \\ 0 & 0 & P_0 \end{bmatrix}.$$

Equation (4.3) may be written as

$$[I + FG][e] = U\,. \tag{4.4}$$

The system (4.1), (4.2) is well-posed if the $\det(I + FG(s))$ is not identically zero for all s. This condition guarantees that for every input signal U (4.1), (4.2) or (4.4) has a unique solution for e given by

$$e = (I + FG)^{-1}[U] \triangleq H[U]. \tag{4.5}$$

Similarly,

$$Y = G(I + FG)^{-1}[U] = GH[U]. \tag{4.6}$$

We can obtain several equivalent expressions for $H(s)$ by using the well-known matrix identities [24]

$$\begin{aligned} (I + PC)^{-1} &= I - P(I + CP)^{-1}C \\ C(I + PC)^{-1} &= (I + CP)^{-1}C \\ (I + P)^{-1} &= I - P(I + P)^{-1}. \end{aligned} \tag{4.7}$$

It can be verified using (4.7) that

$$H(s) = \left[H_1(s)\ H_2(s)\ H_3(s) \right], \tag{4.8}$$

where

$$H_1(s) = \begin{bmatrix} I - C_2P_0(I + C_1C_2P_0)^{-1}C_1 \\ P_0(I + C_1C_2P_0)^{-1}C_1 \\ (I + C_1C_2P_0)^{-1}C_1 \end{bmatrix}$$

$$H_2(s) = \begin{bmatrix} -C_2 + C_2P_0(I + C_1C_2P_0)^{-1}C_1C_2 \\ I - P_0(I + C_1C_2P_0)^{-1}C_1C_2 \\ -(I + C_1C_2P_0)^{-1}C_1C_2 \end{bmatrix}$$

$$H_3(s) = \begin{bmatrix} -C_2P_0(I + C_1C_2P_0)^{-1} \\ P_0(I + C_1C_2P_0)^{-1} \\ (I + C_1C_2P_0)^{-1} \end{bmatrix}.$$

In view of (4.8), it follows that the system is well-posed, i.e., the solution exists and is unique, if $\det(I + C_1C_2P_0)$ is not identically zero for all s.

Definition 4.1 *The feedback system (4.4) is L_p-stable with $p \in [1, \infty]$ if*
(i) it is well-posed
(ii) $U \in L_p$ implies $e, Y \in L_p$
(iii) $\|e\|_p \leq k_1 \|U\|_p$, where $k_1 < \infty$ is independent of U.

Using Definition 4.1, we can establish the following result.

Lemma 4.1 *Assume that the feedback system of Fig. 4 is well posed. Then this system is L_p-stable $\forall p \in [1, \infty]$ iff $H(s)$ is a stable proper transfer matrix.*

Proof. We have

$$e = H(s)[U], \quad Y = GH[U].$$

If we establish that H stable implies that GH is stable, then the proof follows from well-known results [29]. We have $H = (I + FG)^{-1} = I - FG(I + FG)^{-1}$, i.e., $H = I - FGH$. Since F is a nonsingular matrix independent of s, we have $GH = F^{-1} - F^{-1}H$. Hence, GH is stable iff H is stable. □

The stability of $H(s)$ is equivalent to requiring that each of its nine entries be a stable transfer function matrix. It should be noted that this concept of stability is more complete than the usual stability concept of undergraduate textbooks where system stability is checked by examining the roots of the characteristic equation. To further illustrate this fact, we give the following scalar example.

Example 4.1.1. Let $C_1(s) = \frac{-s+1}{s+1}$, $P_0(s) = \frac{1}{-s+1}$, $C_2(s) = 1$. Then the characteristic equation

$$1 + P_0(s)C_1(s)C_2(s) = 1 + \frac{1}{s+1} = 0 \qquad (4.9)$$

has a single root at $s = -2$, which would indicate stability. On the other hand,

$$P_0(s)[1 + C_1(s)C_2(s)P_0(s)]^{-1} = \frac{(s+1)}{(-s+1)(s+2)}$$

is an unstable transfer function so that a bounded disturbance d_2 could produce an unbounded response at e_1. Thus, blindly analyzing the roots of the characteristic equation is no guarantee for closed-loop stability. If, however, $P_0(s)$, $C_1(s)$, $C_2(s)$ are modeled as the ratio of coprime polynomials, i.e., $P_0(s) = \frac{n_0(s)}{d_0(s)}$, $C_1(s) = \frac{n_{c_1}(s)}{d_{c_1}(s)}$, $C_2(s) = \frac{n_{c_2}(s)}{d_{c_2}(s)}$ for some coprime polynomial pairs $(n_0(s), d_0(s))$, $(n_{c_1}(s), d_{c_1}(s))$, $(n_{c_2}(s), d_{c_2}(s))$, then a necessary and sufficient condition for stability is that the equation

$$d_0(s)d_{c_1}(s)d_{c_2}(s) + n_0(s)n_{c_1}(s)n_{c_2}(s) = 0 \qquad (4.10)$$

have all its roots in the open left half plane. For the example just considered, $n_0(s) = 1$, $d_0(s) = (-s+1)$, $n_{c_1}(s) = (-s+1)$, $d_{c_1}(s) = (s+1)$, $n_{c_2}(s) = 1$, $d_{c_2}(s) = 1$. Thus, (4.10) becomes

$$(-s+1)(s+2) = 0\,,$$

which clearly has one root in the right half plane and so the closed-loop system is unstable. □

The above example illustrates that if we have a SISO system, then to determine the closed-loop stability one can either go ahead and blindly check the stability of all the entries of $H(s)$ or, alternatively, analyze the roots of (4.10). If, however, C_1 and C_2 are stable compensators, the stability of $H(s)$ is implied by the stability of only one of its entries, namely, the transfer function matrix between the input d_2 and the output y, i.e., $P_0(I + C_1C_2P_0)^{-1}$, which is the case often emphasized in most books on control systems.

4.2 Uncertainty Characterizations

The first task of a control engineer in designing a control system is to obtain a simplified mathematical model which describes the actual plant to be controlled with a reasonable degree of accuracy over the operating range of interest. While a simple model leads to a simpler control design, such a design must possess a sufficient degree of robustness or insensitivity with respect to the unmodeled plant characteristics. In order to study and improve the robustness properties of the control design, we need to come up with some characterization of the types of plant uncertainties that are likely to be encountered in practice. In the following definitions, we give various descriptions of the plant uncertainty that will be used later for robustness analysis and design.

Definition 4.2 Additive Perturbations: *Suppose that the true plant $P(s)$ and the nominal plant $P_0(s)$ are related by*

$$P(s) = P_0(s) + \Delta_a(s)\,, \tag{4.11}$$

where $\Delta_a(s)$ is stable. Then $\Delta_a(s)$ is called an additive perturbation of $P_0(s)$.

Definition 4.3 Multiplicative Perturbations: *Suppose that the true plant $P(s)$ and the nominal plant $P_0(s)$ are related by*

$$P(s) = (I + \Delta_{\mathrm{m}}(s))P_0(s)\,, \tag{4.12}$$

where $\Delta_{\mathrm{m}}(s)$ is stable. Then $\Delta_{\mathrm{m}}(s)$ is called a multiplicative perturbation of $P_0(s)$.

Definition 4.4 Stable Factor Perturbations: *Suppose that the true plant $P(s)$ and nominal plant $P_0(s)$ have the following right coprime factorizations [30]*

$$\begin{aligned} P(s) &= [N_0(s) + \Delta_1(s)][D_0 + \Delta_2(s)]^{-1} \\ P_0(s) &= N_0(s)D_0^{-1}(s)\,, \end{aligned} \tag{4.13}$$

where $N_0(s)$, $D_0(s)$ are proper stable rational transfer function matrices that are coprime [2] and $[\Delta_1(s)]_{i,j}$, $[\Delta_2(s)]_{i,j}$ are stable. Then $\Delta_1(s)$, $\Delta_2(s)$ are called stable factor perturbations of $P_0(s)$.

---

[2] Two stable rational transfer function matrices are right/left coprime iff every right/left divisor is unimodular in the ring of stable rational transfer function matrices. Two proper rational stable transfer functions are coprime iff they have no finite common zeros in the closed right half s-plane and at least one of them has relative degree zero [30].

Singular Perturbations Another important description of a wide class of plant dynamic uncertainties is given by singular perturbation models. For LTI plants, the following singular perturbation model in the state variable form

$$\dot{x} = A_{11}x + A_{12}z + B_1u\,, \quad x \in \mathbb{R}^n, \quad u \in \mathbb{R}^r \tag{4.14}$$

$$\mu\dot{z} = A_{21}x + A_{22}z + B_2u\,, \quad z \in \mathbb{R}^m \tag{4.15}$$

$$y = C_1x + C_2z\,, \quad y \in \mathbb{R}^q \tag{4.16}$$

can be used to describe the slow (or dominant) and fast (or parasitic) phenomena of the plant. The scalar $\mu > 0$ represents all the small parameters such as small time constants, small masses etc. to be neglected. In most applications, the representation (4.14)–(4.16) with a single parameter μ can be achieved by proper scaling as shown in [31]. All the matrices in (4.14)–(4.16) are assumed to be constant and independent of μ. As explained in [31], this assumption is for convenience only and leads to a minor loss of generality.

Assuming that A_{22} is a stable matrix, an approximation of (4.14) is obtained by setting $\mu = 0$, solving for z in (4.15) and substituting for its value in (4.14) and (4.16), i.e.,

$$\begin{aligned} \dot{x}_0 &= A_0x_0 + B_0u\,, \quad x_0 \in \mathbb{R}^n \\ y_0 &= C_0x_0 + D_0u\,, \end{aligned} \tag{4.17}$$

where $A_0 = A_{11} - A_{12}A_{22}^{-1}A_{21}$, $B_0 = B_1 - A_{12}A_{22}^{-1}B_2$, $C_0 = C_1 - C_2A_{22}^{-1}A_{21}$ and $D_0 = -C_2A_{22}^{-1}B_2$. With μ set to zero, the dimension of the state space of (4.14)–(4.15) reduces from $n+m$ to n because the differential equation (4.15) degenerates into the algebraic equation

$$0 = A_{21}x_0 + A_{22}z_0 + B_2u\,, \quad \text{i.e., } z_0 = -A_{22}^{-1}(A_{21}x_0 + B_2u)\,,$$

where the subscript 0 is used to indicate that the variables belong to a system with $\mu = 0$. The transfer function matrix

$$P_0(s) = C_0(sI - A_0)^{-1}B_0 + D_0 \tag{4.18}$$

represents the nominal or slow or dominant part of the plant. We should emphasize that even though the transfer function matrix $P(s)$ from y to u of the full-order plant given by (4.14)–(4.16) is strictly proper, the nominal $P_0(s)$ may be proper but not strictly proper since $D_0 = -C_2A_{22}^{-1}B_2$ may not be equal to zero. The singular perturbation model (4.14)–(4.16) received considerable attention in the control literature [31], where it is used for control synthesis as well as for robustness analysis.

The state-variable model (4.14)–(4.16) can also be expressed in the s-domain as follows: As in [31], we use the similarity transformation

$$\begin{bmatrix} x \\ z \end{bmatrix} = \begin{bmatrix} I_n & \mu H \\ -L & I_m - \mu LH \end{bmatrix} \begin{bmatrix} x_s \\ z_f \end{bmatrix}, \tag{4.19}$$

where L, H satisfy the algebraic equations

$$\begin{aligned} \mu(A_{11} - A_{12}L)H - H(A_{22} + \mu LA_{12}) + A_{12} &= 0 \\ A_{21} - A_{22}L + \mu LA_{11} - \mu LA_{12}L &= 0\,, \end{aligned} \tag{4.20}$$

to transform (4.14)–(4.16) into the form

$$\dot{x}_s = A_s x_s + B_s u \tag{4.21}$$

$$\mu \dot{z}_f = A_f z_f + B_f u \tag{4.22}$$

$$y = C_s x_s + C_f z_f , \tag{4.23}$$

where

$$A_s = A_{11} - A_{12}L , \quad B_s = B_1 - \mu H B_2 - \mu H L B_1 , \quad C_s = C_1 - C_2 L ,$$

$$A_f = A_{22} + \mu L A_{12} , \quad B_f = B_2 + L B_1 , \quad C_f = C_2 + \mu(C_1 - C_2 L)H .$$

The solution of (4.20), which is of interest, can be shown to be of the form [31]

$$H = A_{12}A_{22}^{-1} + O(\mu) , \quad L = A_{22}^{-1}A_{21} + O(\mu) . \tag{4.24}$$

Omitting initial conditions, the transfer function matrix $P(s, \mu s, \mu)$ from the input $u(s)$ to the output $y(s)$ is given by

$$P(s, \mu s, \mu) = P_s(s, \mu) + P_f(\mu s, \mu) , \tag{4.25}$$

where

$$P_s(s, \mu) = C_s(sI_n - A_s)^{-1}B_s$$

is called the slow transfer function matrix, while

$$P_f(\mu s, \mu) = C_f(\mu s I_m - A_f)^{-1}B_f$$

is called the fast transfer function matrix. It is worth noting that $P_s(s, \mu)$ has the low-pass property, i.e., $P_s(s, \mu) = O(\mu)$ for $s \in [j\omega_0, \infty)$ for some fixed $\omega_0 > 0$. On the other hand, using the identities (4.7) we can express $P_f(\mu s, \mu)$ as

$$P_f(\mu s, \mu) = P_f(0, \mu) + \Delta(\mu s) , \tag{4.26}$$

where $P_f(0, \mu) = -C_f A_f^{-1} B_f$ and

$$\Delta(\mu s) = \mu s C_f A_f^{-1}(\mu s I_m - A_f)^{-1}B_f . \tag{4.27}$$

Hence, for low frequencies, i.e., for $s \in [0, j\omega_1]$ and some fixed $\omega_1 > 0$, we have

$$P_f(\mu s, \mu) = P_f(0, \mu) + O(\mu) , \quad \forall s \in [0, j\omega_1] .$$

In view of (4.26) and the above analysis, we can express $P(s, \mu s, \mu)$ as

$$P(s, \mu s, \mu) = P_0(s, \mu) + \Delta(\mu s) , \tag{4.28}$$

where $P_0(s, \mu) = P_s(s, \mu) + P_f(0, \mu)$ is a low-frequency approximation of $P(s, \mu s, \mu)$ and can be taken to represent the nominal plant. Therefore, $\Delta(\mu s)$ can be treated as an additive perturbation according to Definition 4.2. Using (4.24), it can be shown that $P_0(s) = P(s, 0, 0)$ (given by (4.18)) is also a low-frequency approximation of $P(s, \mu s, \mu)$ since $P_s(s, \mu) = P_0(s) + O(\mu)$ and $P(s, \mu s, \mu) = P_s(s, \mu) + O(\mu)$, $\forall s \in [0, j\omega_1]$. Therefore, $P_0(s)$ can also be taken to be the nominal model. In this case, the corresponding additive perturbation is $\Delta_\alpha(s) =$

$\Delta(\mu s) + \mu\Delta_1(s,\mu)$, where $\Delta_1(s,\mu)$ can be shown to be a proper stable transfer function matrix.

We have shown that the singular perturbation model (4.14)–(4.16) can be treated in the complex domain as a nominal plant P_0 with an additive perturbation Δ_α. We have to emphasize, however, that even though the overall transfer function matrix $P(s,\mu s,s)$ of the plant (4.14)–(4.16) is strictly proper, its nominal part $P_0(s,\mu)$ or $P_0(s)$ may not be so. For example, in (4.28)

$$P_0(s,\mu) = C_s(sI_n - A_s)^{-1}B_s - C_f A_f^{-1} B_f\,,$$

which is strictly proper provided $C_f A_f^{-1} B_f = 0$. The fast dynamics or parasitics in (4.14)–(4.16), which create a throughput $\|-C_f A_f^{-1} B_f\| > O(\mu)$ in $P_0(s,\mu)$, are referred to as strongly controllable and observable parasitics. As discussed in [31,8], if they are not taken into account in the control design, the results may be catastrophic.

One way to eliminate the effect of strongly controllable and strongly observable parasitics is to augment (4.14)–(4.16) or (4.21)–(4.23) with low-pass filters [8] as follows: we pass y through the filter $I_q\frac{f_1}{s+f_0}$ for some fixed f_0, $f_1 > 0$, i.e.,

$$\dot{\hat{y}} = -f_0\hat{y} + f_1 y = -f_0\hat{y} + f_1 C_1 x + f_1 C_2 z\,, \tag{4.29}$$

and augment (4.14)–(4.16) with (4.29) to obtain the $(n+q+m)$ system

$$\dot{\hat{x}} = \hat{A}_{11}\hat{x} + \hat{A}_{12}z + \hat{B}_1 u \tag{4.30}$$

$$\mu\dot{z} = \hat{A}_{21}\hat{x} + A_{22}z + B_2 u \tag{4.31}$$

$$\hat{y} = \hat{C}_1\hat{x}\,, \tag{4.32}$$

where $\hat{x} = [\hat{y}, x^{\mathrm{T}}]^{\mathrm{T}}$ and $\hat{A}_{11}, \hat{A}_{12}, \hat{A}_{21}, \hat{B}_1, \hat{C}_1$ are appropriately defined. Since z does not appear in the filtered output $\hat{y}$, i.e., $\hat{C}_2 = 0$, the parasitics are made weakly observable. In the complex domain, the transfer-function matrix $\hat{P}(s,\mu s,\mu)$ from u to $\hat{y}$ is given by

$$\hat{P}(s,\mu s,\mu) = \hat{P}_0(s,\mu) + \hat{\Delta}(\mu s,s)\,,$$

where $\hat{P}_0(s,\mu) = P_0(s,\mu)\frac{f_1}{s+f_0}$ is now strictly proper and $\hat{\Delta}(\mu s,s) = \Delta(\mu s)\frac{f_1}{s+f_0}$ $= \mu\frac{sf_1}{s+f_0}C_f A_f^{-1}(\mu s I_m - A_f)^{-1}B_f$. Due to the filter and the stability of A_f, $\hat{\Delta}(\mu s,s)$ $= O(\mu)$ for $s = j\omega$ and all ω, whereas without the filter $\Delta(\mu s)$ can be of $O(1)$ at high frequencies.

We should note that in our definition the plant transfer function perturbations are assumed to have stable poles. This is not essential in robust non-adaptive control [30], where the plant perturbations are allowed to have a certain number of unstable poles. Usually the unstable poles in the plant perturbations arise due to parametric uncertainty. Since in adaptive control the parametric uncertainty is not considered as a plant perturbation, the assumption of stable plant perturbations is not restrictive. We illustrate this point with the following example: Consider the plant transfer function

$$G(s) = \frac{1}{s-1+\mu}\,,$$

where μ is a small unknown constant. In nonadaptive control one would express $G(s)$ as

$$G(s) = \frac{1}{s-1}[1+\Delta_{\rm m}(s)], \quad \Delta_{\rm m}(s) = -\frac{\mu}{s-1+\mu},$$

and model it as $G_0(s) = \frac{1}{s-1}$, i.e., $\Delta_{\rm m}(s)$ has one unstable pole for $\mu < 1$. In adaptive control we model $G(s)$ as

$$G(s) = \frac{1}{s+a}$$

for some a which is constant but unknown, i.e., $\Delta_{\rm m}(s) = 0$.

We illustrate the various types of plant uncertainties by the following examples:

Example 4.2.1. Consider the following equations describing the dynamics of a DC motor

$$J\dot{\omega} = ki$$
$$L\frac{di}{dt} = -k\omega - Ri + v\,,$$

where i, v, R and L are the armature current, voltage, resistance and inductance respectively, J is the moment of inertia, ω is the angular speed, and ki and $k\omega$ are the torque and back e.m.f. respectively.

Defining $x = \omega$, $z = i$, we have

$$\begin{aligned} \dot{x} &= \alpha_1 z \\ \mu\dot{z} &= -\alpha_2 x - \alpha_3 z + v \\ y &= x\,, \end{aligned} \tag{4.33}$$

where $\alpha_1 = \frac{k}{J}$, $\alpha_2 = k$, $\alpha_3 = R$ and $\mu = L$, which is in the form of the singular perturbation model (4.14)–(4.16). The transfer function between the input v and the output y is given by

$$\frac{y(s)}{v(s)} = \frac{\alpha_1}{\mu s^2 + \alpha_3 s + \alpha_0} = P(s, \mu s, \mu)\,, \tag{4.34}$$

where $\alpha_0 = \frac{k^2}{J}$. In most DC motors, the inductance $L = \mu$ is small and can be neglected leading to the reduced-order or nominal plant transfer function

$$P_0(s) = P(s, 0, 0) = \frac{\alpha_1}{\alpha_3 s + \alpha_0}\,. \tag{4.35}$$

Using $P_0(s)$ as the nominal transfer function we can express $P(s, \mu s, \mu)$ as

$$P(s, \mu s, \mu) = P_0(s) + \Delta_a(s)\,,$$

where $\Delta_a(s) = -\mu\frac{\alpha_1 s^2}{(\mu s^2+\alpha_3 s+\alpha_0)(\alpha_3 s+\alpha_0)}$ is strictly proper and stable since μ, α_3, $\alpha_0 > 0$, or as

$$P(s, \mu s, \mu) = P_0(s)(1 + \Delta_{\rm m}(s))\,,$$

where $\Delta_m(s) = -\mu\frac{\alpha_1 s^2}{\mu s^2+\alpha_3 s+\alpha_0}$ is proper and stable.

Let us now use definition 4.4 and try to express $P(s,\mu s,\mu)$ in the form of (4.13). We can write

$$P(s,\mu s,s) = \frac{\alpha_1(\mu s^2+\alpha_3 s+\alpha_0)-\mu s^2\alpha_1}{(\alpha_3 s+\alpha_0)(\mu s^2+\alpha_3 s+\alpha_0)}$$

$$= \frac{\alpha_1 - \dfrac{\mu s^2\alpha_1}{\mu s^2+\alpha_3 s+\alpha_0}}{(\alpha_3 s+\alpha_0)}$$

$$= \frac{\dfrac{\alpha_1}{s+\lambda} - \dfrac{\mu s^2\alpha_1}{(\mu s^2+\alpha_3 s+\alpha_0)(s+\lambda)}}{\dfrac{\alpha_3 s+\alpha_0}{s+\lambda}+0}$$

for some $\lambda > 0$, i.e., $N_0(s) = \frac{\alpha_1}{s+\lambda}$, $D_0(s) = \frac{\alpha_3 s+\alpha_0}{s+\lambda}$, $\Delta_1(s) = -\frac{\mu s^2\alpha_1}{(s+\lambda)(\mu s^2+\alpha_3 s+\alpha_0)}$ and $\Delta_2(s) = 0$. □

Example 4.2.2. Consider the system shown in Fig. 5. Here $T > 0$ is a small time delay in the feedback. The transfer function is

$$\frac{y(s)}{u(s)} = P(s) = \frac{1}{s+e^{-Ts}-1} .$$

For low frequencies $e^{-Ts}-1$ is $O(T)$. Therefore, for small T the nominal plant can be taken to be

$$P_0(s) = \frac{1}{s} . \tag{4.36}$$

With $P_0(s) = \frac{1}{s}$, however, $P(s)$ cannot be expressed in the form (4.11) with a stable additive perturbation $\Delta_a(s)$ or in the form (4.12) with a stable multiplicative perturbation $\Delta_m(s)$. We can express $P(s)$, however, in the form (4.13) as follows:

$$P(s) = \frac{\dfrac{1}{s+\lambda}}{\dfrac{s}{s+\lambda}+\dfrac{e^{-Ts}-1}{s+\lambda}} = \frac{N_0(s)+\Delta_1(s)}{D_0(s)+\Delta_2(s)} ,$$

where $N_0(s) = \frac{1}{s+\lambda}$, $D_0(s) = \frac{s}{s+\lambda}$ with $\lambda > 0$ is a coprime factorization of $P_0(s) = \frac{1}{s}$ and $\Delta_1(s) = 0$, $\Delta_2(s) = \frac{e^{-Ts}-1}{s+\lambda}$ are proper stable factor perturbations.

□

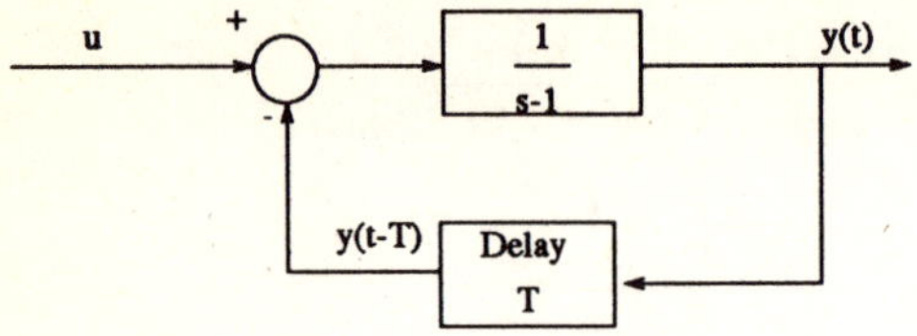

Fig. 5. System for Example 4.2.2.

4.3 Robustness Analysis: Known Parameters

In this subsection, we consider the general feedback system of Fig. 4 subject to the plant perturbations discussed in the last subsection. Our goal is to derive conditions which, if satisfied by the perturbations, will guarantee closed-loop stability. The launching point for any robustness analysis is the stability of the nominal (i.e., unperturbed) plant-controller pair. Accordingly, we assume that in the absence of any perturbations each of the nine entries of $H(s)$ is stable. Using this assumption, we can obtain the following robustness results.

Theorem 4.1 *Let the nominal plant $P_0(s)$ be subjected to an additive perturbation $\Delta_a(s)$. Then, the closed-loop system of Fig. 4 with $P_0(s)$ replaced by $P_0(s) + \Delta_a(s)$ is L_p-stable provided that*

$$\|[I + C_1C_2P_0]^{-1}C_1C_2\Delta_a\|_{ip} < 1.^{3} \tag{4.37}$$

Proof. The perturbed plant-controller configuration for this case is shown in Fig. 6.

We first note that in the absence of Δ_a, we have a stable nominal closed loop consisting of only rational transfer function matrices. As a result, it follows that [24,29] the nominal feedback system is L_p-stable $\forall p \in [1, \infty]$. Thus, the stability of the perturbed closed-loop system will be established once it is shown that r, d_1, $d_2 \in L_p$ implies that $e_2 \in L_p$.

From Fig. 6, using (4.8), it follows that

$$e_2 = [I + C_1C_2P_0]^{-1}C_1r - [I + C_1C_2P_0]^{-1}C_1C_2d_1 + [I + C_1C_2P_0]^{-1}d_2$$
$$-[I + C_1C_2P_0]^{-1}C_1C_2\Delta_a e_2 .$$

[3] In this paper, $\|\cdot\|_{ip}$ denotes the operator norm induced by the L_p norms on the input and output signals

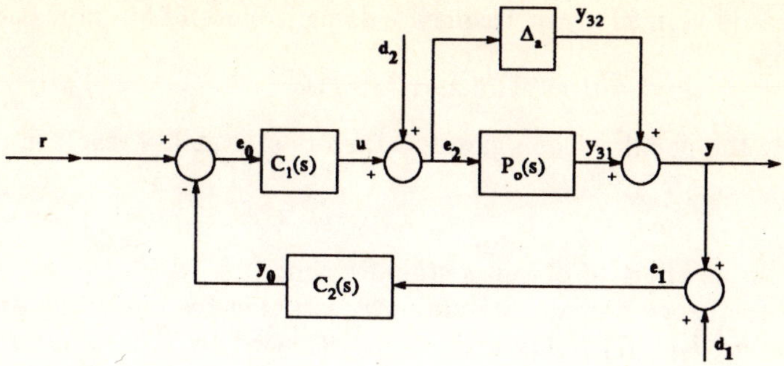

Fig. 6. Perturbed plant-controller configuration (additive perturbation).

Truncating the signals, taking L_p norms on both sides and making use of Minkowski's Inequality [24], it follows that

$$\begin{aligned}\|e_{2t}\|_p \leq & \left\|\left([I+C_1C_2P_0]^{-1}C_1r\right)_t\right\|_p + \left\|\left([I+C_1C_2P_0]^{-1}C_1C_2d_1\right)_t\right\|_p \\ & + \left\|\left([I+C_1C_2P_0]^{-1}d_2\right)_t\right\|_p + \left\|\left([I+C_1C_2P_0]^{-1}C_1C_2\Delta_a e_2\right)_t\right\|_p .\end{aligned}$$

Now from the L_p-stability of the nominal $H(s)$, it follows that each of the operators $[I+C_1C_2P_0]^{-1}C_1$, $[I+C_1C_2P_0]^{-1}C_1C_2$, $[I+C_1C_2P_0]^{-1}$ have finite induced L_p norms. Now, assuming that Δ_a is causal, we obtain

$$\begin{aligned}\|e_{2t}\|_p \leq & \left\|[I+C_1C_2P_0]^{-1}C_1\right\|_{ip} \|r_t\|_p + \left\|[I+C_1C_2P_0]^{-1}C_1C_2\right\|_{ip} \|d_{1t}\|_p \\ & + \left\|[I+C_1C_2P_0]^{-1}\right\|_{ip} \|d_{2t}\|_p + \left\|[I+C_1C_2P_0]^{-1}C_1C_2\Delta_a\right\|_{ip} \|e_{2t}\|_p .\end{aligned}$$

It now follows that if (4.37) holds, then

$$\|e_{2t}\|_p \leq k_1\|r_t\|_p + k_2\|d_{1t}\|_p + k_3\|d_{2t}\|_p$$

for some $k_1,\ k_2,\ k_3 > 0$.
Thus, $r,\ d_1,\ d_2 \in L_p \Rightarrow e_2 \in L_p$.
This completes the proof. □

A few remarks concerning Theorem 4.1 are now in order:

Remark 4.1. It is well known that $\|\cdot\|_{\mathcal{A}}$ [29] of any linear time-invariant operator provides an upper bound on its induced L_p norm $\forall p \in [1,\infty]$. Thus, if (4.37) holds with $\|\cdot\|_{ip}$ replaced by $\|\cdot\|_{\mathcal{A}}$, then the closed loop is guaranteed to be stable $\forall p \in [1,\infty]$. However, $\|\cdot\|_{\mathcal{A}}$ is difficult to compute and also the condition (4.37), with $\|\cdot\|_{ip}$ replaced by $\|\cdot\|_{\mathcal{A}}$, cannot be easily verified in the frequency domain. □

Remark 4.2. If we choose $p = 2$ in Theorem 4.1, then we obtain a condition which can be easily verified in the frequency domain, since (4.37) now becomes

$$\|[I + C_1C_2P_0]^{-1}C_1C_2\Delta_a(s)\|_\infty < 1 .$$

This is precisely the popular "loop-gain-less-than-one" stability result in robust control [32]. □

Remark 4.3. In an effort to obtain a stability condition which can be readily checked in the frequency domain, we are now restricted to concluding only L_2 stability. However, if (4.37) holds with $\|\cdot\|_{ip}$ replaced by the operator norm induced by $\|(\cdot)_t\|_2^\delta$ for some $\delta > 0$, then it can be shown as in [24] that the perturbed system is also L_∞-stable, i.e., L_2 stability with exponential weighting implies L_∞ stability. □

Theorem 4.2 *Let the nominal plant $P_0(s)$ be subjected to a multiplicative perturbation $\Delta_{\rm m}(s)$. Then, the closed-loop system in Fig. 4 with $P_0(s)$ replaced by $[I + \Delta_{\rm m}(s)]P_0(s)$ is L_p-stable provided that*

$$\|[I + P_0C_1C_2]^{-1}P_0C_1C_2\Delta_{\rm m}\|_{ip} < 1 . \tag{4.38}$$

Proof. The proof is very similar to that of Theorem 4.1 and is therefore omitted. □

Remark 4.4. In many controller structures $C_2 = I$. Then, $S \triangleq (I + P_0C_1)^{-1}$ is called the *Sensitivity*, while $T \triangleq (I + P_0C_1)^{-1}P_0C_1$ is called the *Complementary Sensitivity*. Clearly S and T are related by

$$S + T = I . \tag{4.39}$$

Now from (4.38), it is clear that the smaller T is,[4] the larger the robustness margin. In other words, a small T enhances robustness. On the other hand, from the (1,1) entry of $H(s)$, it is clear that for good tracking performance S must be made small. Moreover, it is clear that a small S will lead to better output disturbance rejection. Thus, a smaller S enhances performance. Since S and T are constrained by the relationship (4.39), it follows that in any control design, there is a fundamental tradeoff between performance and robustness. H_2 and H_∞ optimal control [33] have emerged as powerful tools for achieving such a tradeoff.

Now tracking high frequency signals is not very meaningful. Thus, for tracking purposes, S should be made small only in the low-frequency range. On the other hand, at high frequencies the plant is very poorly known so that it is desireable to have T small at high frequencies. These simple observations give the control engineer a qualitative idea of how to achieve such a tradeoff. Of course, the actual design may involve H_2/H_∞ controller synthesis techniques. □

[4] Here the size of T is measured by its H_∞ norm.

Theorem 4.3 *Let the nominal plant $P_0(s)$ have a right coprime factorization $(N_0(s), D_0(s))$ and let $(\Delta_1(s), \Delta_2(s))$ be a stable factor perturbation of $P_0(s)$, so that*

$$P(s) = [N_0(s) + \Delta_1(s)][D_0(s) + \Delta_2(s)]^{-1} \tag{4.40}$$

represents the perturbed plant. Then, the feedback system of Fig. 4 with $P_0(s)$ replaced by $P(s)$ of (4.40) is L_p-stable provided that

$$\left\| \left[-[D_0 + C_1C_2N_0]^{-1}C_1C_2,\ [D_0 + C_1C_2N_0]^{-1} \right] \begin{bmatrix} \Delta_2 \\ \Delta_1 \end{bmatrix} \right\|_{ip} < 1. \tag{4.41}$$

Proof. The proof is essentially the same as that of Theorem 4.1 and is therefore omitted. □

Remarks 4.1, 4.2 and 4.3 apply to Theorems 4.2 and 4.3 also.

4.4 Control Structures for Adaptive Control

In this subsection, we describe three control structures which have become very popular in the adaptive control literature. They are the model reference control structure, the pole placement control structure and the linear quadratic control structure. The controllers themselves are designed on the assumption that there are no disturbances (i.e., $d_1 = d_2 = 0$) or unmodeled dynamics. However, each of these control structures is shown to be a special case of the General Feedback System of Fig. 4 so that the robustness results of the last subsection apply to each one of them. We begin with the model reference control structure.

Model Reference Control Structure. In model reference control, the objective is to stabilize the plant and to force its output y to track the output $y_m(t)$ of a reference model given by:

$$y_m(s) = W_m(s)r(s) = k_m \frac{Z_m(s)}{R_m(s)} r(s) \tag{4.42}$$

for *any* bounded piecewise continuous reference signal $r(t)$, where $k_m > 0$ is a constant and $Z_m(s)$, $R_m(s)$ are monic Hurwitz polynomials of degree m_r, n_r respectively. In order to achieve such an objective the following assumptions [2] are made about the modeled part of the plant transfer function

$$P_0(s) = k_p \frac{Z_0(s)}{R_0(s)} : \tag{4.43}$$

(M1) $Z_0(s)$ is a monic Hurwitz polynomial of degree $m \leq n - 1$,
(M2) $R_0(s)$ is a monic polynomial of degree n,
(M3) the relative degree of $W_m(s)$ is $n_r - m_r = n - m$, and $n_r \leq n$.

The MRC law is given by

$$u = \theta_1^{*\mathrm{T}} \frac{a(s)}{\Lambda(s)} u + \theta_2^{*\mathrm{T}} \frac{a(s)}{\Lambda(s)} y + \theta_3^* y + c_0^* r\,, \tag{4.44}$$

where $a(s) = [s^{n-2}, s^{n-3}, \ldots, 1]^{\mathrm{T}}$, $\Lambda(s) = \Lambda_0(s) Z_{\mathrm{m}}(s)$, $\Lambda_0(s)$ is a monic Hurwitz polynomial of degree $n - m_{\mathrm{r}} - 1$ and θ_i^*, $i = 1, 2, 3$ and c_0^* are the constant controller parameters to be determined so that the control objective is achieved for the modeled part of the plant $P_0(s)$. It can be shown [2] that there exist θ_1^*, θ_2^*, θ_3^*, c_0^* such that the control objective is achieved for the nominal plant $P_0(s)$. Now with

$$C_1(s) = \frac{c_0^* \Lambda(s)}{\Lambda(s) - \theta_1^{*T} a(s)}\,, \quad C_2(s) = -\frac{\theta_2^{*T} a(s) + \theta_3^* \Lambda(s)}{c_0^* \Lambda(s)}\,,$$

it can be easily shown that the general feedback system of Fig. 4 reduces to the model reference control structure of (4.44).

We use the following simple example to illustrate the design analysis and robustness of a model reference control scheme.

Example 4.4.1. Consider the plant

$$y = \frac{1}{s + a}(1 + \Delta_{\mathrm{m}}(s)) u\,, \tag{4.45}$$

where $\Delta_{\mathrm{m}}(s)$ is a multiplicative plant uncertainty which is stable, and a is the parameter of the modeled part of the plant.

The control objective is to choose u such that all the signals in the closed-loop plant are bounded and y is as close as possible to the output y_{m} of the reference model

$$y_{\mathrm{m}} = \frac{b_m}{s + a_{\mathrm{m}}}[r]\,, \tag{4.46}$$

where b_m, $a_{\mathrm{m}} > 0$ are known, for any bounded reference input signal $r(t)$. If a is known, the MRC law (4.44) yields

$$u = \theta_0^* y + b_m r \tag{4.47}$$

with $\theta_0^* = a - a_{\mathrm{m}}$, and this choice of u guarantees that the control objective is met for $\Delta_{\mathrm{m}}(s) = 0$. When $\Delta_{\mathrm{m}}(s) \neq 0$, we have

$$[s + a_{\mathrm{m}} - \theta_0^* \Delta_{\mathrm{m}}(s)] y = b_m (1 + \Delta_{\mathrm{m}}(s)) r\,,$$

so that the closed-loop characteristic equation is

$$(s + a_{\mathrm{m}}) - \theta_0^* \Delta_{\mathrm{m}}(s) = 0\,,$$

or

$$1 - \theta_0^* \frac{\Delta_{\mathrm{m}}(s)}{(s + a_{\mathrm{m}})} = 0\,.$$

Since $\theta_0^* \frac{\Delta_m(s)}{(s+a_m)}$ is a stable transfer function, it follows from the small gain theorem that a sufficient condition for the closed-loop system to be stable is that $\Delta_m(s)$ satisfy

$$\left\| \frac{\theta_0^* \Delta_m(s)}{s+a_m} \right\|_\infty = \left\| \frac{(a-a_m)\Delta_m(s)}{s+a_m} \right\|_\infty < 1 , \tag{4.48}$$

which agrees with the robustness condition (4.38). Furthermore, the tracking error $e_r \triangleq y - y_m$ satisfies

$$e_r = \frac{b_m}{(s+a_m)} \frac{(s+a_m+\theta_0^*)}{(s+a_m-\theta_0^*\Delta_m(s))} \Delta_m(s)[r] .$$

□

The same example can be analyzed in the presence of additive and stable factor perturbations without any significant changes. Condition (4.48) for stability also involves the transfer function of the reference model. One can choose a_m so that $a - a_m$ is as small as possible depending on the flexibility of the control objective.

Pole Placement Control Structure. In pole placement control, the objective is to stabilize the plant $P_0(s)$ by assigning the poles of the closed-loop plant to the roots of some desired polynomial. This stabilization objective can be extended to tackle the tracking problem for a special class of reference input signals by using the so-called *internal model principle.*

Suppose the signal $y_m(t)$ to be tracked satisfies:

$$Q(p)y_m(t) \equiv 0 , \tag{4.49}$$

where $Q(p)$ is a fixed known monic polynomial of the differential operator $p \triangleq \frac{d}{dt}(\cdot)$ of degree q_0 with non-repeated roots on the $j\omega$ axis. In pole placement control, we seek to synthesize the control input $u(t)$ which places the closed-loop poles of the system at the roots of a given Hurwitz polynomial $A^*(s)$ of degree $2n + q_0 - 1$ and guarantees that the output $y(t)$ tracks $y_m(t)$. $Q(s)$ is called the *internal model* of $y_m(t)$.

In order to achieve the pole placement control objective, we make the following assumptions about the modeled part of the plant transfer function $P_0(s) = Z_0(s)/R_0(s)$:

(P1) $R_0(s) = s^n + \bar{R}_0(s, \theta_1^*)$ is a monic polynomial of degree n,
(P2) $Z_0(s) = \bar{Z}_0(s, \theta_2^*)$ is a polynomial of degree $\leq n-1$, and
(P3) $Q(s)R_0(s)$ and $Z_0(s)$ are coprime.

Here θ_1^*, θ_2^* are vectors whose elements are the coefficients of $\bar{R}_0(s)$, $Z_0(s)$ respectively.

The pole placement control law is given by [10,20]

$$u = [\Lambda_1(s) - P(s)Q(s)] \frac{1}{\Lambda_1(s)}[u] + L(s) \frac{1}{\Lambda_1(s)}(y_m - y) , \tag{4.50}$$

where $L(s)$, $P(s)$ are polynomials of degree $q_0 + n - 1$ and $n - 1$ respectively, obtained by solving the Bezout equation

$$P(s)Q(s)R_0(s) + L(s)Z_0(s) = A^*(s), \tag{4.51}$$

where

- q_0 is the degree of $Q(s)$,
- $\Lambda_1(s)$ is a Hurwitz polynomial of degree $n + q_0 - 1$, and
- $A^*(s)$ is the desired closed-loop stable polynomial of degree $2n + q_0 - 1$.

The coprimeness of $Q(s)R_0(s)$ and $Z_0(s)$ guarantees that (4.51) has a unique solution [30]. With $C_1(s) = L(s)/P(s)Q(s)$, $C_2(s) = 1$ it can be easily shown that the general feedback system of Fig. 4 reduces to the pole placement control structure.

From (4.50),

$$P(s)Q(s)u = L(s)(y_m - y)$$
$$P(s)Q(s)Z_0(s)u = L(s)Z_0(s)(y_m - y).$$

Using $Z_0(s)u = R_0(s)y$, we obtain

$$[P(s)Q(s)R_0(s) + L(s)Z_0(s)]y = L(s)Z_0(s)y_m .$$

Using (4.51)

$$y = \frac{L(s)Z_0(s)}{A^*(s)}y_m = \frac{[A^*(s) - P(s)Q(s)R_0(s)]}{A^*(s)}y_m ,$$

i.e.,

$$y - y_m = -\left[\frac{P(s)R_0(s)}{A^*(s)}\right][Q(s)y_m] .$$

Since $Q(s)y_m = 0$, it follows that the tracking error $e_r = y - y_m$ will converge to zero exponentially with time. Since no cancellation of unstable polynomials took place, the closed-loop plant is also internally stable.

In order to implement (4.50)–(4.51), however, we need to know $R_0(s)$ and $Z_0(s)$, i.e., the parameters θ_1^*, θ_2^* of the plant.

Let us now use a simple example to illustrate the design, analysis and robustness of a pole placement control scheme.

Example 4.4.2. Consider the plant

$$y = \frac{b}{s + a}(1 + \Delta_m(s))u , \tag{4.52}$$

where $\Delta_m(s)$ is a multiplicative plant uncertainty which is stable, and a, b are the parameters of the modeled part of the plant.

The control objective is to choose u such that the poles of the closed-loop system are placed at the roots of a given Hurwitz polynomial, say $A^*(s) = (s +$

$1)^2$, and the output $y(t)$ tracks a unit step input $u_s(t)$. Thus, here $y_m(t) = u_s(t)$ and $Q(s) = s$. The control law

$$u = \frac{\lambda_1}{s+\lambda_1}[u] + \frac{l_1 s + l_2}{s+\lambda_1}[y_m - y],$$

where l_1, l_2 satisfy

$$s(s+a) + (l_1 s + l_2)b = (s+1)^2,$$

i.e.,

$$l_1 = \frac{2-a}{b}, \quad l_2 = \frac{1}{b},$$

and $\lambda_1 > 0$ is a design constant, can be shown to meet the control objective when $\Delta_m(s) = 0$. When $\Delta_m(s) \neq 0$, the closed-loop plant is given by

$$y = \frac{b(1+\Delta_m(s))(l_1 s + l_2)}{[(s+1)^2 + b\Delta_m(s)(l_1 s + l_2)]}[y_m].$$

Therefore, for

$$\left\| \frac{b\Delta_m(s)(l_1 s + l_2)}{(s+1)^2} \right\|_\infty < 1, \tag{4.53}$$

the closed-loop plant is stable. Furthermore, the tracking error $e_r = y - y_m$ is given by

$$e_r = -\frac{s+a}{[(s+1)^2 + b\Delta_m(s)(l_1 s + l_2)]}s[y_m] = \frac{s+a}{[(s+1)^2 + b\Delta_m(s)(l_1 s + l_2)]}[0],$$

i.e., if (4.53) is satisfied $e_r(t)$ converges to zero exponentially fast. □

The same result can be established in the presence of additive and stable factor perturbations.

Linear Quadratic Control Structure. In this case the objective is to minimize a particular cost criterion while stabilizing the plant and forcing its output $y(t)$ to track the reference signal $y_m(t)$, which satisfies $Q(p)y_m(t) \equiv 0$ (where $p \triangleq \frac{d}{dt}(\cdot)$) for some monic polynomial $Q(p)$ of degree q_0 with no repeated roots on the $j\omega$ axis. To obtain a solution to this problem, we make the following assumptions about the modeled part of the plant transfer function $P_0(s) = Z_0(s)/R_0(s)$:

(L1) $R_0(s) = s^n + \bar{R}_0(s, \theta_1^*)$ and $Z_0(s) = \bar{Z}_0(s, \theta_2^*)$ are coprime polynomials of degree m and n respectively, with $m < n$, and

(L2) $Z_0(s)$ and $Q(s)$ do not have any common zeros on the $j\omega$ axis.

Here, θ_1^* and θ_2^* are vectors with the coefficients of $\bar{R}_0(s)$, $Z_0(s)$ respectively.

The tracking problem is first transformed into a regulator problem as follows. In Fig. 4, let $C_2(s) = 1$. Then $-e_0 = y - y_m$, which can be rewritten as

$$-e_0 = \frac{Z_0(s)}{R_0(s)}u - y_m$$

$$\Rightarrow -R_0(s)e_0 = Z_0(s)u - R_0(s)y_m.$$

Filtering each side of the above equation by $Q(s)/Q_1(s)$, where $Q_1(s)$ is an arbitrary monic Hurwitz polynomial of degree q_0, and using the fact that $Q(p)y_m(t) \equiv 0$, we obtain

$$-e_0 = P_0'(s)\bar{u}, \tag{4.54}$$

where

$$P_0'(s) = \frac{Z_0(s)}{R_0(s)}\frac{Q_1(s)}{Q(s)}, \quad \bar{u} = \frac{Q(s)}{Q_1(s)}u. \tag{4.55}$$

The LQ controller is designed by minimizing the cost criterion

$$J(\bar{u}) = \int_0^\infty \left[e_0^2(t) + \lambda_w \bar{u}^2(t)\right] dt,$$

where $\lambda_w > 0$ is the control weighting. Let A, b, C be a minimal state-space realization of $P_0'(s)$ in the observable canonical form. Then,

$$A = \left[\begin{array}{c|c} -\tilde{\theta}_1^* & I_{n+q_0-1} \\ & 0 \end{array}\right], \quad b = \tilde{\theta}_2^*, \tag{4.56}$$

where $\tilde{\theta}_1^*$, $\tilde{\theta}_2^*$ are the coefficient vectors of $[s^n + \bar{R}_0(s,\theta_1^*)]Q(s) - s^{n+q_0}$ and $\bar{Z}_0(s,\theta_2^*)Q_1(s)$ respectively, and

$$C = [1\,0\,\cdots\,0]. \tag{4.57}$$

From standard LQR theory for the infinite interval problem [34], it follows that the optimal control $\bar{u}$ is given by

$$\bar{u} = -Ge, \quad G = \frac{1}{\lambda_w}b^{\mathrm{T}}P, \tag{4.58}$$

where e is the state vector corresponding to the realization (A, b, C) and P satisfies the Algebraic Ricatti Equation

$$A^{\mathrm{T}}P + PA + C^{\mathrm{T}}C - \frac{1}{\lambda_w}Pbb^{\mathrm{T}}P = 0. \tag{4.59}$$

When the state vector e is not available, it is replaced in (4.58) by its estimate $\hat{e}$ obtained from the following state observer:

$$\dot{\hat{e}} = A\hat{e} + b\bar{u} + k(C\hat{e} + e_0), \tag{4.60}$$

where k is chosen so that $A + kC$ is a stable matrix. Using standard LQG theory [34], it is possible to obtain an optimal choice for k which involves the solution of a dual Ricatti equation. However, any k that makes $A + kC$ a stable matrix is an acceptable choice. Thus, k is chosen to satisfy

$$A + kC = A^*, \tag{4.61}$$

where A^* is a stable matrix. Without loss of generality, we assume that A^* has the following form

$$A = \left[\begin{array}{c|c} -a^* & \begin{matrix} I_{n+q_0-1} \\ \\ 0 \end{matrix} \end{array} \right], \tag{4.62}$$

so that (4.61) is guaranteed to have a solution for k. The control $u(t)$ is thus given by

$$u = \frac{Q_1(s)}{Q(s)}\bar{u}, \quad \bar{u} = -G\hat{e}, \quad G = \frac{1}{\lambda_w}b^{\mathrm{T}}P, \tag{4.63}$$

where $\hat{e}$ is obtained from (4.60) and P is the unique nonnegative definite solution of (4.59). With

$$C_1(s) = -\frac{Q_1(s)}{Q(s)}G(sI - A - kC + bG)^{-1}k, \quad C_2(s) = 1,$$

it can be easily shown that the general feedback system of Fig. 4 reduces to the linear quadratic control structure.

Let us now use the same example as in the pole placement case to illustrate the design, analysis and robustness of an LQ control scheme.

Example 4.4.3. Consider the plant

$$y = \frac{b_1}{s+a}(1 + \Delta_{\mathrm{m}}(s))u, \tag{4.64}$$

where $\Delta_{\mathrm{m}}(s)$ is a multiplicative plant uncertainty which is stable, and a, b_1 are the parameters of the modeled part of the plant.

The control objective is to choose u to stabilize the plant in a way so as to minimize a particular performance index while making the output $y(t)$ track a unit step input $u_{\mathrm{s}}(t)$. Thus, here $y_{\mathrm{m}}(t) = u_{\mathrm{s}}(t)$ and $Q(s) = s$. Choosing $Q_1(s) = s + q$, $q > 0$ being arbitrary we can convert the tracking problem for (4.64) into an equivalent regulation problem for

$$e_0 = -\frac{(s+q)b_1}{s(s+a)}\left[1 + \Delta_{\mathrm{m}}(s)\right]\bar{u}, \tag{4.65}$$

where

$$\begin{aligned} e_0 &\stackrel{\Delta}{=} y_{\mathrm{m}} - y \\ \bar{u} &\stackrel{\Delta}{=} \frac{s}{s+q}[u]. \end{aligned}$$

Equation (4.65) can be realized in the state-space form as

$$\begin{aligned} \dot{e} &= Ae + b(\bar{u} + \Delta_{\mathrm{m}}(s)\bar{u}) \\ e_0 &= -Ce, \end{aligned} \tag{4.66}$$

where

$$A = \begin{bmatrix} -a & 1 \\ 0 & 0 \end{bmatrix}, \quad b = \begin{bmatrix} b_1 \\ b_1 q \end{bmatrix}, \quad C = [1 \ 0].$$

The LQ control law

$$\bar{u} = -Ge, \quad G = \frac{1}{\lambda_w} b^{\mathrm{T}} P, \tag{4.67}$$

where P is the unique non negative definite solution of the Algebraic Ricatti Equation (ARE)

$$A^{\mathrm{T}}P + PA + C^{\mathrm{T}}C - \frac{1}{\lambda_w} Pbb^{\mathrm{T}}P = 0, \tag{4.68}$$

can be shown to meet the control objective of minimizing the cost criterion $\int_0^\infty [e_0^2(t) + \lambda_w \bar{u}^2(t)]dt$ in the ideal case, i.e., for $\Delta_{\mathrm{m}}(s) = 0$.

Since the state vector e is not available, it is replaced in (4.67) by its estimate $\hat{e}$ obtained from the state observer

$$\dot{\hat{e}} = A\hat{e} + b\bar{u} + k(C\hat{e} + e_0), \tag{4.69}$$

where $k = [a - \alpha_1, \ -\alpha_2]^{\mathrm{T}}$ and $\alpha_1, \alpha_2 > 0$ are chosen such that $s^2 + \alpha_1 s + \alpha_2$ is a Hurwitz polynomial. To summarize, the control input u for (4.64) is given by

$$u = \frac{s+q}{s}[\bar{u}], \quad \bar{u} = -G\hat{e}, \quad G = \frac{1}{\lambda_w} b^{\mathrm{T}} P, \tag{4.70}$$

where $\hat{e}$ is obtained from (4.69) and P is the unique nonnegative definite solution of (4.68). Equation (4.70) along with (4.68), (4.69) guarantee that the LQ control objective is met when $\Delta_{\mathrm{m}}(s) = 0$. When $\Delta_{\mathrm{m}}(s) \neq 0$, we analyze the closed-loop system as follows:

From (4.69), we have

$$\hat{e} = [sI - (A + kC)]^{-1} [b\bar{u} + ke_0].$$

Substituting this expression in (4.70) we obtain

$$\bar{u} = -G[sI - (A + kC)]^{-1} [b\bar{u} + ke_0],$$

or

$$[1 + G[sI - (A + kC)]^{-1} b]\, \bar{u} = -G[sI - (A + kC)]^{-1} ke_0.$$

Combining the above expression with (4.65), we obtain

$$\begin{aligned} &[1 + G(sI - (A + kC))^{-1} b] e_0 \\ &= \frac{(s+q)b_1}{s(s+a)} [1 + \Delta_{\mathrm{m}}(s)] G[sI - (A + kC)]^{-1} ke_0, \end{aligned}$$

or

$$\begin{aligned} \{ s(s+a)[1 + G(sI - (A + kC))^{-1} b] - (s+q) b_1 G[sI - (A + kC)]^{-1} k \\ -(s+q) b_1 \Delta_{\mathrm{m}}(s) G[sI - (A + kC)]^{-1} k \} \, e_0 = 0. \end{aligned} \tag{4.71}$$

Using the small gain theorem as in the earlier examples, we conclude that the closed-loop system is stable provided that

$$\left\| \Delta_{\mathrm{m}}(s) \frac{(s+q)b_1 G[sI-(A+kC)]^{-1}k}{s(s+a)[1+G(sI-(A+kC))^{-1}b]-(s+q)b_1 G[sI-(A+kC)]^{-1}k} \right\|_\infty < 1\,, \tag{4.72}$$

which agrees with the robustness condition (4.38). Also, from (4.71) and the stability of the closed loop it is clear that the tracking error $e_{\mathrm{r}} \triangleq y - y_{\mathrm{m}} = -e_0 \to 0$ exponentially fast. □

The same result can be established in the presence of additive and stable factor perturbations.

5 Robust Adaptive Control Schemes

In Sect. 4.4, we have considered the design of three particular robust controller structures. However, when the parameters of the modeled part of the plant are unknown, the control input u cannot be calculated and therefore none of these controller structures can be implemented. A natural approach to follow in the unknown parameter case is to use the same control law as in the known parameter case, replacing the unknown controller parameters with their estimates obtained by using the identification techniques of Sect. 3. This approach is called Certainty Equivalence and has been widely used in the design of adaptive control schemes. The way the controller parameters are estimated led to two different classes of adaptive control schemes, called direct adaptive control and indirect adaptive control, which are treated in the following subsections.

5.1 Direct Robust Adaptive Control

The appropriate control input u that will force the unknown plant to meet the control objective is obtained by using the following steps:

Step 1: We derive the control law that can meet the control objective when the plant parameters are unknown. This step demonstrates that there is sufficient structural flexibility in the closed-loop plant which allows us to meet the control objective.

Step 2: We use the same control law as in Step 1 but with the controller parameters replaced by their estimates generated by an adaptive law. The adaptive law is designed by first parametrizing the plant in terms of the unknown desired controller parameters and then using any of the methods given in Sect. 3 to derive the adaptive law.

Step 3: We analyze the adaptive control scheme formed in Step 2 and show that it meets the control objective.

The adaptive control scheme developed as in Step 2 is referred to as direct adaptive control for the simple reason that the controller parameters are estimated directly without any explicit information about the plant parameters. The derivation of the adaptive law in Step 2 assumes that the plant equation can be parametrized in terms of the unknown controller parameter vector θ^*, i.e., the unknown plant parameters are expressed in terms of θ^* and no longer appear in the plant equation. In order to eliminate the unknown plant parameters and express the plant equation in terms of θ^* and in the form of the parametric models considered in Sect. 3, we have to cancel the unknown zeros of the modeled part of the plant by filtering. This cancellation restricts the modeled part of the plant to be minimum-phase, i.e., to have stable zeros. As a result, the only class of direct adaptive control for which a complete stability analysis exists, is the one based on the model reference approach [2,9,10,11,13,16,20], which requires the modeled part of the plant to have stable zeros. We illustrate the design and analysis of model reference adaptive control (MRAC) using a simple example.

Example 5.1.1. Consider the plant

$$y = \frac{1}{s+a}(1 + \Delta_{\mathrm{m}}(s))u\,, \tag{5.1}$$

where a is an unknown constant and $\Delta_{\mathrm{m}}(s)$ is a multiplicative plant uncertainty which is assumed to be analytic in $\mathrm{Re}[s] \geq 0$. The control objective is to choose the input $u(t)$ such that all the signals in the closed-loop are bounded and the output y tracks the output of the reference model

$$y_{\mathrm{m}} = \frac{b_m}{s + a_{\mathrm{m}}}[r]\,, \tag{5.2}$$

where $a_{\mathrm{m}} > 0$, b_m are known parameters, as closely as possible for any bounded reference input signal $r(t)$. We design the MRAC scheme by following Steps 1–3.

Step 1: As we have shown in Sect. 4.4 the control law

$$u = \theta_0^* y + b_m r\,, \quad \theta_0^* = a - a_{\mathrm{m}}\,, \tag{5.3}$$

meets the control objective exactly when $\Delta_{\mathrm{m}}(s) = 0$. Furthermore, when $\Delta_{\mathrm{m}}(s) \neq 0$, the control law (5.3) guarantees that for all $\Delta_{\mathrm{m}}(s)$ that satisfy

$$\left\| \frac{\theta_0^* \Delta_{\mathrm{m}}(s)}{s + a_{\mathrm{m}}} \right\|_\infty = \left\| \frac{(a - a_{\mathrm{m}})\Delta_{\mathrm{m}}(s)}{s + a_{\mathrm{m}}} \right\|_\infty < 1\,, \tag{5.4}$$

we have signal boundedness, and the tracking error $e_{\mathrm{r}} = y - y_{\mathrm{m}}$ satisfies

$$e_{\mathrm{r}} = \frac{b_m}{(s + a_{\mathrm{m}})} \frac{(s + a_{\mathrm{m}} + \theta_0^*)}{(s + a_{\mathrm{m}} - \theta_0^* \Delta_{\mathrm{m}}(s))} \Delta_{\mathrm{m}}(s)[r]\,.$$

Step 2: When a is unknown (5.3) cannot be implemented. Therefore, instead of (5.3) we propose the control law

$$u = \theta_0(t) y + b_m r\,, \tag{5.5}$$

where $\theta_0(t)$ is the estimate of θ_0^* at time t. The adaptive law for generating θ_0 is developed as follows:
We substitute $a = \theta_0^* + a_m$ in the plant equation to obtain

$$(s + a_m + \theta_0^*)[y] = (1 + \Delta_m(s))[u] .$$

Filtering by $1/(s + a_m)$, we have

$$z = \frac{1}{s + a_m}[\theta_0^* \omega + \eta_0] , \tag{5.6}$$

where $z = -y + \frac{1}{s+a_m}[u]$, $\omega = y$, $\eta_0 = -\Delta_m(s)[u]$, which is in the form of the linear parametric model (3.3). The adaptive law for estimating θ^* can be developed by using any one of the methods given in Sect. 3. Let us consider the Lyapunov method and the gradient method.

(i) *Lyapunov Method.* Since $\frac{1}{s+a_m}$ is SPR, we choose $L(s) = 1$ and use the results of Sect. 3.1 to obtain the adaptive law

$$\begin{aligned} \dot{\theta}_0 &= -\gamma\epsilon\zeta - \gamma\sigma_s\theta_0 , \quad \zeta = \omega = y \\ \epsilon &= \frac{1}{s + a_m}[\theta_0\zeta - \epsilon n_s^2] - z = \frac{1}{s + a_m}[\phi_0\zeta - \epsilon n_s^2 - \eta] , \end{aligned} \tag{5.7}$$

where $\phi_0 = \theta_0 - \theta_0^*$ and $\eta = \eta_0 = -\Delta_m(s)[u]$

(ii) *Gradient Method.* Using the results of Sect. 3.2, we have

$$\begin{aligned} \dot{\theta}_0 &= -\gamma\epsilon\zeta - \gamma\sigma_s\theta_0 , \quad \zeta = \frac{1}{s + a_m}[y] \\ \epsilon &= \frac{\theta_0\zeta - z}{m^2} = \frac{\phi_0\zeta - \eta}{m^2} , \quad \eta = -\frac{1}{s + a_m}\Delta_m(s)[u] , \end{aligned} \tag{5.8}$$

where $m^2 = \alpha + \beta n_s^2$ for some α, $\beta > 0$. The normalizing signal n_s^2 is generated as $n_s^2 = m_s$, where

$$\dot{m}_s = -\delta_0 m_s + |u|^2 + |y|^2 , \quad m_s(0) = 0 \tag{5.9}$$

for some $a_m > \delta_0 > 0$ and such that $\Delta_m(s - \delta_0/2)$ has stable poles. Hence, the first assumption that $\Delta_m(s)$ has to satisfy in the adaptive case is

(A1) $\Delta_m(s)$ is analytic in $\mathrm{Re}[s] \geq -\delta_0/2$ for some known $\delta_0 > 0$.

Let us now check whether n_s generated from (5.9) has the property $\frac{\zeta}{m}$, $\frac{\eta}{m} \in L_\infty$ in (5.7), (5.8). We first consider (5.7), where $\zeta = y$ and $\eta = -\Delta_m(s)[u]$. We have $\dot{y} = -a_m y + (1 + \Delta_m(s))u - (a - a_m)y$. Using Lemma 2.2, we have

$$|y(t)| \leq c \left\| \frac{s + p}{s + a_m}(1 + \Delta_m(s)) \right\|_\infty^{\delta_0} \|u_t\|_2^{\delta_0} + c \left\| \frac{s + p}{s + a_m} \right\|_\infty^{\delta_0} \|y_t\|_2^{\delta_0}$$

for some $p > \delta_0$ and $c \in \mathbb{R}^+$. Therefore, if $\Delta_m(s)$ is proper, i.e., the H_∞-norms above are finite, we have

$$|y(t)| \leq c \left(\|u_t\|_2^{\delta_0} + \|y_t\|_2^{\delta_0} \right)$$

for some $c \in \mathbb{R}^+$, which implies that $\frac{|y(t)|}{m} \in L_\infty$. Similarly

$$|\eta(t)| \le c\|(s+p)\Delta_{\rm m}(s)\|_\infty^{\delta_0}\,\|u_t\|_2^{\delta_0}\,,$$

which implies that $\frac{|\eta(t)|}{m} \in L_\infty$, provided $\Delta_{\rm m}(s)$ is strictly proper. The requirement that $\Delta_{\rm m}(s)$ be strictly proper can be relaxed by filtering both sides of (5.6) as shown in [17] before deriving the adaptive law (5.7).

Let us now consider (5.8). We have

$$|\zeta(t)| \le c\left\|\frac{(s+p)}{s+a_{\rm m}}\right\|_\infty^{\delta_0}\|y_t\|_2^{\delta_0}\,,\quad |\eta(t)| \le c\left\|\frac{(s+p)}{s+a_{\rm m}}\Delta_{\rm m}(s)\right\|_\infty^{\delta_0}\|u_t\|_2^{\delta_0}\,,$$

which imply that $\frac{|\zeta|}{m}, \frac{|\eta|}{m} \in L_\infty$ provided $\Delta_{\rm m}(s)$ is proper. As we showed in Sect. 3, the adaptive laws (5.7), (5.8) guarantee that (i) θ_0, $\epsilon \in L_\infty$ (ii) ϵ, $\epsilon n_{\rm s}$, $\dot\theta_0 \in \mathcal{S}(\frac{\eta^2}{m^2})$ independent of the boundedness of all the other signals in the closed-loop plant.

Step 3: In this step we use the properties (i) and (ii) of the adaptive law to establish stability and robustness of the overall closed-loop plant, which can be represented by the following equations:

$$\begin{aligned} u &= \theta_0^* y + b_m r + \phi_0 y \\ y &= y_{\rm m} + \frac{1}{s+a_{\rm m}}[\phi_0 y] + \frac{\Delta_{\rm m}(s)}{s+a_{\rm m}}[u]\,. \end{aligned}$$

Using the properties of the L_{2e}^{δ} norm and Lemma 2.2, we have

$$\begin{aligned} \|u_t\|_2^{\delta_0} &\le c\|y_t\|_2^{\delta_0} + \|(\phi_0 y)_t\|_2^{\delta_0} + c \\ \|y_t\|_2^{\delta_0} &\le c\|(\phi_0 y)_t\|_2^{\delta_0} + \Delta_1\|u_t\|_2^{\delta_0} + c\,, \end{aligned}$$

where

$$\Delta_1 = \left\|\frac{\Delta_{\rm m}(s)}{s+a_{\rm m}}\right\|_\infty^{\delta_0}, \tag{5.10}$$

and $c \in \mathbb{R}^+$ is a general bound for the various finite constants which are independent of $\Delta_{\rm m}(s)$. Hence,

$$m \le \sqrt{\alpha + \beta n_{\rm s}^2} \le c + c(\|u_t\|_2^{\delta_0} + \|y_t\|_2^{\delta_0}) \le c\Delta_1\|u_t\|_2^{\delta_0} + c\|(\phi_0 y)_t\|_2^{\delta_0} + c \tag{5.11}$$

for some $c \in \mathbb{R}^+$.

We now need to manipulate the term $\phi_0 y$ in (5.11) using the properties (i) and (ii) of the adaptive law used to generate θ_0.

Let us first use (5.7). We have

$$\phi_0\zeta = \phi_0 y = (s+a_{\rm m})[\epsilon] + \epsilon n_{\rm s}^2 + \eta\,.$$

Using Lemma 2.6, we have

$$\begin{aligned} \phi_0 y &= \left(1 - \frac{\alpha}{s+\alpha}\right)[\phi_0 y] + \frac{\alpha}{s+\alpha}[\phi_0 y] \\ &= \frac{1}{s+\alpha}\left[\dot\phi_0 y + \phi_0\dot y\right] + \frac{\alpha}{s+\alpha}(s+a_{\rm m})[\epsilon] + \frac{\alpha}{s+\alpha}\left[\epsilon n_{\rm s}^2\right] + \frac{\alpha}{s+\alpha}[\eta] \end{aligned}$$

for some $\alpha > \delta_0 > 0$ to be chosen, which implies that

$$\|(\phi_0 y)_t\|_2^{\delta_0} \le \frac{c}{\alpha}\left(\left\|\left(\dot{\phi}_0 y\right)_t\right\|_2^{\delta_0} + \|(\phi_0 \dot{y})_t\|_2^{\delta_0}\right) + c\left\|\left(\epsilon n_s^2\right)_t\right\|_2^{\delta_0} + c\Delta_2\|u_t\|_2^{\delta_0} + c\,,$$

where

$$\Delta_2 = \|\Delta_m(s)\|_\infty^{\delta_0}\,. \tag{5.12}$$

Now $\left|\dot{\phi}_0 y\right| = \left|\dot{\phi}_0 \frac{y}{m} m\right| \le c\left|\dot{\phi}_0 m\right|$ and $\|(\phi_0\dot{y})_t\|_2^{\delta_0} \le c\|\dot{y}_t\|_2^{\delta_0}$ since $\phi_0 \in L_\infty$. Furthermore,

$$\|\dot{y}_t\|_2^{\delta_0} \le c\left\|\frac{s}{s+a_m}\right\|_\infty^{\delta_0} \|(\phi_0 y)_t\|_2^{\delta_0} + c\left\|\frac{s\Delta_m(s)}{s+a_m}\right\|_\infty^{\delta_0} m + c\,,$$

i.e.,

$$\|(\phi_0 y)_t\|_2^{\delta_0} \le \frac{c}{\alpha}\|(\phi_0 y)_t\|_2^{\delta_0} + \frac{c}{\alpha}\Delta_3 m + \frac{c}{\alpha}\left\|\left(\dot{\phi}_0 m\right)_t\right\|_2^{\delta_0} + c\left\|\left(\epsilon n_s^2\right)_t\right\|_2^{\delta_0} + c\Delta_2 m + c\,,$$

where

$$\Delta_3 = \left\|\frac{s\Delta_m(s)}{s+a_m}\right\|_\infty^{\delta_0}\,. \tag{5.13}$$

Choosing $\alpha > 2c$, we have

$$\|(\phi_0 y)_t\|_2^{\delta_0} \le \left(\frac{c}{\alpha}\Delta_3 + c\Delta_2\right) m + \left(\int_0^t e^{-\delta_0(t-\tau)}\gamma^2(\tau) m^2(\tau)d\tau\right)^{\frac{1}{2}} + c\,,$$

where $\gamma^2(\tau) = 2\frac{c}{\alpha}|\dot{\phi}_0|^2 + 2c\epsilon^2 n_s^2$, i.e., $\gamma \in \mathcal{S}(\frac{\eta^2}{m^2})$, which by (5.12) implies that $\gamma \in \mathcal{S}(\Delta_2^2)$. Therefore, from (5.11) we have

$$m \le c\Delta_1 m + c\left(\frac{\Delta_3}{\alpha} + \Delta_2\right) m + c\left(\int_0^t e^{-\delta_0(t-\tau)}\gamma^2(\tau) m^2(\tau)d\tau\right)^{\frac{1}{2}} + c\,.$$

Hence, for

$$c\left(\Delta_1 + \Delta_2 + \frac{\Delta_3}{\alpha}\right) < 1\,, \tag{5.14}$$

we have

$$m \le c\left(\int_0^t e^{-\delta_0(t-\tau)}\gamma^2(\tau) m^2(\tau)d\tau\right)^{\frac{1}{2}} + c\,,$$

or

$$m^2 \le c\int_0^t e^{-\delta_0(t-\tau)}\gamma^2(\tau) m^2(\tau)d\tau + c\,.$$

Using Lemma 2.4, we obtain

$$m^2 \le c\int_0^t e^{-\delta_0(t-\tau)}\gamma^2(\tau) e^{c\int_\tau^t \gamma^2(s)ds} d\tau + c\,.$$

Now $\int_\tau^t \gamma^2(s)ds \le c\Delta_2^2(t-\tau)+c$. Therefore,

$$m^2 \le c\int_0^t e^{-\beta_0(t-\tau)}\gamma^2(\tau)d\tau\,,$$

where

$$\beta_0 = \delta_0 - c\Delta_2^2\,. \tag{5.15}$$

For $c\Delta_2^2 < \delta_0$ we have $\beta_0 > 0$. Since $\gamma(\tau) \in \mathcal{S}(\Delta_2^2)$, it follows from Lemma 2.7 that $m \in L_\infty$. Therefore, $y = \zeta$ and η are bounded, which implies that u and therefore all the signals in the closed loop are bounded. The tracking error $e_r = y - y_m$ is given by

$$e_r = \frac{1}{s+a_m}[\phi_0 y] - \eta = \epsilon + \frac{1}{s+a_m}[\epsilon n_s^2] + \frac{1}{s+a_m}[\eta] - \eta\,,$$

which implies that

$$|e_r| \le |\epsilon(t)| + |\eta(t)| + \|(\epsilon n_s^2)_t\|_2^{a_m} + \|\eta_t\|_2^{a_m}\,.$$

Hence,

$$\int_0^t |e_r(t)|^2 d\tau \le c\Delta_2^2 t + c\,. \tag{5.16}$$

The conditions for stability that $\Delta_m(s)$ has to satisfy when (5.7) is used as the adaptive law are summarized below

(C1) $\Delta_m(s)$ is analytic in $\mathrm{Re}[s] \ge -\delta_0/2$ for some known $\delta_0 > 0$.

(C2) $\Delta_1 = \left\|\frac{\Delta_m(s)}{s+a_m}\right\|_\infty^{\delta_0}$, $\Delta_2 = \|\Delta_m(s)\|_\infty^{\delta_0}$, $\Delta_3 = \left\|\frac{s\Delta_m(s)}{s+a_m}\right\|_\infty^{\delta_0}$ and $\|(s+p)\Delta_m(s)\|_\infty^{\delta_0}$ are finite.

(C3) $c\left(\Delta_1 + \Delta_2 + \frac{\Delta_3}{\alpha}\right) < 1$, where $c \in \mathbb{R}^+$ and α can be arbitrarily large.

(C4) $\Delta_2^2 < \frac{\delta_0}{c}$ for some $c \in \mathbb{R}^+$.

Condition (C2) will be satisfied if $\Delta_m(s)$ is assumed to be strictly proper. The arbitrary constants denoted by c in (C1) to (C4) can be calculated by following the steps given above in a similar way as in [18], where such explicit calculations are performed for a gradient algorithm.

In a similar way we can proceed with the analysis of the MRAC scheme, where (5.8) is used as the adaptive law. In this case, it can be shown that the conditions on $\Delta_m(s)$ for stability are given by

(C1′) $\Delta_m(s)$ is analytic in $\mathrm{Re}[s] \ge -\delta_0/2$ for some known $\delta_0 > 0$.

(C2′) $\Delta_1 = \left\|\frac{\Delta_m(s)}{s+a_m}\right\|_\infty^{\delta_0} \le c$ for some $c > 0$.

(C3′) $\Delta_2 = \left\|\frac{(s+\delta_0)\Delta_m(s)}{(s+a_m)}\right\|_\infty^{\delta_0} \le c\delta_0$ for some $c > 0$.

It is clear that (C1′)–(C3′) allow $\Delta_m(s)$ to be proper, in contrast to (C1)–(C4) for (5.7), where $\Delta_m(s)$ is required to be strictly proper. □

Comparing the stability condition (5.4) in the known parameter case (i.e., the stability condition for robust nonadaptive control) with conditions (C1′)–(C3′) or (C1)–(C4) one may conclude that the conditions in the adaptive case are more restrictive. While this conclusion may be true in some cases, it is not true in others. In particular, it is not true in the case where in robust non-adaptive control $\Delta_m(s)$ is largely due to parametric uncertainty, as shown by the following example.

Example 5.1.2.

$$y = \frac{1}{s+1-\epsilon}[u] \tag{5.17}$$

for some constant ϵ. In robust nonadaptive control we express (5.17) as

$$y = \frac{1}{s+1}[1+\Delta_m(s)],$$

where $\Delta_m(s) = \frac{\epsilon}{s+1-\epsilon}$, i.e., in (5.1) $a = 1$. Then, condition (5.4) becomes

$$\left\| \frac{(1-a_m)\epsilon}{(s+1-\epsilon)(s+a_m)} \right\|_\infty < 1,$$

which requires ϵ to be small, i.e., for $a_m = 0.1$ and $1-\epsilon > 0$ we require $\epsilon < 0.1$. In adaptive control, however, (5.17) is modeled as

$$y = \frac{1}{s+a} u,$$

i.e., $\Delta_m(s) = 0$, which satisfies (C1′)–(C3′) for all ϵ. □

The above example together with our analysis for Example 5.1.1 demonstrate that robust adaptive control can handle a larger class of parametric uncertainties than robust nonadaptive control but a smaller class of non-parametric uncertainties.

5.2 Indirect Robust Adaptive Control

In Sect. 5.1, the control of the unknown plant was carried out by directly estimating the desired controller parameters. An alternative method is to estimate the plant parameters on-line and use them to calculate the controller parameters at each time t. The scheme derived with this method is commonly known as *indirect adaptive control* because the evaluation of the controller parameters is achieved indirectly by using the estimated plant model. The main steps used in the design and analysis of indirect adaptive control are the following:

Step 1: We derive the control law that can be used to meet the control objective as if the plant parameters were known.

Step 2: We propose the same control law as in Step 1 but with the controller parameters calculated at each time t from the estimated plant parameters generated by an adaptive law.

Step 3: We analyze the adaptive control scheme formed in Step 2 and show that it meets the control objective.

In the adaptive control literature, the most frequently encountered indirect schemes are of two types. They are (i) the adaptive pole placement control (APPC) scheme [10,12,13,14,16,20] for which the control objective in Step 1 is pole placement and (ii) the adaptive LQ control (ALQC) scheme [38,39,40,41] for which the control law in Step 1 is obtained by minimizing the LQ cost as discussed in Sect. 4.4. We consider these schemes one by one.

Adaptive Pole Placement Control. We illustrate the design and analysis of an adaptive pole placement control scheme using a simple example.

Example 5.2.1. Consider the plant

$$y = \frac{b}{s+a}(1 + \Delta_{\mathrm{m}}(s))u\,, \tag{5.18}$$

where a, b are unknown constants.

The control objective is to choose u such that the poles of the closed-loop system are placed at the roots of a given Hurwitz polynomial, say $A^*(s) = (s+1)^2$, and the output $y(t)$ tracks a unit step input $u_{\mathrm{s}}(t)$. Thus, here $y_{\mathrm{m}}(t) = u_{\mathrm{s}}(t)$ and $Q(s) = s$.

Step 1: As we have already shown in Sect. 4.4, the control law

$$u = \frac{\lambda_1}{s+\lambda_1}[u] + \frac{l_1 s + l_2}{s+\lambda_1}[y_{\mathrm{m}} - y]\,, \tag{5.19}$$

where

$$l_1 = \frac{2-a}{b}\,, \quad l_2 = \frac{1}{b}\,, \tag{5.20}$$

and $\lambda_1 > 0$ is a design constant, can be used to meet the control objective exactly when $\Delta_{\mathrm{m}}(s) = 0$. Furthermore, if $\Delta_{\mathrm{m}}(s) \neq 0$, then for all $\Delta_{\mathrm{m}}(s)$ satisfying

$$\left\| \frac{b\Delta_{\mathrm{m}}(s)(l_1 s + l_2)}{(s+1)^2} \right\|_\infty < 1\,, \tag{5.21}$$

the closed-loop plant is stable and the tracking error $e_{\mathrm{r}} = y - y_{\mathrm{m}}$ is given by

$$e_{\mathrm{r}} = -\frac{s+a}{[(s+1)^2 + b\Delta_{\mathrm{m}}(s)(l_1 s + l_2)]} s[y_{\mathrm{m}}] = \frac{s+a}{[(s+1)^2 + b\Delta_{\mathrm{m}}(s)(l_1 s + l_2)]}[0]\,,$$

i.e., if (5.21) is satisfied, $e_{\mathrm{r}}(t)$ converges to zero exponentially fast.

Step 2: Since a, b are unknown, (5.20) cannot be used to calculate the controller parameters l_1, l_2. Instead of (5.19)–(5.20) we use

$$u = \frac{\lambda_1}{s+\lambda_1}[u] + \left(\hat{l}_1(t)s + \hat{l}_2(t)\right)\frac{1}{s+\lambda_1}[y_{\mathrm{m}} - y]\,, \tag{5.22}$$

where $\hat{l}_1$, $\hat{l}_2$ are calculated as

$$\hat{l}_1 = \frac{2-\hat{a}}{\hat{b}}, \quad \hat{l}_2 = \frac{1}{\hat{b}}, \tag{5.23}$$

where $\hat{a}$, $\hat{b}$ are the estimates of a, b generated by an adaptive law as follows: From (5.18) we have

$$(s+a)y = bu + b\Delta_{\mathrm{m}}(s)[u]. \tag{5.24}$$

Filtering both sides of (5.24) by $1/(s+\lambda)$, with $\lambda > 0$, and rearranging term,s we obtain

$$z = \theta^{*\mathrm{T}}\zeta + \eta, \tag{5.25}$$

where

$$z = \frac{s}{s+\lambda}[y], \quad \zeta = \frac{1}{s+\lambda}\begin{bmatrix} -y \\ u \end{bmatrix}, \quad \theta^* = [a,\, b]^{\mathrm{T}}$$

$$\eta = \frac{b}{s+\lambda}\Delta_{\mathrm{m}}(s)[u],$$

which is in the form of the linear parametric model used with the gradient and least-squares methods. The adaptive law for estimating θ^* is then given by

$$\begin{aligned} \dot{\theta} &= -\Gamma\epsilon\zeta - \Gamma\sigma_{\mathrm{s}}\theta \\ \epsilon &= \frac{\theta^{\mathrm{T}}\zeta - z}{m^2} = \frac{\phi^{\mathrm{T}}\zeta - \eta}{m^2}, \end{aligned} \tag{5.26}$$

where $m^2 = 1 + n_{\mathrm{s}}^2$ and $n_{\mathrm{s}}^2 = m_{\mathrm{s}}$ is generated from

$$\dot{m}_{\mathrm{s}} = -\delta_0 m_{\mathrm{s}} + u^2 + y^2, \quad m_{\mathrm{s}}(0) = 0.$$

As shown in Sect. 3.2, the adaptive law (5.26) has the following properties (i) ϵ, θ, $\dot{\theta} \in L_\infty$ (ii) ϵ, ϵn_{s}, $\dot{\theta} \in \mathcal{S}(\frac{\eta^2}{m^2})$.

These properties, however, do not guarantee that $\hat{l}_1$, $\hat{l}_2$ are bounded, since (5.26) may generate estimates $\hat{b}$ which are arbitrarily close to zero. If the sign of b and a lower bound $b_0 > 0$ for $|b|$ are known, then (5.26) can be modified so that $|\hat{b}(t)| \geq b_0$ $\forall t \geq 0$ without altering the properties (i) and (ii) [14,11].

Step 3: Let $\bar{x} \triangleq \frac{1}{s+\lambda}[x]$. Since λ and λ_1 are both design constants, we can choose $\lambda = \lambda_1$. Then, using the estimation error $\epsilon m^2 = \theta^{\mathrm{T}}\zeta - z$ and the control law (5.22), we can write

$$\begin{aligned} \dot{\bar{u}} &= -\hat{l}_1\hat{b}\bar{u} - (\hat{l}_2 - \hat{l}_1\hat{a})\bar{y} + \hat{y}_{\mathrm{m}} + \hat{l}_1\epsilon m^2 \\ \dot{\bar{y}} &= -\epsilon m^2 - \hat{a}\bar{y} + \hat{b}\bar{u}, \end{aligned}$$

where $\hat{y}_{\mathrm{m}} = \frac{\hat{l}_1 s + \hat{l}_2}{s+\lambda_1}[y_{\mathrm{m}}]$ is bounded due to $\hat{l}_1$, $\hat{l}_2$, $y_{\mathrm{m}} \in L_\infty$. Defining $x_1 = \bar{y}$, $x_2 = \bar{u}$, we have

$$\dot{x} = A(t)x + \begin{bmatrix} -1 \\ \hat{l}_1 \end{bmatrix}\epsilon m^2 + \begin{bmatrix} 0 \\ \hat{y}_{\mathrm{m}} \end{bmatrix},$$

where $x = [x_1, x_2]^{\mathrm{T}}$ and

$$A(t) = \begin{bmatrix} -\hat{a} & \hat{b} \\ -(\hat{l}_2 - \hat{l}_1\hat{a}) & -\hat{l}_1\hat{b} \end{bmatrix}.$$

For each fixed t, $\det(sI - A(t)) = s^2 + 2s + 1$, and

$$\left\|\dot{A}(t)\right\| \leq c\left(\left|\dot{\hat{a}}\right| + \left|\dot{\hat{b}}\right|\right) \leq c\left\|\dot{\theta}(t)\right\|, \text{ i.e., } \left\|\dot{A}(t)\right\| \in \mathcal{S}\left(\frac{\eta^2}{m^2}\right).$$

Provided $\lambda \geq \delta_0$, using Lemma 2.2, we can show that

$$\frac{\eta^2}{m^2} \leq \frac{1}{\lambda}\left(\|b\Delta_{\mathrm{m}}(s)\|_{\infty}^{\delta_0}\right)^2 = \Delta_0^2,$$

i.e., $\left\|\dot{A}(t)\right\| \in \mathcal{S}(\Delta_0^2)$. Applying Lemma 2.1 it follows that for $\Delta_0 < c$ and some $c > 0$, $A(t)$ is exponentially stable. Hence, from Lemma 2.3 we have

$$\|x_t\|_2^{\delta} \leq c\left\|(\epsilon m^2)_t\right\|_2^{\delta} + c$$

for some $\delta > 0$. Now u and y can be expressed as

$$\begin{bmatrix} y \\ u \end{bmatrix} = (\lambda I + A(t))x + \begin{bmatrix} -1 \\ \hat{l}_1 \end{bmatrix}\epsilon m^2 + \begin{bmatrix} 0 \\ \hat{y}_{\mathrm{m}} \end{bmatrix}.$$

Since $\hat{y}_{\mathrm{m}}$, $A \in L_\infty$, we have

$$\left(\|y_t\|_2^{\delta}\right)^2 + \left(\|u_t\|_2^{\delta}\right)^2 \leq c\left(\|x_t\|_2^{\delta}\right)^2 + c\left(\left\|(\epsilon m^2)_t\right\|_2^{\delta}\right)^2 + c.$$

Therefore,

$$m^2 = 1 + n_{\mathrm{s}}^2 \leq c\left(\left\|(\epsilon m^2)_t\right\|_2^{\delta}\right)^2 + c,$$

i.e.,

$$m^2 \leq c\int_0^t e^{-\delta(t-\tau)}(\epsilon m)^2 m^2 d\tau + c,$$

where $\epsilon m \in \mathcal{S}(\Delta_0^2)$. Hence, by applying Lemma 2.4 it follows that for $\Delta_0 < c$ and some $c > 0$, $m \in L_\infty$, which implies that all signals in the closed loop are bounded.

The following expression for the tracking error

$$\int_0^t |e_{\mathrm{r}}(\tau)|^2 d\tau \leq c\Delta_0^2 t + c$$

can also be obtained in a similar manner as in the direct case. The condition for stability is that

$$\Delta_0 = \frac{1}{\sqrt{\lambda}}\|b\Delta_{\mathrm{m}}(s)\|_{\infty}^{\delta_0} < c$$

for some $c > 0$ which can be calculated if an estimate of the rate of exponential convergence of the transition matrix of $A(t)$ can be obtained. If, instead of leakage, a dead-zone modification is used, the bound for Δ_0 does not depend on the transition matrix of $A(t)$ and is more convenient to calculate [14,18]. □

Adaptive Linear Quadratic Control. We illustrate the design and analysis of an adaptive LQ control scheme using a simple example.

Example 5.2.2. Consider the plant

$$y = \frac{b_1}{s+a}(1 + \Delta_m(s))u\,. \tag{5.27}$$

The control objective is to choose u to stabilize the plant in a way so as to minimize an LQ cost, while making the output $y(t)$ track a unit step input $u_s(t)$. Thus, here $y_m(t) = u_s(t)$ and $Q(s) = s$.

Step 1: As we have already shown in Sect. 4.4, the LQ control law

$$u = \frac{s+q}{s}\bar{u} \tag{5.28}$$

$$\bar{u} = -G\hat{e} \tag{5.29}$$

$$G = \frac{1}{\lambda_w} b^{\mathrm{T}} P \tag{5.30}$$

$$\dot{\hat{e}} = A\hat{e} + b\bar{u} + k(C\hat{e} + e_0)\,, \quad k = [a - \alpha_1,\ -\alpha_2]^{\mathrm{T}} \tag{5.31}$$

$$A^{\mathrm{T}}P + PA + C^{\mathrm{T}}C - \frac{1}{\lambda_w}Pbb^{\mathrm{T}}P = 0\,, \tag{5.32}$$

where

$$A = \begin{bmatrix} -a & 1 \\ 0 & 0 \end{bmatrix}, \quad b = \begin{bmatrix} b_1 \\ b_1 q \end{bmatrix}, \quad C = [1\ 0]$$

can be used to meet the control objective exactly when $\Delta_m(s) = 0$. Furthermore, if $\Delta_m(s) \neq 0$, then for all $\Delta_m(s)$ satisfying

$$\left\| \Delta_m(s) \frac{(s+q)b_1 G[sI - (A+kC)]^{-1}k}{s(s+a)[1 + G(sI - (A+kC))^{-1}b] - (s+q)b_1 G[sI - (A+kC)]^{-1}k} \right\|_\infty < 1\,, \tag{5.33}$$

we have signal boundedness and the tracking error $e_r = y - y_m \to 0$ exponentially fast.

Step 2: Since a, b_1 are unknown, the control law in (5.28)–(5.32) cannot be implemented. Instead, we replace a, b_1 by their estimates $\hat{a}$, $\hat{b}_1$ in the matrices A, b to obtain the matrices $\hat{A}$, $\hat{b}$ and then use the latter instead of A, b in (5.28)–(5.32) to calculate the control input u. Since the design of the adaptive law (5.26) in Example 5.2.1 did not depend on the particular control structure, the same adaptive law can be used here also with the only modification that the parameter estimate $\theta(t)$ now consists of $\hat{a}$ and $\hat{b}_1$. As before, the adaptive law guarantees (i) $\epsilon, \theta, \dot{\theta} \in L_\infty$ (ii) $\epsilon, \epsilon n_s, \dot{\theta} \in \mathcal{S}(\frac{\eta^2}{m^2})$.

The adaptive law (5.26) does not, however, guarantee that the pair $(\hat{A}(t), \hat{b}(t))$ is uniformly controllable for all $t \geq 0$. Such a condition is needed to ensure that the solution $P(t)$ of the ARE (5.32) with A, b replaced by $\hat{A}(t)$, $\hat{b}(t)$ exists and is uniformly bounded. If the sign of b_1 and a lower bound $b_0 > 0$ for $|b_1|$ are known, then (5.26) can be modified to guarantee that $|\hat{b}_1(t)| \geq b_0 \ \forall t \geq 0$ and $\hat{a}(t)$ is strictly bounded away from $q \ \forall t \geq 0$, without altering the properties (i) and (ii) [14,11]. This will guarantee that $\left(\hat{A},\, \hat{b}\right)$ is uniformly controllable $\forall t \geq 0$.

Step 3: The adaptively controlled closed-loop plant can be written as:

$$\begin{aligned}
\dot{\hat{e}} = A_c(t)\hat{e} + \frac{k(s+\lambda)(s+\lambda_1)}{s^2+\alpha_1 s+\alpha_2}\Bigg[&-\phi_1\frac{s}{(s+\lambda)(s+\lambda_1)}[y] + \phi_2\frac{(s+q)}{(s+\lambda)(s+\lambda_1)}[\bar{u}] \\
&+W_{c1}(s)\left\{(W_{b1}(s)[e_0])\,\dot{\phi}_1\right\} + W_{c2}(s)\left\{(W_{b2}(s)[\bar{u}])\,\dot{\phi}_2\right\}\Bigg] \\
&-k\frac{b_1 s\Delta_{\mathrm{m}}(s)}{s^2+\alpha_1 s+\alpha_2}[u] \qquad (5.34)
\end{aligned}$$

$$\begin{aligned}
e_0 = -C\hat{e} + \frac{(s+\lambda)(s+\lambda_1)}{s^2+\alpha_1 s+\alpha_2}\Bigg[&-\phi_1\frac{s}{(s+\lambda)(s+\lambda_1)}[y] + \phi_2\frac{(s+q)}{(s+\lambda)(s+\lambda_1)}[\bar{u}] \\
&+W_{c1}(s)\left\{(W_{b1}(s)[e_0])\,\dot{\phi}_1\right\} + W_{c2}(s)\left\{(W_{b2}(s)[\bar{u}])\,\dot{\phi}_2\right\}\Bigg] \\
&-\frac{b_1 s\Delta_{\mathrm{m}}(s)}{s^2+\alpha_1 s+\alpha_2}[u] \qquad (5.35)
\end{aligned}$$

$$\begin{aligned}
u = &\frac{(s+a)(s+q)M^*(s)}{s^3+\beta_1 s^2+\beta_2 s+\beta_3}[\bar{u}] + \frac{(s+a)(s+q)F^*(s)}{s^3+\beta_1 s^2+\beta_2 s+\beta_3}[y] \\
&-\frac{b_1(s+q)F^*(s)\Delta_{\mathrm{m}}(s)}{s^3+\beta_1 s^2+\beta_2 s+\beta_3}[u] \qquad (5.36)
\end{aligned}$$

$$\bar{u} = -G(t)\hat{e} \qquad (5.37)$$

$$y = y_{\mathrm{m}} - e_0\,, \qquad (5.38)$$

where

- $A_c(t) \stackrel{\Delta}{=} \hat{A}(t) - \hat{b}(t)G(t)$,
- $C = [1\ 0]$,
- $k = [\hat{a} - \alpha_1 \ \ -\alpha_2]^{\mathrm{T}}$,
- $s^2 + \alpha_1 s + \alpha_2$ is an arbitrary monic polynomial of degree 2 with all its roots in $\mathrm{Re}[s] < -\frac{\delta_0}{2}$,
- $(s+\lambda_1)$ is a Hurwitz polynomial with $\lambda_1 > \delta_0$,
- $W_{bi}(s)$, $W_{ci}(s)\,, i = 1,2$, are stable strictly proper transfer function matrices obtained using Lemma 2.5,
- $M^*(s)$ and $F^*(s)$ solve the Diophantine equation

$$M^*(s)s(s+a) + F^*(s)b_1(s+q) = s^3 + \beta_1 s^2 + \beta_2 s + \beta_3\,,$$

  where $s^3 + \beta_1 s^2 + \beta_2 s + \beta_3$ is an arbitrary monic polynomial of degree 3 having all roots in $\mathrm{Re}[s] < -\frac{\delta_0}{2}$.

For each fixed t, it follows from standard LQR Theory [34] that the matrix $A_c(t)$ is stable. Furthermore,

$$\left\|\dot{A}(t)\right\| \leq c\left(\left|\dot{\hat{a}}\right| + \left|\dot{\hat{b}}_1\right|\right) \leq c\left\|\dot{\theta}(t)\right\|\,, \quad \text{i.e.,}\ \left\|\dot{A}_c(t)\right\| \in \mathcal{S}\left(\frac{\eta^2}{m^2}\right).$$

Provided $\lambda \geq \delta_0$, using Lemma 2.2, we can show that

$$\frac{\eta^2}{m^2} \leq \frac{1}{\lambda}\left(\|b_1 \Delta_{\mathrm{m}}(s)\|_\infty^{\delta_0}\right)^2 = \Delta_0^2 ,$$

i.e., $\left\|\dot{A}(t)\right\| \in \mathcal{S}(\Delta_0^2)$. Also, $A(t) \in L_\infty$, $P(t) \in L_\infty$ implies that $A_c(t) \in L_\infty$. Furthermore, since the eigenvalues of $A_c(t)$ are continuous functions of the parameter estimates $\theta(t)$ and $\theta(t) \in L_\infty$ (i.e., $\theta \in K$, Kcompact $\subset \mathbb{R}^2$), it follows that $\exists\, \sigma_s > 0$ such that $\mathrm{Re}\{\lambda_i(A_c(t))\} \leq -\sigma_s \;\forall t \geq 0$. Hence, applying Lemma 2.1, it follows that for $\Delta_0 < c$ and some $c > 0$, $A_c(t)$ is exponentially stable.

Using Lemma 2.5 and the relationship $(s+q)\bar{u} = su$, we obtain

$$\begin{aligned}\frac{s}{s+\lambda_1}[\phi^{\mathrm{T}}\zeta] &= \phi^{\mathrm{T}}\left(\frac{s}{s+\lambda_1}[\zeta]\right) + W_{c3}(s)\left\{\left(W_{b3}(s)\,[\zeta^{\mathrm{T}}]\right)\dot{\phi}\right\} \\ &= \left(-\phi_1\frac{s}{(s+\lambda)(s+\lambda_1)}[y] + \phi_2\frac{(s+q)}{(s+\lambda)(s+\lambda_1)}[\bar{u}]\right) \\ &\quad + W_{c3}(s)\left\{\left(W_{b3}(s)\,[\zeta^{\mathrm{T}}]\right)\dot{\phi}\right\}.\end{aligned}$$

Since $\phi^{\mathrm{T}}\zeta = \epsilon m^2 + \eta$, we can write

$$\begin{aligned}-\phi_1\frac{s}{(s+\lambda)(s+\lambda_1)}[y] + \phi_2\frac{(s+q)}{(s+\lambda)(s+\lambda_1)}[\bar{u}] &= \frac{s}{s+\lambda_1}[\epsilon m^2 + \eta] \\ &\quad - W_{c3}(s)\left\{\left(W_{b3}(s)\,[\zeta^{\mathrm{T}}]\right)\dot{\phi}\right\}.\end{aligned}$$

Using the above expression in (5.34)–(5.35) and making use of (5.36)–(5.38), it is clear that the stability analysis for this example can be completed by essentially duplicating the steps involved in the analysis of Example 5.2.1.

The condition for stability is that

$$\Delta_0 = \frac{1}{\sqrt{\lambda}}\|b_1 \Delta_{\mathrm{m}}(s)\|_\infty^{\delta_0} < c$$

for some $c > 0$ which can be calculated if an estimate of the rate of exponential convergence of the transition matrix of $A_c(t)$ can be obtained. As in Example 5.2.1, the use of a dead-zone instead of a leakage will make the bound for Δ_0 independent of the transition matrix of $A_c(t)$ and hence easier to calculate.

The following expression for the tracking error

$$\int_0^t |e_r(\tau)|^2 d\tau \leq c\Delta_0^2 t + c$$

can also be obtained in a similar manner as in Example 5.1.1. □

An important drawback of indirect adaptive control is what is called the "stabilizability problem": the control law is designed based on the estimated plant, which has to be controllable and observable at each time t. The adaptive laws cannot guarantee such a property alone without additional modifications. One possible modification is to use projection and constrain the estimated parameters to be in a convex set which contains the unknown parameters and is such that for every member of the set the stabilizability condition required for the calculation of the control law is satisfied [14,42]. While such modification seems feasible in theory, in practice the development of such convex sets for higher order plants is complicated and awkward. Other approaches based on persistence of excitation are used in [43,44] to handle the stabilizability problem.

6 Conclusions

In this paper we have given a unified treatment of the design, analysis and robustness of most of the continuous-time robust adaptive control schemes that are currently available in the adaptive control literature. The methodology presented can also be used to synthesize new robust adaptive control schemes by simply combining an adaptive law, developed using the procedure of Sect. 3, with a robust controller structure. Thus, this paper is part of an ongoing effort to develop a systematic and unified theory for the design and analysis of robust adaptive control systems so that the latter cease to be looked upon as a mere collection of assorted tricks and algorithms.

The theory of robust adaptive control is still far from complete. A satisfactory theory for the quantitative analysis and transient performance of robust adaptive control systems is yet to be developed.

The results presented in this paper are applicable to LTI plants. Their extension to linear time-varying plants is attempted with considerable success in [45,46,47,48,49].

References

1. A. S. Morse, "Global stability of parameter adaptive control systems," *IEEE Trans. Aut. Control*, vol. AC-25, pp. 433-439, June 1980.
2. K. S. Narendra, Y. H. Lin, and L. S. Valavani, "Stable adaptive controller design -part II: proof of stability," *IEEE Trans. Aut. Control*, vol. AC-25, pp. 440–448, June 1980.
3. I. D. Landau, *Adaptive Control: The Model Reference Approach*, Marcel Dekker, New York, NY, 1979.
4. G. C. Goodwin, P. J. Ramadge, and P. E. Caines, "Discrete time multivariable adaptive control," *IEEE Trans. Aut. Control*, vol. AC-25, pp. 449–456, June 1980.
5. B. Egardt, *Stability of Adaptive Controllers*, Springer-Verlag, Berlin, 1979.
6. P. A. Ioannou and P. V. Kokotovic, *Adaptive Systems with Reduced Models*, Springer-Verlag, New York, NY, 1983.
7. C. E. Rohrs, L. Valavani, M. Athans, and G. Stein, "Robustness of adaptive control algorithms in the presence of unmodeled dynamics," *Proc. 21st IEEE Conf. Dec. Control*, 1982.
8. P. A. Ioannou and P. V. Kokotovic, "Instability analysis and improvement of robustness of adaptive control," *Automatica*, vol. 20, pp. 583–594, Sept. 1984.
9. P. A. Ioannou and K. S. Tsakalis, "A robust direct adaptive controller," *IEEE Trans. Aut. Control*, vol. AC-31, pp. 1033–1043, Nov. 1986.

10. K. S. Narendra and A. M. Annaswamy, *Stable Adaptive Systems*, Prentice-Hall, Englewood Cliffs, NJ, 1989.
11. L. Praly, "Robust model reference adaptive controllers–part 1: stability analysis," *Proc. 23rd IEEE Conf. Dec. Control*, 1984.
12. G. Kreisselmeier, "A robust indirect adaptive control approach," *Int. J. Control*, vol. 43, pp. 161–175, 1986.
13. P. Ioannou and J. Sun, "Theory and design of robust direct and indirect adaptive control schemes," *Int. J. Control*, vol. 47, pp. 775–813, 1988.
14. R. H. Middleton, G. C. Goodwin, D. J. Hill, and D Q. Mayne, "Design issues in adaptive control," *IEEE Trans. Aut. Control*, vol. AC-33, pp. 50–58, Jan. 1988.
15. B. D. O. Anderson, R. R. Bitmead, C. R. Johnson, Jr., P. V. Kokotovic, R. L. Kosut, I. M. Y. Mareels, L. Praly, and B. D. Riedle, *Stability of Adaptive Systems: Passivity and Averaging Analysis*, MIT Press, Cambridge, MA, 1986.
16. S. S. Sastry and M. Bodson, *Adaptive Control: Stability, Convergence and Robustness*, Prentice-Hall, Englewood Cliffs, NJ, 1989.
17. P. A. Ioannou and K. S. Tsakalis, "Time and frequency domain bounds in robust adaptive control," *Proc. 1988 Amer. Control Conf.*
18. K. S. Tsakalis, "Robustness of model reference adaptive controllers: input-output properties," *Department of Electrical and Computer Engineering, Arizona State University*, Report no. 89-03-01, 1989, to appear in *IEEE Trans. Aut. Control*).
19. J. M. Krause, P. P. Khargonekar and G. Stein, "Robust adaptive control: stability and asymptotic performance," *Proc. 28th IEEE Conf. Dec. Control*, 1989.
20. G. C. Goodwin and K. S. Sin, *Adaptive Filtering, Prediction and Control*, Prentice-Hall, Englewood Cliffs, NJ, 1984.
21. G. Kreisselmeier and D. Joos, "Rate of convergence in model reference adaptive control," *IEEE Trans. Aut. Control*, vol. AC-27, pp. 710–713, June 1982.
22. B. A. Francis, *A Course in H_∞ Control Theory*, Springer-Verlag, Berlin, 1987.
23. T. Kailath, *Linear Systems*, Prentice-Hall, Englewood Cliffs, NJ, 1980.
24. C. A. Desoer and M. Vidyasagar, *Feedback Systems: Input-Output Properties*, Academic Press, New York, NY, 1975.
25. P. J. Gawthrop, *Continuous-Time Self-Tuning Control*, Research Studies Press, Wiley, New York, NY, 1987.
26. R. H. Middleton and G. C. Goodwin, *Digital Control and Estimation: A Unified Approach*, Prentice-Hall, Englewood Cliffs, NJ, 1990.
27. M. Polycarpou and P. Ioannou, "On the existence and uniqueness of solutions in adaptive control systems," *Department of EE-Systems, University of Southern California*, Technical Report no. 90-05-01, 1990.
28. D. G. Luenberger, *Optimization by Vector Space Methods*, Wiley, New York, NY, 1969.
29. M. Vidyasagar, *Nonlinear Systems Analysis*, Prentice-Hall, Englewood Cliffs, NJ, 1978.
30. M. Vidyasagar, *Control System Synthesis: A Factorization Approach*, MIT Press, Cambridge, MA, 1985.
31. P. V. Kokotovic, H. Khalil, and J. O'Reilly, *Singular Perturbation Methods in Control: Analysis and Design*, Academic Press, New York, NY, 1986.
32. M. G. Safonov, A. J. Laub, and G. L. Hartmann, "Feedback properties of multivariable systems: the role and use of the return difference matrix," *IEEE Trans. Aut. Control*, vol. AC-26, pp. 47–65, Feb. 1981.

33. J. C. Doyle, K. Glover, P. P. Khargonekar, and B. A. Francis, "State-space solutions to standard H_2 and H_∞ control problems," *IEEE Trans. Aut. Control*, vol. 34, pp. 831–847, Aug. 1989.
34. H. Kwakernaak and R. Sivan, *Linear Optimal Control Systems*, Wiley, New York, NY, 1972.
35. K. J. Astrom and B. Wittenmark, *Adaptive Control*, Prentice-Hall, Englewood Cliffs, NJ, 1989.
36. G. C. Goodwin and D. Q. Mayne, "A parameter estimation perspective of continuous time model reference adaptive control," *Automatica*, vol. 23, pp. 57-70, Jan. 1987.
37. P. A. Ioannou and G. Tao, "Dominant richness and improvement of performance of robust adaptive control," *Automatica*, vol. 25, pp. 287–291, March 1989.
38. C. Samson, "An adaptive LQ control for nonminimum phase systems," *Int. J. Control*, vol. 35, 1982.
39. D. W. Clarke, P. P. Kanjilal, and C Mohtadi, "A generalized LQG approach to self-tuning control, part I: aspects of design," *Int. J. Control*, vol. 41, 1985.
40. D. W. Clarke, P. P. Kanjilal, and C Mohtadi, "A generalized LQG approach to self-tuning control, part II: implementation and simulations," *Int. J. Control*, vol. 41, 1985.
41. J. Sun and P. Ioannou, "Robust adaptive LQ control schemes," *Proc. 1989 Amer. Control Conf.*
42. Ph. de Larminat, "On the stabilizability condition in indirect adaptive control," *Automatica*, vol. 20, Nov. 1984.
43. G. Kreisselmeier, "An indirect adaptive controller with a self-excitation capability," *IEEE Trans. Aut. Control*, vol. 34, pp. 524–528, May 1989.
44. R. Cristi, "Internal persistency of excitation in indirect adaptive control," *IEEE Trans. Aut. Control*, vol. AC-32, pp. 1101–1103, Dec. 1987.
45. G. Kreisselmeier, "Adaptive control of a class of slowly time-varying plants," *Syst. Control Lett.*, vol. 8, Dec. 1986.
46. K. S. Tsakalis and P. A. Ioannou, "Adaptive control of linear time-varying plants," *Automatica*, vol. 23, pp. 459–468, July 1987.
47. R. H. Middleton and G. C. Goodwin, "Adaptive control of time-varying linear systems," *IEEE Trans. Aut. Control*, vol. AC-33, pp. 150–155, Feb. 1988.
48. K. S. Tsakalis and P. A. Ioannou, "Adaptive control of linear time-varying plants: a new model reference controller structure," *IEEE Trans. Aut. Control*, vol. 34, pp. 1038–1046, Oct. 1989.
49. K. S. Tsakalis and P. A. Ioannou, "A new indirect adaptive control scheme for time-varying plants," *IEEE Trans. Aut. Control*, vol. 35, pp. 697–705, June 1990.

Robust Continuous-Time Adaptive Control by Parameter Projection *

Sanjeev M. Naik,[1] *P. R. Kumar,*[1] *and B. Erik Ydstie*[2]

[1] Coordinated Science Laboratory and
Department of Electrical and Computer Engineering
University of Illinois, Urbana, IL 61801, USA.

[2] Department of Chemical Engineering, Goessmann Laboratory
University of Massachusetts, Amherst, MA 01003, USA.

Abstract. We consider the problem of adaptive control of a continuous-time plant of arbitrary relative degree, in the presence of bounded disturbances as well as unmodeled dynamics. The adaptation law we consider is the usual gradient update law with parameter projection, the latter being the only robustness enhancement modification employed. We show that if the unmodeled dynamics, which consists of multiplicative as well as additive system uncertainty, is small enough, then all the signals in the closed-loop system are bounded. This shows that extra modifications such as, for example, normalization or relative dead zones, are not necessary for robustness with respect to bounded disturbances and small unmodeled dynamics. In the nominal case, where unmodeled dynamics and disturbances are absent, the asymptotic error in tracking a given reference signal is zero. Moreover, the performance of the adaptive controller is also robust in that the mean-square tracking error is quadratic in the magnitude of the unmodeled dynamics and bounded disturbances, when both are present.

1 Introduction

Recently, there have been many attempts to study the adaptive control of plants with bounded disturbances and unmodeled dynamics. In his pioneering work, Egardt [5] showed that even small bounded disturbances can cause instability in adaptively controlled plants. He further demonstrated that modification of the adaptation law by projecting the parameter estimates, at each time instant, onto a compact, convex set known to contain the true parameter vector, provides stability with respect to bounded disturbances.

Proving the stability of an adaptive control system in the presence of unmodeled dynamics is however more difficult, and researchers have proposed various additional modifications to the adaptation law to analyze and bound the effects of unmodeled dynamics. For example, in [8-10], Praly introduced the device of using a normalizing signal in the parameter estimator, while another notable modification is the idea of a normalized dead-zone [11]. Other modifications used in conjunction with normalization include projection [5] or some form

* The research reported here has been supported in part by U.S.A.R.O. under Contract No. DAAL 03-88-K-0046, the JSEP under Contract No. N00014-90-J-1270, and by an International Paper Fellowship for the first author.

of leakage, the concept of which was first introduced in [7], such as switching-σ-modification [12,14]. An excellent unification of these various modifications and results can be found in Tao and Ioannou [18].

In this paper, we show that Egardt's original simple modification of parameter projection is in fact enough to provide stability with respect to small unmodeled dynamics as well as bounded disturbances. Our main results are the following:

(i) A certainty equivalent adaptive controller, using a gradient based parameter estimator with projection, ensures that all closed-loop signals are bounded, when applied to a nominally minimum-phase continuous-time system with bounded disturbances and small unmodeled dynamics (Theorem 9.1).

(ii) In the absence of unmodeled dynamics and disturbances, i.e., in the nominal case, the error in tracking a reference trajectory converges to zero (Theorem 10.1). When unmodeled dynamics as well as bounded disturbances are present, the mean-squared tracking error is quadratic in the magnitude of the unmodeled dynamics and bounded disturbances (Theorem 11.1). Thus, the adaptive controller provides robust performance in addition to robust boundedness.

While our work thus shows that a simple modification is sufficient to ensure robust boundedness and robust performance, and some early modifications may have been proposed due to the limitations of proof techniques used, we feel that it is nevertheless important for future work to compare the various modifications on the basis of the amount of robustness provided, the resulting performance as a function of unmodeled effects, transient response, etc., as well as the complexities of the modifications themselves.

The key stimulus for our work here is the recent paper of Ydstie [19], which showed that parameter projection in a gradient update law is sufficient for ensuring the boundedness of closed loop signals for a nominally minimum-phase, unit delay, discrete-time plant with some types of unmodeled dynamics as well as bounded disturbances.

The continuous-time systems studied here give rise to several additional issues such as filtering of signals, parametrization of systems, differentiability considerations of signals, augmented errors, etc., which motivate various changes, and allow us to establish stability for nominal plants with arbitrary positive relative degree, as well as for a class of unmodeled dynamics which is larger than those considered earlier, for example in [20],[14] or [19]. For instance, unlike Ydstie[19] we do not require the true plant to be stably invertible; only the nominal plant is assumed to be minimum-phase. Additionally, in contrast to [14] we also allow the unmodeled dynamics to be nonlinear or time-varying, and do not require differentiability of either the bounded disturbance, which is lumped together with the unmodeled dynamics in our treatment, or the reference input.

The rest of the paper is organized as follows: Section 2 introduces the system and reference models. In Section 3, we reparametrize these models, and describe, in Section 4, the adaptive control law. Our analysis starts in Section 5, where we show that all closed-loop signals in the system are bounded by a particular signal $m(t)$. In Section 6, we introduce a signal $z(t)$ defined through a "switched system" which overbounds $m(t)$. In Section 7, we show that the filtered signals are comparable to z over certain bounded intervals of time. To apply these results to the stability analysis of the closed-loop system, in Section 8 we present a nonminimal representation of the closed-loop system, which is then used in Section 9 to complete the boundedness analysis by showing that a certain positive definite function of the signal $z(t)$ and the non-minimal system state error $e(t)$ is bounded. In Section 10, we show that asymptotic tracking is achieved in the nominal case, and in Section 11 we establish a mean-square robust performance result. In Section 12 we present simulation examples to illustrate the results. Finally, Section 13 presents some concluding remarks. Some necessary technical results are collected in Appendices A, B, and C. A list of constants appears in Table D. A preliminary version of the results presented here is contained in [20].

2 System and Reference Models

Consider the single-input, single-output system,

$$y(t) = \frac{B(s)}{A(s)}(1 + \mu_{\mathrm{m}} \Delta_{\mathrm{m}}(s))u(t) + v(t)\,, \tag{1}$$

where $B(s)/A(s)$ is the transfer function of the *modeled* part of the plant, $\Delta_{\mathrm{m}}(s)$ represents the multiplicative uncertainty in the plant, and $v(t)$ represents the effect of additional additive unmodeled dynamics as well as bounded disturbances.

We will make the following assumptions on the nominal model of the plant:

(A1) $A(s) = s^n + \sum_{j=0}^{n-1} a_j s^j$, and $B(s) = \sum_{j=0}^{m} b_j s^j$, $0 \leq m < n$, where $A(s)$, $B(s)$ are coprime.

(A2) $B(s - p_0)$ is Hurwitz for some $p_0 > 0$, and $b_m \geq b_{\min} > 0$.

The relative degree of the nominal plant will be denoted by $n_{\mathrm{r}} := n - m$.

We will make the following assumptions on the *unmodeled dynamics* and *bounded disturbance* of the plant:

(A3) The multiplicative uncertainty, $\Delta_{\mathrm{m}}(s)$, is a transfer function with relative degree greater than or equal to $(1 - n_{\mathrm{r}})$ such that $\Delta_{\mathrm{m}}(s - p_0)$ is stable.

(A4) The additive unmodeled dynamics and disturbances give rise to a signal $v(t)$ which satisfies $|v(t)| \leq K_v m(t) + k_v + k_{v0} \exp[-pt]$, where $m(\cdot)$ is defined by,

$$\dot{m}(t) = -d_0 m(t) + d_1(|u(t)| + |y(t)| + 1), \quad m(0) \geq d_1/d_0\,, \tag{2}$$

All the constants are positive. Furthermore, $0 < p \leq d_0$ and $d_0 + d_2 < p_0 < 2d_0 + d_2$, for some $d_2 > 0$. The term $k_{v0} \exp[-pt]$ above allows for the effect of initial conditions.

Example 2.1 The above assumptions allow for the class of linear systems

$$y(t) = \frac{B(s)}{A(s)}\left(1 + \mu_{\mathrm{m}}\frac{N_{\mathrm{m}}(s)}{D_{\mathrm{m}}(s)}\right)u(t) + v(t),$$

where deg $N_{\mathrm{m}}(s) \leq$ deg $D_{\mathrm{m}}(s) + n_{\mathrm{r}} - 1$ (which could be greater than the degree of $D_{\mathrm{m}}(s)$ if $n_{\mathrm{r}} \geq 2$),

$$v(t) = \frac{N_1(s)}{D_1(s)}y(t) + \frac{N_2(s)}{D_2(s)}u(t) + w(t) + c_0 \exp[-pt],$$

where $\frac{N_1(s)}{D_1(s)}, \frac{N_2(s)}{D_2(s)}$ are strictly proper, $w(t)$ is a bounded disturbance, and the polynomials $D_{\mathrm{m}}(s - p_0)$, $D_1(s - p_0)$ and $D_2(s - p_0)$ are Hurwitz. □

Example 2.2 A class of nonlinear unmodeled dynamics satisfying (A4) is

$$v(t) = f(t, g_1(t) \cdot H_1(s)y(t), g_2(t) \cdot H_2(s)u(t)),$$

where f is any nonlinear function such that $|f(t, x_1(t), x_2(t)| \leq k_1|x_1(t)| + k_2|x_2(t)| + k_3$, $g_1(t), g_2(t)$ are bounded time-varying signals, and $H_1(s), H_2(s)$ are strictly proper transfer functions such that $H_1(s - p_0)$ and $H_2(s - p_0)$ are stable. □

The goal of adaptation is to follow the output of a reference model given by

$$y_{\mathrm{m}}(t) = W_{\mathrm{m}}(s)r(t), \tag{3}$$

where $W_{\mathrm{m}}(s)$ is a stable transfer function with relative degree n_{r}, and $r(t)$ is a reference input. We will suppose that $|r(t)| \leq k_{r1}, \forall t \geq 0$ and $|y_{\mathrm{m}}(t)| \leq k_{ym}, \forall t \geq 0$.

3 Parametrization of System and Reference Models

We now reparametrize the system and reference models so that they are in a form more suitable for the development of an adaptive control law. To do this, we need to filter the input and output signals.

Let a be a positive number satisfying $a > d_2 + 2d_0$. We define the "regression" vector,[3]

$$\psi := \left(\frac{1}{(s+a)^{2n-m-1}}y, \ldots, \frac{1}{(s+a)^{n-m}}y, \frac{1}{(s+a)^{2n-m-1}}u, \ldots, \frac{1}{(s+a)^{n-m}}u\right)^{\mathrm{T}}.$$

[3] All of the results of this paper continue to hold if instead of the filter $\frac{1}{(s+a)^{2n-m-1}}$ we use the filter $\frac{1}{(s+a)^{n-m}\lambda_1(s)}$, where $\lambda_1(s) = (s+a)\lambda_1'(s)$ with $\lambda_1'(s)$ being monic, of degree $(n-2)$, and with all roots having real parts less than or equal to $-(2d_0 + d_2)$.

We will presently show that there exists a "parameter vector" $\theta = (\theta_1, \ldots, \theta_{2n})^{\mathrm{T}}$ such that the system (1) can be represented as

$$y(t) = \psi^{\mathrm{T}}(t)\theta + v_{\mathrm{f}}'(t) + \xi_{2d_0}(t)\,,\ \theta_{2n} = b_m\,, \tag{4}$$

where $\psi^{\mathrm{T}}(t)\theta$ represents the nominal part of the system, $\xi_{2d_0}(t)$ represents the effect of initial conditions arising from the filtering operations,[4] and $v_{\mathrm{f}}'(t)$ represents the effect of unmodeled dynamics and bounded disturbances.

We will also reparametrize the reference model (3) as

$$y_{\mathrm{m}}(t) = \frac{1}{(s+a)^{n_{\mathrm{r}}}} r'(t) + \xi_{2d_0}(t)\,, \tag{5}$$

where $r'(t) := (s+a)^{n_{\mathrm{r}}} W_{\mathrm{m}}(s) r(t)$. Note that $r'(t)$ is well defined since the relative degree of $W_{\mathrm{m}}(s)$ is n_{r}. Further, $r'(t)$ is bounded since $r(t)$ is bounded. We shall directly suppose from now onwards that $r'(t) \leq k_r, \forall t \geq 0$.

To see the existence of a θ for which (4) holds, let $F(s)$ be a monic polynomial of degree $(n_{\mathrm{r}} - 1)$ and $G(s)$ a polynomial of degree less than or equal to $(n-1)$, such that $A(s)F(s) + G(s) = \lambda(s)$, where $\lambda(s) := (s+a)^{2n-m-1}$. Then, from (1), taking into consideration the effect of the initial conditions introduced by the filtering operation $1/\lambda(s)$, we have,

$$\begin{aligned} \frac{\lambda(s) - G(s)}{\lambda(s)} y(t) &= F(s) \frac{A(s)}{\lambda(s)} y(t) + \xi_{2d_0}(t) \\ &= F(s) \left[\frac{B(s)}{\lambda(s)} u(t) + \mu_{\mathrm{m}} \frac{B(s)\Delta_{\mathrm{m}}(s)}{\lambda(s)} u(t) + \frac{A(s)}{\lambda(s)} v(t) \right] + \xi_{2d_0}(t)\,. \end{aligned}$$

Thus, if $\theta = (\theta_1, \ldots, \theta_{2n})^{\mathrm{T}}$ is defined from the coefficients of $G(s)$ and $B(s)F(s)$ by, $\theta_n(s+a)^{n-1} + \ldots + \theta_2(s+a) + \theta_1 = G(s)$, $\theta_{2n}(s+a)^{n-1} + \ldots + \theta_{n+2}(s+a) + \theta_{n+1} = B(s)F(s)$, and $v_{\mathrm{f}}'(t) = \frac{A(s)F(s)}{\lambda(s)} v(t) + \mu_{\mathrm{m}} \frac{B(s)F(s)}{\lambda(s)} \Delta_{\mathrm{m}}(s) u(t)$, then (4) is satisfied. We note that since F is monic, $\theta_{2n} = b_m$.

4 The Adaptive Control Law

We will use the control law

$$\phi^{\mathrm{T}}(t)\hat{\theta}(t) = r'(t)\,, \tag{6}$$

to implicitly define the input $u(t)$, where $\phi(t) := (\frac{1}{(s+a)^{n-1}} y, \ldots, y, \frac{1}{(s+a)^{n-1}} u, \ldots, u)^{\mathrm{T}}$, and $\hat{\theta}(t)$ is an estimate of θ that we shall presently specify. Note that the "regression" vector $\psi(t)$ defined earlier satisfies $\psi(t) = \frac{1}{(s+a)^{n-m}} \phi(t)$.

[4] Here and throughout, $\xi_q(t)$ will denote a signal which satisfies the following properties:
(i) $|\xi_q^{(i)}(t)| \leq c_0 \exp[-qt], i = 0, 1$, where $\xi_q^{(i)}(t)$ denotes the i–th derivative of $\xi_q(t)$.
(ii) $\xi_q(t) \equiv 0$ when initial conditions are zero.
When the value of q is unimportant, we will sometimes drop the subscript on ξ.

The adaptive control law (6) is a "certainty equivalent" control law, since if $\hat{\theta}(t) = \theta$ in (6), then the nominal part of the system (4) tracks $y_m(t)$, since

$$y(t) = \psi^T(t)\theta + \xi_{2d_0}(t) = \frac{1}{(s+a)^{n-m}}[\phi^T(t)\theta] + \xi_{2d_0}(t)$$
$$= \frac{1}{(s+a)^{n-m}} r'(t) + \xi_{2d_0}(t) = y_m(t) + \xi_{2d_0}(t) .$$

Let us now turn to the parameter estimator. We shall use the gradient estimator with projection,

$$\dot{\hat{\theta}}(t) = \text{Proj}\left[\hat{\theta}(t), \frac{\alpha\psi(t)e_a(t)}{n(t)}\right], \ ||\hat{\theta}(0)|| \le M, \ \hat{\theta}_{2n}(0) \ge b_{\min} , \tag{7}$$

where $\alpha > 0$, $e_a(t)$ is the "augmented" error,

$$e_a(t) = y(t) - y_m(t) + \frac{1}{(s+a)^{n-m}}\left[\phi^T(t)\hat{\theta}(t)\right] - \psi^T(t)\hat{\theta}(t) ,$$

$$n(t) := \nu + \sum_{i=1}^{2n-m-1}\left(\left[\frac{1}{(s+a)^i}y(t)\right]^2 + \left[\frac{1}{(s+a)^i}u(t)\right]^2\right), \ \nu > 0 ,$$

is a normalizer of the gradient, and $\text{Proj}[\cdot,\cdot]$ is a projection whose i–th component is defined for $i \ne 2n$ by

$$\text{Proj}[p,x]\,|_i = \begin{cases} x_i , & \text{if } ||p|| < M \text{ or } p^T x \le 0 , \\ x_i - \dfrac{p^T x}{||p||^2} p_i , & \text{otherwise} , \end{cases}$$

and for $i = 2n$ by

$$\text{Proj}[p,x]\,|_{2n} = \begin{cases} x_{2n} & \text{if } (||p|| < M \text{ or } p^T x \le 0) \\ & \text{and } (p_{2n} > b_{\min} \text{ or } x_{2n} \ge 0), \\ x_{2n} - \dfrac{p^T x}{||p||^2} p_{2n} , & \text{if } (||p|| \ge M \text{ and } p^T x > 0) \\ & \text{and } (p_{2n} > b_{\min} \text{ or } x_{2n} \ge 0), \\ 0 , & \text{otherwise.} \end{cases}$$

There are several features worth commenting upon. First, the augmented error $e_a(t)$ consists of the tracking error $y(t) - y_m(t)$, as well as the familiar "swapping" term $\frac{1}{(s+a)^{n-m}}[\phi^T(t)\hat{\theta}(t)] - \left[\frac{1}{(s+a)^{n-m}}\phi^T(t)\right]\hat{\theta}(t)$, which would be zero if $\hat{\theta}(t)$ were a constant. Second, note that $n(t)$ is slightly different from the usual $||\psi(t)||^2$ since it additionally contains the lower-order filtered terms $\frac{1}{(s+a)^i}y(t)$ and $\frac{1}{(s+a)^i}u(t)$ for $1 \le i \le n-m-1$, which are absent in $\psi(t)$.

Turning to the "projection" mechanism, it has two features. Without any projection, a gradient scheme would simply consist of

$$\dot{\hat{\theta}}(t) = \frac{\alpha\psi(t)e_a(t)}{n(t)} .$$

However, to keep the estimates inside a sphere of radius M centered at the origin, one would project them according to

$$\dot{\hat{\theta}}(t) = \text{Proj}'\left[\hat{\theta}(t), \frac{\alpha\psi(t)e_a(t)}{n(t)}\right],$$

where $\text{Proj}'[p,x] = x$ if $\|p\| < M$ or $p^T x \leq 0$, i.e., when the estimate is inside the sphere or not about to leave it, and otherwise

$$\text{Proj}'[p,x] = x - \frac{p^T x}{\|p\|^2} p,$$

i.e., one projects the drift term so that it evolves tangentially to the sphere. Our projection $\text{Proj}[p,x]$, however, involves an additional feature to ensure that $\hat{b}_m(t)$, the $2n$–th component of $\hat{\theta}(t)$, is larger than or equal to $b_{\min}$. So we have

$$\text{Proj}\left[\hat{\theta}(t), \frac{\alpha\psi(t)e_a(t)}{n(t)}\right]_{2n} = \text{Proj}'\left[\hat{\theta}(t), \frac{\alpha\psi(t)e_a(t)}{n(t)}\right]_{2n},$$
$$\text{if } \hat{\theta}_{2n}(t) > b_{\min} \text{ or } \text{Proj}'\left[\hat{\theta}(t), \frac{\alpha\psi(t)e_a(t)}{n(t)}\right]_{2n} \geq 0,$$

i.e., if $\hat{b}_m(t)$ is not about to become less than $b_{\min}$, and

$$\text{Proj}\left[\hat{\theta}(t), \frac{\alpha\psi(t)e_a(t)}{n(t)}\right]_{2n} = 0 \text{ otherwise.}$$

The following easily verified consequences of projection are important.

(P0) $\hat{\theta}_{2n}(t) \geq b_{\min}$, $\|\hat{\theta}(t)\| \leq M$.

(P1) $\|\dot{\hat{\theta}}(t)\| \leq |\frac{\alpha\psi(t)e_a(t)}{n(t)}|$

(P2) If $\|\theta\| \leq M$ and $\theta_{2n} = b_m \geq b_{\min}$, i.e., θ lies in the region to which we confine the parameter estimates, then

$$\tilde{\theta}^T(t)\dot{\hat{\theta}}(t) \leq \tilde{\theta}^T(t)\left(\frac{\alpha\psi(t)e_a(t)}{n(t)}\right), \quad \text{where } \tilde{\theta}(t) := \hat{\theta}(t) - \theta.$$

Finally, since the existence of a solution to the above parameter estimator may not be assured due to the discontinuous nature of the projection, we can replace the above projection by the "smooth" projection due to Pomet and Praly [15], for which existence is assured. The key properties (P1) and (P2) continue to hold for the "smooth" projection, and all the results of this paper rigorously hold for this "smooth" projection.

In what follows, we will throughout suppose that the nominal plant described by θ satisfies the assumptions of (P2).

5 Bounding Signals by m

In this section we will show that all the signals in the system can be bounded in terms of $m(t)$ given in (2).[5]

For brevity of notation in the proof, we define

$$\lambda_1(s) = (s+a)^{n-1}\,, \lambda_2(s) = (s+a)^{n_r}\,, \lambda(s) = \lambda_1(s)\lambda_2(s)\,,$$

$$\zeta(t) = \frac{1}{\lambda_2(s)}[\phi^{\mathrm{T}}(t)\hat{\theta}(t)] - \left[\frac{1}{\lambda_2(s)}\phi^{\mathrm{T}}(t)\right]\hat{\theta}(t)\,,$$

$$\psi'(t) = \left[\frac{1}{\lambda(s)}y(t), \ldots, \frac{1}{(s+a)}y(t), \frac{1}{\lambda(s)}u(t), \ldots, \frac{1}{(s+a)}u(t)\right]^{\mathrm{T}}\,, \text{ and}$$

$$e_1(t) = y(t) - y_{\mathrm{m}}(t)\,.$$

Lemma 5.1

$$|u(t)| \leq K_{u\psi'}\|\psi'(t)\| + K_{uv}|v_{\mathrm{f}}'(t)| + k_{uu} + c_0\exp[-2d_0t], \forall t \geq 0\,. \tag{8}$$

Proof. Let $\phi_{\mathrm{I}} := \left(\frac{1}{\lambda_1(s)}y, \ldots, \frac{1}{(s+a)}y, y, \frac{1}{\lambda_1(s)}u, \ldots, \frac{1}{(s+a)}u\right)$ be a sub-vector of ϕ, and $\hat{\theta}_{\mathrm{I}}$ the corresponding sub-vector of $\hat{\theta}$. Then, the control law (6) gives $u = (r' - \phi_{\mathrm{I}}^{\mathrm{T}}\hat{\theta}_{\mathrm{I}}^{\mathrm{T}})/\hat{b}_{\mathrm{m}}$. From (4), we then obtain $|y(t)| \leq M\|\psi'\| + |v_{\mathrm{f}}'| + c_0\exp[-2d_0t]$. Hence $\|\phi_{\mathrm{I}}\| \leq |y| + \|\psi'\| \leq (1+M)\|\psi'\| + |v_{\mathrm{f}}'| + c_0\exp[-2d_0t]$. Since $\hat{b}_{\mathrm{m}} \geq b_{\mathrm{min}}$, and $\|\hat{\theta}_{\mathrm{I}}\| \leq \|\hat{\theta}\| \leq M$, we thus conclude that $|u| \leq \frac{1}{b_{\mathrm{min}}}(k_r + M(1+M)\|\psi'\| + M|v_{\mathrm{f}}'|) + c_0M\exp[-2d_0t]$. □

Theorem 5.1 *For all* $t \geq 0$,

(i) $|v_f(t)| \leq K_{v\mathrm{f}}m(t) + k_{v\mathrm{f}} + c_0\exp[-(d_0 + d_2/2)t] + k_{v\mathrm{f}0}\exp[-pt]$,
(ii) $|v_{\mathrm{f}}'(t)| \leq K_{v\mathrm{f}}'m(t) + k_{v\mathrm{f}} + c_0\exp[-(d_0 + d_2/2)t] + k_{v\mathrm{f}0}\exp[-pt]$,
(iii) $\|\psi'(t)\| \leq K_{\psi'm}m(t) + c_0\exp[-2d_0t]$,
(iv) $|e_{\mathrm{a}}(t)| \leq K_{em}m(t) + k_{v\mathrm{f}} + c_0\exp[-(d_0 + d_2/2)t] + k_{v\mathrm{f}0}\exp[-pt]$ $+c_0\exp[-2d_0t]$,
(v) $|y(t)| \leq K_{ym}m(t) + k_{v\mathrm{f}} + c_0\exp[-(d_0 + d_2/2)t] + k_{v\mathrm{f}0}\exp[-pt]$ $+c_0\exp[-2d_0t]$,
(vi) $|u(t)| \leq K_{um}m(t) + k_{um} + c_0\exp[-(d_0 + d_2/2)t] + K_{uv}k_{v\mathrm{f}0}\exp[-pt]$ $+c_0\exp[-2d_0t]$.

[5] Here and throughout, the values of useful constants used in the bounds are specified in Table D in the Appendix. Certain constants whose exact value is unimportant and which do not depend on $K_v, \mu_{\mathrm{m}}, k_v$ will be denoted generically by C. Any constant whose exact value is unimportant but which depends on $K_v, \mu_{\mathrm{m}}, k_v$, and such that its value decreases as $K_v, \mu_{\mathrm{m}}, k_v$ decrease, will be denoted generically by c. Any positive constant which depends only on initial conditions of some filter, and whose exact value is unimportant will be denoted by c_0. Finally, all constants throughout are positive, unless otherwise noted.

The proof of this theorem is based on the following lemma.

Lemma 5.2 *Let $H(s)$ be a strictly proper, stable transfer function, whose poles $\{p_j\}$ satisfy* $\mathrm{Re}(p_j) \leq -(d_0 + d_2) < 0$ *for all j.*
(i) If $w_{\rm in}(t)$ is the input to a system with transfer function $H(s)$, which satisfies the bound $|w_{\rm in}(t)| \leq k_u|u(t)| + k_y|y(t)| + k_m m(t) + k' + k'' \exp[-p't], \forall t \geq 0$, for $0 < p' \leq d_0$, then the output $w_{\rm out}(t)$ is bounded by $|w_{\rm out}(t)| \leq k_1\|x(0)\|\exp[-(d_0+d_2/2)t] + k_3 m(t) + k_4 + k_5 \exp[-p't]$, where $x(0)$ is the initial state corresponding to a minimal state representation of $H(s)$, and k_1, k_3, k_4 and k_5 are related positive constants specified below.
(ii) Let $H'(s) = k_p + H(s)$. If $w_{\rm in}(t)$ is the input to a system with transfer function $H'(s)$, which satisfies the bound $|w_{\rm in}(t)| \leq k_m m(t) + k' + k'' \exp[-p't]$, $\forall t \geq 0$, for $0 < p' \leq d_0$, then the output $w_{\rm out}(t)$ is bounded by $|w_{\rm out}(t)| \leq k_1\|x(0)\|\exp[-(d_0+d_2/2)t] + k_6 m(t) + k_7 + k_8 \exp[-p't]$, where $x(0)$ is the initial state corresponding to a minimal state representation of $H'(s)$, and k_1, k_6, k_7 and k_8 are related positive constants specified below.[6]

Proof. (i) Let $\dot{x} = Ax + bw_{\rm in}, w_{\rm out} = c_H^{\rm T}x$ be a minimal state representation of $H(s)$. Then there exist constants k_1, k_2 such that $\|c_H^{\rm T}\exp[At]\| \leq k_1 \exp[-(d_0 + d_2/2)t], \forall t \geq 0$, and $|h(t)| \leq k_2 \exp[-(d_0 + d_2/2)t], \forall t \geq 0$, where $h(\cdot)$ is the impulse response of $H(s)$. Thus,

$$
\begin{aligned}
|w_{\rm out}(t)| &\leq \|c_H^{\rm T}\exp[At]\|\,\|x(0)\| + \int_0^t |h(\tau)||w_{\rm in}(t-\tau)|d\tau \\
&\leq k_1\|x(0)\|\exp[-(d_0+d_2/2)t] + k_2k_m\int_0^t \exp[-(d_0+d_2/2)\tau] \\
&\quad m(t-\tau)d\tau + k_2\int_0^t \exp[-(d_0+d_2/2)\tau](k_u|u(t-\tau)| + k_y|y(t-\tau)| \\
&\quad +k' + k''\exp[-p'(t-\tau)])d\tau\,.
\end{aligned}
$$

Using $m(t-\tau) \leq \exp[d_0\tau]m(t)$ gives

$$
\int_0^t \exp[-(d_0+d_2/2)\tau]m(t-\tau)d\tau \leq m(t)\int_0^t \exp[-d_2\tau/2]d\tau \leq 2m(t)/d_2\,.
$$

Also, from $0 < p' \leq d_0$,

$$
\begin{aligned}
&\int_0^t \exp[-(d_0+d_2/2)\tau]\exp[-p'(t-\tau)]d\tau \\
&= \exp[-p't]\int_0^t \exp[-(d_0+d_2/2-p')\tau]d\tau \\
&\leq \frac{\exp[-p't]}{d_0+d_2/2-p'} \leq \frac{2}{d_2}\exp[-p't]\,.
\end{aligned}
$$

[6] k_3, k_4, k_6 and k_7 are independent of k'', which will mean that initial conditions have no influence on the magnitude of the bounds or the size of allowable unmodeled dynamics.

Thus,

$$|w_{\mathrm{out}}(t)| \le k_1 ||x(0)|| \exp[-(d_0 + d_2/2)t] + k_3 m(t) + k_4 + k_5 \exp[-p't] ,$$

where

$$k_3 = k_2\left(\frac{\max\{k_u, k_y\}}{d_1} + \frac{2k_m}{d_2}\right), k_4 = \frac{2k_2 k'}{2d_0 + d_2}, k_5 = \frac{2k_2 k''}{d_2} .$$

(ii) This follows from part (i). □

Proof of Theorem 5.1. The results (i) and (ii) follow from Lemma 5.2 and (A4) by noting that $v_f = \frac{A(s)F(s)}{\lambda(s)} v$ and $v_{\mathrm{f}}' = v_f + \mu_{\mathrm{m}} \frac{B(s)F(s)}{\lambda(s)} \Delta_{\mathrm{m}}(s)u$. The result (iii) follows by applying Lemma 5.2 to each component of ψ', noting that each component is either of the form $\frac{1}{(s+a)^k} y$ or $\frac{1}{(s+a)^k} u$. Since $e_{\mathrm{a}} = -\psi^{\mathrm{T}}\tilde{\theta} + v_{\mathrm{f}}' + \xi_{2d_0}$ the bound for e_{a} in (iv) follows from the bounds for $||\psi|| \le ||\psi'||$ and v_{f}'. The proof of (v) is similar to (iv) since $y = \psi^{\mathrm{T}}\theta + v_{\mathrm{f}}' + \xi_{2d_0}$. Finally, (vi) follows from Lemma 5.1 and (ii), (iii) above. □

6 Bounding by a Switched System

Let us introduce the "switched" system,

$$\dot{z}(t) = I(t)(-d_0 z(t) + d_1(|u(t)| + |e_{\mathrm{a}}(t)| + 1)) + (1 - I(t))(-g_2 z(t) + K_2) , \quad (9)$$

$$z(0) \ge \frac{m(0)}{(1 + \frac{2\alpha K_{u\psi'}}{(a-2d_0)})} ,$$

where

$$I(t) = 1 , \text{ if } (-d_0 z(t) + d_1(|u(t)| + |e_{\mathrm{a}}(t)| + 1)) \ge (-g_2 z(t) + K_2) ,$$
$$= 0 , \text{ otherwise} . \quad (10)$$

In Theorem 6.1(ii) below we will show that m itself is bounded in terms of z, thus showing from Theorem 5.1 that all other signals can also be bounded in terms of z.

Theorem 6.1

$$\text{(i) } \dot{m}(t) \le K_m m(t) + k_m + c_0 \exp[-(d_0 + d_2/2)t] + c_0 \exp[-pt] + c_0 \exp[-2d_0 t] , \ \forall t \ge 0 . \quad (11)$$

$$\text{(ii) } m(t) \le K_{mz} z(t) + k_{mz} , \ \ \forall t \ge 0 . \quad (12)$$

$$\text{(iii) } \dot{z}(t) \le K_z z(t) + k_z + c_0 \exp[-(d_0 + d_2/2)t] + c_0 \exp[-pt] + c_0 \exp[-2d_0 t] , \ \forall t \ge 0 . \quad (13)$$

Proof. (i) This follows from $\dot{m} = -d_0 m + d_1(|u| + |y| + 1)$, by making use of the bounds for $|u|$ and $|y|$ in terms of m given in Theorem 5.1(v),(vi).
(ii) By the definition of the augmented error e_a, we have $y = (e_a + y_m) - \zeta$, which implies $|y| \leq |e_a| + |y_m| + |\zeta|$. By the Swapping lemma (Morse [6]; Goodwin and Mayne [13]), $\zeta(t) = [h^T \exp(Q_1 t)] * (H(t)\dot{\tilde{\theta}}(t))$, where Q_1 is an $(n_r \times n_r)$ stable matrix such that $\det(sI - Q_1) = \lambda_2(s)$, $H^T(t) = (\psi(t), s\psi(t), \ldots, s^{n-m-1}\psi(t))$, h^T is a constant row vector of dimension n_r, and "$*$" denotes convolution. Now,

$$\left\| H\dot{\tilde{\theta}} \right\| = \left\| H \mathrm{Proj}\left[\hat{\theta}, \frac{\alpha\psi e_a}{\nu + \|\psi'\|^2}\right] \right\| \leq \alpha C |e_a| , \tag{14}$$

where the last inequality follows from Property (P1) of the parameter estimator, and since $\|H\| \leq C\|\psi'\|$. Since $\lambda_2(s) = (s+a)^{n_r}$, we have $\|h^T \exp(Q_1 t)\| \leq C \exp(-at/2)$. Therefore $|\zeta(t)| \leq \int_0^t \alpha C \exp[-a(t-\tau)/2] |e_a(\tau)| d\tau$, and so

$$\begin{aligned}
\int_0^t |\zeta| \exp[-d_0(t-\tau)] d\tau &\leq C\alpha \int_0^t \exp[-d_0(t-\tau)] \\
&\qquad \int_0^\tau \exp[-a(\tau - t')/2] |e_a(t')| dt' d\tau \\
&= C\alpha \int_0^t \exp[-d_0 t] \exp[at'/2] \\
&\qquad \int_{t'}^t \exp[-(a/2 - d_0)\tau] d\tau |e_a(t')| dt' \\
&\leq \frac{2C\alpha}{a - 2d_0} \int_0^t \exp[-d_0(t-t')] |e_a(t')| dt'.
\end{aligned}$$

This implies

$$\int_0^t |y| \exp[-d_0(t-\tau)] d\tau \leq \int_0^t \exp[-d_0(t-t')](1 + \frac{2C\alpha}{(a-2d_0)}) |e_a(t')| dt' + \frac{k_{ym}}{d_0}.$$

Now, add $\int_0^t (|u(\tau)| + 1) \exp[-d_0(t-\tau)] d\tau$ to both sides and multiply by d_1. Finally, adding $\exp(-d_0 t) m(0)$ to both sides yields the desired result.
(iii) The result follows from $\dot{z}(t) \leq -(d_0 + g_2) z(t) + d_1(|u| + |e_a| + 1) + K_2$ and the bounds on $|e_a|$ and $|u|$ in Theorem 5.1(iv, vi) by using the bound for m in (ii). □

From Theorem 6.1 it follows that z can grow at most exponentially fast, and therefore does not have a finite escape time. Since z bounds all signals, it follows that no signal in the system has a finite escape time.

In what follows, we choose a large time T_ℓ, and restrict attention to $t \geq T_\ell$, for which the following bounds hold.

(i) $|v(t)| \leq K_v m(t) + k_v + k(T_\ell) \leq K_v K_{mz} z(t) + K_v k_{mz} + k_v + k(T_\ell)$ (15)

(ii) $|v_f'(t)| \leq K_{vf}' K_{mz} z(t) + K_{vf}' k_{mz} + k_{vf} + k(T_\ell)$ (16)

(iii) $\|\psi'(t)\| \le K_{\psi' m} K_{mz} z(t) + K_{\psi' m} k_{mz}$ (17)

(iv) $|y(t)| \le M\|\psi'(t)\| + K'_{vf} K_{mz} z(t) + K'_{vf} k_{mz} + k_{vf} + k(T_\ell)$ (18)

(iv)' $|y(t)| \le K_{yz} z(t) + k_{yz} + k(T_\ell)$ (19)

(v) $|u(t)| \le K_{u\psi'}\|\psi'(t)\| + K'_{vf} K_{uv} K_{mz} z(t) + k_u + k(T_\ell)$ (20)

(v)' $|u(t)| \le K_{uz} z(t) + k_{uz} + k(T_\ell)$ (21)

(vi) $\dot{z}(t) \le K_z z(t) + k_z + k(T_\ell)$ (22)

(vii) $\eta_{\mathrm{m}}(t) := \dfrac{\Delta_{\mathrm{m}}(s)}{q(s)} u(t)$ satisfies $|\eta_{\mathrm{m}}(t)| \le cz(t) + c + k(T_\ell)$, (23)

if $q(s - 2d_0 - d_2)$ is a Hurwitz polynomial such that $\frac{\Delta_{\mathrm{m}}(s)}{q(s)}$ is proper. (The result (vii) follows from Theorem 5.1(vi), Lemma 5.2(ii) and Theorem 6.1(ii)). In all the above inequalities and in what follows, $k(T_\ell)$ denotes a generic constant which decreases with increasing T_ℓ at a rate faster than $\exp[-pT_\ell]$.

7 Comparing ψ' and z

In this section we compare ψ' and z.

Theorem 7.1 *(i) If* $I(t) = 1$, *then* $\dfrac{n(t)}{z(t)^2} \ge K_{nz}$.

(ii) Consider $T' > 0$. *Let* $t_1 \ge T_\ell + T'$ *be any instant such that* $I(t_1) = 1$. *If* $z(t) \ge \frac{L}{2}, \forall t \in (t_1 - T', t_1]$ *where L is a large positive constant (specified in Table D), then there exists a* $K_{v\max} > 0$ *and a* $\mu_{\max} > 0$ *such that for all* $K_v \in [0, K_{v\max}]$ *and for all* $\mu_{\mathrm{m}} \in [0, \mu_{\max}]$, *we have:*

$$\frac{n(t)}{z^2(t)} > \delta(T'), \quad \forall t \in (t_1 - T', t_1]. \tag{24}$$

Proof. (i) Let us fix t such that $I(t) = 1$. By (10), we have $|u| + |e_{\mathrm{a}}| \ge \frac{d_0 - g_2}{d_1} z + \frac{K_2 - d_1}{d_1}$. Using the upper bound (20) on u gives, $K_{u\psi'}\|\psi'\| + |e_{\mathrm{a}}| \ge (\bar{d}_0 - \bar{g}_2)z + \bar{K}_2 =: RHS$. We note that we can take $\bar{d}_0 \ge 0$ and $\bar{K}_2 \ge 0$, by the choices of $K_2, T_\ell, K_{v\max}$, and $\mu_{\max}$ given in Table D. If $\|\psi'\| > \dfrac{RHS}{2K_{u\psi'}}$, then the result clearly holds, so we consider $\|\psi'\| \le \dfrac{RHS}{2K_{u\psi'}}$. Then clearly $|e_{\mathrm{a}}| \ge \dfrac{RHS}{2}$. Therefore using the upper bound (16) for v'_{f}, we obtain by definition of the augmented error,

$$\begin{aligned} \widetilde{M}\|\psi\| &\ge |\tilde{\theta}^{\mathrm{T}}\psi| \\ &\ge |e_{\mathrm{a}}| - |v'_{\mathrm{f}}| - c_0 \exp[-2d_0 T_\ell] \\ &\ge |[(\bar{d}_0 - \bar{g}_2)z + \bar{K}_2]/2| - (K'_{vf} K_{mz} z(t) + K'_{vf} k_{mz} \\ &\quad + k_{vf} + k(T_\ell) + c_0 \exp[-pT_\ell]) . \end{aligned}$$

Note that $(\bar{d}_0 - \bar{g}_2) \ge 4K'_{vf} K_{mz}$ and $\bar{K}_2 \ge 2(K'_{vf} k_{mz} + k_{vf} + k(T_\ell) + k \exp[-pT_\ell])$, again from the choices of $K_2, T_\ell, K_{v\max}$, and $\mu_{\max}$ given in Table D.) This implies

$||\psi'|| \geq ||\psi|| \geq [\frac{(\bar{d}_0 - \bar{g}_2)z}{4\widetilde{M}}]$. Hence we obtain, $\frac{n(t)}{z(t)^2} \geq K_{nz}$.

(ii) First, we will bound the growth rate of $\frac{n(t)}{z^2(t)}$. We do not require z to be large for this part of the proof. Note that $\frac{d}{dt}(\frac{n(t)}{z^2(t)}) = \frac{\dot{n}}{z^2} - 2\frac{n}{z^2}\frac{\dot{z}}{z} \leq \frac{\dot{n}}{z^2} + 2g_2\frac{n}{z^2}$. Also, $\dot{\psi}' = \phi' - a\psi'$, which implies $\dot{n} = 2\psi'^{\mathrm{T}}(\phi' - a\psi') \leq (1-2a)||\psi'||^2 + ||\phi'||^2$, where, $\phi'^{\mathrm{T}} := (\frac{1}{(s+a)^{2n-m-2}}y, \ldots, \frac{1}{(s+a)}y, y, \frac{1}{(s+a)^{2n-m-2}}u, \ldots, \frac{1}{(s+a)}u, u)$, i.e., $\phi' := (s+a)\psi'$. Now, $||\phi'||^2 \leq ||\psi'||^2 + y^2 + u^2$. Using this, we get $\dot{n} \leq 2(1-a)||\psi'||^2 + y^2 + u^2$. Next, using the upper bounds for u and y from (20), (18), and recalling that $n(t) = \nu + ||\psi'(t)||^2$, we get $\frac{d}{dt}(\frac{n(t)}{z^2(t)}) \leq K_{\mathrm{a}}(\frac{n(t)}{z^2(t)}) + K_{\mathrm{b}} + \frac{K_{\mathrm{c}}}{z^2(t)}$. Next, using the fact that $z(t) \geq \frac{L}{2}, \forall t \in (t_1 - T', t_1]$, we get $\frac{d}{dt}(\frac{n(t)}{z^2(t)}) \leq K_{\mathrm{a}}(\frac{n(t)}{z^2(t)}) + K_{\mathrm{d}}$, where $K_{\mathrm{d}} := K_{\mathrm{b}} + \frac{4K_{\mathrm{c}}}{L^2}$. This gives,

$$\begin{aligned}\frac{n}{z^2}(t_1) &\leq \exp[K_{\mathrm{a}}(t_1 - t')][\frac{n(t')}{z^2(t')} + \int_{t'}^{t_1} \exp[K_{\mathrm{a}}(t' - \tau)]K_{\mathrm{d}}d\tau] \\ &\leq \exp[K_{\mathrm{a}}T'][\frac{n(t')}{z^2(t')} + \frac{K_{\mathrm{d}}}{K_{\mathrm{a}}}(1 - \exp[-K_{\mathrm{a}}T'])], \ \forall t' \in [t_1 - T', t_1].\end{aligned}$$

Thus,

$$\begin{aligned}\frac{n(t')}{z^2(t')} &\geq \exp[-K_{\mathrm{a}}T'][\frac{n}{z^2}(t_1) + \frac{K_{\mathrm{d}}}{K_{\mathrm{a}}}] - \frac{K_{\mathrm{d}}}{K_{\mathrm{a}}} \\ &\geq \exp[-K_{\mathrm{a}}T'][K_{nz} + \frac{K_{\mathrm{d}}}{K_{\mathrm{a}}}] - \frac{K_{\mathrm{d}}}{K_{\mathrm{a}}}, \ \forall t' \in [t_1 - T', t_1].\end{aligned}$$

The fact that there exist $K_{v\max} > 0, \mu_{\max} > 0$ such that $\delta(T') > 0$ for all $K_v \in [0, K_{v\max}], \mu_{\mathrm{m}} \in [0, \mu_{\max}]$ follows from the fact that $\frac{K_{\mathrm{d}}}{K_{\mathrm{a}}}$ can be made as small as we wish by making L sufficiently large, and then choosing appropriately small $K_{v\max}$ and $\mu_{\max}$. □

8 A Nonminimal System Representation

Recalling $y = \theta^{\mathrm{T}}\psi + v_{\mathrm{f}}' + \xi_{2d_0}$, $\psi(t) = \frac{1}{\lambda_2(s)}\phi(t)$, $\phi^{\mathrm{T}}(t)\hat{\theta}(t) = r'(t)$, and $v_{\mathrm{m}} = \frac{BF}{\lambda}\Delta_{\mathrm{m}}u$, we note that

$$y = \frac{1}{\lambda_2(s)}(r' - \tilde{\theta}^{\mathrm{T}}\phi) + v_f + \mu_{\mathrm{m}}v_{\mathrm{m}} + \xi_{2d_0}. \tag{25}$$

The system (1) can be represented as

$$\dot{x}_p = A_p x_p + b_p u + \mu_{\mathrm{m}} b_p' \eta_{\mathrm{m}}, \ y = h_p^{\mathrm{T}} x_p + v + \xi_{2d_0}, \tag{26}$$

where $(A_p, b_p, h_p^{\mathrm{T}})$ is a minimal state representation of the *nominal* plant transfer function, $\eta_{\mathrm{m}} := \frac{\Delta_{\mathrm{m}}(s)}{q(s)}u$, with $q(s)$ chosen as some polynomial of degree $n_{\mathrm{r}} - 1$ such that $q(s - 2d_0 - d_2)$ is Hurwitz, thus giving rise to the ξ_{2d_0} term in (26)). In (26) above, $h_p^{\mathrm{T}}(sI - A_p)^{-1}b_p = \frac{B(s)}{A(s)}$ and $h_p^{\mathrm{T}}(sI - A_p)^{-1}b_p' = \frac{B(s)q(s)}{A(s)}$. Note that $\frac{\Delta_{\mathrm{m}}(s)}{q(s)}$ is a proper, stable transfer function. The control law (6) is equivalent to $\theta^{\mathrm{T}}\phi + \widetilde{\theta}^{\mathrm{T}}\phi = r'$. Hence,

$$b_m u = -\theta_u^{\mathrm{T}}\phi_u - \theta_n y - \theta_y^{\mathrm{T}}\phi_y - \widetilde{\theta}^{\mathrm{T}}\phi + r'. \tag{27}$$

Defining $\phi_y = \left(\frac{1}{(s+a)^{n-1}}y, \ldots, \frac{1}{(s+a)}y\right)^{\mathrm{T}}$ and ϕ_u likewise, we note that $\phi(t) = (\phi_y^{\mathrm{T}}(t), y(t), \phi_u^{\mathrm{T}}(t), u(t))^{\mathrm{T}}$, with

$$\dot{\phi}_y(t) = Q\phi_y(t) + qy(t)\,,\; \dot{\phi}_u(t) = Q\phi_u(t) + qu(t), \tag{28}$$

where Q is a stable matrix such that $\det(sI - Q) = (s+a)^{n-1}$, and (Q, q) is a controllable pair.

Define $X_{\mathrm{c}}^{\mathrm{T}} := [x_p^{\mathrm{T}}, \phi_y^{\mathrm{T}}, \phi_u^{\mathrm{T}}]$. Then, using (26), (28), and (27), we get $\dot{X}_{\mathrm{c}} = A_{\mathrm{c}}X_{\mathrm{c}} + b_{\mathrm{c}}(r' - \widetilde{\theta}^{\mathrm{T}}\phi) + b_{\mathrm{c}v}v + \mu_{\mathrm{m}}b_{\mathrm{c}\eta}\eta_{\mathrm{m}}, y = h_{\mathrm{c}}^{\mathrm{T}}X_{\mathrm{c}} + v + \xi_{2d_0}$, where, A_{c} is a stable matrix (as we will prove shortly), $b_{\mathrm{c}}^{\mathrm{T}} := [b_p^{\mathrm{T}}/b_m, 0, q^{\mathrm{T}}/b_m]$, $b_{\mathrm{c}v}^{\mathrm{T}} := [-\theta_n b_p^{\mathrm{T}}/b_m, q^{\mathrm{T}}, -\theta_n q^{\mathrm{T}}/b_m], b_{\mathrm{c}\eta}^{\mathrm{T}} := [b_p'^{\mathrm{T}}, 0, 0]$ and $h_{\mathrm{c}}^{\mathrm{T}} := [h_p^{\mathrm{T}}, 0, 0]$; see [3].

For $\widetilde{\theta} = 0, v \equiv 0$ and $\mu_{\mathrm{m}} = 0$, we have, $\dot{X}_{\mathrm{c}} = A_{\mathrm{c}}X_{\mathrm{c}} + b_{\mathrm{c}}r', y = h_{\mathrm{c}}^{\mathrm{T}}X_{\mathrm{c}} + \xi_{2d_0}$. This implies $y = h_{\mathrm{c}}^{\mathrm{T}}(sI - A_{\mathrm{c}})^{-1}b_{\mathrm{c}}r' + \xi_{2d_0}$. However, we already know (from (25)) that $y = \frac{1}{\lambda_2(s)}r' + \xi_{2d_0}$, in such a case. This means that (after cancellations)

$$h_{\mathrm{c}}^{\mathrm{T}}(sI - A_{\mathrm{c}})^{-1}b_{\mathrm{c}} = \frac{1}{\lambda_2(s)}\,. \tag{29}$$

It can be easily verified (e.g. see [17, p. 136]) that we have only stable pole-zero cancellations because the nominal plant is minimum-phase and Q is a stable matrix. This proves that $(A_{\mathrm{c}}, h_{\mathrm{c}}^{\mathrm{T}})$ is detectable. Since the overall transfer function is stable (by (29)), we conclude that A_{c} is a stable matrix.

Thus we can write the following nonminimal representation of $\frac{1}{\lambda_2(s)}$:

$$\dot{X}_{\mathrm{m}} = A_{\mathrm{c}}X_{\mathrm{m}} + b_{\mathrm{c}}r'\,,\; y_{\mathrm{m}} = h_{\mathrm{c}}^{\mathrm{T}}X_{\mathrm{m}}\,,$$

where $X_{\mathrm{m}}^{\mathrm{T}} := [x_{\mathrm{m}}^{\mathrm{T}}, \phi_{y\mathrm{m}}^{\mathrm{T}}, \phi_{u\mathrm{m}}^{\mathrm{T}}]$. A_{c} being stable and r' being bounded implies that $\|X_{\mathrm{m}}(t)\| \le K_{X\mathrm{m}}, \forall t \ge 0$, for some positive constant $K_{X\mathrm{m}}$. Define the *state error* e as $e := X_{\mathrm{c}} - X_{\mathrm{m}}$. This gives,

$$\dot{e} = A_{\mathrm{c}}e - b_{\mathrm{c}}(\widetilde{\theta}^{\mathrm{T}}\phi) + b_{\mathrm{c}v}v + \mu_{\mathrm{m}}b_{\mathrm{c}\eta}\eta_{\mathrm{m}}\,,\; e_1 = h_{\mathrm{c}}^{\mathrm{T}}e + v\,, \tag{30}$$

where we recall that $e_1 = y - y_{\mathrm{m}}$ is the tracking error.

9 Robust Ultimate Boundedness

Define W, which bounds all signals, by

$$W = k_e e^{\mathrm{T}} P e + \frac{1}{2} z^2 , \tag{31}$$

where $P = P^{\mathrm{T}} > 0$ satisfies $PA_{\mathrm{c}} + A_{\mathrm{c}}^{\mathrm{T}} P = -I$. Such a P exists since A_{c} is a stable matrix.

Our main result on robust ultimate boundedness of the overall system is given by the following theorem. It states that eventually $W(t)$ (which bounds all signals) enters a compact set, the size of which is independent of the initial conditions. Furthermore, the size of the allowable unmodeled dynamics for which this is guaranteed is independent of the initial conditions. It should be noted, however, that the time T_{large} that it takes for $W(t)$ to enter this compact set can depend on the initial conditions.

Choose constants $0 < \gamma < 1$, $\epsilon_0 > 0$ small, and $\epsilon_z > 0$. Let $T, \epsilon, T_\ell, K_{v\max}$, and $\mu_{\max}$ be as in Table D.

Theorem 9.1 (Robust Ultimate Boundedness theorem). *There exist a $T_{\mathrm{large}} \geq T_\ell$, and positive $K_{v\max}$, $\mu_{\max}$, such that for all $K_v \in [0, K_{v\max}]$, and for all $\mu_{\mathrm{m}} \in [0, \mu_{\max}]$,*

$$W(t) \leq 8 K_{Wz}^2 L^2 \exp[4(K_z + \epsilon_z)T], \quad \forall t \geq T_{\mathrm{large}} . \tag{32}$$

Furthermore, $K_{v\max}$, $\mu_{\max}$, T, and L are independent of the initial conditions.

Proof. The idea of the proof, based on the following lemmas, is to show that whenever $W(t) \geq K_{Wz} L^2$ throughout an interval of length $2T$, then at the end of the interval its value is smaller than at the beginning of the interval.

Lemma 9.1

$$W(t) \leq K_{Wz} z^2(t) + k_{Wz}, \forall t \geq T_\ell . \tag{33}$$

Proof. Since $W = k_e e^{\mathrm{T}} P e + \frac{1}{2} z^2, e = X_{\mathrm{c}} - X_{\mathrm{m}}$, $\|X_{\mathrm{m}}\| \leq K_{Xm}$, $X_{\mathrm{c}} = [x_p^{\mathrm{T}}, \phi_y^{\mathrm{T}}, \phi_u^{\mathrm{T}}]^{\mathrm{T}}$, $\|\phi_y\| \leq cz, \|\phi_u\| \leq cz$, it suffices to show that $\|x_p\| \leq cz + c + c_0 \exp[-pt]$.

By (26), defining $u' := \dfrac{1 + \mu_{\mathrm{m}} \Delta_{\mathrm{m}}(s)}{q(s)} u$, $y' = y - v - \xi_{2d_0}(t)$, and b_p'' such that $h_p^{\mathrm{T}}(sI - A_p)^{-1} b_p'' = \dfrac{B(s)q(s)}{A(s)}$, with the (ξ_{2d_0}) term in the above equation possibly different from the (ξ_{2d_0}) term in (26), we get, $\dot{x}_p = A_p x_p + b_p'' u'$, $y' = h_p^{\mathrm{T}} x_p$. Now, from Lemma 5.2(ii), $|u'(t)| \leq cm(t) + c + c_0 \exp[-pt], \forall t \geq 0$, and from Theorem 5.1(v), $|y'(t)| \leq cm(t) + c + c_0 \exp[-pt], \forall t \geq 0$. Noting that $B'(s) := B(s)q(s)$ is such that $B'(s - p)$ is Hurwitz, and applying Lemma A.1, we get $\|x_p(t)\| \leq cm(t) + c + c_0 \exp[-pt], \forall t \geq 0$. Finally applying Theorem 5.2(ii), we get the desired bound on $\|x_p\|$. □

Lemma 9.2 *Consider an interval* $[a, b]$. *Then*

$$\frac{W(b)}{W(a)} \leq \exp\left[\int_a^b g(\tau)d\tau\right]\left(1 + \frac{\beta_4 \exp[\beta(b-a)]}{\beta W(a)}\right), \tag{34}$$

where $g(t) := -\beta + \beta_3 \frac{|\widetilde{\theta}^{\mathrm{T}}(t)\phi(t)|}{z(t)} + \beta_5 \frac{|\widetilde{\theta}^{\mathrm{T}}(t)\psi(t)|}{z(t)}$, β_3, β_4 *and* β_5 *being specified in Appendix C.*

Proof. See Appendix C. □

Lemma 9.3 *Consider a time interval* $[a, b]$ *such that (i)* $b \geq a \geq T_\ell$, *(ii)* $W(t) \geq K_{W_z}L^2, \forall t \in [a, b]$. *Then,*

(i) $\frac{1}{2}z^2(t) \leq W(t) \leq 4K_{W_z}z^2(t), \forall t \in [a, b]$.

(ii) *If* $I(t) = 0, \forall t \in [a, b]$, *then* $W(b) \leq 8K_{W_z}\left(\exp[-g_2(b-a)] + \frac{2K_2}{g_2 L}\right)^2 W(a)$.

(iii) *If* $(b - a) \geq 1$ *and for each* $t \in [a, b]$ *such that* $I(t) = 0$ *there exists a* $t' \in [0, T]$ *such that* $I(t + t') = 1$, *then*
$W(b) \leq k \cdot \exp[-\frac{\beta}{4}(b-a)](1 + \frac{\beta_4}{K_{W_z}\beta L^2}\exp[\beta(b-a)])W(a)$.

(iv) $W(b) \leq 8K_{W_z}\exp[2(K_z + \epsilon_z)(b-a)]W(a)$.

Proof. The result (i) follows by Lemma 9.1 since $W \geq K_{W_z}L^2$ implies $z \geq L/2$. The result (ii) follows from (i) because of $\dot{z} \leq -g_2 z + K_2$, and since $W \geq K_{W_z}L^2$ implies $z \geq L/2$. To prove (iii), note from Lemmas 9.2, B.4, the discussion preceding Lemma B.4, and by our choices of $\epsilon, \mu_{\max}, K_{v\max}$ and L, that

$$\begin{aligned}
\int_a^b g(t)dt &= -\beta(b-a) + \beta_3 \int_a^b \frac{|\widetilde{\theta}^{\mathrm{T}}\phi|}{z}dt + \beta_5 \int_a^b \frac{|\widetilde{\theta}^{\mathrm{T}}\psi|}{z}dt \\
&\leq -\beta(b-a) + \beta_3\nu_3/\epsilon^2 + \beta_3[\mu^2\nu_4(T)/\epsilon^2 + \mu\nu_5(T)/\sqrt{\epsilon} + \nu_6(T)\epsilon^\rho \\
&\quad +4a^{n-m}k(\mu, L)](b-a) + \beta_5\nu_7/\epsilon^2 + \beta_5[\mu^2\nu_8(T)/\epsilon^2 + \sqrt{\epsilon}](b-a) \\
&\leq -\frac{\beta}{4}(b-a) + \frac{(\beta_3\nu_3 + \beta_5\nu_7)}{\epsilon^2},
\end{aligned}$$

which gives us the desired result after defining $k = \exp\left[\frac{(\beta_3\nu_3 + \beta_5\nu_7)}{\epsilon^2}\right]$. Finally, (iv) follows from the bounded growth rate of $z(t)$ shown in (22), and (i). □

Lemma 9.4 (Contraction Property). *If* $t_0 \geq T_\ell$, *and* $W(t) \geq K_{W_z}L^2$ $\forall t \in [t_0, t_0 + 2T]$, *then* $W(t_0 + 2T) \leq \gamma W(t_0)$.

Proof. There are four possibilities:

(1) Suppose $I(t) = 0, \forall t \in [t_0, t_0 + 2T]$. Then from Lemma 9.3(ii), we have $W(t_0 + 2T) \le 8K_{Wz}\left(\exp[-2g_2T] + \frac{2K_2}{g_2L}\right)^2 W(t_0) \le \gamma W(t_0)$, where the last inequality follows by the definitions of T and L given in Table D. Let $t_1 := \min\{t \in [0, 2T] : I(t_0 + t) = 1\}$, and $t_2 := \max\{t \in [0, 2T] : I(t_0 + t) = 1\}$.

(2) Suppose $0 \le t_1 \le t_2 < T$. First, by Lemma 9.3(ii),
$W(t_0 + 2T) \le 8K_{Wz}\left(\exp[-g_2(2T - t_2)] + \frac{2K_2}{g_2L}\right)^2 W(t_0 + t_2)$.

(2A) Suppose $t_2 < 1$. Then, the above inequality implies $W(t_0 + 2T) \le 8K_{Wz}\left(\exp[-g_2(2T - 1)] + \frac{2K_2}{g_2L}\right)^2 W(t_0 + t_2)$. Using Lemma 9.3(iv), we have $W(t_2 + t_0)) \le 8K_{Wz}\exp[2(K_z + \epsilon_z)]W(t_0)$, which implies that $W(t_0 + 2T) \le 64K_{Wz}^2\exp[2(K_z + \epsilon_z)]\left(\exp[-g_2(2T - 1)] + \frac{2K_2}{g_2L}\right)^2 W(t_0)$ so that $W(t_0 + 2T) \le \gamma W(t_0)$, the last inequality following from the definitions of T and L.

(2B) Suppose $t_2 \ge 1$. In this case, we have by Lemma 9.3(iii) and the definitions of T and L, $W(t_0 + t_2) \le \exp[-\beta t_2/4](k + \epsilon_0)W(t_0)$, so that

$$\begin{aligned} W(t_0 + 2T) &\le 8\exp[-\beta/4](k + \epsilon_0)K_{Wz}\left(\exp[-g_2T] + \frac{2K_2}{g_2L}\right)^2 W(t_0) \\ &\le \gamma W(t_0). \end{aligned}$$

Since the rest of the proof is similar, we abbreviate it.

(3) Suppose $0 \le t_1 \le T \le t_2 < 2T$. We apply the definitions of T and L to the following cases just as in cases 1 and 2 above and in each case get $W(t_0 + 2T) \le \gamma W(t_0)$.

(3A) Suppose $I(t_0 + T) = 1$. Then we separately consider the case where $I(t_0 + t) = 1$, for some $t \in [t_2, t_2 + T]$, and where $I(t_0 + t) = 0, \forall t \in [t_2, t_2 + T]$.

(3B) Suppose $I(t_0 + T) = 0$. Define $t_3 := \max\{t \in [0, T] : I(t_0 + t) = 1\}$.

(1) $(t_3 < 1)$

$$\begin{aligned} &(a)\ (t_2 - T < 1)\ (i)\ (I(t_0 + t) = 1,\ \text{for some } t \in [t_2, t_2 + T]) \\ &\qquad\qquad\qquad\ (ii)\ (I(t_0 + t) = 0, \forall t \in [t_2, t_2 + T]) \\ &(b)\ (t_2 - T \ge 1)\ (i)\ (I(t_0 + t) = 1,\ \text{for some } t \in [t_2, t_2 + T]) \\ &\qquad\qquad\qquad\ (ii)\ (I(t_0 + t) = 0, \forall t \in [t_2, t_2 + T]). \end{aligned}$$

(2) $(t_3 \ge 1)$

$$\begin{aligned} &(a)\ (t_2 - T < 1)\ (i)\ (I(t_0 + t) = 1,\ \text{for some } t \in [t_2, t_2 + T]) \\ &\qquad\qquad\qquad\ (ii)\ (I(t_0 + t) = 0, \forall t \in [t_2, t_2 + T]) \\ &(b)\ (t_2 - T \ge 1)\ (i)\ (I(t_0 + t) = 1,\ \text{for some } t \in [t_2, t_2 + T]) \\ &\qquad\qquad\qquad\ (ii)\ (I(t_0 + t) = 0, \forall t \in [t_2, t_2 + T]). \end{aligned}$$

Here one needs to consider the four cases $(t_3 < T/2)/(t_3 \geq T/2)$ and $(t_2 - T < T/2)/(t_2 - T \geq T/2)$. (Because of symmetry, these reduce to just two).

(4) $(T \leq t_1 \leq t_2 < 2T)$. Again, we apply the definitions of T and L as in cases 1 and 2 to get $W(t_0 + 2T) \leq \gamma W(t_0)$ in each case.
(a) $(I(t_0 + t) = 0, \forall t \in [t_2, t_2 + T])$ (i)$(t_2 - T < 1)$ (ii)$(t_2 - T \geq 1)$
(b) $(I(t_0 + t) = 1$, for some $t \in [t_2, t_2 + T])$.

□

Proof of Theorem 9.1. Lemma 9.4 and the fact that z has a bounded growth rate gives us the desired result since $W(t + 2T) \leq 8K_{W_z} \exp[4(K_z + \epsilon_z)T]W(t)$ whenever $W(\tau) \geq K_{W_z} L^2, \forall \tau \in [t, t + 2T]$. Both K_{W_z} and L are independent of initial conditions. However, T_{large} can depend on initial conditions. This concludes the proof. □

Theorem 9.2 *(i)* $||\psi||, ||\psi'|| \in L_\infty$, *(ii)* $v, v_f, v_f' \in L_\infty$, *(iii)* $y \in L_\infty$, *(iv)* $u \in L_\infty$, *(v)* $e_a \in L_\infty$, *(vi)* $\zeta \in L_\infty$, *and (vii)* $\dot{\tilde{\theta}} \in L_\infty$.

Proof. The result (i) follows from Theorem 5.1(iii). The result (ii) follows from assumption (A4) and Theorem 5.1(i),(ii). The result (iii) follows since $y = \psi^T\theta + v_f' + \xi_{2d_0}$, and $||\theta|| \leq M$. Lemma 5.1 implies that (iv) is true, and since $e_a = -\psi^T\tilde{\theta} + v_f' + \xi_{2d_0}$, and $||\tilde{\theta}|| \leq \widetilde{M}$, (v) holds true. The result (vi) follows since $\zeta = h^T \exp[Q_1 t] * (H^T\dot{\tilde{\theta}})$, and $||H^T\dot{\tilde{\theta}}|| \leq C\alpha e_a$. Finally, result (vii) follows from property (P1) of the parameter estimator. □

10 Performance for a Nominal System

We now show that if the system has no unmodeled dynamics and disturbances, then exact asymptotic tracking is achieved.

Theorem 10.1 *If* $K_v = 0, k_v = 0$, *and* $\mu_m = 0$, *then* $|y(t) - y_m(t)| \to 0$ *as* $t \to \infty$.

Proof. It suffices to show $e_a(t) \to 0$ and $\zeta(t) \to 0$ as $t \to \infty$, since $e_1(t) = e_a(t) - \zeta(t)$. Note that since $K_v = 0, k_v = 0, \mu_m = 0$ we have $v_f'(t) \equiv 0$. So, by Lemma A.2, $\frac{e_a^2}{n} \leq -\frac{\dot{V}}{\alpha} + c\exp[-2d_0 t]$. This implies, since $||\tilde{\theta}|| \leq \widetilde{M}$, that

$$\int_0^\infty \frac{e_a^2(\tau)}{n(\tau)} d\tau \leq \frac{\widetilde{M}^2}{\alpha} + \frac{k}{2d_0}. \tag{35}$$

Boundedness of $n(\cdot)$ then implies $e_a \in L_2$. Recall that $e_a \in L_\infty$. $e_a = -\tilde{\theta}^T\psi + \xi_{2d_0}$ implies $|\dot{e}_a| \leq \|\dot{\tilde{\theta}}\|\,\|\psi\| + \|\tilde{\theta}\|\,\|\dot{\psi}\| + c_0\exp[-2d_0t]$, which implies $\dot{e}_a \in L_\infty$. Therefore, by Barbalat's lemma (Popov [1, p. 211], or Sastry and Bodson [17, p. 19]), $e_a(t) \to 0$ as $t \to \infty$. Now, recall that $\zeta(t) = h^T\exp[Q_1t] * (H(t)\dot{\tilde{\theta}}(t))$, and $\|H\dot{\tilde{\theta}}\| \leq \alpha C|e_a|$, which implies,

$$|\zeta(t)| \leq C\alpha\int_0^t \exp[-a(t-\tau)/2]|e_a(\tau)|d\tau\,.$$

Using the just established fact that $e_a \to 0$, we will show that $\zeta \to 0$. Let $e_{am} := \sup\{|e_a(t)| : t \geq 0\}$. Pick a $\delta > 0$. Then, $\exists$ a $T_1 < \infty$ such that (i) $\frac{C\alpha|e_a(t)|}{a} < \delta/4, \forall t > T_1$, and (ii) $\frac{2C\alpha e_{am}\exp[-aT_1/2]}{a} < \delta/2$. Hence, $\forall t > 2T_1$, we have, $|\zeta(t)| \leq C\alpha(e_{am}\int_0^{t/2}\exp[-a(t-t')/2]dt' + \frac{a\delta}{4}\int_{t/2}^t \exp[-a(t-t')/2]dt') < 2e_{am}\exp[-at/4]/a + \delta/2 \leq 2e_{am}\exp[-aT/2]/a + \delta/2 < \delta$, thus proving that $\zeta(t) \to 0$ as $t \to 0$. □

11 Robust Performance

We now show that the performance of the adaptive tracker, as measured by the mean square tracking error, is robust in that it is a quadratic (hence also continuous) in the magnitude of the unmodeled dynamics and bounded disturbances.

Theorem 11.1

$$\limsup_{T_2\to\infty}\frac{1}{T_2}\int_0^{T_2}(y(t) - y_m(t))^2dt \leq c(K_v^2 + k_v^2 + \mu_m^2)\,. \tag{36}$$

Proof. From Lemma A.2, we have $\frac{e_a^2 - v_f'^2}{n} \leq -\frac{\dot{V}}{\alpha} + c_0\exp[-2d_0t]$, which implies

$$\frac{1}{T_2}\int_{T_\ell}^{T_2}\frac{e_a^2 - v_f'^2}{n} \leq \frac{k\exp[-2d_0T_\ell]}{T_2},\ \forall T_2 \geq T_\ell\,.$$

Hence,

$$\limsup_{T_2\to\infty}\frac{1}{T_2}\int_{T_\ell}^{T_2}\frac{e_a^2 - v_f'^2}{n} \leq 0$$

$$\Rightarrow \limsup_{T_2\to\infty}\frac{1}{T_2}\int_{T_\ell}^{T_2}\frac{e_a^2}{n} \leq \limsup_{T_2\to\infty}\frac{1}{T_2}\int_{T_\ell}^{T_2}\frac{v_f'^2}{n}\,.$$

Noting that $z(t) \geq \min\{z(0), K_2/g_2\} =: z_{\min}$,
and defining $z_{\max} := \sup_{t\geq T_\ell} z(t)$, by (16), we get,

$$\frac{v_f'^2}{z^2} \leq 2(K_{vf}'K_{mz})^2 + 2\frac{(K_{vf}'k_{mz} + k_{vf} + k(T_\ell))^2}{z^2}\,,$$

which then gives

$$\frac{1}{T_2}\int_{T_\ell}^{T_2}\frac{v_{\mathrm{f}}'^2}{n} \le \left[2(K_{v\mathrm{f}}'K_{mz})^2 + 2\frac{(K_{v\mathrm{f}}'k_{mz}+k_{v\mathrm{f}}+k(T_\ell))^2}{z_{\min}^2}\right]\frac{z_{\max}^2}{\nu}\,,\ \forall T_2 \ge T_\ell\,.$$

This implies

$$\limsup_{T_2\to\infty}\frac{1}{T_2}\int_{T_\ell}^{T_2}\frac{e_{\mathrm{a}}^2}{n} \le c(K_v^2+k_v^2+\mu_{\mathrm{m}}^2+k^2(T_\ell))\,,$$

which then gives

$$\limsup_{T_2\to\infty}\frac{1}{T_2}\int_{T_\ell}^{T_2} e_{\mathrm{a}}^2 \le c(K_v^2+k_v^2+\mu_{\mathrm{m}}^2+k^2(T_\ell))\,,$$

where we recall that $k(T_\ell)$ depends only on initial conditions and decays exponentially at a rate faster than $\exp[-2d_0T_\ell]$ as T_ℓ increases. Finally, using the fact that e_{a} is bounded and the fact that the expression above is true for all $T_\ell > 0$, we get $\limsup_{T_2\to\infty}\frac{1}{T_2}\int_0^{T_2} e_{\mathrm{a}}^2 \le c(K_v^2+k_v^2+\mu_{\mathrm{m}}^2)$. The proportionality constant c is basically $z_{\max}n_{\max}$ and is therefore $O(z_{\max}^2)$. From the Robust Ultimate Boundedness theorem, we know that it decreases as the unmodeled dynamics and bounded disturbance decrease. This means that the right-hand side in the expression above indeed goes to zero as the unmodeled dynamics and bounded disturbances go to zero, thereby giving us robust performance. Now consider the swapping term. Recall from the proof of Theorem 6.1 that,

$$\begin{aligned}\int_0^t \zeta^2(t')dt' &\le (\alpha C)^2\int_0^t\left(\int_0^{t'}\exp[-a(t'-\tau)/2]|e_{\mathrm{a}}(\tau)|d\tau\right)^2 dt'\\ &\le \frac{4}{a}(\alpha C)^2\int_0^t\int_0^{t'}\exp[-a(t'-\tau)/2]e_{\mathrm{a}}^2(\tau)d\tau dt'\\ &\le \frac{4}{a}(\alpha C)^2\int_0^t\exp[a\tau/2]e_{\mathrm{a}}^2(\tau)\int_\tau^t\exp[-at'/2]dt'd\tau\\ &\le \frac{8}{a^2}(\alpha C)^2\int_0^t e_{\mathrm{a}}^2(\tau)d\tau\,,\end{aligned}$$

where the second inequality follows from Cauchy-Schwartz. This implies that $\limsup_{T_2\to\infty}\frac{1}{T_2}\int_0^{T_2}\zeta^2(t')dt' \le \frac{8}{a^2}(\alpha C)^2(K_v^2+k_v^2+\mu_{\mathrm{m}}^2)c$. Finally, recalling that $e_1(t) = e_{\mathrm{a}}(t) - \zeta(t)$ yields the desired result. □

12 Simulation Examples

We now present two simulation examples to illustrate the results. The first example is the same as the one considered in [12], except that we *do not* assume knowledge of the (nominal) plant gain, and that we also add a bounded disturbance to the output. The actual plant is unstable and has a fast nonminimum-phase zero. This plant is modeled as a nominal plant which is minimum-phase and unstable, with a multiplicative uncertainty.

Example 12.1 The true system is given by

$$y(t) = \frac{(1-\mu s)}{s(s-1)}u(t) + w(t)\,, \tag{37}$$

where $0 < \mu < 1$ and $w(t)$ is a bounded disturbance. The reference model to be matched is

$$y_{\mathrm{m}}(t) = \frac{1}{(s+1)(s+2)}r(t)\,, \tag{38}$$

where $r(t) = 10\sin(0.5t)$. We consider a nominal model of the form $y(t) = \dfrac{1}{s^2 + a_1 s + a_0}u(t)$, which is parameterized using $\lambda_1(s) = (s+1), \lambda_2(s) = (s+1)^2$ (so that $\lambda(s) = (s+1)^3$), which results in $F(s) = (s+4)$, $G(s) = (7s+1)$, and $\theta = (-6, 7, 3, 1)^{\mathrm{T}}$. The simulation results with $\hat{\theta}(0) = (-4, 3, 6, 4)^{\mathrm{T}}$, $\mu = 0.02$, $w(t) =$ unit square wave with period 10, parameter estimator constants $\alpha = 4.0, \nu = 1$ and $b_{\min} = 0.5$, and constants $d_0 = 0.7, d_1 = 1.0, m(0) = 2.0$ for the overbounding signal $m(t)$ are given in Figs. 1–3.

Despite the presence of unmodeled dynamics and a bounded output disturbance, the plant output approximately tracks the given reference model output over the time interval considered. □

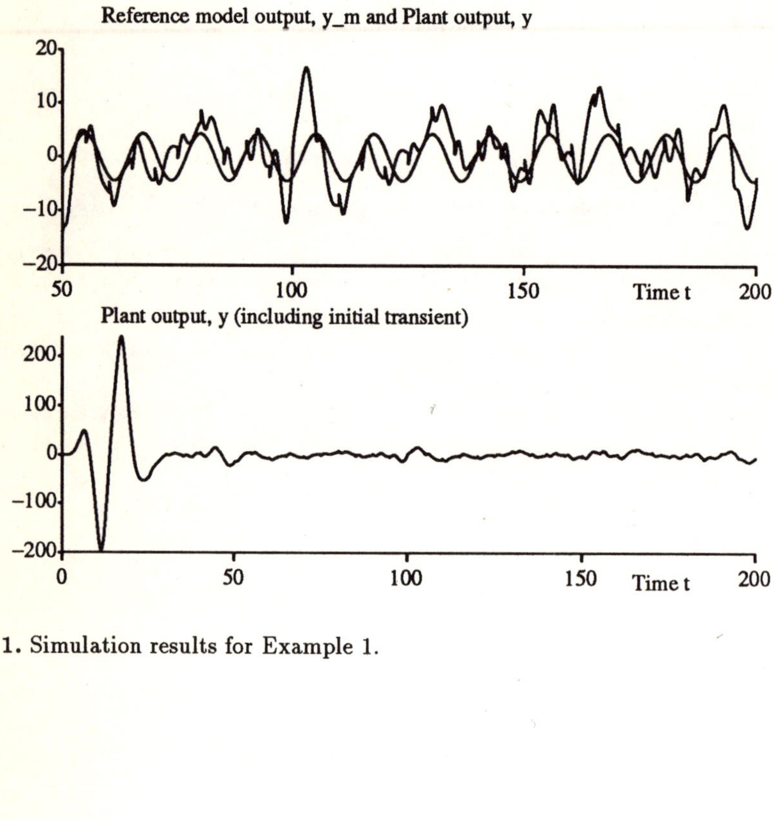

Fig. 1. Simulation results for Example 1.

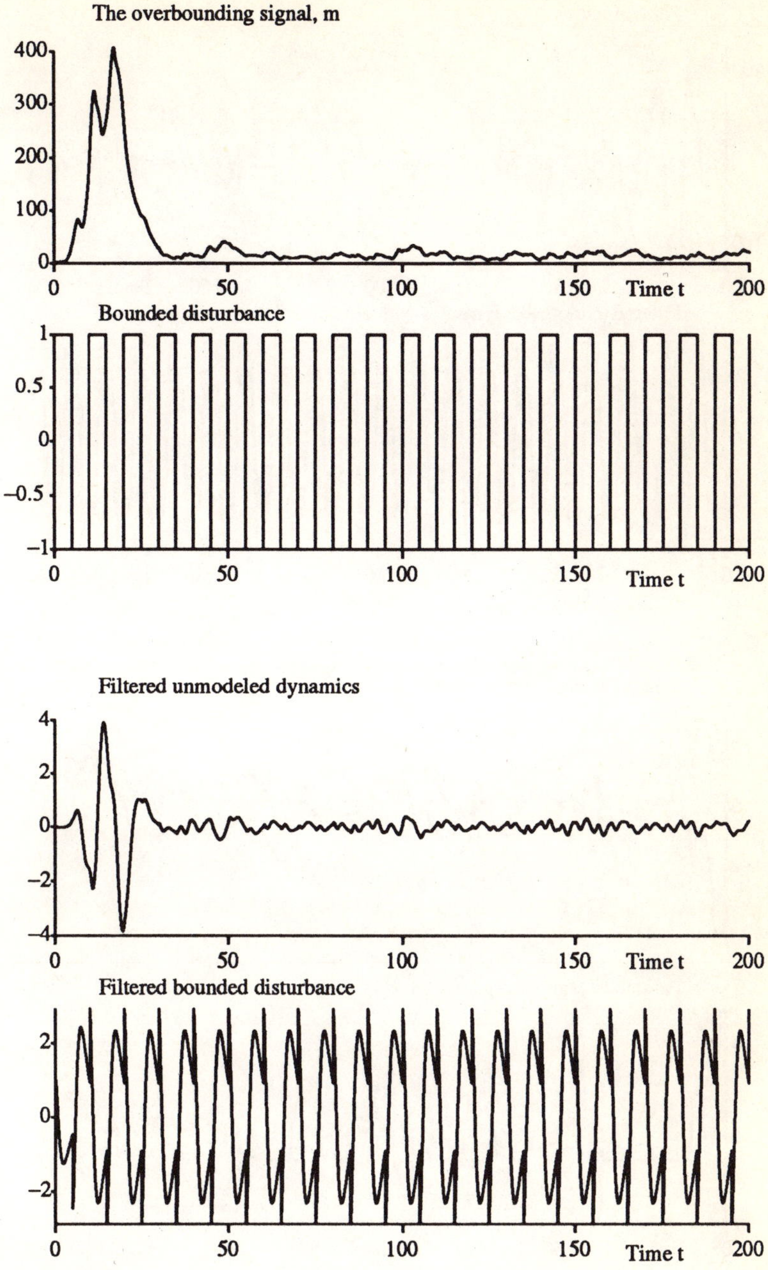

Fig. 2. Simulation results for Example 1.

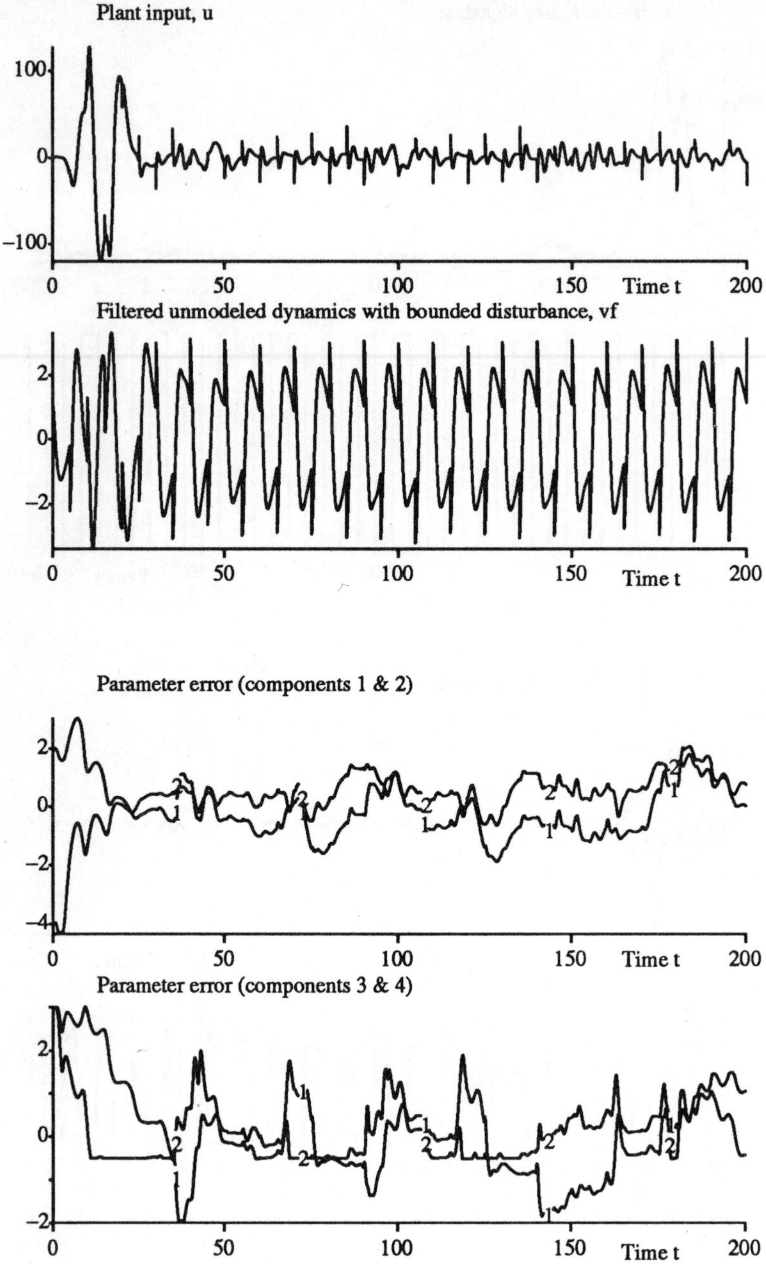

Fig. 3. Simulation results for Example 1.

The next simulation example illustrates a wider class of unmodeled dynamics allowed in our algorithm. Specifically, the true system is a nonlinear, time-varying plant which has a linear time-invariant (LTI) part that is strictly proper and unstable. This LTI part is considered to be the nominal plant, and the nonlinear, time-varying term is treated as unmodeled dynamics. It also includes a constant disturbance term and an exponentially decaying term to allow for initial conditions.

Example 12.2 The true system is given by

$$y(t) = \frac{1}{(s-1)}u(t) + \mu\left[\left|k_y(t)\cdot\frac{1}{(s+2)}y(t)\cdot k_u(t)\cdot\frac{1}{(s+3)}u(t)\right|\right]^{1/2}$$
$$+2+3\exp[-t/2], \tag{39}$$

where $k_y(t) = \sin(t/20), k_u(t) =$ a unit square wave of period 10, and $\mu = 0.05$. The reference model to be matched is

$$y_{\mathrm{m}}(t) = \frac{1}{(s+2)}r(t), \tag{40}$$

where $r(t) = 10\sin(0.5t)$. The nominal model considered is of the form $y(t) = \dfrac{b}{s+a}u(t)$, which is parametrized using $\lambda_1(s) = 1, \lambda_2(s) = (s+1)$ (so that $\lambda(s) = (s+1)$), which results in $F(s) = 1$, $G(s) = 2$, and $\theta = (2,1)^{\mathrm{T}}$.The simulation results with $\hat{\theta}(0) = (-1,2)^{\mathrm{T}}$, and the parameter estimator constants $\alpha = 4.0, \nu = 1$ and, $b_{\min} = 0.2$ are given in Figs. 4 and 5. For the overbounding signal $m(t)$, $d_0 = 0.7$, $d_1 = 1.0$ and $m(0) = 2.0$.

Again, approximate tracking is achieved. The main cause of the lack of tracking accuracy appearing to be due to the constant disturbance of magnitude 2, as is clear by looking at Fig. 6, which displays simulation results with the same $v(t)$ as above, but with $\mu = 0$.

Finally, if both the constant disturbance and unmodeled dynamics are set to zero and $v(t) = 3\exp[-t/2]$, the adaptive control algorithm does indeed result in asymptotically perfect tracking. This is clear by looking at Fig. 7. □

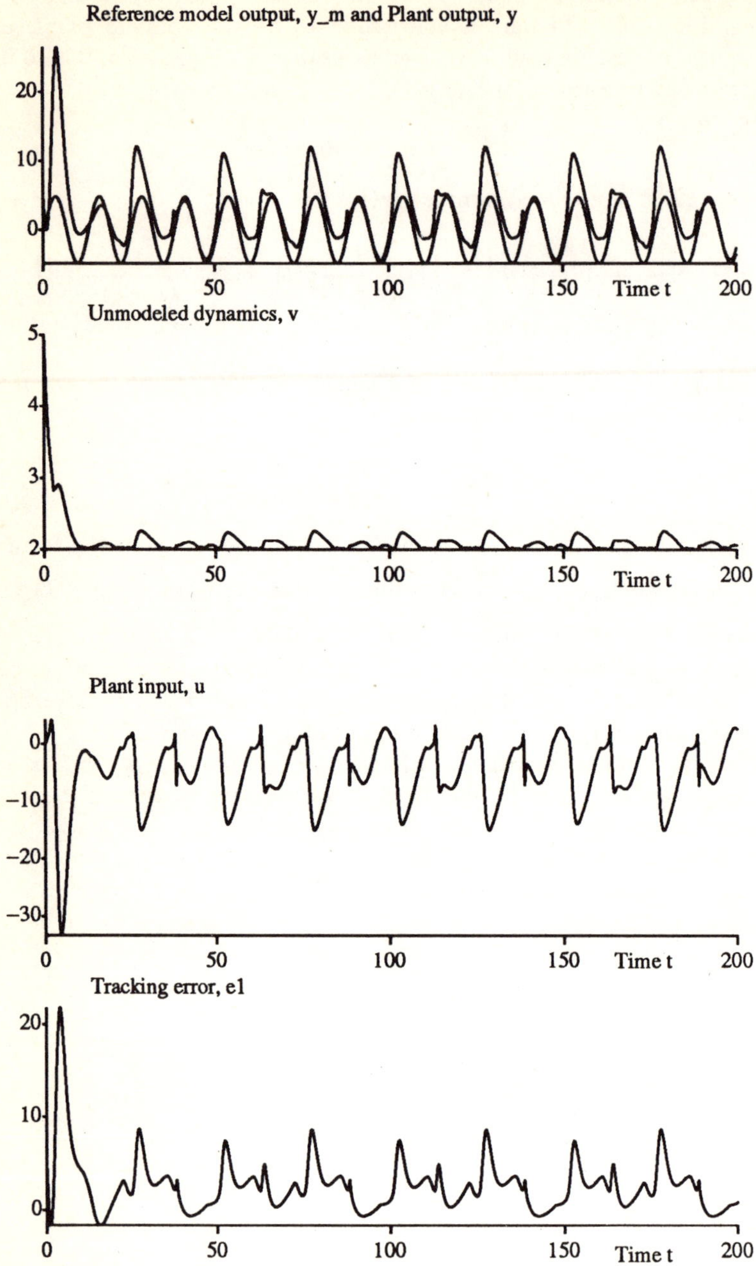

Fig. 4. Simulation results for Example 2.

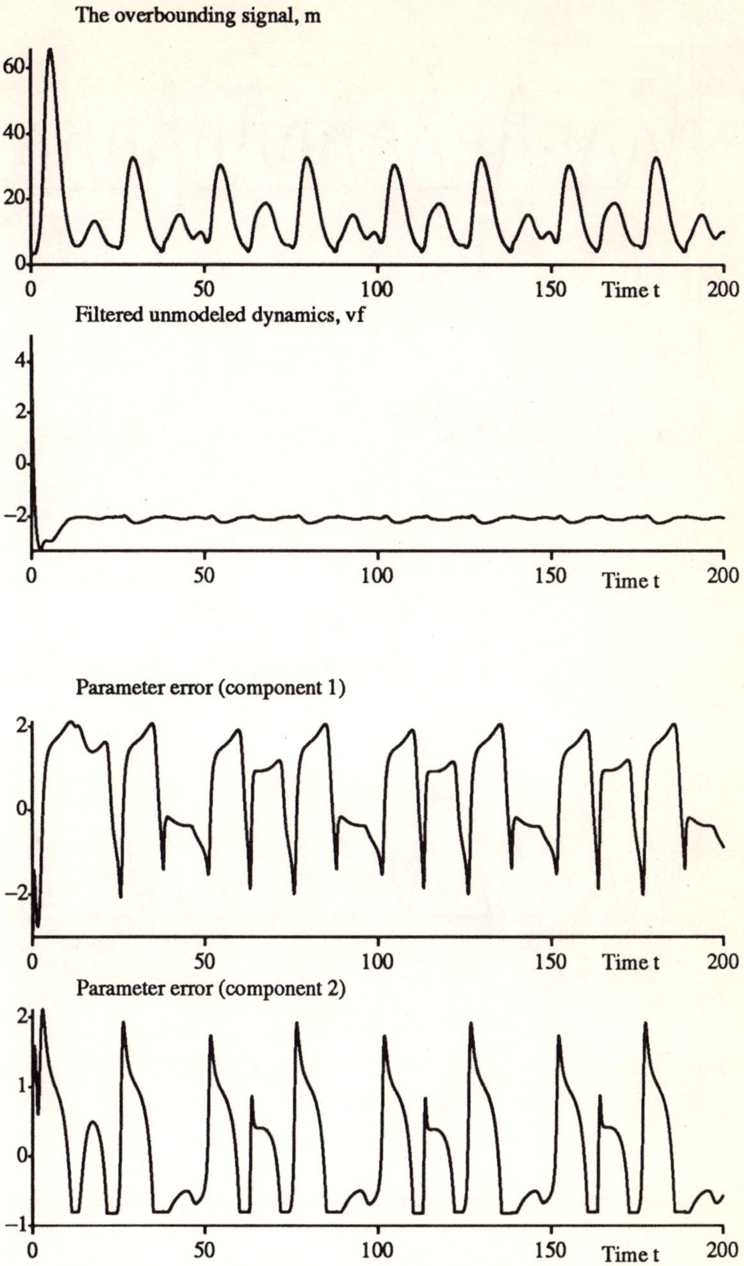

Fig. 5. Simulation results for Example 2.

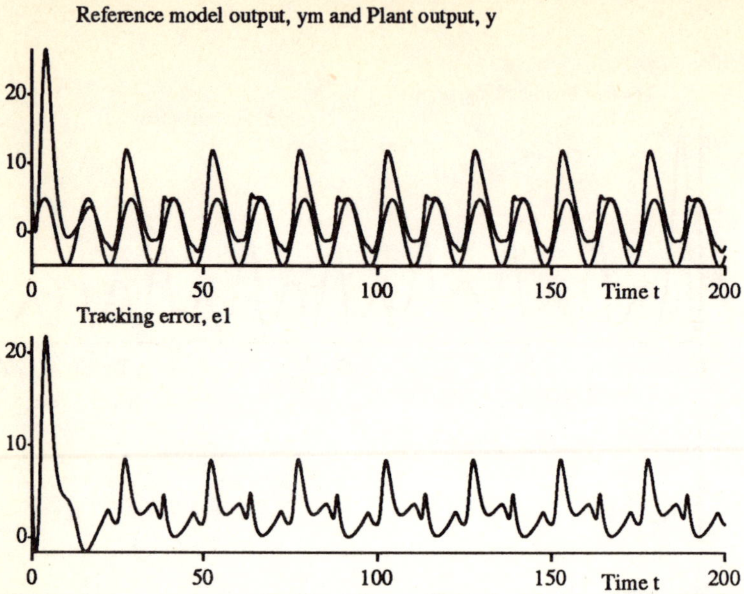

Fig. 6. Simulation results for Example 2 with $v(t) = 2 + 3\exp[-t/2]$.

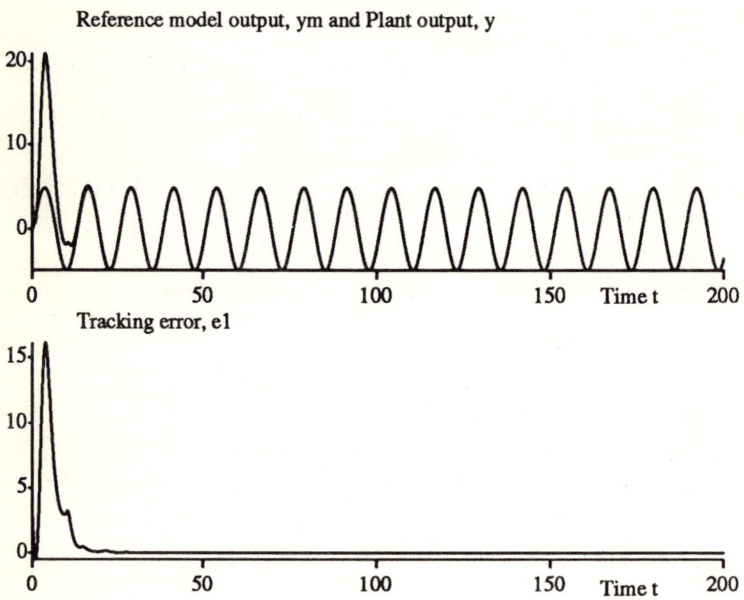

Fig. 7. Simulation results for Example 2 with $v(t) = 3\exp[-t/2]$.

13 Concluding Remarks

In this paper, we have obtained boundedness and performance for continuous-time plants of *arbitrary* relative degree, with a somewhat wider class of unmodeled dynamics than in [12], but without any extra modifications except projection. Unlike [14], we allow non-differentiable bounded disturbances, and non-differentiable reference inputs. We also allow some time-varying and nonlinear uncertainties. The nominal plant is however restricted to be minimum-phase.

We have shown that eventually all the signals enter a compact set, the size of which is independent of initial conditions. Also, the upper-bounds on the size of allowable unmodeled dynamics are independent of initial conditions.

Our results thus show that the projection mechanism alone is sufficient to guarantee robust boundedness and robust performance, at least with respect to small unmodeled dynamics and bounded disturbance. It is important to study the dependence of the bounds on the parameters of the nominal plant, the constants defining the unmodeled dynamics, initial conditions, etc. Also, it is important to reevaluate the various robustness modifications which have earlier been proposed, to examine the amount of robustness they provide, the performance guaranteed in the presence of unmodeled dynamics and disturbances, and to thus determine whether they actually provide some improvements with respect to employing just the projection mechanism.

Acknowledgements.

We thank Petros Ioannou, Jean-Baptiste Pomet, and Laurent Praly for helpful discussions.

References

1. V. M. Popov, *Hyperstability of Control Systems*, Springer-Verlag, Berlin, 1973.
2. C. A. Desoer and M. Vidyasagar, *Feedback Systems: Input-Output Properties*, Academic Press, New York, 1975.
3. K. S. Narendra and L. S. Valavani, "Stable adaptive controller design - Direct control", *IEEE Trans. Aut. Control*, vol. 23, pp. 570–583, 1978.
4. A. Feuer and A. S. Morse, "Adaptive control of SISO linear systems", *IEEE Trans. Aut. Control*, vol. 23, pp. 557–569, 1978.
5. B. Egardt, *Stability of adaptive controllers, Lecture Notes in Control and Info. Sciences*, vol. 20, Springer-Verlag, Berlin, 1979.
6. A. S. Morse, "Global stability of parameter-adaptive control systems", *IEEE Trans. Aut. Control*, vol. 25, pp. 433–439, 1980.
7. G. Kreisselmeier and K. S. Narendra, "Stable MRAC in the presence of bounded disturbances", *IEEE Trans. Aut. Control*, vol. 27, pp. 1169–1175, 1982.
8. L. Praly, "Robustness of model reference adaptive control", K. S. Narendra, ed., *Proc. 3rd Yale Workshop on Adaptive Systems*, pp. 224–226, 1983.
9. L. Praly, "Robust model reference adaptive controllers, part I: stability analysis", *Proc. 23rd IEEE Conf. Dec. Control*, Dec. 1984.
10. L. Praly, "Global stability of a direct adaptive control scheme which is robust w.r.t. a graph topology", *Adaptive and Learning Systems: Theory and Applications*, K. S. Narendra, ed., Plenum Press, New York, 1986.

11. G. Kreisselmeier and B. D.O. Anderson, "Robust model reference adaptive control", *IEEE Trans. Aut. Control*, vol. 31, pp. 127–133, 1986.
12. P. A. Ioannou and K. S. Tsakalis, "A robust direct adaptive controller", *IEEE Trans. Aut. Control*, vol. 31, pp. 1033–1043, 1986.
13. G. C. Goodwin and D. Q. Mayne, "A parameter estimation perspective of continuous MRAC", *Automatica*, vol. 23, pp. 57–70, 1987.
14. P. A. Ioannou and J. Sun, "Theory and design of robust direct and indirect adaptive-control schemes", *Int. J. Control*, vol. 47, pp. 775–813, 1988.
15. J.-B. Pomet and L. Praly, "Adaptive nonlinear regulation: equation error from the Lyapunov equation", *Proc. 28th IEEE Conf. Dec. Control*, pp. 1008–1013, Tampa, FL, 1989.
16. L. Praly, S.-F. Lin and P. R. Kumar, "A robust adaptive minimum variance controller", *SIAM J. Control Optimiz.*, vol.27, pp. 235–266, 1989.
17. S. Sastry and M. Bodson, *Adaptive Control : Stability, Convergence, and Robustness*, Prentice-Hall, Englewood Cliffs, NJ, 1989.
18. G. Tao and P. A. Ioannou, "Robust stability and performance improvement of discrete-time multivariable adaptive control systems", *Int. J. Control*, vol. 50, pp. 1835–1855, 1989.
19. B. E. Ydstie, "Stability of discrete MRAC—revisited", *Syst. Control Lett.*, vol. 13, pp. 429–438, 1989.
20. S. M. Naik and P. R. Kumar, "A robust adaptive controller for continuous-time systems", submitted to the *1991 Amer. Control Conf.*, 1991.

Appendix A

Lemma A.1 *Consider the system:* $\dot{x} = A_w x + b_w w_{in}, w_{out} = h_w^T x$ *with zero initial conditions.* (A_w, b_w, h_w^T) *is a minimal representation of* $H_0(s) = \frac{B'(s)}{A(s)}$, *where* $H_0(s)$ *is strictly proper and of relative degree one, and* $B'(s - p_0)$ *is Hurwitz.*

If $|w_{in}| \leq K_i m + k_i + c_0 \exp[-pt]$, *and* $|w_{out}| \leq K_o m + k_o + c_0 \exp[-pt]$, *then,* $\|x\| \leq K_x m + k_x + c_0 \exp[-pt]$ *for some positive constants* K_x *and* k_x.

Proof. Without loss of generality, suppose

$$H_0(s) = \frac{B'(s)}{\prod_{i=1}^{k}(s^2 + a_{i1}s + a_{i2}) \prod_{j=1}^{n-2k}(s + a_j)} .$$

Since $H_0(s)$ is minimal, the corresponding states are the states corresponding to $\dfrac{w_{in}}{s^2 + a_{i1}s + a_{i2}}, i = 1, \ldots, k$ and $\dfrac{w_{in}}{s + a_j}, j = 1, \ldots, n - 2k$. Now,

$$\frac{1}{s + a_l} w_{in}(t) = \frac{\prod_{i=1}^{k}(s^2 + a_{i1}s + a_{i2}) \prod_{j=1, j \neq l}^{n-2k}(s + a_j)}{B'(s)} w_{out}(t) .$$

Using Lemma 5.2(ii) (since $B'(s - p_0)$ is Hurwitz), we get $|\frac{1}{s+a_l} w_{in}(t)| \leq cm(t) + c + c_0 \exp[-pt], l = 1, \ldots, n - 2k$. Define

$$w_l(t) = \frac{1}{s^2 + a_{l1}s + a_{l2}} w_{in}(t) =: H_l(s) w_{in}(t) .$$

Then, since

$$w_\ell(t) = \frac{\prod_{i=1,i\neq l}^{k}(s^2 + a_{i1}s + a_{i2})\prod_{j=1}^{n-2k}(s + a_j)}{B'(s)} w_{\text{out}}(t) =: H_{\ell\text{out}}(s)w_{\text{out}}(t),$$

using Lemma 5.2(i) (since $B'(s - p_0)$ is Hurwitz), we get $|w_\ell(t)| \leq cm(t) + c + c_0 \exp[-pt], l = 1, \ldots, k$. Further, since $H_{\ell\text{out}}(s)$ is strictly proper, Lemma 5.2(ii) gives $|\dot{w}_\ell(t)| = |sH_{\ell\text{out}}(s)w_{\text{out}}(t)| \leq cm(t)+c+c_0 \exp[-pt]$. Since $w_\ell(t)$ and $\dot{w}_\ell(t)$ are the states corresponding to $H_\ell(s)w_{\text{in}}(t)$, we are done. □

Lemma A.2 *Define* $V(t) := ||\widetilde{\theta}(t)||^2$ *for the parameter estimator with projection. Then, we have*

$$\dot{V}(t) \leq -\alpha\frac{(\widetilde{\theta}^{\text{T}}(t)\psi(t))^2}{n(t)} + \alpha\frac{v_{\text{f}}'^2(t)}{n(t)} + \alpha\frac{c_0\widetilde{M}}{\sqrt{n(t)}}\exp[-2d_0t]. \tag{41}$$

Furthermore, (41) also holds when the term $\widetilde{\theta}^{\text{T}}(t)\psi(t)$ *on the right-hand side is replaced by* $e_{\text{a}}(t)$.

Proof. For notational simplicity, let $c_1(t) := \dfrac{\hat{\theta}(t)}{||\hat{\theta}(t)||}$, and $c_2(t) := \dfrac{\alpha\psi(t)e_{\text{a}}(t)}{n(t)}$. First, consider the parameter estimator when the projection is not used, i.e. $\dot{\hat{\theta}}(t) = c_2(t)$. Then, recalling that $e_{\text{a}}(t) = -\widetilde{\theta}^{\text{T}}(t)\psi(t) + v_{\text{f}}'(t) + \xi_{2d_0}(t)$, we obtain

$$\begin{aligned}
\dot{V}(t) &= 2\widetilde{\theta}(t)^{\text{T}}\frac{\alpha\psi(t)e_{\text{a}}(t)}{n(t)} \\
&= \frac{\alpha}{n(t)}[-2(\widetilde{\theta}^{\text{T}}(t)\psi(t))^2 + 2(\widetilde{\theta}^{\text{T}}(t)\psi(t))v_{\text{f}}'(t) + 2(\widetilde{\theta}^{\text{T}}(t)\psi(t))\xi_{2d_0}(t)] \\
&\leq \frac{\alpha}{n(t)}[-(\widetilde{\theta}^{\text{T}}(t)\psi(t))^2 + v_{\text{f}}'^2(t)] + \alpha\frac{c_0\widetilde{M}}{\sqrt{n(t)}}\exp[-2d_0t].
\end{aligned}$$

Now, consider the parameter estimator with projection. Using property (P2) of the projection, we get

$$\dot{V} = 2\widetilde{\theta}^{\text{T}}\dot{\widetilde{\theta}} = 2\widetilde{\theta}^{\text{T}}Proj(\hat{\theta}, c_2) \leq 2\widetilde{\theta}^{\text{T}}c_2,$$

and so the bound for the estimator without projection still holds. □

Appendix B

In this section, we will assume that (i) $t \geq T_\ell$, and (ii) $W(\cdot) \geq K_{Wz}L^2$, which implies that $z(\cdot) \geq L/2$.

Lemma B.1 (i) $\dfrac{||\psi||}{z}, \dfrac{||\phi_y||}{z}, \dfrac{||\phi_u||}{z}, \dfrac{||\phi||}{z} \leq c.$

(ii) $\dfrac{\|\psi^{(i)}\|}{z} \leq a_i,\ i = 1, \ldots, n-m.$

(iii) $\left|\dfrac{d}{dt}(y - v_f')\right| \leq k_y' z + k(T_\ell)$, *for some* $k_y' > 0$, *and* $k(T_\ell)$ *is a positive constant which decreases exponentially with increasing* T_ℓ.

(iv) $\left|\dfrac{d}{dt}(u - r'/\hat{b}_m + \hat{\theta}_n v_f'/\hat{b}_m)\right| \leq k_u' z + k_{v\mu} + k(T_\ell)$, *for some* $k_u' > 0$, *where* $k(T_\ell)$ *is a positive constant which decreases exponentially with increasing* T_ℓ *and* $k_{v\mu}$ *is a positive constant which is a weighted linear combination of* K_v, k_v *and* μ_m.

Proof. The result (i) follows from Theorems 5.1(iii),(v),(vi), and 6.1(ii), while (ii) follows since $\psi^T = \left(\dfrac{1}{(s+a)^{2n-m-1}} y, \ldots, \dfrac{1}{(s+a)^{n-m}} y, \dfrac{1}{(s+a)^{2n-m-1}} u, \ldots, \dfrac{1}{(s+a)^{n-m}} u\right)$. For (iii) note that $y - v_f' = \psi^T\theta + \xi_{2d_0}$, so that by (ii) above $\left|\dfrac{d}{dt}(y - v_f')\right| \leq a_1 M z + k(T_\ell)$. For (iv) recall that the control law is $u - \dfrac{1}{\hat{b}_m} r' + \dfrac{\hat{\theta}_n}{\hat{b}_m} v_f' = -\dfrac{1}{\hat{b}_m}[\hat{\theta}_u^T \phi_u + \hat{\theta}_y^T \phi_y + \hat{\theta}_n (y - v_f')] =: RHE/\hat{b}_m$. This implies

$$\frac{d}{dt}(u - \frac{1}{\hat{b}_m} r' + \frac{\hat{\theta}_n}{\hat{b}_m} v_f') = -\frac{\dot{\hat{b}}_m}{\hat{b}_m^2}(RHE) - \frac{1}{\hat{b}_m}\left(\dot{\hat{\theta}}_u^T \phi_u + \dot{\hat{\theta}}_y^T \phi_y + \hat{\theta}_u^T \dot{\phi}_u + \hat{\theta}_y^T \dot{\phi}_y + \dot{\hat{\theta}}_n (y - v_f') + \hat{\theta}_n \frac{d}{dt}(y - v_f')\right).$$

Now, $\|\dot{\hat{b}}_m \phi_u\| \leq |\dot{\hat{b}}_m| \|\phi_u\| \leq \|\dot{\hat{\theta}}\| \|\psi'\| \leq \dfrac{\alpha \|\psi'\|^2 |e_a|}{n} \leq \alpha |e_a| \leq cz + k_{v\mu} + k(T_\ell)$. Similar analysis using $\|\hat{\theta}\| \leq M$ and $b_m \geq b_{\min} > 0$, gives $\left|\dfrac{\dot{\hat{b}}_m}{\hat{b}_m^2}(RHE)\right| \leq cz + k_{v\mu} + k(T_\ell)$. Also, $|\phi_u^T \dot{\hat{\theta}}_u| \leq cz + k_{v\mu} + k(T_\ell)$, etc. Finally, using these results, the fact that $\hat{\theta}$ is bounded, and (i), (ii), (iii), etc., we get the desired result. □

For future use, define

$$k_{nz}(T) = \frac{1}{K_{nz}}, \text{ if } I(t) = 1;$$
$$= 1/\delta(T), \text{ if } I(t) = 0, \text{ and } I(t + t') = 1 \text{ for some } t' \in (0, T].$$

Lemma B.2 *For instants* t *such that* $I(t + t') = 1$ *for some* $t' \in [0, T]$,

$$\left|\frac{d}{dt}\left(\frac{\tilde{\theta}^T \psi^{(i)}}{z}\right)(t)\right| \leq k_i(T), \quad i = 0, \ldots, n-m-1. \tag{42}$$

Proof. Note that $|\frac{d}{dt}(\frac{\widetilde{\theta}^{\mathrm{T}}\psi^{(i)}}{z})| \leq \frac{\|\dot{\widetilde{\theta}}\|\,\|\psi^{(i)}\|}{z} + \frac{\|\widetilde{\theta}\|\,\|\psi^{(i+1)}\|}{z} + \frac{\|\widetilde{\theta}\|\,\|\psi^{(i)}\|}{z}\frac{|\dot{z}|}{z}$.
Now, the parameter-update law yields

$$\|\dot{\widetilde{\theta}}\| \leq \frac{\alpha\|\psi\|(|\widetilde{\theta}^{\mathrm{T}}\psi| + |v_{\mathrm{f}}'|)}{n}\frac{z^2}{z^2},$$

and

$$\frac{z^2(t)}{n(t)} \leq k_{nz}(T).$$

This implies that $\|\dot{\widetilde{\theta}}\| \leq k(T)$. Finally, using (16), (22) and Lemma B.1(ii), we get the desired result. □

Lemma B.3 *For instants t such that $I(t+t') = 1$ for some $t' \in [0,T]$,*

$$\left|\frac{d}{dt}\left(\frac{\widetilde{\theta}^{\mathrm{T}}\psi^{(n-m)} - \left(1 - \frac{\theta_{2n}}{\hat{b}_{\mathrm{m}}}\right)r' + \left(\theta_n - \theta_{2n}\frac{\hat{\theta}_n}{\hat{b}_{\mathrm{m}}}\right)v_{\mathrm{f}}'}{z}\right)\right| \leq k_{n-m}(T). \qquad (43)$$

Proof. Since $\phi = (s+a)^{(n-m)}\psi$, we have $\psi^{(n-m)} = \phi - \sum_{i=0}^{n-m-1} c_i\psi^{(i)}$, where c_i are the appropriate constants which depend on a. Using the control law(6), we get,

$$\begin{aligned}
\widetilde{\theta}^{\mathrm{T}}\phi &= \hat{\theta}^{\mathrm{T}}\phi - \theta^{\mathrm{T}}\phi \\
&= r' - \theta^{\mathrm{T}}\phi \\
&= r' - (\theta_u^{\mathrm{T}}\phi_u + \theta_y^{\mathrm{T}}\phi_y) - \theta_n(y - v_{\mathrm{f}}') - \theta_{2n}(u - \frac{1}{\hat{b}_{\mathrm{m}}}r' + \frac{\hat{\theta}_n}{\hat{b}_{\mathrm{m}}}v_{\mathrm{f}}') \\
&\quad -\theta_n v_{\mathrm{f}}' - \frac{\theta_{2n}}{\hat{b}_{\mathrm{m}}}r' + \theta_{2n}\frac{\hat{\theta}_n}{\hat{b}_{\mathrm{m}}}v_{\mathrm{f}}',
\end{aligned}$$

which implies

$$\begin{aligned}
\phi^{\mathrm{T}}\widetilde{\theta} - (1 - \frac{\theta_{2n}}{\hat{b}_{\mathrm{m}}})r' + (\theta_n - \theta_{2n}\frac{\hat{\theta}_n}{\hat{b}_{\mathrm{m}}})v_{\mathrm{f}}' &= -(\theta_u^{\mathrm{T}}\phi_u + \theta_y^{\mathrm{T}}\phi_y) - \theta_n(y - v_{\mathrm{f}}') \\
&\quad -\theta_{2n}(u - \frac{1}{\hat{b}_{\mathrm{m}}}r' + \frac{\hat{\theta}_n}{\hat{b}_{\mathrm{m}}}v_{\mathrm{f}}').
\end{aligned}$$

Applying Lemmas B.1 and B.2 to the above equation yields the desired result, using (22). □

Let $T' \leq T$. Assume that $I(t+t') = 1$ for some $t' \in [T', T]$. Then, from Lemma A.2, we have:

$$\begin{aligned}\frac{1}{T'}\int_t^{t+T'} \frac{(\widetilde{\theta}^{\mathrm{T}}\psi)^2}{z^2}d\tau &\leq \sup \frac{n}{z^2}\left[\frac{V(t)-V(t+T')}{\alpha T'} + \frac{1}{T'}\sup\frac{z^2}{n}\int_t^{t+T'}(\frac{v_{\mathrm{f}}'}{z})^2 d\tau \right. \\ &\quad \left. + \frac{\alpha c_0 \widetilde{M}}{2\nu T' d_0}\exp[-2d_0 t]\right] \\ &\leq \frac{1}{T'}\left(\frac{K_{nn}M^2}{\alpha}\right) + \frac{\alpha c_0\widetilde{M}}{2\nu T' d_0}\exp[-2d_0 T_\ell] \\ &\quad + K_{nn}\left(K_{v\mathrm{f}}' K_{mz} + \frac{K_{v\mathrm{f}}' k_{mz} + k_{v\mathrm{f}} + k(T_\ell)}{L/2}\right)^2 k_{nz}(T),\end{aligned}$$

where the supremum above is taken over the interval $[t, t+T']$. Defining $\mu := K_{v\mathrm{f}}' K_{mz} + \dfrac{K_{v\mathrm{f}}' k_{mz} + k_{v\mathrm{f}} + k(T_\ell)}{L/2}$, the above inequality becomes

$$\begin{aligned}\frac{1}{T'}\int_t^{t+T'} \frac{(\widetilde{\theta}^{\mathrm{T}}\psi)^2}{z^2}d\tau &\leq \frac{1}{T'}\left(\left(\frac{K_{nn}M^2}{\alpha}\right) + \frac{\alpha c_0\widetilde{M}}{2\nu d_0}\exp[-2d_0 T_\ell]\right) \\ &\quad + \mu^2 K_{nn} k_{nz}(T).\end{aligned}$$

Subsequently we will apply Lemma B.4 to this inequality. The important point to note is that μ can be made as small (but positive) as we please by making $K_{v\mathrm{f}}'$ small enough and L large enough, and $K_{v\mathrm{f}}'$ can in turn be made as small as desired by making K_v and μ_{m} appropriately small.

Remark. The necessity of the constraint $1 \leq T'$ in Lemma B.4 will become clear in the proofs which follow. □

Lemma B.4 *(i) If* $1 \leq T' \leq T$, *and*

$$\frac{1}{T'}\int_t^{t+T'}\left(\frac{|\widetilde{\theta}^{\mathrm{T}}\psi|}{z}\right)^2 \leq \frac{c_1}{T'} + \mu^2 c_2(T), \tag{44}$$

then,

$$\begin{aligned}\int_t^{t+T'} \frac{|\widetilde{\theta}^{\mathrm{T}}\phi|}{z} &\leq \frac{\nu_3}{\epsilon^2} + \left[\mu^2\frac{\nu_4(T)}{\epsilon^2} + \mu\frac{\nu_5(T)}{\sqrt{\epsilon}}\right. \\ &\quad \left. + \nu_6(T)\epsilon^{\rho} + 4a^{n-m}k(\mu, L)\right] T'\end{aligned} \tag{45}$$

where $\epsilon \in (0, 1]$, $\rho = 2^{-(n-m+1)}$, *and*

$$k(\mu, L) = \frac{\left(\frac{|\theta_{2n}|}{b_{\min}} + 1\right)k_r}{L/2} + \left(|\theta_n| + \frac{|\theta_{2n}|}{b_{\min}}M\right)\mu.$$

(ii) If $1 \le T' \le T$, *and*

$$\frac{1}{T'}\int_t^{t+T'}\left(\frac{|\widetilde{\theta}^{\mathrm{T}}\psi|}{z}\right)^2 \le \frac{c_1}{T} + \mu^2 c_2(T)\,, \tag{46}$$

then

$$\int_t^{t+T'} \frac{|\widetilde{\theta}^{\mathrm{T}}\psi|}{z} \le \frac{\nu_7}{\epsilon^2} + \left[\mu^2\frac{\nu_8(T)}{\epsilon^2} + \sqrt{\epsilon}\right]T'\,,$$

where $\epsilon \in (0,1]$.

Proof. The result (i) is an extension of the proof of Theorem C1 of [12], which uses Lemmas B.1, B.2, B.3, and B.5, B.6, B.7 below. The result (ii) is a special case of the proof of (i).
The complete proof of (i) is presented following the proof of Lemma B.7. □

Lemma B.5 *Consider t such that* $I(t+t') = 1$ *for some* $t' \in [0,T]$. *If* $\left|\frac{\psi^{\mathrm{T}}\widetilde{\theta}}{z}\right| \le \sqrt{\epsilon}$ *for any* $\epsilon \in [0,1]$, *then*

$$\|\dot{\widetilde{\theta}}\| \le \alpha\sqrt{k_{nz}(T)}(\sqrt{\epsilon}+\mu)\,.$$

Proof.

$$\begin{aligned}
\|\dot{\widetilde{\theta}}\| &\le \alpha\frac{\|\psi\||e_{\mathrm{a}}|}{n} \le \alpha\frac{|e_{\mathrm{a}}|}{\sqrt{n}} \le \alpha\frac{|\psi^{\mathrm{T}}\widetilde{\theta}|}{\sqrt{n}} + \alpha\frac{(v_{\mathrm{f}}' + k(T_\ell))}{\sqrt{n}}\\
&\le \alpha\sqrt{k_{nz}(T)}\left(\frac{|\psi^{\mathrm{T}}\widetilde{\theta}|}{z} + \frac{(v_{\mathrm{f}}' + k(T_\ell))}{z}\right)\\
&\le \alpha\sqrt{k_{nz}(T)}\left(\sqrt{\epsilon} + K_{v\mathrm{f}}'K_{mz} + \frac{2(K_{v\mathrm{f}}'k_{mz} + k_{v\mathrm{f}} + k(T_\ell))}{L}\right),
\end{aligned}$$

which is (letting μ be as in Table D) equivalent to

$$\|\dot{\widetilde{\theta}}\| \le \alpha\sqrt{k_{nz}(T)}\,(\sqrt{\epsilon}+\mu)\,.$$

□

Lemma B.6 *If* $\frac{|\psi^{\mathrm{T}}\widetilde{\theta}|}{z} \le \sqrt{\epsilon}$, $\forall t \in [t_j, t_j + \Delta t]$, $\forall \epsilon \in (0,1]$, *where* $t_j \ge T_\ell$ *and* $\Delta t \ge 1$, *then* $\exists$ *constants* $\lambda_i(T) > 0$ *and continuous functions* $f_i(\epsilon,T)$ *with* $0 < f_i(\epsilon,T) \le O\left(\frac{1}{\sqrt{\epsilon}}\right)$ *such that*

$$\frac{|\widetilde{\theta}^{\mathrm{T}}\psi^{(i)}|}{z} \le \lambda_i(T)\epsilon^{\alpha_i} + \mu f_i(\epsilon,T)\,,\ \forall t \in [t_j, t_j + \Delta t]\,, \tag{47}$$

where

$$\alpha_i = 2^{-(i+1)}\,,\ i = 1, 2, \ldots, n-m-1\,,$$

and

$$\frac{|\widetilde{\theta}^{\mathrm{T}}\psi^{(n-m)}|}{z} \leq \lambda_{n-m}(T)\epsilon^{\alpha_{n-m}} + \mu f_{n-m}(\epsilon, T) + k(\mu, L)\,. \tag{48}$$

Proof (by induction). Consider an instant $t \in [t_j, t_j + \Delta t]$, and $\delta > 0$ such that $t + \delta \in [t_j, t_j + \Delta t]$ (see Fig. 8). Then,

$$\int_t^{t+\delta} \frac{\widetilde{\theta}^{\mathrm{T}}\psi^{(i)}}{z} d\tau = \frac{\widetilde{\theta}^{\mathrm{T}}\psi^{(i-1)}}{z}\bigg|_t^{t+\delta} - \int_t^{t+\delta} \frac{\dot{\widetilde{\theta}}\psi^{(i-1)}}{z} d\tau + \int_t^{t+\delta} \frac{\widetilde{\theta}^{\mathrm{T}}\psi^{(i-1)}}{z}\frac{\dot{z}}{z} d\tau, \quad i = 1, \ldots, n-m\,. \tag{49}$$

Assume (47) holds for $(i-1)$. Next note that $z(\cdot) \geq L/2$ implies $-2K_z z \leq -g_2 z + K_2 \leq \dot{z} \leq K_z z + k_z + k(T_\ell) \leq 2K_z z$,, which further implies that $|\dot{z}/z| \leq 2K_z$. Using this, the fact that $\|\dot{\widetilde{\theta}}\| \leq k_T(\sqrt{\epsilon} + \mu)$(where $k_T := \alpha\sqrt{k_{nz}(T)}$), and Lemma B.1, we have from (49)

$$\left|\int_t^{t+\delta} \frac{\widetilde{\theta}^{\mathrm{T}}\psi^{(i)}}{z}\right| \leq (2 + 2K_z\delta)\left(\lambda_{i-1}(T)\epsilon^{\alpha_{i-1}} + \mu f_{i-1}(\epsilon, T)\right) + \delta k_T(\sqrt{\epsilon} + \mu) a_{i-1} \tag{50}$$

for $i = 1, \ldots, n-m$, with $f_0(\epsilon, T) = 0$, $\lambda_0 = 1$, $a_0 = c$, and $\alpha_0 = \frac{1}{2}$.

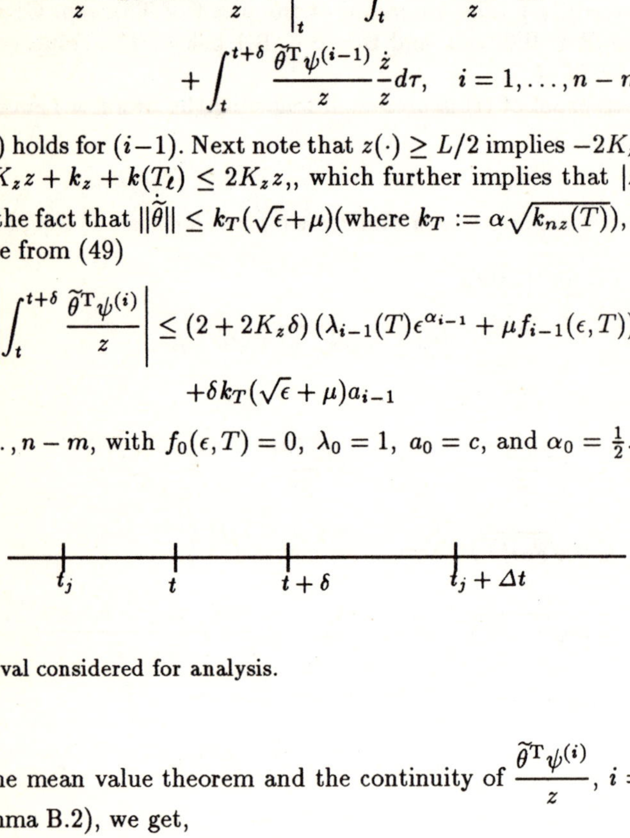

Fig. 8. Interval considered for analysis.

Using the mean value theorem and the continuity of $\dfrac{\widetilde{\theta}^{\mathrm{T}}\psi^{(i)}}{z}$, $i = 1, \ldots, n-m-1$ (Lemma B.2), we get,

$$\int_t^{t+\delta} \frac{\widetilde{\theta}^{\mathrm{T}}\psi^{(i)}}{z} d\tau = \delta \frac{\widetilde{\theta}^{\mathrm{T}}(\tau)\psi^{(i)}(\tau)}{z(\tau)}\,|_{\tau = t+\delta^*}\ , \ \text{some } \delta^* \in [0, \delta]\,. \tag{51}$$

Substituting (51) into (50) and taking $\delta = \epsilon^{\alpha_i}$, we have

$$\frac{|\widetilde{\theta}^{\mathrm{T}}(\tau)\psi^{(i)}(\tau)|}{z(\tau)}\bigg|_{\tau = t+\delta^*} \leq 2(1 + K_z\delta)\left(\lambda_{i-1}(T)\epsilon^{\alpha_i} + \mu\epsilon^{-\alpha_i} f_{i-1}(\epsilon, T)\right) + a_{i-1}k_T(\sqrt{\epsilon} + \mu),\ i = 1, \ldots, n-m-1\,. \tag{52}$$

This follows because $\alpha_i = \frac{1}{2}\alpha_{i-1}$, so that $\epsilon^{\alpha_i} = \epsilon^{\alpha_{i-1}-\alpha_i}$. Since $\epsilon^{\frac{1}{2}} \leq \epsilon^{\alpha_i} \leq 1$, $i = 1, \ldots, n-m-1$, (52) may be written as

$$\left.\frac{|\widetilde{\theta}^{\mathrm{T}}(\tau)\psi^{(i)}(\tau)|}{z(\tau)}\right|_{\tau=t+\delta^*} \leq \overline{\lambda}_i(T)\epsilon^{\alpha_i} + \mu f_i(\epsilon, T)\,, \tag{53}$$

where

$$\begin{aligned} \overline{\lambda}_i(T) &= 2(1+K_z)\lambda_{i-1}(T) + a_{i-1}k_T\,, \\ f_i(\epsilon,T) &= a_{i-1}k_T + 2(1+K_z\delta)\epsilon^{-\alpha_i} f_{i-1}(\epsilon,T)\,, \end{aligned} \quad i = 1, \ldots, n-m-1\,. \tag{54}$$

with $\lambda_0 = 1$, $f_0(\epsilon, T) = 0$.

From Lemma B.2, $\widetilde{\theta}^{\mathrm{T}}\psi^{(i)}/z$ is a uniformly continuous function of time; therefore, its value in the interval $[t, t+\delta]$ will not differ from its value at $t + \delta^*$ by more then $k_i(T)\delta = k_i(T)\epsilon^{\alpha_i}$. Setting $\lambda_i(T) = \overline{\lambda}_i(T) + k_i(T)$, it follows from (53) that

$$\frac{|\widetilde{\theta}^{\mathrm{T}}(\tau)\psi^{(i)}(\tau)|}{z(\tau)} \leq \lambda_i(T)\epsilon^{\alpha_i} + \mu f_i(\epsilon, T), \quad \forall \tau \in [t, t+\delta]\,. \tag{55}$$

Since $\epsilon \in (0, 1]$, it follows from (54) that $f_i(\epsilon, T) \leq O\left(\frac{1}{\sqrt{\epsilon}}\right)$, $i = 1, 2, \ldots, n - m - 1$. Consider now a set of intervals $[t_k, t_k + \delta]$, $k = 0, 1, \ldots, l$ where $l \geq \Delta t/\delta$ such that $t_0 = t_j$, $t_k < t_{k+1} \leq t_k + \delta$ and $t_l + \delta = t_j + \Delta t$. The fact that $\cup_{k=0}^{l}[t_k, t_k + \delta] = [t_j, t_j + \Delta t]$ together with (55) imply (47), and therefore the proof is complete for $i = 1, \ldots, n - m - 1$.

Now, let $i = n - m$. From (50) we have,

$$\left|\int_t^{t+\delta} \frac{\widetilde{\theta}^{\mathrm{T}}\psi^{(n-m)}}{z} d\tau\right| \leq (2 + 2K_z\delta)(\lambda_{n-m-1}(T)\epsilon^{\alpha_{n-m-1}} + \mu f_{n-m-1}(\epsilon, T))$$
$$+\delta k_T(\sqrt{\epsilon} + \mu) \cdot a_{n-m-1}\,.$$

By Lemma B.3,

$$\zeta := \frac{\widetilde{\theta}^{\mathrm{T}}\psi^{(n-m)} - \left(1 - \dfrac{\theta_{2n}}{\widehat{b}_m}\right) r' + \left(\theta_n - \theta_{2n}\dfrac{\widehat{\theta}_n}{\widehat{b}_m}\right) v_{\mathrm{f}}'}{z}$$

is continuous.

Now,

$$\left|\int_t^{t+\delta} \frac{\widetilde{\theta}^{\mathrm{T}}\psi^{(n-m)}}{z} d\tau\right|$$

$$= \left|\int_t^{t+\delta} \left[\zeta + \frac{\left(1 - \frac{\theta_{2n}}{\widehat{b}_m}\right) r' - \left(\theta_n - \theta_{2n}\frac{\widehat{\theta}_n}{\widehat{b}_m}\right) v_{\mathrm{f}}'}{z}\right] d\tau\right|$$

$$\geq \left|\int_t^{t+\delta} \zeta(\tau) d\tau\right| - \int_t^{t+\delta} \frac{\left|\left(1 - \frac{\theta_{2n}}{\widehat{b}_m}\right) r' - \left(\theta_n - \theta_{2n}\frac{\widehat{\theta}_n}{\widehat{b}_m}\right) v_{\mathrm{f}}'\right|}{z} d\tau$$

$$= \delta|\zeta(t+\delta^*)| - \int_t^{t+\delta} \left(\frac{\left|\left(1 - \frac{\theta_{2n}}{\widehat{b}_m}\right) r' - \left(\theta_n - \theta_{2n}\frac{\widehat{\theta}_n}{\widehat{b}_m}\right) v_{\mathrm{f}}'\right|}{z}\right) d\tau\,.$$

Now

$$\frac{\left|\left(1 - \frac{\theta_{2n}}{\widehat{b}_m}\right) r' - \left(\theta_n - \theta_{2n}\frac{\widehat{\theta}_n}{\widehat{b}_m}\right) v_{\mathrm{f}}'\right|}{z} \leq k(\mu, L)\,.$$

This gives

$$\left|\int_t^{t+\delta} \frac{\widetilde{\theta}^{\mathrm{T}}\psi^{(n-m)}}{z} d\tau\right| \geq \delta \frac{|\widetilde{\theta}^{\mathrm{T}}(t+\delta^*)\psi^{(n-m)}(t+\delta^*)|}{z(t+\delta^*)} - 2\delta k(\mu, L)\,.$$

So, we have

$$\delta \frac{|\widetilde{\theta}^{\mathrm{T}}(t+\delta^*)\psi^{(n-m)}(t+\delta^*)|}{z(t+\delta^*)} \leq 2(1+K_z\delta)(\lambda_{n-m-1}(T)\epsilon^{\alpha_{n-m-1}} + \mu f_{n-m-1}(\epsilon, T)) + \delta k_T(\sqrt{\epsilon}+\mu)a_{n-m-1} + 2\delta k(\mu, L)\,. \quad (56)$$

Letting $\delta = \epsilon^{\alpha_{n-m}}$ and dividing both sides by δ, we get

$$\begin{aligned}\left.\frac{|\widetilde{\theta}^{\mathrm{T}}(\tau)\psi^{(n-m)}(\tau)|}{z(\tau)}\right|_{\tau=t+\delta^*} &\leq 2(1+K_z\delta)(\lambda_{n-m-1}(T)\epsilon^{\alpha_{n-m}} \\ &\quad +\mu\epsilon^{-\alpha_{n-m}} f_{n-m-1}(\epsilon, T) \\ &\quad + a_{n-m-1}k_T(\sqrt{\epsilon}+\mu) + 2k(\mu, L) \\ &\leq \overline{\lambda}_{n-m}\epsilon^{\alpha_{n-m}} + \mu f_{n-m}(\epsilon, T) + 2k(\mu, L)\,, \quad (57)\end{aligned}$$

where

$$\overline{\lambda}_{n-m} = 2(1+K_z)\lambda_{n-m-1}() + a_{n-m-1}k_T$$

$$f_{n-m}(\epsilon,T) = a_{n-m-1}k_T + 2(1+K_z)\epsilon^{-\alpha_{n-m}} f_{n-m-1}(\epsilon,T).$$

Next, (57) implies,

$$|\zeta(t+\delta^*)| \le \overline{\lambda}_{n-m}\epsilon^{\alpha_{n-m}} + \mu f_{n-m}(\epsilon,T) + 3k(\mu,L). \tag{58}$$

By Lemma B.3, $\zeta(\tau)$ is uniformly continuous. Therefore its value in $[t, t+\delta]$ will not differ from its value at $t+\delta^*$ by more than $k_{n-m}(T)\delta = k_{n-m}(T)\epsilon^{\alpha_{n-m}}$.

Setting $\lambda_{n-m} = \overline{\lambda}_{n-m} + k_{n-m}(T)$, (58) implies

$$|\zeta(\tau)| \le \lambda_{n-m}\epsilon^{\alpha_{n-m}} + \mu f_{n-m}(\epsilon,T) + 3k(\mu,L), \quad \forall \tau \in [t, t+\delta].$$

This finally gives

$$\left|\frac{\widetilde{\theta}^{\mathrm{T}}(\tau)\psi^{(n-m)}(\tau)}{z(\tau)}\right| \le \lambda_{n-m}\epsilon^{\alpha_{n-m}} + \mu f_{n-m}(\epsilon,T) + 4k(\mu,L), \tag{59}$$

thus concluding the proof. □

Lemma B.7 *If* $t_j \ge T_\ell$, $\Delta t \ge 1$, *and*

$$\frac{|\widetilde{\theta}^{\mathrm{T}}\psi|}{z} \le \sqrt{\epsilon}, \quad \forall t \in [t_j, t_j + \Delta t],$$

then

$$\frac{|\widetilde{\theta}^{\mathrm{T}}\phi|}{z} \le h_0(T)\epsilon^{\rho} + \mu\frac{h_1(T)}{\sqrt{\epsilon}} + 4a^{n-m}k(\mu,L),$$

where $\rho = 2^{-(n-m-1)}$, $\epsilon \in (0,1]$ *and* $h_0(T)$, $h_1(T)$ *are some positive constants which depend on* T.

Proof. $\phi = \psi + \sum_{i=1}^{n-m} c_i\psi^{(i)}$ for some constant coefficients $c_i (c_{n-m} = a^{n-m})$. Therefore by Lemma B.6, we have

$$\begin{aligned}
\frac{|\widetilde{\theta}^{\mathrm{T}}\phi|}{z} &\le \sum_{i=1}^{n-m-1} |c_i|\frac{|\widetilde{\theta}^{\mathrm{T}}\psi^{(i)}|}{z} + |c_{n-m}|\frac{|\widetilde{\theta}^{\mathrm{T}}\psi^{(n-m)}|}{z} + \frac{|\widetilde{\theta}^{\mathrm{T}}\psi|}{z} \\
&\le \sqrt{\epsilon} + \sum_{i=1}^{n-m} |c_i|\left(\lambda_i(T)\epsilon^{\alpha_i} + \mu f_i(\epsilon,T)\right) \\
&\quad + 4|c_{n-m}|\left(\frac{2k_r\left(1+\frac{|\theta_{2n}|}{b_{\min}}\right)}{L} + \mu\left(|\theta_n| + \frac{|\theta_{2n}|}{b_{\min}}M\right)\right).
\end{aligned}$$

Setting $h_0(T) = \sum_{i=1}^{n-m} |c_i|\lambda_i(T) + 1$, $h_1(T) = \sum_{i=1}^{n-m} |c_i|\overline{\lambda}_i(T)$, where $\bar{h}_i(T)$ satisfies $f_i(\epsilon, T) \leq \frac{\bar{h}_i(T)}{\sqrt{\epsilon}}$ and noting that $0 < \epsilon^{\alpha_i} \leq \epsilon^\rho$, $i = 1, 2, \ldots, n-m$, we get

$$\frac{|\widetilde{\theta}^{\mathrm{T}}\phi|}{z} \leq h_0(T)\epsilon^\rho + \mu\frac{h_1(T)}{\sqrt{\epsilon}} + 4a^{n-m}k(\mu, L)\,.$$

□

Proof of Lemma B.4(i). Consider $[t_1, t_1+T']$. Partition this interval into $\lfloor T' \rfloor - 1$ intervals of length 1 and one interval of length ≥ 1. Let $N := \lfloor T' \rfloor$. So we have $\cup_{i=1}^{N}[t_i, t_{i+1}] = [t_i, t_1 + T']$ such that $t_{i+1} - t_i = 1$, $i = 1, \ldots, N-1$ and $t_N - t_{N-1} \geq 1$.

Consider the following sets,

$$\Omega_1 = \left\{ [t_i, t_{i+1}] : \frac{|\psi^{\mathrm{T}}\widetilde{\theta}|}{z} < \sqrt{\epsilon}, \forall t \in [t_i, t_{i+1}] \right\}$$

$$\Omega_2 = [t_1, t_1 + T]\backslash\Omega_1\,.$$

From Lemma B.7 we have that $\forall t \in \Omega_1$, $\frac{|\widetilde{\theta}^{\mathrm{T}}\phi|}{z} \leq h_0(T)\epsilon^\rho + \mu\frac{h_1(T)}{\sqrt{\epsilon}} + 4a^{n-m}k(\mu, L)$.
Now, let $s(t) := \left(\frac{\psi^{\mathrm{T}}(t)\widetilde{\theta}(t)}{z(t)}\right)^2$

$$\begin{aligned} |s(t_b) - s(t_a)| &= \left| \int_{t_a}^{t_b} \frac{d}{dt}\left(\frac{\psi^{\mathrm{T}}(t)\widetilde{\theta}(t)}{z(t)}\right)^2 dt \right| \\ &= 2\left| \int_{t_a}^{t_b} \frac{\psi^{\mathrm{T}}(t)\widetilde{\theta}(t)}{z(t)} \frac{d}{dt}\left(\frac{\psi^{\mathrm{T}}(t)\widetilde{\theta}(t)}{z(t)}\right) dt \right| \\ &\leq 2(t_b - t_a)k_0(T)\widetilde{M}c\,. \end{aligned} \tag{60}$$

If $[t_i, t_{i+1}] \in \Omega_2$, then $s(t) \geq \epsilon$ for some $t^* \in [t_i, t_{i+1}]$. Due to (60) and the fact that $k_s := \max\{1, 2c\widetilde{M}k_0(T)\} \geq 1$ we have (since $t_{i+1} - t_i \geq 1$)

$$\int_{t_i}^{t_{i+1}} s(t)dt \geq \frac{\epsilon^2}{2k_s}, \quad \forall [t_i, t_{i+1}] \in \Omega_2\,. \tag{61}$$

Now let N_1, N_2 be the number of subintervals contained in Ω_1 Ω_2, respectively. Then, from (44) and (61),

$$N_2 \leq \frac{2k_s}{\epsilon^2}\left(c_1 + \mu^2 c_2(T)T'\right), \quad N_1 = \lfloor T' \rfloor - N_2 \leq T'.$$

Hence,

$$\begin{aligned}
\int_{t_1}^{t_1+T'} \frac{|\widetilde{\theta}^{\mathrm{T}}\phi|}{z} d\tau &= \int_{\Omega_1} \frac{|\widetilde{\theta}^{\mathrm{T}}\phi|}{z} d\tau + \int_{\Omega_2} \frac{|\widetilde{\theta}^{\mathrm{T}}\phi|}{z} d\tau \\
&\leq N_1 \left(h_0(T)\epsilon^{\rho} + \mu\frac{h_1(T)}{\sqrt{\epsilon}} + 4a^{n-m}k(\mu, L) \right) + N_2\widetilde{M}C \\
&\leq T' \left(h_0(T)\epsilon^{\rho} + \mu\frac{h_1(T)}{\sqrt{\epsilon}} + 4a^{n-m}k(\mu, L) \right) \\
&\quad + \frac{2k_s}{\epsilon^2}\widetilde{M}C(c_1 + \mu^2 c_2(T)T') \\
&= \frac{2k_s\widetilde{M}Cc_1}{\epsilon^2} + \\
&\quad \left[\mu^2 \frac{2k_s c_2\widetilde{M}C}{\epsilon^2} + \mu\frac{h_1(T)}{\sqrt{\epsilon}} + h_0(T)\epsilon^{\rho} + 4a^{n-m}k(\mu, L) \right] T',
\end{aligned}$$

thereby concluding the proof of Lemma B.4(i), with

$$\nu_3 = 2k_s c_1 \widetilde{M}C$$

$$\nu_4(T) = 2k_s c_2(T)\widetilde{M}C$$

$$\nu_5(T) = h_1(T)$$

$$\nu_6(T) = h_0(T)\,.$$

□

Appendix C

Proof of Lemma 9.2. The proof follows along the lines of Ioannou and Tsakalis [12]. Recalling that $W = k_e e^{\mathrm{T}} P e + \frac{1}{2}z^2$, and that (9) implies $\dot{z} \leq -(d_0 + g_2)z + d_1(|u| + |e_{\mathrm{a}}| + 1) + K_2$, we have

$$\begin{aligned}
\dot{W} &= k_e(\dot{e}^{\mathrm{T}} P e + e^{\mathrm{T}} P \dot{e}) + \dot{z}z && (62)\\
&\leq k_e \left(e^{\mathrm{T}}(A_{\mathrm{c}}^{\mathrm{T}} P + P A_{\mathrm{c}})e + 2\left(-(\widetilde{\theta}^{\mathrm{T}}\phi)b_{\mathrm{c}}^{\mathrm{T}} + b_{\mathrm{c}v}^{\mathrm{T}} v + \mu_m \eta_m b_{\mathrm{c}\eta}^{\mathrm{T}} \right) P e \right) \\
&\quad + (-(d_0 + g_z)z + d_1(|u| + |e_{\mathrm{a}}| + 1) + K_2)z\,. && (63)
\end{aligned}$$

Now,

(i) $e_{\mathrm{a}} = -\psi^{\mathrm{T}}\widetilde{\theta} + v_{\mathrm{f}}' + \xi_{2d_0}$ which implies

$$|e_{\mathrm{a}}| \leq |\psi^{\mathrm{T}}\widetilde{\theta}| + |v_{\mathrm{f}}'| + c_0 \exp[-2d_0 T_\ell]\,. \qquad (64)$$

(ii)

$$y = h_c^{\mathrm{T}} X_c + v + \xi_{2d_0}$$

$$\Rightarrow |y| \le C\|X_c\| + |v| + c_0 \exp[-2d_0 T_\ell]$$

$$\Rightarrow |y| \le C\|e\| + c + |v| + c_0 \exp[-2d_0 T_\ell]\,,$$

and

$$|u| \le \frac{1}{b_{\min}}(|r'| + |\theta_y^{\mathrm{T}}\phi_y| + |\theta_u^{\mathrm{T}}\phi_u| + |\theta_n||y|)\,.$$

Therefore by (30), i.e., the definition of e,

$$|u| \le C\|e\| + c|v| + c + c_0 \exp[-2d_0 T_\ell]\,. \tag{65}$$

Assume T_ℓ large enough to get,

$$|u| + |e_{\mathrm{a}}| \le |\psi^{\mathrm{T}}\widetilde{\theta}| + C\|e\| + C|v| + |v_{\mathrm{f}}'| + 2c\,. \tag{66}$$

Using (66) and (63) we get,

$$\dot{W} \le -k_e\|e\|^2 - (d_0 + g_2)z^2 + 2k_e\|Pb_{\mathrm{c}}\||\widetilde{\theta}^{\mathrm{T}}\phi|\|e\|$$

$$+2k_e\|Pb_{\mathrm{c}v}\||v|\|e\| + 2k_e\mu_m|\eta_m|\|Pb_{\mathrm{c}\eta}\|\,\|e\|$$

$$+\left[d_1\left(|\psi^{\mathrm{T}}\widetilde{\theta}| + C\|e\| + c|v| + |v_{\mathrm{f}}'| + 2c\right) + K_2\right] z\,.$$

Now recall that

$$|v| \le K_v K_{mz} z + K_v k_{mz} + k_v + k(T_\ell)\,,$$

$$|v_{\mathrm{f}}'| \le K_{v\mathrm{f}}' K_{mz} z + K_{v\mathrm{f}}' k_{mz} + k_{vf} + k(T_\ell)\,,$$

and $|\eta_m| \le cz + c + k(T_\ell)$, and let

$$\gamma_1 = 2\|Pb_{\mathrm{c}}\|,\ \gamma_2 = 2\|Pb_{\mathrm{c}v}\|,\ \gamma_3 = 2\|Pb_{\mathrm{c}\eta}\|\,.$$

This gives

$$\begin{aligned}
\dot{W} \le\ & -k_e\|e\|^2 - (d_0 + g_2)z^2 + k_e\gamma_1|\widetilde{\theta}^{\mathrm{T}}\phi|\|e\| + d_1|\psi^{\mathrm{T}}\widetilde{\theta}|z \\
& +\gamma_2 k_e(K_v K_{mz} z + k_v k_{mz} + k_v + k(T_\ell))\|e\| \\
& +\mu_m k_e\gamma_3(cz + c + k(T_\ell))\|e\| \\
& +d_1(C\|e\| + (CK_v K_{mz} + K_{v\mathrm{f}}' K_{mz})z \\
& +CK_v k_{mz} + Ck_v + Ck(T_\ell) + K_{v\mathrm{f}}' k_{mz} + k_{v\mathrm{f}} + k(T_\ell) \\
& +2c + K_2)z \\
=\ & -k_e\|e\|^2 - (d_0 + g_2)z^2 + k_e\gamma_1|\widetilde{\theta}^{\mathrm{T}}\phi|\ \|e\| + d_1|\psi^{\mathrm{T}}\widetilde{\theta}|z \\
& +\ [K_v\gamma_2 k_c K_{mz} + \mu_m\gamma_3 k_e c + d_1 C]\|e\|z \\
& +\ d_1(K_v C + K_{v\mathrm{f}}')K_{mz}z^2 \\
& +k_e[\gamma_2(K_v k_{mz} + k_v + k(T_\ell)) + \mu_m\gamma_3(c + k(T_\ell))]\|e\| +\ d_1[CK_v k_{mz} \\
& +Ck_v + Ck(T_\ell) + K_{v\mathrm{f}}' k_{mz} + k_{v\mathrm{f}} + k(T_\ell) + 2c + \frac{K_2}{d_1}]z\,.
\end{aligned}$$

Letting

$$\begin{aligned}
\beta_0 &= d_1[CK_v k_{mz} + Ck_v + Ck(T_\ell) + K'_{vf}k_{mz} + k_{vf} + k(T_\ell) + 2c + \frac{K_2}{d_1}] \\
\beta_1 &= d_1 C \\
\mu_1 &= (K_v\gamma_2 K_{mz} + \mu_m\gamma_3 C) \\
\mu_2 &= (K_v C + K'_{vf})d_1 K_{mz} \\
\beta'_0 &= [\gamma_2(K_v k_{mz} + k_v + k(T_\ell)) + \mu_m\gamma_3(c + k(T_\ell))]\,,
\end{aligned}$$

we have

$$\dot{W} \le -k_e\|e\|^2 - (d_0+g_2)z^2 + k_e\gamma_1|\widetilde{\theta}^{\mathrm{T}}\phi|\|e\| + d_1|\psi^{\mathrm{T}}\widetilde{\theta}|z$$

$$+\,\beta_1\|e\|z + \mu_1 k_e\|e\|z + \mu_2 z^2 + k_e\beta'_0\|e\| + \beta_0 z\,.$$

Note that μ_1 and μ_2 can be made as small as we desire by making $K_{v\max}$ and $\mu_{\max}$ small enough.

Next, we will choose k_e, $K_{v\max}$ and $\mu_{\max}$ (and hence μ_1, μ_2) appropriately and complete squares.

Using

$$k_e\gamma_1\frac{|\widetilde{\theta}^{\mathrm{T}}\phi|}{\sqrt{W}}\frac{\|e\|}{\sqrt{W}} \le \frac{\sqrt{k_e}\gamma_1\sqrt{2}}{\sqrt{\lambda_{\min}(P)}}\frac{|\widetilde{\theta}^{\mathrm{T}}\phi|}{z}W\,,$$

$$d_1\frac{|\psi^{\mathrm{T}}\widetilde{\theta}|}{z}\frac{z^2}{W}W \le 2d_1\frac{|\psi^{\mathrm{T}}\widetilde{\theta}|}{z}W\,,$$

and defining $\beta_2 = \dfrac{\gamma_1\sqrt{2}}{\sqrt{\lambda_{\min}(P)}}$, we get

$$\begin{aligned}
\dot{W} \le{}& -k_e\|e\|^2 - (d_0+g_2)z^2 + \beta_2\sqrt{k_e}\frac{|\widetilde{\theta}^{\mathrm{T}}\phi|}{z}W + 2d_1\frac{|\psi^{\mathrm{T}}\widetilde{\theta}|}{z}W \\
&+ \beta_0 z + \beta_1\|e\|z + \mu_1 k_e\|e\|z + \mu_2 z^2 + k_e\beta'_0\|e\| \\
={}& -\frac{k_e}{2}\|e\|^2 - \frac{(d_0+g_2)}{2}z^2 + \beta_2\sqrt{k_e}\frac{|\widetilde{\theta}^{\mathrm{T}}\phi|}{z}W + 2d_1\frac{|\psi^{\mathrm{T}}\widetilde{\theta}|}{z}W \\
&- \frac{\|e\|^2}{2}\left[k_e - \frac{4}{(d_0+g_2)}\beta_1^2 - \frac{4}{(d_0+g_2)}k_e^2\mu_1^2 - \frac{(d_0+g_2)}{4}\right] \\
&- \frac{(d_0+g_2)}{8}\left[\left(z - \frac{4\beta_0}{(d_0+g_2)}\right)^2 + \left(z - \frac{4}{(d_0+g_2)}\beta_1\|e\|\right)^2\right. \\
&+ \left(z - \frac{4}{(d_0+g_2)}\mu_1 k_e\|e\|\right)^2 + \left(1 - \frac{8}{(d_0+g_2)}\mu_2\right)z^2 \\
&\left.+ \left(\|e\| - \frac{4}{(d_0+g_2)}p'_0 k_e\right)^2\right] + \frac{2}{(d_0+q_2)}\left(\beta_0^2 + \beta'^2_0 k_e^2\right). \qquad (67)
\end{aligned}$$

Choose $k_e = 2\left(\frac{4}{d_0+g_2}\beta_1^2 + \frac{d_0+g_2}{4}\right)$ and $K_{v\max}$, $\mu_{\max}$ such that $\mu_2 \le \frac{d_0+g_2}{8}$ and $\mu_1 \le \sqrt{\frac{d_0+g_2}{8k_e}}$.

Then, adding and subtracting βW on the right-hand side of (67), we get for each $\mu_m \in [0, \mu_{\max}]$, $K_v \in [0, K_{v\max}]$,

$$\dot{W} \le -\beta W + \beta_2\sqrt{k_e}\frac{|\widetilde{\theta}^{\mathrm{T}}\phi|}{z}W + 2d_1\frac{|\psi^{\mathrm{T}}\widetilde{\theta}|}{z}W$$
$$+\frac{2}{(d_0+g_2)}(\beta_0^2 + \beta'^2_0 k_e^2) - k_e\|e\|^2\left[\frac{1}{2} - \beta\|P\|\right]$$
$$-\frac{1}{2}z^2[(d_0+g_2) - \beta].$$

Choosing $\beta = \min\left\{\frac{1}{2\|P\|}, d_0+g_2\right\}$ and letting

$$\beta_3 = \beta_2\sqrt{k_e}, \beta_4 = \frac{2}{(d_0+g_2)}(\beta_0^2 + \beta'^2_0 k_e^2), \beta_5 = 2d_1\,,$$

we have

$$\dot{W} \le -\beta W + \beta_3\frac{|\widetilde{\theta}^{\mathrm{T}}\phi|}{z}W + \beta_5\frac{|\psi^{\mathrm{T}}\widetilde{\theta}|}{z}W + \beta_4\,.$$

Finally, defining $g(t) := -\beta + \beta_3\frac{|\widetilde{\theta}^{\mathrm{T}}(t)\phi(t)|}{z(t)} + \beta_5\frac{|\widetilde{\theta}^{\mathrm{T}}(t)\psi(t)|}{z(t)}$, we get

$$\frac{W(b)}{W(a)} \le \exp\left[\int_a^b g(\tau)d\tau\right]\left(1 + \left(\beta_4\int_a^b \exp\left[\int_a^{t'} -g(\tau)d\tau\right]dt'\right)/W(a)\right)$$
$$\le \exp\left[\int_a^b g(\tau)d\tau\right]\left(1 + \frac{\beta_4\exp[\beta(b-a)]}{\beta W(a)}\right)\,,\ \text{since } -g(\tau) \le \beta\,.$$

This concludes the proof of Lemma 9.2. □

Table D

$$K_{mz} = 1 + \frac{2C\alpha}{a - 2d_0} \qquad k_{mz} = \frac{d_1 k_{ym}}{d_0}$$

$$K_{u\psi'} = \frac{M(1+M)}{b_{\min}} \qquad K_{uv} = \frac{M}{b_{\min}}$$

$$k_{uu} = \frac{k_r}{b_{\min}} \qquad K_{v\mathrm{f}} = K_v c$$

$$k_{v\mathrm{f}} = k_v c \qquad k_{v\mathrm{f}0} = k_{v0} c$$

$$K'_{v\mathrm{f}} = K_{v\mathrm{f}} + \mu_{\mathrm{m}} c \qquad K_{ym} = M K_{\psi' m} + K'_{v\mathrm{f}}$$

$$K_{um} = K_{u\psi} K_{\psi' m} + K_{uv} K'_{v\mathrm{f}} \qquad k_{um} = K_{uv} k_{v\mathrm{f}} + k_{uu}$$

$$k_u = k_{uu} + K_{uv}(K'_{v\mathrm{f}} k_{mz} + k_{v\mathrm{f}}) \qquad K_m = d_1(K_{um} + K_{ym}) - d_0$$

$$k_m = d_1(1 + k_{v\mathrm{f}} + k_{um} \qquad K_z = d_1 K_{mz}(K_{um} + \widetilde{M} K_{\psi' m} + K'_{v\mathrm{f}}) - (d_0 + g_2)$$

$$k_z = d_1\{[(\widetilde{M} K_{\psi' m} + K'_{v\mathrm{f}}) + K_{um}] k_{mz} + k_{v\mathrm{f}} + k_{um} + 1\} \qquad k''_z = k_z + k(T_\ell)$$

$$\bar{d}_0 = \frac{d_0}{d_1} - K'_{v\mathrm{f}} K_{uv} K_{mz} \qquad \bar{g}_2 = \frac{g_2}{d_1}$$

$$\bar{K}_2 = \frac{K_2}{d_1} - 1 - k - k(T_\ell) \qquad K_{nz} = \frac{(\bar{d}_0 - \bar{g}_2)^2}{4} \min\{\frac{1}{K^2_{u\psi'}}, \frac{1}{4\widetilde{M}^2}\}$$

$$K_{\mathrm{a}} = [2(1 - a + g_2) + 3(M^2 + K^2_{u\psi'})] \qquad K_{\mathrm{b}} = 3K'^2_{v\mathrm{f}}(1 + K^2_{uv}) K^2_{mz}$$

$$K_{\mathrm{c}} = \max\{3[(K'_{v\mathrm{f}} k_{mz} + k_{v\mathrm{f}} + k(T_\ell))^2 + (k + k(T_\ell))^2] - \nu K_{\mathrm{a}}, 0\} \qquad K_{\mathrm{d}} = K_{\mathrm{b}} + \frac{4K_{\mathrm{c}}}{L^2}$$

$$\delta(T) = \exp[-K_{\mathrm{a}} T][K_{nz} + \frac{K_{\mathrm{d}}}{K_{\mathrm{a}}}] - \frac{K_{\mathrm{d}}}{K_{\mathrm{a}}} \qquad K_{Wz} = \tfrac{1}{2} + 2\lambda_{\max}(P)(C + 2g_4)^2$$

$$K_{yz} = (K'_{v\mathrm{f}} + M K_{\psi' m}) K_{mz} \qquad K_{uz} = (K'_{v\mathrm{f}} K_{uv} + K_{u\psi'} K_{\psi' m}) K_{mz}$$

$$k_{yz} = M K_{\psi' m} k_{mz} + K'_{v\mathrm{f}} k_{mz} + k_{v\mathrm{f}} \qquad k_{uz} = k_u + K_{u\psi'} K_{\psi' m} k_{mz}$$

$$K_{nn} = K^2_{\psi' m}(K_{mz} + \frac{2k_{mz}}{L})^2 + \frac{4\nu}{L^2} \qquad \beta = \min\{\frac{1}{2\|P\|}, d_0 + g_2\}$$

$$\beta_1 = d_1 C \qquad k_e = 2(\frac{4}{d_0 + g_2}\beta_1^2 + \frac{d_0 + g_2}{4})$$

$$T = \max\left\{4, 1 + \frac{1}{g_2}\log\left(2\sqrt{\frac{64K^2_{Wz}\exp[2(K_z + \epsilon_z)](k + \epsilon_0)\exp[-\beta/4]}{\gamma}}\right),\right.$$

$$1 + \frac{1}{g_2}\log\left(2\sqrt{\frac{1024K^3_{Wz}\exp[2(K_z + \epsilon_z)](k + \epsilon_0)\exp[-\beta/4]}{\gamma}}\right),$$

$$\frac{1}{2g_2}\log\left(2\sqrt{\frac{8K_{Wz}}{\gamma}}\right), \frac{1}{g_2}\log\left(2\sqrt{\frac{1024K^3_{Wz}\exp[2(K_z + \epsilon_z)]}{\gamma}}\right),$$

$$\frac{2}{\beta}\log\left(\frac{k + \epsilon_0}{\gamma}\right), \frac{1}{g_2}\log\left(2\sqrt{\frac{256K^2_{Wz}(k + \epsilon_0)\exp[-\beta/4]}{\gamma}}\right),$$

$$\frac{1}{2}\left[1+\frac{1}{g_2}\log\left(2\sqrt{\frac{64K_{Wz}^2\exp[2(K_z+\epsilon_z)]}{\gamma}}\right)\right],\frac{8}{\beta}\log\left(\frac{32K_{Wz}(k+\epsilon_0)}{\sqrt{\gamma}}\right),$$

$$\frac{1}{g_2}\log\left(2\sqrt{\frac{8K_{Wz}(k+\epsilon_0)\exp[-\beta/4]}{\gamma}}\right),\frac{2}{g_2}\log\left(2\sqrt{\frac{8K_{Wz}(k+\epsilon_0)}{\sqrt{\gamma}}}\right),$$

$$\frac{4}{\beta}\log\left(\frac{256K_{Wz}^2(k+\epsilon_0)^2\exp[2(K_z+\epsilon_z)]}{\gamma}\right)-1,\frac{4}{\beta}\log\left(\frac{32K_{Wz}(k+\epsilon_0)}{\gamma}\right),$$

$$1+\frac{1}{g_2}\log\left(2\sqrt{\frac{256K_{Wz}^3\exp[4(K_z+\epsilon_z)](k+\epsilon_0)\exp[-\beta/4]}{\gamma}}\right),$$

$$1+\frac{1}{g_2}\log\left(2\sqrt[4]{\frac{4096K_{Wz}^4\exp[4(K_z+\epsilon_z)]}{\gamma}}\right)\Bigg\},$$

$$\epsilon=\min\left\{\sqrt[1/\rho]{\frac{\beta}{8\beta_3\nu_6(T)}},\left(\frac{\beta}{8\beta_5}\right)^2\right\},$$

$$L=\max\left\{\frac{2}{K_z}(k_z+k(T_\ell)),\frac{2K_2}{2K_z-g_2},\frac{2k_z''}{\epsilon_z},\frac{4K_2}{g_2}\sqrt{\frac{64K_{Wz}^2\exp[2(K_z+\epsilon_z)]}{\gamma}},\right.$$

$$\frac{4K_2}{g_2}\sqrt{\frac{8K_{Wz}(k+\epsilon_0)\exp[-\beta/4]}{\gamma}},\sqrt{\frac{\beta_4k\exp[2\beta T]}{\beta K_{Wz}\epsilon_0}},\frac{4K_2}{g_2}\sqrt{\frac{8K_{Wz}(k+\epsilon_0)}{\sqrt{\gamma}}},$$

$$\frac{2K_2}{g_2},\frac{4K_2}{g_2}\sqrt{\frac{256K_{Wz}^3\exp[4(K_z+\epsilon_z)](k+\epsilon_0)\exp[-\beta/4]}{\gamma}},$$

$$\frac{4K_2}{g_2}\sqrt{\frac{8K_{Wz}}{\gamma}},\frac{4K_2}{g_2}\sqrt{\frac{64K_{Wz}^2\exp[2(K_z+\epsilon_z)](k+\epsilon_0)\exp[-\beta/4]}{\gamma}},$$

$$\frac{4K_2}{g_2}\sqrt{\frac{4096K_{Wz}^4\exp[4(K_z+\epsilon_z)]}{\gamma}},\sqrt{\frac{4k_{Wz}}{3K_{Wz}}},$$

$$\left.\frac{4K_2}{g_2}\sqrt{\frac{1024K_{Wz}^3\exp[2(K_z+\epsilon_z)](k+\epsilon_0)\exp[-\beta/4]}{\gamma}}\right\}.$$

Increase L (if necessary) and choose T_ℓ, K_2 large enough and $K_{v\max}, \mu_{\max}$ small enough so that,

$$(\bar{d}_0 - \bar{g}_2) \geq 4K'_{vf}K_{mz}\,, \qquad \bar{K}_2 \geq 2(K'_{vf}k_{mz} + k_{vf} + k(T_\ell))\,,$$

$$\delta(T) \geq k\,, \text{ for some positive scalar } k\,,\ k(\mu_{max}, L) \leq \frac{\beta}{32\beta_3 a^{n-m}}\,,$$

$$\mu_1 \leq \sqrt{\frac{d_0 + g_2}{8k_e}}\,, \qquad \mu_2 \leq \frac{d_0 + g_2}{8}\,,$$

$$\text{and } \mu_{\max} \leq \min\left\{\frac{\beta\sqrt{\epsilon}}{8\beta_3\nu_5(T)}, \epsilon\sqrt{\frac{\beta}{8\beta_3\nu_4(T)}}, \epsilon\sqrt{\frac{\beta}{8\beta_5\nu_8(T)}}\right\},$$

$$\begin{aligned}
\text{where } \mu &:= K'_{vf}K_{mz} + \frac{K'_{vf}k_{mz} + k_{vf} + k(T_\ell)}{L/2} \\
k(\mu, L) &:= \frac{k_r(1 + \frac{|\theta_{2n}|}{b_{\min}})}{L/2} + (|\theta_n| + \frac{|\theta_{2n}|}{b_{\min}}M)\mu \\
\mu_1 &:= K_v\gamma_2 K_{mz} + \mu_{\mathrm{m}}\gamma_3 c \\
\mu_2 &:= (K_v C + K'_{vf})d_1 K_{mz}\,.
\end{aligned}$$

Stability of the Direct Self-Tuning Regulator*

B. Erik Ydstie

Department of Chemical Engineering, Goessmann Laboratory
University of Massachusetts, Amherst, MA 01003, USA.

Abstract. The direct self tuning regulator (STR) with parameter projection is finite gain stable when it is applied to a system where small modeling errors and bounded output perturbations are present. The modeled part of the plant is assumed to be stably invertible; the estimated parameters are projected into a compact convex set which contains a stabilizing controller and excludes the possibility of dividing with small numbers. The convexity assumption implies that we fix the sign and the lower bound for the high frequency gain of the estimated model. The extended plant state converges exponentially fast to an attractor which may be chaotic. The rate of convergence and size of ultimate bounds are estimated. The true plant with the unmodeled dynamics may have unstable zeros and there may be errors in the estimate of the delay.

1 Introduction

The Self-Tuning Regulator (STR) was proposed in [1] as a method to adaptively stabilize a linear system with unknown parameters. The algorithm converges in the ideal case [4]. However, this result is not robust in a classical sense and unmodeled dynamics and disturbances may cause the parameter estimator to become unstable via a mechanism known as parameter drift. In the case that this drift is observable from the parameter estimator it creates a *local* instability which may cause small amplitude chaotic bursting. An infinite parameter drift, which is not observable, causes a *global* instability of the estimator and gives numerical problems. Parameter projection or similar techniques can be applied to prevent the latter type of instability [2]. The aim of the paper is to demonstrate that the first type of instability does not necessarily lead to poor performance.

Indeed, we show that when parameter projection is applied to arrest the infinite drift, then the STR is robust with respect to unmodeled dynamics and bounded perturbations in the sense that we stabilize a compact set which in general contains one or more small amplitude chaotic attractors. It is claimed that the compact set containing these attractors is globally exponentially stable and that the size of the overbounding set scales with design variables, the disturbances and the unmodeled dynamics. In particular, it is shown that the

* This research was supported by the National Science Foundation under contract NSF#CTS-8903160.

algorithm in closed loop with the plant defines a nonlinear system system which is *finite gain stable*. This property gives the algorithm a measure of robustness with respect to nonlinearities and explains why adaptive controllers work well in practical applications.

The results presented in the paper extend quite readily to continuous time adaptive algorithms; algorithms based on indirect pole assignment and the extended horizon approaches can also be analyzed.

2 Setting up the Problem

We consider control of linear stationary plants

$$y(t) = P(q^{-1})u(t-1) + b(t)\,, \tag{1}$$

where $u(t), y(t), b(t)$ are the input, output and disturbance signals respectively. P is a linear shift invariant operator. The objective is to stabilize (1) so that a bounded setpoint $y^*(t)$ is followed using a *certainty equivalent direct adaptive controller*.

To introduce the problem we will first discuss the robust performance problem when we have available a *fixed* reduced order design model

$$A(q^{-1})y(t) = B(q^{-1})u(t-d) + v(t)\,, \tag{2}$$

where

$$A(q^{-1}) = 1 + a_1 q^{-1} + \ldots + a_n q^{-n}$$
$$B(q^{-1}) = b_0 + \ldots + b_m q^{-m}\,.$$

The model orders n, m, the relative degree d and the parameter vector

$$\psi = (a_1, \ldots, a_n, b_0, \ldots, b_m)$$

are design parameters. The signal

$$v(t) = A(q^{-1})(\Delta(q^{-1})u(t-1) + b(t))\,,$$

with

$$\Delta(q^{-1}) = P(q^{-1}) - P_{\mathrm{m}}(q^{-1})q^{-d+1}, \quad P_{\mathrm{m}}(q^{-1}) = \frac{B(q^{-1})}{A(q^{-1})}\,,$$

captures the effect of truncation and parameter errors and bounded noise.

The control will be computed based on the nominal model so that

$$u(t) = \frac{1}{\beta(q^{-1})}\left(y^*(t+d) - \alpha(q^{-1})y(t)\right)\,, \tag{3}$$

where $\alpha(q^{-1})$ and $\beta(q^{-1})$ are polynomials of order $n-1$ and $m+d-1$ respectively determined so that

$$1 = F(q^{-1})A(q^{-1}) + q^{-d}\alpha(q^{-1})$$

and $\beta(q^{-1}) = F(q^{-1})B(q^{-1})$. This gives the closed-loop system shown in Fig. 1. Here the parameters n, m, d are fixed by design, while ψ will be chosen to minimize the sensitivity of the loop. To ensure internal stability we need to perform the minimization subject to the condition that $B(q^{-1})$ is a *Schur-Cohn* polynomial, that is: $B(z^{-1}) \neq 0$ for $|z| \geq \sigma$ where $0 < \sigma < 1$ is a real number. This restriction can be interpreted as a stable invertibility condition for the nominal plant. Note that the plant itself need not be stably invertible. Elements which have an unstable inverse may be "hidden" in the unmodeled term. This is especially important since unmodeled high frequency dynamics invariably are present in practical applications. Also, note that in the adaptive design we will not explicitly enforce the stability property to hold for the estimated parameters. It suffices that there exist parameters such that the nominal plant is stably invertible and that the closed-loop system that results from applying a controller derived from such a model does not have "too high sensitivity". These notions will be made more explicit below.

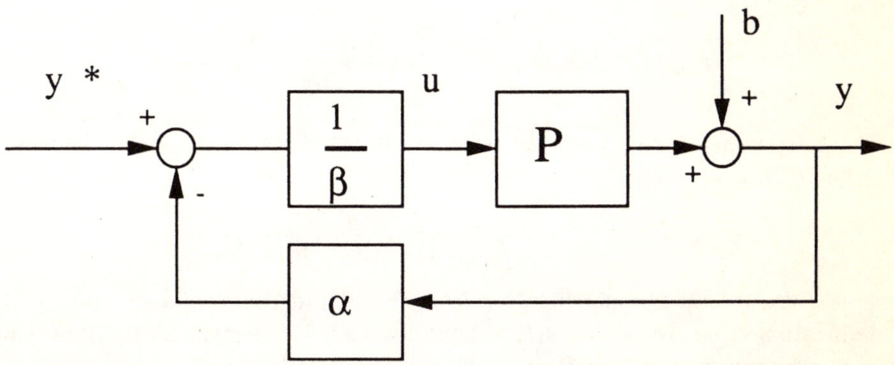

Fig. 1. Block diagram of nominal feedback system.

Using the definitions above we may rewrite (2) as

$$y(t) = \alpha(q^{-1})y(t-d) + \beta(q^{-1})u(t-d) + F(q^{-1})v(t),$$

and the closed-loop expression for the tracking error becomes

$$y(t) - y^*(t) = F(q^{-1})v(t). \tag{4}$$

From the above we then have

$$u(t) = \frac{1}{\beta(q^{-1})}\left(y^*(t+d) - \alpha(q^{-1})(y^*(t) + F(q^{-1})v(t))\right),$$

which we may rewrite as

$$u(t) = \frac{1}{B(q^{-1})}\left(A(q^{-1})y^*(t+d) - \alpha(q^{-1})v(t)\right). \tag{5}$$

It is now clear that the truncation error satisfies

$$v(t) = A(q^{-1})\Delta(q^{-1})\frac{1}{B(q^{-1})}q^{-1}(A(q^{-1})y^*(t+d) - \alpha(q^{-1})v(t)) + A(q^{-1})b(t),$$

which we write as

$$v(t) = H_0(\psi, q^{-1})y^*(t+d) + H_1(\psi, q^{-1})b(t) + \epsilon(t), \tag{6}$$

where $\epsilon(t)$ is introduced to take care of non-zero initial conditions (Fig. 2),

$$H_0(\psi, q^{-1}) = \Delta(q^{-1})\frac{A(q^{-1})q^{-1}}{P_{\rm m}(q^{-1}) + \Delta(q^{-1})\alpha(q^{-1})q^{-1}}$$

is the transfer function from the reference to the disturbance of the closed-loop system, and

$$H_1(\psi, q^{-1}) = \frac{B(q^{-1})}{P_{\rm m}(q^{-1}) + \Delta(q^{-1})\alpha(q^{-1})q^{-1}}$$

is the closed-loop transfer function from the bounded disturbance input to the truncation error. Note that $\epsilon(t)$ is bounded and decays exponentially provided that the transfer function $H_0(\psi, q^{-1})$ is exponentially stable.

From the expressions above it follows that the tracking error satisfies

$$y(t) - y^*(t) = F(q^{-1})(H_0(\psi, q^{-1})y^*(t+d) + H_1(\psi, q^{-1})b(t) + \epsilon(t)),$$

and it is clear that the tracking and disturbance rejection problems can be formulated as optimization problems. In particular, by defining a set Ψ so that $B(q^{-1})^{-1}$ is a stable polynomial for all $\psi \in \Psi$, we solve the tracking problem by minimizing the transfer function $F(q^{-1})H_0(\psi, q^{-1})$ over $\psi \in \Psi$ under a suitable norm. In the multivariable case it is natural to formulate this as an H_∞-type optimization problem. In this paper we work with a SISO system and we will work with an l_2-type objective instead.

We are now in a position to discuss the "optimal centering problem" and we will use the solution to this to define the concept of a best nominal model. Define the "compressed" l_2-gain of a stable linear operator [10][2]

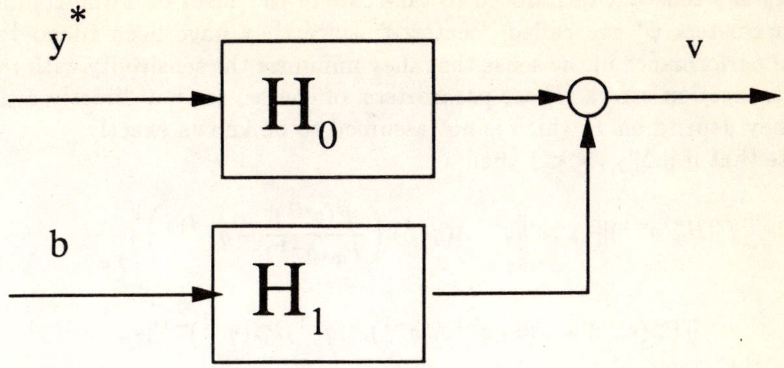

Fig. 2. Block diagram of nominal feedback system.

$$\|H(q^{-1})\|_{2,\sigma} = \left(\oint_{|z|=\sigma^{1/2}} \frac{\sigma}{2\pi z i} |H(z^{-1})|^2 dz \right)^{1/2}$$

and the exponentially weighted (truncated) σ-norm

$$\|x(t)\|_{2,\sigma} = \sqrt{\sum_{i=1}^{t} \sigma^{-i} x(i)^2} \, .$$

Note that in the case $\sigma = 1$ this gives the (truncated) l_2-norm of a function.

From (4), (5) and (6) we can define the optimal set of model parameters with respect to the σ-norm as

$$\psi^* = \arg\min_{\psi \in \Psi} \|H_0(\psi, q^{-1})\|_{2,\sigma} \, .$$

For future reference let $H_0^*(q^{-1}) = H_0(\psi^*, q^{-1})$. The corresponding vector of optimal control parameters becomes

$$\theta^* = (\alpha_1^*, \ldots, \alpha_n^*, \beta_0^*, \ldots, \beta_{m+d-1}^*) \, .$$

---

[2] The argument that follows can equally well be formulated using the "compressed" l_∞ norm

$$|H(q^{-1})|_{\infty,\sigma} = \sup_{|z|=\sqrt{\sigma}} \left(|H(z^{-1})| \right)$$

defined relative to the norm $|x|_\infty = \sup_{0<i<\infty} |x(i)|$. Note that in the case $\sigma = 1$ this gives the standard l_∞-norm which we denote by $|\cdot|_\infty$.

Application of this control gives the performance

$$\|y - y^*\|_2 \leq \epsilon_0 + |F(q^{-1})H_0^*(q^{-1})|_\infty \|y^*\|_2 + |F(q^{-1})H_1^*(q^{-1})|_\infty \|b\|_2 . \quad (7)$$

where ϵ_0 is a constant introduced to take care of the effect of initial conditions. The parameters ψ^* are called "centered" since they have been tuned to give optimal performance in the sense that they minimize the sensitivity with respect to the truncation errors. These parameters, of course, are not directly available since they depend on P which is not assumed to be known exactly.

Note that if $\|\Delta\|_{2,\sigma} << 1$ then

$$\|H_0^*(q^{-1})\|_{2,\sigma} \approx \left\| q^{-1}A(q^{-1}) \left(\frac{P(q^{-1})}{P_{\mathrm{m}}(q^{-1})} - q^{-d+1} \right) \right\|_{2,\sigma} ,$$

or

$$\|H_0^*(q^{-1})\|_{2,\sigma} \approx \|q^{-1}A(q^{-1})\Delta(q^{-1})P_{\mathrm{m}}(q^{-1})^{-1}\|_{2,\sigma} .$$

The first expression indicates that the optimal centering problem is related closely, but is not equivalent to a model matching problem subject to a stability constraint. The second expression indicates that the centering problem can be interpreted as that of finding the stable inverse P_{m}^{-1} that multiplied with the unmodeled dynamics has small gain.

In the certainty equivalent adaptive control strategies, like the one proposed in [1], a linear least-squares-based algorithm "estimates" the parameters of the design model and then the obtained estimates are used to compute the control. In the STR the gains were allowed to go to zero. Here we are interested in studying the dynamics of the feedback system when the gains are finite, but non-zero. To proceed with this design we assume that the disturbance $b(t)$ and the reference $y^*(t)$ are l_∞. That is, there exist constants k_b, k_{y^*} so that $b(t)^2 \leq k_b$ and $y(t)^{*2} \leq k_{y^*}$ for all t. We will also assume that we have available upper and lower bounds for the the control parameters. To be more specific: We know Θ^*, a compact convex set so that $\theta^* \in \Theta^*$ and $|\hat{\beta}_0| \geq \beta_{\min} > 0$ for all vectors $\theta \in \Theta^*$. The parameter estimates will be projected so that they belong to Θ^* at all times. The convexity of Θ^* implies that we fix the sign of the high frequency gain of the modeled part of the plant. The lower bound, $\beta_{\min}$, is introduced to avoid division with small numbers in the control equation and the bounds on θ are introduced to prevent infinite parameter drift.

By noting that we can write the optimized model in the predictive form (2) as

$$y(t) = \phi(t-d)'\theta^* + \gamma(t) ,$$

where $\gamma(t) = F(q^{-1})v(t)$, the following strategy for direct adaptive control becomes obvious

Step 0: Initialize $\theta(0)$ and $\phi(-i), i = 0, 1, \ldots, d-1$ and set $t = 1$

Step 1:

$$\theta(t) = F_{\Theta^*} \left\{ \theta(t-1) + P(t-1)\phi(t-d) \frac{e(t)}{r + \phi(t-d)'P(t-1)\phi(t-d)} \right\} .$$

Step 2: Solve for $u(t)$: $y^*(t+d) - \phi(t)'\theta(t) = 0$
Step 3: Set $t = t+1$ and go to Step 1.

This strategy is called a "certainty equivalent" strategy since the estimated parameters are used directly in the control law. We notice that the control equation (Step 2) is a re-statement of equation (3) using the estimated parameters. In Step 1,

$$e(t) = y(t) - \phi(t-d)'\theta(t-1)$$

is the prediction error, $\theta(t) = (\alpha_1(t), \ldots, \alpha_n(t), \beta_0(t), \ldots, \beta_{m+d}(t))'$ is a vector of estimated parameters and $F_{\Theta^*}\{.\}$ is an orthogonal projection which ensures $\theta(t) \in \Theta^*$ for all t. The parameter $r \geq 0$ controls the (local) speed of adaptation. A modification can be introduced to avoid singularity when $r = 0$ since $\| \phi(t - d)\| = 0$ may give division with zero. [3]

In [1] it was suggested to set $r = \lambda$ where $0 < \lambda \leq 1$ is the forgetting factor, and use recursive least squares to update $P(t)$ so that

$$P(t) = \frac{1}{\lambda}\left(P(t-1) - \frac{P(t-1)\phi(t-d)\phi(t-d)'P(t-1)}{(\lambda + \phi(t-d)P(t-1)\phi(t-d))}\right),$$

The forgetting factor can be updated using the algorithm proposed in [3]. The analysis presented here works for this approach provided $P(t-1)$ is "regularized" so that

$$0 < p_{\min} I \leq P(t-1) \leq p_{\max} I,$$

where $p_{\min} \leq p_{\max}$ are positive numbers. That is, $P(t-1)$ should be bounded and have bounded condition number. However, to simplify the algebra we will take $P(t-1) = P$, a fixed positive definite matrix.

The algorithm described above is known as the Direct Self-tuning Regulator with Parameter Projection. The objective of the paper is to demonstrate that this algorithm is finite gain stable provided that there exist $\sigma < 1$ and a centering so that $\|H_0(\psi^*, q^{-1})\|_{2,\sigma} \leq K_v$, where K_v is a constant determined in Sect. 4. In other words: *The STR with parameter projection solves the optimal centering problem.*

3 Technical Results

The problem with unmodeled dynamics in adaptive control has attracted considerable research attention. It is an interesting problem, mostly because neglected dynamics give rise to disturbance terms whose magnitude grow at the same rate as the process inputs, their effect persists if they have long memory, and quantitative statements about their magnitude cannot be made until the adaptive system has been shown to be stable. The standard analysis, which relies on decoupling and the use of a smallness in the mean property of the squared prediction error which is independent of stability, then fails. Until recently it was thought

[3] In practice a dead-zone is included and we set $\theta(t) = \theta(t-1)$ whenever $\|\phi(t)\| \leq \Delta$ where Δ is an arbitrary small number.

that the algorithm needed to be modified to recover stability and performance. Several approaches have been suggested including data normalization [9] and relative dead-zones [12]. Both of these methods are aimed towards modifying the adaptive algorithm and in this manner the smallness in the mean property can be restored despite the presence of the unmodeled terms. A Bellman-Gronwall argument can then be applied to demonstrate stability of the closed-loop system. A recent survey of these methods is provided in [7].

In this paper we only use parameter projection, and the decoupling property does not play such an explicit role. Instead, we exploit an exponential stabilizing property of the ideal closed-loop system and the bounded rate of growth that results from the parameter projection and then apply the "switched system" approach presented in [14]. Sets of events where the controller destabilizes the closed loop are isolated and bounded rate of growth arguments are applied to show that on these and a certain number of preceding events the regressor overbounds the unmodeled dynamics.

To motivate the development consider an example problem which illustrates, in part, why it is difficult to analyze adaptive control systems.

Example 3.1. (Ydstie and Golden [15]). Consider adaptive control of the system

$$y(t) + a_1 y(t-1) = u(t-1) + v\,,$$

where v is a constant. Furthermore, if the setpoint is a constant different from zero, then the adaptive system can be represented by a mapping in the plane

$$\begin{aligned} y' &\to -\theta' y + \gamma \\ \theta' &\to \theta' + y' \frac{-\theta' y' + \gamma - 1}{c + y_2'^2}\,, \end{aligned} \tag{8}$$

where $\theta' = \hat{\theta}(t) + a_1, y' = y/y^*, c = r/y^{*2}$, and $\gamma = 1 + v/y^*$. A typical result from a simulation is shown in Fig. 3. What we notice here is the persistent oscillatory phenomenon known as "parameter drift and bursting". While the output appears to converge towards the setpoint, the parameter estimate drifts. Eventually the closed-loop system becomes unstable since a closed-loop pole migrates outside the unit circle. The output diverges, this causes excitation and the parameter estimate then retunes. This pattern continues indefinitely and the phase plane plot in Fig. 3 indicates that there is a-periodicity and sensitivity with respect to initial conditions. □

The main issues here are the following:

- It is not possible to give a complete characterization of the dynamic behavior of even very simple adaptive control systems.
- Despite complex dynamics we have a measure of robustness and rapid decay of initial conditions.
- The average tracking error is of the order of the external noise.

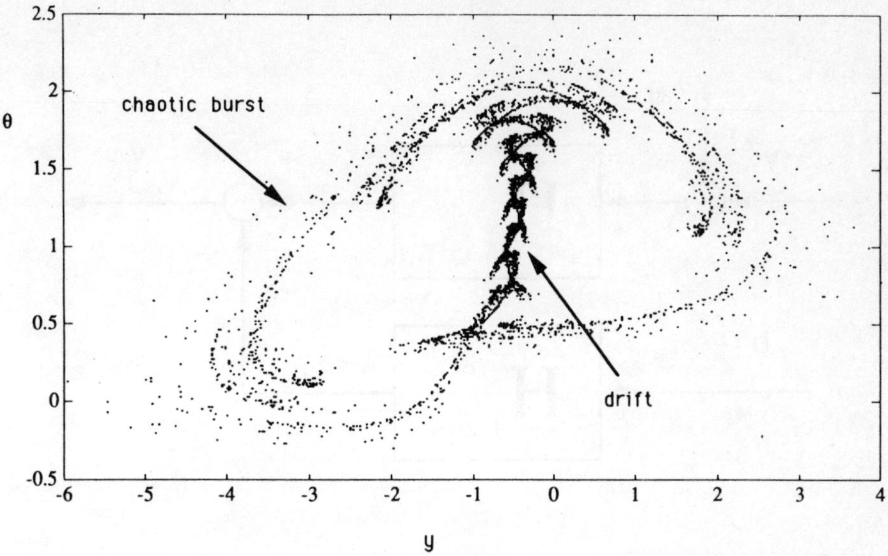

Fig. 3. Simulation result demonstrating persistent parameter drift and bursting. The output y (horizontal axis) converges towards zero while the parameter estimate θ (vertical axis) slowly drifts. Eventually the loop becomes unstable and we observe a burst.

Structural perturbations and noise do not alter the picture described above significantly. The critical property to notice, which we will exploit below, is the following exchange of stability: When the estimator is stable the output diverges and vice versa.

For the purpose of analysis express the adaptive controller in the perturbation form

$$u(t) = \frac{1}{\beta^*(q^{-1})}(y^*(t+d) - \alpha^*(q^{-1})y(t) + \phi(t)'\tilde{\theta}(t)),$$

where $\tilde{\theta}(t) = \theta^* - \theta(t)$ is the parameter error. This result is a direct consequence of (3) and Step 2 of the algorithm. It follows that the effect of parameter errors enters the loop at the same point as the reference. We can now redefine the expression for $v(t)$ to take into account the effect of adaptation and we get

$$v(t) = H_0^*(q^{-1})(y^*(t+d) + \phi(t)'\tilde{\theta}(t)) + H_1^*(q^{-1})b(t) + \epsilon(t).$$

A block diagram for the closed-loop *adaptive* system is shown in Fig. 4 and the analysis proceeds using the machinery set up in the previous section. Let

$$m(t+1) = \sigma m(t) + \left(\phi(t)'\tilde{\theta}(t)\right)^2, \; m(0) = m_0,$$

where m_0 is a constant chosen to overbound the effect of the initial conditions. The existence of this constant is guaranteed by stability of $H_0^*(q^{-1})$. It follows

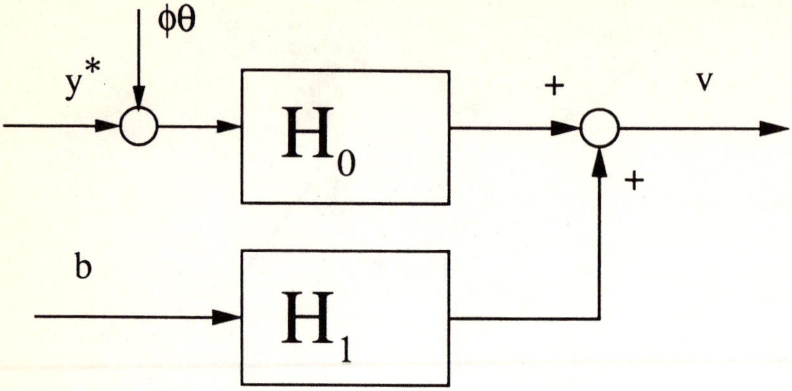

Fig. 4. Block diagram of adaptive feedback system.

that $m(t)$ overbounds the regressor and the perturbation $v(t)$ in the sense that there exist positive constants $\{K_v, K_\phi, k_v, k_\phi\}$ so that

$$v(t)^2 \leq K_v m(t+1) + k_v \tag{9}$$

$$||\phi(t)||^2 \leq K_\phi m(t+1) + k_\phi \, . \tag{10}$$

In particular, we have

$$K_v = 2||H_0^*(q^{-1})||_\sigma \text{ and } k_v = 2\frac{1}{1-\sigma}||H_1^*(q^{-1})||_\sigma k_b \, .$$

We also have

$$\gamma(t)^2 \leq K_\gamma m(t+1) + k_\gamma \, , \tag{11}$$

with $K_\gamma = K_v \; |F(q^{-1})|_\infty$ and $k_\gamma = |F(q^{-1})|_\infty \; k_v$ where $|F(q^{-1})|_\infty = \sup_{|q|=1} F(q^{-1})$.

Another critical property, which follows from the projection $F_{\Theta^*}\{.\}$, is that $||\theta(t)||^2 \leq M_\theta < \infty$ so that $m(t)$ has bounded rate of growth. That is,

$$m(t+1) \leq g_1 m(t) + k_1, \quad \forall t \geq 0 \, , \tag{12}$$

where g_1 and k_1 are positive constants that depend on Θ^*. Note that $k_1 = k_\phi = k_v = k_\gamma = 0$ when $k_b = k_{y^*} = 0$, and that $K_v = K_\gamma = 0$ when there is no mismatch present.

The main idea consists of constructing a comparison signal $a(t)$ which measures all signals in the adaptive loop and in addition captures the drift and burst phenomenon described above. The signal we use is "switched" and defined in the following fashion

$$a(t+1) = A(t)\left(\sigma a(t) + (\phi(t)'\tilde{\theta}(t))^2\right) + (1 - A(t))(g_2 a(t) + k_2), \tag{13}$$

where $A(t) = I_{\{\sigma a(t)+(\phi(t)'\tilde{\theta}(t))^2 \geq (g_2 a(t)+k_2)\}}$ is a zero-one indicator function, $a(0) = m(0)$ and $\sigma < g_2 < 1$ and k_2 are positive constants to be determined. Note that, unlike previous studies, the comparison signal is not used in the algorithm itself–it is only introduced as an instrument to be used in the analysis.

It follows immediately that $a(t)$ overbounds $m(t)$, so that if we can show boundedness of $a(t)$ then boundedness of the adaptive system itself can be inferred from inequality (10). Moreover, $a(t)$ has the bounded rate of growth property

$$a(t+1) \leq g_1 a(t) + k_1, \quad \forall t \geq 0, \tag{14}$$

inherited from (12). (Note that, in general, $k_1 > k_2$ and $g_1 > 1 > g_2$).

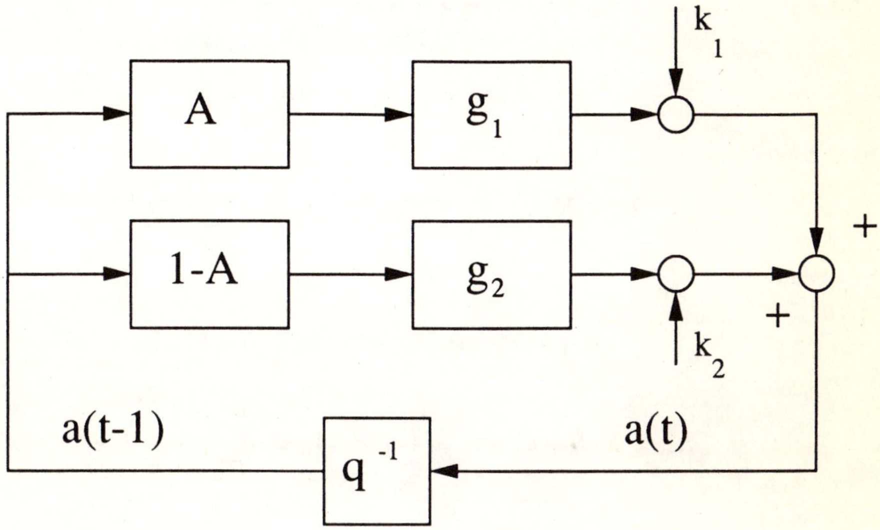

Fig. 5. Block diagram for the "switched" system. The top part has a loop gain larger than one and the lower part is stable. The idea is that the closed-loop system will be stable if the lower path is used sufficiently often.

Equation (13), it turns out, can be interpreted as a nonlinear dynamic system of the form shown in Fig. 5. That is: We have a dynamic feedback system with one exponentially stable component, one exponentially unstable component and a switching function which determines which path is used. In this case a switching function has been defined so that $A(t) = 1$ corresponds (roughly) to the unstable burst phase of the adaptive algorithm and $A(t) = 0$ corresponds to the phase where the closed loop is stable and the parameter estimate may experience drift. It appears reasonable to expect that if the stable path is used sufficiently often then this system may be stable, if not convergent. Indeed this turns out to be the case.

Lemma 3.1 (The Switching Lemma). *Let $A(t)$ be a $(0,1)$ indicator function. Consider the non-negative variable*

$$a(t+1) = A(t)(g_1 a(t) + k_1) + (1 - A(t))(g_2 a(t) + k_2),$$

where k_1 and k_2 are non-negative numbers and $0 < g_2 < 1 < g_1 < \infty$. Define $u_1 = \ln(g_1 + k_1/R)$ and $u_2 = -\ln(g_2 + k_2/R)$ for all t. Let $R > k_2/(1 - g_2)$. It follows that $u_2 > 0$. If for each t, $a(t - i) \geq R$ for $i \in [0, N]$ implies

$$\frac{1}{N} \sum_{i=t-N}^{t} A(i) = U^* < \frac{u_2}{u_1 + u_2}$$

for some integer N,[4] then for every $t > 0$

(i)

$$\sup_{t>1} a(t) \leq \max\{R_m, K_0 a(0) + k_0\},$$

where

$$R_m = K_0 R + k_0, K_0 = g_1^{NU^*}, k_0 = \sum_{i=0}^{NU^*-1} g_1^i k_1$$

and

(ii)

$$a(t+1) \leq \max\{R'_m, e^{-\delta t}(K'_0 a(0) + k'_0\},$$

where

$$R'_m = K'_0 R + k'_0, K'_0 = (\frac{g_1}{g_2})^{NU^*}, k'_0 = \sum_{i=0}^{NU^*-1} \left(\frac{g_1}{g_2}\right)^i k_1$$

and

$$\delta = u_2 - U^*(u_1 - u_2) > 0;$$

(iii)

$$\liminf a(t) \geq R, \quad \limsup a(t) \leq R_m.$$

Proof. Define two variables

$$G_1(t) = g_1 + k_1/a(t), \; G_2(t) = g_1 + k_2/a(t).$$

Then

$$a(t+1) = A(t)G_1(t)a(t) + (1 - A(t))G_2(t)a(t);$$

[4] Here and elsewhere N is an integer.

hence for N = 0,1,2,3,... and $t \geq N$

$$a(t+1) = \prod_{i=t-N}^{t} [A(i)G_1(i) + (1 - A(i))G_2(i)]\, a(t-N)\,.$$

Suppose $a(t-i) \geq R$ for $i = 0, \ldots, N$ and $t \geq N$. Then

$$\ln\left(\max_{t-N \leq i \leq t} G_1(i)\right) \leq u_1,\ \ln\left(\max_{t-N \leq i \leq t} G_2(i)\right) \leq -u_2\,,$$

whence

$$\ln a(t+1) \leq \sum_{i=t-N}^{t} (A(i)u_1 - (1 - A(i))u_2) + \ln a(t-N)\,.$$

However, according to the assumptions,

$$\sum_{i=t-N}^{t} A(i)u_1 - (1 - A(i))u_2 \leq -Nu_2 + N(u_1 + u_2) = -N\delta\,,$$

where $\delta = u_2 - (u_1 - u_2)U^* > 0$. This implies $\ln a(t+1) \leq \ln a(t-N) - N\delta$. Hence,

$$a(t+1) \leq \exp(-N\delta)a(t-N)\,.$$

On the other hand, if there exists an integer $0 < k \leq N$ so that $a(t-k) \leq R$, then due to the bounded rate of growth

$$a(t+1-i) \leq g_1^{NU^*} R + \sum_{i=0}^{NU^*-1} g_1^i k_1, i \in [0, k]\,.$$

Hence, for all $t \geq N$

$$a(t+1) \leq \max\{R_{\mathrm{m}}, e^{-N\delta} a(t-N)\}\,,$$

where

$$R_{\mathrm{m}} = g_1^{NU^*} R + \sum_{i=0}^{NU^*-1} g_1^i k_1\,.$$

Hence, for all t

$$\sup_{t>0} a(t+1) \leq \{R_{\mathrm{m}}, g_1^{NU^*} a(0) + \sum_{i=0}^{NU^*-1} g_1^i k_1\}\,.$$

Having established this, we have

$$a(t+1) \leq \max\left\{R_{\mathrm{m}}, e^{-\delta t}\left(e^{\delta N}\left(g_1^{NU^*} a(0) + \sum_{i=0}^{NU^*-1} g_1^i k_1\right)\right)\right\}\,.$$

Since $a(t) \geq g_2^N a(t-N)$, we have by recursive application of the above

$$a(t+1) \leq \max\left\{ R_{\rm m}, e^{-\delta t}\left(\left(\frac{g_1}{g_2}\right)^{NU^*} a(0) + \sum_{i=0}^{NU^*-1} \left(\frac{g_1}{g_2}\right)^i k_1 \right) \right\}.$$

This establishes the second item and the third follows by taking the limits. □

The statement of the problem sets an effective lower bound on R. Specifically, $R > k_2/(1-g_2)$. This then implies that for initial conditions $a(0) > Rg_2^{-N}$ the interval $[0, R]$ is exponentially attracting with an average exponent δ, i.e., the system decays on the average like a first order system so that

$$z(t+1) = e^{-\delta} z(0)\,, \quad \delta = u_2 - U^*(u_1 - u_2) > 0\,.$$

The program we follow is to show that, under suitable conditions, $A(i)$, defined by (13), satisfies the inequality

$$\frac{1}{N} \sum_{i=t-N}^{t} A(i) \leq U^*\,, \tag{15}$$

where U^* is sufficiently small for the Switching Lemma to apply. The result from applying this type of analysis is that we demonstrate that there exists a compact subset of state space which is globally exponentially attracting. We will not be able to use the this theory to say much about the finer structure of the solutions unless $R = 0$ is allowed. Then we get exponential convergence to zero.

Lemma 3.2 (Switching Frequency for STR with Parameter Projection). *Suppose $a(i) \geq R$ for $i \in [t-N+1,\ t+d]$. Then*

$$\frac{1}{N} \sum_{i=t-N}^{t} A(i) \leq U^*(R, N, \sigma, g_2)\,,$$

where

$$\begin{aligned} U^*(R,N,\sigma,g_2) = \frac{g_2}{g_2-\sigma} \times \Bigg[& c_1 K_\gamma + 2\frac{k_3 + k_\gamma}{R} + 2^{d+1}\left(K_\phi + \frac{k_\phi + r/p_{\max}}{R} \right. \\ & \times d\frac{p_{\max}}{p_{\min}} \left(\frac{M_\theta}{N} + 4\frac{N+d+1}{N} \right. \\ & \left. \left. \times \frac{|F(q^{-1})|_\infty (K_v + k_v/R)}{\left(\frac{g_2}{K_\theta^2}\right)^{N+d+1} \frac{g_1^{-(d+1)}}{M_\theta}(g_2 - \sigma) - 2c_2^2\left(K_v + 2\frac{k_{y^*} + k_v}{R}\right)} \right) \right) \Bigg], \end{aligned}$$

where $k_3 = K_v \sum_{j=0}^{d-1} g_1^j k_1$, $k_2 = k_3(g_2 - \sigma)/g_1^{d+1}$ and $c_1 = 2g_1^d$.

Proof. See Appendix 1. □

Corollary 3.1 *The adaptive control system described in Sect. 2 is stable in the sense that there exist finite positive numbers* δ, R_m, K_0, k_0*, such that*

$$\|\phi(t)\| \le \max\left\{ R_m, K_0 e^{-\delta t}\left(\max_{0\le i<d} \|\phi(-i)\| + K_\phi^{-1} m_0\right) + k_0 \right\},$$

provided there exists $\sigma < 1$ *such that the gain of* $H_0^*(q^{-1})$ *is sufficiently small.*

This result, which is a (significant) generalization of the result presented in [14], follows immediately by inspection. Explicit results are developed in the next section. One obvious conclusion is that that the adaptive controller with parameter projection cannot be destabilized by the addition of bounded disturbances.

4 Stability, Transients and Performance

We focus on establishing four properties of the adaptive system, namely[5]

- Maximum transient ($\sup_{t>0} \|\phi(t)\|^2$).
- Exponential stability of a compact set with radius R_m ($\|\phi(t)\|^2 \le \max\{R_m, K_0 e^{-\delta t}\|\phi(0)\|^2 + k_0\}$).
- Ultimate bounds ($\limsup_{t\to\infty} \|\phi(t)\|^2$).
- Performance ($\lim_{t\to\infty}(y(t) - y^*(t)) = 0$) or finite gain stability ($\|y - y^*\|_2 \le \mu_0 + \mu_1\|b\|_2 + \mu_2\|y^*\|_2$).

The relevant constants are estimated. In addition to slow local adaptation ($r > 0$) we discuss "fast adaptation" ($r = 0$).

In Lemma 3.2, N, R and g_2 are free variables to be optimized. In order to obtain feasible, if not optimal bounds, it is convenient to define the following σ-dependent parameters

$$g_2^* = \arg \min_{\sigma<g_2<1} \frac{g_2}{g_2-\sigma}\left(1 - \frac{\ln g_1}{\ln g_2}\right)$$

(a good approximation is $g_2^* = \sqrt{\sigma}$), and

$$B^* = \frac{-\ln g_2^*}{\ln g_1 - \ln g_2^*}.$$

In addition, we define N^* to be the smallest integer satisfying the inequality

$$N^* > \frac{m}{B^*},$$

where

$$m = 2^{d+1} M_\theta \frac{p_{\max}}{p_{\min}} K_\phi \frac{g_2^*}{g_2^* - \sigma}.$$

Note that we have settled for expedience rather than optimality in some of these definitions.

[5] $\|x\| = \sqrt{\sum_{i=1}^{n} x(i)^2}$ is the Euclidean norm

Proposition 4.1 (Ideal case, fast adaptation). *Consider direct adaptive control of (1) with no model mismatch, $b(t) = y^*(t) = 0$ for all t (ideal case) and $r = 0$ (fast adaptation). Let σ^* be the smallest real number such that $\| H_0^*(q^{-1})\|_{2,\sigma^*} = 0$ and $B(z^{-1}) \neq 0$ for $|z| \geq \sigma^*$. If $\sigma^* < 1$ then the adaptive system has bounded transient so that*

$$\sup_{t>0} \|\phi(t)\|^2 \leq g_1^m \max_{0\leq i<d} \|\phi(-i)\|^2 ,$$

where $m = 2^{d+1} M_\theta \frac{p_{\max}}{p_{\min}} K_\phi \frac{g_2^*}{g_2^* - \sigma^*}$. Moreover, we have exponential decay in the sense that

$$\|\phi(t)\|^2 \leq \left(\frac{g_1}{g_2}\right)^{2m} e^{-\delta t} \max_{0\leq i<d} \|\phi(-i)\|^2 \text{ for } t > 0 ,$$

where $\delta = -0.5 \ln g_2^* > 0$. This implies $\lim y(t) = 0$ and $\lim u(t) = 0$ exponentially fast.

Proof. $r = k_\gamma = k_{y^*} = K_\gamma = 0$ implies $k_3 = k_\phi = 0$. From Lemma 3.2

$$U^*(0, N, \sigma, g_2) = \frac{g_2}{g_2 - \sigma} 2^{d+1} K_\phi \frac{p_{\max}}{p_{\min}} \frac{M_\theta}{N} .$$

It is convenient for now to set $\sigma = \sigma^*, g_2 = g_2^*$ and $N = N^*$. This gives

$$U^*(0, N, \sigma, g_2) < \frac{-\ln g_2^*}{\ln g_1 - \ln g_2^*} = B^* ,$$

and from Lemma 3.1 we conclude boundedness and convergence. In particular, by choosing $a(0) = K_\phi^{-1} \max_{0\leq i<d} \|\phi(-i)\|^2$, the comparison function satisfies

$$\sup_{t>0} a(t) \leq g_1^{NU^*} a(0) ,$$

where, according to the definitions above,

$$NU^* < 2^{d+1} M_\theta \frac{p_{\max}}{p_{\min}} K_\phi \frac{g_2^*}{g_2^* - \sigma} = m .$$

The first part of the result then follows from Lemma 3.1 by application of (10). We also must conclude

$$a(t+1) \leq e^{-t\delta} \left(\frac{g_1}{g_2^*}\right)^m a(0) ,$$

with

$$\delta = -\ln g_2^* - (\ln g_1 - \ln g_2^*) 2^{d+1} K_\phi \frac{M_\theta}{N} \frac{g_2^*}{g_2^* - \sigma} \frac{p_{\max}}{p_{\min}} .$$

Choosing $k = 2N^*$ is convenient in this case and gives the conservative bound

$$\delta = -\ln g_2^* (1 - 1/2) ,$$

which establishes the result. □

The estimate for the rate of convergence has not been optimized and is slower than what we would expect if the parameters of the plant where known exactly. (When the parameters are known we have $\delta = -\ln \sigma^*$. In the adaptive case the bound above ensures $\delta \leq -0.5 \ln \sqrt{\sigma^*}$). We note here that exponential stability of the signals does not necessarily imply convergence of the parameter estimates. It is not even clear whether or not the estimates converge. With the dead-zone we get $\lim \theta(t) = \theta_\infty$ where θ_∞ can be viewed as a random variable and $\limsup ||\phi(t)|| \leq \Delta$.

Indeed, this point has caused some confusion in the field of adaptive control. Several earlier attempts to demonstrate exponential stability of the closed-loop system depended on exponential stability of the parameter estimator and convergence of the parameters. Such an approach relies heavily on the use of the concept of *persistent excitation*. In one form, persistent excitation implies the existence of positive constants c_1, c_2, n such that

$$c_1 I \leq \sum_{t-n}^{t} \phi(i-d)\phi(i-d)' \leq c_2 I .$$

The lower bound is difficult to enforce when there are unmodeled dynamics present. The upper bound can normally be guaranteed through a separate stability analysis.

One contribution of this paper is that we have divorced the question of exponential convergence of the signals from the question of exponential convergence of the parameter estimates and in this way we have obtained the finite gain stability result without using the notion of persistent excitation. We do want to stress that what we show is exponential stability of a set in all cases but in the result above and its counterpart, Proposition 4.4.

Proposition 4.2 (Ideal case, slow adaptation). *Consider adaptive control of (1) with $r > 0$, $b(t) \equiv 0$ and $\sup_{t>0} y(t)^2 \leq k_{y^*}$. Let σ^* be the smallest real number such that $||H_0^*(q^{-1})||_{2,\sigma^*} = 0$ and $B(z^{-1}) \neq 0$ for $|z| \geq \sigma^*$.*[6] *If $\sigma^* < 1$ we have*

$$\sup_{t>0} ||\phi(t)||^2 \leq \max\left\{ g_1^m \max_{0\leq i<d} ||\phi(-i)||^2 + \sum_{i=0}^{m-1} g_1^i k_1, R_\mathrm{m} \right\} + k_\phi ,$$

where

$$R_\mathrm{m} = g_1^{2m} \left(\frac{r}{p_\mathrm{max}} + k_\phi \right) + \sum_{i=0}^{2m-1} g_1^i k_0 .$$

The exponential rate of decay to a compact set is bounded so that

$$||\phi(t)||^2 \leq \max\left\{ R_\mathrm{m}, K_\phi e^{-\delta t} \left(K_0 \max_{0\leq i<d} ||\phi(-i)||^2 + k_0 \right) \right\} + k_\phi \ \mathit{for}\ t \geq 1 ,$$

[6] This condition is stronger than that in [4] where we only need $\sigma^* \leq 1$. We recover the stronger result by treating zeros on the unit circle as unmodeled dynamics, as in Proposition 4.5.

where the constants are defined below and $\delta = -\frac{1}{6}\ln g_2^*$. *In addition,*

$$\lim_{t\to\infty} |y(t) - y^*(t)| = 0\,.$$

Proof. In this case, $k_\gamma = K_\gamma = 0$ which implies $k_3 = 0$. From Lemma 3.2,

$$U^*(R,N,\sigma,g_2) = 2^{d+1}\frac{g_2}{g_2-\sigma}\left(K_\phi + \frac{r/p_{\max} + k_\phi}{R}\right) d\frac{p_{\max}}{p_{\min}}\frac{M_\theta}{N}\,,$$

and stability follows by application of Lemma 3.1, provided R, N and g_2 can be found so that

$$U^*(R,N,\sigma,g_2) < \frac{-\ln g_2}{\ln g_1 - \ln g_2}\,.$$

There is a trade-off between R and N. To obtain stability and convergence to a small set, it is necessary to compensate by using large N. This slows down the estimated rate of convergence.

In the following, set $N = kN^*$ with $k \geq 2, \sigma = \sigma^*$ and $g_2 = g_2^*$ so that

$$\frac{-\ln g_2}{\ln g_1 - \ln g_2} = B^*\,.$$

As in the previous example, the expression for U^* simplifies and we get the condition for stability:

$$U^*(R,N,\sigma,g_2) \leq \left(\frac{1}{k} + \frac{r/p_{\max} + k_\phi}{K_\phi R k}\right) B^*\,,$$

and we guarantee convergence of $a(t)$ to a bounded limit set by choosing

$$R = R^* = \frac{1}{k-1}\frac{r/p_{\max} + k_\phi}{K_\phi}\,.$$

From Lemma 3.1 we now get, by setting $a(0) = \max\left\{K_\phi^{-1}\max_{0\leq i<d}\|\phi(-i)\|^2, R\right\}$:

$$\sup_{t>0} a(t) \leq \max\left\{R_{\mathrm{m}}, g_1^{NU^*}a(0) + \sum_{i=0}^{NU^*-1} g_1^i k_1\right\}\,,$$

where $NU^* = km$, and hence

$$R_{\mathrm{m}} = g_1^{km}\frac{r/p_{\max} + k_\phi}{K_\phi(k-1)} + \sum_{i=0}^{km-1} g_1^i k_1\,, \quad k = 2$$

gives a reasonable bound. The first part of the result follows. The bound for the transient increases with $r/p_{\max}$ large. This is not surprising since a wrong set of initial conditions for the parameter estimates may give a large transient if the rate of adaptation is low. It then takes more time to correct modes leading to instability and the transient performance deteriorates.

We now turn our attention to the exponential decay property. Immediately

$$a(t+1) \leq \max\{R'_{\mathrm{m}}, e^{-\delta t} K'_0 a(0) + k'_0\},$$

where the constants need to be computed. First we have

$$\delta = -\ln g_2^* - U^*(N, R, \sigma, g_2)(\ln g_1 - \ln g_2).$$

With $R = nR^*, n \geq 1$

$$U^*(N, R, \sigma, g_2) = B^* \left(\frac{1}{k} + \frac{k-1}{n}\right)$$

so that, according to Lemma 3.1,

$$\delta = -\ln g_2^* \left(1 - \frac{1}{k} + \frac{k-1}{n}\right).$$

Clearly, to achieve $\delta > 0$ we need

$$\frac{1}{k} + \frac{k-1}{n} < 1$$

or $n > k$. Somewhat arbitrarily, we set $k = 2$ and $n = 3$, which gives

$$\delta = -\frac{1}{6} \ln g_2^*.$$

To finalize, we need to determine NU^*. From the above,

$$NU^* = \frac{5}{6} B^* \leq \frac{5}{4} m.$$

Hence

$$R'_{\mathrm{m}} = 3K'_0 R^* + k'_0, \text{ where } K'_0 = \left(\frac{g_1}{g_2}\right)^m, \quad k'_0 = \sum_{i=0}^{m-1} \left(\frac{g_1}{g_2}\right)^i k_1,$$

and the second part of the result follows. Finally, we show convergence to zero error. First,

$$V(t) \leq V(t-1) - \frac{e(t)^2}{n(t)},$$

where $V(t) = \tilde{\theta}(t)' P^{-1} \tilde{\theta}(t)$. It follows that $\lim V(t) - V(t-1) = 0$. Now from the above

$$e(t)^2 \leq n(t)\left(V(t-1) - V(t)\right).$$

Now $n(t)$ is uniformly bounded, the swapping terms $\phi(t-d)'(\theta(t-1) - \theta(t-d))$ converge to zero and after some simple algebra we have $\lim |y(t) - y^*(t)| = 0$ since $y(t) - y^*(t) = e(t) + \phi(t-d)'(\theta(t-1) - \theta(t-d))$. □

The gist of this is that (at least using this theory) we cannot guarantee exponential convergence to zero error in the absence of persistent excitation when we use slow adaptation and non-zero reference inputs. Instead, we have exponential convergence to an overbounding set – the estimated rate of convergence then slows down as we attempt to decrease the size of this set.

We now move into the part of the paper which concerns the robustness with respect to bounded disturbances. First we have a result on performance when there are l_∞-output perturbations. The main idea is, as in the previous example, to establish ultimate boundedness first and then bring the local analysis to bear on the problem to establish the performance result.

Proposition 4.3 (Bounded disturbances). *Suppose* $\sup_{t>0} b(t)^2 \leq k_b$ *and* $\sup_{t>0} y^*(t)^2 \leq k_{y^*}$. *Let* σ^* *be the smallest real number such that* $\|H_0^*(q^{-1})\|_{2,\sigma^*} = 0$ *and* $B(z^{-1}) \neq 0$ *for* $|z| \geq \sigma^*$. *If* $\sigma^* < 1$, *then*

$$\sup_{t>0} \|\phi(t)\|^2 \leq \max\left\{ R_{\mathrm{m}}, K_0 \max_{0\leq i<d} \|\phi(-i)\|^2 + k_0 \right\} + k_\phi ,$$

where R_{m}, K_0 *and* k_0 *are defined below. Exponential convergence to a compact set is achieved in the sense that*

$$\|\phi(t)\|^2 \leq \max\left\{ R'_{\mathrm{m}}, e^{-\delta t} K'_0 \max_{0\leq i<d} \|\phi(-i)\|^2 + k'_0 \right\} + k_\phi ,$$

where $\delta, R'_{\mathrm{m}}, K'_0$ *and* k'_0 *are positive constants defined below. In addition, we have*[7]

$$\|y - y^*\|_2 \leq \mu_0 + \mu_1 |F(q^{-1})A(q^{-1})|_\infty \|b\|_2 ,$$

where $\mu_i, i = 0, 1$ *are constants independent of the initial conditions.*

Proof. From Lemma 3.2 with $K_\gamma = 0$,

$$U^*(R, N, \sigma, g_2) = \frac{g_2}{g_2 - \sigma}\left[2\frac{k_\gamma}{R} + 2^{d+1}\left(K_\phi + \frac{k_\phi + r/p_{\max}}{R}\right) d\frac{p_{\max}}{p_{\min}} \right.$$
$$\left. \times \left(\frac{M_\theta}{N} + \frac{4}{N} \frac{(N+d+1)k_\gamma/R}{\left(\frac{g_2}{K_\theta^2}\right)^{N+d+1} \frac{g_1^{-(d+1)}}{M_\theta}(g_2 - \sigma) - 2c_2^2 \frac{k_{y^*} + k_\gamma}{R}} \right) \right] .$$

The adaptive system is stable if we can find N, R, σ and g_2 such that

$$U^*(R, N, \sigma, g_2) < \frac{-\ln g_2}{\ln g_1 - \ln g_2} .$$

[7] The l_2 norm of a function x is defined so that

$$\|x\|_2 = \sqrt{\sum_{t\in T} x(t)^2}$$

where $T = [0, 1, 2, \ldots)$, i.e., $\|x\|_2 = \|x\|_{2,1}$. We also note that $|FA|_\infty$ can be improved to $|F|_\infty$ by using algorithms developed for stochastic adaptive control.

Some insight can be gleaned from here by exploring limits. For example, by setting $\sigma = \sigma^*, g_2 = g_2^*$ and letting $R \to \infty$ (large signals) we obtain

$$U(\infty, N, \sigma, g_2) = \frac{g_2}{g_2 - \sigma} 2^{d+1} K_\phi d \frac{p_{\max}}{p_{\min}} \frac{M_\theta}{N}, \tag{16}$$

which coincides with the ideal case using fast adaptation. It follows that bounded disturbances cannot lead to global instability. Moreover, when the signals are large relative to the external disturbances, then the adaptive system behaves roughly like the ideal adaptive controller with no external disturbances or references and fast adaptation, giving similar convergence rate and performance.

Setting $N = kN^*$ and $R = nR^*$ (R^* was defined in the proof of Proposition 4.2) is convenient and gives

$$U^*(R, N, \sigma, g_2) = B^* \left(\frac{1}{k} + \frac{k-1}{n}\right) + \frac{g_2}{g_2 - \sigma} \frac{k_\gamma}{nR^*} \left[2 + 2^{d+3} d \frac{n+k-1}{n} \right.$$
$$\left. \frac{p_{\max}(kN^* + d + 1)}{p_{\min} kN^* \left(\left(\frac{g_2}{K_\theta^2}\right)^{kN^*+d+1} \frac{g_1^{-d-1}}{M_\theta}(g_2 - \sigma) - 2c_2^2 \frac{k_{y^*} + k_\gamma}{nR^*}\right)}\right].$$

By somewhat arbitrarily choosing $k = 2, n \geq 4$ and using the fact that $k_\phi \geq k_{y^*} + k_\gamma$, we get

$$U^*(R, N, \sigma, g_2) \leq \frac{g_2}{g_2 - \sigma} \frac{k_\gamma}{nR^*} \left[2 + 2^{d+3} d \frac{p_{\max}(2N^* + d + 1)}{p_{\min} 2N^*} \right.$$
$$\left. \left(\left(\frac{g_2}{K_\theta^2}\right)^{2N^*+d+1} \frac{g_1^{-d-1}}{M_\theta}(g_2 - \sigma) - 2c_2^2 \frac{k_\phi}{n}\right)^{-1}\right],$$

and we can rewrite the condition for stability as

$$k_\gamma \frac{a}{n}\left[1 + \frac{1}{b - c/n}\right] \leq \frac{1}{4} B^*,$$

where

$$a = 2 \frac{g_2}{g_2 - \sigma} \frac{K_\phi}{(r/p_{\max} + k_\phi)}$$
$$b = 2^{-d-2} \frac{p_{\min}}{p_{\max}} \frac{2N^*}{2N^* + d + 1} \left(\frac{g_2}{K_\theta^2}\right)^{2N^*+d+1} \frac{g_1^{-d-1}}{M_\theta}(g_2 - \sigma)$$
$$c = 2^{-d} c_2^2 k_\phi \frac{p_{\min}}{p_{\max}} \frac{2N^*}{2N^* + d + 1}.$$

After some algebra we get a sufficient condition for stability. In particular, with $R = \max(4, n) R^*$, where

$$n = 2k_\gamma ac \left(\frac{1}{8} B^* c + ab + a - \sqrt{\left(\frac{1}{8} B^* c + ab + a\right)^2 - abcB^*}\right)^{-1},$$

we get $U^* \leq \frac{7}{8}B^*$ and Lemma 3.1 applies. The most important point to notice here is that n, and hence R, scales with k_γ.

We are now in a position to compute the bounds. Substituting into the expressions above gives

$$NU^* \leq 2m\,,$$

so that, with $a(0) = K_\phi^{-1} \max_{0 \leq i < d} \|\phi(-i)\|^2$,

$$\sup_{t>0} a(t) \leq \{R_{\mathrm{m}}, K_0 a(0) + k_0\}\,,$$

where

$$R_{\mathrm{m}} = K_0 R + k_0 \text{ with } K_0 = g_1^{2m}, \quad k_0 = \sum_{i=0}^{2m-1} g_1^i k_1\,.$$

R is as defined above and the first result follows.

We also have

$$a(t) \leq \max\{R'_{\mathrm{m}}, e^{-\delta t} K'_0 a(0) + k'_0\}\,,$$

where the constants are defined so that

$$R'_{\mathrm{m}} = K'_0 R + k'_0 \text{ with } K'_0 = \left(\frac{g_1}{g_2}\right)^{2m}, \quad k'_0 = \sum_{i=0}^{2m-1} \left(\frac{g_1}{g_2}\right),$$

and

$$\delta = -\ln g_2 - U^*(-\ln g_2 + \ln g_1) \geq -\frac{1}{8} \ln g_2^*\,,$$

and exponential stability of the compact set follows.

The expressions above give horrible bounds. This is largely due to the fact that we allow $r = 0$. Sudden changes in the disturbance then cause large excursions in the estimated parameters when $\|\phi(t)\|$ is small. Consider for example the case $y^* = 0$ and $b(t) = 0$ for $t < t_c$ and $b(t) = 1$ for $t \geq t_c$. With $r = 0$ we then get convergence towards zero. The sudden change in the disturbance at $t = t_c$ then causes all the parameters to jump to their bounds and we experience a large (but bounded) transient. One obvious way to remedy this problem is to slow down the (local) speed of adaptation by setting $r > 0$ and average over more samples. This idea is pursued below.

By letting $V(t) = \tilde{\theta}(t)' P^{-1} \tilde{\theta}(t)$ we have

$$\begin{aligned} V(t) &\leq V(t-1) - \frac{e(t)^2}{n(t)} + 2\frac{\gamma(t)e(t)}{n(t)} \\ &= V(t-1) - \frac{e(t)^2}{n(t)} + \frac{\gamma(t)^2}{r} \\ &= V(t-1) - \frac{(\phi(t-d)'\tilde{\theta}(t-1))^2}{n(t)} + \frac{\gamma(t)^2}{n(t)}\,. \end{aligned} \tag{17}$$

Hence,

$$\sum_{i=1}^{t} \frac{e(i)^2}{n(i)} \leq \sum_{i=1}^{t} \frac{\gamma(i)^2}{r} + V(0)$$

and

$$\sum_{i=1}^{t} \frac{(\phi(i-d)'\tilde{\theta}(i-1))^2}{n(i)} \le \sum_{i=1}^{t} \frac{\gamma(i)^2}{n(i)} + V(0).$$

The latter expression is not needed here, but will be used in the proof of Proposition 4.5.

Now the tracking error satisfies

$$y(t) - y^*(t) = r\frac{e(t)}{n(t)} + \phi(t-d)'(\theta(t-d) - \theta(t)).$$

The last term is called the swapping term. From the algorithm we have

$$|\theta(t-d) - \theta(t)| \le \left| \sum_{i=0}^{d-1} \frac{P\phi(t-d-i)e(t-i)/n(t-i)}{n(t-i)} \right| .$$

and it follows that

$$\begin{aligned}(\phi(t-d)'(\theta(t-d) - \theta(t)))^2 &\le \left(\sum_{i=0}^{d-1} \frac{\phi(t-d)'P\phi(t-d-i)e(t-i)}{n(t-i)} \right)^2 \\ &\le 2^{d-1}\|\phi(t-d)\|^2 p_{\max} \sum_{i=0}^{d-1} \frac{e(t-i)^2}{n(t-i)} .\end{aligned}$$

Since

$$\left(\frac{e(t)}{n(t)}\right)^2 \le \frac{1}{r}\frac{e(t)^2}{n(t)},$$

we get, using the expression for the tracking error [8]

$$\begin{aligned}\|y(t) - y^*(t)\|_T &\le \sqrt{r}\|e(t)/\sqrt{n(t)}\|_T + \|\phi(t-d)'(\theta(t-d) - \theta(t))\|_T \\ &\le \sqrt{\|\gamma(t)\|_T^2 + rV(0)} + \|\phi(t-d)'(\theta(t-d) - \theta(t))\|_T .\end{aligned}$$

Since $\sup_{t>-d} \|\phi(t)\|^2 \le \bar{R}_{\mathrm{m}}$, where

$$\bar{R}_{\mathrm{m}} = K_\phi \max \left\{ K_0 R + k_0, g_1^{2m} \max_{0 \le i < d} \|\phi(-i)\|^2 \right\} + k_\phi$$

due to boundedness, it follows that

$$\begin{aligned}\|y(t) - y^*(t)\|_T \le &\left(1 + \sqrt{2^{d-1} d \bar{R}_{\mathrm{m}} \frac{p_{\max}}{r}}\right) \|\gamma(t)\|_T^2 \\ &+ \left(\sqrt{r} + \sqrt{2^{d-1} d \bar{R}_{\mathrm{m}} \frac{p_{\max}}{r}}\right) \sqrt{V(0)} .\end{aligned}$$

[8] $\|\cdot\|_T$ denotes the truncated l_2-norm of a function.

For $r > 0$ (slow adaptation), there exists a constant c_1 so that $\bar{R}_m \le c_1 r/p_{max}$ and we can rewrite this so that

$$||y(t)-y^*(t)||_T \le \left(1+\sqrt{2^{d-1}dc_1}\right)||\gamma(t)||_T^2+\left(\sqrt{r}+\sqrt{2^{d-1}d\bar{R}_m\frac{p_{max}}{r}}\right)\sqrt{V(0)}.$$

The result now follows by taking limits since $\gamma(t) = F(q^{-1})A(q^{-1})b(t)$. The definitions of the gains are obvious. □

Note that it is not critical to use slow adaptation to achieve the finite gain result. We still get a finite gain result for $r = 0$. However, the expressions for the gains become even more cumbersome to develop. More explicit expressions are developed in Proposition 4.5. The size of the gains depend on the speed of adaptation and the performance of the adaptive system improves as the (local) rate of adaptation is decreased. More explicit formulas can be developed to highlight this point and a more detailed analysis will show that it is advantageous to use *time varying* gains. The gains should be high during transients and then they should decrease to zero during steady state performance. This analysis, when carried out in detail, shows why algorithms based on recursive least squares and fixed dead-zones work well in practical applications. Here the algorithms are usually re-initialized with high gains during a transient, using a reset and/or a variable forgetting factor mechanism. Then the gains are allowed to decrease naturally through the "covariance matrix" update, and eventually, when the algorithm reaches steady state, the estimator is simply turned off. It turns out not to be so important how the details of this general approach are implemented.

We now consider an example problem which explains what kind of dynamics can be observed when the gains are non-vanishing.

Example 4.1. (Golden and Ydstie [6]). Consider adaptive control of

$$y(t) = Ku(t-1) + v\,, \tag{18}$$

where K is an unknown gain and v is a constant but unknown perturbation. In this case the adaptive system can be re-parameterized by setting

$$x(t) = \frac{1}{K}\theta(t)\left(1-\frac{v}{y^*}\right)$$

$$\alpha = (1-\frac{v}{y^*})p\,,$$

$$\beta = K^2\frac{r}{(y^*-v)^2} > 0 \text{ for } y^* \ne v\,.$$

(The projection has been omitted). The two design parameters p and r enter into α and β respectively. Note that α represents the noise to signal ratio multiplied with the adaptive gain and that β is non-negative. The closed-loop adaptive system is now described by the mapping $F : R \to R$

$$x \to x + \frac{\alpha(1-x)}{\beta x^2+1}\,, \tag{19}$$

and the system converges so that $\lim_{t\to\infty} x(t) = 1$ if and only if $0 < \alpha/(\beta+1) < 2$. We note that this imposes a strict bound on the noise to signal ratio and the adaptive gain. If this condition is violated then it can be shown that the map period doubles and the dynamics may become complicated. A fairly elaborate theory applies and can be used to trace the evolution of the attractors to a fair degree of accuracy. In particular, it can be shown that for an infinite family of parameters the Schwartzian derivative

$$SF(y) = \frac{D^3F(y)}{DF(y)} - \frac{3}{2}\left(\frac{D^2F(y)}{DF(y)}\right)^2$$

of a transformed mapping equivalent to (19) is negative. We conclude that all points (with respect to Lebesque measure) have the same asymptotic behavior; there is exactly one absolutely continuous invariant measure and the adaptive system is stationary and ergodic. Moreover, there exists a continuous sub-family where (19) has no stable attractor of finite period. It follows that the adaptive system acts somewhat like a Markov chain with a countable (infinite) number of states, the attractor is a closed set which contains no interior or isolated points (a Cantor set) and the dynamics are chaotic [5].

A simulation showing the behavior of the process output is given in Fig. 6. The parameter estimate drifts towards the fixed point, "overshoots" and gives oscillatory behavior, reminiscent of the drift and burst phenomenon. The amplitude of the oscillation is small as seen at the process output, and it is clear that aperiodicity and lack of convergence do *not* lead to poor performance. By averaging over the sample paths we have, due to ergodicity,

$$E\{y - y^*\} = \lim_{t\to\infty} \frac{1}{t}\sum_{i=1}^{t}(y(i) - y^*), \quad E\left\{(y - y^*)^2\right\} = \lim_{t\to\infty} \frac{1}{t}\sum_{i=1}^{t}(y(i) - y^*)^2 .$$

The simulation indicates that $E\left\{(y - y^*)^2\right\} \le v^2$ when the projection is included. □

Example 4.2. (Praly and Pomet [11]). Consider the system

$$y(t) = ay(t-1) + u(t-1) + v .$$

This is the same example treated in the beginning of Sect. 3. However, the set point is equal to zero. In this case the parameter drift and bursting is transient and the adaptive system converges either to a stable (elliptic) period-two attractor, which is optimal, *or* a period-ten attractor which yields slightly higher variance. Further discussions of this problem are given in [6] and [11]. □

Finally, we introduce the model-order mismatch. We consider the case with fast adaptation, no output disturbance and zero setpoint first. Define the σ-dependent parameter

$$K_v^* = \frac{1}{2b}\left(1 + bc + a\epsilon^{2N^*} - \sqrt{(1 + bc + a\epsilon^{2N^*+d+1})^2 - 4ac\epsilon^{2N^*+d+1}}\right),$$

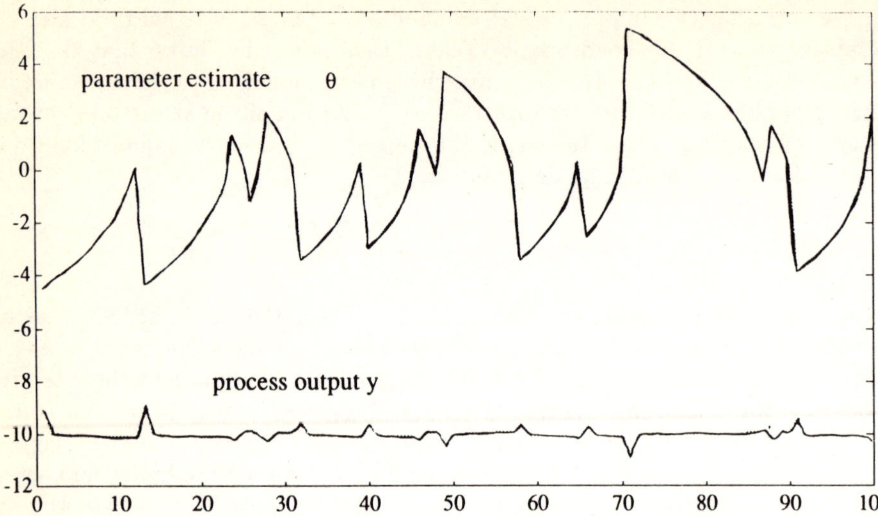

Fig. 6. Simulation showing parameter drift and bursting in a simple adaptive feedback system.

where

$$a = \frac{2N^* c_1 p_{\min} g_1^{-(d+1)}(g_2 - \sigma)}{(2N^* + d + 1)2^{d+3} p_{\max} M_\theta K_\phi}$$

$$b = \frac{c_1 c_2^2 p_{\min}}{(2N^* + d + 1)K_\phi p_{\max} 2^{d+2}}$$

$$c = \frac{1}{2c_1} B^*(1 - \frac{\sigma}{g_2})/|F(q^{-1})|_\infty$$

$$\epsilon = \frac{g_2}{K_\theta^2} .$$

We have the following result on robust performance of the adaptive system when there are no external perturbations.

Proposition 4.4. *Suppose $b(t) \equiv 0, y^*(t) \equiv 0, r = 0$. Let σ^* be the smallest number such that $||H_0^*(q^{-1})||_{2,\sigma^*} \leq K_v^*$ and $B(z^{-1}) \neq 0$ for $|z| \geq \sigma^*$. Provided that $\sigma^* < 1$, we have*

$$\sup_{t>0} \| \phi(t)\|^2 \leq g_1^{2m} \left(\max_{0 \leq i < d} ||\phi(-i)||^2 + K_\phi^{-1} m_0 \right) .$$

Moreover, the convergence is exponential, so that

$$||\phi(t)||^2 \leq \left(\frac{g_1}{g_2}\right)^{2m} e^{-\delta t}(\max_{0 \leq i < d} ||\phi(-i)||^2 + K_\phi^{-1} m_0)$$

for $t > 0$, *where*

$$\delta = -\frac{1}{6}\ln g_2^* \,.$$

Consequently, $\lim \|\phi(t)\| = 0$, *which implies* $\lim y(t) = \lim u(t) = 0$.

Proof. Let $K_\gamma = |F(q^{-1})|_\infty K_v^*$. From Lemma 3.2 with $k_\gamma = k_{y^*} = r = 0$

$$U^*(R,N,g_2) = \frac{g_2}{g_2-\sigma}\left(c_1 K_\gamma + 2^{d+1} K_\phi d \frac{p_{\max}}{p_{\min}}\right.$$
$$\left.\times\left(\frac{M_\theta}{N} + \frac{4(N+d+1)K_\gamma}{N\left(\left(\frac{g_2}{K_\theta^2}\right)^{N+d+1}\frac{g_1^{-(d+1)}}{M_\theta}(g_2-\sigma) - 2c_2^2 K_v\right)}\right)\right).$$

As before, we require

$$U^*(R,N,g_2) < \frac{-\ln g_2}{\ln g_1 - \ln g_2}\,.$$

In this case the analysis works with $R = 0$. Setting $g_2 = g_2^*$ and $N = 2N^*$ yields the condition

$$\frac{g_2}{g_2-\sigma}\left(c_1 K_\gamma + 2^{d+3} K_\phi d \frac{p_{\max}}{p_{\min}}\right.$$
$$\left.\times\frac{K_\gamma(2N^*+d+1)}{2N^*\left(\left(\frac{g_2}{K_\theta^2}\right)^{2N^*+d+1}\frac{g_1^{-(d+1)}}{M_\theta}(g_2-\sigma) - 2c_2^2 K_v\right)}\right) \le \frac{1}{2}B^*,$$

which, for the purpose of developing explicit bounds, we rewrite as

$$K_v\left(1 + \frac{1}{a\epsilon^{2N^*+d+1} - K_v b}\right) \le c\,,$$

where a, b, c and ϵ are defined above. It now follows from Lemma 3.1 that

$$\sup_{t>0} a(t) \le g_1^{NU^*} a(0)\,,$$

where $NU^* = 2m$ and $a(0) = K_\phi \max_{0\le i<d} \|\phi(-i)\| + m_0$. Similarly,

$$a(t) \le \left(\frac{g_1}{g_2}\right)^{NU^*} e^{-\delta t} a(0)\,.$$

After substituting back we find $U^* \le \frac{5}{6}B^*$. Hence, $\delta = \ln g_2(1-\frac{5}{6})$, and the result follows. □

This result is remarkable in the sense that if there are only unmodeled dynamics present, then we can guarantee exponential stability and convergence to zero error when the mismatch is not too serious. Since bounded perturbations cannot cause a global instability, this implies that the adaptive algorithm can handle some plants with zeros on or even outside the unit circle provided there is a time scale separation. (The unmodeled unstable zeros should be sufficiently fast or slow relative to the ones we model). This result does not say anything about the properties of the estimates. This result was anticipated in [8].

Example 4.3. (Mareels and Bitmead [8]). Consider adaptive control of the system

$$y(t) = ay(t-1) + u(t) + cy(t-2)\,.$$

The parameter a is estimated and the term $cy(t-2)$ is left as unmodeled dynamics. The setpoint equals zero. Mareels and Bitmead show that under certain conditions the parameter estimate undergoes period doubling bifurcations and behaves erratically. Convincing evidence has been presented that the parameter estimate displays at least transient chaos. In particular, it has been shown that the fourth iterate of the mapping satisfies a set of conditions leading up to the application of the Smale-Birkhoff "Homoclinic Theorem". (It is shown that there exists a transversal intersection between the $W^s(p)$ and $W^u(p)$ at a point $q \neq p$, where p is a hyperbolic fixed point. A similar type of analysis is used in [6] to indicate the presence of chaos in the mapping described in Example 3.1). The Homoclinic Theorem then guarantees the existence of a hyperbolic invariant set topologically similar to a horseshoe. This again gives a strong indication, if not a proof, of the existence of a strange attractor. □

The example discussed by Mareels and Bitmead must be treated with some caution since the transformation used to define the mapping in question may be singular in the limit $t = \infty$. In fact, whether or not the adaptive system will display persistent chaos in this example depends on how this particular problem is dealt with. Nevertheless, it is safe to say that there is transient chaos present here. Examples 3.1 and 4.1 should dispel any further doubt whether or not chaos is present asymptotically. In fact, it appears that chaos will almost always be present when adaptive controllers are applied to practical control systems. The overriding and important point to notice is that, if the algorithm is implemented properly, then the erratic behavior of the parameter estimates does *not* translate into poor performance. This is eminently clear in the case discussed above, since the adaptive system converges, so that $\lim y(t) = 0$ exponentially fast [8]. Proposition 4.4 generalizes this to the case where several parameters are estimated.

Finally, we tie everything together.

Proposition 4.5. (Ultimate Boundedness and Robustness). *Suppose $y^*(t)$ and $b(t) \in l_\infty$; that is, there exists constants so that $\sup_{t>0} y^*(t)^2 \leq k_{y^*}$ and $\sup_{t>0} b(t)^2 \leq k_b$. Let σ^* be the smallest constant so that $\|H_0^*(q^{-1})\|_{2,\sigma^*} \leq 0.5K_v^*$ and $B(z^{-1}) \neq 0$ for $|z| \geq \sigma^*$. If $\sigma^* < 1$ then the STR with parameter projection is stable in the sense that*

$$\sup_{t>0} \|\phi(t)\|^2 \leq K_\phi \max\{R_{\mathrm{m}}, K_0 \max_{0 \leq i < d} \|\phi(-i)\|^2 + k_0 + K_\phi^{-1} m_0\} + k_\phi\,,$$

where R_{m}, K_0 and k_0 are positive constants and m_0 overbounds the initial conditions. Convergence to a compact set which scales with respect to k_{y^*} and k_b is exponential in the sense that

$$\|\phi(t)\|^2 \leq K_\phi \max\{R'_{\mathrm{m}}, e^{-\delta t} K'_0 \max_{0 \leq i < d} \|\phi(-i)\|^2 + k'_0 + K_\phi^{-1} m_0\} + k_\phi\,,$$

where $\delta = -\frac{1}{6}\ln g_2^*$, R'_{m}, K'_0 and k'_0 are positive constants, independent of initial conditions. In addition, we have the following finite gain result

$$\|y - y^*\|_2 \le \mu_0 + \mu|F(q^{-1})H_1^*(q^{-1})|_\infty \|b\|_2 + \mu|F(q^{-1})H_0^*(q^{-1})|_\infty \|y^*\|_2 ,$$

where the gains μ_0 and μ are defined below.[9]

Proof. It is going to be convenient to choose $g_2 = g_2^*$. In addition, define a parameter

$$R^* = \max\left\{2\frac{k_3 + k_\gamma}{c_1 K_\gamma}, \frac{k_\phi + r/p_{\max}}{K_\phi}, \frac{k_{y^*} + k_v}{K_v}\right\},$$

where $K_\gamma = 0.5|F(q^{-1})|_\infty K_v^* \neq 0$. By choosing $R = R^*$ we can (after some algebra and hindsight) rewrite the expression for U^* so that the following inequality is satisfied

$$U^*(R, N, \sigma, g_2) \le \frac{g_2}{g_2 - \sigma}\left[2c_1 K_\gamma + 2^{d+2} K_\phi d \frac{p_{\max}}{p_{\min}} \times \left(\frac{M_\theta}{N} + \frac{4}{N}\frac{(N+d+1)K_\gamma}{\left(\frac{g_2}{K_\theta^2}\right)^{N+d+1} \frac{g_1^{-(d+1)}}{M_\theta}(g_2 - \sigma) - 2c_2^2 K_v}\right)\right].$$

By dividing through with 2 this problem is essentially the same as that leading up to the result in Proposition 4.4. We get the bounds for transient performance by defining the constants as in Proposition 4.3 and replacing R by R^* as defined above. We then have

$$\sup_{t>0} a(t) \le \bar{R} \text{ with } \bar{R} = K_0 R^* + k_0$$

and

$$\sup_{t>0} \| \phi(t) \|^2 \le \bar{R}_{\mathrm{m}} \text{ with } \bar{R}_{\mathrm{m}} = K_\phi \bar{R} + k_\phi$$

Following a procedure similar to the one developed in Proposition 4.3 we get the finite gain result. The difference is that we will simplify the expressions by assuming that the (local) adaptation is slow enough to have

$$R^* = \frac{k_\phi + r/p_{\max}}{K_\phi}$$

(this is not critical, however it leads to more economical representations).

As before, we start out by defining the tracking error, then we introduce the swapping error and relate the obtained expressions to the comparison function $V(t)$. We get, using the same development as before (see proof of Proposition 4.3)

$$\|y(t) - y^*(t)\|_T \le \left(1 + \sqrt{2^{d-1} d \bar{R}_{\mathrm{m}} \frac{p_{\max}}{r}}\right) \|\gamma(t)\|_T + \sqrt{rV(0)}.$$

[9] Ideally we would like $\mu_0 = \epsilon_0$, $\mu = 1$ to recover the performance bound (7).

There is now an added catch since $\gamma(t)$ contains the effect of the unmodeled terms and is not defined explicitly. First note that

$$\begin{aligned}\|\gamma(t)\|_T \leq& \left|F(q^{-1})H_0^*(q^{-1})\right|_\infty \left(\|y^*(t+d)\|_T + \|\phi(t)'(\tilde{\theta}(t))\|_T\right) \\ &+|F(q^{-1})H_1^*(q^{-1})|_\infty \|b(t)\|_T + m_0/(1-\sigma)\,.\end{aligned}$$

From the definition of $V(t)$, the swapping term, and the fact that the signals are uniformly bounded we now have

$$\begin{aligned}\|\phi(t)'\tilde{\theta}(t)\|_T \leq& \sqrt{1+\bar{R}\frac{p_{\max}}{r}}\|\gamma(t+d)\|_T + \|\phi(t)'(\theta(t)-\theta(t+d-1))\|_T \\ &+\sqrt{(r+p_{\max}\bar{R}_{\mathrm{m}})V(0)} \\ \leq& \sqrt{1+\bar{R}\frac{p_{\max}}{r}}\|\gamma(t)\|_T + d\sqrt{K_\gamma \bar{R} + k_\gamma} \\ &+\|\phi(t)'(\theta(t)-\theta(t+d-1))\|_T + \sqrt{(r+p_{\max}\bar{R}_{\mathrm{m}})V(0)}\,.\end{aligned}$$

By using the expression for the swapping error from Proposition 4.3,

$$(\phi(t-d)'(\theta(t-d)-\theta(t-1)))^2 \leq 2^{d-1}\|\phi(t-d)\|^2 p_{\max}\sum_{i=0}^{d-1}\frac{e(t-i)^2}{n(t-i)}\,.$$

Using the ultimate bounds and the slow adaptation,

$$\bar{R}_{\mathrm{m}} = K_\phi\left(K_0\left(\frac{k_\phi + r/p_{\max}}{K_\phi}\right) + k_0\right) + k_\phi\,.$$

Hence

$$\|\phi(t)'(\theta(t)-\theta(t+d-1))\|_T \leq \sqrt{2^{d-1}d\bar{R}_{\mathrm{m}}p_{\max}}\|\gamma(t)\|_T/\sqrt{r} + \sqrt{V(0)}\,.$$

We can then rewrite the inequality for $\gamma(t)$ as

$$\begin{aligned}\|\gamma(t)\|_T \leq& |F(q^{-1})H_0^*(q^{-1})|_\infty \Big[\|y^*(t+d)\|_T \\ &+\left(\sqrt{1+\bar{R}\frac{p_{\max}}{r}} + \sqrt{2^{d-1}d\bar{R}_{\mathrm{m}}\frac{p_{\max}}{r}}\right)\|\gamma(t)\|_T + d\sqrt{K_\gamma\bar{R}+k_\gamma} \\ &+\sqrt{(r+p_{\max}\bar{R})V(0)} + \sqrt{2^{d-1}d\bar{R}_{\mathrm{m}}\frac{p_{\max}}{r}}\sqrt{V(0)}\Big] \\ &+|F(q^{-1})H_1^*(q^{-1})|_\infty\|b(t)\|_T + m_0/(1-\sigma)\,,\end{aligned}$$

which gives

$$\begin{aligned}(1-g)\|\gamma(t)\|_T \leq& c_0 + |F(q^{-1})H_0^*(q^{-1})|_\infty\|y^*(t+d)\|_T \\ &+|F(q^{-1})H_1^*(q^{-1})|_\infty\|b(t)\|_T\,,\end{aligned}$$

where

$$g = |F(q^{-1})H_0^*(q^{-1})|_\infty \left(\sqrt{1+\bar{R}_m\frac{p_{\max}}{r}} + \sqrt{2^{d-1}d\bar{R}_m\frac{p_{\max}}{r}}\right),$$

$$c_0 = d\sqrt{K_\gamma \bar{R} + k_\gamma} + \sqrt{(r+\frac{p_{\max}}{r})V(0)} + \sqrt{2^{d-1}d\bar{R}_m\frac{p_{\max}}{r}}\sqrt{V(0)} + \frac{m_0}{1-\sigma}.$$

This we then substitute back into the expression for the tracking error developed in the proof of Proposition 4.3. However, to proceed we will need $g < 1$. This is automatically satisfied by the bound imposed, since K_v^* will have to be quite small for this approach to work at all.

From the expressions above we then finally get

$$\begin{aligned}\|y(t)-y(t)^*\|_T \le \mu_0 &+ \mu|F(q^{-1})H_1^*(q^{-1})|_\infty\|b(t)\|_T \\ &+ \mu|F(q^{-1})H_0^*(q^{-1})|_\infty\|y^*(t+d)\|_T,\end{aligned}$$

where

$$\mu_0 = \frac{c_0 + \sqrt{rV(0)} + \sqrt{2^{d-1}d\bar{R}_m\frac{p_{\max}}{r}}\sqrt{V(0)}}{1-g}$$

and

$$\mu = \frac{1+\sqrt{d2^{d-1}\bar{R}_m\frac{p_{\max}}{r}}}{1-g}$$

and the result follows by extending the norms. □

A few comments are now in order.

It is clear from Proposition 4.5 that the design parameters influence the gains and that we can improve performance by tuning. It appears particularly effective to slow down the rate adaptation after a transient. In the limit ($p_{\max}/r \to 0$) we have

$$\mu = \frac{1+\sqrt{K_0}}{1-|FH_0|_\infty}.$$

$|FH_0|_\infty$ is quite small and K_0 can be made close to one by using a separate backwards in time argument. It follows that $\mu \approx 2$ and the performance of the adaptive system is almost as good as that of the optimized non-adaptive system. This performance is achieved at the cost of a very long transient due to the small gains. One thing which is clear is that during a transient the gains should be high and the condition number of the gain matrix should be close to one.

The amount of mismatch that can be tolerated by the algorithm discussed in Section 2 is very small. We believe that this is due to an overly conservative analysis and current work is aimed towards tightening up the robustness result as well as the performance bounds.

We recover the ideal case (Propositions 4.1 and 4.2) since $|H_0^*|_\infty = 0$ and $b \in l_2 \cap l_\infty$ implies $(y-y^*) \in l_2 \cap l_\infty$. If $|H_0^*|_\infty \neq 0$ and $b \in l_\infty$ only, then we recover Proposition 4.3 and the dynamics become complicated. All we can say is that that $(y-y^*) \in l_\infty$. If $y \in l_2$ and $b \in l_2$ then we recover Proposition 4.4 and all signals converge exponentially fast. The latter result best describes the transient behavior of the algorithm.

Example 4.5. (Ydstie [13], Golden and Ydstie [5]) Consider adaptive control of the system

$$y(t) + ay(t-1) = bu(t-1) + cy(t-2) + v\,.$$

The parameter a is estimated, v is here a constant perturbation and c represents an unmodeled pole. There are two sources of mismatch in the adaptive control system and the dynamics become complicated, but constrained to a small subset of the phase space. □

5 Conclusions and Discussion

In this paper it has been shown that the discrete self tuning regulator with non-vanishing gain can be applied to linear systems that are undermodeled, provided that the estimated parameters are projected into a compact convex set which contains the stabilizing parameters and excludes the possibility of dividing with zero. The latter problem implies that we know the sign and a lower bound for the high frequency gain of the modeled part of the system. The performance of the adaptive system is related to the size of the external perturbations and we obtain the "ideal result" as these tend to zero.

We have established the following design guidelines:

- The condition number of P should be close to one during transients.
- Fast adaptation should be used (r small and/or $p_{\min}$ large) initially to minimize the transient.
- Slow adaptation should be used to improve local performance.
- Prior knowledge of the parameter values, in particular $\beta_{\min}$, should be built into Θ^*.
- The data vector should be filtered so that $\|\phi(t)\| = 0$ at steady state.

This points towards the use of one of the robustified Recursive Least Squares Algorithms. These usually are initialized so that the gains are high in the transient phase and then the gains decrease rapidly as the system approach steady state. It can be shown that this not only improves the transient performance, but it also gives the algorithm the ability to handle more unmodeled dynamics.

The results extend readily to indirect adaptive laws like pole assignment, optimal and predictive controllers. Such control laws are used in practice, since they can be applied even when the stable invertibility condition is violated. Some care will have to be taken to ensure stabilizability of the estimated control model when such laws are applied.

Acknowledgements

The author is grateful for numerous discussions with Iven Mareels at the University of Newcastle, Australia, Laurent Praly at Ecole des Mines de Paris, France, and Chris Hollot at the University of Massachusetts. Finally, I would to thank

Melinda Golden, who is currently working for Simulation Sciences in Los Angeles, for many inspiring discussions while she was working on her PhD on the Dynamics of Adaptive Control Systems.

References

1. K. J. Åström and B. Wittenmark, "On self-tuning regulators," *Automatica*, vol. 9, pp. 195–199, 1973.
2. B. Egard, *Stability of adaptive controllers*, Springer-Verlag, New York, NY, 1979.
3. T. R. Fortescue, L. S. Kershenbaum, and B. E. Ydstie, "Self-tuning regulators with variable forgetting factors," *Automatica*, vol. 17, pp. 831–835, 1981.
4. G. C. Goodwin, P. J. Ramadge, and P. E. Caines, "Discrete time multivariable adaptive control," *IEEE Trans. Aut. Control*, vol 25, pp. 449–456, 1990.
5. M. P. Golden and B. E. Ydstie, "Parameter drift in adaptive control systems: Bifurcation analysis of a simple case," *Tech. Report, University of Massachusetts at Amherst*, 1990.
6. M. P. Golden and B. E. Ydstie, "Small amplitude chaos and ergodicity in adaptive control systems," *Tech. Report, University of Massachusetts at Amherst*, 1990. To be presented at the *IFAC Symposium on Identification and System Parameter Estimation*, Budapest, Hungary, 1991.
7. P. Ioannou and J. Sun, "Theory and design of robust direct and indirect adaptive control systems," *Int J. Control*, vol. 47, pp. 775–813, 1988.
8. I. M. Y. Mareels and R. R. Bitmead, "Nonlinear dynamics in adaptive control: chaotic and periodic stabilization," *Automatica*, vol. 22, pp. 641–665, 1986.
9. L. Praly, "Robustness of model reference adaptive control," *Proc. 2nd Yale Workshop on Adaptive Systems*, 1983.
10. L. Praly, *Commande lineaire adaptative: Solutions bornees et leurs proprietes.* PhD thesis, Universite Paris IX, Dauphine, Paris, France, 1988.
11. L. Praly and J.-B. Pomet, "Periodic solutions in adaptive systems: the regular case," *Proc. 10th IFAC World Congress*, Munich, vol. 10, pp. 40–45, 1987.
12. C. Samson, *Problemes en identification et commande de systemes dynamique*, PhD thesis, Universite de Rennes, France, 1983.
13. B. E. Ydstie, "Bifurcations and complex dynamics in adaptive control systems," *Proc. 25th IEEE Conf. Dec. Control*, Athens, Greece, 1986.
14. B. E. Ydstie, "Stability of discrete model reference control–revisited," *Syst. Control Lett.*, vol. 13, pp. 429–438, 1989.
15. B. E. Ydstie and M. P. Golden, "Chaos and strange attractors in adaptive control systems," *Proc. 10th IFAC World Congress*, Munich, vol. 10, pp. 127–132, 1987.

Appendix 1: Proof of Lemma 3.2

Multiplying through with $A(t)$ in equation (13) we obtain

$$A(t)a(t+1) \leq A(t)\left(\sigma a(t) + (\phi(t)'\tilde{\theta}(t))^2\right).$$

And by dividing through with $a(t+1)$ and using $a(t+1) \geq g_2 a(t)$,

$$\left(1 - \frac{\sigma}{g_2}\right) A(t) \leq \frac{1}{a(t+1)} \left(\phi(t)'\tilde{\theta}(t)\right)^2 . \tag{20}$$

To determine the relative occurrence of the events $A(t) = 1$ we need to relate this equation to a very specific gradient property of the parameter estimator.

The swapping error $\eta(t) = \phi(t-d)'(\theta(t-1) - \theta(t-d))$ satisfies

$$|\eta(t)| \leq \|\phi(t-d)\| \sum_{i=1}^{d-1} \|P\phi(t-d-i)e(t-i)/n(t-i)\|,$$

where we used the fact that $F_{\Theta^*}\{.\}$ projects orthogonally onto Θ^* so that $\|F_{\Theta^*}\{\theta^* - \theta\}\| \leq \|\theta^* - \theta\|$ for all $\theta \in \Theta^*$.

Since $\|P\phi(t-d)\|^2/n(t) \leq p_{\max}$ we then have, using (10)

$$|\eta(t)| \leq (K_\phi a(t-d+1) + k_\phi)^{1/2} p_{\max}^{1/2} \sum_{i=1}^{d-1} \left| \frac{e(t-i)}{n(t-i)^{1/2}} \right| .$$

Here we used the fact that $a(t) \geq m(t)$. By noting that $\phi(t-d)'\tilde{\theta}(t-d) = e(t) - \gamma(t) + \eta(t)$ we have the inequality

$$\begin{aligned} \left|\phi(t)'\tilde{\theta}(t)\right| &\leq |e(t+d)| + |K_\gamma a(t+d+1) + k_\gamma|^{1/2} \\ &\quad + |K_\phi a(t+1) + k_\phi|^{1/2} p_{\max}^{1/2} \sum_{i=1}^{d-1} \left| \frac{e(t+d-i)}{n(t+d-i)^{1/2}} \right| , \end{aligned}$$

so that from (20)

$$\begin{aligned} A(t)\left(1 - \frac{\sigma}{g_2}\right) \leq \frac{1}{a(t+1)} &\left[n(t+d)^{1/2} \frac{|e(t+d)|}{n(t+d)^{1/2}} + |K_\gamma a(t+d+1) + k_\gamma|^{1/2} \right. \\ &\left. + (K_\phi a(t+1) + k_\phi)^{1/2} p_{\max}^{1/2} \sum_{i=1}^{d-1} \frac{|e(t+i)|}{n(t+i)^{1/2}} \right]^2 . \end{aligned}$$

Now, $a(t)$ enjoys the same bounded rate of growth property as $m(t)$ (Equation (12)), so that for all t and $i > 0$ we have

$$a(t+i) \leq g_1^i a(t) + k_3, \quad k_3 = K_\gamma \sum_{j=0}^{i-1} g_1^j k_1 .$$

Since $n(t+d) \leq r + p_{\max}\|\phi(t)\|^2 \leq r + p_{\max}(K_\phi a(t+1) + k_\phi)$ this gives, by re-arranging the expression above

$$\begin{aligned} A(t)\left(1 - \frac{\sigma}{g_2}\right) &\leq \frac{1}{a(t+1)} \left[|g_1^d K_\gamma a(t+1) + k_3 + k_\gamma|^{1/2} \right. \\ &\quad \left. + |r + p_{\max}(K_\phi a(t+1) + k_\phi)|^{1/2} \sum_{i=1}^{d} \frac{|e(t+i)|}{n(t+i)^{1/2}} \right]^2 \\ &\leq c_1 K_\gamma + 2\frac{k_3 + k_\gamma}{a(t+1)} + 2^d \left(p_{\max} K_\phi + \frac{r + k_\phi p_{\max}}{a(t+1)} \right) \\ &\quad \times \sum_{i=1}^{d} \frac{e(t+i)^2}{n(t+i)} , \end{aligned} \tag{21}$$

where $c_1 = 2g_1^d$. From this we then have by summing over the interval $[t-N,t]$

$$\left(1-\frac{\sigma}{g_2}\right)\sum_{i=t-N}^{t} A(i) \leq \sum_{i=t-N}^{t}\left(c_1 K_\gamma + 2\frac{k_3+k_\gamma}{a(i+1)}\right.$$
$$\left. + 2^d\left(p_{\max}K_\phi + \frac{r+p_{\max}k_\phi}{a(i+1)}\right)\sum_{j=1}^{d}\frac{e(i+j)^2}{n(i+j)}\right). \quad (22)$$

The last term is the sum of squared a posteriori errors. These are known to satisfy a smallness in the mean condition when the parameter estimates and the unmodeled terms are bounded. Below we show that $e(t)^2/n(t)$ is small in the mean on the sets where the adaptive controller is diverging. We use this property to demonstrate that the regressor measures $a(t)$ and from this we get ultimate boundedness of the extended state when there are unmodeled dynamics. To elaborate on this point we introduce another comparison function, namely $V(t) = \tilde{\theta}(t)'P^{-1}\tilde{\theta}(t)$. A bit of algebra shows that with the projection we have

$$V(t) \leq V(t-1) - \frac{e(t)^2 - 2e(t)\gamma(t)}{n(t)} \leq V(t-1) - \frac{1}{2}\frac{e(t)^2}{n(t)} + 2\frac{\gamma(t)^2}{n(t)}.$$

Let R be an arbitrary positive number. By assuming $a(i) \geq R$ for $i \in [t-N+1,\ t+d]$, we obtain from inequality (22)

$$\left(1-\frac{\sigma}{g_2}\right)\sum_{i=t-N}^{t} A(i) \leq N\left(c_1K_\gamma + 2\frac{k_3+k_\gamma}{R}\right) + 2^{d+1}\left(p_{\max}K_\phi + \frac{r+k_\phi p_{\max}}{R}\right)$$
$$\times\, d\sum_{i=t-N+1}^{t+d}\left(V(i-1) - V(i) + 2\frac{\gamma(i)^2}{n(i)}\right).$$

Hence,

$$\left(1-\frac{\sigma}{g_2}\right)\sum_{i=t-N}^{t} A(i) \leq N\left(c_1K_\gamma + 2\frac{k_\gamma+k_3}{R}\right)$$
$$+ 2^{d+1}\left(K_\phi p_{\max} + \frac{r+p_{\max}k_\phi}{R}\right)$$
$$\times\, d\left[V(t-N) - V(t+d) + 2(N+d+1)p_{\min}^{-1}\right.$$
$$\left.\times \max_{t-N+1\leq i\leq t+d}\frac{K_\gamma a(i+1)+k_\gamma}{\|\phi(i-d)\|^2 + r/p_{\min}}\right]. \quad (23)$$

What is required now is to show that the last term, under suitable conditions, is small. In other words, we need to show that the regressor overbounds the unmodeled dynamics. We use a "backwards in time" argument which goes as follows: First we show that if $A(t_k) = 1$ for one or more events $t_k \in [t-N+1, t+d]$, then the regressor overbounds the measuring signal $a(t)$

on this event. This follows from equation (13) since if $A(t_k) = 1$ then from its definition $[\phi(t_k)'\tilde{\theta}(t_k)]^2 \geq (g_2 - \sigma)a(t_k) + k_2$, so that

$$\|\phi(t_k)\|^2 \geq \frac{1}{M_\theta}\left((g_2 - \sigma)a(t_k) + k_2\right) . \tag{24}$$

We then use a bounded rate of growth argument to show that if R is chosen large and the model mismatch is small then $\|\phi(t-d)\|$ overbounds $a(t+1)$ on a limited number of preceding events, due to a bounded rate of growth property of $\|\phi(t)\|^2$ and the fact that $a(t)$ decreases when $A(t) = 0$.

The expression

$$\phi(t) = A(\theta(t))\phi(t-1) + b_1 v(t) + b_2 y^*(t+d) ,$$

derived from equation (2) and the control law (Step 2), gives an alternative description of the closed loop of the adaptive system. Here $|A(\theta(t))| \leq K_\theta < \infty$ because $\theta(t) \in \Theta^*$. This gives

$$\|\phi(t)\| \leq K_\theta^N \|\phi(t-N)\| + \sum_{i=0}^{N-1} K_\theta^i \left[|y^*(t+d-i)| + |v(t-i)|\right]$$

for all $N > 0$. This bound is obviously very conservative and presents us with a serious problem when it comes to establishing tight bounds.

Dividing through with $a(t+d+1)^{1/2}$ and using $a(t) \geq g_2^i a(t-i)$ and $a(t+d+1) \leq g_1^{d+1}a(t) + k_3$ gives

$$\begin{aligned}\frac{\|\phi(t)\|}{\left(g_1^{d+1}a(t)+k_3\right)^{1/2}} \leq {} & \left(\frac{K_\theta}{g_2^{1/2}}\right)^N \frac{\|\phi(t-N)\|}{a(t+d+1-N)^{1/2}} \\ & + \sum_{i=0}^{N-1}\left[\left(\frac{K_\theta}{g_2^{1/2}}\right)^i \frac{1}{g_2^{(d-1)/2}} \right. \\ & \left. \times \left(\left(\frac{k_{y^*}}{a(t+1-i)}\right)^{1/2} + \left|K_v + \frac{k_v}{a(t+1-i)}\right|^{1/2}\right)\right] .\end{aligned}$$

This obviously gives what is needed, since we now have for all N

$$\begin{aligned}\frac{\|\phi(t-N)\|}{a(t+1+d-N)^{1/2}} \geq {} & \left(\frac{g_2^{1/2}}{K_\theta}\right)^N \frac{\|\phi(t)\|}{\left(g_1^{d+1}a(t)+k_3\right)^{1/2}} \\ & - \sum_{i=0}^{N-1}\left[\left(\frac{g_2^{1/2}}{K_\theta}\right)^{N-i} \frac{1}{g_2^{(d-1)/2}} \left(\left(\frac{k_{y^*}}{a(t+1-i)}\right)^{1/2} \right.\right. \\ & \left.\left. + \left|K_v + \frac{k_v}{a(t+1-i)}\right|^{1/2}\right)\right] .\end{aligned}$$

The last term defines a geometric series which converges as $N \to \infty$ (quite fast, as it turns out). It follows, by still assuming $a(i) \geq R$,

$$\frac{\|\phi(t-N)\|}{a(t+1+d-N)^{1/2}} \geq \left(\frac{g_2^{1/2}}{K_\theta}\right)^N \frac{\|\phi(t)\|}{\left(g_1^{d+1}a(t)+k_3\right)^{1/2}} - c_2\left(K_v + 2\frac{k_{y^*}+k_v}{R}\right)^{1/2},$$

where

$$c_2 = \frac{1}{\left(1 - g_2^{1/2}/K_\theta\right) g_2^{(d-1)/2}} .$$

At this point it is convenient to set $k_2 = k_3(g_2 - \sigma)/g_1^{d+1}$. From the above and equation (24) this implies that if $A(t_k) = 1$ for some $t_k \in [t-N, t+d]$, then

$$\frac{\|\phi(t_k - d - i)\|}{a(t_k + 1 - i)^{1/2}} \geq \left(\frac{g_2^{1/2}}{K_\theta}\right)^N \sqrt{\frac{g_1^{-(d+1)}}{M_\theta}(g_2-\sigma)} - c_2\left(K_v + 2\frac{k_{y^*}+k_v}{R}\right)^{1/2}$$

for all $i > 0$. The expression on the right hand side is positive and makes sense if K_v is small and R is large. We now have, using expression (23) and the inequality derived above, still assuming that $a(t+d-i) \geq R$:

$$\begin{aligned}
\left(1 - \frac{\sigma}{g_2}\right)\frac{1}{N}\sum_{i=t-N}^{t} A(i) &\leq c_1 K_\gamma + 2\frac{k_3+k_\gamma}{R} + 2^{d+1}\left(K_\phi + \frac{k_\phi + r/p_{\max}}{R}\right) \\
&\quad \times d\left(p_{\max}\frac{V(t-N)-V(t+d)}{N} + 2\frac{N+d+1}{N}\frac{p_{\max}}{p_{\min}}\right. \\
&\quad \left.\times \max_{t-N+1\leq i\leq t+d}\left(\frac{K_\gamma a(i+1)+k_\gamma}{\|\phi(i-d)\|^2 + r/p_{\min}}\right)\right) \\
&\leq c_1 K_\gamma + 2\frac{k_3+k_\gamma}{R} + 2^{d+1}\left(K_\phi + \frac{k_\phi + r/p_{\max}}{R}\right) \\
&\quad \times d p_{\max}\left[\frac{V(t-N)-V(t+d)}{N}\right. \\
&\quad + \frac{2(N+d+1)(K_\gamma + k_\gamma/R)}{p_{\min}N} \\
&\quad \times \left[\left(\frac{g_2^{1/2}}{K_\theta}\right)^{N+d+1}\sqrt{\frac{g_1^{-(d+1)}}{M_\theta}(g_2-\sigma)}\right. \\
&\quad \left.\left. - c_2\left(K_v + 2\frac{k_{y^*}+k_v}{R}\right)^{1/2}\right]^{-2}\right. .
\end{aligned}$$

We now use the inequality $(a-b)^2 \geq 0.5a^2 - b^2$, which gives

$$\begin{aligned}
&\left[\left(\frac{g_2^{1/2}}{K_\theta}\right)^{N+d+1}\sqrt{\frac{g_1^{-(d+1)}}{M_\theta}(g_2-\sigma)} - c_2\left(K_v + 2\frac{k_{y^*}+k_v}{R}\right)^{1/2}\right]^2 \\
&\quad \geq 0.5\left(\frac{g_2}{K_\theta^2}\right)^{N+d+1}\frac{g_1^{-(d+1)}}{M_\theta}(g_2-\sigma) - c_2^2\left(K_v + 2\frac{k_{y^*}+k_v}{R}\right),
\end{aligned}$$

and this establishes Lemma 3.2. □

Adaptive-Invariant Discrete Control Systems

Ya. Z. Tsypkin

Institute of Control Sciences, Moscow, USSR.

Abstract. The paper investigates the structure and properties of selective-invariant and adaptive-invariant discrete control systems in which the effects of external regular disturbances are eliminated and those of external stochastic disturbances are essentially weakened. Possibilities of eliminating the constraints caused by multiple delays and nonminimum phase of a dynamic plant are described.

1 Introduction

A great number of works [1] is devoted to the problem of invariance or, in other words, the problem of compensating the influence of external disturbances both in continuous and discrete control systems.

Conditions of absolute or complete invariance, providing accurate compensation of arbitrary disturbances are, as a rule, physically unrealizable. The need to relax these invariance conditions, and ensure compensation of a known class of disturbances, has led to the notion of selective invariance in continuous systems which was introduced by V. S. Kulebakin [2,3].

Availability of a priori information from various disturbance models made it possible to design systems capable of resisting broader classes of disturbances. The theory of such control systems has been developed by Johnson [4,5] for both continuous [4] and discrete systems [5]. With an insufficient level of a priori information on disturbances it is expedient to use the adaptive approach which provides an estimate of the class of disturbances and uses it for disturbance compensation. Systems which use adaptation to achieve invariance are called adaptive-invariant systems.

This paper presents design principles of σ^2-adaptive-invariant discrete-time control systems. After a brief review of discrete-time control systems the character of a priori and current information about regular invariant systems is considered and structures of selective- and adaptive-invariant systems are established. Then it is shown that an insignificant modification of system structures allows us to remove constraints caused by delays and nonminimum-phase plants. Properties of selective- and adaptive-invariant systems are analyzed when the compensation condition is accurately satisfied. Finally, possibilities of constructing systems with stochastic disturbances which are approximately adaptive-invariant, that is σ^2-invariant, are analyzed and new problems are formulated.

2 Properties of Discrete Control Systems

Discrete-time control of continuous plants leads to difference equations which can be written in the following form

$$Q(q)y(n) = q^{1+k} P_u(q)u(n) + f(n) . \tag{1}$$

At n-th time instant, $y(n)$ is the output, $u(n)$ is the control, $f(n)$ is an external disturbance and q is the unit lag operator:

$$q^m x(n) = x(n-m), \qquad m = 1, 2 \dots .$$

The polynomials $Q(q)$ and $P_u(q)$ with degrees N and N_1, respectively, are such that $Q(0) = 1$, $P_u(0) \neq 0$. A polynomial satisfying the condition $Q(0) = 1$ is referred to as a monic polynomial. To simplify the notation the lag operator and the polynomial variable are denoted by the same letter. The integer $k > 0$ indicates a multiple lag. The minimum possible lag of a plant is equal to 1, and corresponds to $k = 0$ [6]. This fact should be taken into account in the analysis and synthesis of discrete-time control systems.

Furthermore, we shall assume that the plant is stable and minimum-phase, i.e., that all the zeros of the polynomials $P_u(q)$ and $Q(q)$ lie outside the unit circle $|q| \leq 1$. For brevity such polynomials are said to be "external" [7] or stable. The requirement of plant stability is not an essential constraint, since a known unstable plant can be always stabilized by feedback.

A nonminimum-phase plant imposes some definite limitations to controller synthesis [6,7]. The possibility of overcoming these constraints will be discussed in Section 7. Meanwhile, if not stated otherwise, a plant is assumed to be minimum-phase and stable. The block-scheme of a plant is given in Fig. 1.

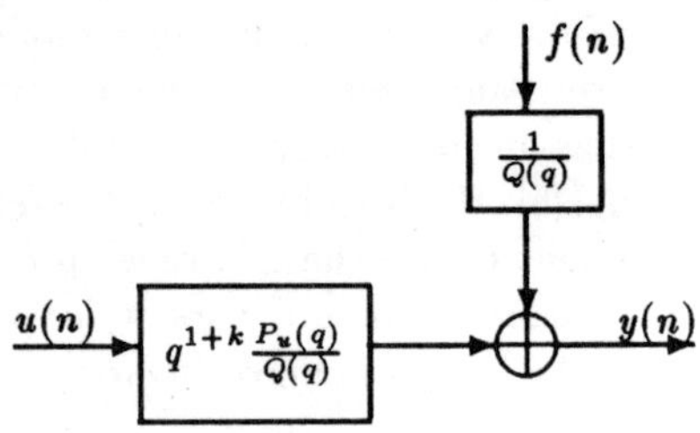

Fig. 1. Plant to be controlled.

Without external disturbances, $f(n) = 0$, the plant equation (1) becomes

$$Q(q)y(n) = q^{1+k} P_u(q)u(n) . \tag{2}$$

As an optimal controller for such a plant we define the controller which satisfies the following condition:

$$T^0(q)(y(n) - y_0(n)) = 0, \quad n \geq 1 + k , \tag{3}$$

where $T^0(q)$ is an external monic polynomial of degree $N_2 > 0$, with the assigned distribution of zeros, and $y_0(n)$ is the output of a reference model,

$$G_0(q)y_0(n) = q^{1+k}H_0(q)r(n)\,, \tag{4}$$

where $r(n)$ is a given reference input, and $G_0(q)$ and $H_0(q)$ are external polynomials of degrees N_3 and N_4, respectively. It is known from [6,7] that an optimal controller for plant (1) is described by the equation

$$P_u(q)S(q)u(n) = T^0(q)\bar{r}(n) - P(q)y(n)\,. \tag{5}$$

Here $S(q)$ is a monic polynomial of degree k and $P(q)$ is a polynomial of degree $\max(N-1, N_2-k-1)$ satisfying the polynomial equation

$$Q(q)S(q) + q^{1+k}P(q) = T^0(q) \tag{6}$$

and $\bar{r}(n)$ is defined by

$$\bar{r}(n) = \frac{H_0(q)}{G_0(q)}r(n)\,. \tag{7}$$

The block-scheme of the optimal system described by equations (2) and (5) is shown in Fig. 2a, and its equivalent form is shown in Fig. 2b. With $k = 0$, which

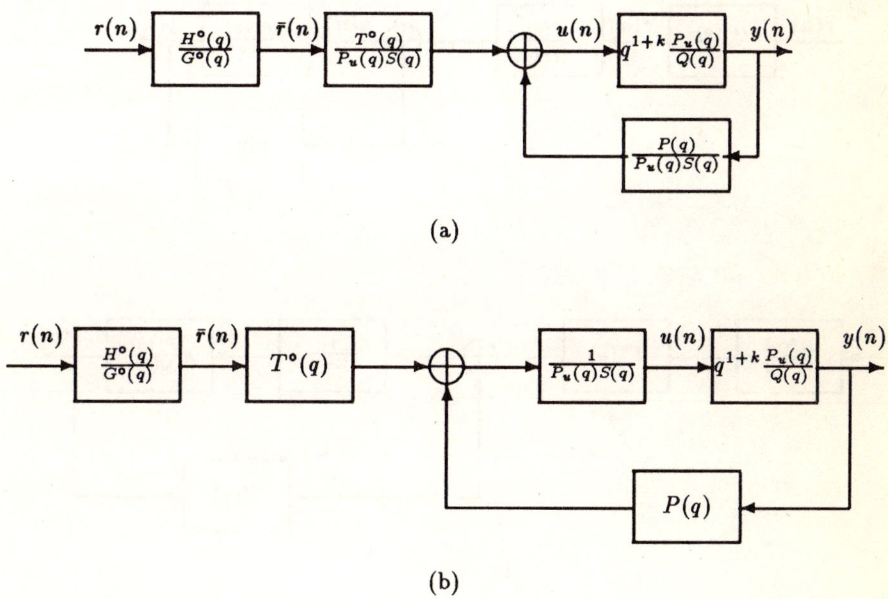

Fig. 2. Two equivalent forms of the optimal system.

corresponds to $S(q) = 1$, we obtain from (2) and (5)

$$Q(q)y(n) = qP_u(q)u(n)\,, \tag{8}$$

$$P_u(q)u(n) = T^0(q)\bar{r}(n) - P(q)y(n), \tag{9}$$

From (6) it follows that

$$P_u(q) = q^{-1}(T^0(q) - Q(q)) \tag{10}$$

and (3) reduces to

$$T^0(q)(y(n) - y_0(n)) = 0, \quad n \geq 1, \tag{11}$$

where

$$y_0(n) = q\frac{H_0(q)}{G_0(q)}r(n). \tag{12}$$

In this case it is not necessary to solve (6) for $S(q)$ and $P(q)$, nor to include a multiple lag in the reference model. The block-scheme of the optimal system (8) and (9) is shown in Fig. 3a and its equivalent form is given in Fig. 3b. Note that for a nonminimum-phase plant the system becomes non-robust [8] due to cancellation of the polynomials $P_u(q)$ in the plant and the controller. Thus a change in the structure of the controller is necessary. Nonminimum-phase of a plant and its multiple lag impose additional constraints upon the choice of a reference model [6,8].

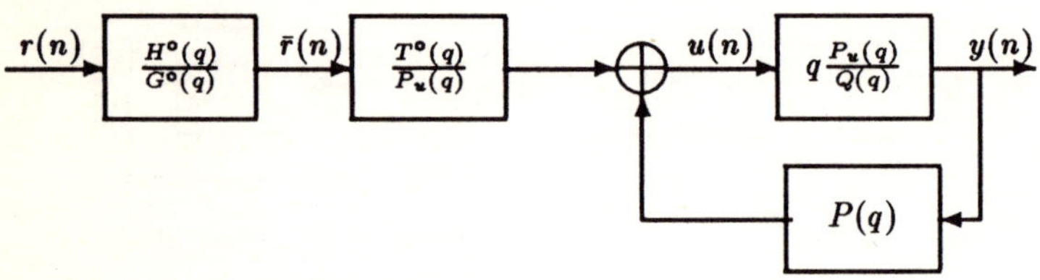

(a)

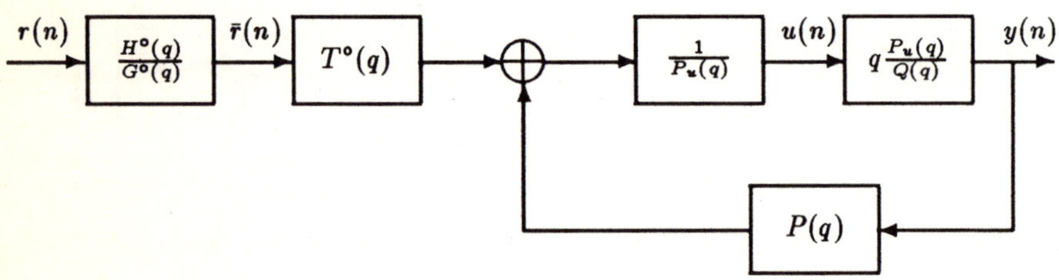

(b)

Fig. 3. Optimal system forms in the case $k = 0$.

For completely deterministic perturbations $f(n) \neq 0$ the synthesis problem of optimal control for a plant described by (1), with respect to various optimality criteria, leads to the solution of corresponding variational problems [8].

For example, if the disturbances satisfy a known bound, then the minimax or game approaches can be used to determine the "best" result for the "worst" disturbance [9–11].

Absolutely invariant discrete systems under any arbitrary disturbances are, as a rule, unrealizable, because they require infinite feedback gain [12] which is not allowed because of the lag inherent in the discrete-time plants.

The problem of synthesizing an adaptive-invariant discrete system consists in determining the structure and parameters of a controller, such that the control system response coincides with the response of the optimal system in the absence of external regular disturbances.

Further problems consist in consideration of possibilities to synthesize the adaptive-invariant systems, where the constraints caused by multiple lag and plant nonminimum-phase would be removed. Solution of these problems will be based on a combination of the adaptive approach [7] with the theory of selective invariance [2,3] for discrete control systems.

3 A Priori and Current Information on Disturbance

A priori information on the disturbance plays an important role in synthesis of the adaptive-invariant systems. Assume that disturbance satisfies a difference equation

$$f(n) = D(q)f(n-k-1)\,, \tag{13}$$

where $D(q)$ is a polynomial of degree M. In the sequel (13) is called the equation of extrapolation or prediction, while the polynomial $D(q)$ is a prediction polynomial. The prediction equation is easily reduced to the compensation equation

$$(1 - q^{1+k}D(q))f(n) = 0\,. \tag{14}$$

Let us call the polynomial $1 - q^{1+k}D(q)$ a compensation polynomial.

Equation (14) indicates the conditions under which the function $f(n)$ is being compensated. Theorems providing the possibilities to represent a wide class of continuous functions $f(n)$ in the form of solutions of homogeneous differential equations have been formulated by C. Shannon in his investigations of differential analyzers [13] and by V. S. Kulebakin for problems of electromechanics and the development of selective-invariant systems [2,3]. Representation of functions $f(n)$ in the form of homogeneous difference equations was considered in [14]. In the presence of sufficiently complete information on disturbance $f(n)$ a prediction polynomial is quite easily determined. Thus, if $f(n) \equiv d$ is constant then $D(q) \equiv 1$, if $f(n) = d_1 n + d_0$ is linear, then $D(q) = 2 - q$, if $f(n) = a^n$ is an exponential function then $D(q) = a$, etc.

In some cases it is expedient to use polynomial splines [15] instead of a conventional prediction polynomial $D(q)$. With incomplete a priori information a prediction polynomial is known only up to its coefficients which can be reconstructed from disturbance observations. Thus, there is a need in obtaining information about the disturbance and its estimate. As a rule, direct measurement of a disturbance is impossible. Therefore an indirect adjustable model method will be employed which is well known in identification [6,7,16]. However, the adjustable model will now play a different role. First, its parameters are not to be identified but rather determined from known values of the plant coefficients. Second, this model is a two-input finite impulse response (FIR) model [16]. Third, it is introduced here to identify a deterministic disturbance, rather than a stochastic one. The equation of this model has the following form:

$$\hat{y}(n) = (1 - Q(q))y(n) + q^{1+k} P_u(q)u(n) . \tag{15}$$

We define the error

$$\epsilon(n) = y(n) - y_0(n) \tag{16}$$

and obtain from (15) that

$$\epsilon(n) = Q(q)y(n) - q^{1+k} P_u(q)u(n) . \tag{17}$$

By virtue of the plant equation (1) the error is equal to the disturbance:

$$\epsilon(n) = f(n) . \tag{18}$$

Thus, the two-input FIR model provides an indirect measurement of the disturbance. It would seem that only one step is left to achieve invariance – to send a signal of the measured disturbance $\epsilon(n)$ with the oppostie sign in order to cancel the effect of the disturbance. But this signal can be sent to the plant only by the controller, and the lag in the plant will cause the difference of disturbances $f(n) - f(n-1)$ to appear. This difference for an arbitrary disturbance will differ from zero even in the absence of a multiple lag, $k = 0$. This fact most evidently testifies to the unrealizability of absolute invariance.

4 Selective-Invariant Systems

Let us use the above indirect method of measuring disturbances for synthesis of a selective-invariant optimal system. Consider a controller equation which differs from equation (5) in having an additional signal, depending on the error $\epsilon(n)$:

$$P_u(q)S(q)u(n) = T^0(q)\bar{r}(n) - P(q)y(n) - D(q)\epsilon(n) . \tag{19}$$

Here $\epsilon(n)$ is found from equation (17) and the polynomial $D(q)$ is a prediction polynomial, so that

$$\epsilon(n) = D(q)\epsilon(n-k-1) . \tag{20}$$

Now, taking into account that $\epsilon(n) = f(n)$, we obtain the compensation condition

$$(1 - q^{1+k} D(q))f(n) = 0 . \tag{21}$$

Let us show that controller (19) provides selective invariance for the plant

$$Q(q)y(n) = q^{1+k} P_u(q)u(n) + f(n) . \tag{22}$$

Indeed, substituting $u(n)$ from (19) into (22) and taking into account polynomial equation (6) and equations of reference model (4) and (7), we obtain

$$T^0(q)y(n) = T^0(q)y_0(n) - (1 - q^{1+k}D(q))S(q)f(q)\,. \tag{23}$$

But, by virtue of the *compensation condition* (21), the last term in (23) vanishes and the optimality condition (3) follows from (23). The block-scheme of the selective-invariant system, described by equation (22), (14), (17) is shown in Fig. 4. There the processes do not depend on the disturbance $f(n)$. The system

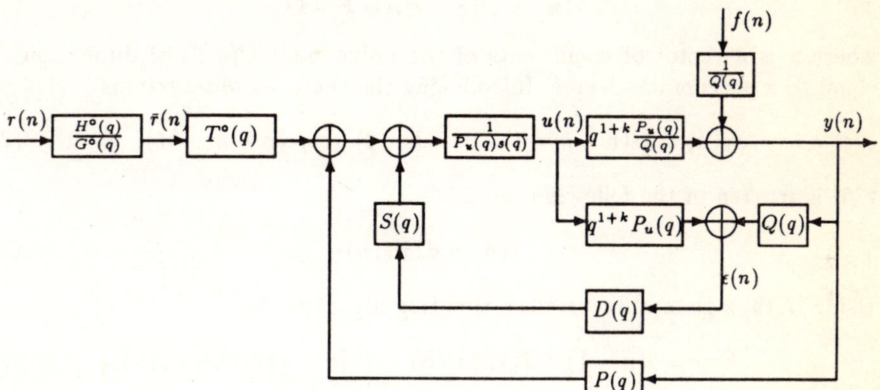

Fig. 4. Selective-invariant system.

is equivalent to the optimal system in Fig. 2. Comparing the controller equation (5) of the optimal system in Fig. 2 with equation (19) of the selective-invariant optimal system in Fig. 4, we conclude that it differs only in the additional signal $D(q)S(q)\epsilon(n)$ where $D(q)$ is a prediction polynomial and $S(q)$ is the solution of polynomial equation (6). This signal is generated by the FIR model (15) and a predictor described by equation (20). Equation (19) with $\epsilon(n)$ substituted from (17) can be written in the explicit form as

$$(1 - q^{1+k}D(q))P_u(1)u(n) = T^0(q)\bar{r}(n) - [P(q) + D(q)S(q)Q(q)]y(n)\,. \tag{24}$$

This equation reveals the role of the polynomial $D(q)$ which is closely connected with the main ideas of the $K(D)$-transform used by V. S. Kulebakin for selective invariance [3]. Equation (24) is also close to the one used by C. Johnson [5] to construct control systems which accommodate disturbances. However instead of a two-input FIR model, [5] uses a reduced-order Luenberger observer and does not point to the connection with selective-invariance results.

5 Adaptive-Invariant Systems

If the a priori information on disturbances is not sufficient to determine in advance all predictor polynomial coefficients, then (19) will represent a polynomial with unknown coefficients θ, and so instead of (19) we shall have

$$P_u(q)S(q)u(n) = T^0(q)\bar{r}(n) - P(q)y(n) - D(q,\theta)S(q)\epsilon(n)\,. \tag{25}$$

An estimator of the coefficients $\theta = \theta(n)$ from the observations $\epsilon(n) = f(n)$ can be realized using the adaptive approach [19,7], by means of a recurrent algorithm. To construct an algorithm, we consider a prediction equation similar to (20), namely

$$\hat{\epsilon}(n) = D(q,\theta)\epsilon(n-k-1)\,, \tag{26}$$

where θ is a vector of coefficients of the polynomial $D(q,\theta)$ of dimension M, equal to a polynomial degree. Introducing the vector of observations

$$\phi_k(n) = (\epsilon(n-k-1), \epsilon(n-k-2), \ldots, \epsilon(n-k-M))^{\mathrm{T}}\,, \tag{27}$$

(26) is written in the following form:

$$\hat{\epsilon}(n) = \theta^{\mathrm{T}}\psi_k(n)\,. \tag{28}$$

Using [7,19] a general recurrent estimation algorithm is

$$\theta(n) = \theta(n-1) + \Gamma_0(n)\left[\epsilon(n) - \theta^{\mathrm{T}}(n-1)\psi_k(n)\right]\psi_k(n)\,, \tag{29}$$

where $\Gamma_0(n)$ is a gain matrix. In the simplest Kaczmarz algorithm [6,20] this matrix is

$$\Gamma_0(n) = \left[\psi_k^{\mathrm{T}}(n)\psi_k(n)\right]^{-1} I\,, \tag{30}$$

where I is a unit matrix. For the orthogonal projection algorithm [7,16,17] we have

$$\Gamma_0(n) = \left[\psi_k^{\mathrm{T}}(n)\Gamma_0(n)\psi_k(n)\right]^{-1}\Gamma(n) \tag{31}$$

$$\Gamma(n) = \Gamma(n-1) - \frac{\Gamma(n-1)\psi_k(n)\psi_k^{\mathrm{T}}(n)\Gamma(n-1)}{\psi_k^{\mathrm{T}}(n)\Gamma(n-1)\psi_k(n)}\,, \tag{32}$$

if $\psi_k^{\mathrm{T}}(n)\Gamma(n-1)\psi_k(n) \neq 0$. Whenever $\psi_k^{\mathrm{T}}(n)\Gamma(n-1)\psi_k(n) = 0$, then instead of (31) and (28) we let

$$\theta(n) = \theta(n-1), \quad \Gamma(n) = \Gamma(n-1)\,. \tag{33}$$

In practice the condition $\psi_k^{\mathrm{T}}(n)\Gamma(n-1)\psi_k(n) \neq 0$ is usually replaced by the condition

$$\psi_k^{\mathrm{T}}(n)\Gamma(n-1)\psi_k(n) > \delta_0\,,$$

where δ_0 is some positive number [7,20]. The estimation speed of the Kaczmarz algorithm can be increased by using a modification [21,22]. The orthogonal projection algorithm provides the estimate of coefficients in a finite number of steps

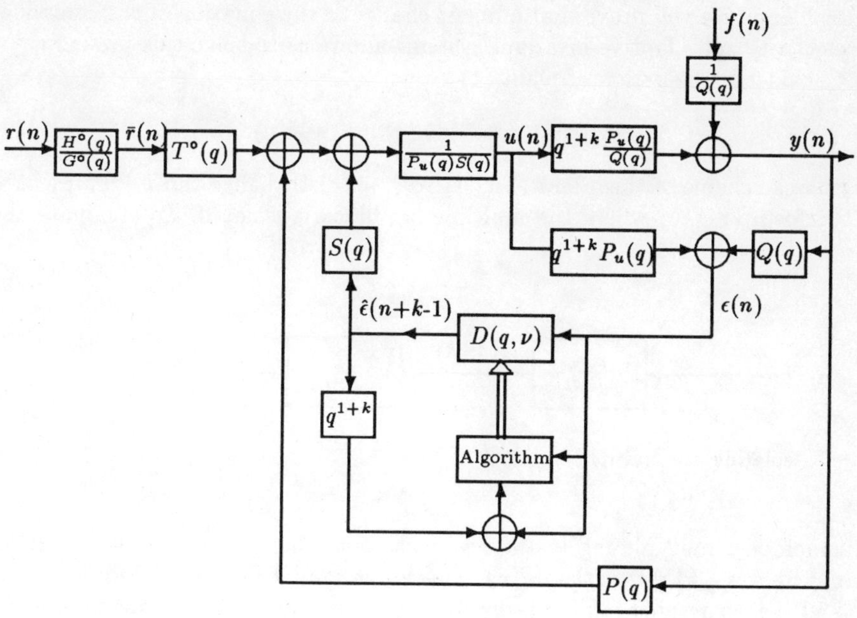

Fig. 5. Adaptive-invariant system.

[7,20]. The block-scheme of the adaptive-invariant system, described by equations (22), (25) and (24) is presented in Fig. 5.

In contrast to the selective-invariant system, the system in Fig. 5 has an adaptive loop which accomplishes parameter estimation in the part of the controller which compensates the disturbance. The adaptive loop identifies the disturbance or, to be more precise, implements a prediction of the disturbance. The adaptive-invariant system becomes equivalent to the optimal system by the end of the adaptation process. Robustness of the adaptive-invariant system coincides with the robustness of the optimal system to which it is asymptotically equivalent. As it is well known [23,24], these optimal systems are robust, i.e., their properties change slightly for small changes in the plant and (or) controller. Hence the adaptive-invariant systems also possess the property of robustness.

6 Elimination of Constraints Caused by the Time Lag

The presence of a multiple lag in the plant, $k > 0$, causes the controller to be more complex. Its order, defined by the degree of polynomial $P_u(q)S(q)$, is equal to $(N_1 + 1 + k)$, and it increases with the growth of k, while for $k = 0$, we saw that (9) acquires an extremely simple form. In this case there is no necessity in solving the polynomial equation with respect to $S(q)$ and $P(q)$, since $S(q) \equiv 1$ and $P(q)$ is explicitly expressed in terms of $T^0(q)$ and $Q(q)$. Let us now consider the problem of eliminating the influence of a multiple lag on the processes in the control loop. We will prove that a minor change in the equation of the controller of selective- and adaptive-invariant systems allows us to solve this problem.

Consider the equation of plant (1)

$$Q(q)y(n) = q^{1+k}P_u(q)u(n) + f(n)\,. \tag{34}$$

The block scheme of this plant (Fig. 1) with no disturbance can be represented in the form (Fig. 6) where the multiple lag block is isolated. To eliminate the

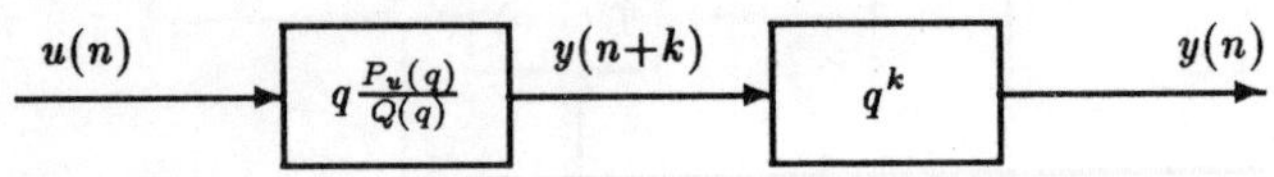

Fig. 6. Isolating the multiple lag block.

influence of a multiple lag in the feedback loop, the input into the controller should be $y(n+k)$ rather than $y(n)$. This input cannot be obtained directly, but instead it is generated indirectly by means of the two-input FIR model used in the selective-invariant system. Denoting the difference of the two signals by

$$z(n) = y(n) - y(n+k)\,, \tag{35}$$

and the error of the simplified model by

$$\epsilon^0(n) = Q(q)y(n) - qP_u(q)u(n)\,, \tag{36}$$

we obtain $z(n)$ from

$$Q(q)z(n) = \epsilon^0(n)\,. \tag{37}$$

The equation of an optimal controller of the selective-invariant system, in which the influence of a multiple lag is eliminated, is represented in the following form:

$$P_u(q)u(n) = T^0(q)\bar{r}(n) - P(q)(y(n) - z(n)) - D(q)\epsilon(n) \tag{38}$$

where, as it has been seen earlier in (17),

$$\epsilon(n) = Q(q)y(n) - q^{1+k}P_u(q)u(n)\,. \tag{39}$$

The prediction polynomial $D(q)$ is such that

$$(1 - q^{1+k}D(q))f(n) = 0\,, \tag{40}$$

and $z(n)$ is found from (36) and (37). Here we have in mind that the polynomial $P(q)$ is determined by (10). Now the feedback signal is equal to $y(n) - z(n) = y(n+k)$ as is seen from (36). Eliminating $u(n)$ from (34) and (38) and taking into account (36), (37), (33) and (40), we obtain the optimality conditions (11), (12) for the optimal system without disturbance and multiple lag.

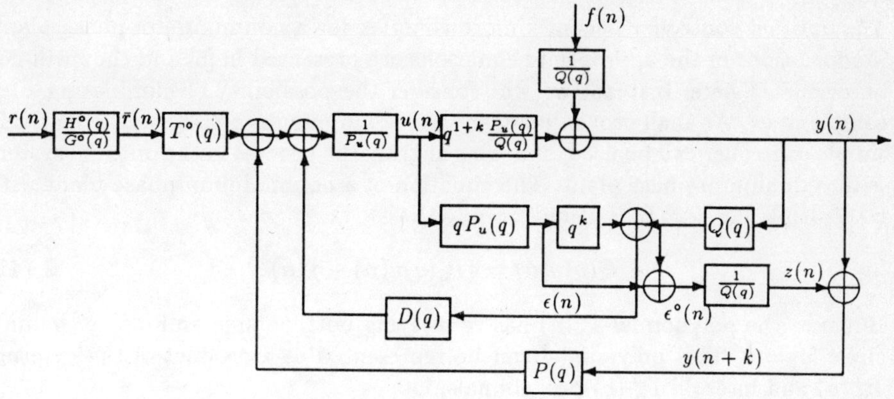

a)

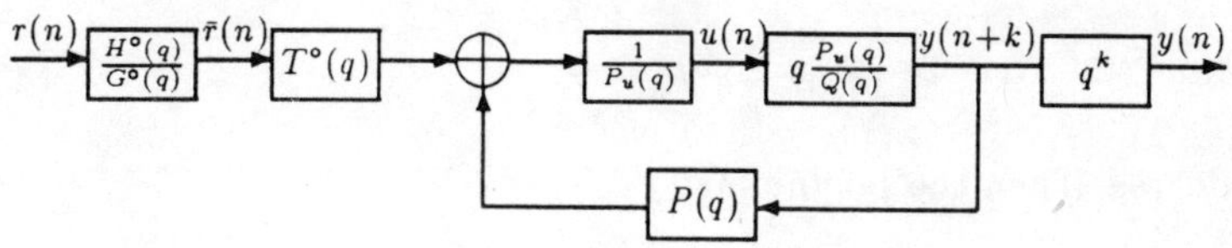

b)

Fig. 7. Selective-invariant system with multiple lag.

The block-scheme of a selective-invariant system without a multiple lag influence is shown in Fig. 7a, where the two previous models are combined. The block-scheme in Fig. 7b is equivalent except for the location of the lag element outside of the feedback loop. The processes in these systems are shifted relatively to another for an interval equal to the multiple lag. The idea of eliminating the lag influence in continuous systems on the basis of introducing dynamic models of a plant with and without lag was proposed by O. Smith [25,26]. It was also used in discrete systems [27–29].

The adaptive-invariant system which eliminates the influence of multiple lag differs from the selective-invariant system (Fig. 7a) by the presence of an adaptation loop similar to that shown in Fig. 5.

7 Elimination of Constraints Caused by Nonminimum-Phase Plant

The optimal controller becomes more complex for a nonminimum-phase plant. Modifications of the appropriate equations are presented in [6] and they will not be discussed here. Instead, we will consider the possibility of eliminating such complexities. We shall prove that even for a nonminimum-phase plant a relatively simple controller can be used, changing slightly the process in the optimal system with a minimum-phase plant. The equation of a nonminimum-phase plant with no multiple lag, $k = 0$, is found from (34):

$$Q(q)y(n) = qP_u(q)u(n) + f(n) . \tag{41}$$

But now the polynomial $P_u(q)$ has zeros lying both outside and inside the unit circle $|q| = 1$. This polynomial can be represented as a product of the external $P_u^+(q)$ and internal $P_u^-(q)$ polynomials, i.e.,

$$P_u(q) = P_u^+(q)P_u^-(q) . \tag{42}$$

Introduce the notation of a normalized inverse polynomial [6]:

$$\tilde{P}_u^-(q) = \frac{1}{\kappa} q^\ell P_u(q^{-1}) , \tag{43}$$

where ℓ is its degree. Then the polynomial

$$\tilde{P}_u(q) = P_u^+(q)\tilde{P}_u^-(q) \tag{44}$$

will be external. From (44) and (43) there follows the equality

$$\frac{P_u(q)}{\tilde{P}_u(q)} \frac{P_u(q^{-1})}{\tilde{P}_u(q^{-1})} = \kappa^2 . \tag{45}$$

The division of the polynomial $\tilde{P}_u(q)$ by the polynomial $P_u(q)$ is realized with the help of a single iteration of factorization [30] with no need for explicitly determining polynomials zeros. Making use of the notation $\tilde{P}_u^-(q)$ in (43), we represent the block-scheme of a nonminimum-phase plant described by (41) without disturbances in the form shown in Fig. 8. In this case the nonminimum-phase

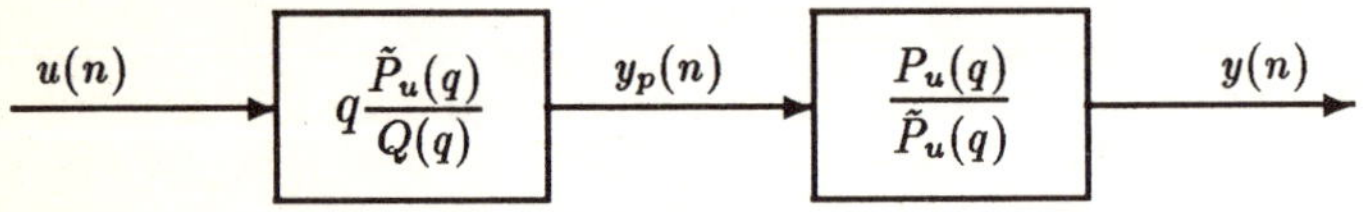

Fig. 8. Nonminimum-phase plant without disturbances.

element is isolated. Its transfer function is equal to $P_u(q) \Big/ \tilde{P}_u(q)$. With $q = e^{-j\omega}$ it follows from (45) that the amplitude frequency characteristic of this element is constant, that is it does not depend on the frequency ω [6]:

$$y_p(n) = \frac{\tilde{P}_u(q)}{Q(q)} u(n) . \tag{46}$$

It can be isolated by changing appropriately a two-input model, and then used as in Section 6 to eliminate the lag influence. In this case the equation of an optimal controller in the selective-invariant system, where the influence of nonminimum phase is eliminated, has the form

$$\tilde{P}_u(q)u(q) = T^0(q)\bar{r}(n) - P(q)(y(n) - z(n)) - D(q)\epsilon(q) , \tag{47}$$

where $\epsilon(n)$, $z(n)$ and $Q(q)z(n) = \epsilon^0(n)$, are defined from the equations

$$\epsilon(n) = Q(q)y(n) - qP_u(q)u(n) + f(n) . \tag{48}$$

$$z(n) = y(n) - y_p(n) = y(n) - \frac{\tilde{P}_u(q)}{Q(q)} u(n) \tag{49}$$

$$\epsilon^0(n) = Q(q)y(n) - q\tilde{P}_u(q)u(n) + f(n) . \tag{50}$$

The compensation condition (40) is fulfilled and the polynomial $P_u(n)$ is found from equality (10).

The block-scheme of a selective-invariant system with the eliminated influence of nonminimum-phase is shown in Fig. 9a, and the block-scheme which is equivalent to it is given in Fig. 9b. It differs from the optimal control system of a minimum-phase plant in that the phase change is outside the control loop. Of course in this case $y_p(n)$ will differ from $y(n)$. All we can say is that amplitude-frequency spectra of the processes $y_p(n)$ and $y(n)$ will coincide. This is the price we paid for nonminimum-phase influence elimination. Moreover, elimination of the influence of nonminimum phase and plant lag can be combined if necessary changes are entered in the utilized two-input model.

By introduction of the adaptive loop into the selective-invariant system we obtain the adaptive-invariant system to which the optimal system will be equivalent after the completion of the adaptation process.

8 Approximate Realizations of Adaptive-Invariant Systems

Assume that a disturbance $f(n)$ is generated by a piecewise continuous function $f(t)$ but not by a continuous one. In this case compensation equation (14) acquires the form

$$(1 - q^{1+k} D(q)) f(n) = x(n) , \tag{51}$$

where $x(n)$ are jumps caused by the change of intervals of the continuous functions $f(t)$.

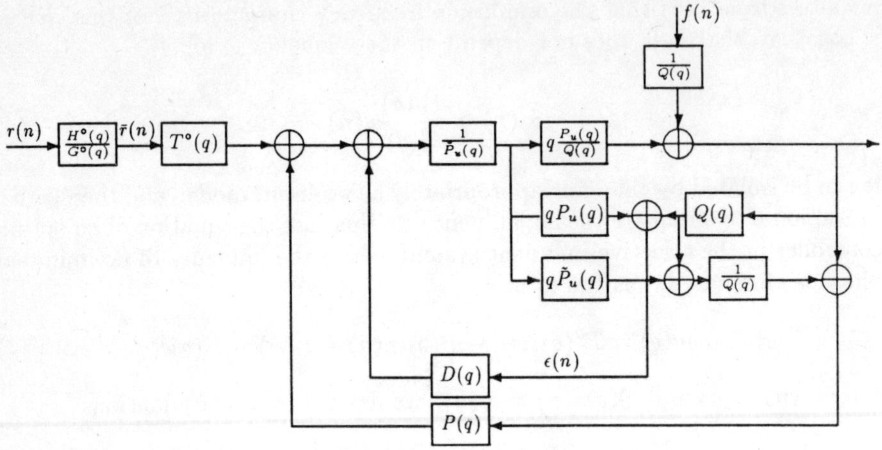

a)

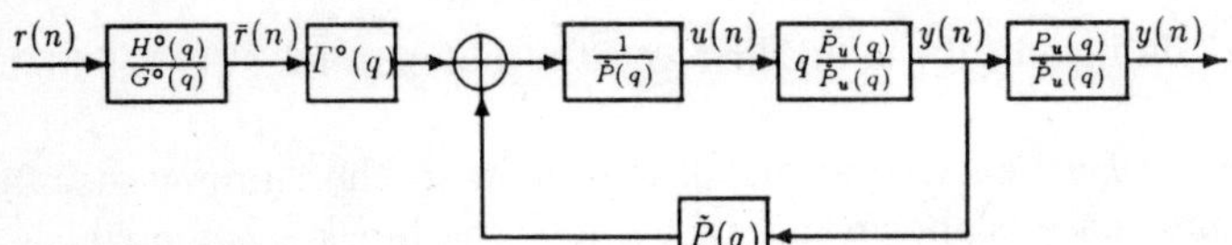

b)

Fig. 9. Selective-invariant system with a nonminimum-phase plant.

In this case the optimality condition (3) is replaced by

$$T^0(q)(y(n) - y_0(n)) = x(n). \tag{52}$$

If the change of intervals of $f(t)$ occurs very seldom then the influence of $x(n)$ on the processes in the system is insignificant.

In those cases where the disturbance $f(n)$ is changing so slowly that the first difference $\Delta f(n) = f(n) - f(n-1)$ is sufficiently small, it is sufficient to assume $D(q) \equiv 1$. In this case there will be no need for adaptation and we come to the realization of the simplest selective-invariant system.

If the first difference is not very small but the value $\Delta f(n) - \alpha \Delta f(n-1)$ is small, which is with $\alpha = 1$ equal to the second difference, then we should assume a prediction polynomial of the following form

$$D(q) = (1+\alpha) - \alpha q\,. \tag{53}$$

Tuning of the parameter α is done by the adaptation loop. In this case we realize the simplest adaptive-invariant system.

9 σ^2-Adaptive-Invariant Systems

Up to now it was assumed that external disturbances $f(n)$ are determinatistic. Let us now consider that the disturbances $\zeta(n)$ are stochastic and they satisfy the conditions

$$E\{\zeta(n)\} = 0, \quad E\{\zeta(n)^2\} = \sigma_\zeta^2\,. \tag{54}$$

For simplicity we consider a minimum-phase, stable plant without multiple lag ($k = 0$) and write its equation in the form

$$Q(q)y(n) = qP_u(q)u(n) + \zeta(n)\,. \tag{55}$$

As it follows from [8,20,31] an optimal controller, minimizing the criterion of the generalized mean square error

$$J = E\left\{\left[T^0(q)(y(n) - y_0(n))\right]^2\right\}, \tag{56}$$

coincides with the optimal controller (9) for the same plant but with no disturbance. However, in contrast to (11) we now have

$$T^0(q)(y(n) - y_0(n)) = \zeta(n)\,, \tag{57}$$

and, consequently, a minimum value of criterion (56) is equal to the variance of the random disturbance:

$$J_{\min} = E\left\{\left[T^0(q)(y(n) - y_0(n))\right]^2\right\} = E\{\zeta(n)^2\} = \sigma_\zeta^2\,. \tag{58}$$

The block scheme of the optimal system with stochastic disturbance is shown in Fig. 10.

It should be noted that, as opposed to systems with deterministic disturbances, where the transient processes were of main interest, in the systems with stochastic disturbances the steady-state processes are in the center of attention. With the stochastic disturbances the notion of invariance loses its meaning. However, the problem of synthesizing the system, that is adaptive-invariant up to σ^2 can be formulated, i.e., the system for which the optimality criterion $J_{\min}$ reaches the smallest possible value.

We assume a priori that the disturbance $\zeta(n)$ is determined by the autoregression (AR) equation [32,16]:

$$\zeta(n) = C(q)\zeta(n-1) + \varepsilon(n)\,. \tag{59}$$

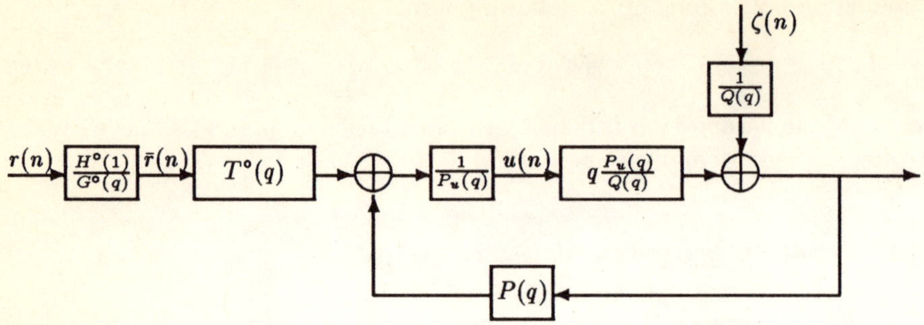

Fig. 10. Optimal system with stochastic disturbance.

Here $\varepsilon(n)$ is a sequence of independent identically distributed random values, satisfying the conditions

$$E\{\varepsilon(n)\} = 0, \qquad E\{\varepsilon(n)\varepsilon(n-m)\} = \begin{cases} \sigma^2 & \text{with } m = 0 \\ 0 & \text{with } m \neq 0, \end{cases} \tag{60}$$

and $C(q)$ is an external polynomial of degree M, with unknown coefficients.

This polynomial is actually a prediction polynomial. The variance σ_ζ^2 of $\zeta(n)$ is equal to the minimum of the optimality criterion (58). Substituting $\zeta(n)$ from (59) in (58) and taking into account (60) and equalities $E\{\varepsilon(n)\varepsilon(n-m)\} = 0$ with $m \geq 0$ we obtain

$$\sigma_\zeta^2 = E\left\{[C(q)\zeta(n)]^2\right\} + \sigma^2 , \tag{61}$$

whence it follows that always

$$J_{\min} = \sigma_\zeta^2 > \sigma^2 . \tag{62}$$

AR-equation (59) leads to the form

$$(1 - qC(q))\zeta(n) = \varepsilon(n) . \tag{63}$$

This is the equation of incomplete compensation while the polynomial $1 - qC(q)$ of degree $M + 1$ is referred to as a polynomial of incomplete compensation. The problem of synthesizing the σ^2-adaptive-invariant system consists in determining the structure and parameters of a controller which reduces $J_{\min} = \sigma_\zeta^2$ to smallest possible value $J_{\min} = \sigma^2$.

The equation of an optimal controller of the σ^2-adaptive-invariant system for the plant

$$Q(q)y(n) = qP_u(q)u(n) + \zeta(n) \tag{64}$$

can be represented in the form

$$P_u(q)u(n) = T^0(q)\bar{r}(n) - P(q)y(n) - C(q,\theta)\epsilon(n), \tag{65}$$

where $\epsilon(n)$ is the error

$$\epsilon(n) = Q(q)y(n) - qP_u(q)u(n), \tag{66}$$

which, as it becomes obvious from the comparison of (64) and (66), is equal to the disturbance

$$\epsilon(n) = \zeta(n). \tag{67}$$

The prediction polynomial $C(q,\theta)$ in (65) is not fully determined. The vector θ of its coefficients is unknown. To obtain the estimate $\theta(n)$ we shall use the estimation algorithms in the presence of noise [16].

Introducing the observation vector

$$\phi(n) = (\epsilon(n-1), \dots, \epsilon(n-M))^{\mathrm{T}}, \tag{68}$$

prediction value of $\epsilon(n)$ denoted by $\hat{\epsilon}(n)$ is

$$\hat{\epsilon}(n) = C(q,\theta)\epsilon(n-1) = \theta^{\mathrm{T}}\psi(n). \tag{69}$$

The best prediction will be realized with minimization of the prediction criterion

$$J_{\mathrm{pr}}(\theta) = E\left\{[\epsilon(n) - \hat{\epsilon}(n)]^2\right\} = E\left\{\left[\epsilon(n) - \theta^{\mathrm{T}}\psi(n)\right]^2\right\}, \tag{70}$$

or, which is equivalent, satisfying the following condition:

$$\nabla J_{\mathrm{pr}}(\theta) = -E\left\{\left[\epsilon(n) - \theta^{\mathrm{T}}\psi(n)\right]\psi(n)\right\} = 0. \tag{71}$$

On the basis of the adaptive approach [19,16] we can select an estimation algorithm, for example, of the least squares type:

$$\begin{aligned} &\theta(n) = \theta(n-1) + \Gamma(n)\left[\epsilon(n) - \theta^{\mathrm{T}}\psi(n)\right]\psi(n), \qquad \theta(0) = \theta_0 \\ &\Gamma(n) = \Gamma(n-1) - \frac{\Gamma(n-1)\psi(n)\psi(n)\Gamma(n-1)}{1 + \psi^{\mathrm{T}}(n)\Gamma(n-1)\psi(n)}, \ \Gamma(0) = \alpha I, \alpha \gg 1. \end{aligned} \tag{72}$$

This estimation algorithm is optimal on the class of disturbances $\varepsilon(n)$ with bounded variance [16]. It realizes identification of the disturbance $\zeta(n)$. The block-scheme of the σ^2-adaptive-invariant system, described by equations (64)-(66), (72) is shown in Fig. 11. The adaptation loop corrects the controller in order to decrease the influence of the disturbance. An equivalent block-scheme for the σ^2-adaptive-invariant system has a similar form shown in Fig. 10 after a disturbance $\zeta(n)$ was replaces by $\varepsilon(n)$. If $\zeta(n)$ is a sequence with independent increments [32] then

$$(1-q)\zeta(n) = \varepsilon(n),$$

which corresponds to $C(q) = 1$ in (59). In this case the optimal system is unfit for operation since $\sigma_\zeta^2 = \infty$. Assuming that $J_{\min} = \infty$ in (65) we obtain $C(q,\theta)$. The

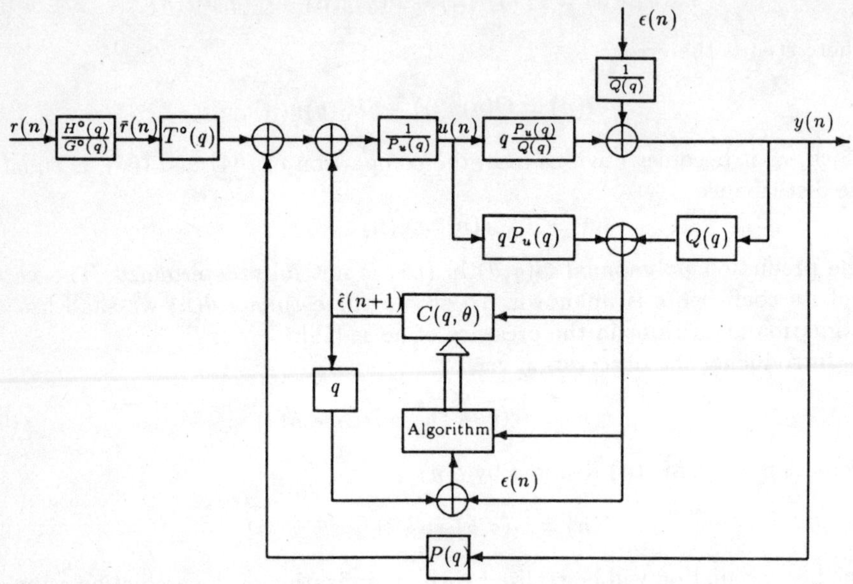

Fig. 11. σ^2-adaptive-invariant system.

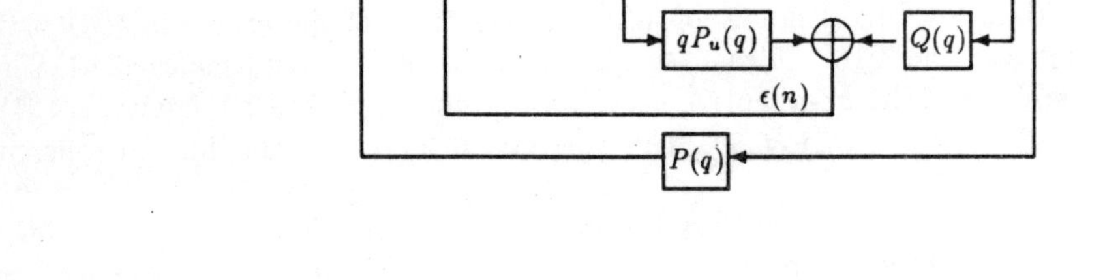

Fig. 12. Special case: $\zeta(n)$ with independent increments.

block-scheme of the selective-invariant up to σ^2 system is shown in Fig. 12. There is no adaptation loop in it and the prediction equation realization is extremely simplified. The equivalent block-scheme for the selective-invariant system has the form similar to that shown in Fig. 10 after a disturbance $\zeta(n)$ was replaced by $\varepsilon(n)$. In this case the selective invariance up to σ^2 turns the unfit original system into an optimal system with a minimum error variance.

If $\zeta(n)$ is a stochastic disturbance, determined by a moving average autoregression equation (ARMA) [16,32]

$$\zeta(n) = C(q)\zeta(n-1) + \varepsilon(n) + B(q)\varepsilon(n-1),$$

where $B(q)$ is an external polynomial, then instead of (63) we obtain the equation of incomplete compensation:

$$\left(1 - q\frac{C(q) + B(q)}{1 + qB(q)}\right)\zeta(n) = \varepsilon(n), \tag{73}$$

and the prediction equation is

$$\zeta(n) = \frac{C(q) + B(q)}{1 + qB(q)}\zeta(n-1).$$

Now, instead of the prediction polynomial there appears a prediction transfer function. The block-scheme of such a σ^2-adaptive-invariant system coincides with the one shown in Fig. 11, if in the adaptation loop we use $\dfrac{C(q) + B(q)}{1 + qB(q)}$ instead of $C(q, \theta)$. Surely, in this case the algorithms become complex, because it is necessary to estimate the coefficients of both $C(q, \theta)$ and $B(q, \theta)$.

The results obtained can be generalized to the case when the plant possesses a multiple lag. However, we shall not spend time on describing this procedure.

Due to incomplete compensation of a stochastic disturbance, eliminating the constraints caused by multiple lag and nonminimum phase will lead to a transformation of the disturbance $\varepsilon(n)$ and the variance of this transformed disturbance will be greater than σ^2. This is a price to be paid for constraints elimination.

10 Some Generalizations

Assume that the disturbance is the sum of a deterministic, $f(n)$, and a stochastic disturbance, $\zeta(n)$. In this case the compensation equation (14) with $\zeta(n)$ is replaced by the equation of incomplete compensation

$$(1 - D(q))(f(n) + \zeta(n)) = (1 - D(q))\zeta(n). \tag{74}$$

Thus, it follows that in a selective-invariant system a stochastic disturbance is transformed and the variance of the transformed disturbance is increased because

$$E\left\{[(1 - qD(q))\zeta(n)]^2\right\} > E\left\{\zeta(n)^2\right\} = \sigma_\zeta^2. \tag{75}$$

Let us find another prediction polynomial with which the variance of the transformed stochastic disturbance would be smaller:

$$\begin{aligned}(1-qD(q))(1-qC(q))(f(n)+\zeta(n)) &= (1-qC(q))\,[(1-qD(q))f(n)] \\ &= (1-qD(q))\,[(1-qC(q))f(n)]\,. \end{aligned} \quad (76)$$

Taking into consideration the compensation equation of a deterministic disturbance (14) with $k=0$ and the condition of incomplete compensation of the stochastic disturbance (63) we obtain from the equation of incomplete compensation

$$(1-qD(q))(1-qC(q))(f(n)+\zeta(n)) = (1-D(q))\zeta(n)\,. \quad (77)$$

It is obvious that

$$E\left\{[(1-qD(q))\zeta(n)]^2\right\} < E\left\{[(1-qD(q))\zeta(n)]^2\right\} \quad (78)$$

and, hence:

$$\sigma^2 < E\left\{[(1-qD(q))\varepsilon(n)]^2\right\} < \sigma_\zeta^2\,. \quad (79)$$

Thus, using the equation of incomplete compensation (76), the variance of the transformed stochastic disturbance is decreased in comparison with σ_ζ^2, but it does not reach the smallest possible value σ^2. Noting that

$$(1-qD(q))(1-qC(q)) = (1-q(D(q)+C(q))) + q^2D(q)C(q)\,,$$

it follows that the prediction polynomial in this case is

$$A(q) = D(q) + C(q) - qD(q)C(q)\,. \quad (80)$$

In the adaptive-invariant system we must estimate the coefficients of this polynomial. They enter (80) nonlinearly, or if overparameterized, they become dependent, which makes the algorithm more complex.

Essential simplification is attained in the approximate realization of the invariant systems described in Sections 6 and 10. Thus, assuming $D(q)\equiv 1$ in (80) we obtain

$$A(q) = 1 + (1-q)C(q)\,. \quad (81)$$

If, in addition, $f(n)$ is a slowly changing deterministic disturbance, then we can assume $A(q)\equiv 1$. The processes in this system will be weakly dependent on $f(n)$.

11 Examples

Consider a dynamic plant described by the equation

$$Q(q)y(n) = qP_u(q)u(n) + \zeta(n) + f(n),$$

where

$$Q(q) = 1 + 0.5q + 0.2q^2; \quad P_u(q) = 1.1 + 0.3q, \tag{82}$$

to which a regular disturbance $f(n)$ and (or) a stochastic disturbance $\zeta(n)$ are applied. The regular disturbance is a periodic saw-tooth signal

$$f_1(t) = \begin{cases} 0.4(n-50m) & 50k - 12.5 \leq n < 12.5 + 50k, \\ -0.4(n-25-50k) & 12.5 + 50k \leq n < 37.5 + 50k, \end{cases} \tag{83}$$

where k is an integer; or the sum of a periodical saw-tooth signal and a harmonic disturbance

$$f(n) = f_1(n) + f_2(n), \tag{84}$$

where

$$f_2(n) = 5 \sin \frac{2\pi n}{25}. \tag{85}$$

Alternatively, the disturbance $f_2(n)$ can be stochastic, $f_2(n) = \zeta(n)$, with $\zeta(n)$ described by the equation of the form (63), or

$$\zeta(n) = \frac{1}{1 - qC(q)}\varepsilon(n), \quad C(q) = 2 - q, \tag{86}$$

where $\varepsilon(n)$ satisfies conditions (60) with $\sigma = 1$.

The reference model is determined by equation (4) with

$$H^0(q) = G^0(q) = 1, \quad k = 0, \quad y_0(n) = qr(n) = r(n-1).$$

The signals $\bar{r}(n) = r(n)$ and $y_0(n)$ are presented in Fig. 13.

For selective-invariant systems (as it follows from (19)) with $k = 0$, $S(q) = 1$, the equation of a controller takes the following form

$$P_u(q)u(n) = T^0(q)\bar{r}(n) - P(q)y(n) - D(q)\epsilon(n) \tag{87}$$

$$\epsilon(n) = Q(q)u(n) - qP_u(q)u(n), \tag{88}$$

where $T^0(q) = 1 + 0.5q$, $Q(q)$ and $P_u(q)$ are assigned by equation (82):

$$\begin{aligned} P(q) &= q^{-1}(T^0(q) - Q(q)) = 0.3 - 0.2q \\ D(q) &= 2 - q. \end{aligned} \tag{89}$$

Under the control law (87) the responses of the plant $y(n)$ and the reference model $y_0(n)$ to $f(n)$ (83), which is equal to the sum of regular disturbances (83), (85) (Fig. 14) are presented in Fig. 15. With $D(q) = 0$, the error of a conventional system, shown in Fig. 16, is increased when compared with Fig. 15.

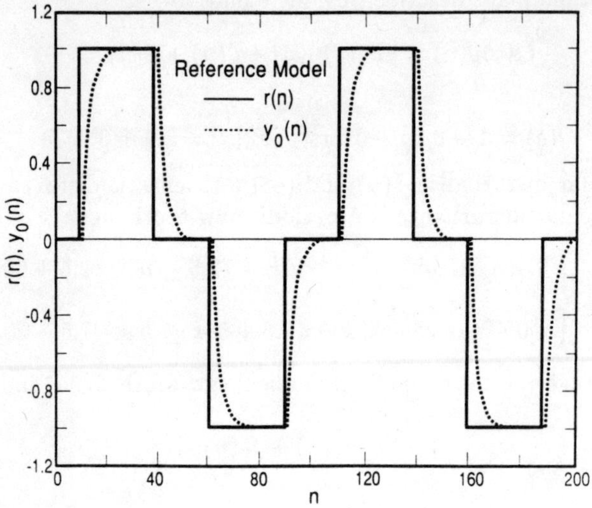

Fig. 13. Reference model signals $r(n)$ and $y_0(n)$.

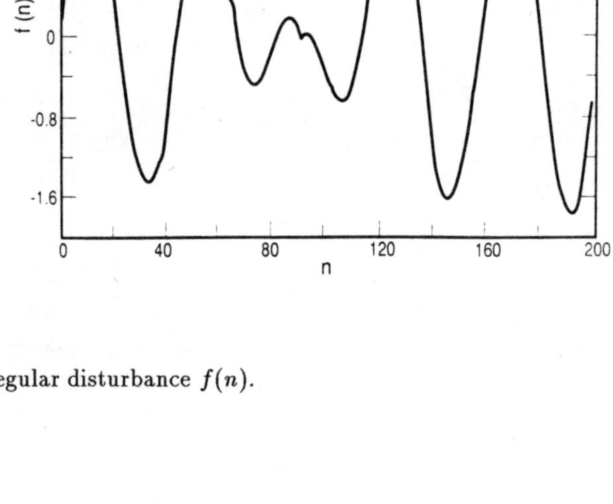

Fig. 14. Regular disturbance $f(n)$.

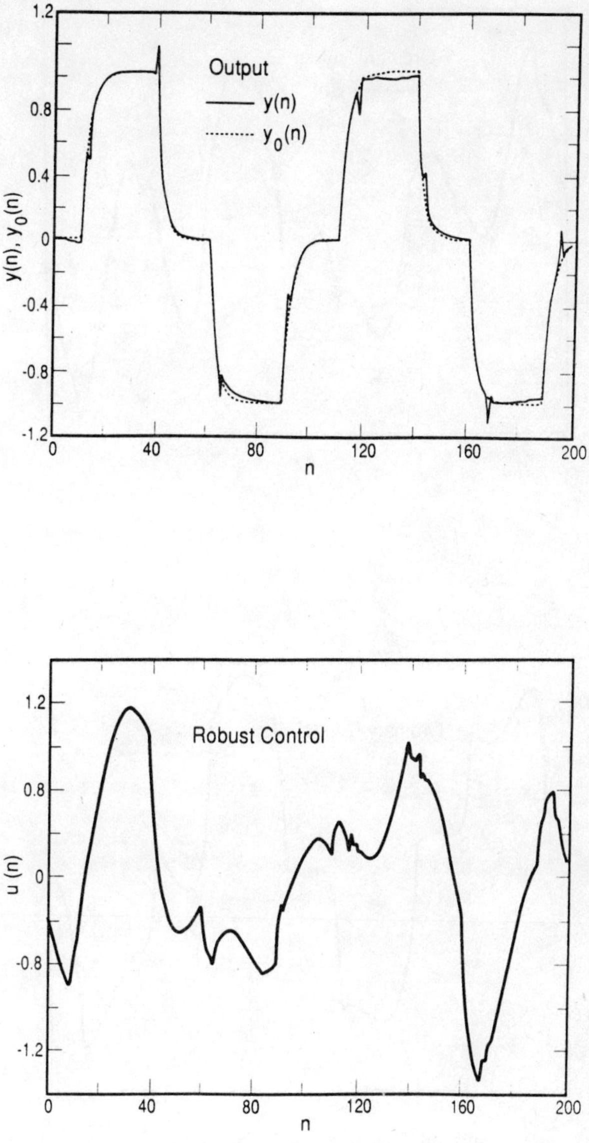

Fig. 15. Selective-invariant system output and control.

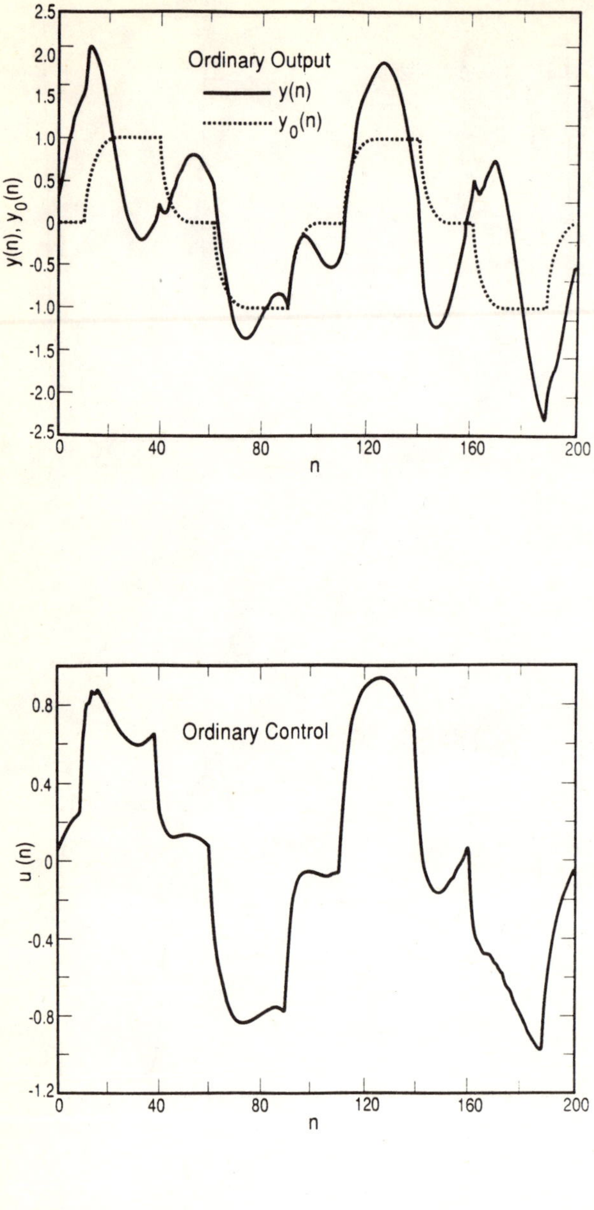

Fig. 16. Conventional system output and control.

Finally, we assume that an additional stochastic disturbance (86) (Fig. 17) was applied along with regular disturbance (84) to a plant. In this case the error of the selective-invariant system, as shown in Fig. 18, is quite close to that of the nominal system, i.e., of the system in which regular disturbances are absent and instead of the stochastic disturbance $\zeta(n)$ we apply to the plant the disturbance $\varepsilon(n)$ whose variance is close to 1. A conventional system (with $D(q) = 1$), as it can be seen in Fig. 19, is not applicable. Control actions in selective-invariant and nominal systems essentially differ from one another (see Fig. 18 and Fig. 19). The reason is that the variance $\zeta(n)$ in the course of time tends to infinity. We shall limit ourselves to these simple examples as illustrations of the principle of selective invariance.

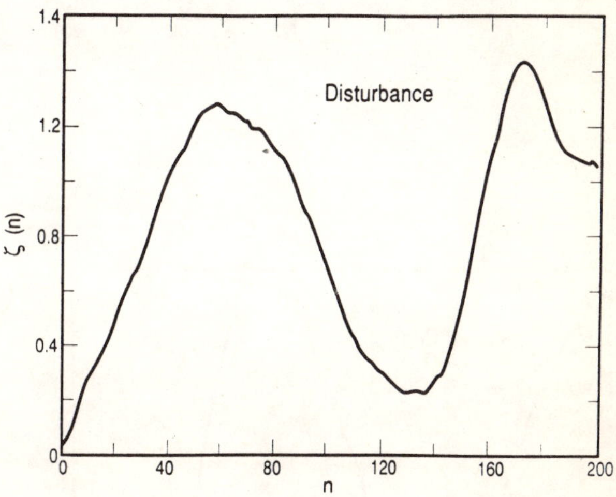

Fig. 17. Stochastic disturbance $\zeta(n)$.

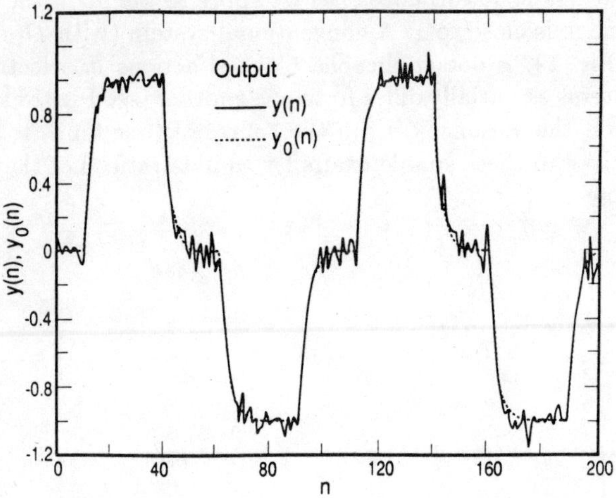

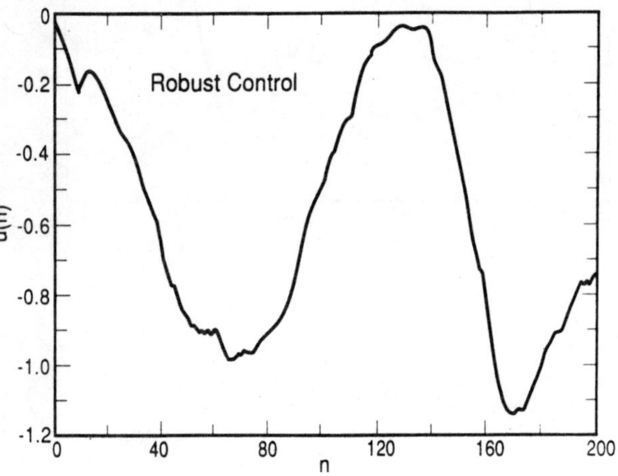

Fig. 18. Selective-invariant system output and control.

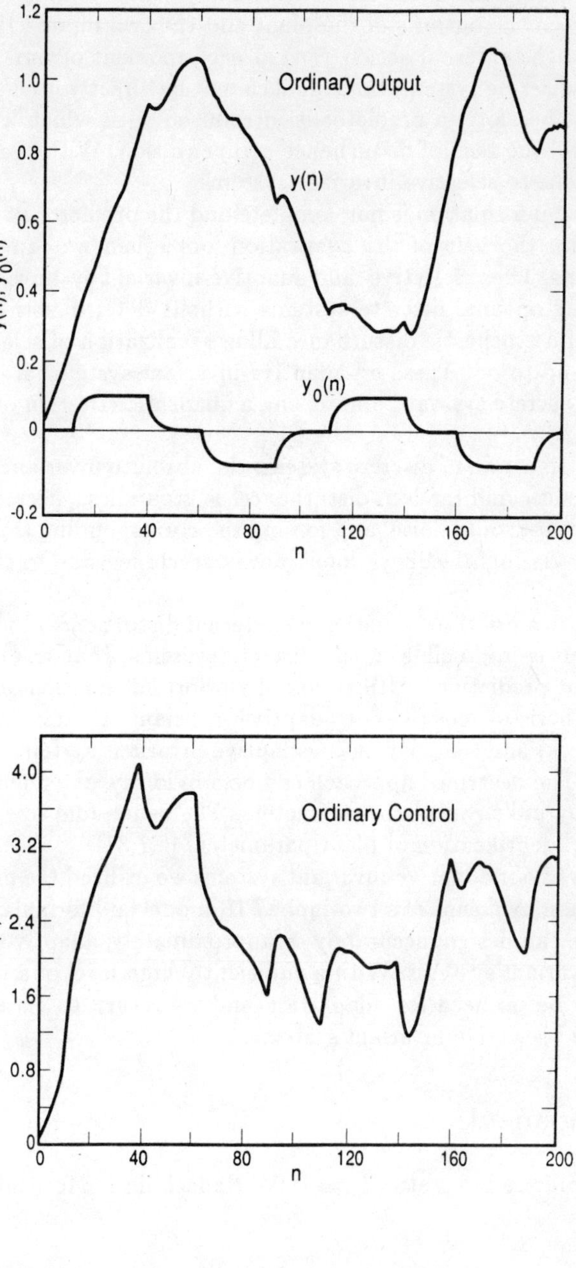

Fig. 19. Conventional system output and control.

12 Concluding Remarks

The main idea of designing adaptive-invariant systems consists in utilizing the error $\epsilon(n)$ between the outputs of the plant and the two-input FIR model. This error is equal to the external action $f(n)$ at each moment of time n. Because of the plant lag in discrete systems this signal cannot be directly used to compensate a disturbance. Therefore, a predictor is introduced with which a control law a designed to fulfill the task of disturbance compensation. With the deterministic predictor we come to selective-invariant systems.

If the a priori information is not complete and the predictor parameters have to be tuned with the help of the adaptation loop, then we come to adaptive-invariant systems. These selective- and adaptive-invariant systems are equivalent to corresponding optimal discrete systems without deterministic disturbances. The presence of a stochastic disturbance allows realization of selective or adaptive invariance up to σ^2. These σ^2-adaptive-invariant systems are equivalent to corresponding discrete systems, minimizing a quadratic criterion of a generalized error.

An attempt to attain in discrete systems the absolute invariance with respect to any deterministic and random disturbances is groundless, because it is impossible to compensate white noise, and forcing the corresponding transfer function to tend to zero via high feedback loop gain is precluded due to the time lag in the plant.

Thus, without a priori information on external disturbances their elimination or compensation is impossible in the discrete systems, that is: "there is no invariance without prediction". Utilization of a priori information on disturbances allows the synthesis of selective- or adaptive-invariant systems (with deterministic disturbances) and selective or σ^2-adaptive-invariant systems (with random disturbances). The described approach can be, obviously, extended to the uncertainty caused by unknown plant parameters. The usual adaptive approach was connected with identification of plant parameters [6,7,31].

As the basis of our adaptive-invariant systems we utilized the principle of disturbance isolation by means of a two-input FIR model and formulated conditions under which we can design, accurately or approximately, adaptive-invariant and σ^2-adaptive-invariant systems. With a sufficiently high level of a priori information there may be no need for adaptation and we return to classical selective-invariant and σ^2-selective-invariant systems.

Acknowledgement

The author expresses his gratitude to P. V. Nadezhdin for fruitful discussions.

References

1. A. I. Kukhtenko, "The main stages of the invariance theory formation, Parts 1 and 2," *Automatika*, 1984, no. 2, pp. 3–13; 1985, no. 2, pp. 3-14.
2. V. S. Kulebakin, "On the behavior of the continuously disturbed automatic linear systems," *Doklady of the USSR Academy of Sciences*, vol. 68, no. 5, pp. 73–79, 1949.
3. V. Kulebakin, "*TheK(D)* transform and its practical application," *Trudy of the N. Zhukovsky VVIA*, vol. 695, pp. 59, 1958.
4. C. D. Johnson, "Theory of disturbance accommodating controllers," in *Advances in Control and Dynamic Systems*, Chap. 12, New York, NY: Academic Press, 1976.
5. C. D. Johnson, "Discrete-time disturbance accommodating control theory with applications to missile digital control," *J. Guidance and Control*, vol. 4, no. 2, pp. 116–125, 1981.
6. Ya. Z. Tsypkin, "Nonminimum phase in discrete control systems," *Itogi nauki i tekhniki.*, pp. 3–40, 1989. Ser. Tekhn. kibernetiki. Moscow: VINITI, T. 18.
7. Ya. Z. Tsypkin and E. D. Aved'yan, "Discrete adaptive control systems for deterministic plants," *Itogi nauki i tekhniki.*, pp. 45–78, 1985. Ser. Tekhn. kibernetiki. Moscow: VINITI, vol. 18.
8. L. N. Volgin, *Optimal Discrete Control of Dynamic Plants.* Moscow: Nauka, 1986.
9. E. D. Yakubovich, "Solution of an optimal control problem by means of discrete linear system," *Avtomatika i telemekhanika*, vol. 9, pp. 73–79, 1975.
10. A. B. Kurzhansky, *Control and Observation under Uncertainty.* Moscow: Nauka, 1977.
11. V. M. Kuntsevich and M. M. Lytchak, *Synthesis of Optimal and Adaptive Control Systems. Game Approach.* Kiev: Naukova dumka, 1985.
12. A. I. Kukhtenko, *Problem of Invariance in Automatics.* Kiev: Uk. SSR: GITL, 1963.
13. C. Shannon, "Mathematical theory of the differential analyzer," *Journal of Mathematics and Physics*, vol. 20, no. 4, pp. 337–352, 1941.
14. V. A. Nikol'sky and I. P. Sevastyanov, "$q(e)$-transformation of lattice functions discrete systems," in *Automatics and Electromechanics*, pp. 30–36, Moscow: Nauka, 1973.
15. J. H. Ahlberg, E. N. Nilson, and J. L. Walsh, *The Theory of Splines and Their Applications.* New York and London: Academic Press, 1967.
16. J. Zypkin, *Grundlagen der Informationellen Theorie der Identification.* Berlin: VEB Verlag Technik, 1987.
17. A. G. Ivakhnenko, *Electrical Automatics.* Kiev: Uk. SSR: Gostekhizdat, 1957.
18. B. M. Mensky, *Invariance Principle in Automatic Regulation and Control.* Moscow: Mashinostrojenije, 1972.
19. Ya. Z. Tsypkin, *Adaptation and Learning in Automatic Systems.* Moscow: Nauka, 1968.
20. G. C. Goodwin and K. S. Sin, *Adaptive Filtering, Prediction and Control.* Englewood Cliffs, NJ: Prentice Hall, 1984.
21. E. D. Aved'yan and Ya. Z. Tsypkin, "The generalized Kaczmarz algorithm," *Avtomatika i telemekhanika*, vol. 3, pp. 71–84, 1979.
22. Y. Markl, "Convergence acceleration of the Kaczmarz algorithm in the case of input process correlation," *Avtomatika i telemekhanika*, vol. 8, pp. 70–73, 1980.
23. K. J. Åström, *Introduction to Stochastic Control Theory.* New York, NY: Academic Press, 1970.

24. V. A. Yakubovich, "Optimization and invariance of linear stationary control systems," *Avtomatika i telemekhanika*, vol. 8, pp. 5–45, 1984.
25. O. Smith, "Closing control of loops with dead time," *Chemical Engineering Progress*, vol. 53, no. 3, pp. 217–219, 1957.
26. O. Smith, *Feedback Control Systems.* New York, NY: McGraw-Hill, 1958.
27. Ya. Z. Tsypkin, "Lag influence compensation in sampled-data systems," *Theory and Applications of Discrete Automatic Systems*, pp. 157–171, 1960. Moscow: Izd. of the USSR Academy of Sci.
28. H. Gorecki, *Analiza i synteza ukladow regulacji z opoznienie.* Warszawa: Wydawnictwa Naucowotechniczne, 1971.
29. E. F. Vogel and T. E. Edgar, "A new dead-time compensator for digital control," in *Proceedings of the ISA, Annual Conference*, pp. 29–46, 1980.
30. Z. Vostry, "New algorithms for polynomial spectral factorization with quadratic convergence," *Kibernetika*, vol. 11, no. 6, pp. 415–422, 1975.
31. Ya. Z. Tsypkin, *Optimality in adaptive control systems. Uncertainty and Control, Lecture Notes in Control and Information Sciences*, vol. 70, Berlin: Springer Verlag, 1985.
32. T. W. Anderson, *The Statistical Analysis of Time Series.* New York, NY: John Wiley, 1971.

Stochastic Adaptive System Theory: Recent Advances and a Reappraisal *

W. Ren and P. R. Kumar

Coordinated Science Laboratory
and
Department of Electrical and Computer Engineering
University of Illinois, Urbana, IL 61801, USA.

Abstract. Progress has been made in the past year towards the solution of several long standing open problems in stochastic adaptive systems for identification, signal processing and control. We provide an account of these recent advances and a fresh reappraisal of the field.

This paper divides itself naturally into two parts. Part I considers identification, adaptive prediction and control based on the ARMAX model. Recent results on the self-optimality of adaptive minimum variance prediction and model reference adaptive control for general delay systems are presented. Both direct and indirect approaches based on non-interlaced extended least squares as well as stochastic gradients algorithms are considered. We emphasize the use of a generalized certainty equivalence approach where the estimates of disturbance as well as parameters are utilized. We also show that self-optimality in the mean square sense in general implies self-tuning, by exhibiting the convergence of the parameter estimates to the null space of a certain covariance matrix.

Part II considers stochastic parallel model adaptation problems, which include output error identification, adaptive IIR filtering, adaptive noise and echo cancelling, and adaptive feedforward control with or without input contamination. Recent results on the convergence of these parallel model adaptation schemes in the presence of nonstationary colored noise are presented.

1 Introduction

1.1 What is Stochastic Adaptive System Theory?

The problems of recursive identification, adaptive signal processing, and adaptive control of linear stochastic systems have been an active area of research for more than two decades. In this paper, we provide a fresh appraisal of these fields, based on many recent advances, some occurring in the last year alone. [23,26,20,17,18]

We focus on analyzing of the *performance* of various discrete time *parameter*

* The research reported here has been partially supported by the U.S. Army Research Office under Contract No. DAAL-03-88-K0046, by the Joint Services Electronics Program under Contract No. N00014-90-J-1270, and by an International Paper Company Fellowship for the first author.

adaptation algorithms in the presence of *stochastic* disturbances.[2] This is due to our belief that adaptive systems are used in many applications to improve system performance. In many such applications, the key to improving performance is to exploit any statistical properties of the disturbances and signals, which can realistically and profitably be assumed in many practical situations, to obtain sharp performance results. One of the main contributions of stochastic adaptive system theory is to show how one may do this.

By the "performance" of a stochastic *adaptive* system, we mean the asymptotic properties of relevant signals and parameters such as the following:

i) Stability: Is the system stable in a mean square sense? Does it have bounded states?
ii) Self-optimality, or more generally the convergence of signals: Do the signals of interest converge to what they would be if the parameters of the system were known? For example, in the case of prediction, does the adaptive prediction converge to the optimal prediction, say in a mean-square or some other sense?
iii) Parameter convergence: Does the parameter estimate converge, and if so to what set? Does it converge to the "true" parameter vector, i.e., is it strongly consistent?
iv) Self-tuning property: Does the adaptive filter, predictor or controller converge to what we would have designed had we known the parameters? Note that even if the parameter estimates do not converge to the true parameter, the limit set to which they converge may still yield the self-tuning property.

The probabilistic properties that we will exploit in this stochastic adaptive system theory are the following; i) the auto-correlation property of a signal; ii) the cross-correlation property between signals; and iii) any persistence of excitation property of a signal. Property i) characterizes the "predictability" of a signal from its own past. By exploiting such an auto-correlation property, we may be able to estimate the innovation process of a signal, and the signal model which generates it. This could then be employed either for adaptive prediction of a signal, or to identify a system model based on say the prediction error identification method. Or, we could use it to attenuate the effect of the predictable portion of an unmeasured but partially predictable disturbance on a system, using adaptive feedback control. By exploiting property ii), we may be able to define "instruments" to obtain asymptotically unbiased estimates of the parameters in the presence of disturbances, by an instrumental variable or output error identification scheme. Alternatively, by making use of an auxiliary correlated measurement, we could extract a signal from noise, as in adaptive noise or echo cancelling, or to cancel the effect of a measurable disturbance, as in adaptive feedforward control. Finally concerning property iii), the persistency of disturbance provides a "natural" excitation to the system, thus allowing us to identify the system better and to establish stronger parameter convergence

---

[2] A disturbance or noise in identification or control of dynamical systems may very well be the signal of interest in a signal processing context. In this paper we mostly adopt a control viewpoint.

results. Thus, we could, for example, conclude the self-tuning property of an adaptive controller, once we know that it is self-optimal.

1.2 A Deterministic Reduction

In the stochastic adaptive system theory that we develop below, the key concern is to deal with the stochastically modeled disturbances and signals. However, it is possible to fully dispense with all probabilistic assumptions, and adopt a completely deterministic model of such signals and disturbances. Some readers may find that this is a more convenient point of view to adopt.

To start, let us consider the key properties satisfied by a "white noise" sequence $w(t)$. A probabilistic model for "white noise" is a stochastic process which satisfies the following three assumptions.

i) $w(t)$ is a martingale difference sequence with respect to an increasing sequence of σ-algebras $\{\mathcal{F}_t\}$, i.e.,

$$E[w(t) \mid \mathcal{F}_{t-1}] = 0 \text{ a.s., } \forall t .$$

ii) $E[w^2(t)|\mathcal{F}_{t-1}] = \sigma^2 > 0$ a.s., $\forall t$.
iii) $\sup_t E[|w(t)|^\alpha|\mathcal{F}_{t-1}] < +\infty$ a.s. for some $\alpha > 2$.

The key consequences of these assumptions are given in the following lemma, see [1-4].

Lemma 1 *Let f_t be a $\mathcal{F}_t$-measurable sequence. If assumptions i) and iii) above are satisfied, then*

a) $\sum_{t=1}^N f_{t-1}w(t) = o\left(\sum_{t=1}^N f_{t-1}^2\right) + O(1)$ *a.s.*[3]
b) $\sum_{t=1}^N f_{t-1}w(t)$ *converges a.s. on the event* $\{\sum_{t=1}^\infty f_{t-1}^2 < \infty\}$,
c) $\sum_{t=1}^N |f_{t-1}|w_t^2 = O\left(\sum_{t-1}^N |f_{t-1}|\right)$ *a.s. on the event* $\{\sup_t |f_t| < \infty\}$.

If, in addition, $w(t)$ satisfies assumption ii) above, then

d) $\frac{1}{N}\sum_{t=1}^N(w^2(t) - \sigma^2) = O(N^{-\delta})$ *a.s., where* $0 < \delta < \frac{\alpha-2}{\alpha}$,
e) $\lim \frac{1}{N}\sum_{t=1}^N f_{t-1}^2 = 0$ *a.s. on the event*

$$\{\lim \frac{1}{N}\sum_{t=1}^N f_{t-1}^2 w_t^2 = 0 \textit{ and } \sup_t |f_{t-1}| < \infty\}.$$

In the above, f_t is a signal which depends only on $w(k)$, $k \leq t$, and on any other signal supposed to be "uncorrelated " with $\{w(t)\}$. Properties a) and b) essentially capture the fact that $w(t)$ is mean zero, and moreover is "uncorrelated" with f_{t-1}. Since the value of $f_{t-1}w(t)$ is therefore as "likely" to be positive as

[3] Here and in the sequel, we write $\alpha_t = o(\beta_t)$ if $\alpha_t/\beta_t \to 0$, $\alpha_t = O(\beta_t)$ if $\sup_t \alpha_t/\beta_t < \infty$, and $\alpha_t \sim \beta_t$ if $\alpha_t = O(\beta_t)$ and $\beta_t = O(\alpha_t)$.

negative, in forming the sum $\sum_{t=1}^{N} f_{t-1}w(t)$ there are likely to be many cancellations, thus rendering it small in comparison with the energy, $\sum_{t=1}^{N} f_{t-1}^2$, in f. Property c) essentially states that since $w^2(t)$ has a bounded average value, by assumption iii), the sum $\sum_{t=1}^{N} |f_{t-1}|w^2(t)$ grows at the same rate as $\sum_{t=1}^{N} |f_{t-1}|$. Finally properties d) and e) show that w is a natural non-negligible unpredictable excitation.

The above sample path properties capture the essence of a white noise. Thus, instead of starting with a probabilistic set of assumptions, one could simply forego all stochastic assumptions, and simply suppose that the noise has the properties (a-e) above, and all the results in this paper would continue to hold. Readers more familiar with deterministic theory may well find this viewpoint to be more enlightening with respect to understanding the goals and results of stochastic adaptive system theory.

In this paper, we will mostly model "stochastic" disturbances and signals as a moving average of such a "white noise" sequence. Thus we will consider disturbances or signals $v(t)$ which can be represented as,

$$v(t) = \sum_{i=-\infty}^{t} c_t(t-i)w(i)\,, \tag{1.1}$$

where the weights $\{c_t(i)\}_{i=0}^{\infty}$ are deterministic.

The shift structure contained in the above moving average representation (1.1), and Lemma 1 a)-c), will provide us with certain auto-correlation and cross-correlation properties which we can exploit. On the other hand, Lemma 1 d) and e) will provide us with certain "persistency of excitation" properties which we will find useful.

The rest of this paper divides itself naturally into two parts. Section 2 treats identification, adaptive prediction and adaptive control based on the ARMAX model, using the so called "equation error" approach. Recent results on the self-optimality of adaptive prediction and control schemes for general delay systems are presented. Both direct and indirect approaches based on non-interlaced extended least squares (ELS) as well as stochastic gradient (SG) algorithms are considered. We motivate the use of a generalized certainty equivalence approach whereby the estimates of disturbance as well as parameters are utilized. We also show that self-optimality implies self-tuning in general.

Section 3 considers stochastic parallel model adaptation problems which consist of a unknown linear time invariant system and a partially or wholly tunable system connected in parallel, with a common input to each. The goal of adaptation is to tune the latter system so that its output matches that of the unknown system, despite the presence of an arbitrary disturbance which is stochastically uncorrelated with the common input. Such a general formulation includes the problems of output error identification, adaptive IIR filtering, adaptive noise and echo cancelling, and adaptive feedforward control with or without "input contamination." The convergence of such schemes will be established for fairly general nonstationary colored noise.

2 Identification, Adaptive Prediction and Control Based on an ARMAX Model

There has been a great deal of interest in establishing the convergence properties of recursive identification, adaptive prediction and control schemes based on an ARMAX model. Despite the notable contributions, among others, of Åström and Wittenmark [5], Ljung [6,7], Solo [8], Goodwin, Ramadge, and Caines [9], Goodwin, Sin and Saluja [10], Fuchs [11], Lai and Wei [2], Becker, Kumar and Wei [3], Chen and Caines [12], Kumar and Praly [13], Lai and Wei [14], Chen and Guo [15,4], and Radenković and Stanković [16], several significant open problems remain. Among them are: i) the self-optimality of adaptive prediction and control based on an ELS algorithm; ii) the self-optimality of adaptive prediction and control schemes for systems with general delay using non-interlaced algorithms; iii) a general self-tuning result.

Recently in [21], a fairly complete solution to these problems is obtained for generically all systems under the restrictive condition of Gaussian white noise. Early this year Guo and Chen [23] have succeeded brilliantly in establishing the optimality of adaptive minimum variance (MV) control based on the ELS algorithm, for systems with unit delay. The convergence of non-interlaced algorithms for the case of general delay has been resolved in Ren and Kumar [17]. Very recently, a general self-tuning result has been established by Ren and Kumar [18]. It is shown that for the SG as well as the ELS algorithms, the parameter estimate converges almost surely to the null space of a certain covariance matrix. Moreover, each point in the null space gives rise to, for example, the optimal controller in the case of adaptive MV control.

2.1 Identification of an ARMAX model

Consider the single input, single output linear system described by the following ARMAX model

$$A(q^{-1})y(t) = q^{-d}B(q^{-1})u(t) + C(q^{-1})w(t)\,, \tag{2.1}$$

where q^{-1} is the backward shift operator, i.e., $q^{-1}y(t) = y(t-1)$,

$$\begin{aligned} A(q^{-1}) &= 1 + a_1 q^{-1} + \cdots + a_p q^{-p} \\ B(q^{-1}) &= b_0 + b_1 q^{-1} + \cdots + b_h q^{-h} \\ C(q^{-1}) &= 1 + c_1 q^{-1} + \cdots + c_r q^{-r}\,, \end{aligned} \tag{2.2}$$

$\{y(t)\}$ and $\{u(t)\}$ are the output and input of the system, and $\{w(t)\}$, the disturbance, is a martingale difference sequence with respect to an increasing sequence of σ-fields $\{\mathcal{F}_t\}$ which satisfies

$$E[w^2(t)|\mathcal{F}_{t-1}] = \sigma^2, \qquad \text{a.s., for all } t \geq 1$$

$$\sup_t E[|w(t)|^\alpha|\mathcal{F}_{t-1}] < \infty, \qquad \text{a.s., for some } \alpha > 2\,.$$

The system (2.1) can be also parametrized as [4]

$$y(t) = R(q^{-1})u(t-d) + S(q^{-1})y(t-d) + q^{-d}D(q^{-1})v(t) + v(t)\,, \tag{2.3}$$

where $v(t) = F(q^{-1})w(t)$, and $F(q^{-1}) = \sum_{i=0}^{d-1} f_i q^{-1}$, $f_0 = 1$. This alternative model has certain advantages when used for adaptive prediction and control, which will become clear later.

In this subsection we consider the parameter estimation of both models (2.1) and (2.3) . The following stochastic regression model encompasses both models,

$$y(t) = x^{\mathrm{T}}(t-s)\beta + d_1 v(t-s) + \cdots + d_\ell v(t-s-\ell+1) + v(t)\,, \tag{2.4}$$

where $v(t) = \sum_{i=0}^{s-1} f_i w(t-i)$, $s \geq 1$, $x(t)$ and β are column vectors, and $x(t)$ is $\mathcal{F}_t$-measurable and is available at time t. The correspondence between (2.3) and (2.4) with $s = d$ is clear, while for the model (2.1), $s = 1$, $v(t) = w(t)$, $x(t-1) = [y(t-1), \ldots, y(t-p), u(t-d), \ldots, u(t-d-h)]^{\mathrm{T}}$, $\beta = [a_1, \ldots, a_p, b_0, \ldots, b_h]^{\mathrm{T}}$, $\ell = r$ and $d_i = c_i, i = 1, \ldots, r$.

Note that the regression model (2.4) is a generalization of that considered in [14], which corresponds to the case $s = 1$ here.

Let $\theta_0 := [\beta^{\mathrm{T}}, d_1, \ldots, d_\ell]^{\mathrm{T}}$. To recursively estimate θ_0, a natural extension of the algorithm considered in [14], i.e., for the case $s = 1$, is as follows: Let $\theta(t-1)$ be the estimate of θ_0 at time $t-1$, and construct an estimate $\widehat{v}(t-1)$ of the unobservable $v(t-1)$,

$$\widehat{v}(t-1) := y(t-1) - \phi^{\mathrm{T}}(t-s-1)\theta(t-1)\,,$$

where $\phi^{\mathrm{T}}(t-s-1) := [x^{\mathrm{T}}(t-s-1), \widehat{v}(t-s-1), \cdots, \widehat{v}(t-s-\ell)]$. Then we update the estimate to $\theta(t)$ using either the recursive least squares algorithm,

$$\theta(t) = \theta(t-1) + P(t-s)\phi(t-s)(y(t) - \phi^{\mathrm{T}}(t-s)\theta(t-1)) \tag{2.5}$$

$$P(t-s)^{-1} = P(t-s-1)^{-1} + \phi(t-s)\phi^{\mathrm{T}}(t-s); \quad P(-s) = I \tag{2.6}$$

or its stochastic gradient counterpart,

$$\theta(t) = \theta(t-1) + \frac{\phi(t-s)}{r(t-s)}(y(t) - \phi^{\mathrm{T}}(t-s)\theta(t-1)) \tag{2.7}$$

$$r(t-s) = r(t-s-1) + \|\phi(t-s)\|^2, \quad r(-s) = 1\,. \tag{2.8}$$

In contrast to the above algorithms, for $s > 1$, interlaced algorithms of the form

$$\theta(t) = \theta(t-s) + P(t-s)\phi(t-s)(y(t) - \phi^{\mathrm{T}}(t-s)\theta(t-s)) \tag{2.9}$$

$$P^{-1}(t-s) = P^{-1}(t-2s) + \phi(t-s)\phi^{\mathrm{T}}(t-s)\,, \tag{2.10}$$

have usually been considered in the literature [22]. Due to the interlacing of s separate recursions, the algorithm (2.9, 2.10) requires the additional storage in

[4] Let polynomials F and G satisfy $AF + q^{-d}G = C$, where $\deg F \leq d-1$, and let polynomials $\overline{F}$ and $\overline{G}$ satisfy $C\overline{F} + q^{-d}\overline{G} = 1$, with $\deg \overline{F} \leq d-1$. Then $R = BF\overline{F}, D = \overline{G}, S = G\overline{F} + \overline{G}$. See [22].

memory of $(\theta(t-2),\ldots,\theta(t-s))$ and $(P(t-s-1),\ldots,P(t-2s))$ at each time t.

Since each interlaced recursion effectively sees $v(t)$ as a "white noise", it is clear that the results for $s=1$ can be easily extended to the case $s>1$ by the use of interlaced algorithms. The consideration of such interlaced algorithms have been purely an artifact of the theoretical inability to demonstrate the convergence of non-interlaced algorithms.

In Theorems 2.1 and 2.2 below we resolve the convergence properties of the non-interlaced SG and ELS algorithms defined by (2.7, 2.8) and (2.5, 2.6), respectively.

The following preliminary fact is useful.

Lemma 2.1 *Consider the regression model (2.4). Let* $D(z)=d_1+d_2z+\cdots+d_\ell z^\ell$. *Then*

$$\left(1+q^{-s}D(q^{-1})\right)\left(\widehat{v}(t)-v(t)\right)=-\phi^{\mathrm{T}}(t-s)\widetilde{\theta}(t), \tag{2.11}$$

where $\widetilde{\theta}(t):=\theta(t)-\theta_0$.

Proof.

$$\begin{aligned}
&(1+q^{-s}D(q^{-1}))(\widehat{v}(t)-v(t))\\
&=\widehat{v}(t)+q^{-s}D(q^{-1})\widehat{v}(t)-y(t)+x^{\mathrm{T}}(t-s)\beta\\
&=-\phi^{\mathrm{T}}(t-s)\theta(t)+q^{-s}D(q^{-1})\widehat{v}(t)+x^{\mathrm{T}}(t-s)\beta\\
&=-\phi^{\mathrm{T}}(t-s)(\theta(t)-\theta_0)\,.
\end{aligned}$$

□

For convenience, let us define

$$\begin{aligned}
R(N) &:= I+\sum_{1}^{N}\phi(t)\phi^{\mathrm{T}}(t)\\
\phi_0(t) &:= [x^{\mathrm{T}}(t),v(t),\cdots,v(t-\ell+1)]^{\mathrm{T}}\\
R_0(N) &:= I+\sum_{1}^{N}\phi_0(t)\phi_0^{\mathrm{T}}(t)\\
r_0(N) &:= \text{ trace }(R_0(N))\,.
\end{aligned}$$

Theorem 2.1 *Consider the SG algorithm defined by (2.7) and (2.8), and applied to the stochastic regression model (2.4). Assume that* $1+q^{-d}D(q^{-1})$ *is Hurwitz* [5] *and is strictly positively real (SPR), i.e.,*

$$\mathrm{Re}[1+z^dD(z)]>0, \quad \textit{for } z=e^{j\omega},\forall\omega\,. \tag{2.12}$$

[5] Here and in the sequel, we shall say that a polynomial $P(q^{-1})$ is Hurwitz if its zeroes are inside the open unit disk, and that a rational transfer function is stable if its denominator polynomial is Hurwitz.

Then,

$$\|\theta(t)\| \text{ converges a.s.} \tag{2.13}$$

$$\sum_{t=1}^{\infty} \|\theta(t) - \theta(t-k)\|^2 < \infty, \text{ a.s., } \forall k < \infty \tag{2.14}$$

$$\sum_{t=1}^{+\infty} \frac{(\widehat{v}(t) - v(t))^2}{r(t-s-1)} < \infty, \text{ a.s.} \tag{2.15}$$

$$r(t) \sim r_0(t) \text{ a.s.} \tag{2.16}$$

$$\sum_{t=1}^{\infty} \frac{(y(t) - \phi^{\mathrm{T}}(t-s)\theta(t-k) - v(t))^2}{r(t-s)} < \infty, \text{ a.s., } \quad \forall k < \infty. \tag{2.17}$$

Moreover, if

$$r_0(N) \to \infty \quad \text{and} \quad \sup_N \left(\frac{\lambda_{\max}(R_0(N))}{\lambda_{\min}(R_0(N))} \right) < \infty, \text{ a.s.,} \tag{2.18}$$

then

$$\theta(t) \to \theta_0 \text{ a.s.}.$$

Proof. Premultiplying (2.7) by $\phi^{\mathrm{T}}(t-s)$ and subtracting from $y(t)$, we can rewrite (2.7) as

$$\widetilde{\theta}(t) - \frac{\phi(t-s)}{r(t-s-1)}\widehat{v}(t) = \widetilde{\theta}(t-1).$$

Taking squares yields

$$\|\widetilde{\theta}(t)\|^2 - \frac{2\phi(t-s)\widetilde{\theta}(t)\widehat{v}(t)}{r(t-s-1)} + \frac{\|\phi(t-s)\|^2\widehat{v}^2(t)}{r^2(t-s-1)} = \|\widetilde{\theta}(t-1)\|^2.$$

Let $V(t){:=}\widetilde{\theta}(t)\|^2$, and for convenience define the operator $\mathcal{H}_t(\cdot)$ on a random variable z, as follows,

$$\mathcal{H}_t(z) := z - E[z|\mathcal{F}_{t-s}]. \tag{2.19}$$

Making use of (2.19) to rewrite (2.7), We obtain,

$$\begin{aligned} V(t) &- \frac{2\phi^{\mathrm{T}}(t-s)\widetilde{\theta}(t)(\widehat{v}(t) - v(t))}{r(t-s-1)} - \frac{2\phi^{\mathrm{T}}(t-s)E[\widetilde{\theta}(t)|\mathcal{F}_{t-s}]v(t)}{r(t-s-1)} \\ &+ \frac{\|\phi(t-s)\|^2\widehat{v}^2(t)}{r^2(t-s-1)} = V(t-1) + \frac{2\phi^{\mathrm{T}}(t-s)\mathcal{H}_t(\theta(t))v(t)}{r(t-s-1)}. \end{aligned} \tag{2.20}$$

We will now show that the last term above is absolutely summable, almost surely. Applying the operator $\mathcal{H}_t(\cdot)$ on both sides of (2.7) and noting that $y(t) - v(t)$, $\phi(t-s)$, $r(t-s)$ and $\theta(t-s)$ are all $\mathcal{F}_{t-s}$-measurable, we have, for $i = 0, \ldots, s-2$,

$$\begin{aligned} \mathcal{H}_t(\theta(t-i)) &= \frac{r(t-s-i)I - \phi(t-s-i)\phi^{\mathrm{T}}(t-s-i)}{r(t-s-i)}\mathcal{H}_t(\theta(t-i-1)) \\ &+ \frac{\phi(t-s-i)}{r(t-s-i)}\mathcal{H}_t(v(t-i)), \end{aligned} \tag{2.21}$$

and

$$\mathcal{H}_t(\theta(t-s+1)) = \frac{\phi(t-2s+1)}{r(t-2s+1)}\mathcal{H}_t(v(t-s+1)). \tag{2.22}$$

Now using (2.21) with $i=0$ and $\mathcal{H}_t(v(t)) = v(t)$,

$$\frac{\phi^{\mathrm{T}}(t-s)\mathcal{H}_t(\theta(t))v(t)}{r(t-s-1)} = \frac{\phi^{\mathrm{T}}(t-d)\mathcal{H}_t(\theta(t-1))v(t)}{r(t-s)} + \frac{\|\phi(t-s)\|^2}{r(t-s)r(t-s-1)}v^2(t). \tag{2.23}$$

Since $\dfrac{\|\phi(t-s)\|^2}{r(t-s)r(t-s-1)}$ is summable, it follows from Lemma 1 c) that the last term on the RHS of (2.23) is summable. To show that the first term on the RHS of (2.23) is also summable, note that

$$\sum_{t=1}^{\infty}\left|\frac{\phi^{\mathrm{T}}(t-s)\mathcal{H}_t(\theta(t-1))v(t)}{r(t-s)}\right| \leq \left(\sum_{t=1}^{\infty}\frac{\|\phi(t-s)\|^2 v^2(t)}{r^2(t-s)}\right)^{\frac{1}{2}}\left(\sum_{t=1}^{\infty}\|\mathcal{H}_t(\theta(t-1))\|^2\right)^{\frac{1}{2}}. \tag{2.24}$$

As earlier, the first summation on the RHS above is finite a.s.. To show that the second summation is also finite, note that from (2.21), for $i = 1, \ldots, s-2$,

$$\|\mathcal{H}_t(\theta(t-i))\| \leq \|\mathcal{H}_t(\theta(t-i-1))\| + \frac{\|\phi(t-s-i)\|}{r(t-s-i)}|\mathcal{H}_t(v(t-i))|, \tag{2.25}$$

and from (2.22),

$$\|\mathcal{H}_t(\theta(t-s+1))\| \leq \frac{\|\phi(t-2s+1)\|}{r(t-2s+1)}|\mathcal{H}_t(v(t-s+1))|. \tag{2.26}$$

So,

$$\|\mathcal{H}_t(\theta(t-1))\| \leq \sum_{i=1}^{s-1}\frac{\|\phi(t-s-i)\|}{r(t-s-i)}|\mathcal{H}_t(v(t-s-i))|. \tag{2.27}$$

Hence,

$$\sum_{t=1}^{\infty}\|\mathcal{H}_t(\theta(t-1))\|^2 < \infty. \tag{2.28}$$

From (2.23), (2.24) and (2.28), it then follows that

$$\sum_{t=1}^{\infty}\left|\frac{\phi^{\mathrm{T}}(t-s)\mathcal{H}_t(\theta(t))v(t)}{r(t-s)}\right| < \infty, \text{ a.s..}$$

We now bound the third term on the LHS of (2.20). Applying Lemma 1 a), we can obtain that

$$\sum_{t=1}^{N} \frac{\phi^{\mathrm{T}}(t-s)E[\widetilde{\theta}(t)|\mathcal{F}_{t-s}]v(t)}{r(t-s-1)} = o\left(\sum_{t=1}^{N}\left(\frac{\phi^{\mathrm{T}}(t-s)E[\widetilde{\theta}(t)|\mathcal{F}_{t-s}]}{r(t-s-1)}\right)^2\right) + O(1)\,.$$

Using the decomposition $\widetilde{\theta}(t) = E[\widetilde{\theta}(t)|\mathcal{F}_{t-s}] + \mathcal{H}_t(\theta(t))$, we can further write

$$\begin{aligned}\sum_{t=1}^{N} \frac{\phi^{\mathrm{T}}(t-s)E[\widetilde{\theta}(t)\mid\mathcal{F}_{t-s}]v(t)}{r(t-s-1)} &= o\left(\sum_{t=1}^{N}\left(\frac{\phi^{\mathrm{T}}(t-s)\widetilde{\theta}(t))}{r(t-s-1)}\right)^2\right) \\ &\quad + o\left(\sum_{t=1}^{N}\left(\frac{\phi^{\mathrm{T}}(t-s)\mathcal{H}_t(\theta(t))}{r(t-s-1)}\right)^2\right) + O(1)\,.\end{aligned} \tag{2.29}$$

Now, from (2.23), we have

$$\begin{aligned}\sum_{t=1}^{N}\left(\frac{\phi^{\mathrm{T}}(t-s)\mathcal{H}_t(\theta(t))}{r(t-s-1)}\right)^2 &= O\left(\sum_{t=1}^{N}\left(\frac{\phi^{\mathrm{T}}(t-s)\mathcal{H}_t(\theta(t-1))}{r(t-s)}\right)^2\right) \\ &\quad + O\left(\sum_{t=1}^{n}\left(\frac{\|\phi(t-s)\|^2}{r(t-s)r(t-s-1)}\right)^2 v^2(t)\right) \\ &= O(1)\,,\end{aligned} \tag{2.30}$$

where in the last equality we have used (2.28) and the fact that $\left(\frac{\|\phi(t-s)\|^2}{r(t-s)r(t-s-1)}\right)^2$ is summable. It then follows from (2.29) and (2.30) that

$$\sum_{t=1}^{N} \frac{\phi^{\mathrm{T}}(t-s)E[\widetilde{\theta}(t)|\mathcal{F}_{t-s}]v(t)}{r(t-s-1)} = o\left(\sum_{t=1}^{N}\left(\frac{\phi^{\mathrm{T}}(t-s)\widetilde{\theta}(t)}{r(t-s-1)}\right)^2\right) + O(1)\,. \tag{2.31}$$

Finally, from (2.11) and the SPR assumption (2.12), we have

$$S(N) = \sum_{t=1}^{N}[-\phi^{\mathrm{T}}(t-s)\widetilde{\theta}(t)\,(\widehat{v}(t)-v(t)) - \epsilon(\widehat{v}(t)-v(t))^2] + S(0) \geq 0\,, \tag{2.32}$$

for some constant $\epsilon > 0$, and random variable $S(0) < \infty$ a.s.. Summing by parts, we have

$$\sum_{t=1}^{N} \frac{S(t)-S(t-1)}{r(t-s-1)}$$
$$= \frac{S(N)}{r(N-s-1)} + \sum_{t=1}^{N-1} S(t)\left(\frac{1}{r(t-s-1)} - \frac{1}{r(t-s)}\right) - \frac{S(0)}{r(-s)}. \quad (2.33)$$

Since $r(t)$ is monotonically nondecreasing, it follows from (2.32), (2.33) that

$$\sum_{t=1}^{N} \frac{[-\phi^{\mathrm{T}}(t-s)\widetilde{\theta}(t)\,(\widehat{v}(t)-v(t)) - \epsilon(\widehat{v}(t)-v(t))^2]}{r(t-s-1)} \geq -\frac{S(0)}{r(-s)}. \quad (2.34)$$

Using (2.34), (2.31) and (2.28) in (2.20), we can immediately conclude (2.15) and (2.14). (2.16) follows from (2.15). (2.13) can be then shown using Lemma 1 b). The following series of inequalities proves (2.17).

$$\sum_{t=1}^{\infty} \frac{[y(t)-\phi^{\mathrm{T}}(t-d)\theta(t-s)-v(t)]^2}{r(t-s)}$$
$$\leq \sum_{t=1}^{\infty} \frac{2[y(t)-\phi(t-s)\theta(t)-v(t)]^2}{r(t-s)} + \sum_{t=1}^{\infty} \frac{2[\phi^{\mathrm{T}}(t-s)(\theta(t)-\theta(t-k))]^2}{r(t-s)}$$
$$\leq \sum_{t=1}^{\infty} \frac{2[\widehat{v}(t)-v(t)]^2}{r(t-s-1)} + 2\left(\sum_{t=1}^{\infty} \frac{\|\phi(t-s)\|^2}{r^2(t-s)}\right)^{\frac{1}{2}} \left(\sum_{t=1}^{\infty} \|\theta(t)-\theta(t-k)\|^2\right)^{\frac{1}{2}}$$
$$< \infty. \quad (2.35)$$

We now turn to the issue of parameter consistency when (2.18) holds. Multiplying (2.7) by$R(t-s)$, we have

$$R(t-s)\widetilde{\theta}(t) = R(t-s-1)\widetilde{\theta}(t-1) + \phi(t-s)\phi^{\mathrm{T}}(t-s)\widetilde{\theta}(t-1)$$
$$+R(t-s)\frac{\phi(t-s)(y(t)-\phi^{\mathrm{T}}(t-s)\theta(t-1))}{r(t-s)}.$$

So,

$$\frac{1}{r(N-s)}\left\|R(N-s)\widetilde{\theta}(N)\right\|$$
$$\leq \frac{1}{r(N-s)}\left\|\sum_{t=1}^{N}\phi(t-s)\phi(t-s)^1\widetilde{\theta}(t-1)\right\|$$
$$+\frac{1}{r(N-s)}\left\|\sum_{t=1}^{N}\frac{R(t-s)}{r(t-s)}\phi(t-s)(y(t)-\phi^{\mathrm{T}}(t-s)\theta(t-1)-v(t))\right\|$$
$$+\frac{1}{r(N-s)}\left\|\sum_{t=1}^{N}\frac{R(t-s)}{r(t-s)}\phi(t-s)v(t)\right\| + o(1).$$

The terms on the right hand side above converge to zero by the Schwarz inequality, (2.17) and Lemma 1. Hence,

$$\frac{1}{r(N-s)}\|R(N-s)\widetilde{\theta}(N)\| \to 0 \quad \text{a.s.}. \tag{2.36}$$

From (2.15),

$$\frac{1}{r(N-s)}(R(N-s) - R_0(N-s)) \to 0 \quad \text{a.s.}. \tag{2.37}$$

Combining (2.36), (2.37), (2.16) and (2.18) completes the proof. □

Theorem 2.2 *Consider the ELS algorithm defined by (2.5) and (2.6), and applied to the stochastic regression model (2.4). Assume that* $1 + q^{-s}D(q^{-1})$ *is Hurwitz and that*

$$\operatorname{Re}\left[(1 + z^s D(z))^{-1} - \frac{1}{2}\right] > 0, \text{ for } z = e^{j\omega}, \quad \forall\omega\,. \tag{2.38}$$

Then,

$$\sum_{t=1}^{N}(\widehat{v}(t) - v(t))^2 = 0(\log r(N-s)) \text{ a.s.} \tag{2.39}$$

$$r(t) \sim r_0(t) \text{ a.s.} \tag{2.40}$$

$$\|\widetilde{\theta}(N)\|^2 = 0\left(\frac{\log r(N-s)}{\lambda_{\min}(R(N-s))}\right) \text{ a.s.} \tag{2.41}$$

$$\sum_{t=1}^{N}\frac{[y(t) - \phi^{T}(t-s)\theta(t-k) - v(t)]^2}{1 + \phi^{T}(t-s)P(t-s-k)\phi(t-s)} = 0(\log r(N-s)) \text{ a.s.}, \tag{2.42}$$

$\forall k < \infty$.

Moreover, if

$$r_0(N) \to \infty \text{ and } \frac{\log \lambda_{\max}(R_0(N))}{\lambda_{\min}(R_0(N))} \to 0 \text{ a.s.}, \tag{2.43}$$

then

$$\theta(t) \to \theta_0 \text{ a.s.}.$$

The proof of Theorem 2.2 below is presented in a manner which follows the steps of the proof of Theorem 2.1, so that the similarities and differences are transparent.

Proof. Rewrite (2.5) as

$$\widetilde{\theta}(t) - P(t-s-1)\phi(t-s)\widehat{v}(t) = \widetilde{\theta}(t-1)\,, \tag{2.44}$$

and also as,

$$P(t-s-1)^{-1}\widetilde{\theta}(t) - \phi(t-s)\widehat{v}(t) = P(t-s-1)^{-1}\widetilde{\theta}(t-1)\,. \tag{2.45}$$

Multiplying (2.44) and (2.45), we have

$$V(t) - [\phi^{\mathrm{T}}(t-s)\widetilde{\theta}(t)]^2 + \phi(t-s)^{\mathrm{T}}P(t-s-1)\phi(t-s)\widehat{v}^2(t) \\ -2\phi^{\mathrm{T}}(t-s)\widetilde{\theta}(t)\widehat{v}(t) = V(t-1)\,,$$

where $V(t) := \widetilde{\theta}^{\mathrm{T}}(t)P(t-s)^{-1}\widetilde{\theta}(t)$. Hence,

$$V(t-1) = V(t) - 2\phi^{\mathrm{T}}(t-s)\widetilde{\theta}(t)\left[\frac{1}{2}\phi^{\mathrm{T}}(t-s)\widetilde{\theta}(t) + (\widehat{v}(t) - v(t))\right] \\ +\phi(t-s)^{\mathrm{T}}P(t-s-1)\phi(t-s)\widehat{v}^2(t) - 2\phi^{\mathrm{T}}(t-s)\widetilde{\theta}(t)v(t)\,. \tag{2.46}$$

We will now show that

$$\sum_{t=1}^{N}\phi^{\mathrm{T}}(t-s)\widetilde{\theta}(t)v(t) = o\left(\sum_{t=1}^{N}\left(\phi^{\mathrm{T}}(t-s)\widetilde{\theta}(t)\right)^2\right) + O(\log r(N-s))\,. \tag{2.47}$$

We first decompose it as

$$\sum_{t=1}^{n}v(t)\phi^{\mathrm{T}}(t-s)\widetilde{\theta}(t) = \sum_{t=1}^{n}v(t)\phi^{\mathrm{T}}(t-s)E[\widetilde{\theta}(t)|\mathcal{F}_{t-s}] \\ +\sum_{t=1}^{n}v(t)\phi(t-s)\mathcal{H}_t(\theta(t))\,. \tag{2.48}$$

We will first bound the second term on the RHS above. From (2.5), we have for $i = 0, \ldots, s-1$,

$$P^{-1}(t-s-i)\theta(t-i) = \phi(t-s-i)y(t-i) + P^{-1}(t-s-i-1)\theta(t-i-1)\,.$$

Noting that $y(t) - v(t)$, $\phi(t)$ and $P(t)$ are all $\mathcal{F}_t$- measurable, we have for $i = 0, \ldots, s-2$

$$P^{-1}(t-s-i)\mathcal{H}_t(\theta(t-i)) = \phi(t-s-i)\mathcal{H}_t(v(t-i)) \\ +P^{-1}(t-s-i-1)\mathcal{H}_t(\theta(t-i-1))\,. \tag{2.49}$$

For $i = s-1$, we have

$$\mathcal{H}_t(\theta(t-s+1)) = P(t-2s+1)\phi(t-2s+1)\mathcal{H}_t(v(t-s+1))\,. \tag{2.50}$$

From (2.49) and (2.50), it is easy to see that

$$\mathcal{H}_t(\theta(t)) = P(t-s)\sum_{i=0}^{s-1}\phi(t-s-i)\mathcal{H}_t(v(t-i))\,.$$

So,

$$\begin{aligned}
&\sum_{t=1}^{N} v(t)\phi^{\mathrm{T}}(t-s)\mathcal{H}_t(\theta(t)) \\
&= \sum_{t=1}^{N} \phi^{\mathrm{T}}(t-s)P^{-1}(t-s)\sum_{i=0}^{s-1}\phi(t-s-i)v(t)\mathcal{H}_t(v(t-i)) \\
&= O\left(\sum_{i=0}^{s-1}\left(\sum_{t=1}^{N}\phi^{\mathrm{T}}(t-s)P(t-s)\phi(t-s)\right)^{\frac{1}{2}}\right. \\
&\qquad\left.\left(\sum_{t=1}^{N}\phi^{\mathrm{T}}(t-s-i)P(t-s)\phi(t-s-i)\right)^{\frac{1}{2}}\right) + O(1) \\
&= O(\log r(N-s)), \qquad (2.51)
\end{aligned}$$

where in the last equality we have used the fact

$$\sum_{t=1}^{N}\phi^{\mathrm{T}}(t-i)P(t)\phi(t-i) = O(\log r(N)), \forall i \geq 0. \qquad (2.52)$$

We now bound the first term on the RHS of (2.48). Applying Lemma 1 a), we can obtain that

$$\sum_{t=1}^{N}\phi^{\mathrm{T}}(t-s)E[\widetilde{\theta}(t)|\mathcal{F}_{t-s}]v(t) = o\left(\sum_{t=1}^{N}\left(\phi^{\mathrm{T}}(t-s)E[\widetilde{\theta}(t)|\mathcal{F}_{t-s}]\right)^2\right) + O(1).$$

Using the decomposition $\widetilde{\theta}(t) = E[\widetilde{\theta}(t)|\mathcal{F}_{t-s}] + \mathcal{H}_t(\theta(t))$, we can further write

$$\begin{aligned}
&\sum_{t=1}^{N}\phi^{\mathrm{T}}(t-s)E[\widetilde{\theta}(t) \mid \mathcal{F}_{t-s}]v(t) \\
&= o\left(\sum_{t=1}^{N}(\phi^{\mathrm{T}}(t-s)\widetilde{\theta}(t))^2\right) + o\left(\sum_{t=1}^{N}\left[\phi^{\mathrm{T}}(t-s)\mathcal{H}_t(\theta(t))\right]^2\right) + O(1). \quad (2.53)
\end{aligned}$$

Now, from (2.50), we have

$$\begin{aligned}
&\sum_{t=1}^{N}\left[\phi^{\mathrm{T}}(t-s)\mathcal{H}_t(\theta(t))\right]^2 \\
&= O\left(\sum_{i=1}^{s-1}\sum_{t=1}^{N}(\phi^{\mathrm{T}}(t-s)P(t-s)\phi(t-s-i))^2[\mathcal{H}_t(v(t-i))]^2\right) \\
&= O\left(\sum_{i=1}^{s-1}\sum_{t=1}^{n}\phi^{\mathrm{T}}(t-s)P(t-s)\phi(t-s)\phi^{\mathrm{T}}(t-s-i)P(t-s)\phi(t-s-i)\right) \\
&= O(\log r(N-s)). \qquad (2.54)
\end{aligned}$$

Combining (2.48), (2.51), (2.53) and (2.54) then establishes (2.47).

Finally, from (2.11) and the SPR assumption (2.38), we have

$$\sum_{t=1}^{N} -\phi^{\mathrm{T}}(t-s)\widetilde{\theta}(t)\left[\frac{1}{2}\phi^{\mathrm{T}}(t-s)\widetilde{\theta}(t)+\widehat{v}(t)-v(t)\right]$$
$$\geq \epsilon_1\sum_{t=1}^{N}[\phi^{\mathrm{T}}(t-s)\widetilde{\theta}(t)]^2+\epsilon_2\sum_{t=1}^{N}(\widehat{v}(t)-v(t))^2+O(1)\,, \tag{2.55}$$

for some constants $\epsilon_1 > 0$, $\epsilon_2 > 0$. Using (2.55) and (2.47) in (2.46), we can immediately conclude (2.39) and (2.41). (2.40) follows from (2.39).

To show (2.42), we have the following chain of inequalities,

$$\begin{aligned}
&\sum_{t=1}^{N}\frac{(y(t)-\phi^{\mathrm{T}}(t-s)\theta(t-k)-v(t))^2}{1+\phi^{\mathrm{T}}(t-s)P(t-s-k)\phi(t-s)}\\
&\leq 2\sum_{t=1}^{N}(\widehat{v}(t)-v(t))^2+2\sum_{t=1}^{N}\frac{[\phi^{\mathrm{T}}(t-s)(\theta(t)-\theta(t-k))]^2}{1+\phi^{\mathrm{T}}(t-s)P(t-s-k)\phi(t-s)}\\
&\leq O(\log r(N-s))\\
&\quad+2\sum_{t=1}^{N}\frac{\left[\sum_{i=0}^{k-1}\phi^{\mathrm{T}}(t-s)P(t-s-i-1)\phi(t-s-i)\widehat{v}(t-i))\right]^2}{1+\phi^{\mathrm{T}}(t-s)P(t-s-k)\phi(t-s)}\\
&\leq O(\log r(N-s))\\
&\quad+O\left(\sum_{i=1}^{k-1}\sum_{t=1}^{N}\phi^{\mathrm{T}}(t-s-i)P(t-s-i-1)\phi(t-s-i)\widehat{v}^2(t-i)\right)\\
&= O(\log r(N-s))\,.
\end{aligned}$$

The parameter consistency when (2.43) holds can be shown as in [14]. □

2.2 Adaptive prediction

We now consider the prediction of a stochastic process $y(t)$ which satisfies the ARMAX model (2.1). Such a model is appropriate, for example, when $y(t)$ is the output of a dynamical system with a random disturbance and an $\mathcal{F}_t$-measurable input $u(t)$, or when $y(t)$ contains seasonal fluctuations and $u(t)$ is the appropriate known periodic function.

For simplicity, we consider only d-step ahead prediction. Let $F'(q^{-1}) := 1+f_1'(q^{-1})+\cdots+f_{d-1}'q^{-(d-1)}$, and $G'(q^{-1}) := \sum_{i=0}^{p} g_i'q^{-i}$ be the solution of the following "long division",

$$A(q^{-1})F'(q^{-1})+q^{-d}G'(q^{-1})=1\,. \tag{2.56}$$

Also let $F(q^{-1}) = 1 + f_1(q^{-1}) + \cdots + f_{d-1}q^{-(d-1)}$, and $G(q^{-1}) = \sum_{i=0}^{p} g_i q^{-i}$ be the solution of

$$A(q^{-1})F(q^{-1}) + q^{-d}G(q^{-1}) = C(q^{-1}). \tag{2.57}$$

Then (2.1) can be reparameterized as

$$y(t+d) - F(q^{-1})w(t+d) = B(q^{-1})F'(q^{-1})u(t) + G'(q^{-1})y(t) + M(q^{-1})w(t), \tag{2.58}$$

where

$$M(q^{-1}) = q^d(C(q^{-1})F'(q^{-1}) - F(q^{-1})).$$

From (2.56) and (2.57),

$$A(q^{-1})M(q^{-1}) = G(q^{-1}) - G'(q^{-1})C(q^{-1}).$$

Hence, it is clear that

$$M(q^{-1}) = \text{ Truncation } (F'(q^{-1})C(q^{-1}), d), \tag{2.59}$$

where

$$\text{Truncation } \left(\sum_{i=0}^{n} \alpha_i q^{-i}, d\right) := \sum_{i=0}^{n-d} \alpha_{i+d} q^{-i}.$$

From (2.59), it follows that the RHS of (2.58) is $\mathcal{F}_t$-measurable. Hence the conditional mean, and thus also the minimum variance prediction,

$$\bar{y}(t+d) := E[y(t+d)|\mathcal{F}_t],$$

is given by

$$\bar{y}(t+d) = B(q^{-1})F'(q^{-1})u(t) + G'(q^{-1}) + M(q^{-1})w(t). \tag{2.60}$$

Note that though $w(t)$ is not an observed variable, we could estimate w by $\bar{w}$ using the following "observer",

$$C(q^{-1})\bar{w}(t) = B(q^{-1})u(t-d) - A(q^{-1})y(t). \tag{2.61}$$

An *indirect approach* to adaptive prediction is based on the following procedure: Estimate the parameters A, B, C and the innovation sequence $w(t)$; solve the Diophantine equation (2.56) with the estimated parameters to obtain estimates of F' and G'; obtain an estimate of M based on (2.59); replace all the unknowns in (2.60), including $w(t)$, by their estimates to generate adaptive prediction $\widehat{y}(t+d)$, i.e.,

$$\widehat{y}(t+d) = (\widehat{B}(q^{-1},t)\widehat{F}'(q^{-1},t))u(t) + \widehat{G}'(q^{-1},t)y(t) + \widehat{M}(q^{-1},t)\widehat{w}(t), \tag{2.62}$$

where $\widehat{F}'$ and $\widehat{G}'$ are obtained from

$$\widehat{A}(q^{-1},t)\widehat{F}'(q^{-1},t) + q^{-d}\widehat{G}'(q^{-1},t) = 1, \tag{2.63}$$

and $\widehat{M}$ is defined as in (2.59) using $\widehat{F}'$ and $\widehat{C}'$.[6]

It should be noted that since $w(t)$ is estimated along with parameters, the linear observer (2.61) based on the estimated parameters need not be used. The use of the disturbance estimate as well as parameter estimate results in a more general certainty equivalence approach than commonly advocated.

Alternatively, one could use a *direct* approach to adaptive prediction based on the reparameterized model (2.3), since then the conditional estimate satisfies

$$\bar{y}(t+d) = R(q^{-1})u(t-d) + S(q^{-1})y(t-d) + D(q^{-1})v(t)\,. \qquad (2.64)$$

One can therefore simply estimate R, S, D and $v(t)$ in the model (2.3) and use the estimates in (2.64), to give d-step ahead prediction,

$$\widehat{y}(t+d) = \phi^{\mathrm{T}}(t)\theta(t)\,. \qquad (2.65)$$

Theorems 2.3-2.6 below establish the self-optimality of both the direct and indirect approaches to adaptive predictions.

Theorem 2.3 *Consider the adaptive prediction based on the direct approach (2.65) using the SG algorithm (2.7-2.8) with $s = d$. Suppose that $1+q^{-d}D(q^{-1})$ is Hurwitz and the following SPR condition holds*

$$\min_{|z|=1} \mathrm{Re}\left[1 + z^d D(z)\right] > 0,$$

and

$$\sup_N \frac{1}{N}\sum_1^N \left(y^2(t) + u^2(t)\right) < \infty \ a.s.. \qquad (2.66)$$

Then, the adaptive predictor is self-optimizing in the sense

$$\lim_{N\to\infty} \frac{1}{N}\sum_{t=1}^N \left(\widehat{y}(t+d) - \bar{y}(t+d)\right)^2 = 0 \ a.s.. \qquad (2.67)$$

Proof. When $r(t)$ is finite, (2.67) holds trivially. When $r(t) \to \infty$, the optimality (2.67) follows easily from (2.65), (2.66), (2.16-2.17) and Kronecker Lemma. □

Theorem 2.4 *Consider adaptive prediction based on the direct approach (2.65) using the ELS algorithm (2.5-2.6) with $s = d$. Suppose that $1 + q^{-d}D(q^{-1})$ is Hurwitz and the following SPR condition holds*

$$\min_{|z|=1} \mathrm{Re}\left[1 + z^d D(z)\right]^{-1} > \frac{1}{2}\,,$$

and

$$y^2(t) + u^2(t) = O\left(\frac{t}{\log t}\right) \ a.s.. \qquad (2.68)$$

[6] The multiplication of two time varying polynomials of q^{-1} is to be understood as ordinary polynomial multiplication disregarding the time dependence of their coefficients.

Then the adaptive predictor is self-optimizing in the sense of (2.67). Moreover, if $y(t)$, $u(t)$ and $w(t)$ are uniformly bounded, then the "accumulated regret" satisfies,

$$\sum_{t=1}^{N} (\widehat{y}(t+d) - \bar{y}(t+d))^2 = o\left(\log^2 N\right) \quad a.s.. \tag{2.69}$$

Proof. As in the proof of Theorem 2.3, it is enough to consider the case $r(t) \to \infty$.
Following [23], let

$$\alpha_t := \frac{(\widehat{y}(t) - \bar{y}(t))^2}{1 + \phi^{\mathrm{T}}(t-d)P(t-2d)\phi(t-d)}$$

$$\delta_t := \text{ trace } \left(P(t-2d) - P(t-d)\right).$$

Then from (2.65), (2.42), (2.40) and (2.39), we have

$$\begin{aligned}
\sum_{t=r}^{N} (\widehat{y}(t) - \bar{y}(t))^2 &\leq \sum_{t=1}^{N} 2\alpha_t + \sum_{t=1}^{N} \alpha_t \phi^{\mathrm{T}}(t-d)\left(P(t-2d) - P(t-d)\right)\phi(t-d) \\
&= O(\log r_0(N)) + o(\log r_0(N) \cdot \max_{1 \leq t \leq N} \|\phi(t-d)\|^2 \\
&= O(\log r_0(N)) + o\left(\log r_0(N) \max_{1 \leq t \leq N} \|\phi(t) - \phi_0(t)\|\right)^2 \\
&\quad + o\left(\log r_0(N) \max_{1 \leq t \leq N} \|\phi_0(t)\|^2\right) \\
&= o(\log^2 r_0(N)) + o(\log r_0(N)) \max_{1 \leq t \leq N} \|\phi_0(t)\|^2.
\end{aligned}$$

From (2.68), we have

$$\log r_0(N) = O(\log(N)).$$

The optimality (2.67) and the rate of the regret (2.69) then follow. □

Remark: It is worthwhile noting that adaptive prediction based on the least squares algorithm is capable of self-optimizing even for processes with a growing trend as in (2.68). □

Theorem 2.5 *Consider adaptive prediction based on the indirect approach (2.62) using the SG algorithm (2.7-2.8) with $s = 1$. Suppose that $C(q^{-1})$ is Hurwitz and the following SPR condition holds:*

$$\min_{|z|=1} \mathrm{Re}[C(z)] > 0.$$

Then the adaptive predictor is self-optimizing in the sense of (2.67).

Proof. From (2.1), we have

$$\begin{aligned}&\widehat{A}(q^{-1},t)y(t+d-j)+(A(q^{-1})-\widehat{A}(q^{-1},t))y(t+d-j)\\&=\widehat{B}(q^{-1},t)u(t-j)+(B(q^{-1})-\widehat{B}(q^{-1},t))u(t-j)\\&\quad+\widehat{C}(q^{-1},t)\widehat{w}(t+d-j)+C(q^{-1})(w(t+d-j)-\widehat{w}(t+d-j))\\&\quad+(C(q^{-1})-\widehat{C}(q^{-1},t)\widehat{w}(t+d-j),\end{aligned}$$

and hence

$$\begin{aligned}&\widehat{A}(q^{-1},t)y(t+d-j)=\widehat{B}(q^{-1},t)u(t-j)+\widehat{C}(q^{-1},t)\widehat{w}(t+d-j)\\&+C(q^{-1})(w(t+d-j)-\widehat{w}(t+d-j))-\phi^{\mathrm{T}}(t+d-j-1)\widetilde{\theta}(t).\end{aligned}\tag{2.70}$$

So,

$$\begin{aligned}&(\widehat{A}(q^{-1},t)\widehat{F}'(q^{-1},t))y(t+d)=(\widehat{B}(q^{-1},t)\widehat{F}'(q^{-1},t))u(t)\\&+(\widehat{F}'(q^{-1},t)\widehat{C}(q^{-1},t))\widehat{w}(t+d)+\sum_{i=0}^{d-1}\widehat{f}_i'(t)\cdot\phi^{\mathrm{T}}(t+d-i-1)\widetilde{\theta}(t)\\&-(\widehat{F}'(q^{-1},t)C(q^{-1}))(w(t+d)-\widehat{w}(t+d)).\end{aligned}$$

From (2.62), (2.63) and the above, we have

$$\begin{aligned}&y(t+d)-\widehat{F}(q^{-1},t)w(t+d)-\widehat{y}(t+d)\\&=\left(\widehat{F}'(q^{-1},t)C(q^{-1})\right)(w(t+d)-\widehat{w}(t+d))+\ \widehat{F}(q^{-1},t)(\widehat{w}(t+d)-(w(t+d))\\&-\sum_{i=0}^{d-1}\widehat{f}_i'(t)\phi^{\mathrm{T}}(t+d-i-1)\widetilde{\theta}(t),\end{aligned}\tag{2.71}$$

where $\widehat{F}(q^{-1},t)$ satisfies

$$\widehat{A}(q^{-1},t)\widehat{F}(q^{-1},t)+q^{-d}\widehat{G}(q^{-1},t)=\widehat{C}(q^{-1},t).$$

Since $\|\widetilde{\theta}(t)\|$ is bounded and the mappings from $\theta(t)$ to $\widehat{F}(q^{-1},t)$ and $\widehat{F}'(q^{-1},t)$ are continuous, the coefficients of $\widehat{F}(q^{-1},t)$ and $\widehat{F}'(q^{-1},t)$ are also bounded. Hence it follows from (2.17) that

$$\lim_{N\to\infty}\frac{1}{N}\sum_{t=1}^{N}\left[y(t+d)-\widehat{F}(q^{-1},t)w(t+d)-\widehat{y}(t+d)\right]^2=0\ \text{a.s.}.\tag{2.72}$$

So

$$\lim_{N\to\infty}\frac{1}{N}\sum_{t=1}^{N}\left[\bar{y}(t+d)-\widehat{y}(t+d)+(F(q^{-1})-\widehat{F}(q^{-1},t))w(t+d)\right]^2=0\ \text{a.s.}.\tag{2.73}$$

Noting that both $\bar{y}(t+d)$ and $\widehat{y}(t+d)$ are $\mathcal{F}_t$-measurable, and that $\deg \widehat{F}(q^{-1},t) = \deg F(q^{-1}) = d-1$, we have

$$\sum_{t=1}^{N}\left[\bar{y}(t+d)-\widehat{y}(t+d)+(F(q^{-1})-\widehat{F}(q^{-1},t))w(t+d)\right]^2$$
$$=\sum_{t=1}^{N}[\bar{y}(t+d)-\widehat{y}(t+d)]^2+\sum_{t=1}^{N}\left[(F(q^{-1})-\widehat{F}(q^{-1},t))w(t+d)\right]^2$$
$$+2\sum_{t=1}^{N}(\bar{y}(t+d)-\widehat{y}(t+d))\cdot(F(q^{-1})-\widehat{F}(q^{-1},t))w(t+d)$$
$$=(1+o(1))\sum_{t=1}^{N}(\bar{y}(t+d)-\widehat{y}(t+d))^2+\sum_{t=1}^{N}((F(q^{-1})-\widehat{F}(q^{-1},t))w(t+d))^2.$$

The optimality (2.72) then follows from the above equation and (2.73). □

Theorem 2.6 *Consider adaptive prediction based on the indirect approach (2.62) using the the ELS algorithm (2.5-2.6) with $s=1$. Let $C(q^{-1})$ be Hurwitz and the following SPR condition hold:*

$$\min_{|z|=1}[\mathrm{Re}C^{-1}(z)] > \frac{1}{2},$$

and

$$y^2(t)+u^2(t)=O\left(\frac{t}{\log^d t}\right),$$

then the adaptive predictor is self-optimizing in the sense of (2.67). Moreover, if $y(t)$, $u(t)$ and $w(t)$ are uniformly bounded, then

$$\sum_{t=1}^{N}(\widehat{y}(t+d)-\bar{y}(t-d))=o(\log^{d+1}N).$$

Proof. It follows from (2.41) and (2.63) that

$$|\widehat{f}_i'(t)|^2=O(\log^{d-1}r(t)).$$

The rest of the proof consists of applying the technique of the proof of Theorem 2.4 to (2.71). The details are omitted. □

2.3 Adaptive feedback control

We consider adaptive feedback control of plants represented by the ARMAX model (2.1).

As for adaptive prediction, certainty equivalence adaptive control can also make use of the extra degree of freedom afforded by the availability of the disturbance estimate. Thus, we consider the general control law of the form

$$R(q^{-1})u(t) = S(q^{-1})y(t) + T(q^{-1})z(t) + M(q^{-1})w(t), \tag{2.74}$$

where R, S, T and M are polynomials, and $z(t)$ is the command signal.

In this section, we restrict our attention to a class of certainty equivalence adaptive control which may be regarded as a model reference adaptive control with disturbance attenuation capability. The minimum variance (MV) control may be regarded as a special case with a unity reference model and optimal disturbance attenuation. In order for the closed loop transfer function from the command signal $z(t)$ to the output $y(t)$ to match an arbitrary reference model, we assume that the plant is minimum phase, i.e., $B(q^{-1})$ is Hurwitz, and we consider the general control law (2.74) with the restriction that $R(q^{-1})$ presumes $B(q^{-1})$ as a factor, i.e.,

$$R(q^{-1}) = R'(q^{-1})B(q^{-1}),$$

for some polynomial $R'(q^{-1})$.

With this control law, it is easy to calculate that

$$y(t) = \frac{q^{-d}T(q^{-1})z(t)}{A(q^{-1})R'(q^{-1}) - q^{-d}S(q^{-1})} + \frac{R'(q^{-1})C(q^{-1}) + q^{-d}M(q^{-1})}{A(q^{-1})R'(q^{-1}) - q^{-d}S(q^{-1})}w(t),$$

where the first term on the RHS above is the command response, and the second term is the disturbance response.

Suppose that the desired trajectory is given by a reference model

$$y^*(t) = \frac{q^{-d}B_m(q^{-1})}{A_m(q^{-1})}z(t),$$

then for the command response of $y(t)$ to match that of the reference model, we require that

$$\begin{aligned} A(q^{-1})R'(q^{-1}) - q^{-d}S(q^{-1}) &= H(q^{-1}) \\ T(q^{-1}) &= A_0(q^{-1})B_m(q^{-1}), \end{aligned} \tag{2.75}$$

where $A_0(q^{-1})$ is some "observer" polynomial, and $H(q^{-1}) = A_m(q^{-1})A_0(q^{-1})$.

We now consider the optimization of the disturbance response by making use of the remaining freedom in the control law. Let $(R'(q^{-1}), S(q^{-1}))$ be a particular solution of (2.75), then $(R'(q^{-1})+q^{-d}\Lambda(q^{-1}), S(q^{-1})+A(q^{-1})\Lambda(q^{-1}))$ is the class of all possible solutions, where $\Lambda(q^{-1})$ is a free polynomial. Thus by choosing $M(q^{-1})$ one exploits the freedom given by $\Lambda(q^{-1})$in shaping the disturbance response. Hence we need to only consider the appropriate choice of $M(q^{-1})$ to "optimize" the disturbance response.

A simple choice is to set

$$M(q^{-1}) = -\text{ Truncation } (R'(q^{-1})C(q^{-1}), d), \tag{2.76}$$

which minimizes the variance of $H(q^{-1})(y(t) - y^*(t))$, and yields

$$H(q^{-1})(y(t) - y^*(t)) = F_1(q^{-1})w(t),$$

where

$$F_1(q^{-1}) = R'(q^{-1})C(q^{-1}) + M(q^{-1})q^{-d}.$$

To minimize the variance of $(y(t) - y^*(t))$, note that since $\{w(t)\}$ is an uncorrelated sequence, the variance of

$$\frac{R'(q^{-1})C(q^{-1}) + q^{-d}M(q^{-1})}{H(q^{-1})} w(t)$$

is minimized with respect to $M(q^{-1})$ if for some polynomial $F_2(q^{-1})$, $\deg F_2(q^{-1}) \leq d-1$, the following holds,

$$\frac{R'(q^{-1})C(q^{-1}) + q^{-d}M(q^{-1})}{H(q^{-1})} = F_2(q^{-1}).$$

Hence the optimum $M(q^{-1})$ is the minimal degree solution with respect to F_2 of the following Diophantine equation,

$$H(q^{-1})F_2(q^{-1}) - q^{-d}M(q^{-1}) = R'(q^{-1})C(q^{-1}). \tag{2.77}$$

It is easy to see that in fact $F_2(q^{-1}) = F(q^{-1})$, where $F(q^{-1})$ is as in (2.57). Hence, with this choice of $M(q^{-1})$, we have

$$y(t) - y^*(t) = F(q^{-1})w(t).$$

The difference between the model reference control law described above and a MV control law with $y^*(t)$ as the command signal should be noted. The former may have the advantage of resulting in a smaller feedback controller gain, and hence may be more robust in the presence of unmodeled dynamics.

The corresponding *indirect* adaptive control law is to estimate the parameters A, B and C, and the innovation process $w(t)$ of the model (2.1), solve (2.75), (2.76) or (2.77), and use the control law (2.74), with the estimated quantities.

Theorems 2.7 and 2.8 below establish the self-optimality of such adaptive control laws based on either the SG or ELS algorithms.

Theorem 2.7 *Consider the indirect model reference adaptive control based on the SG algorithm (2.7-2.8) with* $s = 1$. *Assume*

A2.1) $H(q^{-1})$ *is stable, and* $z(t)$ *is bounded.*
A2.2) $B(q^{-1})$ *is Hurwitz,* $b_0 \neq 0$.
A2.3) $\min_{|z|=1} \mathrm{Re}[C(z)] > 0$.
A2.4) The distribution of $(w(0), \ldots, w(t))$ *is absolutely continuous with respect to Lebesque measure for* $\forall t \geq 0$.

Then, with the design (2.76), we have

$$\lim_{N\to\infty}\frac{1}{N}\sum_{t=0}^{N}\left(H(q^{-1})(y(t)-y^*(t))-F_1(q^{-1})w(t)\right)^2=0 \ a.s., \tag{2.78}$$

and with the design (2.77), we have

$$\lim_{N\to\infty}\frac{1}{N}\sum_{t=1}^{N}\left(y(t)-y^*(t)-F(q^{-1})w(t)\right)^2=0 \ a.s.. \tag{2.79}$$

Proof. Substituting the adaptive control law into (2.71), and proceeding as the proof of Theorem 2.5, one can show that

$$\frac{1}{r_0(N-1)}\sum_{t=1}^{N}\Big(H(q^{-1})(y(t)-y^*(t))$$
$$+(\widehat{R}(q^{-1},t)\widehat{C}(q^{-1},t)+q^{-d}\widehat{M}(q^{-1},t))w(t)\Big)^2\to 0 \text{ a.s.},$$

Because of the minimum phase assumption (A2.2),

$$r_0(N-1)=O(N)+O\left(\sum_{t=0}^{N}y^2(t)\right).$$

Hence,

$$\sum_{t=0}^{N}\left(H(q^{-1})\,(y(t)-y^*(t))+\left(\widehat{R}(q^{-1},t)\widehat{C}(q^{-1},t)+q^{-d}\widehat{M}(q^{-1},t)\right)w(t)\right)^2$$
$$=o(N)+o\left(\sum_{t=0}^{N}y^2(t)\right) \text{ a.s..}$$

Because $y^*(t)$ is bounded, $H(q^{-1})$ is stable, and $\widehat{R}(q^{-1},t)$, $\widehat{C}(q^{-1},t)$ and $\widehat{M}$ (q^{-1},t) are bounded, we have

$$\sum_{t=0}^{N}y^2(t)=O(N) \text{ a.s..}$$

Hence,

$$r_0(N)=O(N) \text{ a.s..}$$

So

$$\frac{1}{N}\sum_{t=0}^{N}\Big[H(q^{-1})(y(t)-y^*(t))$$
$$+(\widehat{R}(q^{-1},t)\widehat{C}(q^{-1},t)+q^{-d}\widehat{M}(q^{-1},t))w(t)\Big]^2\to 0 \text{ a.s..}$$

With $\widehat{M}$ designed through (2.76), we have

$$\frac{1}{N}\sum_{t=0}^{N}\left[H(q^{-1})(y(t)-y^*(t))+\widehat{F}_1(q^{-1},t)w(t)\right]^2 \to 0 \text{ a.s.}.$$

It is clear that $H(q^{-1})y(t) - F_1(q^{-1})w(t)$ is $\mathcal{F}_{t-d}$ measurable. The rest of the proof for (2.78) then follows the proof of Theorem 2.5. The proof for (2.79) is similar and is omitted. □

We now consider the ELS algorithm based adaptive control. The following theorem is essentially due to Guo and Chen [23].

Theorem 2.8 *Consider the indirect model reference adaptive control for unit delay systems based on the ELS algorithm. Let the assumptions (A2.1), (A2.2) of Theorem 2.7, and the following conditions hold,*

A2.5) $\min_{|z|=1} \mathrm{Re}[C^{-1}(z)] > \frac{1}{2}$.

A2.6) $\lim_{t\to\infty} \inf |\widehat{b}_0(t)| > 0$.

Then with $\widehat{M}$ designed through (2.76), we have

$$\sum_{t=0}^{N}\left[H(q^{-1})(y(t)-y^*(t))-w(t)\right]^2 = O(N^{\delta}), \quad a.s.\ , \forall\delta\in\left(\frac{2}{\alpha},1\right).$$

With $\widehat{M}$ designed through (2.77), we have

$$\sum_{t=0}^{N}(y(t)-y^*(t)-w(t))^2 = O(N^{\delta}), \qquad a.s., \forall\delta\in\left(\frac{2}{\alpha},1\right).$$

Proof. First note that with the delay equal to one, $F(q^{-1})$, $F_1(q^{-1})$, $F'(q^{-1})$ and their estimates are all equal to 1. Without loss of generality, we can also take $\widehat{R}'$ to be 1. Substituting the control law into (2.70), we have

$$\begin{aligned} H(q^{-1})(y(t+1)-y^*(t+1)) &= M(q^{-1},t)\widehat{w}(t)+\widehat{R}'(q^{-1},t)\widehat{C}(q^{-1},t)\widehat{w}(t+1) \\ &\quad +C(q^{-1})(w(t+1)-\widehat{w}(t+1))+\phi^{\mathrm{T}}(t)\widetilde{\theta}(t)\,, \end{aligned}$$

with $\widehat{M}$ designed via (2.76),

$$H(q^{-1})(y(t+1)-y^*(t+1)) = \widehat{w}(t+1)+C(q^{-1})(w(t+1)-\widehat{w}(t+1))+\phi^{\mathrm{T}}(t)\widetilde{\theta}(t)\,.$$

The rest of the proof then follows Guo and Chen [23]. □

It is also possible to reparameterize the model in the form of the controller, so that the controller parameters can be estimated directly. This is the *direct* approach.

Let us consider the direct approach to adaptive MV control for systems with general delay. With the plant parameterized in the form of (2.3), the MV controller for tracking $y^*(t)$ is simply given by

$$y^*(t+d) = R(q^{-1})u(t) + S(q^{-1})y(t) + D(q^{-1})v(t)\,.$$

The corresponding adaptive control law is to replace R, S, D and $v(t)$ by their estimates. If the algorithms (2.5-2.6) and (2.7-2.8) in Section 2.1 are employed for estimation, the adaptive MV tracker simply is given by

$$y^*(t+d) = \phi^{\mathrm{T}}(t)\theta(t)\,. \tag{2.80}$$

The self-optimality of the direct MV adaptive control (2.80) can be established as for adaptive prediction. For details, we refer the reader to [18].

2.4 The convergence of adaptive controllers

In the previous section, we have established the self-optimality, i.e., the optimal performance with respect to a certain cost criterion on the output y of the system, of a class of certainty equivalence adaptive control. In this section we consider the issue of parameter convergence and show that self-optimality generally implies self-tuning in some sense.

Let us consider the indirect adaptive MV control using the stochastic gradient algorithm. In the previous subsection, we have shown that

$$\lim_{N\to\infty} \frac{1}{N}\sum_{t=0}^{N}(y(t) - y^o(t))^2 = 0 \text{ a.s.},$$

where $y^o(t) := y^*(t) + F(q^{-1})w(t)$. Let $u^o(t)$ be the input corresponding to $y^o(t)$, i.e., it satisfies

$$A(q^{-1})y^o(t) = q^{-d}B(q^{-1})u^o(t) + C(q^{-1})w(t)\,. \tag{2.81}$$

Since $B(q^{-1})$ is Hurwitz, it follows that

$$\lim_{N\to\infty} \frac{1}{N}\sum_{t=0}^{N}(u(t) - u^o(t))^2 = 0 \text{ a.s.}.$$

Define $\phi^o(t)$ and $R^o(t)$ as $\phi_0(t)$ and $R_0(t)$, using $y^o(t)$ and $u^o(t)$ in place of $y(t)$ and $u(t)$, respectively. Then

$$\frac{1}{N}\sum \|\phi^o(t) - \phi(t)\|^2 \to 0 \text{ a.s.}$$

and

$$\frac{1}{N}\|R^o(t) - R(t)\| \to 0 \text{ a.s.}.$$

It then follows from (2.36) that

$$\frac{1}{N}\|R^o(N-1)\widetilde{\theta}(N)\| \to 0 \text{ a.s.}.$$

For simplicity, assume that the ensemble correlations of the command signal $y^*(t)$ exists. Then $\frac{R^o(N-1)}{N}$ converges. Let

$$\Phi := \lim_{N\to\infty} \frac{R^o(N-1)}{N} .$$

It can then be shown that $\widetilde{\theta}(t)$ converges almost surely to the null space of Φ. The following theorem summarizes the above development and further characterizes the null space of Φ.

Theorem 2.11 *Consider the indirect adaptive MV control using the SG algorithm. Let the assumptions of Theorem 2.7 hold, and suppose that the ensemble correlations of $\{y^*(t)\}$ exist. Then*

$$\widetilde{\theta}(t) \to \mathcal{N}(\Phi) \ a.s..$$

Further $\forall\widetilde{\theta} \in \mathcal{N}(\Phi)$, using the obvious notation, let $\widehat{A}$, $\widehat{B}$, $\widehat{C}$, $\widehat{F}$ and $\widehat{G}$ be the appropriate polynomials corresponding to the parameter vector $\widetilde{\theta} + \theta_o$. Then

$$F(q^{-1}) = \widehat{F}(q^{-1}), \tag{2.82}$$

$$B(q^{-1})\widehat{G}(q^{-1}) = \widehat{B}(q^{-1})G(q^{-1}), \tag{2.83}$$

and

$$\frac{1}{N}\sum_{t=0}^{N}\left((\hat{B}C - B\hat{C})y^*(t)\right)^2 \to 0 . \tag{2.84}$$

Moreover, if $B(q^{-1})$ and $G(q^{-1})$ do not have a common factor, and $\{y^*(t)\}$ is persistently exciting of order $\ell_p \geq \min(\deg C(q^{-1}), \deg A(q^{-1})) + 1$, then

$$\theta(t) \to \theta_0 \ a.s..$$

Proof. Since $\widetilde{\theta} \in \mathcal{N}(\Phi)$, we have

$$\frac{1}{N}\sum_{t=1}^{N}((A-\widehat{A})y^o(t) + q^{-d}(\widehat{B}-B)u^o(t) + (\widehat{C}-C)w(t))^2 \to 0 \text{ a.s.}.$$

So,

$$\frac{1}{N}\sum_{t=1}^{N}((A-\widehat{A})y^*(t)+q^{-d}(\widehat{B}-B)u^o(t)+((A-\widehat{A})F+\widehat{C}-C)w(t))^2 \to 0 . \tag{2.85}$$

Since $u^o(t)$ is $\mathcal{F}_t$-measurable, it follows as in [3,13] that the first d coefficients of $(A-\widehat{A})F + \widehat{C} - C$ have to be zero, which yields (2.82).

Multiplying (2.85) by $B(q^{-1})$ and substituting (2.81) into (2.85), we obtain

$$\frac{1}{N}\sum_{t=0}^{N}\left[(\widehat{B}A - B\widehat{A})y^*(t) + (B(\widehat{C} - \widehat{A}F) - \widehat{B}(C - AF))w(t)\right]^2 \to 0 \text{ a.s.}.$$

Hence,

$$\frac{1}{N}\sum_{t=0}^{N}\left((\widehat{B}A - B\widehat{A})y^*(t)\right)^2 \to 0 \tag{2.86}$$

and

$$\frac{1}{N}\sum_{t=0}^{N}\left((B(\widehat{C} - \widehat{A}F) - \widehat{B}(C - AF))w(t)\right) \to 0 \text{ a.s.}.$$

So

$$B(\widehat{C} - \widehat{A}F) = \widehat{B}(C - AF), \tag{2.87}$$

which is equivalent to (2.83). If B and G do not have a common factor, then

$$\widehat{B} = \lambda B, \quad \widehat{G} = \lambda B$$

for some scalar λ.

If $\ell_p \geq \deg A + 1$, from the above and (2.86), we obtain $\lambda A - \widehat{A} = 0$. It then follow from (2.87) that $\lambda = 1$, $\widehat{A} = A$, and $\widehat{C} = C$.

From (2.87), we obtain

$$(B\widehat{A} - \widehat{B}A)F = \widehat{B}C - B\widehat{C}.$$

Hence if $\ell_p \geq \deg C + 1$, we can proceed similarily. This completes the proof. □

The parameter convergence results above can be easily extended to more general adaptive control laws for which signal convergence in the mean square sense can be established, e.g., model reference adaptive control. The extension to the ELS algorithm is also possible. For these extensions, we refer the readers to Ren and Kumar [18].

In the case of direct adaptive MV *regulation* of systems with general delay, based on the SG algorithm using a-priori estimates, the parameters of $C(q^{-1})$ need not be estimated, and it can then be shown that if there is no overparameterization, the limit set $\Phi + \theta_o$ of $\theta(t)$, is a line passing through the origin and the true parameter. The fact that $||\widetilde{\theta}(t)||$ converges can then be used to establish the convergence of $\theta(t)$ to a particular point on the line. This result then generalizes to the case of general delay the result of Becker, Kumar and Wei [3], which establishes the convergence of the parameter estimates to a random multiple of the true parameter for the unit delay case.

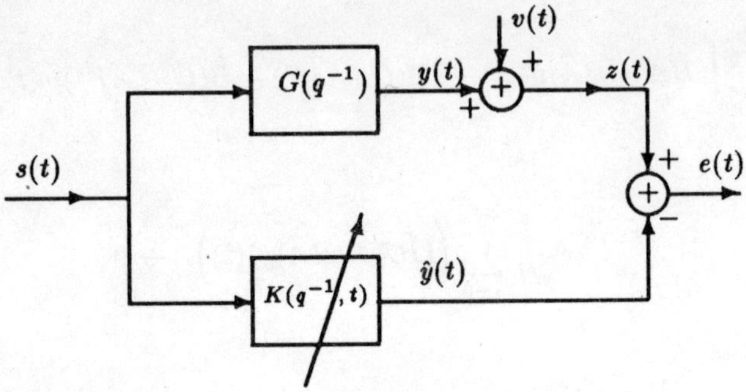

Fig. 1. Problem A: Output error identification and adaptive IIR filtering.

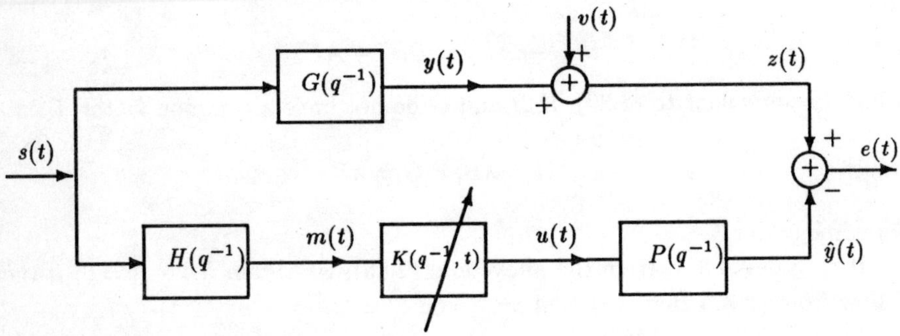

Fig. 2. Problem B: Adaptive feedforward control.

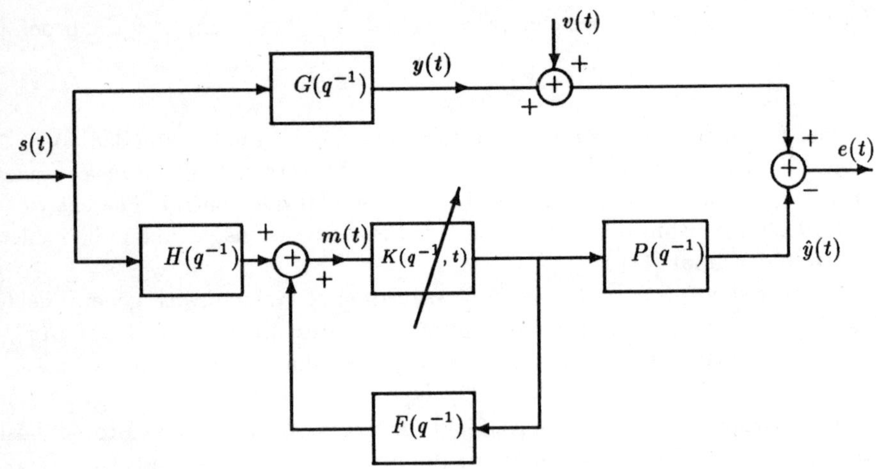

Fig. 3. Problem C: Adaptive feedforward control with input contamination.

3 Stochastic Parallel Model Adaptation Problems

In this section, we consider three successively more complicated parallel model adaptation problems which are all inspired by practical problems and are illustrated in Figs. 1, 2 and 3. All the fixed systems in Figs. 1–3 are unknown but stable, linear and time invariant. In all three figures, $G(q^{-1})$ is the system whose output $y(t)$ is to be matched by $\hat{y}(t)$, the output of the (partially) tunable system in the lower channels of the figures. The signal $s(t)$ is the common input to both the upper and lower channels, while $v(t)$ is the disturbance.

Problem A, which is the simplest of the three and is depicted in Fig. 1, encompasses both output error identification as well as adaptive filtering. In the *identification* context, $G(q^{-1})$ is the unknown dynamical system to be identified, and $z(t)$, $s(t)$ and $v(t)$ are the output, input, and disturbance plus measurement noise of the system, respectively. In the *filtering* context, $v(t)$ is the signal to be extracted, $y(t)$ is the noise corrupting the measured signal $z(t)$, and $s(t)$ can be regarded as the source of the noise. In both cases, $v(t)$ can be quite arbitrary, except for the reasonable and practical assumption that $\{v(t)\}$ and $\{s(t)\}$ are stochastically uncorrelated. The challenge here is to exploit this correlational property to provide unbiased estimation of the dynamical system $G(q^{-1})$, in the context of identification, or undistorted extraction of the signal $v(t)$, in the context of filtering.

Problem B, depicted in Fig. 2, is clearly a generalization of Problem A, and arises in practice as an adaptive feedforward control scheme for disturbance cancelling, which has the following physical interpretation. $P(q^{-1})$ is an unknown "plant", for which $u(t)$ is the input, and $e(t)$ is the output, which is subject to additive disturbances $v(t)$ and $y(t)$ which are mutually uncorrelated. A signal $m(t)$ related to $s(t)$, the source of the disturbance $y(t)$, is measured. The goal of adapting the feedforward controller $K(q^{-1},t)$ is to eliminate the part due to $y(t)$ from the output $e(t)$, leaving only the uncancelable disturbance $v(t)$.

Problem C, depicted in Fig. 3, represents a further complication that is inspired by practical applications of adaptive feedforward control in the areas of active acoustic noise and vibration control [24]. The feedback $F(q^{-1})$ in Fig. 3 models the contamination of the measurable signal $m(t)$ by the control input $u(t)$. In active acoustic noise control this effect is called acoustic feedback. More generally, it has been termed as *input contamination* in [24].

Section 3.1 considers Problem A and develops tools that allow us to exploit the uncorrelatedness between signals. The extension to Problems B and C is considered in Section 3.2.

3.1 The convergence of output error recursions

In this subsection, we consider Problem A, the output error identification and adaptive IIR filtering problem. Referring to Fig. 1, let

$$G(q^{-1}) = \frac{B(q^{-1})}{A(q^{-1})},$$

where

$$A(q^{-1}) = 1 + a_1 q^{-1} + \cdots + a_n q^{-n}$$
$$B(q^{-1}) = b_0 + b_1 q^{-1} + \cdots + b_m q^{-m}.$$

Hence,

$$A(q^{-1})y(t) = B(q^{-1})s(t), \tag{3.1}$$

and

$$z(t) = y(t) + v(t).$$

As mentioned, the goal is to estimate the parameters of A and B, and the signal $v(t)$. We consider the following algorithm which is a slight variation of Landau's output error identification algorithm [25] in that a projection of the parameter estimate onto a compact convex set $\mathcal{M}$ known to contain $\theta_0 = [a_1, \ldots, a_n, b_0, \ldots, b_m]$ is employed. We estimate θ_0 by $\theta(t) := [\widehat{a}_1(t), \ldots, \widehat{a}_n(t), \widehat{b}_0(t), \ldots, \widehat{b}_m(t)]^{\mathrm{T}}$, as follows,

$$\theta'(t) = \theta(t-1) + \frac{\phi(t-1)}{r(t-1)} e(t), \quad \theta(0) \in \mathcal{M} \tag{3.2}$$
$$\theta(t) = F[\theta'(t)],$$

where $F[\cdot]$ denotes the projection onto $\mathcal{M}$,

$$\begin{aligned} r(t-1) &= r(t-2) + \|\phi(t-1)\|^2, \quad r(-1) = 1 \\ e(t) &:= z(t) - \phi^{\mathrm{T}}(t-1)\theta(t-1) \\ \phi(t-1) &:= [-y'(t-1), \ldots, -y'(t-n), s(t), \ldots, s(t-m)], \end{aligned} \tag{3.3}$$

and

$$y'(t) = \phi^{\mathrm{T}}(t-1)\theta'(t).$$

Because the estimate of $v(t)$ is given by $e(t)$, we shall say that the algorithm (3.2-3.3) is self-optimizing if,

$$\lim_{N\to\infty} \frac{1}{N} \sum_{t=1}^{N} (e(t) - v(t))^2 = 0 \text{ a.s.}. \tag{3.4}$$

Theorem 3.1 below establishes this optimality under the following set of assumptions.

Assumptions

A3.1) The disturbance is of the form $v(t) = \sum_{i=0}^{+\infty} c_i(t) w(t-i)$, where $\{w(t)\}$ is a martingale difference sequence with respect to the increasing sequence of σ-fields $\mathcal{F}_t$ generated by $(w(0), \ldots, w(t), \{u(k)\}_0^\infty)$, and $\{c_i(t)\}$ is deterministic and satisfies $|c_i(t)| \leq K_c \alpha^{-i}$, for some $\alpha > 1$, $K_c < \infty$, and $\forall t \geq 0$.

A3.2) The sequences $\{|y(t)|\}$, $\{|s(t)|\}$, $\{|v(t)|\}$, and $\{|w(t)|\}$, are uniformly bounded by a finite number Γ, almost surely.

A3.3) $A(q^{-1})$ is Hurwitz and satisfies the following strictly positive real (SPR) condition,

$$\mathrm{Re}[A(e^{j\omega})] > 0, \forall\omega .$$

A3.4) $\theta_0 \in \mathcal{M}$.

A3.5) $\lim_{N\to\infty} \inf \frac{1}{N}\sum_{t=1}^{N} s^2(t) > 0$.

Note that the set $\mathcal{M}$ can be constructed from minimal a-priori information about the transfer function $\frac{B(q^{-1})}{A(q^{-1})}$, for example, from the knowledge of a finite upper bound for its H^∞-norm.

Theorem 3.1 *Let the assumptions (A3.1) to (A3.5) hold. Then for the algorithm described by (3.2 - 3.3), the self-optimality result (3.4), as well as the following results hold:*

$$\sup_N \frac{r(N)}{N} < \infty \quad a.s.,$$

$$\sum_{t=1}^{\infty} \|\theta(t) - \theta(t-1)\|^2 < \infty \quad a.s..$$

Proof. Let us denote

$$v'(t) := z(t) - y'(t)$$

$$\Delta(t) := \theta(t) - \theta(t-1), \quad \Delta'(t) := \theta'(t) - \theta(t-1),$$

$$\widetilde{\theta}(t) = \theta(t) - \theta_0, \text{ and } \widetilde{\theta}'(t) := \theta'(t) - \theta_0 .$$

Because of the projection facility, $\theta(t), \theta'(t), \Delta(t)$ and $\Delta'(t)$ are all uniformly bounded. As a consequence, $\|\phi(t)\|$ has a bounded growth rate, and

$$\frac{r(t)}{r(t-k)}$$

is uniformly bounded in t, for finite k.

Let us consider the "stochastic Lyapunov function" $V(N) := \|\widetilde{\theta}(N)\|^2$. It can be shown to satisfy the recursion

$$V(N) + \sum_{t=1}^{N}\|\Delta'(t)\|^2 + \sum_{t=1}^{N}\frac{2[A(q^{-1})(v'(t) - v(t))][v'(t) - v(t)]}{r(t-2)} \le V(0) + \sum_{t=1}^{N}\frac{2\phi^{\mathrm{T}}(t-1)\widetilde{\theta}'(t)v(t)}{r(t-2)} . \tag{3.5}$$

From the positive real assumption (A3.3), the third term on the left hand side above satisfies,

$$\sum_{t=1}^{N}\frac{[A(q^{-1})(v'(t) - v(t))][v'(t) - v(t)]}{r(t-2)} \ge \sum_{t=1}^{N}\frac{\delta_1[(v'(t) - v(t))]^2 + \delta_2[\phi^{\mathrm{T}}(t-1)\widetilde{\theta}'(t)]^2}{r(t-2)} - S_0 ,$$

for some finite positive constants δ_1 and δ_2, and a random variable $S_0 < +\infty$ a.s..

Turning next to the last term on the right hand side of (3.5), our goal is to show that for $\forall \epsilon > 0$,

$$\left|\sum_{t=1}^{N} \frac{\phi^{\mathrm{T}}(t-1)\widetilde{\theta}'(t)v(t)}{r(t-2)}\right| \leq O(1) + o\left(\sum_{t=1}^{N}\left(\frac{\phi^{\mathrm{T}}(t-1)\widetilde{\theta}'(t)}{r(t-2)}\right)^2\right) + \epsilon \sum_{t=1}^{N} \|\Delta'(t)\|^2 . \tag{3.6}$$

To exhibit this let us consider a decomposition of the signal $v(t)$ as

$$v(t) = v_1(t) + v_2(t)$$

where

$$v_1(t) = \begin{cases} \sum_{i=0}^{d(t)-1} c_i(t)w(t-i), & d(t) \geq 1, \\ 0\,, & d(t) = 0, \end{cases}$$

$$v_2(t) = \sum_{i=d(t)}^{+\infty} c_i(t)w(t-i)\,,$$

and $d(t) := \max\{d \in Z^+ | \epsilon_d \log_\alpha r(t-d) \geq d\}$, where ϵ_d is a sufficiently small positive number.

Because of the monotonicity of $\log_\alpha r(t)$, $d(t)$ is clearly uniquely defined and depends only on $r(0), \ldots, r(t-d(t))$, hence is $\mathcal{F}_{t-d(t)}$-measurable. Moreover,

$$\epsilon_d \log_\alpha r(t-d(t)-1) - 1 \leq d(t) \leq \epsilon_d \log_\alpha r(t-d(t)) \leq \epsilon_d \log r(t)\,.$$

The basic idea for proving (3.6) is to bound

$$\left|\sum_{t=1}^{N} \frac{\phi^{\mathrm{T}}(t-1)\widetilde{\theta}'(t)v_2(t)}{r(t-2)}\right|$$

by exploiting the vanishing magnitude of $|v_2(t)|$, and

$$\left|\sum_{t=1}^{N} \frac{\phi(t-1)\widetilde{\theta}'(t)v_1(t)}{r(t-2)}\right|$$

by using a backward recursion developed in [26], and Lemma 1 a). By backward recursion we approximate the regression vector $\phi(t-1)$ by $f(t-1)$, which is the fictitious state of the parallel adaptation model at time $t-1$ with the parameter fixed at $\theta(t-d(t))$, and hence is $\mathcal{F}_{t-d(t)}$-measurable.

For the rest of the details, we refer the reader to [19,20]. □

Having established the self-optimality, we turn next to the issue of parameter consistency.

Theorem 3.2 *In addition to the assumptions of Theorem 3.1, assume that $A(q^{-1})$ and $B(q^{-1}))$ do not have a common factor, and that θ_0 is an interior point of the set $\mathcal{M}$. If $\{u(t)\}$ satisfies one of the following two conditions,*

i) $\{u(t)\}$ is sufficiently rich of order greater than or equal to $n+m+1$, i.e., $\exists$ an $\epsilon > 0$, an integer $T < \infty$, such that

$$\sum_{i=t}^{t+T}[u(t),\ldots,u(t-n-m)][u(t),\ldots,u(t-n-m)]^{\mathrm{T}} \geq \epsilon I_{n+m+1}\,,$$

for all t sufficiently large; or

ii) $u(t)$ is an ARMA process, i.e., $u(t) = C(q^{-1})\epsilon(t)$, where $C(q^{-1})$ is a stable rational transfer function, and $\epsilon(t)$ is a martingale difference sequence with finite but nonzero variance.

Then, $\theta(t) \to \theta_0$ a.s..

Proof. As in Section 2.1, it can be shown that

$$\begin{aligned}
&\frac{1}{N}\left\|R(N-1)\widetilde{\theta}(N)\right\| \\
&\leq \frac{1}{N}\left\|\sum_1^N \phi(t-1)\phi(t-1)^{\mathrm{T}}\widetilde{\theta}(t-1)\right\| + \frac{1}{N}\left\|\sum_1^N \frac{R(t-1)}{(t-1)}\phi(t-1)(e(t)-v(t))\right\| \\
&\quad + \frac{1}{N}\left\|\sum_1^N \frac{R(t-1)}{r(t-1)}\phi(t-1)v(t)\right\| + \frac{1}{N}\sum_1^N \|R(t-1)\|\,\|\theta'(t)-\theta(t)\| + o(1)
\end{aligned}$$

where $R(N) := \sum_1^N \phi(t)\phi^{\mathrm{T}}(t)$.

The first two terms on the RHS above converge to zero by Schwarz inequality and the self-optimality established earlier. The third term can be shown to converge to zero by the techniques in the proof of Theorem 3.1, i.e., the decomposition of $v(t)$, the backward recursion for $\phi(t-1)$, and Lemma 1 a).

We turn next to the fourth term on the RHS above. Let $\overline{I}_D(x)$ denote the indicator function of the complement of a set D. Clearly

$$\|\widetilde{\theta}'(t) - \widetilde{\theta}(t)\| \leq \overline{I}_{\mathcal{M}}(\theta'(t))\|\Delta'(t)\|\,.$$

Hence,

$$\begin{aligned}
&\frac{1}{N}\sum_1^N \|R(t-1)\|\,\|\theta'(t)-\theta(t)\| \\
&\leq \frac{1}{N}\sum_1^N \|\phi(t-1)e(t)\|\,\overline{I}_{\mathcal{M}}(\theta'(t)) \\
&\leq \frac{1}{N}\sum_1^N \|\phi(t-1)(e(t)-v(t))\| + \frac{1}{N}\sum_1^N \|\phi(t-1)v(t)\|\,\overline{I}_{\mathcal{M}}(\theta'(t)) \\
&\leq o(1) + \Gamma\frac{1}{N}\left(\sum_1^N \|\phi(t-1)\|^2\right)^{\frac{1}{2}}\left(\sum_1^N \overline{I}_{\mathcal{M}}(\theta'(t))\right)^{\frac{1}{2}}.
\end{aligned}$$

From either condition i) or ii) of the theorem, by using the techniques of [3,13] we can establish that $\theta(t)$ converges to θ_0 in the Cesaro sense, i.e., for any open set D containing θ_0,

$$\frac{1}{N}\sum_{1}^{N} \overline{I}_D(\theta(t)) \to 0 \quad \text{a.s.}.$$

Since θ_0 is an interior point of $\mathcal{M}$, and $\|\theta'(t) - \theta(t)\| \to 0$, it follows that

$$\frac{1}{N}\sum_{1}^{N} \overline{I}_{\mathcal{M}}(\theta'(t)) \to 0 \quad \text{a.s.}.$$

Therefore,

$$\frac{1}{N}\|R(N-1)\widetilde{\theta}(N)\| \to 0 \quad \text{a.s.}.$$

From either i) or ii), and the self-optimality, it can be shown that

$$\liminf_{N\to\infty} \frac{1}{N}\lambda_{\min}(R(N)) > 0\,.$$

Hence, $\|\widetilde{\theta}(t)\| \to 0$ a.s.. □

3.2 Adaptive feedforward control

Let us consider Problems B and C depicted in Figs. 2 and 3. Clearly,

$$e(t) = G(q^{-1})s(t) - P(q^{-1})u(t) + v(t)\,. \tag{3.7}$$

We make the following assumptions.

Assumptions

A3.6) $G(q^{-1})$, $P(q^{-1})$, $H(q^{-1})$ and $F(q^{-1})$ are stable rational transfer functions, and $P(q^{-1})$ is minimum phase.

A3.7) The pure delay in $G(q^{-1})$ is larger than or equal to that in $H(q^{-1})P(q^{-1})$, and $H(q^{-1})$ is minimum phase.

The model (3.7) can then be equivalently represented by

$$A(q^{-1})(e(t) - v(t)) = q^{-d}B(q^{-1})u(t) + q^{-k}D(q^{-1})m(t)\,, \tag{3.8}$$

where $A(q^{-1})$, $B(q^{-1})$ and $D(q^{-1})$ are polynomials, $A(q^{-1})$ is Hurwitz, and $d \leq k$.

With the model parameterized as in (3.8), it is clear that the minimum variance feedforward control is given by

$$B(q^{-1})u(t) + q^{-(k-d)}D(q^{-1})m(t) = 0\,.$$

Let us now consider the adaptation procedure. Let $B(q^{-1}) = b_0 + \cdots + b_\ell q^{-\ell}, D(q^{-1}) = d_0 + \cdots + d_n q^{-n}, \theta_0 = [b_0, \ldots, b_\ell, d_0, \ldots, d_n]$, and

$$\phi(t) := [u(t), \ldots, u(t-\ell-1), m(t-k+d), \ldots, m(t-k+d-n-1)]^{\mathrm{T}}.$$

Then

$$A(q^{-1})(e(t)-v(t)) = \phi^{\mathrm{T}}(t-d)\theta_0 .$$

We propose the following adaptive feedforward control algorithm. The control $u(t)$ is generated so as to satisfy

$$\phi^{\mathrm{T}}(t)\theta(t) = 0 ,$$

and $\theta(t)$, the estimate of θ_0, is updated by

$$\theta(t) = F\left(\theta(t-1) + \frac{a\phi(t-d)}{r(t-d)}e(t)\right), \quad a > 0$$

$$r(t-d) = r(t-d-1) + \|\phi(t-d)\|^2, \quad r(-d) = 1 ,$$

where $F(\cdot)$ denotes the projection onto a convex compact set $\mathcal{M}$. The following theorem establishes the optimality of the above adaptive control law.

Theorem 3.3 *Let the following conditions hold:*

i) $\{v(t)\}$ *and* $\{s(t)\}$ *satisfy the assumptions (A3.1), (A3.2) and (A3.5).*
ii) Assumptions (A3.6) and (A3.7) hold.
iii)

$$\min_{|z|=1} \mathrm{Re}[A(z)] > \left(d - \frac{1}{2}\right) a \tag{3.9}$$

iv) $\theta_0 \in \mathcal{M}$, *and* $\widehat{b}_0 \neq 0, \forall \widehat{\theta} \in \mathcal{M}$.

Then

$$\lim_{N\to\infty} \frac{1}{N} \sum_{t=1}^{N} [e(t) - v(t)]^2 = 0 \quad a.s.. \tag{3.10}$$

Proof. Let us denote

$$V(t) := \|\widetilde{\theta}(t)\|^2 ,$$

and

$$\Delta(t) := \frac{\|a\phi(t-d)e(t)\|}{r(t-d)} .$$

Then $V(t)$ satisfies the recursion

$$\begin{aligned} V(N) \leq V(0) + \sum_{t=1}^{N} \Delta^2(t) + 2a \sum_{t=1}^{N} \frac{\phi^{\mathrm{T}}(t-d)\widetilde{\theta}(t-d)e(t)}{r(t-d)} \\ +2a \sum_{t=1}^{N} \frac{\phi^{\mathrm{T}}(t-d)(\widetilde{\theta}(t-1) - \widetilde{\theta}(t-d))e(t)}{r(t-d)} . \end{aligned} \tag{3.11}$$

The last term on the RHS above can be bounded as follows,

$$
\begin{aligned}
2a\sum_{t=1}^{N}&\frac{\phi^{\mathrm{T}}(t-d)(\widetilde{\theta}(t-1)-\widetilde{\theta}(t-d))e(t)}{r(t-d)} \\
&\leq 2\sum_{t=1}^{N}\Delta(t)\sum_{i=1}^{d-1}\Delta(t-i) \\
&\leq \sum_{t=1}^{N}\sum_{i=1}^{d-1}(\Delta^2(t)+\Delta^2(t-i)) \\
&\leq 2(d-1)\sum_{t=1}^{N}\Delta^2(t)+O(1)\,.
\end{aligned}
\tag{3.12}
$$

As in the proof of Theorem 3.1, we can show that

$$
\begin{aligned}
&\left|\sum_{t=1}^{N}\frac{\phi^{\mathrm{T}}(t-d)\widetilde{\theta}(t-d)v(t)}{r(t-d)}\right| \\
&\leq O(1)+o\left(\sum_{t=1}^{N}\left(\frac{\phi^{\mathrm{T}}(t-d)\widetilde{\theta}(t-d)}{r(t-d)}\right)^2\right)+\epsilon\sum_{t=1}^{N}\Delta^2(t)\,,
\end{aligned}
\tag{3.13}
$$

for $\forall\epsilon>0$. Substituting (3.12) and (3.13) into (3.11), we obtain

$$
\begin{aligned}
V(N)\leq (2d-1+\epsilon)\sum_{t=1}^{N}\Delta^2(t)+2a\sum_{t=1}^{N}\frac{\phi^{\mathrm{T}}(t-d)\widetilde{\theta}(t-d)(e(t)-v(t))}{r(t-d)} \\
+o\left(\sum_{t=1}^{N}\left(\frac{\phi^{\mathrm{T}}(t-d)\widetilde{\theta}(t-d)}{r(t-d)}\right)^2\right)+O(1)\,.
\end{aligned}
\tag{3.14}
$$

Now,

$$
\begin{aligned}
\sum_{t=1}^{N}\Delta^2(t)&=a^2\sum_{t=1}^{N}\frac{\|\phi(t-d)\|^2(e(t)-v(t))^2+v^2(t)+2e(t)v(t)}{r^2(t-d)} \\
&\leq a^2\sum_{t=1}^{N}\frac{(e(t)-v(t))^2}{r(t-d)}+O(1)\,.
\end{aligned}
$$

Substituting the above into (3.14), noting that

$$A(q^{-1})(e(t)-v(t))=-\phi^{\mathrm{T}}(t-d)\widetilde{\theta}(t-d)\,,$$

and using (3.9), we obtain

$$\sum_{t=1}^{\infty}\frac{(e(t)-v(t))^2}{r(t-1)}<\infty. \quad \text{a.s.}.$$

Finally, since $F(q^{-1})$ is stable and $P(q-1)$ is minimum phase, we have

$$\begin{aligned} r(N-1) &= O\left(\sum_{t=1}^{N}(m^2(t)+u^2(t))\right) \\ &= O\left(\sum_{t=1}^{N}u^2(t)\right) \\ &= O\left(N+\sum_{t=1}^{N}(e(t)-v(t))\right) \quad \text{a.s..} \end{aligned}$$

The optimality (3.10) then follows as in the proof of Theorem 2.7. □

Remarks: i) It is worth noting that the presence of input contamination $F(q^{-1})$ *does not* require any algorithmic modification. ii) Unlike the output recursions for filtering or identification, it is *not* possible to use a posteriori estimates in the regression vector for adaptive feedforward control. iii) When $s(t)$ is periodic, the assumption (A3.7) can be removed from Theorem 3.3. □

4 Concluding Remarks

As we have noted at the outset, stochastic adaptive system theory has as its goal the improvement of the performance of an adaptive system by exploiting any stochastic, typically correlational, properties of disturbances and signals. As a step towards providing such a coherent view of stochastic adaptive system theory, we have provided a somewhat unified treatment of several problems in recursive identification, adaptive signal processing, and adaptive control, based both on the equation error method as well as the parallel model method.

There is yet another dimension towards a more encompassing stochastic adaptive system theory which at the current time remains largely an open area. This is to use any prior information about the *likely* values of the true parameters and treat them also as random variables. Such an approach leads us back essentially to stochastic control, and what is commonly termed "dual" control. While the determination of "optimal" algorithms for such a model seems intractable, there are interesting issues that arise. For example, will an "optimal" or "good" algorithm be stable for a fixed value of the parameters? How "robust" are these algorithms with respect to the assumption on the underlying probability distribution for θ_0? Such a theory suggests the promise of a better treatment and analysis of *transient* properties of adaptive algorithms, in contrast to the largely asymptotic theory currently existing.

We have also not dealt here with the "robustness" issues in adaptive control. Much progress has been made on this problem, and is surveyed elsewhere in this volume. However, many *performance* issues relating to the choice of both adaptation as well as control designs remain open.

A genuinely comprehensive theory of adaptive systems, it seems, will require a unification of all these considerations, and still remains to be developed.

References

1. Y. S. Chow, "Local convergence of martingales and the law of large numbers," *Ann. Math. Statist.*, vol. 36, pp. 552–558, 1965.
2. T. L. Lai and C. Z. Wei, "Least squares estimate in stochastic regression with applications to identification and control of dynamic systems," *Ann. Math. Statist.*, vol. 10, pp. 154–166, 1982.
3. A. Becker, P. R. Kumar, and C. Z. Wei, "Adaptive control with the stochastic approximation algorithm: Geometry and convergence," *IEEE Trans. Aut. Control*, vol. AC-30, pp. 330–338, April 1985.
4. H. F. Chen and L. Guo, "Asymptotically optimal adaptive control with consistent parameter estimates," *SIAM J. Control Optimiz.*, vol. 25, no. 3, pp. 558–575, 1987.
5. K. J. Åström and B. Wittenmark, "On self-tuning regulators," *Automatica*, vol. 9, pp. 185–199, 1973.
6. L. Ljung, "Analysis of recursive stochastic algorithms," *IEEE Trans. Aut. Control*, vol. AC-22, pp. 551–575, 1977.
7. L. Ljung, "On positive real transfer functions and the convergence of some recursive schemes," *IEEE Trans. Aut. Control*, vol. AC-22, pp. 539–551, 1977.
8. V. Solo, "The convergence of AML," *IEEE Trans. Aut. Control*, vol. AC-24, pp. 958–962, 1979.
9. G. C. Goodwin, P. J. Ramadge, and P. E. Caines, "Discrete time stochastic adaptive control," *SIAM J. Control Optimiz.*, vol. 19, pp. 829–853, 1981.
10. G. C. Goodwin, K. S. Sin, and K. K. Saluja, "Stochastic adaptive control and prediction–the general delay-colored noise case," *IEEE Trans. Aut. Control*, vol. AC-25, pp. 946–949, 1980.
11. J. J. Fuchs, "Indirect stochastic adaptive control: The general delay–white noise case," *IEEE Trans. Aut. Control*, vol. AC-27, pp. 219–223, 1982.
12. H. F. Chen and P. E. Caines, "Strong consistency of the stochastic gradient algorithm of adaptive control," *IEEE Trans. Aut. Control*, vol. AC-30, pp. 159–192, 1985.
13. P. R. Kumar and L. Praly, "Self-tuning trackers," *SIAM J. Control Optimiz.*, vol. 25, pp. 1053–1071, July 1987.
14. T. L. Lai and C. Z. Wei, "Extended least squares and their applications to adaptive control and prediction in linear systems," *IEEE Trans. Aut. Control*, vol. AC-31, pp. 898–906, 1986.
15. H. F. Chen and L. Guo, "Convergence rate of least squares identification and adaptive control for stochastic systems," *Int. J. Control*, vol. 44, pp. 1459–1476, 1986.
16. M. Radenković and S. Stanković, "Strong consistency of parameter estimates in direct self-tuning control algorithms based on stochastic approximations," *Automatica*, vol. 26, pp. 533–544, May 1990.
17. W. Ren and P. R. Kumar, "Direct stochastic minimum variance control with noninterlaced algorithms." Submitted to *1991 Amer. Control Conf.*, September 1990.

18. W. Ren and P. R. Kumar, "On stochastic adaptive control," to appear as a Technical Report, 1990.
19. W. Ren and P. R. Kumar, "The Convergence of Output Error Recursions in Infinite Order Moving Average Noise," to appear in *New Directions in Time Series Analysis, IMA Volumes in Mathematics and its Applications*, Springer-Verlag, 1991.
20. W. Ren and P. R. Kumar, "On stochastic parallel model adaptation problems," to be submitted, 1990.
21. P. R. Kumar, "Convergence of adaptive control schemes using least-squares parameter estimates," *IEEE Trans. Aut. Control*, vol. AC-35, pp. 416–424, 1990.
22. K. S. Sin, G. C. Goodwin, and R. R. Bitmead, "An adaptive *d*-step ahead predictor based on least squares," *IEEE Trans. Aut. Control*, vol. AC-25, pp. 1161–1164, 1980.
23. L. Guo and H. F. Chen, "Revisit to Åström-Wittenmark's self-tuning regulator and ELS-based adaptive trackers," Research Report, Academia Sinica, Institute of Systems Sciences, Beijing, 1990.
24. W. Ren and P. R. Kumar, "Adaptive active noise control: Structures, algorithms and convergence analysis," in *Proc. Inter-Noise 89*, pp. 435–440, Newport Beach, CA, Dec. 1989.
25. I. D. Landau, "Unbiased recursive identification using model reference adaptive techniques," *IEEE Trans. Aut. Control*, vol. AC-21, pp. 194–202, 1976.
26. W. Ren and P. R. Kumar, "The convergence of output error identification and adaptive IIR filtering algorithms in the presence of colored noise," *Proc. 29th Conf. Dec. Control*, pp. 3534–3539, Honolulu, HI, Dec. 1990.

Part II

Adaptive Nonlinear Control

Adaptive Feedback Linearization of Nonlinear Systems*

P. V. Kokotović,[1] *I. Kanellakopoulos,*[1] *and A. S. Morse*[2]

[1] Coordinated Science Laboratory
University of Illinois, Urbana, IL 61801, USA.
[2] Department of Electrical Engineering
Yale University, New Haven, CT 06520-1968, USA.

Abstract. After an examination of the restrictive assumptions that limit the applicability of existing adaptive nonlinear control schemes, new adaptive regulation and tracking schemes are developed for a class of feedback linearizable nonlinear systems. The coordinate-free geometric conditions, which characterize this class of systems, neither restrict the location of the unknown parameters, nor constrain the growth of the nonlinearities. Instead, they require that the nonlinear system be transformable into the so-called pure-feedback form. When this form is "strict", the proposed scheme guarantees global regulation and tracking properties, and substantially enlarges the class of nonlinear systems with unknown parameters for which global stabilization can be achieved. The new design procedure is systematic and its stability proofs use simple analytical tools, familiar to most control engineers.

1 Introduction

Until a few years ago, adaptive linear [1,2] and geometric nonlinear [3,4] methods belonged to two separate areas of control theory. They were helpful in the design of controllers for plants containing either unknown parameters or known nonlinearities, but not both. In the last few years the problem of *adaptive nonlinear control* was formulated to deal with the control of plants containing both unknown parameters and known nonlinearities. A realistic plan of attack for this challenging new problem led through a series of simpler problems, each formulated under certain restrictive assumptions. The two most common assumptions are those of linear parametrization and full-state feedback.

Linear parametrization. This assumption, adopted by all the researchers in the field [5–18], requires that in a nonlinear plant the unknown parameters either appear, or can be made to appear, linearly. For example, if the plant model contains not only θ_1 and θ_2, but also $e^{\theta_1\theta_2}$, it is to be "overparametrized" by

* The work of the first two authors was supported in part by the National Science Foundation under Grant ECS-87-15811 and in part by the Air Force Office of Scientific Research under Grant AFOSR 90-0011. The work of the third author was supported by the National Science Foundation under Grants ECS-88-05611 and ECS-90-12551.

introducing $\theta_3 = e^{\theta_1\theta_2}$ as an additional parameter. For linear plants, an analogous linear parametrization is to consider the coefficients of a transfer function as "plant parameters", although they may be nonlinear functions of physical parameters such as resistance, inertia etc. Most adaptive nonlinear control results have been obtained for linearly parametrized systems of the form

$$\dot{\zeta} = f_0(\zeta) + \sum_{i=1}^{p} \theta_i f_i(\zeta) + \left[g_0(\zeta) + \sum_{i=1}^{p} \theta_i g_i(\zeta) \right] u, \tag{1.1}$$

where $\zeta \in \mathbb{R}^n$ is the state, $u \in \mathbb{R}$ is the input, $\theta = [\theta_1, \ldots, \theta_p]^{\mathrm{T}}$ is the vector of unknown constant parameters, and f_i, g_i, $0 \leq i \leq p$, are smooth vector fields in a neighborhood of the origin $\zeta = 0$ with $f_i(0) = 0$, $0 \leq i \leq p$, $g(0) \neq 0$.

Full-state feedback. With the exception of some very recent papers [17,18], all the adaptive nonlinear control literature [5–16] deals with the full-state feedback problem. The assumption of full-state feedback is motivated by the geometric conditions for input-output or full-state linearization of (1.1).

Parameter-dependent forms of the feedback linearization conditions for (1.1) and their "certainty-equivalence" implementation require that further restrictions be imposed either on the location of the unknown parameters or on the type of nonlinearities. According to these additional restrictions, the existing adaptive schemes can be classified into *uncertainty-constrained schemes* and *nonlinearity-constrained schemes.*

Uncertainty-constrained schemes , surveyed in Section 2, impose restrictions (*matching conditions*) on the location of unknown parameters, but can handle all types of nonlinearities. References dealing with this class of schemes are [5–11,14].

Nonlinearity-constrained schemes , surveyed in Section 3, do not restrict the location of unknown parameters. Instead, they impose restrictions on the nonlinearities of the original system, as well as on those appearing in the transformed error system. References dealing with this class of schemes are [12–16].

The brief survey of the existing adaptive schemes in Sections 2 and 3 stresses their major limitations. One of the main results of this paper is that these limitations can be removed for the so-called *pure-feedback systems.* This is now the broadest class of nonlinear systems for which adaptive controllers can be systematically designed without imposing any growth constraints on system nonlinearities.

Pure-feedback systems are introduced in Section 4. Their geometric characterization identifies the *level of uncertainty* and *nonlinear complexity* as structural obstacles to adaptive feedback linearization. For an unknown parameter, the level of uncertainty is its "distance", in terms of the number of integrators, from the control input. The larger this distance is, the smaller is the number of state variables on which the nonlinearity multiplying this parameter is allowed to depend (nonlinear complexity).

Systematic design procedure. The new adaptive scheme for pure-feedback systems is designed by a step-by-step procedure which interlaces, at each step, the change of coordinates required for feedback linearization, and the construction of parameter update laws required for adaptation. The main idea of this new nonlinear procedure evolved from an earlier linear result of Feuer and Morse [19]. In Section 5, the procedure is developed for single-input systems. Its feasibility and the stability of the resulting adaptive system are established in Section 6. The extension to the multi-input case is given in Section 7.

One of the most important stability and robustness properties of every adaptive system is the size of its region of attraction, relative to the size of the region that would have been achieved if all the parameters were known. When with the known parameters the stability and tracking properties are global, but the same properties of the adaptive scheme are only local, then the loss of globality is due to adaptation. To avoid this loss, the schemes of [12,13,15,16] require that the nonlinearities and some of their derivatives satisfy a linear growth condition, which severely limits the applicability of these schemes. The class of systems for which the new adaptive scheme guarantees global regulation and tracking is much wider.

Global regulation and tracking. It is shown is Section 8 that the region of attraction for the new adaptive scheme is global if the feedback linearization is global. A subclass of pure-feedback systems for which this global property is easy to establish are *strict feedback systems*. For these systems the new adaptive scheme achieves both global regulation and, as shown in Section 9, global tracking of smooth bounded reference inputs. In contrast to the earlier schemes, these global results are obtained without any growth constraints on system nonlinearities. This is illustrated by two examples in Section 10.

2 Uncertainty-Constrained Schemes

The uncertainty-constrained schemes developed by Kanellakopoulos, Kokotović and Marino [6,7,8] and Campion and Bastin [5], use the so-called *extended matching condition* (EMC), which was introduced in [6,7] and reformulated in [5] as a "strong linearizability" condition. For the nonlinear system (1.1), the EMC has the following geometric expression:

$$\begin{aligned} f_i &\in \text{ sp }\{g_0, \text{ad}_{f_0} g_0\} \\ g_i &\in \text{ sp }\{g_0\}\,. \end{aligned} \tag{2.1}$$

This condition extends the *strict matching condition* (SMC)

$$f_i, g_i \in \text{ sp }\{g_0\}\,, \tag{2.2}$$

which was used by Taylor, Kokotović, Marino and Kanellakopoulos [10,11], and, in a special case, by Slotine and Coetsee [9]. For full-state feedback linearizable systems, the EMC is necessary and sufficient for the existence of a parameter-independent diffeomorphism $x = \phi(\zeta)$ which transforms the system (1.1) into

$$\begin{aligned} \dot{x}_i &= x_{i+1}\,, \qquad i = 1, \ldots, n-2 \\ \dot{x}_{n-1} &= x_n + \sum_{i=1}^{\ell} \theta_i \gamma_i(x) \\ \dot{x}_n &= \alpha_0(x) + \sum_{i=1}^{\ell} \theta_i \alpha_i(x) + \left[\beta_0(x) + \sum_{i=1}^{\ell} \theta_i \beta_i(x)\right] u\,. \end{aligned} \tag{2.3}$$

To illustrate the properties of the EMC-based schemes of [6,7] we use the system

$$\begin{aligned} \dot{x}_1 &= x_2 \\ \dot{x}_2 &= x_3 + \theta\gamma(x_1) \\ \dot{x}_3 &= u\,, \end{aligned} \tag{2.4}$$

with $\gamma(x_1)$ a smooth nonlinear function of x_1. This system is already in the form (2.3). The design procedure of [6,7] employs an estimate $\hat{\theta}$ of the unknown parameter θ and replaces x_3 with the new state

$$\hat{x}_3 = x_3 + \hat{\theta}\gamma(x_1)\,. \tag{2.5}$$

The system (2.4) is then rewritten as

$$\begin{aligned} \dot{x}_1 &= x_2 \\ \dot{x}_2 &= \hat{x}_3 + \gamma(x_1)(\theta - \hat{\theta}) \\ \dot{\hat{x}}_3 &= u + \hat{\theta}\frac{\partial\gamma(x_1)}{\partial x_1}x_2 + \dot{\hat{\theta}}\gamma(x_1)\,. \end{aligned} \tag{2.6}$$

It is now possible to design the control as a function of the state and both $\hat{\theta}$ and $\dot{\hat{\theta}}$, because the derivative $\dot{\hat{\theta}}$ will be explicitly known from the update law. The control

$$u = -\hat{\theta}\frac{\partial\gamma(x_1)}{\partial x_1}x_2 - \dot{\hat{\theta}}\gamma(x_1) - k_1x_1 - k_2x_2 - k_3[x_3 + \hat{\theta}\gamma(x_1)] \tag{2.7}$$

renders the system (2.6) linear in the parameter error $\theta - \hat{\theta}$:

$$\begin{aligned} \dot{x}_1 &= x_2 \\ \dot{x}_2 &= \hat{x}_3 + (\theta - \hat{\theta})\gamma(x_1) \\ \dot{\hat{x}}_3 &= -k_1x_1 - k_2x_2 - k_3\hat{x}_3\,. \end{aligned} \tag{2.8}$$

The gains k_1, k_2, k_3 are chosen to place the eigenvalues of the system matrix

$$A = \begin{bmatrix} 0 & 1 & 0 \\ 0 & 0 & 1 \\ -k_1 & -k_2 & -k_3 \end{bmatrix} \tag{2.9}$$

at some desired stable locations. The general form of (2.8) is

$$\dot{\hat{x}} = A\hat{x} + W(\hat{x})(\theta - \hat{\theta}), \quad \hat{x} = \begin{bmatrix} x_1 \\ x_2 \\ \hat{x}_3 \end{bmatrix}, \quad W(\hat{x}) = \begin{bmatrix} 0 \\ \gamma(x_1) \\ 0 \end{bmatrix}. \tag{2.10}$$

This form is desirable, because it suggests that a parameter update law can be designed using a result of adaptive linear control [1]. This update law is

$$\dot{\hat{\theta}} = W^T(\hat{x})P\hat{x}\,, \tag{2.11}$$

where $P = P^T > 0$ is chosen to satisfy $PA + A^TP = -I$. The stability of the equilibrium $\hat{x} = 0$, $\hat{\theta} = \theta$ of the closed-loop system (2.10)–(2.11) is then established using the *quadratic* Lyapunov function $V = \hat{x}^T P\hat{x} + (\theta - \hat{\theta})^2$, whose derivative along the solutions of (2.10)–(2.11) is $\dot{V} = -||\hat{x}||^2 \leq 0$. Since the feedback linearization of (2.4) can be achieved for all x and θ, this stability result is global. The largest invariant set of (2.10)–(2.11) contained in the set where $\dot{V} = 0$ is

$$M = \{(\hat{x}, \hat{\theta}) : \hat{x} = 0, (\theta - \hat{\theta})\gamma(0) = 0\}\,. \tag{2.12}$$

Furthermore, if $\gamma(0) \neq 0$, then $(0, \theta)$ can be shown to be an *exponentially stable* equilibrium of (2.10)–(2.11). Thus, the stability of this adaptive scheme is *robust* with respect to both fast stable unmodeled dynamics and small disturbances. In the case when $\gamma(0) = 0$, or, in the multi-input case, when the number of unknown parameters is more than twice the number of control inputs, only stability, rather than exponential stability, can be achieved.

The above example demonstrates the main advantage of EMC-based schemes: their stability properties can be established *independently of the type of nonlinearities*. Clearly, the nonlinear function $\gamma(x_1)$ in (2.4) can be any smooth function, and is not restricted by global Lipschitz or sector-type conditions.

Even though the EMC is quite restrictive, it is satisfied by many systems of practical importance, such as most types of electric motors [7] and autocatalyzed chemical reactions [5]. The robustness of EMC-based schemes with respect to unmodeled dynamics [6,7] can be exploited to extend their applicability. For systems that do not satisfy the EMC, a partial high-gain feedback control can be used to induce a two-time-scale property, such that the slow subsystem satisfies the EMC, and the fast stable subsystem is treated as unmodeled dynamics. This possibility was exploited by Taylor [10] and is illustrated on the system

$$\begin{aligned} \dot{x}_1 &= x_2 + \theta\gamma(x_1) \\ \dot{x}_2 &= x_3 \\ \dot{x}_3 &= u\,, \end{aligned} \tag{2.13}$$

which does not satisfy the EMC (2.1). Using the control

$$u = \frac{1}{\mu}(v - x_3)\,, \tag{2.14}$$

where v is the new control variable and μ is a small positive constant, and replacing x_3 with the new state $\eta = x_3 - v$, we transform (2.13) into the standard singular perturbation form

$$\begin{aligned} \dot{x}_1 &= x_2 + \theta\gamma(x_1) \\ \dot{x}_2 &= v + \eta \\ \mu\dot{\eta} &= -\eta + \mu\dot{v}\,. \end{aligned} \tag{2.15}$$

Neglecting the fast unmodeled dynamics by setting $\eta \equiv 0$, we obtain the reduced-order system

$$\begin{aligned} \dot{x}_1 &= x_2 + \theta\gamma(x_1) \\ \dot{x}_2 &= v\,, \end{aligned} \tag{2.16}$$

for which an EMC-based scheme is:

$$\hat{x}_2 = x_2 + \hat{\theta}\gamma(x_1) \tag{2.17}$$

$$v = -k_1 x_1 - k_2\hat{x}_2 - \hat{\theta}\frac{\partial\gamma(x_1)}{\partial x_1}\hat{x}_2 - \dot{\hat{\theta}}\gamma(x_1) \tag{2.18}$$

$$\dot{\hat{\theta}} = \left[\gamma(x_1)\ \ \hat{\theta}\frac{\partial\gamma(x_1)}{\partial x_1}\gamma(x_1)\right] P \begin{bmatrix} x_1 \\ \hat{x}_2 \end{bmatrix}. \tag{2.19}$$

When applied to (2.15), this scheme results in the closed-loop system

$$\begin{aligned} \dot{x}_1 &= \hat{x}_2 + \gamma(x_1)(\theta - \hat{\theta}) \\ \dot{\hat{x}}_2 &= -k_1 x_1 - k_2\hat{x}_2 - \hat{\theta}\frac{\partial\gamma(x_1)}{\partial x_1}\gamma(x_1)(\theta - \hat{\theta}) + \eta \\ \mu\dot{\eta} &= -\eta + \mu\dot{v} \\ \dot{\hat{\theta}} &= \left[\gamma(x_1)\ \ \hat{\theta}\frac{\partial\gamma(x_1)}{\partial x_1}\gamma(x_1)\right] P \begin{bmatrix} x_1 \\ \hat{x}_2 \end{bmatrix}. \end{aligned} \tag{2.20}$$

As shown in [7], this system has $x_1 = \hat{x}_2 = \eta = 0$, $\hat{\theta} = \theta$ as a stable equilibrium with a guaranteed region of attraction.

The EMC-based schemes of [6,7,8] and [5] may result in implicit expressions for the control and the update law. The implicit form will arise if and only if there is at least one unknown parameter θ_i which enters the control vector field (so that its derivative $\dot{\hat{\theta}}_i$ is defined in terms of the control u) and whose derivative has to be cancelled by the control (so that u is in turn defined in terms of $\dot{\hat{\theta}}_i$). For the system (1.1), the *explicit extended matching condition* (EEMC), which is necessary and sufficient for u and $\dot{\hat{\theta}}$ to be explicitly defined, is

$$\begin{aligned} &f_i, g_i \in \operatorname{sp}\{g_0\} \\ &\qquad\text{or} \qquad\qquad\qquad\qquad \forall\, i = 1,\ldots,\ell\,. \\ &f_i \in \operatorname{sp}\{g_0,\ \mathrm{ad}_{f_0}g_0\},\ g_i \equiv 0 \end{aligned} \tag{2.21}$$

To illustrate the implicit form of the scheme of [6,7,8], we use the system

$$\begin{aligned} \dot{x}_1 &= x_2 \\ \dot{x}_2 &= x_3 + \theta(1 + x_1^2) \\ \dot{x}_3 &= (1 + \theta x_1^2)u\,, \end{aligned} \tag{2.22}$$

for which the new state is

$$\hat{x}_3 = x_3 + \hat{\theta}(1 + x_1^2). \tag{2.23}$$

The control and the parameter update law are

$$u = \frac{-(1 + x_1^2)\dot{\hat{\theta}} - 2x_1 x_2 \hat{\theta} - k_1 x_1 - k_2 x_2 - k_3 \hat{x}_3}{1 + \hat{\theta} x_1^2}, \tag{2.24}$$

$$\dot{\hat{\theta}} = \begin{bmatrix} 0 & 1 + x_1^2 & x_1^2 u \end{bmatrix} P \begin{bmatrix} x_1 \\ x_2 \\ \hat{x}_3 \end{bmatrix}. \tag{2.25}$$

The key point is that (2.24) and (2.25) define u and $\dot{\hat{\theta}}$ as implicit functions of x and $\hat{\theta}$. Eliminating $\dot{\hat{\theta}}$ we obtain

$$\left[1 + \hat{\theta} x_1^2 + (1 + x_1^2) x_1^2 (P_{13} x_1 + P_{23} x_2 + P_{33} \hat{x}_3)\right] u =$$

$$= -(1 + x_1^2)^2 (P_{12} x_1 + P_{22} x_2 + P_{23} \hat{x}_3) - 2x_1 x_2 \hat{\theta} - k_1 x_1 - k_2 x_2 - k_3 \hat{x}_3. \tag{2.26}$$

To express u explicitly, the term multiplying it must be nonzero, which is true in a region S around the equilibrium $x_1 = x_2 = \hat{x}_3 = 0$, $\hat{\theta} = \theta$. The boundaries of S are the manifolds on which the term multiplying u in (2.26) is zero.

For the scheme of Campion and Bastin [5], the counterpart of (2.26) is equation (20) of [5], assumed to be solvable in a region $D_x \times D_\theta$.

An *overparametrization* approach that always avoids the implicit form is due to Pomet and Praly [14]. To illustrate this approach, the system (2.22) is rewritten as

$$\begin{aligned} \dot{x}_1 &= x_2 \\ \dot{x}_2 &= x_3 + \theta_1(1 + x_1^2) \\ \dot{x}_3 &= (1 + \theta_2 x_1^2) u, \end{aligned} \tag{2.27}$$

where the same unknown parameter θ is treated as two parameters θ_1 and θ_2, which are estimated separately:

$$\begin{bmatrix} \dot{\hat{\theta}}_1 \\ \dot{\hat{\theta}}_2 \end{bmatrix} = \begin{bmatrix} 0 & 1 + x_1^2 & 0 \\ 0 & 0 & x_1^2 u \end{bmatrix} P \begin{bmatrix} x_1 \\ x_2 \\ \hat{x}_3 \end{bmatrix}, \tag{2.28}$$

and the implicit form is avoided because

$$\hat{x}_3 = x_3 + \hat{\theta}_1(1 + x_1^2) \tag{2.29}$$

$$u = \frac{-(1 + x_1^2)\dot{\hat{\theta}}_1 - 2x_1 x_2 \hat{\theta}_1 - k_1 x_1 - k_2 x_2 - k_3 \hat{x}_3}{1 + \hat{\theta}_2 x_1^2}. \tag{2.30}$$

A price paid in this overparametrization is the loss of exponential stability. However, the restriction to the region S has been removed and the only requirement is that $1 + \hat{\theta}_2 x_1^2$ be bounded away from zero. Hence, the designer is faced with a *tradeoff*: if the solvability region S is large enough, the implicit scheme is preferable. If, on the other hand, S is too small, then the overparametrized explicit scheme must be used, and additional measures will have to be taken to ensure robustness.

3 Nonlinearity-Constrained Schemes

If a system, such as (2.13), does not satisfy the EMC, one has to resort to one of the nonlinearity-constrained schemes developed in [12–16]. To illustrate these schemes we again use the system (2.13).

The scheme of Sastry and Isidori [15] is developed for input-output linearizable systems with exponentially stable zero dynamics. When applied to (2.13), this scheme employs the change of coordinates

$$\begin{aligned} y_1 &= x_1 \\ y_2 &= \dot{y}_1 = x_2 + \theta\gamma(x_1) \\ y_3 &= \dot{y}_2 = x_3 + \theta\frac{\partial\gamma(x_1)}{\partial x_1}(x_2 + \theta\gamma(x_1)) \end{aligned} \tag{3.1}$$

to transform the system (2.13) into the normal form:

$$\begin{aligned} \dot{y}_1 &= y_2 \\ \dot{y}_2 &= y_3 \\ \dot{y}_3 &= u + \theta\left[\frac{\partial\gamma(x_1)}{\partial x_1}x_3 + \frac{\partial^2\gamma(x_1)}{\partial x_1^2}x_2^2\right] \\ &\quad + \theta^2\left[2\frac{\partial^2\gamma(x_1)}{\partial x_1^2}\gamma(x_1)x_2 + \left(\frac{\partial\gamma(x_1)}{\partial x_1}\right)^2 x_2\right] \\ &\quad + \theta^3\left[\frac{\partial^2\gamma(x_1)}{\partial x_1^2}\gamma^2(x_1) + \left(\frac{\partial\gamma(x_1)}{\partial x_1}\right)^2\gamma(x_1)\right], \end{aligned} \tag{3.2}$$

which is no longer linear in θ. One way to circumvent this obstacle is to introduce two new parameters θ_2 and θ_3:

$$\theta_1 = \theta,\ \theta_2 = \theta^2,\ \theta_3 = \theta^3. \tag{3.3}$$

This overparametrization makes (3.2) linear in θ_1, θ_2 and θ_3, but it also increases the dimension and complexity of the adaptive controller. A "certainty-equivalence" control is implemented as

$$u = -k_1y_1 - k_2\hat{y}_2 - k_3\hat{y}_3 - \hat{\theta}_1w_{11}(x) - \hat{\theta}_2w_{21}(x) - \hat{\theta}_3w_3(x), \tag{3.4}$$

where

$$w_{11}(x) = \frac{\partial\gamma(x_1)}{\partial x_1}x_3 + \frac{\partial^2\gamma(x_1)}{\partial x_1^2}x_2^2 \tag{3.5}$$

$$w_{21}(x) = 2\frac{\partial^2\gamma(x_1)}{\partial x_1^2}\gamma(x_1)x_2 + \left(\frac{\partial\gamma(x_1)}{\partial x_1}\right)^2 x_2 \tag{3.6}$$

$$w_3(x) = \frac{\partial^2\gamma(x_1)}{\partial x_1^2}\gamma^2(x_1) + \left(\frac{\partial\gamma(x_1)}{\partial x_1}\right)^2\gamma(x_1). \tag{3.7}$$

The control (3.4) uses the estimated states

$$\hat{y}_2 = x_2 + \hat{\theta}_1\gamma(x_1) \tag{3.8}$$

$$\hat{y}_3 = x_3 + \hat{\theta}_1\frac{\partial\gamma(x_1)}{\partial x_1}x_2 + \hat{\theta}_2\frac{\partial\gamma(x_1)}{\partial x_1}\gamma(x_1), \tag{3.9}$$

since the states y_2 and y_3, defined by (3.1), depend on the unknown parameters and thus cannot be computed from the measurements of x_1, x_2 and x_3. Substitution of (3.4) into (3.2) and use of the notation

$$w_1(x) = w_{11}(x) + k_2\gamma(x_1) + k_3\frac{\partial\gamma(x_1)}{\partial x_1}x_2 \tag{3.10}$$

$$w_2(x) = w_{21}(x) + k_3\frac{\partial\gamma(x_1)}{\partial x_1}\gamma(x_1), \tag{3.11}$$

results in the error system

$$\begin{aligned} \dot{y}_1 &= y_2 \\ \dot{y}_2 &= y_3 \\ \dot{y}_3 &= -k_1y_1 - k_2y_2 - k_3y_3 + w_1(x)(\theta_1 - \hat{\theta}_1) \\ &\quad + w_2(x)(\theta_2 - \hat{\theta}_2) + w_3(x)(\theta_3 - \hat{\theta}_3), \end{aligned} \tag{3.12}$$

which is linear in $\theta_i - \hat{\theta}_i$, $i = 1, 2, 3$. The gradient-type update law of [15] consists of fifteen dynamic equations (three for the update law and twelve for the four third-order filters), and is not presented here for brevity.

The main limitation of this scheme is that it guarantees global stability and tracking only under the assumption, made in Theorem 3.3 of [15], that the change of coordinates (3.1) is globally Lipschitz in x. This forces $\gamma(x_1)$ to be *linear*, since if $\frac{\partial\gamma(x_1)}{\partial x_1}$ were *any* function of x_1 other than a constant, the term $\theta\frac{\partial\gamma(x_1)}{\partial x_1}x_2$ would *not* be globally Lipschitz in x_1. A further nonlinearity constraint used in [15], is that the derivatives of the regressor vector with respect to both x and $\hat{\theta}$ be bounded. In the above example this requirement is satisfied by the regressor $[w_1(x)\ w_2(x)\ w_3(x)]^T$ only when $\gamma(x_1)$ is linear.

For input-output linearizable systems of relative degree n, the scheme of Nam and Arapostathis [12] replaces globally Lipschitz conditions with less restrictive sector-type bounds. However, its stability proof requires the initial parameter estimates to be close enough to the true parameter values. In the case of the system (2.13), this scheme employs the change of coordinates

$$\begin{aligned} \hat{y}_1 &= x_1 \\ \hat{y}_2 &= x_2 + \hat{\theta}\gamma(x_1) \\ \hat{y}_3 &= x_3 + \hat{\theta}\frac{\partial\gamma(x_1)}{\partial x_1}x_2 + \hat{\theta}^2\frac{\partial\gamma(x_1)}{\partial x_1}\gamma(x_1) \end{aligned} \tag{3.13}$$

and the control

$$\begin{aligned} u = {} & k_1\hat{y}_1 - k_2\hat{y}_2 - k_3\hat{y}_3 \\ & -\hat{\theta}\frac{\partial\gamma(x_1)}{\partial x_1}x_3 - \hat{\theta}\frac{\partial^2\gamma(x_1)}{\partial x_1^2}x_2(x_2 + \hat{\theta}\gamma(x_1)) \\ & -\hat{\theta}^2\left[\left(\frac{\partial\gamma(x_1)}{\partial x_1}\right)^2 + \frac{\partial^2\gamma(x_1)}{\partial x_1^2}\gamma(x_1)\right](x_2 + \hat{\theta}\gamma(x_1)) \end{aligned} \tag{3.14}$$

to obtain the error system

$$\dot{\hat{y}} = A\hat{y} + \Psi(x,\hat{\theta})(\theta - \hat{\theta}) + D(x,\hat{\theta})\dot{\hat{\theta}}\,, \tag{3.15}$$

where

$$\hat{y} = \begin{bmatrix} \hat{y}_1 \\ \hat{y}_2 \\ \hat{y}_3 \end{bmatrix}, \quad A = \begin{bmatrix} 0 & 1 & 0 \\ 0 & 0 & 1 \\ -k_1 & -k_2 & -k_3 \end{bmatrix}, \tag{3.16}$$

$$\Psi(x,\hat{\theta}) = \begin{bmatrix} \gamma(x_1) \\ \hat{\theta}\dfrac{\partial\gamma}{\partial x_1}\gamma(x_1) \\ \hat{\theta}\dfrac{\partial^2\gamma}{\partial x_1^2}\gamma(x_1)[x_2 + \hat{\theta}\gamma(x_1)] + \hat{\theta}^2\left(\dfrac{\partial\gamma}{\partial x_1}\right)^2\gamma(x_1) \end{bmatrix},$$

$$D(x,\hat{\theta}) = \begin{bmatrix} 0 \\ \gamma(x_1) \\ \dfrac{\partial\gamma(x_1)}{\partial x_1}x_2 + 2\hat{\theta}\dfrac{\partial\gamma}{\partial x_1}\gamma(x_1) \end{bmatrix}. \tag{3.17}$$

Proposition 4.1 in [12] assumes the following sector-type bounds (cf. (4.9)–(4.10) of [12]) to be valid for all $\hat{\theta} \in B(\theta,\rho)$, a ball of radius ρ around θ:

$$\|\Psi(x,\hat{\theta})\| \le c_1 + c_2\|\hat{y}\| \tag{3.18}$$

$$\|D(x,\hat{\theta})\| \le d_1 + d_2\|\hat{y}\|\,, \tag{3.19}$$

where c_1, c_2, d_1, d_2 are positive constants. Furthermore, the initial parameter estimate is required to satisfy

$$\|\hat{\theta}(0) - \theta\| \le \rho' = \min\left\{\rho, \frac{1}{d_2}\right\}. \tag{3.20}$$

Upon closer inspection of (3.18), one realizes that $\gamma(x_1)$, although not required to be linear as in the scheme of Sastry and Isidori [15], is still severely restricted. For example, if

$$\gamma(x_1) = x_1 \sin x_1\,, \tag{3.21}$$

then $\left(\dfrac{\partial\gamma}{\partial x_1}\right)^2\gamma(x_1)$ and $\dfrac{\partial^2\gamma}{\partial x_1^2}\gamma^2(x_1)$ grow like cubic functions of x_1, and cannot be bounded by $\|\hat{y}\|$, which can grow at most quadratically with x_1.

The above two nonlinearity-constrained schemes employ parameter update laws analogous to those used in *direct* adaptive linear control. In contrast, the "*semi-indirect*" scheme of Teel, Kadiyala, Kokotović and Sastry [16], developed for the same class of systems as that of Sastry and Isidori, combines parameter estimation elements from both the direct and the indirect approaches. In the case of the system (2.13), this scheme uses the change of coordinates (3.13) and the control (3.14), and results in the same error system (3.15). In addition, following [5,13], the scheme employs the state observer (cf. (75) in [16])

$$\dot{\bar{y}} = A\hat{y} + D(x,\hat{\theta})\dot{\hat{\theta}} + \Omega(\bar{y} - \hat{y})\,, \tag{3.22}$$

where Ω is a Hurwitz matrix. Subtracting (3.22) from (3.15) provides the parameter estimation error system:

$$\dot{\varepsilon} = \Omega\varepsilon + \Psi(x,\hat{\theta})(\theta - \hat{\theta}), \quad \varepsilon \equiv \bar{y} - \hat{y}\,. \tag{3.23}$$

Assumption (A9) of Theorem 4.2 in [16] requires that the initial parameter estimate satisfy the bound

$$|\hat{\theta}(0) - \theta| \leq \kappa\,, \tag{3.24}$$

where κ is a constant which depends on various gains and Lipschitz constants. In addition, Assumption (A5) requires that the change of coordinates (3.13) be globally Lipschitz in $\hat{\theta}$, uniformly in x. This means that the derivative with respect to $\hat{\theta}$ of (3.13):

$$\begin{bmatrix} 0 \\ \gamma(x_1) \\ \dfrac{\partial\gamma(x_1)}{\partial x_1}x_2 + 2\hat{\theta}\dfrac{\partial\gamma(x_1)}{\partial x_1}\gamma(x_1) \end{bmatrix}, \tag{3.25}$$

is required to be uniformly bounded, which is true if and only if $\gamma(x_1) \equiv$ const. Finally, Assumption (A7) requires that

$$\left\| D(x,\hat{\theta})\Psi^T(x,\hat{\theta}) \right\| \leq c\|x\|, \quad c \geq 0\,. \tag{3.26}$$

When $\gamma(x_1) \equiv \gamma \equiv$ const, this inequality becomes $|\gamma| \leq c\|x\|$, and thus $\gamma(x_1) \equiv 0$, which implies that this scheme is not applicable to the system (2.13). Let us note, however, that the proof of Theorem 4.2 in [16]—in particular, equation (92)—is still valid under the slightly less restrictive assumption

$$\left\| D(x,\hat{\theta})\Psi^T(x,\hat{\theta}) \right\| \leq c_1 + c_2\|x\|\,, \tag{3.27}$$

under which γ can be any constant and not necessarily zero.

All the schemes we have examined so far assume that the nonlinear system is either full-state or input-output feedback linearizable. This enables them to borrow stability results obtained with quadratic Lyapunov functions from adaptive linear control.

In contrast, the approach of Pomet and Praly [13,14] assumes the existence of a more general Lyapunov function with prespecified growth properties achievable by an implementable feedback control. While this approach offers some significant advantages, it also has a major drawback: the designer is required to *construct* such a Lyapunov function, although its existence cannot be ascertained *a priori*. In spite of this difficulty, this approach has produced [14] the only available scheme that can stabilize the system (2.13) with $\gamma(x_1) = x_1^2$.

4 Pure-Feedback Systems

In order to remove some of the limitations of the existing adaptive nonlinear schemes, we need to examine whether they are structural, that is, caused by the nature of the problem, or methodological, that is, caused by the design approach. Within the framework of feedback linearization, a quick qualitative examination is now made using the following "benchmark" examples with nonlinearities which do not satisfy linear growth constraints:

nonlinear complexity ↓

$$\begin{aligned} \dot{z}_1 &= z_2 \\ \dot{z}_2 &= z_3 + \theta z_1^2 \\ \dot{z}_3 &= u \end{aligned} \qquad \begin{aligned} \dot{z}_1 &= z_2 + \theta z_1^2 \\ \dot{z}_2 &= z_3 \\ \dot{z}_3 &= u \end{aligned}$$

$$\begin{aligned} \dot{z}_1 &= z_2 \\ \dot{z}_2 &= z_3 + \theta z_2^2 \\ \dot{z}_3 &= u \end{aligned} \qquad \begin{aligned} \dot{z}_1 &= z_2 + \theta z_2^2 \\ \dot{z}_2 &= z_3 \\ \dot{z}_3 &= u \end{aligned}$$

$$\begin{aligned} \dot{z}_1 &= z_2 \\ \dot{z}_2 &= z_3 + \theta z_3^2 \\ \dot{z}_3 &= u \end{aligned} \qquad \begin{aligned} \dot{z}_1 &= z_2 + \theta z_3^2 \\ \dot{z}_2 &= z_3 \\ \dot{z}_3 &= u\,. \end{aligned}$$

uncertainty level →

Fig. 1. Systems with different levels of uncertainty and nonlinear complexity.

The two columns in Fig. 1 differ in their *uncertainty level.* In the left column the unknown parameter θ is separated from the control u by only one integrator and in the right column by two integrators. Hence, the uncertainty levels of these systems are one and two, respectively. The level of uncertainty and, hence, the design difficulties increase from the left to the right, because the systems on the left satisfy the EMC and those on the right don't.

The design difficulties also increase going from top to bottom, even when the parameter θ is known. The *nonlinear complexity* is greater when the nonlinearity multiplying the unknown parameter depends on state variables which are "fed forward". An examination of the systems in the right column reveals that with z_1^2 feedback linearization is global, with z_2^2 it is only local, and with z_3^2, which is fed forward over one integrator, feedback linearization is impossible.

These benchmark examples motivate our interest in *pure-feedback systems,* defined as systems of the form

$$\dot{\zeta} = f_0(\zeta) + \sum_{i=1}^{p} \theta_i f_i(\zeta) + \left[g_0(\zeta) + \sum_{i=1}^{p} \theta_i g_i(\zeta) \right] u\,, \tag{4.1}$$

which can be transformed via a parameter-independent diffeomorphism $z = \phi(\zeta)$ into the *pure-feedback form*:

$$\begin{aligned} \dot{z}_1 &= z_2 + \theta^{\mathrm{T}} \gamma_1(z_1, z_2) \\ \dot{z}_2 &= z_3 + \theta^{\mathrm{T}} \gamma_2(z_1, z_2, z_3) \\ &\ \vdots \\ \dot{z}_{n-1} &= z_n + \theta^{\mathrm{T}} \gamma_{n-1}(z_1, \ldots, z_n) \\ \dot{z}_n &= \gamma_0(z) + \theta^{\mathrm{T}} \gamma_n(z) + \left[\beta_0(z) + \theta^{\mathrm{T}} \beta(z)\right] u\,, \end{aligned} \tag{4.2}$$

with

$$\gamma_0(0) = 0\,, \gamma_1(0) = \cdots = \gamma_n(0) = 0\,,\ \beta_0(0) \neq 0\,. \tag{4.3}$$

Necessary and sufficient conditions for the existence of such a diffeomorphism are given in the following proposition.

Proposition 4.1 *A diffeomorphism $z = \phi(\zeta)$, with $\phi(0) = 0$, transforming (4.1) into (4.2), exists in a neighborhood $B_z \subset U$ of the origin if and only if the following conditions are satisfied in U:*

(i) Feedback linearization condition. *The distributions*

$$\mathcal{G}^i = \operatorname{span}\left\{g_0, \operatorname{ad}_{f_0} g_0, \ldots, \operatorname{ad}_{f_0}^i g_0\right\}, \quad 0 \leq i \leq n-1 \tag{4.4}$$

are involutive and of constant rank $i+1$.

(ii) Pure-feedback condition.

$$\left.\begin{aligned} & g_i \in \mathcal{G}^0\,, \\ & [X, f_i] \in \mathcal{G}^{j+1}\,,\ \forall X \in \mathcal{G}^j\,, \quad 0 \leq j \leq n-2\,, \end{aligned}\right\} \quad 1 \leq i \leq p\,. \tag{4.5}$$

Proof. Sufficiency. As proved in [20], condition (i) is sufficient for the existence of a diffeomorphism $z = \phi(\zeta)$ that transforms the system

$$\dot{\zeta} = f_0(\zeta) + g_0(\zeta) u\,,\ f_0(0) = 0\,,\ g_0(0) \neq 0 \tag{4.6}$$

into the system

$$\begin{aligned} \dot{z}_i &= z_{i+1}\,,\quad 1 \leq i \leq n-1 \\ \dot{z}_n &= \gamma_0(z) + \beta_0(z) u\,, \end{aligned} \tag{4.7}$$

with

$$\gamma_0(0) = 0\,,\ \beta_0(0) \neq 0\,. \tag{4.8}$$

Hence, in the coordinates of (4.7) we have

$$f_0(\phi^{-1}(z)) = [z_2 \cdots z_n\, \gamma_0(z)]^{\mathrm{T}} \tag{4.9}$$

$$g_0(\phi^{-1}(z)) = [0 \cdots 0\, \beta_0(z)]^{\mathrm{T}} \tag{4.10}$$

$$\mathcal{G}^i = \operatorname{span}\left\{\frac{\partial}{\partial z_n}, \ldots, \frac{\partial}{\partial z_{n-i}}\right\}, \quad 0 \le i \le n-1\,. \tag{4.11}$$

Because of (4.11), the pure-feedback condition (4.5), expressed in the z-coordinates, states that

$$\left.\begin{aligned} g_i &\in \operatorname{span}\left\{\frac{\partial}{\partial z_n}\right\}, \\ \left[\frac{\partial}{\partial z_j}, f_i\right] &\in \operatorname{span}\left\{\frac{\partial}{\partial z_n}, \ldots, \frac{\partial}{\partial z_{j-1}}\right\}, \quad 2 \le j \le n\,, \end{aligned}\right. \quad 1 \le i \le p\,. \tag{4.12}$$

But (4.12) can be equivalently rewritten as

$$g_i(\phi^{-1}(z)) = \begin{pmatrix} 0 \\ \vdots \\ 0 \\ \beta_i(z) \end{pmatrix}, \quad f_i(\phi^{-1}(z)) = \begin{pmatrix} \gamma_1(z_1) \\ \gamma_2(z_1, z_2) \\ \vdots \\ \gamma_{n-1,i}(z_1, \ldots, z_n) \\ \gamma_{n,i}(z_1, \ldots, z_n) \end{pmatrix}, \quad 1 \le i \le p\,. \tag{4.13}$$

Furthermore, since $\phi(0) = 0$ and $f_i(0) = 0, 1 \le i \le p$, we conclude from (4.13) that

$$\gamma_1(0) = \cdots = \gamma_n(0) = 0\,. \tag{4.14}$$

Combining (4.9), (4.10), (4.13) and (4.14), we see that in the z-coordinates the system (4.1) becomes (4.2).

Necessity. If there exists a diffeomorphism $z = \phi(\zeta)$ that transforms (4.1) into (4.2), one can directly verify that the coordinate-free conditions (i) and (ii) are satisfied for the system (4.2), and hence for the system (4.1). □

Remark 4.2. The EMC (2.1) is a special case of the "pure-feedback" condition (4.5). This is clear because if the system (4.1) satisfies the feedback linearization condition (4.4) and (2.1), then it is transformable into the pure-feedback form (4.2) with $\gamma_1 \equiv 0, \ldots, \gamma_{n-2} \equiv 0$. □

5 Adaptive Scheme Design

Since the diffeomorphism $z = \phi(\zeta)$ does not depend on the unknown parameter vector θ, Proposition 4.1 gives an *a priori verifiable* characterization of the class of nonlinear systems to which the new adaptive scheme is applicable. Assuming that the transformation of (4.1) into (4.2) has been performed, the new adaptive scheme is designed for the pure-feedback system:

$$\begin{aligned} \dot{z}_i &= z_{i+1} + \theta^{\mathrm{T}}\gamma_i(z_1,\ldots,z_{i+1}), \quad 1 \le i \le n-1 \\ \dot{z}_n &= \gamma_0(z) + \theta^{\mathrm{T}}\gamma_n(z) + \left[\beta_0(z) + \theta^{\mathrm{T}}\beta(z)\right] u, \end{aligned} \tag{5.1}$$

with

$$\gamma_0(0) = 0, \gamma_1(0) = \cdots = \gamma_n(0) = 0, \ \beta_0(0) \neq 0. \tag{5.2}$$

Recall that γ_0, β_0, and the components of β and γ_i, $1 \le i \le n$, are smooth nonlinear functions in B_z, a neighborhood of the origin $z = 0$.

The following step-by-step procedure interlaces, at each step, a change of coordinates with the construction of a parameter update law:

Step 0: Define $x_1 = z_1$, and denote by $c_1, c_2, \ldots, c_n$ constant coefficients to be chosen later.

Step 1: Starting with

$$\dot{x}_1 = z_2 + \theta^{\mathrm{T}}\gamma_1(z_1, z_2), \tag{5.3}$$

let ϑ_1 be an estimate of θ and define the new state x_2 as

$$x_2 = c_1 x_1 + z_2 + \vartheta_1^{\mathrm{T}}\gamma_1(z_1, z_2). \tag{5.4}$$

Substitute (5.4) into (5.3) to obtain

$$\begin{aligned} \dot{x}_1 &= -c_1 x_1 + x_2 + (\theta - \vartheta_1)^{\mathrm{T}}\gamma_1(z_1, z_2) \\ &= -c_1 x_1 + x_2 + (\theta - \vartheta_1)^{\mathrm{T}} w_1(x_1, x_2, \vartheta_1). \end{aligned} \tag{5.5}$$

Then, let the update law for the parameter estimate ϑ_1 be

$$\dot{\vartheta}_1 = x_1\, w_1(x_1, x_2, \vartheta_1). \tag{5.6}$$

Step 2: Using the definitions for x_1, x_2 and $\dot{\vartheta}_1$, write $\dot{x}_2$ as

$$\begin{aligned} \dot{x}_2 &= c_1[-c_1 x_1 + x_2 + (\theta - \vartheta_1)^{\mathrm{T}} w_1(x_1, x_2, \vartheta_1)] + z_3 + \theta^{\mathrm{T}}\gamma_2(z_1, z_2, z_3) \\ &\quad + x_1 w_1(x_1, x_2, \vartheta_1)^{\mathrm{T}}\gamma_1(z_1, z_2) + \vartheta_1^{\mathrm{T}}\left[\frac{\partial\gamma_1}{\partial z_1}(z_2 + \theta^{\mathrm{T}}\gamma_1) + \frac{\partial\gamma_1}{\partial z_2}(z_3 + \theta^{\mathrm{T}}\gamma_2)\right] \\ &= \left(1 + \vartheta_1^{\mathrm{T}}\frac{\partial\gamma_1}{\partial z_2}\right)[z_3 + \theta^{\mathrm{T}}\gamma_2(z_1, z_2, z_3)] + \varphi_2(x_1, x_2, \vartheta_1) + \theta^{\mathrm{T}}\psi_2(x_1, x_2, \vartheta_1). \end{aligned} \tag{5.7}$$

Let ϑ_2 be a *new* estimate of θ and define the new state x_3 as

$$\begin{aligned} x_3 &= c_2 x_2 + \left(1 + \vartheta_1^{\mathrm{T}}\frac{\partial\gamma_1}{\partial z_2}\right)[z_3 + \vartheta_2^{\mathrm{T}}\gamma_2(z_1, z_2, z_3)] \\ &\quad + \varphi_2(x_1, x_2, \vartheta_1) + \vartheta_2^{\mathrm{T}}\psi_2(x_1, x_2, \vartheta_1). \end{aligned} \tag{5.8}$$

Substitute (5.8) into (5.7) to obtain

$$\begin{aligned}\dot{x}_2 &= -c_2 x_2 + x_3 \\ &\quad + (\theta - \vartheta_2)^{\mathrm{T}} \left[\psi_2(x_1, x_2, \vartheta_1) + \left(1 + \vartheta_1^{\mathrm{T}} \frac{\partial \gamma_1(z_1, z_2)}{\partial z_2}\right) \gamma_2(z_1, z_2, z_3) \right] \\ &= -c_2 x_2 + x_3 + (\theta - \vartheta_2)^{\mathrm{T}} w_2(x_1, x_2, x_3, \vartheta_1, \vartheta_2) . \end{aligned} \tag{5.9}$$

Then, let the update law for the new estimate ϑ_2 be

$$\dot{\vartheta}_2 = x_2\, w_2(x_1, x_2, x_3, \vartheta_1, \vartheta_2) . \tag{5.10}$$

Step i ($2 \le i \le n-1$): Using the definitions for $x_1, \ldots, x_i$ and $\dot{\vartheta}_1, \ldots, \dot{\vartheta}_{i-1}$, express the derivative of x_i as

$$\begin{aligned}\dot{x}_i &= \left(1 + \vartheta_1^{\mathrm{T}} \frac{\partial \gamma_1}{\partial z_2}\right) \cdots \left(1 + \vartheta_{i-1}^{\mathrm{T}} \frac{\partial \gamma_{i-1}}{\partial z_i}\right) \left[z_{i+1} + \theta^{\mathrm{T}} \gamma_i(z_1, \ldots, z_{i+1})\right] \\ &\quad + \varphi_i(x_1, \ldots, x_i, \vartheta_1, \ldots, \vartheta_{i-1}) + \theta^{\mathrm{T}} \psi_i(x_1, \ldots, x_i, \vartheta_1, \ldots, \vartheta_{i-1}) . \end{aligned} \tag{5.11}$$

Let ϑ_i be a *new* estimate of θ and define the new state x_{i+1} as

$$\begin{aligned} x_{i+1} &= c_i x_i + \left(1 + \vartheta_1^{\mathrm{T}} \frac{\partial \gamma_1}{\partial z_2}\right) \cdots \left(1 + \vartheta_{i-1}^{\mathrm{T}} \frac{\partial \gamma_{i-1}}{\partial z_i}\right) \left[z_{i+1} + \vartheta_i^{\mathrm{T}} \gamma_i(z_1, \ldots, z_{i+1})\right] \\ &\quad + \varphi_i(x_1, \ldots, x_i, \vartheta_1, \ldots, \vartheta_{i-1}) + \vartheta_i^{\mathrm{T}} \psi_i(x_1, \ldots, x_i, \vartheta_1, \ldots, \vartheta_{i-1}) . \end{aligned} \tag{5.12}$$

Substitute (5.12) into (5.11) to obtain

$$\begin{aligned}\dot{x}_i &= -c_i x_i + x_{i+1} + (\theta - \vartheta_i)^{\mathrm{T}} \left[\psi_i + \left(1 + \vartheta_1^{\mathrm{T}} \frac{\partial \gamma_1}{\partial z_2}\right) \cdots \left(1 + \vartheta_{i-1}^{\mathrm{T}} \frac{\partial \gamma_{i-1}}{\partial z_i}\right) \gamma_i \right] \\ &= -c_i x_i + x_{i+1} + (\theta - \vartheta_i)^{\mathrm{T}} w_i(x_1, \ldots, x_{i+1}, \vartheta_1, \ldots, \vartheta_i) . \end{aligned} \tag{5.13}$$

Then, let the update law for ϑ_i be

$$\dot{\vartheta}_i = x_i\, w_i(x_1, \ldots, x_{i+1}, \vartheta_1, \ldots, \vartheta_i) . \tag{5.14}$$

Step n: Using the definitions for $x_1, \ldots, x_n$ and $\dot{\vartheta}_1, \ldots, \dot{\vartheta}_{n-1}$, express the derivative of x_n as

$$\begin{aligned}\dot{x}_n &= \left(1 + \vartheta_1^{\mathrm{T}} \frac{\partial \gamma_1}{\partial z_2}\right) \cdots \left(1 + \vartheta_{n-1}^{\mathrm{T}} \frac{\partial \gamma_{n-1}}{\partial z_n}\right) \left[\beta_0(z) + \theta^{\mathrm{T}} \beta(z)\right] u \\ &\quad + \varphi_n(x, \vartheta_1, \ldots, \vartheta_{n-1}) + \theta^{\mathrm{T}} \psi_n(x, \vartheta_1, \ldots, \vartheta_{n-1}) . \end{aligned} \tag{5.15}$$

Let ϑ_n be a *new* estimate of θ and define the control u as

$$u = \frac{1}{\bar{\beta}(z, \vartheta_1, \ldots, \vartheta_n)} \left[-c_n x_n - \varphi_n - \vartheta_n^{\mathrm{T}} \psi_n\right] , \tag{5.16}$$

where

$$\bar{\beta}(z, \vartheta_1, \ldots, \vartheta_n) = \left(1 + \vartheta_1^{\mathrm{T}} \frac{\partial \gamma_1}{\partial z_2}\right) \cdots \left(1 + \vartheta_{n-1}^{\mathrm{T}} \frac{\partial \gamma_{n-1}}{\partial z_n}\right) \left[\beta_0(z) + \vartheta_n^{\mathrm{T}} \beta(z)\right] . \tag{5.17}$$

Substitute (5.16) into (5.15) to obtain

$$\begin{aligned}\dot{x}_n &= -c_n x_n + (\theta - \vartheta_n)^{\mathrm{T}} \left[\psi_n + \left(1 + \vartheta_1^{\mathrm{T}} \frac{\partial \gamma_1}{\partial z_2}\right) \cdots \left(1 + \vartheta_{n-1}^{\mathrm{T}} \frac{\partial \gamma_{n-1}}{\partial z_n}\right) \beta(z) u\right] \\ &= -c_n x_n + (\theta - \vartheta_n)^{\mathrm{T}} w_n(x, \vartheta_1, \ldots, \vartheta_n),\end{aligned} \tag{5.18}$$

where (5.16) is used in the definition of w_n. Finally, let the update law for the estimate ϑ_n be

$$\dot{\vartheta}_n = x_n\, w_n(x, \vartheta_1, \ldots, \vartheta_n). \tag{5.19}$$

Feasibility of this design procedure and the stability of the resulting closed-loop adaptive system are analyzed in the next section.

6 Feasibility and Stability

The above design procedure has introduced a control law defined by (5.16)–(5.17) and a set of new coordinates $x_1, \ldots, x_n$ defined by (5.12). In order to ensure that the procedure is *feasible*, we construct in Proposition 6.1 an estimate $\mathcal{F} \subset \mathrm{IR}^{n(1+p)}$ of the *feasibility region* such that for all $(z, \vartheta_1, \ldots, \vartheta_n) \in \mathcal{F}$ the denominator in (5.16) is nonzero and the coordinate change (5.12) is one-to-one, onto, continous and has a continuous inverse.

Proposition 6.1 *Suppose the pure-feedback form (5.1) of the system (4.1) exists in B_z, and let $B_\vartheta \subset \mathrm{IR}^p$ be an open set such that*

$$\left|1 + \vartheta_i^{\mathrm{T}} \frac{\partial \gamma_i(z)}{\partial z_{i+1}}\right| > 0\,,\ \forall z \in B_z\,,\ \forall \vartheta_i \in B_\vartheta\,,\ 1 \le i \le n-1 \tag{6.1}$$

$$\left|\beta_0(z) + \vartheta_n^{\mathrm{T}} \beta(z)\right| > 0\,,\ \forall z \in B_z\,,\ \forall \vartheta_n \in B_\vartheta\,. \tag{6.2}$$

Then, the set $\mathcal{F} = B_z \times B_\vartheta^n$ is a subset of the region in which the design procedure of Section 5 is feasible.

Proof. Obvious, since (6.1) and (6.2) guarantee that in $B_z \times B_\vartheta^n$ the denominator in (5.16) is nonzero and (5.12) is *uniquely* solvable for z_i. □

Remark 6.2. In general, the feasibility region is not global. However, this is not due to the adaptive scheme, because, even when the parameters θ are known, the feedback linearization of the system (5.1) can only be guaranteed for $\theta \in B_\theta \subset \mathrm{IR}^p$, an open set such that

$$\left|1 + \theta^{\mathrm{T}} \frac{\partial \gamma_i(z)}{\partial z_{i+1}}\right| > 0\,,\ \forall z \in B_z\,,\ \forall \theta \in B_\theta\,,\ 1 \le i \le n-1 \tag{6.3}$$

$$\left|\beta_0(z) + \theta^{\mathrm{T}} \beta(z)\right| > 0\,,\ \forall z \in B_z\,,\ \forall \theta \in B_\theta\,. \tag{6.4}$$

□

In the feasibility region, the adaptive system resulting from the design procedure can be expressed in the x-coordinates as

$$\begin{aligned} \dot{x}_1 &= -c_1 x_1 + x_2 + (\theta - \vartheta_1)^{\mathrm{T}} w_1(x_1, x_2, \vartheta_1) \\ &\vdots \\ \dot{x}_{n-1} &= -c_{n-1} x_{n-1} + x_n + (\theta - \vartheta_{n-1})^{\mathrm{T}} w_{n-1}(x_1, \ldots, x_n, \vartheta_1, \ldots, \vartheta_{n-1}) \\ \dot{x}_n &= -c_n x_n + (\theta - \vartheta_n)^{\mathrm{T}} w_n(x, \vartheta_1, \ldots, \vartheta_n) \\ \dot{\vartheta}_i &= x_i\, w_i(x_1, \ldots, x_{i+1}, \vartheta_1, \ldots, \vartheta_i)\,, \quad 1 \leq i \leq n\,. \end{aligned} \tag{6.5}$$

A nice property of this system is that its stability can be established using the quadratic Lyapunov function

$$V(x, \vartheta_1, \ldots, \vartheta_n) = \frac{1}{2} x^{\mathrm{T}} x + \frac{1}{2} \sum_{i=1}^{n} (\theta - \vartheta_i)^{\mathrm{T}} (\theta - \vartheta_i)\,. \tag{6.6}$$

The derivative of $V(x, \vartheta_1, \ldots, \vartheta_n)$ along the solutions of (6.5) is

$$\begin{aligned} \dot{V} &= -\sum_{i=1}^{n} \left[c_i x_i^2 + (\theta - \vartheta_i)^{\mathrm{T}} (x_i w_i - \dot{\vartheta}_i) \right] + \sum_{i=1}^{n-1} x_i x_{i+1} \\ &= -\sum_{i=1}^{n} c_i x_i^2 + \sum_{i=1}^{n-1} x_i x_{i+1}\,. \end{aligned} \tag{6.7}$$

At this point we can choose the coefficients $c_1, \ldots, c_n$ to guarantee that $\dot{V}$ is negative semidefinite. The choice $c_i \geq 2$, for all $i = 1, \ldots, n$, yields:

$$\dot{V} \leq -\|x\|^2\,. \tag{6.8}$$

This proves the uniform stability of the equilibrium

$$x = 0\,, \; \vartheta_1 = \theta\,, \ldots, \; \vartheta_n = \theta \tag{6.9}$$

of the adaptive system (6.5). To give an estimate Ω of the region of attraction of this equilibrium, we note that Ω must be a subset of our estimate $\mathcal{F}$ of the feasibility region. Let $\Omega(c)$ be the invariant set of (6.5) defined by $V < c$, and let c^* be the largest constant c such that $\Omega(c) \subset \mathcal{F}$. Then an estimate Ω of the region of attraction is

$$\Omega = \Omega(c^*) = \{(x, \vartheta_1, \ldots, \vartheta_n) : V(x, \vartheta_1, \ldots, \vartheta_n) < c^*\}\,, \; c^* = \arg \sup_{\Omega(c) \subset \mathcal{F}} \{c\}\,. \tag{6.10}$$

Remark 6.3. It can be expected that the above estimate is not tight because the choice of the unity gains in the update laws was made for simplicity. With some *a priori* knowledge about the shape of $\mathcal{F}$, different adaptation gains can be found so that Ω is maximized by a better fit of $\mathcal{F}$. □

Next, using the invariance theorem of LaSalle, we establish that for all initial conditions $(x, \vartheta_1, \ldots, \vartheta_n)_{t=0} \in \Omega$, the adaptive system (6.5) has the following regulation properties:

$$\lim_{t\to\infty} x(t) = 0\,, \quad \lim_{t\to\infty} \dot{x}(t) = 0\,, \quad \lim_{t\to\infty} \dot{\vartheta}_i(t) = 0\,,\ 1 \le i \le n\,. \tag{6.11}$$

In order to return to the original coordinates ζ, we note that (6.1) guarantees, first, that the solution $z_2 = \cdots = z_n = 0$ of the system of equations

$$z_{i+1} + \theta^{\mathrm{T}}\gamma_i(0, z_2, \ldots, z_{i+1}) = 0\,, \quad 1 \le i \le n-1\,, \tag{6.12}$$

is *unique* in $B_z \times B_\vartheta$, and, second, that $z_1, \ldots, z_n$ can be expressed as smooth functions of $x, \vartheta_1, \ldots, \vartheta_n$ using (5.12). Combining these two facts with (6.11), we obtain

$$\lim_{t\to\infty} z_1(t) = 0\,, \quad \lim_{t\to\infty} \dot{z}_i(t) = 0\,, \quad 1 \le i \le n\,. \tag{6.13}$$

Using an induction argument, it is now shown that $z_i(t) \to 0$ as $t \to \infty$, $1 \le i \le n$:

- For $i = 1$, we have $z_1(t) \to 0$ as $t \to \infty$.
- For $i = k$, $2 \le k \le n$, we assume that $z_j(t) \to 0$ as $t \to \infty$, $1 \le j \le k-1$. Then, from (6.13) we have

$$\lim_{t\to\infty} \dot{z}_{k-1}(t) = \lim_{t\to\infty} \left\{ z_k + \theta^{\mathrm{T}}\gamma_{k-1}(z_1, \ldots, z_{k-1}, z_k) \right\} = 0\,, \tag{6.14}$$

  and from the uniqueness of solutions of (6.12) we conclude that $z_k(t) \to 0$ as $t \to \infty$.

Hence, $z(t) \to 0$ as $t \to \infty$. Finally, since $z = \phi(\zeta)$ is a diffeomorphism with $\phi(0) = 0$, regulation is achieved in the original coordinates ζ, namely

$$\lim_{t\to\infty} \zeta(t) = 0\,. \tag{6.15}$$

The above facts prove the following result:

Theorem 6.4 *Suppose that the system (4.1) satisfies Proposition 4.1 and that the design procedure of Section 5 is applied to its pure-feedback form (5.1). Then the equilibrium (6.9) of the resulting adaptive system (6.5) is uniformly stable and its region of attraction includes the set Ω defined in (6.10). Furthermore, regulation of the state $\zeta(t)$ is achieved for all initial conditions in Ω.* □

7 Multi-input Systems

The design procedure of Section 5 can be easily extended to multi-input nonlinear systems of the form

$$\dot{\zeta} = f_0(\zeta) + \sum_{i=1}^{p} \theta_i f_i(\zeta) + \sum_{j=1}^{m} \left[g_0^j(\zeta) + \sum_{i=1}^{p} \theta_i g_i^j(\zeta) \right] u_j \,, \tag{7.1}$$

with

$$f_0(0) = f_1(0) = \cdots = f_p(0) = 0 \,, \ \operatorname{rank} G_0(0) = m \,, \ G_0 = [g_0^1 \cdots g_0^m] \,, \tag{7.2}$$

that can be transformed into

$$\begin{aligned} \dot{z}_i^j &= z_{i+1}^j + \theta^{\mathrm{T}} \gamma_i^j \left(z_1^1, \ldots, z_{k_1-k_j+2}^1, \ldots, z_1^m, \ldots, z_{k_m-k_j+2}^m \right) \\ & \qquad 1 \le i \le k_j - 1 \,, \ 1 \le j \le m \\ \dot{z}_{k_j}^j &= \gamma_0^j(z) + \theta^{\mathrm{T}} \gamma_{k_j}^j(z) + \left[\beta_0^j(z) + \sum_{\ell=1}^{p} \theta_\ell \beta_\ell^j(z) \right]^{\mathrm{T}} u \,, \ 1 \le j \le m \,, \end{aligned} \tag{7.3}$$

with

$$\gamma_\ell^i(0) = 0 \,, \ 0 \le i \le k_j \,, \ 1 \le j \le m \,, \ \det B_0(0) \ne 0 \,, \tag{7.4}$$

where $B_0 = [\beta_0^1, \ldots, \beta_0^m]^{\mathrm{T}}$, and $\sum_{j=1}^{m} k_j = n$.

Proposition 7.1 *A parameter-independent diffeomorphism $z = \phi(\zeta)$, with $\phi(0) = 0$, transforming (7.1) into (7.3), exists in a neighborhood $B_z \subset U$ of the origin if and only if the following conditions are satisfied in U:*

(i) Feedback linearization condition. *The distributions*

$$\mathcal{G}^i = \operatorname{span} \left\{ g_0^j, \operatorname{ad}_{f_0} g_0^j, \ldots, \operatorname{ad}_{f_0}^i g_0^j, \ 1 \le j \le m \right\}, \quad 0 \le i \le n-1 \tag{7.5}$$

are involutive and of constant rank r_i, with $r_{n-1} = n$.

(ii) Pure-feedback condition.

$$\left. \begin{aligned} & g_i^j \in \mathcal{G}^0 \,, \quad 1 \le j \le m \,, \\ & [X, f_i] \in \mathcal{G}^{k+1} \,, \ \forall X \in \mathcal{G}^k \,, \ 0 \le k \le n-2 \,, \end{aligned} \right\} \quad 1 \le i \le p \,. \tag{7.6}$$

Proof. As proved in [24,25], condition (i) is necessary and sufficient for the existence of a diffeomorphism $z = \phi(\zeta)$ such that in the z-coordinates we have

$$f_0(\phi^{-1}(z)) = \left[z_1^1 \cdots z_{k_1-1}^1 \, \gamma_0^1(z) \cdots z_1^m \cdots z_{k_m-1}^m \, \gamma_0^m(z) \right]^{\mathrm{T}} \tag{7.7}$$

$$G_0(\phi^{-1}(z)) = \left[0 \cdots 0 \, \beta_0^1(z) \cdots 0 \cdots 0 \, \beta_0^m(z) \right]^{\mathrm{T}} \tag{7.8}$$

$$\mathcal{G}^i = \operatorname{span} \left\{ \frac{\partial}{\partial z_{k_j}}, \ldots, \frac{\partial}{\partial z_{k_j - i}}, 1 \le j \le m \right\}, \ 0 \le i \le n-1 \,. \tag{7.9}$$

It is now a tedious but straightforward task to verify that condition (ii) is equivalent to

$$g_i^j(\phi^{-1}(z)) = \left[0 \cdots 0\, \beta_{j,i}^1(z) \cdots 0 \cdots 0\, \beta_{j,i}^m(z)\right]^{\mathrm{T}}, \quad 1 \le i \le p, \quad 1 \le j \le m \tag{7.10}$$

$$f_i(\phi^{-1}(z)) = \begin{bmatrix} \gamma_{1,i}^1(z_1^1, z_2^1, \ldots, z_1^m, \ldots, z_{k_m-k_1+2}^m) \\ \vdots \\ \gamma_{k_1,i}^1(z) \\ \vdots \\ \gamma_{1,i}^m(z_1^1, \ldots, z_{k_1-k_m+2}^1, \ldots, z_1^m, z_2^m) \\ \vdots \\ \gamma_{k_m,i}^m(z) \end{bmatrix}, \quad 1 \le i \le p. \tag{7.11}$$

□

The design procedure for the system (7.3) is the following:

Steps 0 *through* $(n-m)$: Apply steps 0 through $(k_j - 1)$ of the single-input procedure to the first $(k_j - 1)$ equations of each of the m subsystems of (7.3), to obtain the system:

$$\begin{aligned} \dot{x}_i^j &= -c_i^j x_i^j + x_{i+1}^j + (\theta - \vartheta_\ell)^{\mathrm{T}} w_i^j(x, \vartheta_1, \ldots, \vartheta_{\ell-1}), \ c_i^j \ge 2, \\ &\qquad \ell = \sum_{\rho=1}^{j-1}(k_\rho - 1) + i, \quad 1 \le i \le k_j, \quad 1 \le j \le m \\ \dot{\vartheta}_\ell &= x_i^j\, w_i^j(x, \vartheta_1, \ldots, \vartheta_\ell), \quad 1 \le \ell \le n-m \end{aligned} \tag{7.12}$$

$$\frac{d}{dt}\begin{bmatrix} x_{k_1}^1 \\ \vdots \\ x_{k_m}^m \end{bmatrix} = \left[\bar{B}_0(z, \vartheta_1, \ldots, \vartheta_{n-m}) + \sum_{i=1}^p \bar{B}_i(z, \vartheta_1, \ldots, \vartheta_{n-m})\theta_i\right] u$$
$$+\Phi(x, \vartheta_1, \ldots, \vartheta_{n-m}) + W^{\mathrm{T}}(x, \vartheta_1, \ldots, \vartheta_{n-m})\theta\,,$$

where $\bar{B}_0, \bar{B}_1, \ldots, \bar{B}_p$ are defined as

$$\bar{B}_i(z, \vartheta_1, \ldots, \vartheta_{n-m}) =$$
$$= \begin{bmatrix} \left(1 + \vartheta_1^{\mathrm{T}} \dfrac{\partial \gamma_1^1}{\partial z_2^1}\right) \cdots \left(1 + \vartheta_{k_1-1}^{\mathrm{T}} \dfrac{\partial \gamma_{k_1-1}^1}{\partial z_{k_1}^1}\right) \beta_i^{1\,\mathrm{T}}(z) \\ \vdots \\ \left(1 + \vartheta_{n-m-k_m+2}^{\mathrm{T}} \dfrac{\partial \gamma_1^m}{\partial z_2^m}\right) \cdots \left(1 + \vartheta_{n-m}^{\mathrm{T}} \dfrac{\partial \gamma_{k_m-1}^m}{\partial z_{k_m}^m}\right) \beta_i^{m\,\mathrm{T}}(z) \end{bmatrix}. \tag{7.13}$$

Step $n-m+1$: Let ϑ_{n-m+1} be a new estimate of θ and define the control u as

$$u = \left[\bar{B}_0(z,\vartheta_1,\ldots,\vartheta_{n-m}) + \sum_{i=1}^{p} \bar{B}_i(z,\vartheta_1,\ldots,\vartheta_{n-m})\vartheta_{n-m+1,i}\right]^{-1}$$
$$\times\left\{-\left[c^1_{k_1}x^1_{k_1}\cdots c^m_{k_m}x^m_{k_m}\right]^{\mathrm{T}} - \Phi(x,\vartheta_1,\ldots,\vartheta_{n-m})\right.$$
$$\left.-W^{\mathrm{T}}(x,\vartheta_1,\ldots,\vartheta_{n-m})\vartheta_{n-m+1}\right\}, \quad c^j_{k_j} \geq 2,\ 1 \leq j \leq m. \tag{7.14}$$

Substitute (7.14) into (7.12) and rewrite the last m equations of (7.12) as

$$\frac{d}{dt}\begin{bmatrix} x^1_{k_1} \\ \vdots \\ x^m_{k_m} \end{bmatrix} = -\begin{bmatrix} c^1_{k_1}x^1_{k_1} \\ \vdots \\ c^m_{k_m}x^m_{k_m} \end{bmatrix} + \left\{W + \left[\bar{B}_1 u\cdots\bar{B}_p u\right]\right\}(\theta - \vartheta_{n-m+1})$$
$$= -\begin{bmatrix} c^1_{k_1}x^1_{k_1} \\ \vdots \\ c^m_{k_m}x^m_{k_m} \end{bmatrix} + W^{\mathrm{T}}_{n-m+1}(x,\vartheta_1,\ldots,\vartheta_{n-m+1})(\theta - \vartheta_{n-m+1}), \tag{7.15}$$

where (7.14) was used in the definition of W_{n-m+1}. Finally, let the update law for the estimate ϑ_{n-m+1} be

$$\dot{\vartheta}_{n-m+1} = W_{n-m+1}(x,\vartheta_1,\ldots,\vartheta_{n-m+1})\begin{bmatrix} x^1_{k_1} \\ \vdots \\ x^m_{k_m} \end{bmatrix}. \tag{7.16}$$

Note that this procedure is feasible only in the region in which the matrix $\bar{B} = \bar{B}_0 + \sum_{i=1}^{p} \bar{B}_i\vartheta_{n-m+1,i}$ is invertible. The stability properties of the resulting closed-loop system are analogous to those listed in Theorem 6.4, and can be similarly established using the Lyapunov function

$$V(x,\vartheta_1,\ldots,\vartheta_{n-m+1}) = \frac{1}{2}x^{\mathrm{T}}x + \frac{1}{2}\sum_{i=1}^{n-m+1}(\theta - \vartheta_i)^{\mathrm{T}}(\theta - \vartheta_i). \tag{7.17}$$

8 Global Regulation

There are strong theoretical and practical reasons for investigating whether the stability properties of an adaptive system can be made global in the space of the states and parameter estimates. Systems with a finite region of attraction may not possess a wide enough robustness margin for disturbances and unmodeled dynamics. Furthermore, it is usually hard to find nonconservative estimates of finite regions of attraction.

Another aspect of the global stability issue is the comparison of the proposed adaptive controller with its deterministic counterpart, that is, the controller that would be used if the parameter vector θ were known. Suppose that for all values of θ there exists a deterministic controller which achieves global stabilization and regulation of the system. If, with θ unknown, the proposed adaptive controller does not achieve the same global stability, this loss is clearly due to adaptation.

The stability result of Theorem 6.4 is not global. However, as pointed out in Remark 6.2, this is not due to adaptation, because for pure-feedback systems global stability may not be achievable even with θ known. In Proposition 8.3 we define the class of "strict-feedback" systems, for which a globally stabilizing controller exists when θ is known. We then prove that for this class of systems our adaptive scheme guarantees global stability when θ is unknown.

In order to characterize the class of "strict-feedback" systems, we use the following assumption about the part of the system (4.1) that does not contain unknown parameters:

Assumption 8.1 *There exists a* global *diffeomorphism* $z = \phi(\zeta)$, *with* $\phi(0) = 0$, *transforming the system*

$$\dot{\zeta} = f_0(\zeta) + g_0(\zeta)u\,, \tag{8.1}$$

into the system

$$\begin{aligned} \dot{z}_i &= z_{i+1}\;, 1 \le i \le n-1 \\ \dot{z}_n &= \gamma_0(z) + \beta_0(z)u\,, \end{aligned} \tag{8.2}$$

with

$$\gamma_0(0) = 0\,,\; \beta_0(z) \neq 0 \;\forall z \in \mathbb{R}^n\,. \tag{8.3}$$

□

Remark 8.2. The *local* existence of such a diffeomorphism is equivalent to the feedback linearization condition (4.4). At present there are no necessary and sufficient conditions verifying the *global* validity of this assumption. Sufficient conditions for Assumption 8.1 are given in [21], while necessary and sufficient conditions for the case where $\beta_0(z) \equiv$ const. can be found in [22,23]. □

Proposition 8.3 *Under Assumption 8.1, the system (4.1) is globally diffeomorphically equivalent to the "strict-feedback" system*

$$\begin{aligned} \dot{z}_i &= z_{i+1} + \theta^{\mathrm{T}}\gamma_i(z_1,\ldots,z_i)\,, \quad 1 \le i \le n-1 \\ \dot{z}_n &= \gamma_0(z) + \theta^{\mathrm{T}}\gamma_n(z) + \beta_0(z)u \end{aligned} \tag{8.4}$$

if and only if the following condition holds globally:

Strict-feedback condition.

$$\left.\begin{aligned} &g_i \equiv 0\,, \\ &[X, f_i] \in \mathcal{G}^j\,, \quad \forall X \in \mathcal{G}^j\,, \quad 0 \le j \le n-2\,, \end{aligned}\right\} \quad 1 \le i \le p\,, \tag{8.5}$$

with $\mathcal{G}^j, 0 \le j \le n-1$, *as defined in (4.4).*

Proof. The proof is very similar to that of Proposition 4.1. First, because of the assumptions that the diffeomorphism $z = \phi(\zeta)$ is global and $\beta_0(z) \neq 0 \ \forall z \in \mathbb{R}^n$, the distributions $\mathcal{G}^j$, $0 \leq j \leq n-1$, are globally defined and can be expressed in the z-coordinates as

$$\mathcal{G}^j = \mathrm{span}\left\{\frac{\partial}{\partial z_n}, \dots, \frac{\partial}{\partial z_{n-j}}\right\}, \quad 0 \leq j \leq n-1. \tag{8.6}$$

To prove sufficiency, note that if the pure-feedback condition (4.5) of Proposition 4.1 is replaced by the strict-feedback condition (8.5), then (4.12) is replaced by

$$\left.\begin{array}{l} g_i \equiv 0\,, \\ \left[\dfrac{\partial}{\partial z_j}, f_i\right] \in \mathrm{span}\left\{\dfrac{\partial}{\partial z_n}, \dots, \dfrac{\partial}{\partial z_j}\right\}, \quad 2 \leq j \leq n\,, \end{array}\right. \quad 1 \leq i \leq p. \tag{8.7}$$

Thus, the expression for $f_i(\phi^{-1}(z))$ in (4.13) becomes

$$f_i(\phi^{-1}(z)) = \begin{pmatrix} \gamma_{1,i}(z_1) \\ \gamma_{2,i}(z_1, z_2) \\ \vdots \\ \gamma_{n-1,i}(z_1, \dots, z_{n-1}) \\ \gamma_{n,i}(z_1, \dots, z_n) \end{pmatrix}, \quad 1 \leq i \leq p. \tag{8.8}$$

The necessity is again straightforward. □

The above proposition gives a geometric characterization of the class of systems for which the following global properties can be achieved.

Theorem 8.4 *Suppose that the system (4.1) satisfies Proposition 8.3 and that the design procedure of Section 5 is applied to its strict-feedback form (8.4). Then the equilibrium*

$$x = 0\,,\ \vartheta_1 = \theta\,, \dots,\ \vartheta_n = \theta$$

of the resulting adaptive system is globally uniformly stable. Furthermore, regulation of the state $\zeta(t)$ is achieved:

$$\lim_{t\to\infty} \zeta(t) = 0\,, \tag{8.9}$$

for all initial conditions in $\Omega = \mathbb{R}^{n(1+p)}$.

Proof. When the adaptive design procedure (5.3)–(5.19) is applied to the system (8.4), then for all $\vartheta_i \in \mathbb{R}^p$, $1 \leq i \leq n$, the change of coordinates (5.12) is one-to-one, onto, continuous and has a continuous inverse, and the control (5.16) is well-defined, since

$$\frac{\partial \gamma_i}{\partial z_{i+1}}(z) = 0\,, \quad \beta(z) = 0\,, \quad \beta_0(z) \neq 0 \quad \forall z \in \mathbb{R}^n. \tag{8.10}$$

Hence (6.1)–(6.2) are trivially satisfied on $\mathcal{F} = B_z \times B_\vartheta^n = \mathbb{R}^{n(1+p)}$, and from (6.10) we conclude that $\Omega = \mathbb{R}^{n(1+p)}$. □

9 Global Tracking

Every regulation result in Sections 4–6 has its tracking counterpart. For brevity, we restrict our presentation to the tracking version of the global regulation result in Section 8. The counterparts of nonglobal regulation results can be obtained using the same Lyapunov function argument as in this section to determine an invariant set in which asymptotic tracking is guaranteed.

Consider the nonlinear system

$$\begin{aligned} \dot{\zeta} &= f_0(\zeta) + sum_{i=1}^{p}\theta_i f_i(\zeta) + g_0(\zeta)u \\ y &= h(\zeta), \end{aligned} \tag{9.1}$$

where $\zeta \in \mathbb{R}^n$ is the state, $u \in \mathbb{R}$ is the input, $y \in \mathbb{R}$ is the output, $\theta = [\theta_1, \ldots, \theta_p]^{\mathrm{T}}$ is the vector of unknown constant parameters, h is a smooth function on $\mathbb{R}^n$ with $h(0) = 0$, and the vector fields g_0, f_i, $0 \leq i \leq p$, are smooth on $\mathbb{R}^n$ with $g(\zeta) \neq 0, \forall \zeta \in \mathbb{R}^n$, $f_i(0) = 0$, $0 \leq i \leq p$. We first formulate the input-output counterpart of Assumption 8.1:

Assumption 9.1 *There exist $n-\rho$ smooth functions $\phi_i(\zeta)$, $\rho+1 \leq i \leq n$, such that the change of coordinates*

$$\begin{aligned} z_1 &= h(\zeta) \\ z_2 &= L_{f_0}h(\zeta) \\ z_3 &= L_{f_0}^2 h(\zeta) \\ &\vdots \\ z_\rho &= L_{f_0}^{\rho-1}h(\zeta) \\ z_i &= \phi_i(\zeta), \quad \rho+1 \leq i \leq n \end{aligned} \tag{9.2}$$

is a global *diffeomorphism $z = \phi(\zeta)$ transforming the system*

$$\begin{aligned} \dot{\zeta} &= f_0(\zeta) + g_0(\zeta)u \\ y &= h(\zeta) \end{aligned} \tag{9.3}$$

into the special normal form

$$\begin{aligned} \dot{z}_1 &= z_2 \\ &\vdots \\ \dot{z}_{\rho-1} &= z_\rho \\ \dot{z}_\rho &= \gamma_0(z) + \beta_0(z)u \\ \dot{z}^{\mathrm{r}} &= \Phi_0(y, z^{\mathrm{r}}) \\ y &= z_1, \end{aligned} \tag{9.4}$$

with

$$\gamma_0(0) = L_{f_0}^{\rho}h(0) = 0, \; \Phi_0(0,0) = 0 \tag{9.5}$$

$$\beta_0(z) = L_{g_0}L_{f_0}^{\rho-1}h(\zeta) \neq 0 \; \forall z \in \mathbb{R}^n . \tag{9.6}$$

□

Remark 9.2. In order for (9.3) to be *locally* equivalent to (9.4), it is necessary and sufficient that the following conditions hold in a neighborhood of the origin $\zeta = 0$:

$$L_{g_0} L_{f_0}^i h \equiv 0 \,, \; 0 \leq i \leq \rho - 2 \tag{9.7}$$

$$L_{g_0} L_{f_0}^{\rho-1} h(0) \neq 0 \tag{9.8}$$

$$\mathcal{G}^i \text{ is involutive and of constant rank } i+1 \,, \quad 0 \leq i \leq \rho - 1 \,. \tag{9.9}$$

The sufficiency of these conditions is a consequence of Proposition 10 in [26]. The necessity can be easily established by verifying that (9.7)–(9.9) hold in the coordinates of (9.4). However, at present there are no necessary and sufficient conditions that can verify the *global* validity of this assumption. □

We are now ready to formulate the input-output counterpart of Proposition 8.3:

Proposition 9.3 *Under Assumption 9.1, the system (9.1) is globally diffeomorphically equivalent to the "strict-feedback" normal form*

$$\begin{aligned}
\dot{z}_i &= z_{i+1} + \theta^{\mathrm{T}} \gamma_i(z_1, \ldots, z_i, z^{\mathrm{r}}) \,, \quad 1 \leq i \leq \rho - 1 \\
\dot{z}_\rho &= \gamma_0(z) + \theta^{\mathrm{T}} \gamma_\rho(z) + \beta_0(z) u \\
\dot{z}^{\mathrm{r}} &= \Phi_0(y, z^{\mathrm{r}}) + \sum_{i=1}^{p} \theta_i \Phi_i(y, z^{\mathrm{r}}) \\
y &= z_1 \,,
\end{aligned} \tag{9.10}$$

if and only if the following condition holds globally:

Strict-feedback condition.

$$[X, f_i] \in \mathcal{G}^j \,, \quad \forall X \in \mathcal{G}^j, \quad 0 \leq j \leq \rho - 2 \,, \quad 1 \leq i \leq p, \tag{9.11}$$

with $\mathcal{G}^j, 0 \leq j \leq \rho - 1$, *as defined in (4.4).*

Proof. The proof follows closely that of Proposition 8.3. First, because of the assumptions that the diffeomorphism $z = \phi(\zeta)$ defined in (9.2) is global and that $\beta_0(z) \neq 0 \; \forall z \in \mathbb{R}^n$, the distributions $\mathcal{G}^j, 0 \leq j \leq \rho - 1$, are globally defined and can be expressed in the z-coordinates as

$$\mathcal{G}^j = \operatorname{span}\left\{ \frac{\partial}{\partial z_\rho}, \ldots, \frac{\partial}{\partial z_{\rho-j}} \right\}, \quad 0 \leq j \leq \rho - 1 \,. \tag{9.12}$$

The sufficiency follows from the fact that, by (9.11) and (9.12),

$$\left[\frac{\partial}{\partial z_j}, f_i \right] \in \operatorname{span}\left\{ \frac{\partial}{\partial z_\rho}, \ldots, \frac{\partial}{\partial z_j} \right\}, \quad 2 \leq j \leq \rho, \quad 1 \leq i \leq p. \tag{9.13}$$

Thus, the expression for $f_i(\phi^{-1}(z))$ is

$$f_i(\phi^{-1}(z)) = \begin{pmatrix} \gamma_{1,i}(z_1, z^{\mathrm{r}}) \\ \gamma_{2,i}(z_1, z_2, z^{\mathrm{r}}) \\ \vdots \\ \gamma_{\rho-1,i}(z_1, \ldots, z_{\rho-1}, z^{\mathrm{r}}) \\ \gamma_{\rho,i}(z_1, \ldots, z_\rho, z^{\mathrm{r}}) \\ \Phi_i(z_1, z^{\mathrm{r}}) \end{pmatrix}, \quad 1 \le i \le p. \tag{9.14}$$

The necessity is again straightforward. □

Remark 9.4. To obtain the input-output counterpart of Proposition 4.1, one just needs to replace the feedback linearization condition (4.4) with conditions (9.7)–(9.9), and the pure-feedback condition (4.5) with

$$\left. \begin{aligned} & g_i \in \mathcal{G}^0 , \\ & [X, f_i] \in \mathcal{G}^{j+1} , \ \forall X \in \mathcal{G}^j , \quad 0 \le j \le \rho - 2 , \end{aligned} \right. \qquad 1 \le i \le p . \tag{9.15}$$

□

As in most tracking problems, we need an assumption about the stability of the zero-dynamics of (9.10):

Assumption 9.5 *The z^{r}-subsystem of (9.10) has the bounded-input-bounded-state (BIBS) property with respect to y as its input.* □

It was shown in [15, Proposition 2.1] that the following conditions are sufficient for Assumption 9.5 to be satisfied:

(i) the zero dynamics of (9.1) are globally exponentially stable, and

(ii) the vector field $\Phi = \Phi_0 + \sum_{i=1}^{p} \theta_i \Phi_i$ in (9.10) is globally Lipschitz in z.

These conditions are more convenient for nonglobal results, where (i) can be used to estimate the region of attraction via a converse Lyapunov theorem. However, they are too restrictive for global results. For example, the system $\dot{z}^{\mathrm{r}} = -(z^{\mathrm{r}})^3 + y^2$ violates both (i) and (ii), but is easily seen to satisfy Assumption 9.5.

The control objective is to force the output y of the system (9.1) to *asymptotically track* a known reference signal $y_{\mathrm{r}}(t)$, while keeping all the closed-loop signals bounded.

Assumption 9.6 *The reference signal $y_{\mathrm{r}}(t)$ and its first ρ derivatives are known and bounded.* □

To achieve the asymptotic tracking objective, the design procedure presented in Section 5 is modified as follows:

Step 0: Define

$$x_1 = z_1 - y_\mathrm{r}\,, \tag{9.16}$$

and denote by $c_1, c_2, \ldots, c_\rho$ constant coefficients to be chosen later.

Step 1: Starting with

$$\dot{x}_1 = z_2 + \theta^\mathrm{T}\gamma_1(z_1, z^\mathrm{r}) - \dot{y}_\mathrm{r}\,, \tag{9.17}$$

let ϑ_1 be an estimate of θ and define the new state x_2 as

$$\begin{aligned} x_2 &= c_1x_1 + z_2 + \vartheta_1^\mathrm{T}\gamma_1(z_1, z^\mathrm{r}) - \dot{y}_\mathrm{r} \\ &= c_1x_1 + z_2 + \vartheta_1^\mathrm{T}w_1(x_1, z^\mathrm{r}, y_\mathrm{r}) - \dot{y}_\mathrm{r}\,. \end{aligned} \tag{9.18}$$

Substitute (9.18) into (9.17) to obtain

$$\dot{x}_1 = -c_1x_1 + x_2 + (\theta - \vartheta_1)^\mathrm{T}w_1(x_1, z^\mathrm{r}, y_\mathrm{r})\,. \tag{9.19}$$

Then, let the update law for the parameter estimate ϑ_1 be

$$\dot{\vartheta}_1 = x_1\, w_1(x_1, z^\mathrm{r}, y_\mathrm{r})\,. \tag{9.20}$$

Step 2: Using the definitions for x_1, x_2 and $\dot{\vartheta}_1$, write $\dot{x}_2$ as

$$\begin{aligned} \dot{x}_2 &= c_1[-c_1x_1 + x_2 + (\theta - \vartheta_1)^\mathrm{T}w_1(x_1, z^\mathrm{r}, y_\mathrm{r})] + z_3 + \theta^\mathrm{T}\gamma_2(z_1, z_2, z^\mathrm{r}) \\ &\quad + x_1w_1(x_1, z^\mathrm{r}, y_\mathrm{r})^\mathrm{T}\gamma_1(z_1, z^\mathrm{r}) + \vartheta_1^\mathrm{T}\left[\frac{\partial\gamma_1(z_1, z^\mathrm{r})}{\partial z_1}\left(z_2 + \theta^\mathrm{T}\gamma_1(z_1, z^\mathrm{r})\right)\right. \\ &\quad \left. + \frac{\partial\gamma_1(z_1, z^\mathrm{r})}{\partial z^\mathrm{r}}\left(\Phi_0(z_1, z^\mathrm{r}) + \sum_{i=1}^{p}\theta_i\Phi_i(z_1, z^\mathrm{r})\right)\right] - \ddot{y}_\mathrm{r} \\ &= z_3 + \varphi_2(x_1, x_2, z^\mathrm{r}, \vartheta_1, y_\mathrm{r}, \dot{y}_\mathrm{r}, \ddot{y}_\mathrm{r}) + \theta^\mathrm{T}w_2(x_1, x_2, z^\mathrm{r}, \vartheta_1, y_\mathrm{r}, \dot{y}_\mathrm{r})\,. \end{aligned} \tag{9.21}$$

Let ϑ_2 be a *new* estimate of θ and define the new state x_3 as

$$x_3 = c_2x_2 + z_3 + \varphi_2(x_1, x_2, z^\mathrm{r}, \vartheta_1, y_\mathrm{r}, \dot{y}_\mathrm{r}, \ddot{y}_\mathrm{r}) + \vartheta_2^\mathrm{T}w_2(x_1, x_2, z^\mathrm{r}, \vartheta_1, y_\mathrm{r}, \dot{y}_\mathrm{r})\,. \tag{9.22}$$

Substitute (9.22) into (9.21) to obtain

$$\dot{x}_2 = -c_2x_2 + x_3 + (\theta - \vartheta_2)^\mathrm{T}w_2(x_1, x_2, z^\mathrm{r}, \vartheta_1, y_\mathrm{r}, \dot{y}_\mathrm{r})\,, \tag{9.23}$$

Then, let the update law for the new estimate ϑ_2 be

$$\dot{\vartheta}_2 = x_2\, w_2(x_1, x_2, z^\mathrm{r}, \vartheta_1, y_\mathrm{r}, \dot{y}_\mathrm{r})\,. \tag{9.24}$$

Step i $(2 \le i \le \rho - 1)$: Using the definitions for $x_1, \ldots, x_i$ and $\dot{\vartheta}_1, \ldots, \dot{\vartheta}_{i-1}$, express the derivative of x_i as

$$\begin{aligned} \dot{x}_i &= z_{i+1} + \varphi_i(x_1, \ldots, x_i, z^\mathrm{r}, \vartheta_1, \ldots, \vartheta_{i-1}, y_\mathrm{r}, \ldots, y_\mathrm{r}^{(i)}) \\ &\quad + \theta^\mathrm{T}w_i(x_1, \ldots, x_i, z^\mathrm{r}, \vartheta_1, \ldots, \vartheta_{i-1}, y_\mathrm{r}, \ldots, y_\mathrm{r}^{(i-1)})\,. \end{aligned} \tag{9.25}$$

Let ϑ_i be a *new* estimate of θ and define the new state x_{i+1} as

$$\begin{aligned} x_{i+1} &= c_ix_i + z_{i+1} + \varphi_i(x_1, \ldots, x_i, z^\mathrm{r}, \vartheta_1, \ldots, \vartheta_{i-1}, y_\mathrm{r}, \ldots, y_\mathrm{r}^{(i)}) \\ &\quad + \vartheta_i^\mathrm{T}w_i(x_1, \ldots, x_i, z^\mathrm{r}, \vartheta_1, \ldots, \vartheta_{i-1}, y_\mathrm{r}, \ldots, y_\mathrm{r}^{(i-1)})\,. \end{aligned} \tag{9.26}$$

Substitute (9.26) into (9.25) to obtain

$$\dot{x}_i = -c_i x_i + x_{i+1} + (\theta - \vartheta_i)^{\mathrm{T}} w_i(x_1, \ldots, x_i, z^{\mathrm{r}}, \vartheta_1, \ldots, \vartheta_{i-1}, y_{\mathrm{r}}, \ldots, y_{\mathrm{r}}^{(i-1)}) . \tag{9.27}$$

Then, let the update law for ϑ_i be

$$\dot{\vartheta}_i = x_i \, w_i(x_1, \ldots, x_i, z^{\mathrm{r}}, \vartheta_1, \ldots, \vartheta_{i-1}, y_{\mathrm{r}}, \ldots, y_{\mathrm{r}}^{(i-1)}) . \tag{9.28}$$

Step ρ: Using the definitions for $x_1, \ldots, x_n$ and $\dot{\vartheta}_1, \ldots, \dot{\vartheta}_{\rho-1}$, express the derivative of x_n as

$$\begin{aligned} \dot{x}_\rho = {} & \beta_0(z)u + \varphi_\rho(x_1, \ldots, x_\rho, z^{\mathrm{r}}, \vartheta_1, \ldots, \vartheta_{\rho-1}, y_{\mathrm{r}}, \ldots, y_{\mathrm{r}}^{(\rho)}) \\ & + \theta^{\mathrm{T}} w_\rho(x_1, \ldots, x_\rho, z^{\mathrm{r}}, \vartheta_1, \ldots, \vartheta_{\rho-1}, y_{\mathrm{r}}, \ldots, y_{\mathrm{r}}^{(\rho-1)}) . \end{aligned} \tag{9.29}$$

Let ϑ_ρ be a *new* estimate of θ and define the control u as

$$u = \frac{1}{\beta_0(z)} \left[-c_\rho x_\rho - \varphi_\rho - \vartheta_\rho^{\mathrm{T}} w_\rho \right] . \tag{9.30}$$

Substitute (9.30) into (9.29) to obtain

$$\dot{x}_\rho = -c_\rho x_\rho + (\theta - \vartheta_\rho)^{\mathrm{T}} w_\rho(x_1, \ldots, x_\rho, z^{\mathrm{r}}, \vartheta_1, \ldots, \vartheta_{\rho-1}, y_{\mathrm{r}}, \ldots, y_{\mathrm{r}}^{(\rho-1)}) . \tag{9.31}$$

Finally, let the update law for the estimate ϑ_ρ be

$$\dot{\vartheta}_\rho = x_\rho \, w_\rho(x_1, \ldots, x_\rho, z^{\mathrm{r}}, \vartheta_1, \ldots, \vartheta_{\rho-1}, y_{\mathrm{r}}, \ldots, y_{\mathrm{r}}^{(\rho-1)}) . \tag{9.32}$$

As was the case in the regulation result of Section 8, the assumptions of Proposition 9.3 guarantee that the design procedure (9.16)–(9.32) is globally feasible. The resulting closed-loop adaptive system is given by

$$\begin{aligned} \dot{x}_1 &= -c_1 x_1 + x_2 + (\theta - \vartheta_1)^{\mathrm{T}} w_1(x_1, z^{\mathrm{r}}, y_{\mathrm{r}}) \\ &\ \vdots \\ \dot{x}_{\rho-1} &= -c_{\rho-1} x_{\rho-1} + x_\rho + (\theta - \vartheta_{\rho-1})^{\mathrm{T}} \\ &\qquad \times w_{\rho-1}(x_1, \ldots, x_{\rho-1}, z^{\mathrm{r}}, \vartheta_1, \ldots, \vartheta_{\rho-1}, y_{\mathrm{r}}, \ldots, y_{\mathrm{r}}^{(\rho-2)}) \\ \dot{x}_\rho &= -c_\rho x_\rho + (\theta - \vartheta_\rho)^{\mathrm{T}} w_\rho(x_1, \ldots, x_\rho, z^{\mathrm{r}}, \vartheta_1, \ldots, \vartheta_{\rho-1}, y_{\mathrm{r}}, \ldots, y_{\mathrm{r}}^{(\rho-1)}) \\ \dot{z}^{\mathrm{r}} &= \Phi_0(y, z^{\mathrm{r}}) + \sum_{i=1}^{p} \theta_i \Phi_i(y, z^{\mathrm{r}}) \\ \dot{\vartheta}_i &= x_i \, w_i(x_1, \ldots, x_i, z^{\mathrm{r}}, \vartheta_1, \ldots, \vartheta_{i-1}, y_{\mathrm{r}}, \ldots, y_{\mathrm{r}}^{(i-1)}), \quad 1 \le i \le \rho \\ y &= x_1 + y_{\mathrm{r}} . \end{aligned} \tag{9.33}$$

The stability and tracking properties of (9.33) will be established using the quadratic function

$$V_\rho(x_1, \ldots, x_\rho, \vartheta_1, \ldots, \vartheta_\rho) = \frac{1}{2} \sum_{i=1}^{\rho} \left[x_i^2 + (\theta - \vartheta_i)^{\mathrm{T}} (\theta - \vartheta_i) \right] . \tag{9.34}$$

The derivative of V_ρ along the solutions of (9.33), with $c_i \geq 2$, $1 \leq i \leq \rho$, is

$$\begin{aligned}\dot{V}_\rho &= -\sum_{i=1}^{\rho}\left[c_i x_i^2 + (\theta - \vartheta_i)^{\mathrm{T}}(x_i w_i - \dot{\vartheta}_i)\right] + \sum_{i=1}^{\rho-1} x_i x_{i+1} \\ &= -\sum_{i=1}^{\rho} c_i x_i^2 + \sum_{i=1}^{\rho-1} x_i x_{i+1} \\ &\leq -\sum_{i=1}^{\rho} x_i^2 \leq 0 \,. \end{aligned} \tag{9.35}$$

This proves that V_ρ is bounded. Hence $x_1, \ldots, x_\rho$ and $\vartheta_1, \ldots, \vartheta_\rho$ are bounded, and $\dot{V}_\rho$ is bounded and integrable. The boundedness of x_1 and y_{r} implies that y is bounded. Combining this with Assumption 9.5 proves that z^{r} is bounded. Therefore, the state vector of (9.33) is bounded. This fact, combined with Assumption 9.6, implies the boundedness of z, ζ and u. Thus, the derivatives $\dot{x}_1, \ldots, \dot{x}_\rho$ are bounded, which implies that $\ddot{V}_\rho$ is bounded. Hence, $\dot{V}_\rho \to 0$ as $t \to \infty$, which, combined with (9.35), proves that

$$\lim_{t\to\infty} x_i(t) = 0\,, \quad 1 \leq i \leq \rho\,. \tag{9.36}$$

In particular, this means that asymptotic tracking is achieved:

$$\lim_{t\to\infty} x_1(t) = \lim_{t\to\infty} [y(t) - y_{\mathrm{r}}(t)] = 0\,. \tag{9.37}$$

These results are summarized as:

Theorem 9.7 *Suppose that the system (9.1) satisfies Proposition 9.3 and Assumption 9.5, and that the design procedure (9.16)–(9.32) is applied to its strict-feedback normal form (9.10). Furthermore, suppose that the reference signal y_{r} satisfies Assumption 9.6. Then, all the signals in the resulting closed-loop adaptive system (9.33) are bounded and asymptotic tracking (9.37) is achieved for all initial conditions in $\mathbb{R}^{n+\rho p}$.* □

10 Discussion and Examples

With the help of two examples, we now discuss some of the main features of the new adaptive scheme. The first example illustrates the systematic nature of the design procedure, while the second one compares the stability properties of the new scheme with those of the nonlinearity-constrained scheme of [15].

Example 10.1 (Regulation). We first consider a "benchmark" example of adaptive nonlinear regulation:

$$\begin{aligned}\dot{z}_1 &= z_2 + \theta z_1^2 \\ \dot{z}_2 &= z_3 \\ \dot{z}_3 &= u\,,\end{aligned} \tag{10.1}$$

where θ is an unknown constant parameter. This system violates both the geometric conditions of [5,6,7] and the growth assumptions of [12,13,15,16]. In fact, the only available global result for this example was obtained in [14].

The system (10.1) is already in the form of (8.4) with $\beta_0 \equiv 1$. Hence, this system satisfies the conditions of Theorem 8.4, which guarantees that the point $z = 0$, $\vartheta_1 = \vartheta_2 = \vartheta_3 = \theta$ is a globally stable equilibrium of the adaptive system. Moreover, for any initial conditions $z(0) \in \mathbb{R}^3$, $(\vartheta_1(0), \vartheta_2(0), \vartheta_3(0)) \in \mathbb{R}^3$, the regulation of the state $z(t)$ is achieved:

$$\lim_{t\to\infty} z(t) = 0\,. \tag{10.2}$$

The design procedure of Section 5, applied to (10.1), is as follows:

Step 0: Define $x_1 = z_1$.

Step 1: Let ϑ_1 be an estimate of θ and define the new state x_2 as

$$x_2 = 2x_1 + z_2 + \vartheta_1 x_1^2\,. \tag{10.3}$$

Substitute (10.3) into (10.1) to obtain

$$\dot{x}_1 = -2x_1 + x_2 + x_1^2(\theta - \vartheta_1)\,. \tag{10.4}$$

Then, let the update law for ϑ_1 be

$$\dot{\vartheta}_1 = x_1^3\,. \tag{10.5}$$

Step 2: Using (10.3) and (10.5), write $\dot{x}_2$ as

$$\dot{x}_2 = 2(z_2 + \theta z_1^2) + z_3 + \vartheta_1 2x_1(z_2 + \theta z_1^2) + x_1^5\,. \tag{10.6}$$

Let ϑ_2 be a new estimate of θ, and define the new state

$$x_3 = 2x_2 + 2(z_2 + \vartheta_2 z_1^2)(1 + \vartheta_1 x_1) + x_1^5 + z_3\,. \tag{10.7}$$

Substitute (10.7) into (10.6) to obtain

$$\dot{x}_2 = -2x_2 + x_3 + 2x_1^2(1 + \vartheta_1 x_1)(\theta - \vartheta_2)\,. \tag{10.8}$$

Then, let the update law for ϑ_2 be

$$\dot{\vartheta}_2 = 2x_2 x_1^2(1 + \vartheta_1 x_1)\,. \tag{10.9}$$

Step 3: Using (10.3), (10.5), (10.7) and (10.8), write $\dot{x}_3$ as

$$\begin{aligned}\dot{x}_3 = {} & 2\left[-2x_2 + x_3 + 2x_1^2(1 + \vartheta_1 x_1)(\theta - \vartheta_2)\right] + 2\left[z_3 + 2z_1\vartheta_2(z_2 + \theta z_1^2)\right. \\ & \left. + 2z_1^2 x_2 x_1^2(1 + \vartheta_1 x_1)\right](1 + \vartheta_1 x_1) + 2(z_2 + \vartheta_2 z_1^2)\left[x_1^4 + \vartheta_1(z_2 + \theta z_1^2)\right] \\ & + 5x_1^4(z_2 + \theta z_1^2) + u\,. \end{aligned} \tag{10.10}$$

Let ϑ_3 be a new estimate of θ, and define the control u as

$$u = -2x_3 - 2\left[-2x_2 + x_3 + 2x_1^2(1+\vartheta_1 x_1)(\theta - \vartheta_2)\right] - 2\left[z_3 + 2z_1\vartheta_2(z_2 + \theta z_1^2)\right.$$
$$\left. + 2z_1^2 x_2 x_1^2(1+\vartheta_1 x_1)\right](1+\vartheta_1 x_1) - 2(z_2 + \vartheta_2 z_1^2)\left[x_1^4 + \vartheta_1(z_2 + \theta z_1^2)\right]$$
$$-5x_1^4(z_2 + \theta z_1^2)\,. \tag{10.11}$$

Substitute (10.11) into (10.10) to obtain

$$\dot{x}_3 = -2x_3 + \left[2x_1^2(1+2\vartheta_1 x_1) + 4z_1^3\vartheta_2 + 2\vartheta_1(z_2 + \vartheta_2 z_1^2)z_1^2 + 5x_1^6\right](\theta - \vartheta_3)\,. \tag{10.12}$$

Finally, let the parameter update law for ϑ_3 be

$$\dot{\vartheta}_3 = x_3\left[2x_1^2(1+2\vartheta_1 x_1) + 4z_1^3\vartheta_2 + 2\vartheta_1(z_2 + \vartheta_2 z_1^2)z_1^2 + 5x_1^6\right]. \tag{10.13}$$

The resulting adaptive system is

$$\begin{aligned}
\dot{x}_1 &= -2x_1 + x_2 + x_1^2(\theta - \vartheta_1)\\
\dot{x}_2 &= -2x_2 + x_3 + 2x_1^2(1+\vartheta_1 x_1)(\theta - \vartheta_2)\\
\dot{x}_3 &= -2x_3 + \left[2x_1^2(1+\vartheta_1 x_1) + 4z_1^3\vartheta_2 + 2\vartheta_1(z_2 + \vartheta_2 z_1^2)z_1^2 + 5x_1^6\right](\theta - \vartheta_3)\\
\dot{\vartheta}_1 &= x_1^3\\
\dot{\vartheta}_2 &= 2x_2 x_1^2(1+\vartheta_1 x_1)\\
\dot{\vartheta}_3 &= x_3\left[2x_1^2(1+\vartheta_1 x_1) + 4z_1^3\vartheta_2 + 2\vartheta_1(z_2 + \vartheta_2 z_1^2)z_1^2 + 5x_1^6\right].
\end{aligned} \tag{10.14}$$

Using the Lyapunov function

$$V = \frac{1}{2}\left[x_1^2 + x_2^2 + x_3^2 + (\theta - \vartheta_1)^2 + (\theta - \vartheta_2)^2 + (\theta - \vartheta_3)^2\right], \tag{10.15}$$

it is straightforward to establish the global stability and regulation properties of (10.14). □

Example 10.2 (Tracking). Consider now the problem in which the output y of the nonlinear system

$$\begin{aligned}
\dot{z}_1 &= z_2 + \theta z_1^2\\
\dot{z}_2 &= u + z_3\\
\dot{z}_3 &= -z_3 + y\\
y &= z_1\,,
\end{aligned} \tag{10.16}$$

is required to asymptotically track the reference signal $y_r = 0.1\sin t$.

For the sake of comparison, let us first solve this problem using the scheme of [15]. This scheme employs the control

$$u = -z_3 + k_1(z_1 - y_r) + k_2(z_2 + \hat{\theta}_1 z_1^2 - \dot{y}_r) + \ddot{y}_r - 2\hat{\theta}_1 z_1 z_2 - 2\hat{\theta}_2 z_1^3\,, \tag{10.17}$$

where $\hat{\theta}_1$, $\hat{\theta}_2$, the estimates of θ, θ^2, respectively, are obtained from the update laws:

$$\dot{\hat{\theta}}_1 = \frac{e_1\xi_1}{1+\xi_1^2+\xi_2^2}\,, \quad \dot{\hat{\theta}}_2 = \frac{e_1\xi_2}{1+\xi_1^2+\xi_2^2}\,. \tag{10.18}$$

Using a relative-degree-two stable filter $M(s)$, the variables e_1, ξ_1, ξ_2 in (10.18) are defined as

$$
\begin{aligned}
e_1 &= y - y_r + \omega - \hat{\theta}_1\xi_1 - \hat{\theta}_2\xi_2 && (10.19)\\
\xi_1 &= M(s)\left[2z_1z_2 + k_2z_1^2\right] && (10.20)\\
\xi_2 &= M(s)\left[2z_1^3\right] && (10.21)\\
\omega &= M(s)\left[\hat{\theta}_1\left(2z_1z_2 + k_2z_1^2\right) + \hat{\theta}_2\left(2z_1^3\right)\right]. && (10.22)
\end{aligned}
$$

Simulations of this system were performed with

$$M(s) = \frac{1}{s^2 + 5s + 6}, \; \theta = 1, \; k_1 = -6, \; k_2 = -5, \tag{10.23}$$

and all the initial conditions zero, except for $z_1(0)$, which was varied between 0 and 0.45. The results of these simulations, shown in Fig. 2, indicate that the response of the closed-loop system is bounded for $z_1(0)$ sufficiently small, that is, for $z_1(0) < 0.45$. However, for $z_1(0) \geq 0.45$, the response is unbounded. This behavior is consistent with the proof of Theorem 3.3 in [15], which guarantees boundedness for all initial conditions only under a global Lipschitz assumption. In the above system, the presence of the term z_1^2 leads to the violation of this assumption, and, as the simulations show, to unbounded response.

The unbounded behavior in Fig. 2 is avoided by the new scheme, which results in globally stable tracking. The design procedure of Section 9, applied to the system (10.16), results in the control

$$u = -z_3 - 3x_2 - 2(z_2 + \vartheta_2 z_1^2)(1 + \vartheta_1 z_1) - x_1 z_1^4 + 2\dot{y}_r + \ddot{y}_r, \tag{10.24}$$

and the update laws

$$\dot{\vartheta}_1 = x_1 z_1^2, \quad \dot{\vartheta}_2 = 2x_2 z_1^2(1 + \vartheta_1 x_1), \tag{10.25}$$

where

$$
\begin{aligned}
x_1 &= z_1 - y_r\\
x_2 &= 2(z_1 - y_r) + z_2 + \vartheta_1 z_1^2 - \dot{y}_r.
\end{aligned} \tag{10.26}
$$

Theorem 9.7 establishes that uniform stability and asymptotic tracking are achieved for all $z_1(0), z_2(0), z_3(0), \vartheta_1(0), \vartheta_2(0)$. This is illustrated by simulations in Fig. 3. □

The above example illustrates an obvious advantage of the new scheme when applied to strict-feedback systems: it guarantees global stability for all types of smooth nonlinearities. For pure-feedback systems, when the feedback linearization is not global, the new scheme provides an estimate of the region of attraction. An advantage of the schemes in [5,12,13,14,15,16] is that they provide local results without assuming the pure-feedback form. However, estimates of the region of attraction are given only in [5,13,14]. A quantitative comparison of the regions of attraction and robustness properties guaranteed by different schemes is a topic of future research.

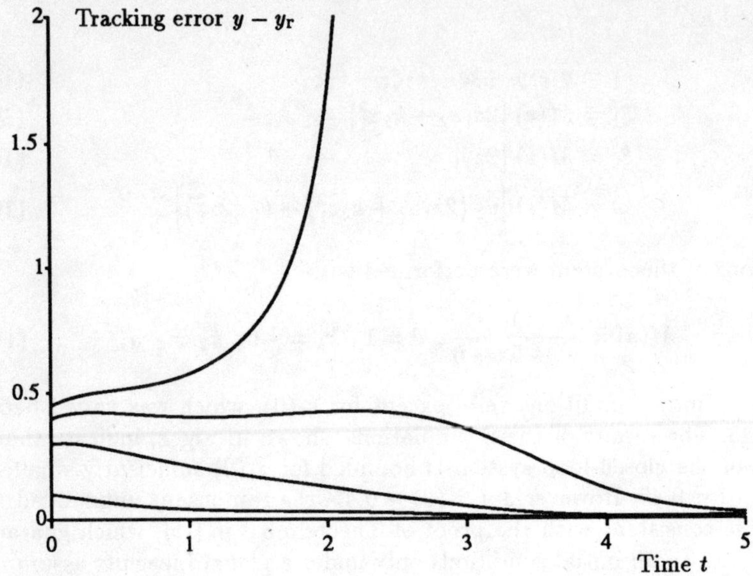

Fig. 2. Locally stable tracking with the adaptive scheme of [15].

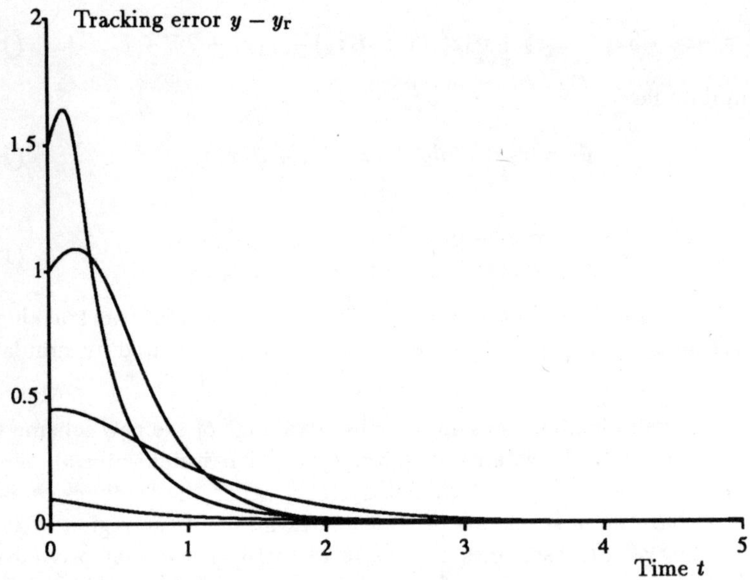

Fig. 3. Globally stable tracking with the new adaptive scheme.

11 Conclusions

The results of this paper have advanced in several directions our ability to control nonlinear systems with unknown constant parameters. The most significant progress has been made in solving the *global* adaptive regulation and tracking problems. The class of nonlinear systems for which these problems can be solved systematically has been substantially enlarged. The strict-feedback condition precisely characterizes the class of systems for which the global results hold with any type of smooth nonlinearities. For the broader class of systems satisfying the pure-feedback condition, the regulation and tracking results may not be global, but are guaranteed in regions for which *a priori* estimates are given. It is crucial that the loss of globality, when it occurs, is not due to adaptation, but is inherited from the deterministic part of the problem. All these results are obtained using a step-by-step procedure which, at each step, interlaces a change of coordinates with the construction of an update law. Apart from the geometric conditions, this paper uses simple analytical tools, familiar to most control engineers.

Acknowledgements

The authors are grateful to Professors Riccardo Marino and Laurent Praly for their insightful comments and helpful suggestions.

References

1. K. S. Narendra and A. M. Annaswamy, *Stable Adaptive Systems*, Prentice-Hall, Inc., Englewood Cliffs, NJ, 1989.
2. S. S. Sastry and M. Bodson, *Adaptive Control: Stability, Convergence and Robustness*, Prentice-Hall, Inc., Englewood Cliffs, NJ, 1989.
3. A. Isidori, *Nonlinear Control Systems*, 2nd ed., Springer-Verlag, Berlin, 1989.
4. H. Nijmeijer and A. J. van der Shaft, *Nonlinear Dynamical Control Systems*, Springer-Verlag, Berlin, 1990.
5. G. Campion and G. Bastin, "Indirect adaptive state feedback control of linearly parametrized nonlinear systems," *Int. J. Adapt. Control Sig. Proc.*, vol. 4, pp. 345–358, Sept. 1990.
6. I. Kanellakopoulos, P. V. Kokotović, and R. Marino, "Robustness of adaptive nonlinear control under an extended matching condition," *Prepr. IFAC Symp. Nonlinear Control Syst. Design*, pp. 192–197, Capri, Italy, June 1989.
7. I. Kanellakopoulos, P. V. Kokotović, and R. Marino, "An extended direct scheme for robust adaptive nonlinear control," *Automatica*, to appear, March 1991.
8. R. Marino, I. Kanellakopoulos, and P. V. Kokotović, "Adaptive tracking for feedback linearizable SISO systems," *Proc. 28th IEEE Conf. Dec. Control*, pp. 1002–1007, Tampa, FL, Dec. 1989.
9. J.-J. E. Slotine and J. A. Coetsee, "Adaptive sliding controller synthesis for nonlinear systems," *Int. J. Control*, vol. 43, pp. 1631–1651, June 1986.
10. D. G. Taylor, *Feedback Control of Uncertain Nonlinear Systems with Applications to Electric Machinery and Robotic Manipulators*, Ph.D. Thesis, University of Illinois, Urbana, IL, May 1988.

11. D. G. Taylor, P. V. Kokotović, R. Marino, and I. Kanellakopoulos, "Adaptive regulation of nonlinear systems with unmodeled dynamics," *IEEE Trans. Aut. Control*, vol. 34, pp. 405–412, April 1989.
12. K. Nam and A. Arapostathis, "A model-reference adaptive control scheme for pure-feedback nonlinear systems," *IEEE Trans. Aut. Control*, vol. 33, pp. 803–811, Sept. 1988.
13. J.-B. Pomet and L. Praly, *Adaptive nonlinear control: an estimation-based algorithm*, in *New Trends in Nonlinear Control Theory*, J. Descusse, M. Fliess, A. Isidori and D. Leborgne Eds., Springer-Verlag, Berlin, 1989.
14. J.-B. Pomet and L. Praly, "Adaptive nonlinear regulation: equation error from the Lyapunov equation," *Proc. 28th IEEE Conf. Dec. Control*, pp. 1008–1013, Tampa, FL, Dec. 1989.
15. S. S. Sastry and A. Isidori, "Adaptive control of linearizable systems," *IEEE Trans. Aut. Control*, vol. 34, pp. 1123–1131, Nov. 1989.
16. A. Teel, R. Kadiyala, P. V. Kokotović, and S. S. Sastry, "Indirect techniques for adaptive input output linearization of nonlinear systems," *Int. J. Control*, vol. 53, pp. 193–222, Jan. 1991.
17. I. Kanellakopoulos, P. V. Kokotović, and R. H. Middleton, "Observer-based adaptive control of nonlinear systems under matching conditions," *Proc. 1990 Amer. Control Conf.*, pp. 549–555, San Diego, CA, May 1990.
18. I. Kanellakopoulos, P. V. Kokotović, and R. H. Middleton, "Indirect adaptive output-feedback control of a class of nonlinear systems," *Proc. 29th IEEE Conf. Dec. Control*, pp. 2714–2719, Honolulu, HI, Dec. 1990.
19. A. Feuer and A. S. Morse, "Adaptive control of single-input single-output linear systems," *IEEE Trans. Aut. Control*, vol. AC-23, pp. 557–569, Aug. 1978.
20. R. Su, "On the linear equivalents of nonlinear systems," *Syst. Control Lett.*, vol. 2, pp. 48–52, July 1982.
21. L. R. Hunt, R. Su, and G. Meyer, "Global transformations of nonlinear systems," *IEEE Trans. Automat. Control*, vol. AC-28, pp. 24–31, Jan. 1983.
22. W. Dayawansa, W. M. Boothby and D. L. Elliott, "Global state and feedback equivalence of nonlinear systems," *Syst. Control Lett.*, vol. 6, pp. 229–234, 1985.
23. W. Respondek, *Global aspects of linearization, equivalence to polynomial forms and decomposition of nonlinear control systems*, in *Algebraic and Geometric Methods in Nonlinear Control Theory*, M. Fliess and M. Hazewinkel, eds., pp. 257–284, D. Reidel Publishing Co., Dordrecht, 1986.
24. B. Jakubczyk and W. Respondek, "On linearization of control systems," *Bull. Acad. Pol. Science, Ser. Science Math.*, vol. 28, no. 9-10, pp. 517–522, 1980.
25. L. R. Hunt, R. Su, and G. Meyer, *Design for multi-input nonlinear systems*, in *Differential Geometric Control Theory*, R. W. Brockett, R. S. Millman and H. S. Sussmann, eds., pp. 268–297, Birkhäuser, Boston, MA, 1983.
26. R. Marino, W. M. Boothby and D. L. Elliott, "Geometric properties of linearizable control systems," *Math. Syst. Theory*, vol. 18, pp. 97–123, 1985.

Adaptive Stabilization of Nonlinear Systems

L. Praly,[1] *G Bastin,*[2] *J.-B. Pomet,*[3] *and Z. P. Jiang*[1]

[1] Centre d'Automatique et Informatique – Section Automatique,
École des Mines de Paris, 35 rue St Honoré, 77305 Fontainebleau cédex, FRANCE.
[2] Laboratoire d'Automatique, de Dynamique et d'Analyse des Systèmes,
Université Catholique de Louvain, Place du Levant 3, B-1348 Louvain-La-Neuve, BELGIUM.
[3] Department of Mathematics and Statistics,
Queens University, Kingston, Ontario, K7L 3N6, CANADA.

Abstract. An overview of the various parametric approaches which can be adopted to solve the problem of adaptive stabilization of nonlinear systems is presented. The Lyapunov design and two estimation designs –equation error filtering and regressor filtering– are revisited. This allows us to unify and generalize most of the available results on the topic and to propose a classification depending on the required extra assumptions – matching conditions or growth conditions.

1 Problem Statement and Assumptions

1.1 Problem Statement

We consider a dynamic system which admits a finite state-space representation and whose dynamics are described by an equation which involves uncertain constant parameters. We are concerned with the design of a dynamic state-feedback controller which ensures, in spite of that uncertainty, that solutions of the closed-loop system are bounded and their x-components converge to a desired set point.

Example: **(1)**
Consider the following one-dimensional system:

$$\dot{x} = p^* x^2 + u, \tag{2}$$

where p^* is a constant parameter. Would the value of p^* be known, we could use the following linearizing control law to globally stabilize the origin of the closed-loop system:

$$u = -p^* x^2 - x. \tag{3}$$

When only an approximate value $\widehat{p}$ of p^* is known and is used in the control law (3), we obtain the following closed-loop system:

$$\dot{x} = -x + (p^* - \widehat{p})\, x^2. \tag{4}$$

This system (4) has two equilibrium points:

1. $x = 0$, which is exponentially stable,
2. $x = \dfrac{1}{p^* - \widehat{p}}$, which is exponentially unstable.

Hence, as long as $\widehat{p}$ is not exactly equal to p^*, the global asymptotic stability of $x = 0$ is lost. We notice also that the simple linear control

$$u = -x \tag{5}$$

gives exactly the same qualitative behavior.

Assume now that we apply the following dynamic controller:

$$\begin{aligned} \dot{\widehat{p}} &= x^3 \\ u &= -\widehat{p}\, x^2 - x\,. \end{aligned} \tag{6}$$

This is the linearizing controller where instead of the true value p^* of the parameter, we use an on-line updated estimate $\widehat{p}$. The corresponding closed-loop system we get is:

$$\begin{aligned} \dot{\widehat{p}} &= x^3 \\ \dot{x} &= -x + (p^* - \widehat{p})\, x^2\,. \end{aligned} \tag{7}$$

To study the stability, we consider the function:

$$W(x, \widehat{p}) = \frac{1}{2}\left(x^2 + (\widehat{p} - p^*)^2\right). \tag{8}$$

Its time derivative along the solutions of (7) is:

$$\dot{W} = -x^2\,. \tag{9}$$

It follows that any solution of the closed-loop system (7) is bounded and:

$$\lim_{t \to \infty} x(t) = 0\,. \tag{10}$$

Therefore, the convergence of x to 0 is restored. □

In this example, the parameter p^* enters linearly in the dynamic equation (2). This assumption is fundamental throughout the paper. It is formalized as follows:

Assumption Λ-LP (Λ-Linear Parameterization) (11)

We can find a set of measured *coordinates x in $\mathbb{R}^n$ such that, given an integer k and a C^1 function $\Lambda : \mathbb{R}^n \times \mathbb{R}_+ \to \mathcal{M}_{kn}(\mathbb{R})$, there exist an integer l, two C^1 functions:*

$$a : \mathbb{R}^n \times \mathbb{R}^m \to \mathbb{R}^n\,, \qquad A : \mathbb{R}^n \times \mathbb{R}^m \to \mathcal{M}_{nl}(\mathbb{R})\,,$$

and an unknown *parameter vector p^* in $\mathbb{R}^l$ such that:*

1. *the functions $\Lambda(x,t)a(x,u)$ and $\Lambda(x,t)A(x,u)$ are* known, *i.e., can be evaluated,*
2. *the dynamics of the system to be controlled are described by:*

$$\dot{x} = a(x, u) + A(x, u)\, p^*\,, \tag{12}$$

where u is the input vector in $\mathbb{R}^m$.

To deal with the case where a and A are affine in u, it is useful to introduce the following notation:

$$a(x,u) = a_0(x) + \sum_{i=1}^{m} u_i\, a_i(x) \qquad A(x,u) = A_0(x) + \sum_{i=1}^{m} u_i\, A_i(x) \tag{13}$$
$$\stackrel{\text{def}}{=} a_0(x) + u \odot b(x)\,, \qquad\qquad \stackrel{\text{def}}{=} A_0(x) + u \odot B(x)\,.$$

Example: Introduction to System (17) **(14)**

Clearly, in the case of the system (2), if we choose:

$$\Lambda(x,t) = 1\,, \tag{15}$$

assumption Λ-LP is satisfied with the known functions:

$$a(x,u) = u\,, \quad A(x,u) = x^2\,. \tag{16}$$

Let us now consider the following two-dimensional system:

$$\begin{aligned} \dot{x}_1 &= x_1^2\,(x_2^2 + c_1 + c_2\,u) \\ \dot{x}_2 &= -c_2\,x_2^2\,(x_2 + c_3\,x_1)\,, \end{aligned} \tag{17}$$

where c_1, c_2 and c_3 are three unknown real numbers. Depending on our choice for the function Λ, we get different parameterizations.

- A first possible choice is, with $k = 2$:

$$\Lambda(x_1,x_2,t) = \begin{pmatrix} 1 & 0 \\ 0 & 1 \end{pmatrix}. \tag{18}$$

  Then Λ-LP is met with $l = 3$, the parameter vector:

$$p^* = (c_1\,,\ c_2\,,\ c_2c_3)^{\mathrm{T}}\,, \tag{19}$$

  and the known functions:

$$a(x_1,x_2) = \begin{pmatrix} x_1^2\,x_2^2 \\ 0 \end{pmatrix}, \qquad A(x_1,x_2,u) = \begin{pmatrix} x_1^2 & x_1^2\,u & 0 \\ 0 & -x_2^3 & x_2^2 x_1 \end{pmatrix}. \tag{20}$$

- Another possibility is, with $k = 1$,

$$\Lambda(x_1,x_2,t) = (x_1\ 0)\,. \tag{21}$$

  Assumption Λ-LP holds with $l = 2$, the parameter vector:

$$p^* = (c_1 \quad c_2)^{\mathrm{T}}\,, \tag{22}$$

  and the functions:

$$a(x_1,x_2) = \begin{pmatrix} x_1^2\,x_2^2 \\ 0 \end{pmatrix}, \qquad A(x_1,x_2,u) = \begin{pmatrix} x_1^2 & x_1^2\,u \\ 0 & -x_2^2\,(x_2 + c_3\,x_1) \end{pmatrix}. \tag{23}$$

  Here the function A is unknown since it involves the unknown constant c_3. But both $\Lambda\,a$ and $\Lambda\,A$ are known. In particular, the choice (21) for the function Λ implies that we pay attention to the dynamics of x_1 only and disregard the dynamics of x_2.

□

This example illustrates that in order to find a state-space representation and a parameterization and to choose a function Λ satisfying assumption Λ-LP, it may be useful not to work with the a priori given, say physical, coefficients. First, the knowledge of these coefficients may not be relevant to the Stabilization problem. As emphasized by Sastry and Kokotovic [28], this is in particular the case if, by looking at these coefficients as disturbances, they can be rejected. For the system (17) in Example (14), we will see in Example (49) that the coefficient c_3 is irrelevant to the stabilization. Second, it may also be useful to change the a priori given coordinates and define the parameters as functions of the coefficients. For instance, in Example (14), we may choose $p_3 = c_2 c_3$. This trick is classically used in robotics (see [14]). It has also been used by Sastry and Isidori [29] for input-output linearization of systems with relative degree larger than one and by Kanellakopoulos, Kokotovic and Middleton [10] for adaptive dynamic output feeeback.

Example: Introduction to System (25) (24)
Consider a system whose dynamics is described by the following second-order differential equation:

$$\ddot{y} = u + L(y)\, p^* , \tag{25}$$

where y is a measured output, u is the input and L is a known smooth function. A straightforward state-space representation with $(y\,,\,\dot{y})$ as the state vector is not appropriate for satisfying assumption Λ-LP, since $\dot{y}$ is not measured. Instead, let us introduce the following filtered quantities:

$$\begin{aligned} &\ddot{y}_f + \dot{y}_f + y_f = y \qquad && y_f(0) = \dot{y}_f(0) = 0 \\ &\ddot{u}_f + \dot{u}_f + u_f = u \qquad && u_f(0) = \dot{u}_f(0) = 0 \\ &\ddot{L}_f + \dot{L}_f + L_f = L \qquad && L_f(0) = \dot{L}_f(0) = 0\,. \end{aligned} \tag{26}$$

Then, (25) can be rewritten with the following non-minimal state-space representation, with $\dot{-}$ denoting differentiation with respect to time:

$$\dot{x} = \begin{pmatrix} 0 & 1 & 0 & 0 & 0 & 0 \\ 0 & 0 & 1 & 0 & p^* & 0 \\ 0 & 0 & 0 & 1 & 0 & 0 \\ 0 & 0 & -1 & -1 & 0 & 0 \\ 0 & 0 & 0 & 0 & 0 & 1 \\ 0 & 0 & 0 & 0 & -1 & -1 \end{pmatrix} x + \begin{pmatrix} 0 \\ \delta(t) \\ 0 \\ u \\ 0 \\ L(y) \end{pmatrix} \tag{27}$$

$$\begin{aligned} y &= \dot{y}_f + y_f + u_f + L_f p^* + \delta(t) \\ \dot{y} &= \dot{y}_f + u_f + \dot{u}_f + \left(L_f + \dot{L}_f\right) p^* + \overline{\dot{\delta}(t)} + \delta(t) , \end{aligned} \tag{28}$$

where x is the following state vector:

$$x = \left(y_f \ \dot{y}_f \ u_f \ \dot{u}_f \ L_f \ \dot{L}_f\right)^{\mathrm{T}} , \tag{29}$$

and $\delta(t)$ is the solution of:

$$\ddot{\delta} + \dot{\delta} + \delta = 0 , \qquad \text{with} \quad \delta(0) = y(0) , \ \dot{\delta}(0) = \dot{y}(0) - y(0) . \tag{30}$$

Since x can be obtained from the knowledge of y and u only, assumption Λ-LP is satisfied up to the presence of the exponentially decaying time function δ by choosing:

$$\Lambda(x,t) = (0\ 1\ 0\ 0\ 0\ 0) . \tag{31}$$

The presence of δ implies that our forthcoming results will not apply in a straightforward manner. In each case we will have to study the effects of this term. □

According to assumption Λ-LP, the only knowledge we have about the system to be controlled is that it is a member of a linearly parameterized family of systems whose dynamics satisfy the following equation, denoted by (S_p) in the sequel:

$$\dot{x} = a(x,u) + A(x,u)\, p . \tag{S_p}$$

In this case, we formalize the Adaptive Stabilization problem as follows:

Adaptive Stabilization Problem: Find an integer ν and two functions

$$\mu_1 : \mathbb{R}^n \times \mathbb{R}^\nu \to \mathbb{R}^\nu , \qquad \mu_2 : \mathbb{R}^n \times \mathbb{R}^\nu \to \mathbb{R}^m ,$$

such that there exists an open subset $\mathcal{D}$ of $\mathbb{R}^n \times \mathbb{R}^\nu$ with the following property: The solutions $(x(t), \chi(t))$ of the system composed of the system (S_{p^*}) to be controlled and the following dynamic *state-feedback* controller:

$$\begin{aligned} \dot{\chi} &= \mu_1(x,\chi) \\ u &= \mu_2(x,\chi) , \end{aligned} \tag{32}$$

with $(x(0), \chi(0))$ in $\mathcal{D}$,

(AS1) are well-defined, unique and bounded on $[0, +\infty)$,
(AS2) have the property

$$\lim_{t\to\infty} x(t) = \mathcal{E} , \tag{33}$$

where $\mathcal{E}$ is a desired set point for x, which may depend on p^*.

A typical illustration of this problem has been given in Example (1). It turns out that, in all the solutions to this problem which are proposed in this paper, part of the state χ of the controller can be considered as an estimate $\widehat{p}$ of the unknown parameter vector p^*. This motivates the adjective "adaptive".

Several solutions to the Adaptive Stabilization problem have been proposed in the literature under particular assumptions. Extending the work of Pomet [19], we present here a framework allowing us to unify and generalize most of these solutions.

1.2 Connection with the Error Feedback Regulator Problem

When, in equation (S_p), the function A does not depend on u, i.e., $A(x,u) = A_o(x)$, and a is affine in u (see (13)), the Adaptive Stabilization problem described above has similarities with the Error Feedback Regulator problem as stated by Isidori for systems of the form [6, Sect. 7.2]:

$$\begin{aligned} \dot{x} &= a_0(x) + u \odot b(x) + A_o(x)\, p \\ \dot{p} &= s(p) \\ y &= h(x) + q(p)\,, \end{aligned} \tag{34}$$

where p in $\mathbb{R}^l$ is an unmeasured disturbance while y in $\mathbb{R}^k$ is a measured output signal.

Error Feedback Regulator Problem: Find an integer ν and two functions

$$\mu_1 : \mathbb{R}^k \times \mathbb{R}^\nu \to \mathbb{R}^\nu\,, \qquad \mu_2 : \mathbb{R}^\nu \to \mathbb{R}^m\,,$$

such that:

(EFR1) the equilibrium point of

$$\begin{aligned} \dot{x} &= a_0(x) + \mu_2(\chi) \odot b(x) + A_0(x)\, p^* \\ \dot{\chi} &= \mu_1(h(x), \chi) \end{aligned} \tag{35}$$

is asymptotically stable in the first approximation,

(EFR2) there exists an open subset $\mathcal{D}$ of $\mathbb{R}^n \times \mathbb{R}^\nu$ such that the solutions of the system (34) controlled by the following dynamic *output* feedback:

$$\begin{aligned} \dot{\chi} &= \mu_1(y, \chi) \\ u &= \mu_2(\chi)\,, \end{aligned} \tag{36}$$

with initial condition in this set $\mathcal{D}$, have the property

$$\lim_{t \to \infty} y(t) = \mathcal{E}\,, \tag{37}$$

where $\mathcal{E}$ is a desired set point for y, independent of p.

It would appear that the family of systems (S_p) dealt with in our Adaptive Stabilization problem is a subclass of the systems (34) of the Error Feedback Regulator problem. This family is obtained by imposing a constant disturbance, i.e.,

$$\dot{p} = s(p) = 0\,, \tag{38}$$

and a measured state, i.e.,

$$h(x) = x \qquad \text{and} \qquad q(p) = 0\,. \tag{39}$$

Isidori gives a solution to the Error Feedback Regulator problem in [6, Theorem 7.2.10]. It applies to systems (S_p) under the following assumptions, using the notation (13):

(A1) There exists a C^2 function $u_0(p)$ defined on an open neighborhood of p^* such that:

$$a_0(\mathcal{E}) + u_0(p) \odot b(\mathcal{E}) + A_0(\mathcal{E})\,p = 0\,, \tag{40}$$

(A2) The $n \times l$ matrix $A_0(\mathcal{E})$ has rank l,

(A3) The pair $\left(\frac{\partial a_0}{\partial x}(\mathcal{E}) + u_0(p^*) \odot \frac{\partial b}{\partial x}(\mathcal{E}) + \frac{\partial A_0}{\partial x}(\mathcal{E})\,p^*\,,\; b(\mathcal{E})\right)$ is stabilizable.

Assumption (A1) is quite natural if $\mathcal{E}$ is not allowed to depend on p. It states only the existence of a control, smoothly depending on p, making $\mathcal{E}$ a set point of (S_p) for all p close enough to p^*. Assumption (A2) is much more restrictive since it imposes that the number l of parameters do not exceed the dimension n of the state and that the matrix $A_0(x)$ cannot degenerate at $x = \mathcal{E}$. Finally, assumption (A3) excludes systems whose linearization has uncontrollable modes associated with pure imaginary eigenvalues and, in particular, systems which are not feedback linearizable.

Example: System (17) Continued **(41)**

Assumptions (A2) and (A3) are not satisfied by the following system (i.e., system (17) with $c_2 = 1$ to make A independent of u):

$$\begin{aligned} \dot{x}_1 &= x_1^2\left(x_2^2\,p_1 + u\right) \\ \dot{x}_2 &= -\,x_2^2\left(x_2 + p_2\,x_1\right). \end{aligned} \tag{42}$$

Indeed, $A(\mathcal{E})$, $\frac{\partial a}{\partial x}(\mathcal{E}, u)$ and $\frac{\partial A}{\partial x}(\mathcal{E})$ are all zero. □

In fact, assumptions (A2) and (A3) follow from the strong requirement EFR1 of asymptotic stability. In this paper, it is precisely in order to be able to deal with systems which may not satisfy assumptions (A2) or (A3), such as system (42), that, instead of requirement EFR1, we ask, in the Adaptive Stabilization problem, for the less stringent Lagrange stability requirement AS1.

1.3 Assumptions

Our counterparts of assumptions (A1), (A2) and (A3) are the following assumptions on the particular system (S_{p^*}):

Let Π be an open subset of $\mathbb{R}^l$ and Ω be an open neighborhood of $\mathcal{E}$ in $\mathbb{R}^n$. There exist two known *functions:*

$$V : \Omega \times \Pi \to \mathbb{R}_+ \text{ of class } C^2 , \qquad u_n : \Omega \times \Pi \to \mathbb{R}^m \text{ of class } C^1 ,$$

such that:

Assumption BO (Boundedness Observability) (43)
There exist an open neighborhood Ω_0 of $\mathcal{E}$ in Ω and a strictly positive constant α_0 such that for all real numbers α, $0 < \alpha < \alpha_0$, all compact subsets $\mathcal{K}$ of Π and all vectors $x_0 \in \Omega_0$, we can find a compact subset Γ of Ω such that, for any C^1-time functions $\widehat{p} : \mathbb{R}_+ \to \Pi$ and $u : \mathbb{R}_+ \to \mathbb{R}^m$ and any solution $x(t)$ of:

$$\dot{x} = a(x,u(t)) + A(x,u(t))\,p^* , \qquad x(0) = x_0 \in \Omega_0 \tag{44}$$

defined on $[0,T)$, we have the following implication:

$$V(x(t),\widehat{p}(t)) \leq \alpha \text{ and } \widehat{p}(t) \in \mathcal{K} \;\; \forall t \in [0,T) \quad \Longrightarrow \quad x(t) \in \Gamma \quad \forall t \in [0,T) . \tag{45}$$

Assumption PRS (Pointwise Reduced-Order Stabilizability) (46)
For all (x,p) in $\Omega \times \Pi$, we have:

$$\frac{\partial V}{\partial x}(x,p)\left[a(x,u_n(x,p)) + A(x,u_n(x,p))\,p\right] \leq 0 , \tag{47}$$

where the inequality is strict iff $V(x,p) \neq 0$.

In the sequel, the case where

$$\Omega = \Omega_0 = \mathbb{R}^n \qquad \text{and} \qquad \alpha_0 = +\infty \tag{48}$$

is called the global case.

Example: System (17) Continued (49)
Consider the two-dimensional system (17) in Example (14) which, according to the second parameterization we mentioned, is rewritten as:

$$\begin{aligned} \dot{x}_1 &= x_1^2\left(x_2^2 + p_1 + p_2\,u\right) \\ \dot{x}_2 &= -p_2\,x_2^2\left(x_2 + c_3\,x_1\right) , \end{aligned} \tag{50}$$

where p_2 is known to be strictly positive. We choose:

$$V(x_1,x_2,p_1,p_2) = \frac{1}{2}\,x_1^2 , \qquad u_n(x_1,x_2,p_1,p_2) = -\frac{x_2^2 + p_1 + x_1}{p_2} \tag{51}$$

$$\Omega_0 = \Omega = \mathbb{R}^2 , \qquad \Pi = \mathbb{R} \times (\mathbb{R}_+ - \{0\}) , \qquad \alpha_0 = +\infty . \tag{52}$$

Notice that the constant c_3 is not involved. This justifies a posteriori the choice of the parameterization.

Assumption PRS is satisfied. Indeed, we get:

$$\frac{\partial V}{\partial x}(x,p)\,[a(x,u_n(x,p)) + A(x,u_n(x,p))\,p] = -x_1^4 . \tag{53}$$

Assumption BO holds also if, for any C^1 time function $x_1(t)$, we have:

$$|x_1(t)| \le \alpha \quad \forall t \in [0,T) \qquad \Longrightarrow \qquad |x_2(t)| \le \gamma \quad \forall t \in [0,T) , \tag{54}$$

where $x_2(t)$ is a solution of:

$$\dot{x}_2 = -p_2^* x_2^2 (x_2 + c_3 x_1(t)) , \tag{55}$$

and γ is a positive function of α and $x_2(0)$. To prove this implication, we notice that $p_2{}^* > 0$ implies:

$$|x_2(t)| > |c_3|\,|x_1(t)| \qquad \Longrightarrow \qquad \dot{\overline{x_2^2(t)}} < 0 . \tag{56}$$

Hence, with (54), we can choose:

$$\gamma = \max\{|x_2(0)| , |c_3|\,\alpha\} . \tag{57}$$

□

This example illustrates two points of our assumptions:

1. The parameter vector p must in some cases be constrained to lie in Π, an open set strictly contained in $\mathbb{R}^l$. Indeed, here, for $p_2 = 0$, u_n is not defined.
2. The function V looks like a Lyapunov function in x, but actually it may not be radially unbounded in x. For instance, in Example (49), x_2 may go to infinity without V going to infinity. The radial unboundedness of a Lyapunov function $V(x(t),t)$ guarantees that the magnitude of $V(x(t),t)$ at time t gives a bound on the norm of the *full* state vector $x(t)$ for the same time t. In contrast, as illustrated by Example (49), the magnitude of $V(x(t),p)$, given by assumption BO, at time t gives a bound on the norm of only a *part* of the state vector $x(t)$ at the same time t, namely $x_1(t)$. What assumption BO actually guarantees is the "observability" of the boundedness of the full state vector x from the "output" function $V(x,p)$, i.e., that if the trajectory $\{V(x(t),p)\}_{t\in[0,T)}$ of the "output" is bounded, then so is the trajectory $\{x(t)\}_{t\in[0,T)}$ of the full state. Then, assumption PRS guarantees the existence of a control law which forces the part of the state vector mentioned above to converge to the corresponding part of the equilibrium point $\mathcal{E}$. Later on, we will add more constraints on the function V, e.g. assumption MC. In order to allow for more possibilities to find such a function V meeting all these requirements, it is useful at this stage to have assumption BO instead of the more standard but more restrictive radial unboundedness.

To guarantee convergence to $\mathcal{E}$ of the whole state vector x and not only of its reduced-order part, we will need also:

Assumption CO (Convergence Observability) (58)

For any bounded C^1 time functions $\widehat{p} : \mathbb{R}_+ \to \Pi$ and $u : \mathbb{R}_+ \to \mathbb{R}^m$ with $\dot{\widehat{p}}$ also bounded and for any solution $x(t)$ of (44) defined on $[0, +\infty)$, we have the following implication:

$$\left.\begin{array}{r} \lim_{t\to\infty} V(x(t), \widehat{p}(t)) \text{ exists and is zero,} \\ x(t) \text{ is bounded on } [0, +\infty) \\ \text{and} \quad x(t) \in \Omega \quad \forall t \in [0, +\infty) \end{array}\right\} \implies \lim_{t\to\infty} x(t) \text{ exists and is equal to } \mathcal{E}.$$

Example: System (17) Continued (59)

Assumption CO is satisfied for the system (17) rewritten as (50) with the choice (51). This is a consequence of the uniform asymptotic stability of the zero solution of (see Lakshmikantham and Leela [11, Theorem 3.8.3]):

$$\dot{x}_2 = -p_2\, x_2^3 \,. \tag{60}$$

Indeed, since x_1 is a component of a solution of an ordinary differential equation, the convergence of $x_1^2(t)$ to 0, implies the boundedness:

$$|x_1(t)| \leq \alpha \quad \forall t \,. \tag{61}$$

Then, since assumption BO is satisfied,

$$|x_2(t)| \leq \gamma = \max\{|x_2(0)|\,,\, |c_3|\,\alpha\} \quad \forall t \,. \tag{62}$$

Now, the convergence of $x_1^2(t)$ to 0 implies also that for all $\varepsilon > 0$ there exists a time T such that:

$$|x_1(t)| \leq \varepsilon \qquad \forall t \geq T \,. \tag{63}$$

And, with (55), we have:

$$\dot{\overline{x_2^2(t)}} \leq -p_2 \left(x_2^4(t) - |c_3|\,\varepsilon\, |x_2(t)|^3 \right) \quad \forall t \geq T \,. \tag{64}$$

It follows that:

$$|x_2(t)| \geq \varepsilon\,(1 + |c_3|) \qquad \to \qquad \dot{\overline{x_2^2(t)}} \leq -\varepsilon^4 \,. \tag{65}$$

Therefore, with (62), defining $\tau(\varepsilon)$ by:

$$\tau = \frac{1}{\varepsilon^4} \left(\gamma^2 - \varepsilon^2\,(1 + |c_3|)^2 \right), \tag{66}$$

we have established:

$$\forall \varepsilon > 0\ \exists T, \exists \tau \,:\, \forall t \geq T + \tau\,,\, |x_2(t)| \leq \varepsilon\,(1 + |c_3|)\,. \tag{67}$$

This is CO. □

This Example illustrates a typical fact about assumption CO: it is usually difficult to prove that it holds. Above we used a Total Stability argument, here is another example invoking Barbălat's Lemma [26, p.211]:

Example: **(68)**

Consider the following system:

$$\begin{aligned} \dot{x}_1 &= x_2 + p^* \\ \dot{x}_2 &= u\,. \end{aligned} \tag{69}$$

We wish to stabilize the equilibrium point $x_1 = 0, x_2 = -p^*$. Assumptions BO and PRS are satisfied when we choose:

$$V(x_1, x_2, p) = x_1^2 + (x_2 + p + x_1)^2\,, \quad u_n(x_1, x_2, p) = -2\,(x_2 + p + x_1)\,. \tag{70}$$

To check that CO is also met, we follow the same steps as the ones proposed by Kanellakopoulos, Kokotovic and Marino for the proof of [9, Theorem 1]. Clearly, if $V(x_1(t), x_2(t), \widehat{p}(t))$ tends to 0, the same holds for $x_1(t)$ and $x_2(t) + x_1(t) + \widehat{p}(t)$. To obtain our conclusion, it is sufficient to prove that $\dot{\overline{x_1(t)}}$ tends to 0. Indeed, in such a case, the first equation in (69) implies in this case that $x_2(t) + p^*$ tends to 0. Since we have:

$$\lim_{t\to+\infty} x_1(t) \;=\; 0 \;=\; x_1(0) + \lim_{t\to+\infty} \int_0^t \dot{\overline{x_1(s)}}\, ds\,, \tag{71}$$

from Barbălat's Lemma, $\dot{\overline{x_1(t)}}$ tends to 0 if this time function is uniformly continuous. This is indeed the case since its time derivative is:

$$\ddot{\overline{x_1(t)}} \;=\; \dot{\overline{x_2(t)}} \;=\; u(t)\,, \tag{72}$$

where, by assumption, the time function $u(t)$ is bounded. □

Assumptions BO, PRS and CO are weaker than (A1), (A2) and (A3). Indeed, it follows from linear systems theory [7] and Total Stability theorems [11] that assumptions (A1) and (A3) imply the existence of:

- Ω, an open neighborhood of $\mathcal{E}$,
- Π, an open neighborhood of p^*,
- P, an $n \times n$ positive definite matrix,
- C, an $n \times m$ matrix, and
- k, a strictly positive constant,

such that, by letting:

$$V(x,p) = V(x) = (x - \mathcal{E})^{\mathrm{T}}\, P\,(x - \mathcal{E})\,, \quad u_n(x,p) = -C\,(x - \mathcal{E}) + u_0(p)\,, \tag{73}$$

we have:

$$\frac{\partial V}{\partial x}(x)\,[a_0(x) + u_n(x,p) \odot b(x) + A_0(x)\,p] \le k\,V(x)\,, \quad \forall (x,p) \in \Omega \times \Pi\,. \tag{74}$$

This implies that assumption PRS is satisfied. Also, the function V is positive definite and radially unbounded. Hence, assumptions BO and CO hold with $\Omega_0 = \Omega$ and α_0 the largest positive real number α such that:

$$V(x) < \alpha \qquad \Longrightarrow \qquad x \in \Omega . \tag{75}$$

Finally, note that (A2) is not needed.

If the value of the parameter vector p^* were known, assumptions BO, PRS and CO would be sufficient to guarantee the stabilizability of $\mathcal{E}$. This is made precise as follows:

Proposition **(76)**
Let assumptions BO and PRS hold, p^* be in Π and the control $u_n(x,p^*)$ be applied to the system (S_{p^*}). Under these conditions, all the solutions $x(t)$ with initial condition $x(0)$ in Ω_0 and satisfying $V(x(0),p^*) < \alpha_0$ are well-defined on $[0,+\infty)$, unique, bounded and:

$$\lim_{t\to\infty} V(x(t),p^*) = 0 . \tag{77}$$

If, moreover, assumption CO holds, then $x(t)$ converges to $\mathcal{E}$.

Proof. The system we consider is:

$$\dot{x} = a(x,u_n(x,p^*)) + A(x,u_n(x,p^*))p^* . \tag{78}$$

With p^* fixed in Π, this is an autonomous system with its right-hand side continuously differentiable in the open subset Ω of $\mathbb{R}^n$. Hence, for any initial condition $x(0)$ in $\Omega_0 \subset \Omega$, there exists a unique solution $x(t)$ of (78) in Ω. It is a continuously differentiable time function (and so is u_n) defined on a right maximal interval $[0,T)$, with T a strictly positive (possibly infinite) real number. Let us prove by contradiction that $T = +\infty$ if $V(x(0),p^*) < \alpha_0$. Assume the contrary. From the theorem on continuation of solutions [5, Theorem I.2.1], $x(t)$ tends to the boundary of Ω as t tends to T. But since:

$$x(t) \in \Omega , \ \forall t \in [0,T) \qquad \text{and} \qquad p^* \in \Pi , \tag{79}$$

we may use (47) in assumption PRS and conclude:

$$\dot{\overline{V(x(t),p^*)}} \leq 0 \quad \forall t \in [0,T) . \tag{80}$$

This yields:

$$V(x(t),p^*) \leq V(x(0),p^*) < \alpha_0 \quad \forall t \in [0,T) . \tag{81}$$

Then, from assumption BO, we know there exists a compact subset Γ of Ω, depending on p^*, $x(0)$ and $V(x(0),p^*)$, such that:

$$x(t) \in \Gamma , \ \forall t \in [0,T) . \tag{82}$$

Since the set Γ is compact, it is strictly contained in the open set Ω. This establishes the contradiction and the fact that the time functions $x(t)$ and $u(t) = u_n(x(t), p^*)$ are bounded on $[0, +\infty)$.

Now, from (47), we have:

$$\overline{V(x(t), p^*)}^{\cdot} < 0 \quad \forall t : V(x(t), p^*) \neq 0 \qquad (83)$$
$$= 0 \quad \forall t : V(x(t), p^*) = 0\,.$$

Since the function V is nonnegative, this implies (77).

Finally, convergence to $\mathcal{E}$ of $x(t)$ is a straightforward consequence of assumption CO. □

In the following we will show that even when the value of p^* is unknown, in which case we cannot implement $u_n(x, p^*)$, the Adaptive Stabilization problem can be solved if extra assumptions are added. In Section 2, a solution will be obtained from the Lyapunov design with the assumption that a so-called matching condition is satisfied. In Section 3, other solutions will be obtained from an estimation approach. They will require a stronger version of assumption PRS and that either a matching condition or some growth conditions on the nonlinearities be satisfied.

2 Lyapunov Design

In a famous paper, Parks [18] suggested a very efficient way of getting a controller for the linearly parameterized family of systems (S_p). The idea is to use the control:

$$u = u_n(x, \widehat{p}) \qquad (84)$$

with the time function $\widehat{p}$ selected so that a positive definite radially unbounded function of x and $\widehat{p}$ be decaying. Even though here the function V is not radially unbounded, let us pursue this idea and compute the time derivative of

$$W(x, \widehat{p}) = V(x, \widehat{p}) + \frac{1}{2}\, \|\widehat{p} - p^*\|^2 \qquad (85)$$

along the solutions of (S_{p^*})-(84). Using (47) in assumption PRS, we get:

$$\begin{aligned}
\dot{W} &= \frac{\partial V}{\partial x}(x, \widehat{p})\, [a(x, u_n(x, \widehat{p})) + A(x, u_n(x, \widehat{p}))\, p^*] \\
&\quad + \left(\frac{\partial V}{\partial p}(x, \widehat{p}) + [\widehat{p} - p^*]^{\mathrm{T}}\right) \dot{\widehat{p}} \\
&= \frac{\partial V}{\partial x}(x, \widehat{p})\, [a(x, u_n(x, \widehat{p})) + A(x, u_n(x, \widehat{p}))\, \widehat{p}] \\
&\quad + \left[-\frac{\partial V}{\partial x}(x, \widehat{p})\, A(x, u_n(x, \widehat{p})) + \dot{\widehat{p}}^{\mathrm{T}}\right] [\widehat{p} - p^*] + \frac{\partial V}{\partial p}(x, \widehat{p})\dot{\widehat{p}} \\
&\leq \left[-\frac{\partial V}{\partial x}(x, \widehat{p})\, A(x, u_n(x, \widehat{p})) + \dot{\widehat{p}}^{\mathrm{T}}\right] [\widehat{p} - p^*] + \frac{\partial V}{\partial p}(x, \widehat{p})\dot{\widehat{p}}\,. \qquad (86)
\end{aligned}$$

2.1 Case: V Independent of p, i.e., $V(x,p) = V(x)$

It follows that in the particular case where the function V does not depend on p, by choosing:

$$\dot{\widehat{p}} = \left[\frac{\partial V}{\partial x}(x)\, A(x, u_{\mathrm{n}}(x, \widehat{p}))\right]^{\mathrm{T}} , \qquad \widehat{p}(0) \in \Pi , \tag{87}$$

we are guaranteed that W remains bounded and in particular:

$$V(x(t)) \leq V(x(0)) + \frac{1}{2} \|\widehat{p}(0) - p^*\|^2 . \tag{88}$$

However, this boundedness property is not sufficient, since to use assumptions BO and PRS, we must check that for all t:

1. the following inequality is satisfied:

$$V(x(t)) < \alpha_0 , \tag{89}$$

2. the following membership property is satisfied:

$$\widehat{p}(t) \in \Pi . \tag{90}$$

It is straightforward to see that, if $x(t)$ and $\widehat{p}(t)$ are continuous functions of t and the following function W:

$$W(x(t), \widehat{p}(t)) = \frac{\alpha_1 V(x(t))}{\alpha_1 - V(x(t))} + \frac{1}{2} \|\widehat{p}(t) - p^*\|^2 \tag{91}$$

is positive and bounded for all t, then, necessarily, $V(x(t))$ is strictly smaller than α_1 for all t. Consequently, (89) is satisfied for all t if we can ensure that the modified function $W(x(t), \widehat{p}(t))$ (91) is positive and bounded for all t and the constant α_1 is chosen smaller than or equal to α_0.

In order to meet the membership property (90), we shall constrain $\widehat{p}$ to remain in a closed convex subset of Π by projection of $\dot{\widehat{p}}$. For this, we need the following property of the set Π:

Assumption ICS (Imbedded Convex Sets) (92)
There exists a known *convex C^2 function $\mathcal{P}$ from $\mathbb{R}^l$ to $\mathbb{R}$ such that:*

1. *for each real number λ in $[0, 1]$, the set:*

$$\Pi_\lambda = \{p \mid \mathcal{P}(p) \leq \lambda\} \tag{93}$$

is contained in Π,

2. *there exists a strictly positive constant d such that:*

$$\left\|\frac{\partial \mathcal{P}}{\partial p}(p)\right\| \geq d \qquad \forall p \in \{p \mid 0 \leq \mathcal{P}(p)\} , \tag{94}$$

3. the parameter vector p^* of the system to be actually controlled satisfies:

$$\mathcal{P}(p^*) < 0 \qquad \textit{and} \qquad D^* \stackrel{\text{def}}{=} \operatorname{dist}(p^*, \{p \mid \mathcal{P}(p) = 0\}) > 0 . \tag{95}$$

Example: System (17) Continued **(96)**

For the system (17) rewritten as (50) in Example (49), the set Π is defined by $p_2 > 0$. Then, by choosing:

$$\mathcal{P}(p_1, p_2) = 2\left(1 - \frac{p_2}{\varepsilon}\right), \tag{97}$$

with $\varepsilon > 0$, this set Π satisfies assumption ICS if $p_2^* > \varepsilon$.

More generally, consider the case where the set Π is:

$$p = (p_1, \ldots, p_l)^{\mathrm{T}} \in \Pi \quad \Longleftrightarrow \quad |p_i - \rho_i| < \sigma_i , \quad \forall i \in \{1, \ldots, l\}, \tag{98}$$

with ρ_i and σ_i some given real numbers. To meet assumption ICS we may choose the function $\mathcal{P}$ as:

$$\mathcal{P}(p) = \frac{2}{\varepsilon}\left[\sum_{i=1}^{l} \left|\frac{p_i - \rho_i}{\sigma_i}\right|^q - 1 + \varepsilon\right], \tag{99}$$

with $0 < \varepsilon < 1$ and $q \geq 2$ two real numbers. In this case, we get:

$$\Pi_\lambda = \left\{p \;\middle|\; \sum_{i=1}^{l} \left|\frac{p_i - \rho_i}{\sigma_i}\right|^q \leq 1 - \varepsilon\left(1 - \frac{\lambda}{2}\right)\right\}, \tag{100}$$

and the set Π_λ, for $\lambda = 1$, approaches Π when ε decreases and q increases. □

Assumption ICS allows us to define the closed convex subset Π_1 of Π as:

$$\Pi_1 = \{p \mid \mathcal{P}(p) \leq 1\} , \tag{101}$$

and the function Proj as:

$$\operatorname{Proj}(M, p, y) = \begin{cases} y & \text{if } \mathcal{P}(p) \leq 0 \text{ or } \frac{\partial \mathcal{P}}{\partial p}(p)\, y \leq 0 \\ y - \frac{\mathcal{P}(p) \frac{\partial \mathcal{P}}{\partial p}(p) y}{\frac{\partial \mathcal{P}}{\partial p}(p) M \frac{\partial \mathcal{P}}{\partial p}(p)^{\mathrm{T}}} M \frac{\partial \mathcal{P}}{\partial p}(p)^{\mathrm{T}} & \text{if } \mathcal{P}(p) > 0 \text{ and } \frac{\partial \mathcal{P}}{\partial p}(p)\, y > 0 , \end{cases} \tag{102}$$

where M is a symmetric positive definite $l \times l$ matrix. Namely, $\operatorname{Proj}(M, p, y)$ is equal to y if p belongs to the set $\{\mathcal{P}(p) \leq 0\}$. In the set $\{0 \leq \mathcal{P}(p) \leq 1\}$, it substracts a vector M-normal to the boundary $\{\mathcal{P}(p) = \lambda\}$, so that we get a smooth transformation from the original vector field for $\lambda = 0$ to an inward or tangent vector field for $\lambda = 1$. We have the following technical properties proved in Appendix A:

Lemma **(103)**

Let $\mathcal{M}$ be the open set of symmetric positive definite $l \times l$ matrices. If assumption ICS holds, then:

1. *The function* $\mathrm{Proj}(M,p,y) \;:\; \mathcal{M} \times \Pi \times \mathbb{R}^l \;\to\; \mathbb{R}^l$ *is locally Lipschitz-continuous.*
2. $\mathrm{Proj}(M,p,y)^{\mathrm{T}} M^{-1} \mathrm{Proj}(M,p,y) \leq y^{\mathrm{T}} M^{-1} y \qquad \forall p \in \Pi_1$
3. $\frac{\partial \mathcal{P}}{\partial p}(p)(p-p^*) \geq D^* \left\| \frac{\partial \mathcal{P}}{\partial p}(p) \right\| \qquad \forall p : \mathcal{P}(p) \geq 0$ *with D^* defined in (95),*
4. $(p-p^*)^{\mathrm{T}} M^{-1} \mathrm{Proj}(M,p,y) \leq (p-p^*)^{\mathrm{T}} M^{-1} y$
5. *Let $(M,y) \;:\; \mathbb{R}_+ \;\to\; \mathcal{M} \times \mathbb{R}^l$ be a C^1 time function. On their domain of definition, the solutions of:*

$$\dot{\widehat{p}} = \mathrm{Proj}\left(M(t)\,,\,\widehat{p}\,,\,y(t)\right) \quad \widehat{p}(0) \in \Pi_1 \tag{104}$$

satisfy $\widehat{p}(t) \in \Pi_1$.

We are now ready to propose the following dynamic controller to solve the Adaptive Stabilization problem when the function V does not depend on p:

$$\begin{aligned} \dot{\widehat{p}} &= \mathrm{Proj}\left(I\,,\,\widehat{p}\,,\,\frac{{\alpha_1}^2}{(V(x)-\alpha_1)^2}\left[\frac{\partial V}{\partial x}(x)\,A(x,u_{\mathrm{n}}(x,\widehat{p}))\right]^{\mathrm{T}}\right) \\ u &= u_{\mathrm{n}}(x\,,\,\widehat{p})\,, \end{aligned} \tag{105}$$

where $\widehat{p}(0)$ is selected in Π_1 and the matrix M used in the function Proj is I, the identity matrix. We have:

Proposition **(106)**

Let assumptions BO, PRS and ICS hold with a function V not depending on p. Assume also that assumption Λ-LP is satisfied with:

$$\Lambda(x,t) \;=\; \frac{\partial V}{\partial x}(x)\,. \tag{107}$$

If, in (105), α_1 is chosen smaller than or equal to α_0, then all the solutions $(x(t),\widehat{p}(t))$ of (S_{p^*})-(105) with $x(0) \in \Omega_0$ and $V(x(0)) < \alpha_1$ are well-defined on $[0,+\infty)$, unique, bounded and:

$$\lim_{t\to\infty} V(x(t)) \;=\; 0\,. \tag{108}$$

It follows that the Adaptive Stabilization problem is solved if assumption CO also holds.

Proof. The system we consider is:

$$\begin{aligned} \dot{x} &= a(x,u_{\mathrm{n}}(x\,,\,\widehat{p})) + A(x,u_{\mathrm{n}}(x\,,\,\widehat{p}))p^* \\ \dot{\widehat{p}} &= \mathrm{Proj}\left(I\,,\,\widehat{p}\,,\,\frac{{\alpha_1}^2}{(V(x)-\alpha_1)^2}\left[\frac{\partial V}{\partial x}(x)\,A(x,u_{\mathrm{n}}(x,\widehat{p}))\right]^{\mathrm{T}}\right). \end{aligned} \tag{109}$$

From our smoothness assumptions on the functions a, A, $u_{\rm n}$ and V and with Point 1 of Lemma (103), this system has a locally Lipschitz-continuous right-hand side in the open set defined by:

$$(x, \widehat{p}) \in \Omega \times \Pi \qquad \text{and} \qquad V(x) < \alpha_1 . \tag{110}$$

It follows that, for any initial condition in this open set, there exists a unique solution $(x(t), \widehat{p}(t))$, defined on a right maximal interval $[0, T)$, with T possibly infinite, and satisfying (110) for all t in $[0, T)$. Applying point 5 of Lemma (103), we also know that $\widehat{p}(t) \in \Pi_1$ for all t in $[0, T)$. Then, let us compute the time derivative of W defined in (91) along such a solution. With assumption PRS, we get as in (86):

$$\dot{W} \leq \left[-\frac{\alpha_1^2}{(V(x(t)) - \alpha_1)^2} \left(\frac{\partial V}{\partial x}(x(t))\, A(x, u_n(x(t), \widehat{p}(t)) \right)^{\rm T} + \dot{\widehat{p}}^{\rm T}(t) \right] (\widehat{p}(t) - p^*) , \tag{111}$$

with a strict inequality if $V(x(t)) \neq 0$. But, with the expression of $\dot{\widehat{p}}$ and Point 4 of Lemma (103), we get readily:

$$\begin{aligned} \dot{W} &\leq 0 \ \text{ if } V(x(t)) = 0 \\ &< 0 \ \text{ if } V(x(t)) \neq 0 . \end{aligned} \tag{112}$$

It follows that for all t in $[0, T)$:

$$\begin{aligned} & V(x(t)) < \alpha_1 \\ & \frac{\alpha_1 V(x(t))}{\alpha_1 - V(x(t))} \leq \frac{\alpha_1 V(0)}{\alpha_1 - V(0)} + \frac{1}{2} \|\widehat{p}(0) - p^*\|^2 \stackrel{\rm def}{=} \beta \\ & \|\widehat{p}(t) - p^*\|^2 \leq 2\beta . \end{aligned} \tag{113}$$

Hence, we get:

$$V(x(t)) \leq \frac{\alpha_1 \beta}{\alpha_1 + \beta} \stackrel{\rm def}{=} \alpha < \alpha_1 \leq \alpha_0 , \tag{114}$$

and we know that $\widehat{p}(t) \in \mathcal{K}$, where $\mathcal{K}$ is the following compact subset of Π:

$$\mathcal{K} = \left\{ p \,\middle|\, \|p - p^*\|^2 \leq 2\beta \right\} \bigcap \Pi_1 . \tag{115}$$

Then, from assumption BO, we know the existence of a compact subset Γ of Ω such that:

$$x(t) \in \Gamma \quad \forall t \in [0, T) . \tag{116}$$

Hence, the solution remains in a compact subset of the open set defined in (110). It follows by contradiction that $T = +\infty$ and, in particular, that the time functions $x(t)$, $\widehat{p}(t)$, $u(t) = u_{\rm n}(x(t), \widehat{p}(t))$ and $\dot{\widehat{p}}(t)$ are bounded on $[0, +\infty)$. Then (108) is a straightforward consequence of (112) and LaSalle's Theorem [5, Theorem X.1.3]. The conclusion follows readily from assumption CO. □

Example: (117)
Consider the following system:

$$\dot{x} = p^* x^3 + (1 - x^2)\, u \tag{118}$$

with p^* positive. We wish to stabilize the equilibrium point $x = 0$. Clearly, even if p^* were known, this would not be possible globally, but only for $x \in (-1\,,\,1)$. Then let $\Omega = (-1\,,\,1)$ and choose:

$$V(x) = x^2 \quad \text{and} \quad u_{\mathrm{n}}(x,p) = -\frac{p\,x^3 + x}{1 - x^2}\,. \tag{119}$$

Assumptions Λ-LP, BO, PRS and CO are satisfied with $\alpha_0 = 1$. According to Proposition (106), the following dynamic controller guarantees the convergence of $x(t)$ to 0 for all initial conditions $\widehat{p}(0)$ and $x(0)$, with $|x(0)| < 1$:

$$\begin{aligned} \dot{\widehat{p}} &= \frac{2x^4}{(1-x^2)^2} \\ u &= -\frac{\widehat{p}\,x^3 + x}{1 - x^2}\,. \end{aligned} \tag{120}$$

□

Compared with Proposition (76), this Proposition (106) states that, for solving the Adaptive Stabilization problem when p^* is unknown, the existence of a function V independent of p is a sufficient condition. This shows that we should look for a control law u_{n} for which we can find such a function V to meet assumption BO and PRS (see [19, Chapter 2]). This is where the fact that V need not be radially unbounded proves to be useful, as we now illustrate:

Example: System (17) Continued (121)
For the system (17) rewritten as (50) in Example (49), we have established that assumptions Λ-LP, BO, PRS and ICS hold when Λ, V and u_{n} are chosen as:

$$\Lambda(x_1, x_2, t) = (x_1\ 0)\,, \qquad V(x_1, x_2, p_1, p_2) = \frac{1}{2}\,x_1^2\,, \tag{122}$$

and:

$$u_{\mathrm{n}}(x_1, x_2, p_1, p_2) = -\frac{x_2^2 + p_1 + x_1}{p_2}\,. \tag{123}$$

By specializing the dynamic controller (105) to this system (50) and choosing $\alpha_1 = +\infty$ and $\mathcal{P}$ as in (97), the following controller solves the Adaptive Stabilization problem:

$$\begin{aligned} \dot{\widehat{p}}_1 &= x_1^3 \\ \dot{\widehat{p}}_2 &= -\frac{x_1^3(x_2^2 + \widehat{p}_1 + x_1)}{\widehat{p}_2} && \text{if } \widehat{p}_2 \geq \varepsilon \ \text{ or } \ \frac{x_1^3(x_2^2 + \widehat{p}_1 + x_1)}{\widehat{p}_2} \leq 0 \\ &= \left(1 - 2\frac{\widehat{p}_2}{\varepsilon}\right) \frac{x_1^3(x_2^2 + \widehat{p}_1 + x_1)}{\widehat{p}_2} && \text{if } \widehat{p}_2 < \varepsilon \text{ and } \frac{x_1^3(x_2^2 + \widehat{p}_1 + x_1)}{\widehat{p}_2} < 0 \\ u &= -\frac{x_2^2 + \widehat{p}_1 + x_1}{\widehat{p}_2}\,. \end{aligned} \tag{124}$$

□

Proposition (106) generalizes results established by Sastry and Isidori [29] for systems that are input-output linearizable via state feedback, and by Taylor et al. [31] for state-feedback linearizable systems:

Corollary [Sastry and Isidori [29, Relative Degree One]] (125)
Let a and A in equation (S_p) be affine in u, let the system to be controlled have a single input, i.e., $m = 1$ and, finally, let Π be an open subset of $\mathbb{R}^l$ which satisfies assumption ICS. Assume the existence of two C^2 functions $h : \mathbb{R}^n \to \mathbb{R}$ and $\varphi : \mathbb{R}^n \to \mathbb{R}^{n-1}$ such that:

1. *$h(\mathcal{E}) = 0$,*
2. *the functions $\frac{\partial h}{\partial x}(x)a(x,u)$ and $\frac{\partial h}{\partial x}(x)A(x,u)$ are* known,
3. *we have for all (x,p) in $\mathbb{R}^n \times \Pi$ (with notation (13)):*

$$\frac{\partial h}{\partial x}(x)\,(b(x) + B(x)p) \neq 0\,, \tag{126}$$

4. *$(h(x), \varphi(x))$ is a diffeomorphism and defines new coordinates with which the system (S_{p^*}) can be rewritten as:*

$$\dot{h} = \frac{\partial h}{\partial x}(x)\,[a(x,u) + A(x,u)p^*] \tag{127}$$

$$\dot{\varphi} = Z(\varphi, h)\,, \tag{128}$$

where Z is a function which is further assumed to be globally Lipschitz.

Assume also that $\varphi(\mathcal{E})$ is an exponentially stable equilibrium point in $\mathbb{R}^{n-1}$ of:

$$\dot{\varphi} = Z(\varphi, 0)\,. \tag{129}$$

Under these conditions, we can find functions V, u_{n} and $\mathcal{P}$ such that the corresponding dynamic controller (105) solves the Adaptive Stabilization problem.

Assumption (126) means that each system (S_p), with $p \in \Pi$, is of relative degree one with respect to the output function h (see Isidori [6]). And the exponential stability of (129) implies that (S_{p^*}) is globally exponentially minimum phase.

Proof. Assumption Λ-LP is met with the choice:

$$\Lambda(x,t) = \frac{\partial h}{\partial x}(x)\,. \tag{130}$$

Then, let:

$$\Omega = \Omega_0 = \mathbb{R}^n\,, \qquad \alpha_0 = +\infty\,, \tag{131}$$

and choose the function V independent of p as simply:

$$V(x) = h(x)^2\,. \tag{132}$$

According to [29, Proposition 2.1], assumption BO is satisfied. An argument similar to the one used in Example (59) (replace K by ε in [29, 2.27]) proves that assumption CO holds also.

Now from (126), there exists a C^1 function $u_n : \mathbb{R}^n \times \Pi \to \mathbb{R}$ such that:

$$\frac{\partial h}{\partial x}(x)\,[a(x,u_n(x,p)) + A(x,u_n(x,p))p] = -c\,h(x)\,, \tag{133}$$

with c a strictly positive constant. With (127), this implies that assumption PRS holds.

In conclusion, the controller (105) may be employed, with the function $\mathcal{P}$ given by assumption ICS and $\alpha_1 = +\infty$. □

Corollary [Taylor et al. [31]] **(134)**
Let, in equation (S_p)*, the functions* a *and* A *be* known *and affine in* u*, let* Π *be an open subset of* $\mathbb{R}^l$ *which satisfies assumption ICS and, finally, let* p_0 *be a* known *vector in* Π*. Assume there exist an open neighborhood* Ω *of* $\mathcal{E}$ *in* $\mathbb{R}^n$ *and three* known *functions:*

$$\begin{array}{lll} \Phi & : \Omega \to \mathbb{R}^n & \text{of class } C^2 \text{ which is a diffeomorphism,} \\ w_1 & : \Omega \to \mathbb{R}^m & \text{of class } C^1\,, \text{ and} \\ w_2 & : \Omega \to \mathcal{GL}(m,\mathbb{R}) & \text{of class } C^1\,, \end{array}$$

such that:

1. *by letting:*
$$\varphi = \Phi(x) \qquad \text{and} \qquad u = w_1(x) + w_2(x)\,\vartheta\,, \tag{135}$$
the time derivative of φ *along the solutions of* (S_{p_0}) *satisfies, for all* ϑ *in* $\mathbb{R}^m$*:*
$$\dot{\varphi} = C\varphi + D\vartheta\,, \tag{136}$$
where D *is an* $n \times m$ *matrix and* C *is an* $n \times n$ *matrix satisfying:*
$$PC + C^{\mathrm{T}}P = -I\,, \tag{137}$$
with P *a symmetric positive definite matrix,*
2. *for all* (x,p,u) *in* $\Omega \times \Pi \times \mathbb{R}^m$*, we have, with notation (13):*
$$\operatorname{rank}\{b(x) + B(x)p\} = m \tag{138}$$
$$[A_0(x) + u \odot B(x)]\,p :\in \operatorname{span}\{b(x) + B(x)p_0\}\,. \tag{139}$$

Under these conditions, we can find functions V*,* u_n *and* $\mathcal{P}$ *and a constant* α_1 *such that the corresponding dynamic controller (105) solves the Adaptive Stabilization problem.*

Note that (136) simply implies that $u = w_1 + w_2\vartheta$ is a feedback linearizing control for the system (S_{p_0}) in the coordinates $\varphi = \Phi(x)$.

Proof. Under the above assumptions, it is proved in [31, Proposition S] that:

$$\Phi^{-1}(0) = \mathcal{E} \tag{140}$$

and there exists a known C^1 function $u_n : \Omega \times \Pi \to \mathbb{R}^m$ such that, for all (x,p) in $\Omega \times \Pi$, we have:

$$\frac{\partial \Phi}{\partial x}(x)\left(a(x,p) + A(x,p)u_n(x,p)\right) = C\Phi(x). \tag{141}$$

It follows that assumption PRS is satisfied if we choose:

$$V(x) = \Phi(x)^{\mathrm{T}} P \Phi(x). \tag{142}$$

To check that assumption BO is satisfied, let us define α_0 as the largest (possibly infinite) real number such that the set:

$$\{\varphi \mid \varphi^{\mathrm{T}} P \varphi < \alpha_0\}$$

is contained in $\Phi(\Omega)$. Since $\Phi(\mathcal{E}) = 0$, Φ is a diffeomorphism, and P is a positive definite matrix, the so-defined α_0 is strictly positive. Then the sets:

$$\Gamma_\alpha = \Phi^{-1}\{\varphi \mid \varphi^{\mathrm{T}} P \varphi \leq \alpha\} \tag{143}$$

are compact subsets of Ω for all $\alpha < \alpha_0$. It follows that:

$$V(x) \leq \alpha \quad \Longrightarrow \quad x \in \Gamma_\alpha, \tag{144}$$

which implies that BO is satisfied with $\Omega_0 = \Omega$.

Assumption CO holds since:

$$V(x(t)) \to 0 \implies \Phi(x(t)) \to 0 \implies x(t) = \Phi^{-1}(\Phi(x(t))) \to \mathcal{E}. \tag{145}$$

Since assumption Λ-LP is also satisfied with the function Λ equal to the $n \times n$ identity matrix, the controller (105) may be employed, with the function $\mathcal{P}$ given by assumption ICS and any α_1 such that $0 < \alpha_1 \leq \alpha_0$. □

One of the features of the result established by Taylor et al. [31] is that it gives a sufficient condition which guarantees the existence of a function V not depending on p while satisfying assumptions BO, PRS and CO. This condition is assumption (139), called the strict matching condition in [31, Assumption L]. It turns out that (with notation (13)), when $B \equiv 0$ and $b(\mathcal{E})$ has full rank m, this condition is also the necessary and sufficient condition for the existence of a regular static state-feedback law:

$$u = c(x) + d(x)\,w + e(x)p, \tag{146}$$

which decouples the state vector x from p seen as a measured disturbance (see [6, Proposition 5.5.1 and Proposition 7.3.1]). As pointed out by Pomet in [19, Sect. 1.2.3], it follows that, by applying the control:

$$u = e(x)\,(p - p_0) + w \tag{147}$$

to any system (S_p) (with input u), $p \in \Pi$, we obtain the particular system (S_{p_0}) (with input w). In fact the same result holds if $B \not\equiv 0$:

Lemma [19, Théorème 2.2] **(148)**
Let, in equation (S_p), the functions a and A be known *and affine in u, let Π be an open subset of $\mathbb{R}^l$, let Ω be an open neighborhood of $\mathcal{E}$ in $\mathbb{R}^n$ and, finally, let p_0 be a* known *vector in Π. Assume:*

$$\operatorname{rank}\{b(x) + B(x)p\} = m \qquad \forall (x,p) \in \Omega \times \Pi . \tag{149}$$

Under these conditions, assumption (139), i.e.,

$$[A_0(x) + u \odot B(x)]\, p :\in \operatorname{span}\{b(x) + B(x)p_0\} \qquad \forall (x,p,u) \in \Omega \times \Pi \times \mathbb{R}^m , \tag{150}$$

is equivalent to the following proposition:
There exist two known *C^1 functions:*

$$c : \Omega \times \Pi \to \mathbb{R}^m , \qquad d : \Omega \times \Pi \to \mathcal{GL}(m, \mathbb{R}),$$

such that, for all (x,p,w) in $\Omega \times \Pi \times \mathbb{R}^m$:

$$a(x,w) + A(x,w)\, p_0 = a\,(x\,,\, c(x,p) + d(x,p)w) + A\,(x\,,\, c(x,p) + d(x,p)w)\, p\, . \tag{151}$$

A straightforward consequence is:

Corollary [19, Proposition 8.13] **(152)**
Let, in equation (S_p), the functions a and A be known *and affine in u, let Π be an open subset of $\mathbb{R}^l$ which satisfies assumption ICS and, finally, let p_0 be a* known *vector in Π. Assume there exists an open neighborhood Ω of $\mathcal{E}$ in $\mathbb{R}^n$ such that:*

1. $\operatorname{rank}\{b(x) + B(x)p\} = m \qquad \forall (x,p) \in \Omega \times \Pi$,
2. $[A_0(x) + u \odot B(x)]\, p \in \operatorname{span}\{b(x) + B(x)p_0\} \qquad \forall (x,p,u) \in \Omega \times \Pi \times \mathbb{R}^m$,
3. *there exist two* known *functions:*

$$V : \Omega \to \mathbb{R}_+ \text{ of class } C^2 , \qquad u_0 : \Omega \to \mathbb{R}^m \text{ of class } C^1 ,$$

such that the functions $\frac{\partial V}{\partial x}(x)\, a(x,u)$ and $\frac{\partial V}{\partial x}(x)\, A(x,u)$ are known, *assumptions BO and CO hold and we have, for all x in Ω:*

$$\frac{\partial V}{\partial x}(x)\, [a(x, u_0(x)) + A(x, u_0(x))\, p_0] \leq 0 , \tag{153}$$

where the inequality is strict iff $V(x) \neq 0$.

Under these conditions, we can find functions V, u_n and $\mathcal{P}$ and a constant α_1 such that the corresponding dynamic controller (105) solves the Adaptive Stabilization problem.

Proof. Assumption Λ-LP holds with:

$$\Lambda(x,t) = \frac{\partial V}{\partial x}(x). \tag{154}$$

Then, since assumptions BO, CO and ICS are satisfied, it is sufficient to define a function u_n meeting assumption PRS. According to Lemma (148), let:

$$u_n(x,p) = c(x,p) + d(x,p)\, u_0(x). \tag{155}$$

This is a C^1 function, and inequality (47) in assumption PRS is satisfied since (151) and (153) hold. □

Assumption (139) is very restrictive. Fortunately, (139) is not necessary for the existence of a function V independent of p, as illustrated below.

Example: **(156)**
Consider the following system:

$$\begin{aligned} \dot{x}_1 &= u \\ \dot{x}_2 &= x_1 \\ \dot{x}_3 &= x_2 + p^* \,(x_3 + x_2)\,(x_1 + 2x_2 + 2x_3). \end{aligned} \tag{157}$$

We have:

$$\begin{aligned} a_0(x) = \begin{pmatrix} 0 \\ x_1 \\ x_2 \end{pmatrix}, \qquad & A_0(x) = \begin{pmatrix} 0 \\ 0 \\ (x_3 + x_2)\,(x_1 + 2x_2 + 2x_3) \end{pmatrix}, \\ b(x) = \begin{pmatrix} 1 \\ 0 \\ 0 \end{pmatrix} \qquad & \text{and} \qquad B(x) = 0. \end{aligned} \tag{158}$$

Hence, even though b is of full rank 1, assumption (139) does not hold. Nevertheless, assumptions BO, PRS and CO are satisfied if we choose V independent of p as:

$$V(x_1, x_2, x_3) = \frac{x_3^2}{2} + \frac{(x_3 + x_2)^2}{2} + \frac{(x_1 + 2x_2 + 2x_3)^2}{2} \tag{159}$$

and u_n as:

$$u_n(x_1, x_2, x_3, p) = -3x_3 - 5x_2 - 3x_1 - p\,(x_3 + x_2)\,(6x_3 + 5x_2 + 2x_1). \tag{160}$$

Indeed, V is positive definite and radially unbounded and a straightforward computation leads to the following expression for the time derivative of V along the solutions of (157):

$$\begin{aligned} \dot{V} = & - x_3^2 - (x_3 + x_2)^2 \\ & + (x_1 + 2x_2 + 2x_3)\,[u + 2x_1 + 3x_2 + x_3 + p\,(x_3 + x_2)\,(6x_3 + 5x_2 + 2x_1)]. \end{aligned} \tag{161}$$

□

2.2 Case: V Dependent on p with a "Matching Condition"

When V depends on p, choosing $\dot{\widehat{p}}$ as in (87) is not sufficient to guarantee that $\dot{W}$ is negative when W is given by (85). This follows from the disturbing term $\frac{\partial V}{\partial p}(x,\widehat{p})\dot{\widehat{p}}$ present in (86). To overcome this difficulty, a very fruitful idea proposed by Kanellakopoulos, Kokotovic and Marino [8] (see also Middleton and Goodwin [15]) is to compensate this measurable disturbing term by modifying the control to:

$$u = u_{\mathrm{n}}(x,\widehat{p}) + v\,. \tag{162}$$

Indeed, in this case, (86) becomes:

$$\begin{aligned}\dot{W} \leq \Big[&-\frac{\partial V}{\partial x}(x,\widehat{p})\,A(x,u_{\mathrm{n}}(x,\widehat{p})+v)+\dot{\widehat{p}}^{\mathrm{T}}\Big]\,[\widehat{p}-p^*] + \frac{\partial V}{\partial p}(x,\widehat{p})\dot{\widehat{p}}\\ &+\frac{\partial V}{\partial x}(x,\widehat{p})\,[a(x,u_{\mathrm{n}}(x,\widehat{p})+v)-a(x,u_{\mathrm{n}}(x,\widehat{p}))\\ &+(A(x,u_{\mathrm{n}}(x,\widehat{p})+v)-A(x,u_{\mathrm{n}}(x,\widehat{p})))\,\widehat{p}]\,.\end{aligned} \tag{163}$$

Hence, taking:

$$\dot{\widehat{p}} = \left[\frac{\partial V}{\partial x}(x,\widehat{p})\,A(x,u_{\mathrm{n}}(x,\widehat{p})+v)\right]^{\mathrm{T}}, \qquad \widehat{p}(0) \in \Pi\,, \tag{164}$$

and choosing v as a solution of:

$$\begin{aligned}0 = \frac{\partial V}{\partial p}(x,\widehat{p})\,\dot{\widehat{p}} + \frac{\partial V}{\partial x}(x,\widehat{p})\,&[a(x,u_{\mathrm{n}}(x,\widehat{p})+v)-a(x,u_{\mathrm{n}}(x,\widehat{p}))\\ &+(A(x,u_{\mathrm{n}}(x,\widehat{p})+v)-A(x,u_{\mathrm{n}}(x,\widehat{p})))\,\widehat{p}]\,,\end{aligned} \tag{165}$$

we guarantee that $\dot{W}$ is negative. However, a new difficulty may arise from the fact that (164) and (165) is a system of implicit equations to be solved in $\dot{\widehat{p}}$ and v. A solution smoothly depending on x and $\widehat{p}$ may not exist. Pomet [19] (see also [21]) has proposed the following way to avoid this difficulty, when a and A are affine in u:
With the control u as in (162), i.e.,

$$u = u_{\mathrm{n}}(x,p) + v\,, \tag{166}$$

we may embed the family of systems (S_p) into the following larger family $(S_{p,q})$ (with notation (13)):

$$\dot{x} = a_0 + (u_{\mathrm{n}}+v)\odot b + (A_0 + u_{\mathrm{n}}\odot B)\,p + v\odot B\,q\,, \quad (p,q)\in\Pi\times\Pi\,. \tag{$S_{p,q}$}$$

Then, we modify the function W to:

$$W(x,\widehat{p}) = V(x,\widehat{p}) + \frac{1}{2}\|\widehat{p}-p^*\|^2 + \frac{1}{2}\|\widehat{q}-p^*\|^2 . \tag{167}$$

As above, the time derivative of this function W is made negative if we choose:

$$\begin{aligned}
\dot{\widehat{p}} &= \left[\frac{\partial V}{\partial x}(x,\widehat{p})\,(A_0(x)+u_{\rm n}(x,\widehat{p})\odot B(x))\right]^{\rm T}, \qquad \widehat{p}(0)\in\Pi\\
\dot{\widehat{q}} &= \left[\frac{\partial V}{\partial x}(x,\widehat{p})\,v\left(x,\widehat{p},\widehat{q},\dot{\widehat{p}}\right)\odot B(x)\right]^{\rm T}, \qquad \widehat{q}(0)\in\Pi\\
u &= u_{\rm n}(x,\widehat{p}) + v\left(x,\widehat{p},\widehat{q},\dot{\widehat{p}}\right).
\end{aligned} \tag{168}$$

This dynamic controller is well-defined with unique solutions if we assume:

Assumption MC (Matching Condition) (169)
The functions a and A are affine in u and there exists a known C^1 *function $v(x,p,q,\partial)$ from $\Omega\times\Pi\times\Pi\times\mathbb{R}^l$ to $\mathbb{R}^m$ satisfying:*

$$\frac{\partial V}{\partial p}(x,p)\,\partial + \frac{\partial V}{\partial x}(x,p)\,v\odot[b(x)+B(x)\,q] = 0 . \tag{170}$$

We remark that, if V is independent of p, this assumption is trivially satisfied (by $v=0$). Also, the way this assumption is stated is too restrictive for our purpose. In order to make $\dot{W}$ negative it is sufficient that (170) be satisfied for the particular case $\partial=\dot{\widehat{p}}$ and not for all $\partial\in\mathbb{R}^l$. We have stated assumption MC is these terms to allow us independent choices for the function V and for $\dot{\widehat{p}}$. Indeed, relaxing (170) by replacing ∂ by $\dot{\widehat{p}}$ implies that V, $u_{\rm n}$ and $\dot{\widehat{p}}$ must be designed all together so that BO, CO, PRS and MC are satisfied at the same time.

Example: (171)
Consider the system, with $n=2$ and $l=1$:

$$\begin{aligned}
\dot{x}_1 &= u\\
\dot{x}_2 &= x_2^2\,(x_1+p\,x_2^2).
\end{aligned} \tag{172}$$

Assumptions BO, PRS and CO are satisfied, with:

$$V(x_1,x_2,p) = x_2^2 + (x_1+x_2+p\,x_2^2)^2 \tag{173}$$

$$u_{\rm n}(x_1,x_2,p) = -x_2 - x_2^3 - (x_1+p\,x_2^2)\left(1+x_2^2\,(1+2\,p\,x_2)\right). \tag{174}$$

Assumption Λ-LP is also satisfied with:

$$\Lambda(x_1,x_2,t) = \frac{\partial V}{\partial x}(x_1,x_2,\widehat{p}(t)). \tag{175}$$

Finally, assumption MC holds since in this case equation (170) is:

$$2\,x_2^2\,(x_1+x_2+p\,x_2^2)\,\partial + 2\,(x_1+x_2+p\,x_2^2)\,v = 0, \tag{176}$$

which is satisfied by the choice

$$v = -x_2^2\,\partial\,. \tag{177}$$

□

Finally, to guarantee that $\widehat{p}(t)$ and $\widehat{q}(t)$ remain in Π and that $V(x(t),\widehat{p}(t))$ remains smaller than α_0, we have to modify further the controller (168) as in the previous section:

$$\begin{aligned}
\dot{\widehat{p}} &= \operatorname{Proj}\left(I, \widehat{p}, \frac{{\alpha_1}^2}{(V(x,\widehat{p})-\alpha_1)^2}\left[\frac{\partial V}{\partial x}(x,\widehat{p})\,(A_0(x) + u_{\mathrm{n}}(x,\widehat{p}) \odot B(x))\right]^{\mathrm{T}}\right)\\
\dot{\widehat{q}} &= \operatorname{Proj}\left(I, \widehat{q}, \frac{{\alpha_1}^2}{(V(x,\widehat{p})-\alpha_1)^2}\left[\frac{\partial V}{\partial x}(x,\widehat{p})\,v(x,\widehat{p},\widehat{q},\dot{\widehat{p}}) \odot B(x)\right]^{\mathrm{T}}\right)\\
u &= u_{\mathrm{n}}(x,\widehat{p}) + v\left(x,\widehat{p},\widehat{q},\dot{\widehat{p}}\right),
\end{aligned} \tag{178}$$

with $\widehat{p}(0)$ and $\widehat{q}(0)$ in Π_1. We have:

Proposition **(179)**
Let assumptions BO, PRS, ICS and MC hold. Assume also that assumption Λ-LP is satisfied with:

$$\Lambda(x,t) = \frac{\partial V}{\partial x}(x,\widehat{p}(t))\,. \tag{180}$$

If, in (178), α_1 is chosen smaller than or equal to α_0, then all the solutions $(x(t),\widehat{p}(t),\widehat{q}(t))$ of (S_{p^*})-(178), with $x(0)$ in Ω_0 and $V(x(0),\widehat{p}(0)) < \alpha_1$, are well-defined on $[0,+\infty)$, unique, bounded and:

$$\lim_{t\to\infty} V(x(t),\widehat{p}(t)) = 0\,. \tag{181}$$

It follows that the Adaptive Stabilization problem is solved if assumption CO holds also.

Proof. The system we consider is:

$$\begin{aligned}
\dot{x} &= a\left(x\,,\,u_{\mathrm{n}}(x,\widehat{p}) + v(x,\widehat{p},\widehat{q},\dot{\widehat{p}})\right) + A\left(x\,,\,u_{\mathrm{n}}(x,\widehat{p}) + v(x,\widehat{p},\widehat{q},\dot{\widehat{p}})\right)\,p^*\\
\dot{\widehat{p}} &= \operatorname{Proj}\left(I, \widehat{p}, \frac{{\alpha_1}^2}{(V(x,\widehat{p})-\alpha_1)^2}\left[\frac{\partial V}{\partial x}(x,\widehat{p})\,(A_0(x) + u_{\mathrm{n}}(x,\widehat{p}) \odot B(x))\right]^{\mathrm{T}}\right)\\
\dot{\widehat{q}} &= \operatorname{Proj}\left(I, \widehat{q}, \frac{{\alpha_1}^2}{(V(x,\widehat{p})-\alpha_1)^2}\left[\frac{\partial V}{\partial x}(x,\widehat{p})\,v(x,\widehat{p},\widehat{q},\dot{\widehat{p}}) \odot B(x)\right]^{\mathrm{T}}\right),
\end{aligned} \tag{182}$$

with $\widehat{p}(0)$ and $\widehat{q}(0)$ in Π_1. From our smoothness assumptions on the functions a, A, u_{n} and V and with Point 1 of Lemma (103), this system has a locally Lipschitz-continuous right-hand side in the open set defined by:

$$(x,\widehat{p},\widehat{q}) \in \Omega \times \Pi \times \Pi \qquad \text{and} \qquad V(x,\widehat{p}) < \alpha_1\,. \tag{183}$$

It follows that, for any initial condition in this open set, there exists a unique solution $(x(t), \widehat{p}(t), \widehat{q}(t))$, defined on a right maximal interval $[0, T)$, with T possibly infinite, and satisfying (183) for all t in $[0, T)$. Applying Point 5 of Lemma (103), we also know that $\widehat{p}(t) \in \Pi_1$ and $\widehat{q}(t) \in \Pi_1$ for all t in $[0, T)$. Then, let us compute the time derivative, along such a solution, of the function W defined by:

$$W(x, \widehat{p}) = \frac{\alpha_1 V(x, \widehat{p})}{\alpha_1 - V(x, \widehat{p})} + \frac{1}{2} \|\widehat{p} - p^*\|^2 + \frac{1}{2} \|\widehat{q} - p^*\|^2 . \tag{184}$$

With assumption PRS, Point 4 of Lemma (103) and equation (170) with $\partial = \dot{\widehat{p}}$ satisfied by $v\left(x, \widehat{p}, \widehat{q}, \dot{\widehat{p}}\right)$, we get:

$$\begin{aligned} \dot{W} &\leq 0 \ \text{ if } V(x(t), \widehat{p}(t)) = 0 \\ &< 0 \ \text{ if } V(x(t), \widehat{p}(t)) \neq 0 . \end{aligned} \tag{185}$$

From there, we conclude exactly as in the proof of Proposition (106). □

Example: (186)
The assumptions of Proposition (179) are satisfied by the system (172) of Example (171). For this system, the Adaptive Stabilization problem is solved by the following dynamic controller (using (164), (173) and (177)):

$$\begin{aligned} \dot{\widehat{p}} &= 2 \left[x_2 + \left(x_1 + x_2 + \widehat{p} x_2^2\right)(1 + 2\widehat{p} x_2)\right] x_2^4 \\ u &= -x_2 - x_2^3 - \left(x_1 + p\, x_2^2\right)\left(1 + x_2^2 (1 + 2\, p\, x_2)\right) \\ &\quad - 2\, x_2^2 \left[x_2 + \left(x_1 + x_2 + \widehat{p} x_2^2\right)(1 + 2\widehat{p} x_2)\right] x_2^4 . \end{aligned} \tag{187}$$

□

Proposition (179) generalizes to the case where V is not radially unbounded a result established by Kanellakopoulos, Kokotovic and Marino for state-feedback linearizable systems [8]:

Corollary [Kanellakopoulos, Kokotovic and Marino [8]] (188)
Let, in equation (S_p), the functions a and A be known *and affine in u, let p_0 be a* known *vector in $\mathbb{R}^l$. Assume there exist a bounded open subset Ω of $\mathbb{R}^n$, an open neighborhood $\mho$ of p_0 in $\mathbb{R}^l$ and three* known *functions:*

$$\begin{aligned} &\Psi : \Omega \to \mathbb{R}^n && \textit{of class } C^2 \textit{ which is a diffeomorphism,} \\ &w_1 : \Omega \to \mathbb{R}^m && \textit{of class } C^1 , \textit{ and} \\ &w_2 : \Omega \to \mathcal{GL}(m, \mathbb{R}) && \textit{of class } C^1 , \end{aligned}$$

such that:

1. *by letting:*

$$\varphi = \Psi(x) \qquad \text{and} \qquad u = w_1(x) + w_2(x)\,\vartheta\,, \tag{189}$$

the time derivative of φ along the solutions of (S_{p_0}) satisfies, for all ϑ in $\mathbb{R}^m$:

$$\dot{\varphi} = C\varphi + D\vartheta\,, \tag{190}$$

where D is an $n \times m$ matrix and C is an $n \times n$ matrix satisfying:

$$PC + C^{\mathrm{T}}P = -I\,, \tag{191}$$

with P a symmetric positive definite matrix,

2. *for all (x,p,u) in $\Omega \times \mho \times \mathbb{R}^m$, we have, with notation (13):*

$$\operatorname{rank}\{b(x) + B(x)p\} = m \tag{192}$$

$$u \odot B(x)\,p \in \operatorname{span}\{b(x) + B(x)p_0\}$$

$$A_0(x)\,p :\in \operatorname{span}\{b(x) + B(x)p_0\,,\ [a_0 + A_0p_0\,,\ \operatorname{span}\{b + Bp_0\}]\,(x)\}\,, \tag{193}$$

where $[\cdot\,,\,\cdot]$ denotes the Lie bracket.

Under these conditions and if p^* is close enough to p_0, we can find functions V, u_n and $\mathcal{P}$ and a constant α_1 such that the corresponding dynamic controller (178) solves the Adaptive Stabilization problem.

Proof. From the sections "State diffeomorphism" and "Ideal feedback control" in [8] (see also [9]), there exist an open neighborhood Π_d of p_0 and three known functions:

$\Phi \;: \Omega \times \Pi_d \to \mathbb{R}^n$ of class C^2, a diffeomorphism for each p,
$u_n : \Omega \times \Pi_d \to \mathbb{R}^m$ of class C^1, and
$v \;: \Omega \times \Pi_d \times \Pi_d \times \mathbb{R}^l \to \mathbb{R}^m$ of class C^1,

such that $\mathcal{E}$ belongs to Ω,

$$\Phi(\mathcal{E},p^*) = 0\,, \quad \Phi(x,p_0) = \Psi(x) \quad \forall x \in \Omega\,, \tag{194}$$

and, for all (x,p,q,∂) in $\Omega \times \Pi_d \times \Pi_d \times \mathbb{R}^l$, we have:

$$\frac{\partial\Phi}{\partial x}(x,p)\,[a(x,p) + A(x,p)u_n(x,p)] = C\,\Phi(x,p) \tag{195}$$

$$\frac{\partial\Phi}{\partial p}(x,p)\,\partial + \frac{\partial\Phi}{\partial x}(x,p)\,v \odot [b(x) + B(x)\,q] = 0\,. \tag{196}$$

Then, let us choose the function V as:

$$V(x,p) = \Phi(x,p)^{\mathrm{T}}P\,\Phi(x,p)\,. \tag{197}$$

In order to choose the set Π and the scalar α_0, let us define a function F by:

$$\begin{array}{rcl} F : \Omega \times \Pi_d & \rightarrow & \mathbb{R}^n \times \Pi_d \,. \\ (x,p) & & (\Phi(x,p)\,,\,p) \end{array} \tag{198}$$

This function is a diffeomorphism satisfying $F(\Psi^{-1}(0), p_0) = (0, p_0)$. Since $\Omega \times \Pi_d$ is an open neighborhood of $(\Psi^{-1}(0), p_0)$ and p^* is assumed to be close enough from p_0, there exist strictly positive real numbers α_0 and π such that:

$$\|p^* - p_0\|^2 < \frac{\pi}{1+\varepsilon}\,, \tag{199}$$

with $0 < \varepsilon < 1$, and the set:

$$\left\{(\varphi,p) \,\middle|\, \|p - p_0\|^2 < \pi \quad \text{and} \quad \varphi^{\mathrm{T}} P \varphi < \alpha_0\right\}$$

is contained in $F(\Omega \times \Pi_d)$ and contains $\mathcal{E}$. Then, let us define the function $\mathcal{P}$ by:

$$\mathcal{P}(p) = \frac{2}{\varepsilon}\left[\frac{1+\varepsilon}{\pi}\|p - p_0\|^2 - 1\right], \tag{200}$$

and the set Π by:

$$\Pi = \{p \,|\, \mathcal{P}(p) < 1+\varepsilon\}\,. \tag{201}$$

Assumption ICS is satisfied.

From (195) and (196), it is clear that assumptions PRS and MC are satisfied. To check that assumption BO is satisfied, we remark that for all compact subsets $\mathcal{K}$ of Π and all $\alpha < \alpha_0$, the sets:

$$\{(x,p) \,|\, \Phi(x,p)^{\mathrm{T}} P \Phi(x,p) \leq \alpha \quad \text{and} \quad p \in \mathcal{K}\}$$

are compact subsets of $\Omega \times \Pi$ and therefore their projections:

$$\Gamma_{\alpha,\mathcal{K}} = \{x \,|\, \exists p \in \mathcal{K} : \Phi(x,p)^{\mathrm{T}} P \Phi(x,p) \leq \alpha \quad\} \tag{202}$$

are compact subsets of Ω. It follows that:

$$V(x,p) \leq \alpha \quad \text{and} \quad p \in \mathcal{K} \qquad \Longrightarrow \qquad x \in \Gamma_{\alpha,\mathcal{K}}\,, \tag{203}$$

which implies that BO is satisfied with $\Omega_0 = \Omega$.

The proof that assumption CO holds follows from the remark:

$$V(x(t), \widehat{p}(t)) \rightarrow 0 \quad \Longrightarrow \quad \Phi(x(t, \widehat{p}(t))) \rightarrow 0 = \Phi(\mathcal{E}, p^*) \tag{204}$$

and the proof of [9, Theorem 1] (see also Example (68)).

Since assumption Λ-LP is also satisfied with the function Λ equal to the $n \times n$ identity matrix, the controller (178) applies. □

One of the nice results proved by Kanellakopoulos, Kokotovic and Marino is that for feedback linearizable systems, assumption (193), called the extended matching condition in [8, Assumption E] is necessary and sufficient for assumption MC to hold. In fact, as for the case where V does not depend on p, this assumption (193) relies on the fact that we can transform any system (S_p) into a particular one (S_{p_0}) but this time by using both feedback and diffeomorphism. Precisely, we have:

Lemma [19, Théorème 2.9 and Lemme 8.14] **(205)**
Let the functions a and A in equation (S_p) be affine in u and let $\mathcal{U}$ be an open neighborhood of $\mathcal{E}$ in $\mathbb{R}^n$. Assume that, in $\mathcal{U}$ (with notation (13)):

1. *The distribution* $\operatorname{span}\{b + Bp^*\}$ *is involutive with constant full rank m,*
2. *The distribution* $\operatorname{span}\{b + Bp^*\,,\, [a_0 + A_0 p^*\,,\, \operatorname{span}\{b + Bp^*\}]\}$ *has constant rank.*

Under these conditions, assumption (193), i.e.,

$$\begin{aligned} u \odot B(x)\, p &:\in \operatorname{span}\{b(x) + B(x)p_0\} \\ A_0(x)\, p &:\in \operatorname{span}\{b(x) + B(x)p_0\,,\, [a_0 + A_0 p_0\,,\, \operatorname{span}\{b + Bp_0\}]\,(x)\}\,, \end{aligned} \tag{206}$$

is equivalent to the following proposition:
There exist an open neighborhood Ω of $\mathcal{E}$, an open neighborhood Π of p^* of $\mathbb{R}^l$, a vector p_0 in Π and four smooth functions:

$$\begin{array}{lll} \Phi : \Omega \times \Pi \;\to\; \mathbb{R}^n & \textit{of class } C^2, \textit{ a diffeomorphism for each } p, \\ c\ : \Omega \times \Pi \;\to\; \mathbb{R}^m & \textit{of class } C^1\,, \\ d\ : \Omega \times \Pi \;\to\; \mathcal{GL}(m, \mathbb{R}) & \textit{of class } C^1\,, \textit{ and} \\ v\ : \Omega \times \Pi \times \Pi \times \mathbb{R}^l \;\to\; \mathbb{R}^m & \textit{of class } C^1\,, \end{array}$$

such that, for all (x, p, q, ∂) in $\Omega \times \Pi \times \Pi \times \mathbb{R}^l$, we have:

$$\frac{\partial \Phi}{\partial p}(x,p)\,\partial + \frac{\partial \Phi}{\partial x}(x,p)\, v \odot [b(x) + B(x)\, q] = 0\,, \tag{207}$$

and, for all (x, p, w) in $\Omega \times \Pi \times \mathbb{R}^m$, we have:

$$\begin{aligned} &a(\Phi(x,p), w) + A(\Phi(x,p), w)\, p_0 \\ &\quad = \frac{\partial \Phi}{\partial x}(x,p)\,[a\,(x\,,\, c(x,p) + d(x,p)w) + A\,(x\,,\, c(x,p) + d(x,p)w)\, p]\,. \end{aligned} \tag{208}$$

Comparing (207) and (170) in assumption MC, we understand the importance of this result. It gives us a possible route for finding functions V and u_n satisfying all our assumptions BO, PRS, CO and MC. Indeed, if

1. the conditions of Lemma (205) are satisfied, and
2. for the particular system (S_{p_0}), with p_0 given by Lemma (205), there exist V_0 and u_0 satisfying assumptions BO, PRS and CO but this time with the family (S_p) reduced to the single element (S_{p_0}), i.e., $p = p^* = p_0$,

then, by choosing:

$$\begin{aligned} V(x,p) &= V_0(\Phi(x,p)) \\ u_n(x,p) &= c(x,p) + d(x,p)\, u_0\,(\Phi(x,p))\,, \end{aligned} \tag{209}$$

assumptions BO, PRS, CO and MC are necessarily satisfied.

However, it is important to notice that finding a solution v in $\mathbb{R}^m$ to the n equations (207) is more difficult than finding a solution v in $\mathbb{R}^m$ to the single equation (170). For single-input two-dimensional affine in u systems with $b(\mathcal{E}) \neq 0$ and $B(x) \in \operatorname{span}\{b(x)\}$, the assumptions of Lemma (205) are "generically" satisfied. Indeed, we can expect that, for almost all x close to $\mathcal{E}$, we have:

$$\operatorname{span}\{b(x), [a_0, b](x)\} = \mathbb{R}^2. \tag{210}$$

However, the set of such x's may not be a neighborhood of $\mathcal{E}$, implying that (207) may not hold. Nevertheless, d'Andréa-Novel, Pomet and Praly [1] have shown that, for this two-dimensional case, there are explicit expressions for the functions V, u_n and v satisfying assumptions BO, PRS and MC.

Compared with Proposition (76), Proposition (179) states that, when p^* is unknown, the solution to the Adaptive Stabilization problem given by the Lyapunov design requires the Matching Condition (MC). And nothing is known if this condition does not hold.

Another but very particular possibility to handle the case where V depends on p is to choose:

$$W = \frac{\alpha_1 V(x,p^*)}{\alpha_1 - V(x,p^*)} + \frac{1}{2}\|\widehat{p} - p^*\|^2 \qquad \text{and} \qquad u = u_n(x,\widehat{p}). \tag{211}$$

This gives:

$$\dot{W} = \dot{\widehat{p}}^{\mathrm{T}}\,[\widehat{p} - p^*] + \frac{{\alpha_1}^2}{(\alpha_1 - V(x,p^*))^2}\frac{\partial V}{\partial x}(x,p^*)\,[a(x,u_n(x,\widehat{p})) + A(x,u_n(x,\widehat{p}))p^*]. \tag{212}$$

Then, using assumption PRS, we get:

$$\begin{aligned}\dot{W} \leq{}& \dot{\widehat{p}}^{\mathrm{T}}\,[\widehat{p} - p^*] + \frac{{\alpha_1}^2}{(\alpha_1 - V(x,p^*))^2}\frac{\partial V}{\partial x}(x,p^*)\\ &\times [a(x,u_n(x,\widehat{p})) - a(x,u_n(x,p^*)) + (A(x,u_n(x,\widehat{p})) - A(x,u_n(x,p^*)))\,p^*].\end{aligned} \tag{213}$$

There is no general expression of $\dot{\widehat{p}}$ not depending on p^* and making the right-hand side of this inequality negative. However, Slotine and Li [30] have shown that in the particular case of rigid robot arms, it is possible to find functions V and u_n which satisfy assumption BO, PRS and CO and are such that:

There exists a C^1 function $Z : \Omega \times \Pi \to \mathbb{R}^l$ such that, for the particular value p^* but for all $(x,p) \in \Omega \times \Pi$, we have:

$$\begin{aligned}[p^* - p]^{\mathrm{T}}\, Z(x,p) ={}& \frac{{\alpha_1}^2}{(\alpha_1 - V(x,p^*))^2}\frac{\partial V}{\partial x}(x,p^*)\\ &\times [a(x,u_n(x,p)) - a(x,u_n(x,p^*)) + (A(x,u_n(x,p)) - A(x,u_n(x,p^*)))\,p^*].\end{aligned} \tag{214}$$

Indeed, with such a property, the Adaptive Stabilization problem is solved by choosing:

$$\dot{\widehat{p}} = \operatorname{Proj}(I, \widehat{p}, Z(x, \widehat{p})). \tag{215}$$

3 Estimation Design

Another way to obtain dynamic controllers to solve the Adaptive Stabilization problem is to hope that a separation principle holds, i.e., that we can get an estimate $\widehat{p}$ of p^* by using a parameter estimator and simultaneously we apply the control u_n with the unknown parameter vector p^* replaced by its estimate $\widehat{p}$, i.e.,

$$u = u_n(x, \widehat{p}). \tag{216}$$

For the estimation, we note that, thanks to assumption Λ-LP, equation (S_p) is linear in p, i.e., it can be rewritten as:

$$z = Z\,p, \tag{217}$$

with:

$$z = \dot{x} - a(x, u) \qquad \text{and} \qquad Z = A(x, u). \tag{218}$$

Also, the vector p^* we want to estimate is constant, i.e., it is a solution of:

$$\dot{p} = 0. \tag{219}$$

Estimating p^* is then equivalent to observing, through the observation equation (217), the state vector p which obeys the (trivial) dynamic equation (219). From linear observer theory [7], we know that an observer can be written as:

$$\dot{\widehat{p}} = -K\,(Z\,\widehat{p} - z), \tag{220}$$

with K the observer gain. Unfortunately, such an observer cannot be implemented, since it makes use of the unknown quantities $\dot{x}$, $a(x, u)$ and $A(x, u)$. This difficulty is handled as follows:

About the unmeasured time derivative $\dot{x}$, it is quite clear that an integration should help us. It turns out that two ways of implementing this integration are fruitful:

1. equation error filtering, and
2. regressor filtering.

These techniques will be presented hereafter.

To deal with the fact that only the functions $\Lambda\, a$ and $\Lambda\, A$ are known, we select an integer k and a C^2 function $h : \Omega \times \Pi \to \mathbb{R}^k$ such that the functions

$\frac{\partial h}{\partial x}(x,p)\, a(x,u)$ and $\frac{\partial h}{\partial x}(x,p)\, A(x,u)$ are known functions of (x,u,p), i.e., assumption Λ-LP is met with:

$$\Lambda(x,t) \;=\; \frac{\partial h}{\partial x}(x,\widehat{p}(t))\,. \tag{221}$$

This function h is called the *observation function*. For any C^1 time functions $\widehat{p} : \mathbb{R}_+ \rightarrow \Pi$ and $u : \mathbb{R}_+ \rightarrow \mathbb{R}^m$, the time derivative of $h(x(t),\widehat{p}(t))$ with $x(t)$ a solution of (S_{p^*}) satisfies:

$$\dot{h} \;=\; \frac{\partial h}{\partial x}(x,\widehat{p})\,(a(x,u) + A(x,u)\,p^*) \;+\; \frac{\partial h}{\partial p}(x,\widehat{p})\,\dot{\widehat{p}}\,. \tag{222}$$

This is again an equation linear in p^* and equality (217) is again satisfied but with the following definitions of z and Z:

$$\begin{aligned} z &= \dot{h} \;-\; \frac{\partial h}{\partial x}(x,\widehat{p})\,a(x,u) \;-\; \frac{\partial h}{\partial p}(x,\widehat{p})\,\dot{\widehat{p}} \\ Z &= \frac{\partial h}{\partial x}(x,\widehat{p})\,A(x,u)\,. \end{aligned} \tag{223}$$

Example: System (17) Continued **(224)**

For the system (17) rewritten as (50) in Example (49), let us choose the observation function h as:

$$h(x_1,x_2,p_1,p_2) \;=\; \frac{1}{2}\,x_1^2\,. \tag{225}$$

We get:

$$\dot{\overline{h(x_1(t),x_2(t),p_1(t),p_2(t))}} \;=\; x_1^3(t)\,\left(x_2^2(t) + p_1^* + p_2^*\,u(t)\right). \tag{226}$$

By letting:

$$\begin{aligned} z(t) &= \dot{\overline{h(x_1(t),x_2(t),p_1(t),p_2(t))}} \;-\; x_1^3(t)\,x_2^2(t) \\ Z(t) &= \begin{pmatrix} x_1^3(t) & x_1^3(t)\,u(t) \end{pmatrix}, \end{aligned} \tag{227}$$

we obtain equation (217), i.e.,

$$z(t) \;=\; Z(t)\begin{pmatrix} p_1^* \\ p_2^* \end{pmatrix}. \tag{228}$$

□

3.1 Equation Error Filtering

The estimator is obtained from the observer (220) where the so called equation error $Z\widehat{p} - z$ is replaced by a filtered version. Namely, let e in $\mathbb{R}^k$ be defined as follows:

$$\dot{e} + r(e, x, \widehat{p})\, e = Z\, \widehat{p} - z\,, \tag{229}$$

or equivalently, using (223):

$$\begin{aligned} \dot{\widehat{h}} &= -r(e, x, \widehat{p})\, e + \frac{\partial h}{\partial x}(x, \widehat{p})\, A(x, u)\, \widehat{p} + \frac{\partial h}{\partial x}(x, \widehat{p})\, a(x, u) + \frac{\partial h}{\partial p}(x, \widehat{p})\, \dot{\widehat{p}} \\ e &= \widehat{h} - h(x, \widehat{p})\,, \end{aligned} \tag{230}$$

with r a positive C^1 function defined in $\mathbb{R}^k \times \Omega \times \Pi$. Definitely, e can be obtained without knowing $\dot{h}$ and the estimate $\widehat{p}$ is then given by:

$$\dot{\widehat{p}} = \mathrm{Proj}\,(I\,, \widehat{p}\,, -K\, e)\,, \qquad \widehat{p}(0) \in \Pi_1\,, \tag{231}$$

where the matrix M used in the function Proj is the identity matrix and, typically:

$$K = Z^{\mathrm{T}} = \left(\frac{\partial h}{\partial x}(x, \widehat{p})\, A(x, u)\right)^{\mathrm{T}}. \tag{232}$$

Unfortunately, as in the Lyapunov design case, if, instead of (216), we implement the control:

$$u = u_{\mathrm{n}}(x, \widehat{p}) + v\,, \tag{233}$$

where v depends on $\dot{\widehat{p}}$, we have again an implicit definition of $\dot{\widehat{p}}$ when v is explicitly involved in the right-hand side of (231). In such a case, if the functions a and A are affine in u, we derive the estimator from equation $(S_{p,q})$ instead of equation (S_p). Using notation (13), this leads to the following observer:

$$\begin{aligned} \dot{\widehat{h}} &= -r(e, x, \widehat{p})\, e + \frac{\partial h}{\partial x}(x, \widehat{p})\, [a_0(x) + A_0(x)\widehat{p} + u_{\mathrm{n}}(x, \widehat{p}) \odot (b(x) + B(x)\widehat{p})] \\ &\quad + \frac{\partial h}{\partial x}(x, \widehat{p})\, v\left(x, \widehat{p}, \widehat{q}, \dot{\widehat{p}}\right) \odot (b(x) + B(x)\widehat{q}) + \frac{\partial h}{\partial p}(x, \widehat{p})\, \dot{\widehat{p}} \\ e &= \widehat{h} - h(x, \widehat{p}) \\ \dot{\widehat{p}} &= \mathrm{Proj}\left(I, \widehat{p}, -\left[\frac{\partial h}{\partial x}(x, \widehat{p})\,(A_0(x) + u_{\mathrm{n}}(x, \widehat{p}) \odot B(x))\right]^{\mathrm{T}} e\right), \ \widehat{p}(0) \in \Pi_1 \\ \dot{\widehat{q}} &= \mathrm{Proj}\left(I, \widehat{q}, -\left[\frac{\partial h}{\partial x}(x, \widehat{p})\, v(x, \widehat{p}, \widehat{q}, \dot{\widehat{p}}) \odot B(x)\right]^{\mathrm{T}} e\right), \ \widehat{q}(0) \in \Pi_1\,. \end{aligned} \tag{234}$$

In the sequel, such a modified estimator will be implicitly assumed to be used whenever the Matching Condition holds.

We have:

Lemma **(235)**

Assume ICS is satisfied. For any C^1 time function $u : \mathbb{R}_+ \to \mathbb{R}^m$, all the solutions $(x(t), \widehat{p}(t), \widehat{h}(t))$ of (S_{p^*})-(230)-(231)-(232) defined on $[0,T)$ with $x(t)$ remaining in Ω satisfy for all t in $[0,T)$:

1. $\widehat{p}(t) \in \Pi_1$
2. $$\|\widehat{p}(t) - p^*\|^2 + \|e(t)\|^2 + 2\int_0^t r\,\|e\|^2 \leq \|\widehat{p}(0) - p^*\|^2 + \|e(0)\|^2 .$$

Similarly for the solutions of (S_{p^*,p^*})-(234), we have, for all t in $[0,T)$:

3. $\widehat{p}(t) \in \Pi_1$ *and* $\widehat{q}(t) \in \Pi_1$
4. $$\left\| \begin{matrix} \widehat{p}(t) - p^* \\ \widehat{q}(t) - p^* \end{matrix} \right\|^2 + \|e(t)\|^2 + 2\int_0^t r\,\|e\|^2 \leq \left\| \begin{matrix} \widehat{p}(0) - p^* \\ \widehat{q}(0) - p^* \end{matrix} \right\|^2 + \|e(0)\|^2 .$$

Proof. Point 1 is a straightforward consequence of Point 5 of Lemma (103). For Point 2, let us denote η the input of the dynamic system (229), i.e.,

$$\eta = Z\widehat{p} - z . \tag{236}$$

This system with output e is passive, namely, it satisfies for all t in $[0,T)$:

$$\int_0^t e^{\mathrm{T}}\eta = \int_0^t e^{\mathrm{T}}(\dot{e} + re) \tag{237}$$

$$= \frac{1}{2}\|e(t)\|^2 - \frac{1}{2}\|e(0)\|^2 + \int_0^t r\,\|e\|^2 \geq -\frac{1}{2}\|e(0)\|^2 . \tag{238}$$

On the other hand, the dynamic system (231) with input e and output y defined by:

$$y = Z\,(p^* - \widehat{p}) \tag{239}$$

is also passive. Indeed, we have thanks to Point 4 of Lemma (103):

$$\int_0^t y^{\mathrm{T}} e = \int_0^t (\widehat{p} - p^*)^{\mathrm{T}}(-Z^{\mathrm{T}})e \tag{240}$$

$$\geq \int_0^t (\widehat{p} - p^*)^{\mathrm{T}}\,\mathrm{Proj}\,(I, \widehat{p}, -Z^{\mathrm{T}}e) \tag{241}$$

$$\geq \int_0^t (\widehat{p} - p^*)^{\mathrm{T}}\,\dot{\widehat{p}} \tag{242}$$

$$\geq \frac{1}{2}\|\widehat{p}(t) - p^*\|^2 - \frac{1}{2}\|\widehat{p}(0) - p^*\|^2 \tag{243}$$

$$\geq -\frac{1}{2}\|\widehat{p}(0) - p^*\|^2 . \tag{244}$$

Noting that p^* satisfies:

$$z = Z p^* , \tag{245}$$

the two passive systems are interconnected with:

$$y = -\eta . \tag{246}$$

Then, Point 2 follows directly from standard passivity theorems (see Landau [12]) or more directly by comparing (238) and (243). The proof of Points 3 and 4 is similar. □

In fact, as emphasized by Landau [12] and expected from this proof, a similar Lemma would be obtained if, instead of the filter (229), we would have used any passive operator. In particular, as already noticed for adaptive control of linear systems by Narendra and Valavani [17], Pomet [19] has shown that, by choosing the identity matrix for the observation function h and a copy of the controlled system itself as the filter (229), we can rederive the adaptation law (178) of the Lyapunov design. Indeed, assume that BO, PRS and MC hold and define e as the output of, instead of (229), the following system with input η and state χ, with notation (13):

$$\begin{aligned} \dot{\chi} &= a_0(\chi) + A_0(\chi)\widehat{p} + u_{\mathrm{n}}(\chi, \widehat{p}) \odot (b(\chi) + B(\chi)\widehat{p}) \\ &\qquad + v\left(\chi, \widehat{p}, \widehat{q}, \dot{\widehat{p}}\right) \odot (b(\chi) + B(\chi)\widehat{q}) + \eta \\ e &= \frac{\partial V}{\partial \chi}(\chi, \widehat{p})^{\mathrm{T}} . \end{aligned} \tag{247}$$

To see that this system is passive, we look at the time derivative of $V(\chi(t), \widehat{p}(t))$. We get:

$$\dot{V} = \frac{\partial V}{\partial \chi}\left[a_0 + A_0 \widehat{p} + u_{\mathrm{n}} \odot (b + B\widehat{p})\right] + \frac{\partial V}{\partial \chi} v \odot (b + B\widehat{q}) + \frac{\partial V}{\partial \widehat{p}}\dot{\widehat{p}} + \frac{\partial V}{\partial \chi}\eta . \tag{248}$$

Hence, the definition of e, (47) in assumption PRS and (170) in assumption MC give readily:

$$\dot{V} \leq e^{\mathrm{T}}\eta . \tag{249}$$

Therefore, for any solution of (247) defined on $[0, T)$, we have:

$$\int_0^t e^{\mathrm{T}}\eta \geq V(\chi(t), \widehat{p}(t)) - V(\chi(0), \widehat{p}(0)) \qquad \forall t \in [0, T) . \tag{250}$$

In the proof of Lemma (235), we have seen that if equation $(S_{p,q})$ is satisfied with $p = p^*$ and $q = p^*$, the following system with input e, state $\begin{pmatrix} \widehat{p} \\ \widehat{q} \end{pmatrix}$ and ouput y:

$$\begin{aligned} \dot{\widehat{p}} &= \operatorname{Proj}\left(I, \widehat{p}, -\left[(A_0(\chi) + u_{\mathrm{n}}(\chi, \widehat{p}) \odot B(\chi))\right]^{\mathrm{T}} e\right) , \quad \widehat{p}(0) \in \Pi_1 \\ \dot{\widehat{q}} &= \operatorname{Proj}\left(I, \widehat{q}, -\left[v(\chi, \widehat{p}, \widehat{q}, \dot{\widehat{p}}) \odot B(\chi)\right]^{\mathrm{T}} e\right) , \quad \widehat{q}(0) \in \Pi_1 \\ y &= (A_0(\chi) + u_{\mathrm{n}}(\chi, \widehat{p}) \odot B(\chi))(\widehat{p} - p^*) + v\left(\chi, \widehat{p}, \widehat{q}, \dot{\widehat{p}}\right) \odot B(\chi)(\widehat{q} - p^*) \end{aligned} \tag{251}$$

is passive, namely:

$$\int_0^t e^{\mathrm{T}} y \geq \frac{1}{2} \left\| \begin{matrix} \widehat{p}(t) - p^* \\ \widehat{q}(t) - p^* \end{matrix} \right\|^2 - \frac{1}{2} \left\| \begin{matrix} \widehat{p}(0) - p^* \\ \widehat{q}(0) - p^* \end{matrix} \right\|^2 \qquad \forall t \in [0, T). \tag{252}$$

It follows that the observer defined by equations (247) and (251) is such that $V(\mathcal{X}(t), \widehat{p}(t))$, $\widehat{p}(t)$ and $\widehat{q}(t)$ are bounded if η is chosen such that:

$$\begin{aligned} \eta &= -y \\ &= (A_0(\mathcal{X}) + u_{\mathrm{n}}(\mathcal{X}, \widehat{p}) \odot B(\mathcal{X}))\,(p^* - \widehat{p}) \\ &\qquad + v\left(\mathcal{X}, \widehat{p}, \widehat{q}, \dot{\widehat{p}}\right) \odot B(\mathcal{X})\,(p^* - \widehat{q}). \end{aligned} \tag{253}$$

It remains to check that the observer (247)–(251) can be implemented while satisfying this equality, i.e., with (247) rewritten as:

$$\begin{aligned} \dot{\mathcal{X}} &= a_0(\mathcal{X}) + A_0(\mathcal{X})p^* + u_{\mathrm{n}}(\mathcal{X}, \widehat{p}) \odot (b(\mathcal{X}) + B(\mathcal{X})p^*) \\ &\qquad + v\left(\mathcal{X}, \widehat{p}, \widehat{q}, \dot{\widehat{p}}\right) \odot (b(\mathcal{X}) + B(\mathcal{X})p^*) + \eta. \end{aligned} \tag{254}$$

The right-hand side of this equation is nothing but a copy of the system (S_{p^*}) with the control:

$$u = u_{\mathrm{n}}(x, \widehat{p}) + v\left(x, \widehat{p}, \widehat{q}, \dot{\widehat{p}}\right). \tag{255}$$

It follows that if, in (247), we choose the initial condition

$$\mathcal{X}(0) = x(0), \tag{256}$$

then, necessarily, for all t:

$$\mathcal{X}(t) = x(t). \tag{257}$$

This implies that, in fact, in the observer, the $\dot{\mathcal{X}}$ equation does not need to be implemented, but that we simply have to replace $\mathcal{X}$ by x in the definition of e, $\dot{\widehat{p}}$ and $\dot{\widehat{q}}$. This gives exactly the adaptation law in (178) provided by the Lyapunov design.

3.2 Regressor Filtering

To overcome the difficulty of having $\dot{x}$ or more precisely $\dot{h}$ unmeasured, another way to implement an integration follows from the following remark:
Let z_{f} and Z_{f} be the following filtered quantities:

$$\begin{aligned} \dot{z}_{\mathrm{f}} + \rho(x, \widehat{p}, u)\, z_{\mathrm{f}} &= z \;, \quad z_{\mathrm{f}}(0) = 0 \\ \dot{Z}_{\mathrm{f}} + \rho(x, \widehat{p}, u)\, Z_{\mathrm{f}} &= Z \;, \quad Z_{\mathrm{f}}(0) = 0, \end{aligned} \tag{258}$$

or equivalently, using (223):

$$
\begin{aligned}
\dot{\widehat{z}}_{\mathrm{f}} &= \rho(x,\widehat{p},u)\, z_{\mathrm{f}} + \frac{\partial h}{\partial x}(x,\widehat{p})\, a(x,u) + \frac{\partial h}{\partial p}(x,\widehat{p})\, \dot{\widehat{p}} \ , \quad \widehat{z}_{\mathrm{f}}(0) = h(x(0),\widehat{p}(0)) \\
z_{\mathrm{f}} &= h(x,\widehat{p}) - \widehat{z}_{\mathrm{f}} \\
\dot{Z}_{\mathrm{f}} &= -\rho(x,\widehat{p},u)\, Z_{\mathrm{f}} + \frac{\partial h}{\partial x}(x,\widehat{p})\, A(x,u) \qquad , \quad Z_{\mathrm{f}}(0) = 0 \,,
\end{aligned}
\tag{259}
$$

where ρ is a C^1 positive function. Clearly, the knowledge of $\dot{h}$ is not needed and equation (S_{p^*}) gives:

$$
z_{\mathrm{f}} = Z_{\mathrm{f}}\, p^* \,. \tag{260}
$$

This equation is again linear in p^*. This yields the following linear observer for p^*:

$$
\begin{aligned}
e &= Z_{\mathrm{f}}\, \widehat{p} - z_{\mathrm{f}} \\
\dot{\widehat{p}} &= \mathrm{Proj}\,(M\,,\, \widehat{p}\,,\, -K\,e)\,, \qquad \widehat{p}(0) \in \Pi_1\,,
\end{aligned}
\tag{261}
$$

where, typically, the observer gain K and the matrix M, used in the function Proj, are given by:

$$
\begin{aligned}
K &= M\, Z_{\mathrm{f}}^{\mathrm{T}} r(x,\widehat{p},u,e) \\
\dot{M} &\geq -(2-\varepsilon_1)\, M\, Z_{\mathrm{f}}^{\mathrm{T}} Z_{\mathrm{f}}\, M\, r(x,\widehat{p},u,e)\,, \quad \varepsilon_2\, M \leq I\,, \quad I \leq \varepsilon_3\, M(0)\,,
\end{aligned}
\tag{262}
$$

where $0 < \varepsilon_1 < 2$, $0 < \varepsilon_3$, $0 < \varepsilon_2$, r is a strictly positive C^1 function and, if the observer gain K is allowed to decay to zero with time:

$$
\dot{M} \leq -\varepsilon_4\, M\, Z_{\mathrm{f}}^{\mathrm{T}} Z_{\mathrm{f}}\, M\, r(x,\widehat{p},u,e)\,, \quad 2-\varepsilon_1 \geq \varepsilon_4 > 0\,. \tag{263}
$$

Note that, in (261), $\dot{\widehat{p}}$ depends on u via the dependence of the observer gain K on r. Unfortunately, in this case, it is useless to extend the parameterization by embedding (S_p) into $(S_{p,q})$, since r will remain a factor in the extended observer gain. It follows that to make sure that $\dot{\widehat{p}}$ is well-defined, we shall impose that r does not depend on u whenever u is allowed to depend on $\dot{\widehat{p}}$.

We have:

Lemma: (264)

Assume ICS is satisfied. There exists a positive continuous function k_1 such that, for any C^1 time function $u : \mathbb{R}_+ \to \mathbb{R}^m$, all $(x(t),\widehat{p}(t),\widehat{z}_{\mathrm{f}}(t),Z_{\mathrm{f}}(t),M(t))$, solutions of (S_{p^*})-(259)-(261)-(262), defined on $[0,T)$ with $(x(t),M(t))$ remaining *in $\Omega \times \mathcal{M}$ satisfy for all t in $[0,T)$:*

1. $\widehat{p}(t) \in \Pi_1$
2. $\varepsilon_2 \|\widehat{p}(t) - p^*\|^2 + \varepsilon_1 \int_0^t r\, \|e\|^2 \le \varepsilon_3 \|\widehat{p}(0) - p^*\|^2$
3. *if (263) holds, then* $\int_0^t \left\|\dot{\widehat{p}}\right\| \le k_1(p^*, \widehat{p}(0))$.

Moreover, for all constant k_2, the property:

$$\frac{\|Z(t)\|}{\rho(x(t), \widehat{p}(t), u(t))} \le k_2 \quad \forall t \in [0,T) \tag{265}$$

implies the property:

$$\|Z_{\mathrm{f}}(t)\| \le k_2 \quad \textit{and} \quad \|e(t)\| \le k_2 \sqrt{\frac{\varepsilon_3}{\varepsilon_2}}\, \|\widehat{p}(0) - p^*\| \quad \forall t \in [0,T). \tag{266}$$

Note that a trivial way of meeting the condition (265) is to choose:

$$\rho(x, \widehat{p}, u) = 1 + \|Z\|. \tag{267}$$

Proof. Point 1 is a straightforward consequence of Point 5 of Lemma (103). For Point 2, first note that (260) and (261) imply:

$$e = Z_{\mathrm{f}}\,(\widehat{p} - p^*). \tag{268}$$

Then, let us consider the time derivative of:

$$W_1(t) = \frac{1}{2}\,(\widehat{p}(t) - p^*)^{\mathrm{T}}\, M^{-1}(t)\,(\widehat{p}(t) - p^*) \tag{269}$$

along the solutions of (261). This is a well-defined time function since by assumption $M(t) \in \mathcal{M}$. First note that, from (262), we get:

$$\dot{\overline{M^{-1}}} \le (2 - \varepsilon_1)\, Z_{\mathrm{f}}^{\mathrm{T}} Z_{\mathrm{f}}\, r\,, \tag{270}$$

then, with Point 4 of Lemma (103) and (268), we get successively:

$$\begin{aligned}
\dot{W}_1 &= (\widehat{p} - p^*)^{\mathrm{T}} \left[\frac{1}{2}\dot{\overline{M^{-1}}}\,(\widehat{p} - p^*) + M^{-1}\,\dot{\widehat{p}}\right] \\
&\le \left(1 - \frac{\varepsilon_1}{2}\right) r\, \|Z_{\mathrm{f}}\,(\widehat{p} - p^*)\|^2 \\
&\qquad + (\widehat{p} - p^*)^{\mathrm{T}}\, M^{-1}\, \mathrm{Proj}\left(M\,,\, \widehat{p}\,,\, -r\,M\,Z_{\mathrm{f}}^{\mathrm{T}} e\right) \\
&\le \left(1 - \frac{\varepsilon_1}{2}\right) r\, \|Z_{\mathrm{f}}\,(\widehat{p} - p^*)\|^2 - r\,(\widehat{p} - p^*)^{\mathrm{T}}\, Z_{\mathrm{f}}^{\mathrm{T}} e \\
&\qquad + (\widehat{p} - p^*)^{\mathrm{T}}\, M^{-1}\, \mathrm{Proj}\left(M\,,\, \widehat{p}\,,\, -r\,M\,Z_{\mathrm{f}}^{\mathrm{T}} e\right) - (\widehat{p} - p^*)^{\mathrm{T}} \left(-r Z_{\mathrm{f}}^{\mathrm{T}} e\right) \\
&\le \left(1 - \frac{\varepsilon_1}{2}\right) r\, \|Z_{\mathrm{f}}\,(\widehat{p} - p^*)\|^2 - r\,(\widehat{p} - p^*)^{\mathrm{T}}\, Z_{\mathrm{f}}^{\mathrm{T}} e \\
&\le -\frac{\varepsilon_1}{2}\, r\, \|e\|^2\,.
\end{aligned} \tag{271}$$

On the other hand, from (262), we have:

$$\frac{\varepsilon_2}{2} \|\widehat{p}(t) - p^*\|^2 \leq W_1(t) \leq \frac{\lambda_{\max}\{M^{-1}(t)\}}{2} \|\widehat{p}(t) - p^*\|^2 , \qquad (272)$$

where $\lambda_{\max}\{\cdot\}$ denotes the maximum eigenvalue. Point 2 follows readily and also with (271):

$$\int_0^t (\widehat{p} - p^*)^{\mathrm{T}} (-r\, Z_{\mathrm{f}}^{\mathrm{T}} e) - \int_0^t (\widehat{p} - p^*)^{\mathrm{T}} M^{-1} \operatorname{Proj}(M\,,\, \widehat{p}\,,\, -rM\, Z_{\mathrm{f}}^{\mathrm{T}} e) \leq \frac{\varepsilon_3}{2} \|\widehat{p}(0) - p^*\|^2 . \qquad (273)$$

To prove Point 3, let us define $\mathcal{X}$ as follows:

$$\begin{aligned} \mathcal{X} &= 0 \quad \text{if } \mathcal{P}(\widehat{p}) \leq 0 \ \text{ or } \ \tfrac{\partial \mathcal{P}}{\partial p}(\widehat{p})\, M\, Z_{\mathrm{f}}^{\mathrm{T}} e \geq 0 \\ &= \mathcal{P}(\widehat{p}) \text{ if } \mathcal{P}(\widehat{p}) > 0 \ \text{ and } \ \tfrac{\partial \mathcal{P}}{\partial p}(\widehat{p})\, M\, Z_{\mathrm{f}}^{\mathrm{T}} e < 0 . \end{aligned} \qquad (274)$$

With the definition (102) of Proj and Point 3 in Lemma (103), we get:

$$\begin{aligned} &(\widehat{p} - p^*)^{\mathrm{T}} (-Z_{\mathrm{f}}^{\mathrm{T}} e) - (\widehat{p} - p^*)^{\mathrm{T}} M^{-1} \operatorname{Proj}(M\,,\, \widehat{p}\,,\, -M\, Z_{\mathrm{f}}^{\mathrm{T}} e) \\ &\qquad = \mathcal{X} \frac{\left(-\frac{\partial \mathcal{P}}{\partial p}(\widehat{p}) M Z_{\mathrm{f}}^{\mathrm{T}} e\right) \left(\frac{\partial \mathcal{P}}{\partial p}(\widehat{p})\,(\widehat{p} - p^*)\right)}{\frac{\partial \mathcal{P}}{\partial p}(\widehat{p}) M \frac{\partial \mathcal{P}}{\partial p}(\widehat{p})^{\mathrm{T}}} && (275) \\ &\qquad = \mathcal{X} \frac{\left|\frac{\partial \mathcal{P}}{\partial p}(\widehat{p}) M Z_{\mathrm{f}}^{\mathrm{T}} e\right| \left(\frac{\partial \mathcal{P}}{\partial p}(\widehat{p})\,(\widehat{p} - p^*)\right)}{\frac{\partial \mathcal{P}}{\partial p}(\widehat{p}) M \frac{\partial \mathcal{P}}{\partial p}(\widehat{p})^{\mathrm{T}}} && (276) \\ &\qquad \geq D^* \mathcal{X} \frac{\left|\frac{\partial \mathcal{P}}{\partial p}(\widehat{p}) M Z_{\mathrm{f}}^{\mathrm{T}} e\right| \left\|\frac{\partial \mathcal{P}}{\partial p}(\widehat{p})\right\|}{\frac{\partial \mathcal{P}}{\partial p}(\widehat{p}) M \frac{\partial \mathcal{P}}{\partial p}(\widehat{p})^{\mathrm{T}}} . && (277) \end{aligned}$$

Then, the expression of $\dot{\widehat{p}}$ in (261) and the inequality on $\dot{M}$ in (263) give with (262):

$$\begin{aligned} \left\|\dot{\widehat{p}}\right\| &= \left\|\operatorname{Proj}(M\,,\, \widehat{p}\,,\, -r\, M\, Z_{\mathrm{f}}^{\mathrm{T}} e)\right\| && (278) \\ &\leq r \left\|M Z_{\mathrm{f}}^{\mathrm{T}} e\right\| + r\, \mathcal{X} \frac{\left|\frac{\partial \mathcal{P}}{\partial p}(\widehat{p}) M Z_{\mathrm{f}}^{\mathrm{T}} e\right| \left\|M \frac{\partial \mathcal{P}}{\partial p}(\widehat{p})\right\|}{\frac{\partial \mathcal{P}}{\partial p}(\widehat{p}) M \frac{\partial \mathcal{P}}{\partial p}(\widehat{p})^{\mathrm{T}}} . && (279) \end{aligned}$$

And, with the Cauchy Schwarz inequality and (262), we get:

$$\begin{aligned}\int_0^t \left\|\dot{\widehat{p}}\right\| &\leq \sqrt{\int_0^t r\,\|e\|^2}\,\sqrt{\int_0^t \operatorname{tr}\left\{rMZ_{\mathrm{f}}^{\mathrm{T}}Z_{\mathrm{f}}M\right\}} \\ &\qquad + \int_0^t \frac{r\chi}{\varepsilon_2}\,\frac{\left|\frac{\partial\mathcal{P}}{\partial p}(\widehat{p})MZ_{\mathrm{f}}^{\mathrm{T}}e\right|\left\|\frac{\partial\mathcal{P}}{\partial p}(\widehat{p})\right\|}{\frac{\partial\mathcal{P}}{\partial p}(\widehat{p})M\frac{\partial\mathcal{P}}{\partial p}(\widehat{p})^{\mathrm{T}}} \\ &\leq \sqrt{\int_0^t r\,\|e\|^2}\,\sqrt{\int_0^t \operatorname{tr}\left\{\frac{-\dot{M}}{\varepsilon_4}\right\}} \\ &\qquad + \int_0^t \frac{r\chi}{\varepsilon_2}\,\frac{\left|\frac{\partial\mathcal{P}}{\partial p}(\widehat{p})MZ_{\mathrm{f}}^{\mathrm{T}}e\right|\left\|\frac{\partial\mathcal{P}}{\partial p}(\widehat{p})\right\|}{\frac{\partial\mathcal{P}}{\partial p}(\widehat{p})M\frac{\partial\mathcal{P}}{\partial p}(\widehat{p})^{\mathrm{T}}}\,, \end{aligned} \tag{280}$$

where $\operatorname{tr}\{\cdot\}$ denotes the trace of a matrix. The proof of Point 3 is now straightforward. We use the inequalities given by Point 2, (273) and (277) and get:

$$\int_0^t \left\|\dot{\widehat{p}}\right\| \leq \sqrt{\frac{\varepsilon_3}{\varepsilon_2}}\,\|\widehat{p}(0)-p^*\|\,\sqrt{\frac{\operatorname{tr}\{M(0)\}}{\varepsilon_4}} + \frac{\varepsilon_3}{2\,D^*\,\varepsilon_2}\,\|\widehat{p}(0)-p^*\|^2\,. \tag{281}$$

Finally, we prove the implication stated in the Lemma. To simplify the notations, let:

$$\rho(t) = \rho\left(x(t),\widehat{p}(t),u(t)\right). \tag{282}$$

The solution $Z_{\mathrm{f}}(t)$ of (258) is for all t in $[0,T)$:

$$Z_f(t) = \int_0^t \exp\left(-\int_s^t \rho(\tau)\,d\tau\right)\,Z(s)\,ds\,. \tag{283}$$

Therefore, if for all t:

$$\|Z(t)\| \leq k_2\,\rho(t)\,, \tag{284}$$

we have:

$$\begin{aligned}\|Z_{\mathrm{f}}(t)\| &\leq \int_s^t \exp\left(-\int_s^t \rho(\tau)\,d\tau\right)\|Z(s)\|\,ds && (285)\\ &\leq k_2\int_0^t \exp\left(-\int_0^t \rho(\tau)\,d\tau\right)\,\rho(s)\,ds && (286)\\ &\leq k_2\,. && (287)\end{aligned}$$

Finally, the inequality on $\|e\|$ is a straightforward consequence of this inequality, Point 2 and (268). □

An important drawback of this result is that it requires a particular initial condition for the initial condition $\widehat{z}_{\mathrm{f}}(0)$. Choosing another initial condition may create difficulties if at the same time r is not guaranteed to be bounded.

3.3 Estimation Design

The estimation design of a dynamic controller for solving the Adaptive Stabilization problem consists in choosing the observation function h and either the equation error filtering or the regressor filtering technique. This provides an estimate $\widehat{p}$ which we use in the nominal control u_{n} to obtain the control:

$$u = u_{\mathrm{n}}(x, \widehat{p}) + v\,, \tag{288}$$

where, as in the Lyapunov design case, v may be introduced to counteract the effects of updating $\widehat{p}$, so that v may depend on $\dot{\widehat{p}}$. In both the equation error filtering case and the regressor filtering case, the estimator is trying to find a function $\widehat{p}$ fitting as well as possible the $\dot{h}$ equation (222). On the other hand, to show that we are solving the Adaptive Stabilization problem, we see, from the proof of Proposition (76), that both the estimation and the control should be such that $\dot{V}$ is negative. With assumption PRS, such an objective is met by the control u_{n} as long as the $\dot{V}$ equation is well-fitted by $\widehat{p}$. Consequently, the fact that the Adaptive Stabilization problem is solved depends crucially on the fact that, when the $\dot{h}$ equation is well-fitted, the same holds for the $\dot{V}$ equation. More precisely, this fact relies on the properties of the following set value maps:

$$\begin{aligned} F(p,\eta) &= \{(p,\nu)\ |\ \nu = V(x,p) \quad \text{and} \quad h(x,p) = \eta\} \\ F^{\dagger}(p,\nu) &= \{(p,\eta)\ |\ \eta = h(x,p) \quad \text{and} \quad V(x,p) = \nu\}\,. \end{aligned} \tag{289}$$

4 Estimation Design with the Observation Function $h = \frac{\alpha_1 V}{\alpha_1 - V}$

When we choose the observation function h as:

$$h(x,p) = \frac{\alpha_1 V(x,p)}{\alpha_1 - V(x,p)}\,, \tag{290}$$

the above set value maps F and $F^{\dagger}$ are in fact standard applications defined by:

$$F(p,\eta) = \left(p\,, \frac{\eta\,\alpha_1}{\eta + \alpha_1}\right) \qquad \text{and} \qquad F^{\dagger}(p,\nu) = \left(p\,, \frac{\nu\,\alpha_1}{\alpha_1 - \nu}\right). \tag{291}$$

As we will show this is a very favorable situation compared with other possible choices for the observation function h.

4.1 Equation Error Filtering

Let $\widehat{p}$ be obtained from the equation error filtering technique (230)–(232). If we implement the control:

$$u = u_{\mathrm{n}}(x, \widehat{p}) , \tag{292}$$

then, with (47) in assumption PRS, equation (230) gives:

$$\dot{\widehat{h}} \leq -r(e, x, \widehat{p})\, e + \frac{{\alpha_1}^2}{(V(x, \widehat{p}) - \alpha_1)^2} \frac{\partial V}{\partial p}(x, \widehat{p})\, \dot{\widehat{p}} , \tag{293}$$

where:

$$\widehat{h} = \frac{\alpha_1 V}{\alpha_1 - V} + e . \tag{294}$$

If, instead, assumption MC holds, let $\widehat{p}$ and $\widehat{q}$ be obtained from (234) and use the control:

$$u = u_{\mathrm{n}}(x, \widehat{p}) + v\left(x, \widehat{p}, \widehat{q}, \dot{\widehat{p}}\right) , \tag{295}$$

with v given by (170) in MC with $\partial = \dot{\widehat{p}}$. Then, inequality (293) reduces to:

$$\dot{\widehat{h}} \leq -r(e, x, \widehat{p})\, e . \tag{296}$$

Since V is positive and e is bounded (see Lemma (235)), $\widehat{h}$ is lower bounded. To conclude that $\widehat{h}$ is also upper bounded and therefore V is strictly smaller than α_1, we would need to know that re is in L^1 but we only know, from Lemma (235), that $\sqrt{r}\, e$ is in L^2. In fact to prove boundedness of $\widehat{h}$, we need to strengthen assumption PRS:

Assumption URS (Uniform Reduced-order Stabilizability) (297)
There exists a positive constant c such that, for all (x, p) in $\Omega \times \Pi$, we have:

$$\frac{\partial V}{\partial x}(x, p) \left[a(x, u_{\mathrm{n}}(x, p)) + A(x, u_{\mathrm{n}}(x, p))\, p\right] \leq -c\, V(x, p) . \tag{298}$$

Example: System (17) Continued (299)
Consider the system (17) rewritten as (50). Assumptions BO, CO and URS are satisfied if we choose:

$$u_{\mathrm{n}}(x_1, x_2, p_1, p_2) = -\frac{x_2^2 + p_1 + x_1}{p_2} , \tag{300}$$

and:

$$\begin{aligned} V(x_1, x_2, p_1, p_2) = V(x_1) &= \left(x_1 + \frac{2}{3}\right)^2 && \text{if } \; x_1 \leq -1 \\ &= \frac{1}{9} \exp\left(3\left(1 - \frac{1}{x_1^2}\right)\right) && \text{if } \; -1 < x_1 < 1 \\ &= \left(x_1 - \frac{2}{3}\right)^2 && \text{if } \; 1 \leq x_1 . \end{aligned} \tag{301}$$

Indeed:

1. The function V is of class C^2 with:

$$\begin{aligned} \frac{d^2V}{dx_1{}^2}(x_1) &= 2 && \text{if} \quad |x_1| \geq 1 \\ &= 2\,\frac{2 - x_1^2}{x_1^6}\,\exp\left(3\left(1 - \frac{1}{x_1^2}\right)\right) && \text{if} \quad |x_1| < 1\,. \end{aligned} \tag{302}$$

2. If $V(x_1)$ is bounded, so is x_1, and if $V(x_1)$ tends to 0, so does x_1. Hence, as in Example (49), BO and CO are satisfied.
3. Finally, assumption URS is met, since we get:

$$\begin{aligned} \frac{\partial V}{\partial x}(x,p)\,[a(x,u_{\mathrm{n}}(x,p)) + A(x,u_{\mathrm{n}}(x,p))\,p] \\ = -2\left(x_1 + \frac{2}{3}\right) x_1^3 && \text{if} \quad x_1 \leq -1 \\ -\frac{2}{3}\exp\left(3\left(1 - \frac{1}{x_1^2}\right)\right) && \text{if} \quad -1 < x_1 < 1 \\ -2\left(x_1 - \frac{2}{3}\right) x_1^3 && \text{if} \quad 1 \leq x_1 \\ \leq -2\,V(x,p)\,. \end{aligned} \tag{303}$$

□

Example: System (25) Continued **(304)**

For the system (25), assume the function L satisfies the following growth condition: there exist a positive constant γ and an integer j such that:

$$|L(y)| \leq \gamma\left(|y| + |y|^j\right). \tag{305}$$

Under this condition, we can find functions V and u_{n} such that assumptions BO, CO and URS are satisfied by the non-minimal state-space representation (27) of (25). Indeed, let $\Phi(p)$ be the following invertible matrix:

$$\Phi(p) = \begin{pmatrix} 1&0&0&0&0&0 \\ 0&1&0&0&0&0 \\ 0&0&1&0&p&0 \\ 0&0&0&1&0&p \\ 0&0&0&0&1&0 \\ 0&0&0&0&0&1 \end{pmatrix}. \tag{306}$$

It allows us to define new coordinates:

$$\chi = \Phi(p)\,x = \begin{pmatrix} \chi_1 \\ \chi_2 \end{pmatrix}, \tag{307}$$

and to rewrite (27) in the following block form:

$$\dot{\chi} = \begin{pmatrix} F_1 & 0 \\ 0 & F_2 \end{pmatrix} \chi + \begin{pmatrix} G_1 \\ 0 \end{pmatrix} [u + L(y)] + \begin{pmatrix} 0 \\ G_2 \end{pmatrix} L(y) \tag{308}$$

$$y = (H_1\ 0)\chi + \delta(t)\,, \tag{309}$$

where the pair $(F_1\,,\,G_1)$ is controllable and the matrix F_2 is Hurwitz. It follows that there exist matrices C_1, P_1 and P_2 and strictly positive constants α_1 and α_2 such that:

$$\begin{aligned} P_1\,(F_1 - G_1C_1) + (F_1 - G_1C_1)^{\mathrm{T}}\,P_1 &= -I \le -2\alpha_1\,P_1 \\ P_1\,F_2 + F_1^{\mathrm{T}}P_2 &= -I \le -2\alpha_2\,P_2\,. \end{aligned} \tag{310}$$

Let us now define a function U by:

$$U(\chi) = \frac{(\chi_1^{\mathrm{T}}P_1\chi_1)^j}{2j} + \frac{(\chi_1^{\mathrm{T}}P_1\chi_1)}{2} + \beta\,\frac{(\chi_2^{\mathrm{T}}P_2\chi_2)}{2}\,, \tag{311}$$

where β is a strictly positive constant. By letting:

$$V(x,p) = U\,(\Phi(p)\,x)\,, \tag{312}$$

assumptions BO and CO are clearly satisfied. To check that assumption URS holds also, we choose u_{n} as:

$$u_{\mathrm{n}}(x,p) = -C_1\,\chi_1 - L(y)\,, \qquad \begin{pmatrix}\chi_1\\ \chi_2\end{pmatrix} = \Phi(p)\,x\,. \tag{313}$$

Then along the solutions of:

$$\dot{\chi} = \begin{pmatrix} F_1 - G_1C_1 & 0 \\ 0 & F_2 \end{pmatrix}\chi + \begin{pmatrix} 0 \\ G_2 \end{pmatrix} L\,(H_1\chi_1 + \delta(t))\,, \tag{314}$$

the time derivative of U satisfies:

$$\begin{aligned} \dot{U} &= \left[(\chi_1^{\mathrm{T}}P_1\chi_1)^{j-1} + 1\right]\chi_1^{\mathrm{T}}P_1\,(F_1 - G_1C_1)\,\chi_1 \\ &\quad + \beta\,\left[\chi_2^{\mathrm{T}}P_2F_2\chi_2 + \chi_2^{\mathrm{T}}P_2G_2L\,(H_1\chi_1 + \delta(t))\right] \\ &\le -\alpha_1\left[(\chi_1^{\mathrm{T}}P_1\chi_1)^j + (\chi_1^{\mathrm{T}}P_1\chi_1)\right] - \alpha_2\beta\,(\chi_2^{\mathrm{T}}P_2\chi_2) \\ &\quad + \beta\,(\chi_2^{\mathrm{T}}P_2\chi_2)^{\frac{1}{2}}\,|L|\,\sqrt{G_2{}^{\mathrm{T}}P_2G_2}\,. \end{aligned} \tag{315}$$

But the growth condition (305), satisfied by the function L, and Young's inequality imply the existence of four positive constants γ_1 to γ_4 such that, for all χ_1, χ_2 and δ, we have:

$$\begin{aligned} &(\chi_2^{\mathrm{T}}P_2\chi_2)^{\frac{1}{2}}\,|L(H_1\chi_1 + \delta)|\,\sqrt{G_2{}^{\mathrm{T}}P_2G_2} \\ &\le \frac{\alpha_2}{2}\,(\chi_2^{\mathrm{T}}P_2\chi_2) + \gamma_1\,(\chi_1^{\mathrm{T}}P_1\chi_1) + \gamma_2\,|\delta|^2 + \gamma_3\,(\chi_1^{\mathrm{T}}P_1\chi_1)^j + \gamma_4\,|\delta|^{2j} \end{aligned} \tag{316}$$

We have established:

$$\dot{U} \le -cU + \beta\left[\gamma_2\,|\delta|^2 + \gamma_4\,|\delta|^{2j}\right], \tag{317}$$

where:

$$c = \min\{2(\alpha_1 - \beta\gamma_1)\,,\, 2j(\alpha_1 - \beta\gamma_3)\,,\, \beta\alpha_2\}\,. \tag{318}$$

It follows that assumption URS holds, up to the presence of the exponentially decaying terms in δ, if β is chosen sufficiently small. □

Example: Introduction to System (320) **(319)**

Consider the system

$$\begin{aligned}\dot{x}_1 &= x_2 + p^* x_1^2\\ \dot{x}_2 &= x_3\\ \dot{x}_3 &= u\,.\end{aligned} \tag{320}$$

Following the Lyapunov design proposed in [27], we choose:

$$\begin{aligned}u_{\mathrm{n}}(p,x) = &-\left(\frac{1}{2k}+\frac{1}{2}+2p\chi_1\right)\left(\chi_3-\frac{\chi_2}{2}-\chi_1^{2k-1}\right)\\ &+\left[\frac{1}{4k^2}+\frac{2p\chi_1}{k}-(2k-1)\chi_1^{2k-2}-2p\chi_2\right]\left(\chi_2-\frac{\chi_1}{2k}\right)\\ &-\frac{\chi_3}{2}-\chi_2\left(\frac{\chi_2^2}{2}+\frac{\chi_1^{2k}}{2k}\right)^{j-1},\end{aligned} \tag{321}$$

and:

$$V(p,x) \;=\; U(\chi) \;=\; \frac{\chi_3^2}{2}+\frac{1}{j}\left(\frac{\chi_2^2}{2}+\frac{\chi_1^{2k}}{2k}\right)^j, \tag{322}$$

where k and j are integers larger or equal to 1 and χ is given by:

$$\chi \;=\; \begin{pmatrix}\chi_1\\ \chi_2\\ \chi_3\end{pmatrix} \;=\; \begin{pmatrix} x_1\\ x_2+\dfrac{x_1}{2k}+p\,x_1^2\\ x_3+\dfrac{1}{2}\left(x_2+\dfrac{x_1}{2k}+px_1^2\right)\\ \quad+\left(\dfrac{1}{2k}+2px_1\right)(x_2+px_1^2)+x_1^{2k-1}\end{pmatrix}. \tag{323}$$

Note that, if $k=j=1$, u_{n} is a linearizing feedback.

Since V is positive definite and radially unbounded, assumptions BO and CO are satisfied. Assumption URS holds also, since a straightforward computation shows that the time derivative of V along the solutions of (320) with p instead of p^* satisfies:

$$\dot{V} \;=\; -V\,. \tag{324}$$

□

Proposition **(325)**

Let assumptions BO, URS and ICS hold and choose:

$$h(x,p) \;=\; \frac{\alpha_1 V(x,p)}{\alpha_1 - V(x,p)} \qquad \textit{and} \qquad r(e,x,p) \;=\; 1+\left(\frac{\alpha_1 V(x,p)}{\alpha_1 - V(x,p)}\right)^2. \tag{326}$$

If assumption Λ-LP is satisfied with:

$$\Lambda(x,t) \;=\; \frac{\partial h}{\partial x}(x,\widehat{p}(t))\,, \tag{327}$$

and:

either *assumption MC holds,* α_1 *is chosen smaller than or equal to* α_0 , $x(0)$ *belongs to* Ω_0 *and* $V(x(0), \widehat{p}(0)) < \alpha_1$,

or *we are in the global case, i.e.,* $\Omega_0 = \Omega = \mathbb{R}^n$, $\alpha_1 = \alpha_0 = +\infty$ *and there exists a* C^0 *function* $d_1 : \Pi \to \mathbb{R}_+$ *such that, for all* (x,p) *in* $\mathbb{R}^n \times \Pi$*:*

$$\left\| \frac{\partial V}{\partial p}(x,p) \right\| \cdot \left\| \left[\frac{\partial V}{\partial x}(x,p)\, A(x, u_{\mathrm{n}}(x,p)) \right]^{\mathrm{T}} \right\| \leq d_1(p) \max\{1\,,\, V(x,p)^2\}\,, \tag{328}$$

then all the corresponding solutions $(x(t), \widehat{p}(t), \widehat{h}(t))$ *of* (S_{p^*})*-(230)-(231)-(232) are well-defined on* $[0, +\infty)$*, unique, bounded and:*

$$\lim_{t\to\infty} V(x(t), \widehat{p}(t)) = 0\,. \tag{329}$$

It follows that the Adaptive Stabilization problem is solved if assumption CO holds also.

Proposition (325) is an extension to the case when V is not radially unbounded of a result established by Pomet and Praly [23,21].

Proof.
Case: Assumption MC holds: The system we consider is, with notation (13):

$$\begin{aligned}
\dot{x} &= a\left(x\,,\, u_{\mathrm{n}}(x,\widehat{p}) + v(x,\widehat{p},\widehat{q},\dot{\widehat{p}})\right) + A\left(x\,,\, u_{\mathrm{n}}(x,\widehat{p}) + v(x,\widehat{p},\widehat{q},\dot{\widehat{p}})\right) p^* \\
\dot{\widehat{h}} &= -r(e,x,\widehat{p})\, e + \frac{\partial h}{\partial x}(x,\widehat{p})\, [a_0(x) + A_0(x)\widehat{p} + u_{\mathrm{n}}(x,\widehat{p}) \odot (b(x) + B(x)\widehat{p})] \\
&\quad + \frac{\partial h}{\partial x}(x,\widehat{p})\, v\left(x,\widehat{p},\widehat{q},\dot{\widehat{p}}\right) \odot (b(x) + B(x)\widehat{q}) + \frac{\partial h}{\partial p}(x,\widehat{p})\, \dot{\widehat{p}} \\
e &= \widehat{h} - h(x,\widehat{p}) \\
\dot{\widehat{p}} &= \mathrm{Proj}\left(I, \widehat{p}, -\left[\frac{\partial h}{\partial x}(x,\widehat{p})\, (A_0(x) + u_{\mathrm{n}}(x,\widehat{p}) \odot B(x)) \right]^{\mathrm{T}} e \right)\,,\ \widehat{p}(0) \in \Pi_1 \\
\dot{\widehat{q}} &= \mathrm{Proj}\left(I, \widehat{q}, -\left[\frac{\partial h}{\partial x}(x,\widehat{p})\, v(x,\widehat{p},\widehat{q},\dot{\widehat{p}}) \odot B(x) \right]^{\mathrm{T}} e \right)\,,\ \widehat{q}(0) \in \Pi_1\,,
\end{aligned} \tag{330}$$

with:

$$h(x,p) = \frac{\alpha_1\, V(x,p)}{\alpha_1 - V(x,p)}\,, \qquad r(e,x,p) = 1 + h(x,p)^2\,, \tag{331}$$

and, with notation (13):

$$\frac{\partial V}{\partial p}(x,\widehat{p})\, \dot{\widehat{p}} + \frac{\partial V}{\partial x}(x,\widehat{p})\, v(x,\widehat{p},\widehat{q},\dot{\widehat{p}}) \odot [b(x) + B(x)\, \widehat{q}] = 0\,. \tag{332}$$

From our smoothness assumptions on the various functions and with Point 1 of Lemma (103), this system has a locally Lipschitz-continuous right-hand side in the open set defined by:

$$(x,\widehat{p},\widehat{q},\widehat{h}) \in \Omega \times \Pi \times \Pi \times \mathbb{R} \qquad \text{and} \qquad V(x,\widehat{p}) < \alpha_1\,. \tag{333}$$

It follows that, for any initial condition in this open set, there exists a unique solution $(x(t), \widehat{p}(t), \widehat{q}(t), \widehat{h}(t))$, defined on a right maximal interval $[0, T)$, with T possibly infinite, and satisfying (333) for all t in $[0, T)$ and in particular:

$$V(x(t), \widehat{p}(t)) < \alpha_1 \quad \forall t \in [0, T). \tag{334}$$

Applying Points 3 and 4 of Lemma (235), we also know that, for all t in $[0, T)$:

$$\begin{gathered}\widehat{p}(t) \in \Pi_1 \quad \text{and} \quad \widehat{q}(t) \in \Pi_1 \\ \left\| \begin{matrix} \widehat{p}(t) - p^* \\ \widehat{q}(t) - p^* \end{matrix} \right\|^2 + \|e(t)\|^2 + 2\int_0^t r\,\|e\|^2 \leq \left\| \begin{matrix} \widehat{p}(0) - p^* \\ \widehat{q}(0) - p^* \end{matrix} \right\|^2 + \|e(0)\|^2 \stackrel{\text{def}}{=} \beta^2 .\end{gathered} \tag{335}$$

Then, with assumption URS, (332), (334) and (331), we get successively from the $\dot{\widehat{h}}$ equation in (330):

$$\dot{\widehat{h}} \leq -r\,e - c\frac{{\alpha_1}^2 V}{(\alpha_1 - V)^2} \tag{336}$$

$$\leq \sqrt{r}\,\left(\sqrt{r}\,|e|\right) - c\frac{\alpha_1}{\alpha_1 - V}\,h \tag{337}$$

$$\leq (1 + h)\,\left(\sqrt{r}\,|e|\right) - c\,h \tag{338}$$

$$\leq -\left(c - \left(\sqrt{r}\,|e|\right)\right) h + \sqrt{r}\,|e| \tag{339}$$

$$\leq -\left(c - \left(\sqrt{r}\,|e|\right)\right) \widehat{h} + (1 + |e|)\sqrt{r}\,|e| + c\,|e| . \tag{340}$$

Now, inequality (335) implies that the assumption of Lemma (583) in Appendix B is satisfied with:

$$X = \widehat{h}, \tag{341}$$

and, by using the fact that $r \geq 1$:

$$\begin{aligned} \vartheta_1 &= \sqrt{r}\,|e|, & \sigma_1 &= 2,\ S_{11} = \frac{\beta^2}{2} \\ \varpi_1 &= (1+\beta)\sqrt{r}\,|e| + c\,|e|, & \zeta_1 &= 2,\ S_{21} = 2\left((1+\beta)^2 + c^2\right)\frac{\beta^2}{2} . \end{aligned} \tag{342}$$

It follows that there exists a constant Υ, depending only on the initial condition, such that, for all t in $[0, T)$, we have:

$$0 \leq \frac{\alpha_1 V(x(t), \widehat{p}(t))}{\alpha_1 - V(x(t), \widehat{p}(t))} = h(x(t), \widehat{p}(t)) = \widehat{h}(t) - e(t) \leq \Upsilon + \beta . \tag{343}$$

Hence, we have established, for all t in $[0,T)$:

$$\begin{aligned} V(x(t),\widehat{p}(t)) &\leq \frac{\alpha_1(\Upsilon+\beta)}{\alpha_1+\Upsilon+\beta} < \alpha_1 \\ -\beta &\leq \widehat{h}(t) \leq \Upsilon \\ \|\widehat{p}(t)-p^*\| \leq \beta &\quad \text{and} \quad \widehat{p}(t) \in \Pi_1 \\ \|\widehat{q}(t)-p^*\| \leq \beta &\quad \text{and} \quad \widehat{q}(t) \in \Pi_1 \,. \end{aligned} \tag{344}$$

Then, from assumption BO, we know the existence of a compact subset Γ of Ω such that:

$$x(t) \in \Gamma \quad \forall t \in [0,T)\,. \tag{345}$$

Hence, the solution remains in a compact subset of the open set defined in (333). It follows by contradiction that $T=+\infty$ and that $x(t)$ and $\widehat{p}(t)$, $u(t)$ and $\dot{\widehat{p}}(t)$ are bounded on $[0,+\infty)$. Then, using the second conclusion of Lemma (583), we have:

$$\limsup_{t\to+\infty} \widehat{h}(t) \leq 0\,. \tag{346}$$

Also, from (229) and the fact that the solution is bounded, we deduce that $\dot{e}$ is bounded. Since, from (335), e is in $L^2([0,+\infty))$, we have established:

$$\lim_{t\to+\infty} e(t) = 0\,. \tag{347}$$

This yields:

$$0 \leq \limsup_{t\to+\infty} h(x(t),\widehat{p}(t)) = \limsup_{t\to+\infty} \widehat{h}(t) - \lim_{t\to+\infty} e(t) \leq 0\,. \tag{348}$$

With the definition of h, (334), and assumption CO, this implies:

$$\lim_{t\to+\infty} x(t) = \mathcal{E}\,. \tag{349}$$

Case: Inequality (328) holds: The system we consider is:

$$\begin{aligned} \dot{x} &= a\,(x\,,\,u_{\mathrm{n}}(x,\widehat{p})) + A\,(x\,,\,u_{\mathrm{n}}(x,\widehat{p}))\,p^* \\ \dot{\widehat{h}} &= -r(e,x,\widehat{p})\,e + \frac{\partial V}{\partial x}(x,\widehat{p})\,A(x,u_{\mathrm{n}}(x,\widehat{p}))\,\widehat{p} + \frac{\partial V}{\partial x}(x,\widehat{p})\,a(x,u_{\mathrm{n}}(x,\widehat{p})) \\ &\quad + \frac{\partial V}{\partial p}(x,\widehat{p})\,\dot{\widehat{p}} \\ e &= \widehat{h} - V(x,\widehat{p}) \\ \dot{\widehat{p}} &= \mathrm{Proj}\left(I\,,\,\widehat{p}\,,\,-\left(\frac{\partial V}{\partial x}(x,\widehat{p})\,A(x,u_{\mathrm{n}}(x,\widehat{p}))\right)^{\mathrm{T}} e\right), \qquad \widehat{p}(0) \in \Pi_1\,, \end{aligned} \tag{350}$$

with:

$$r(e, x, \widehat{p}) = 1 + V(x, \widehat{p})^2 . \tag{351}$$

From our smoothness assumptions on the various functions and with Point 1 of Lemma (103), this system has a locally Lipschitz-continuous right-hand side for all $(x, \widehat{p}, \widehat{h})$ in $\mathbb{R}^n \times \Pi \times \mathbb{R}$. It follows that, for any initial condition in this open set, there exists a unique solution $(x(t), \widehat{p}(t), \widehat{h}(t))$, defined on a right maximal interval $[0, T)$, with T possibly infinite, and remaining in this set. Applying Points 1 and 2 of Lemma (235), we also know that, for all t in $[0, T)$:

$$\begin{aligned}
&\widehat{p}(t) \in \Pi_1 \\
&\|\widehat{p}(t) - p^*\|^2 + \|e(t)\|^2 + 2\int_0^t r\,\|e\|^2 \;\le\; \|\widehat{p}(0) - p^*\|^2 + \|e(0)\|^2 \stackrel{\text{def}}{=} \beta^2 .
\end{aligned} \tag{352}$$

Then, as in the previous case, with assumption URS, we get from the $\dot{\widehat{h}}$ equation in (350):

$$\dot{\widehat{h}} \le -r\,e - c\,V + \frac{\partial V}{\partial p}\dot{\widehat{p}} . \tag{353}$$

But, by using inequality (328), Point 2 of Lemma (103), (351) and the expression of $\dot{\widehat{p}}$ in (350), we get successively:

$$\begin{aligned}
\left\|\frac{\partial V}{\partial p}\dot{\widehat{p}}\right\| &\le \left\|\frac{\partial V}{\partial p}\right\| \left\|\text{Proj}\left(I, \widehat{p}, -\left(\frac{\partial V}{\partial x} A\right)^{\text{T}} e\right)\right\| \\
&\le \left\|\frac{\partial V}{\partial p}\right\| \left\|\left[\frac{\partial V}{\partial x} A\right]^{\text{T}}\right\| |e| \\
&\le d_1(\widehat{p}(t)) \max\{1, V^2\} |e| \\
&\le d_1(\widehat{p}(t)) \left(1 + V^2\right) |e| \\
&\le d_1(\widehat{p}(t))\, r\, |e| .
\end{aligned} \tag{354}$$

Since d_1 depends continuously on $\widehat{p}(t)$ which satisfies (352), there exists a constant k depending only on $\widehat{p}(0)$ and $e(0)$ such that (353), (354) and the expression of e in (350) give:

$$\begin{aligned}
\dot{\widehat{h}} &\le (1 + k)\sqrt{r}\,(\sqrt{r}\,|e|) - c\,V \\
&\le (1 + k)(1 + \widehat{h} - e)\,(\sqrt{r}\,|e|) - c\,(\widehat{h} - e) \\
&\le -\left(c - (1 + k)\sqrt{r}|e|\right)\widehat{h} + (1 + k)(1 + |e|)(\sqrt{r}\,|e|) + c\,|e| .
\end{aligned} \tag{355}$$

Hence, with Point 2 of Lemma (235), the assumption of Lemma (583) in Appendix B is satisfied with (see (352)):

$$X = \widehat{h} \tag{356}$$

and

$$\begin{aligned} &\vartheta_1 = (1+k)\sqrt{r}|e|\,, && \sigma_1 = 2,\ S_{11} = (1+k)^2\frac{\beta^2}{2} \\ &\varpi_1 = (1+k)(1+\beta)\left(\sqrt{r}|e|\right) + c|e|,\ \zeta_1 = 2, && S_{21} = 2\left((1+k)^2(1+\beta)^2 + c^2\right)\frac{\beta^2}{2}\,. \end{aligned} \tag{357}$$

From here, we conclude the proof exactly as in the previous case. □

Compared with the Lyapunov design, we see that, when the Matching Condition (MC) holds, the equation error filtering technique, with V as the observation function, requires the more restrictive Uniform Reduced-order Stabilizability (URS) assumption. However, if assumption MC does not hold, nothing is known for the Lyapunov design, whereas here the Adaptive Stabilization problem is solved in the global case by the equation error filtering technique if the quadratic growth condition (328) is satisfied.

Example: System (320) Continued **(358)**
For the system (320), we have shown that assumption BO, CO and URS are satisfied. For assumption MC, we get (see (322)):

$$\frac{\partial V}{\partial p} = \chi_2\chi_1^2\left(\frac{\chi_2^2}{2} + \frac{\chi_1^{2k}}{2k}\right)^{j-1} + \chi_3\left[\frac{\chi_1^2}{2} + 2\chi_1\left(\chi_2 - \frac{\chi_1}{2k}\right) + \left(\frac{1}{2k} + 2p\chi_1\right)\chi_1^2\right], \tag{359}$$

and:

$$\frac{\partial V}{\partial x} b = \chi_3\,. \tag{360}$$

Hence, (170) cannot be satisfied, since, when $\frac{\partial V}{\partial x} b$ is zero, $\frac{\partial V}{\partial p}$ is not necessarily zero. Then, let us see if the growth condition (328) holds. We have to compare the product of the norms of (see (322)):

$$\frac{\partial V}{\partial x} A = \frac{\partial U}{\partial \chi}\frac{\partial \chi}{\partial x_1} x_1^2 \qquad \text{and} \qquad \frac{\partial V}{\partial p} = \frac{\partial U}{\partial \chi}\frac{\partial \chi}{\partial p}$$

to a power of $V = U$. We have, writing everything in the χ coordinates in which U has a simpler expression:

$$\frac{\partial U}{\partial \chi} = \left(\left(\frac{\chi_2^2}{2} + \frac{\chi_1^{2k}}{2k}\right)^{j-1}\chi_1^{2k-1},\quad \left(\frac{\chi_2^2}{2} + \frac{\chi_1^{2k}}{2k}\right)^{j-1}\chi_2\,,\quad \chi_3\right) \tag{361}$$

$$\frac{\partial \chi}{\partial x_1} = \begin{pmatrix} 1 \\ \frac{1}{2k} + 2p\chi_1 \\ \left(\frac{1}{2k} + 2p\chi_1\right)\left(\frac{1}{2} + 2p\chi_1\right) + 2p\left(\chi_2 - \frac{\chi_1}{2k}\right) + (2k-1)\,\chi_1^{2k-2} \end{pmatrix} \tag{362}$$

$$\frac{\partial \chi}{\partial p} = \begin{pmatrix} 0 \\ \chi_1^2 \\ \frac{\chi_1^2}{2} + 2\chi_1\left(\chi_2 - \frac{\chi_1}{2k}\right) + \left(\frac{1}{2k} + 2p\chi_1\right)\chi_1^2 \end{pmatrix}. \tag{363}$$

To obtain our inequalities, we note that:

$$|\chi_1| \leq (\gamma U)^{\frac{1}{2kj}}\,, \quad |\chi_2| \leq (\gamma U)^{\frac{1}{2j}}\,, \quad |\chi_3| \leq (\gamma U)^{\frac{1}{2}}\,, \tag{364}$$

with:

$$\gamma = \sup\left\{j(2k)^j\,,\, j2^j\,,\, 2\right\}\,, \tag{365}$$

and, for any positive α:

$$|a + b\,U^\alpha| \leq (|a| + |b|)\sup\{1\,,\, U^\alpha\}\,. \tag{366}$$

We get:

$$\left\|\frac{\partial V}{\partial x}A\right\| \leq d_1(p)\,\sup\{1\,,\, U^{\alpha_1}\} \quad \text{and} \quad \left\|\frac{\partial V}{\partial p}\right\| \leq d_2(p)\,\sup\{1\,,\, U^{\alpha_2}\}\,, \tag{367}$$

with:

$$\begin{aligned} d_1(p) &= \left[(c_1 + 2\,|p|) + (c_1c_2 + 2c_2\,|p| + 4p^2) + 2\,|p| + (2k-1)\right]\gamma^{\alpha_1} \\ \alpha_1 &= \sup\left\{1 + \frac{1}{2kj}\,,\, 1 - \frac{1}{2j} + \frac{3}{2kj}\,,\, \frac{1}{2} + \frac{2}{kj}\,,\, \frac{1}{2} + \frac{1}{2j} + \frac{1}{kj}\,,\, \frac{1}{2} + \frac{k}{kj}\right\} \\ d_2(p) &= \left[2 + |c_1 - c_2| + 2\,|p|\right]\gamma^{\alpha_2} && (368) \\ \alpha_2 &= \sup\left\{1 - \frac{1}{2j} + \frac{1}{kj}\,,\, \frac{1}{2} + \frac{3}{2kj}\,,\, \frac{1}{2} + \frac{k+1}{2kj}\right\}. && (369) \end{aligned}$$

It follows that:

$$\left\|\frac{\partial V}{\partial x}(p,x)\,A\left(x, u_n(p,x)\right)\right\|\,\left\|\frac{\partial V}{\partial p}(p,x)\right\| \;\leq\; d_1(p)\,d_2(p)\,\left(1 \,+\, V(p,x)^\alpha\,\right), \tag{370}$$

where α depending on j and k is given in Table 1. Hence, for this example, (328) is satisfied if we choose $k > 2$ and $j > 1$. □

Table 1. $\alpha(k,j)$

k \ j	1	2	3	4	5
1	9/2	11/4	5/2	19/8	23/10
2	11/4	17/8	25/12	33/16	41/20
3	8/3	2	2	2	2
4	21/8	31/16	47/24	63/32	79/40
5	13/5	19/10	29/15	39/20	49/25

4.2 Regressor Filtering

When the regressor filtering (258)–(261) is used, using (259) and (261) we get:

$$\begin{aligned}\dot{\widehat{h}} &= -\rho(x,\widehat{p},u)\,e + Z_{\mathrm{f}}\dot{\widehat{p}} \\ &\quad + \frac{{\alpha_1}^2}{(\alpha_1 - V(x,\widehat{p}))^2}\left(\frac{\partial V}{\partial x}(x,\widehat{p})\,(a(x,u) + A(x,u)\widehat{p}) + \frac{\partial V}{\partial p}(x,\widehat{p})\dot{\widehat{p}}\right), \end{aligned} \tag{371}$$

where:

$$\widehat{h} = h(x,\widehat{p}) + e = Z_{\mathrm{f}}\widehat{p} + \widehat{z}_{\mathrm{f}}\,. \tag{372}$$

Hence, if:

$$u = u_{\mathrm{n}}(x,\widehat{p})\,, \tag{373}$$

and (47) in assumption PRS holds, we have:

$$\dot{\widehat{h}} \leq -\rho(x,\widehat{p},u)\,e + \left(\frac{{\alpha_1}^2}{(\alpha_1 - V(x,\widehat{p}))^2}\,\frac{\partial V}{\partial p}(x,\widehat{p}) + Z_{\mathrm{f}}\right)\dot{\widehat{p}}\,. \tag{374}$$

Compared with equation (293) for the equation error filtering case, we have the extra term $Z_{\mathrm{f}}\dot{\widehat{p}}$. But, thanks to Lemma (264), we know that we may choose ρ and $\dot{M}$ in order to guarantee that Z_{f} is bounded and $\dot{\widehat{p}}$ is absolutely integrable. It follows that the following counterpart to Proposition (325) can be established:

Proposition **(375)**
Let assumptions BO, URS and ICS hold. Choose:

$$\begin{aligned} h(x,p) &= \frac{\alpha_1\,V(x,p)}{\alpha_1 - V(x,p)} \\ \rho(x,p,u) &= 1 + \left\|\frac{{\alpha_1}^2}{(\alpha_1 - V(x,p))^2}\,\frac{\partial V}{\partial x}(x,p)\,A(x,u)\right\| \\ r(x,p,u,e) &= \rho(x,p,u)^2\,, \end{aligned} \tag{376}$$

and:

$$\dot{M} = G(M\,,Z_{\mathrm{f}}\,,r)\,, \qquad \varepsilon_3\,M(0) > I\,, \tag{377}$$

where G is a negative symmetric matrix, depending locally-Lipschitz-continuously on its arguments and satisfying (see (262) and (263)):

$$-\varepsilon_4 M Z_f^T Z_f M r \geq G \geq -(2-\varepsilon_1) M Z_f^T Z_f M r . \tag{378}$$

If assumption Λ-LP is satisfied with:

$$\Lambda(x,t) = \frac{\partial h}{\partial x}(x, \widehat{p}(t)), \tag{379}$$

and:

either *assumption MC holds and, with notation (13), $B(x) \equiv 0$, α_1 is chosen smaller than or equal to α_0, $x(0)$ belongs to Ω_0 and $V(x(0), \widehat{p}(0)) < \alpha_1$,*

or *we are in the global case, i.e., $\Omega_0 = \Omega = \mathbb{R}^n$, $\alpha_1 = \alpha_0 = +\infty$ and there exists a C^0 function $d_2 : \Pi \to \mathbb{R}_+$ such that, for all (x,p) in $\mathbb{R}^n \times \Pi$:*

$$\left\| \frac{\partial V}{\partial p}(x,p) \right\| \leq d_2(p) \max\{1, V(x,p)\}, \tag{380}$$

then all the corresponding solutions $(x(t), \widehat{p}(t), \widehat{z_f}(t), Z_f(t)), M(t)$ of (S_{p^*}), (259), (261), (262) and (377) are well-defined on $[0, +\infty)$, unique, bounded and:

$$\lim_{t\to\infty} V(x(t), \widehat{p}(t)) = 0 . \tag{381}$$

It follows that the Adaptive Stabilization problem is solved if assumption CO holds also.

Proof.
Case: Assumption MC holds: In this case, we assume also that $A(x,u)$ does not depend on u, i.e., with notation (13), we have:

$$A(x,u) = A_0(x) . \tag{382}$$

The system we consider is:

$$\begin{aligned}
\dot{x} &= a(x, u) + A_0(x)\, p^* \\
\dot{\widehat{z_f}} &= \rho(x,\widehat{p})\, z_f + \frac{\partial h}{\partial x}(x,\widehat{p})\, a(x,u) + \frac{\partial h}{\partial p}(x,\widehat{p})\, \dot{\widehat{p}}, \quad \widehat{z_f}(0) = h(x(0), \widehat{p}(0)) \\
z_f &= h(x,\widehat{p}) - \widehat{z_f} \\
\dot{Z}_f &= -\rho(x,\widehat{p})\, Z_f + \frac{\partial h}{\partial x}(x,\widehat{p})\, A_0(x), \quad Z_f(0) = 0 \\
e &= Z_f \widehat{p} - z_f \\
\dot{\widehat{p}} &= \mathrm{Proj}\left(M, \widehat{p}, -M Z_f^T r(x,\widehat{p})\, e\right), \quad \widehat{p}(0) \in \Pi_1 \\
\dot{M} &= G(M, Z_f, r(x,\widehat{p})), \quad M(0) > 0,
\end{aligned} \tag{383}$$

with:

$$u = u_{\mathrm{n}}(x,\widehat{p}) + v(x,\widehat{p},\dot{\widehat{p}}) \,, \quad h(x,p) = \frac{\alpha_1 V(x,p)}{\alpha_1 - V(x,p)} \,, \quad r(x,p) = \rho(x,p)^2 \,, \tag{384}$$

and

$$\rho(x,p) = 1 + \left\| \frac{{\alpha_1}^2}{(\alpha_1 - V(x,p))^2} \frac{\partial V}{\partial x}(x,p) A_0(x)) \right\| \,, \tag{385}$$

where:

$$\frac{\partial V}{\partial p}(x,\widehat{p})\, \dot{\widehat{p}} + \frac{\partial V}{\partial x}(x,\widehat{p})\, v(x,\widehat{p},\dot{\widehat{p}}) \odot b(x) = 0 \,. \tag{386}$$

Since r does not depend on u, $\dot{\widehat{p}}$ is explicitly defined. Then, from our smoothness assumptions on the various functions and with Lemma (103), this system (383) has a locally Lipschitz-continuous right-hand side in the open set defined by:

$$(x, \widehat{z_{\mathrm{f}}}, Z_{\mathrm{f}}, \widehat{p}, M) \in \Omega \times \mathbb{R} \times \mathbb{R}^l \times \Pi \times \mathcal{M} \qquad \text{and} \qquad V(x,\widehat{p}) < \alpha_1 \,. \tag{387}$$

It follows that, for any initial condition in this open set, there exists a unique solution $(x(t), \widehat{z_{\mathrm{f}}}(t), Z_{\mathrm{f}}(t), \widehat{p}(t), M(t))$, defined on a right maximal interval $[0,T)$, with T possibly infinite, and satisfying (387). Applying Points 1 and 2 of Lemma (264), we also know that, for all t in $[0,T)$:

$$\begin{aligned} &\widehat{p}(t) \in \Pi_1 \\ &\varepsilon_2 \left\|\widehat{p}(t) - p^*\right\|^2 + \varepsilon_1 \int_0^t r \left\|e\right\|^2 \leq \varepsilon_3 \left\|\widehat{p}(0) - p^*\right\|^2 \stackrel{\text{def}}{=} \beta^2 \,. \end{aligned} \tag{388}$$

Moreover, our choice for ρ implies with the last statement of Lemma (264) that, for all t in $[0,T)$:

$$\|Z_{\mathrm{f}}\| \leq 1 \qquad \text{and} \qquad \|e\| \leq \sqrt{\frac{\varepsilon_3}{\varepsilon_2}} \, \|\widehat{p}(0) - p^*\| \,. \tag{389}$$

Then, since we have (see (260)):

$$z_{\mathrm{f}} = Z_{\mathrm{f}}\, p^* \,, \tag{390}$$

we conclude also that for all t in $[0,T)$:

$$\|z_{\mathrm{f}}\| \leq \|p^*\| \,. \tag{391}$$

Finally, since $M(t) \in \mathcal{M}$ for all t in $[0,T)$, (377) and (378) imply that $\dot{M}$ is negative and:

$$\dot{\overline{M^{-1}}} \leq (2 - \varepsilon_1)\, Z_{\mathrm{f}}^{\mathrm{T}} Z_{\mathrm{f}}\, r \,. \tag{392}$$

With (389) and (377), this implies, for all t in $[0,T)$:

$$\frac{I}{\varepsilon_3 + (2 - \varepsilon_1) \int_0^T r} \leq M(t) \leq M(0) \,. \tag{393}$$

Then, for $\widehat{h}$ defined in (372), we get by (371) and (298) in assumption URS:

$$\dot{\widehat{h}} \leq -\rho e - c\frac{\alpha_1{}^2 V}{(\alpha_1 - V)^2} + Z_f \dot{\widehat{p}} \tag{394}$$

$$\leq -(\sqrt{r}\,|e|) - c\,h + Z_f \dot{\widehat{p}} \tag{395}$$

$$\leq -c\widehat{h} + \sqrt{r}\,|e| + c\,|e| + Z_f \dot{\widehat{p}}. \tag{396}$$

With (388) and Point 3 of Lemma (264), the assumption of Lemma (583) in Appendix B is satisfied with:

$$X = \widehat{h} \tag{397}$$

and:

$$\begin{aligned} \varpi_1 &= \sqrt{r}\,|e| + c\,|e|\,, \zeta_1 = 2,\, S_{21} = 2\left(1 + c^2\right)\frac{\beta^2}{\varepsilon_1} \\ \varpi_2 &= \left\|\dot{\widehat{p}}\right\|, \qquad \zeta_2 = 1,\, S_{22} = k_1(p^*, \widehat{p}(0))\,. \end{aligned} \tag{398}$$

It follows that there exists a constant Υ depending only on the initial conditions such that, for all t in $[0, T)$, we have:

$$0 \leq \frac{\alpha_1 V(x(t), \widehat{p}(t))}{\alpha_1 - V(x(t), \widehat{p}(t))} = h(x(t), \widehat{p}(t)) \tag{399}$$

$$= \widehat{h}(t) - e(t) \tag{400}$$

$$\leq \Upsilon + \sqrt{\frac{\varepsilon_3}{\varepsilon_2}}\,\|\widehat{p}(0) - p^*\|\,, \tag{401}$$

where we have used (389). Hence, we have established, for all t in $[0, T)$:

$$\begin{aligned} V(x(t), \widehat{p}(t)) &\leq \frac{\alpha_1 (\Upsilon + \sqrt{\frac{\varepsilon_3}{\varepsilon_2}}\,\|\widehat{p}(0) - p^*\|)}{\alpha_1 + \Upsilon + \sqrt{\frac{\varepsilon_3}{\varepsilon_2}}\,\|\widehat{p}(0) - p^*\|} < \alpha_1 \\ \|\widehat{p}(t) - p^*\| &\leq \frac{\beta}{\sqrt{\varepsilon_2}} \qquad \text{and} \qquad \widehat{p}(t) \in \Pi_1 \\ &\|Z_f(t)\| \leq 1 \\ &\|z_f(t)\| \leq \|p^*\|\,. \end{aligned} \tag{402}$$

Then, from assumption BO, we know the existence of a compact subset Γ of Ω such that:

$$x(t) \in \Gamma \quad \forall t \in [0, T)\,. \tag{403}$$

We have also:

$$\frac{I}{\varepsilon_3 + (2 - \varepsilon_1)\,T\,R} \leq M(t) \leq M(0)\,, \tag{404}$$

where R is an upper bound for $\int_0^T r$ whose existence follows from the continuity of the function r and the boundedness of x and $\widehat{p}$. Finally, with the continuity of the function h and the fact that:

$$\widehat{z}_{\rm f} = h(x,\widehat{p}) - z_{\rm f}, \tag{405}$$

we have proved that the solution remains in a compact subset of the open set defined in (387). It follows by contradiction that $T = +\infty$ and in particular that $x(t)$, $\widehat{p}(t)$, $u(t)$ and $\dot{\widehat{p}}(t)$ are bounded on $[0,+\infty)$. From here, we conclude the proof exactly as in the proof of Proposition (325).

Case: Inequality (380) holds: The system we consider is the system (383) with:

$$u = u_{\rm n}(x,\widehat{p}), \quad h(x,p) = V(x,p), \tag{406}$$

$$\rho(x,p,u) = 1 + \left\|\frac{\partial V}{\partial x}(x,p)\,A(x,u))\right\| \quad \text{and} \quad r(x,p,u,e) = \rho(x,p,u)^2. \tag{407}$$

From our smoothness assumptions on the various functions and with Lemma (103), this system has a locally Lipschitz-continuous right-hand side in the open set defined by:

$$(x,\widehat{z}_{\rm f},Z_{\rm f},\widehat{p},M) \in \mathbb{R}^n \times \mathbb{R} \times \mathbb{R}^l \times \Pi \times \mathcal{M}. \tag{408}$$

It follows that, for any initial condition in this open set, there exists a unique solution $(x(t),\widehat{z}_{\rm f}(t),Z_{\rm f}(t),\widehat{p}(t),M(t))$, defined on a right maximal interval $[0,T)$, with T possibly infinite and satisfying (408). Then, as in the previous case, we have, for all t in $[0,T)$:

$$\begin{aligned}
&\widehat{p}(t) \in \Pi_1 \\
&\varepsilon_2\,\|\widehat{p}(t)-p^*\|^2 + \varepsilon_1 \int_0^t r\,\|e\|^2 \le \varepsilon_3\,\|\widehat{p}(0)-p^*\|^2 \stackrel{\rm def}{=} \beta^2 \\
&\|Z_{\rm f}\| \le 1 \qquad \text{and} \qquad \|e\| \le \sqrt{\tfrac{\varepsilon_3}{\varepsilon_2}}\,\|\widehat{p}(0)-p^*\| \\
&\|z_{\rm f}\| \le \|p^*\| \\
&\frac{I}{\varepsilon_3 + (2-\varepsilon_1)\int_0^T r} \le M(t) \le M(0).
\end{aligned} \tag{409}$$

By taking care of the fact that we are not using v to cancel the term $\frac{\partial V}{\partial p}\dot{\widehat{p}}$, we get also:

$$\dot{\widehat{h}} \le -\rho\,e - c\,V + \left(\frac{\partial V}{\partial p} + Z_{\rm f}\right)\dot{\widehat{p}}. \tag{410}$$

And, by using (409), inequality (380), and the continuity of the function d_2, we have:

$$\left\|\frac{\partial V}{\partial p}\right\| \le d_2(\widehat{p}(t))\,\max\{1\,,\,V\} \tag{411}$$

$$\le k\,(1+V) \tag{412}$$

for some constant k depending only on $\widehat{p}(0)$. Hence, with our choice for ρ and r, we have established:

$$\dot{\widehat{h}} \leq -\rho e - cV + (k(1+V)+1)\left\|\dot{\widehat{p}}\right\| \tag{413}$$

$$\leq -\left(c - k\left\|\dot{\widehat{p}}\right\|\right)\widehat{h} + \sqrt{r}\,|e| + c\,|e| + (k(1+|e|)+1)\left\|\dot{\widehat{p}}\right\| . \tag{414}$$

Hence, with (409), the assumption of Lemma (583) in Appendix B is satisfied with:

$$X = \widehat{h} \tag{415}$$

and

$$\begin{aligned} \vartheta_1 &= k\left\|\dot{\widehat{p}}\right\|, & \sigma_1 &= 1,\ S_{11} = k\,k_1(p^*, \widehat{p}(0)) \\ \varpi_1 &= \sqrt{r}\,|e| + c\,|e|, & \zeta_1 &= 2,\ S_{21} = 2\left(1+c^2\right)\frac{\beta^2}{\varepsilon_1} \\ \varpi_2 &= \left[k\left(1+\sqrt{\tfrac{\varepsilon_3}{\varepsilon_2}}\,\|\widehat{p}(0)-p^*\|\right)+1\right]\left\|\dot{\widehat{p}}\right\|, & \zeta_2 &= 1,\ S_{22} = k_1(p^*, \widehat{p}(0)) . \end{aligned} \tag{416}$$

From here, we conclude the proof exactly as in the previous case. □

In practice, to apply this Proposition, we must be allowed to choose the particular value for the initial condition $\widehat{z}_f(0)$ and a vanishing observer gain K ((263) is assumed). Compare with Proposition (325) where we have no constraint on the initial conditions and the observer gain is not forced to decay to zero. However, Proposition (375) shows that the regressor filtering technique with V as observation function has also the property of providing a solution to the Adaptive Stabilization problem in the case where the Matching Condition (MC) does not hold. Then, besides this question of initial conditions and observer gains, the choice between the two estimation techniques depends on which of the linear growth condition (380) or quadratic growth condition (328) holds.

Example: System (320) Continued **(417)**
For the system (320), let us see if the growth condition (380) holds. From (367), we have:

$$\left\|\frac{\partial V}{\partial p}\right\| \leq d_2(p)\,\sup\left\{1\,,\,U^{\alpha_2}\right\}, \tag{418}$$

where d_2 is given by (368) and α_2 satisfying (369) depends on j and k and is given in Table 2. Hence, for this example (380) is satisfied if we choose $k > 1$ and $j > 1$. This is different from what we obtained in Example (358) where the equation error filtering is to be used. □

5 Estimation Design with an Observation Function h not Directly Related to V

When the observation function h is not directly related to V, the set value maps F and $F^\dagger$ defined in (289) do not have any particular properties. Consequently

Table 2. $\alpha_2(k,j)$

k \ j	1	2	3	4	5
1	2	5/4	7/6	9/8	11/10
2	5/4	1	1	1	1
3	7/6	11/12	17/18	23/24	29/30
4	9/8	7/8	11/12	15/16	19/20
5	11/10	17/20	9/10	37/40	47/50

assumption PRS or URS does not give any information on the dynamics of h. Similarly, the estimation does not necessarily provide a better fit of the $\dot{V}$ equation. To overcome this difficulty and guarantee some properties for F and $F^\dagger$, we have to refine our assumptions. Namely, we have to relate V to h. In the global case, i.e., for $\Omega = \mathbb{R}^n$, this yields:

Assumption RBO (Refined Boundedness Observability) (419)
For all positive real numbers α, all compact subsets $\mathcal{K}$ of Π and all vectors $x_0 \in \mathbb{R}^n$, we can find a compact subset Γ of $\mathbb{R}^n$ such that for any C^1 time functions $\widehat{p} : \mathbb{R}_+ \to \Pi$ and $u : \mathbb{R}_+ \to \mathbb{R}^m$ and any solution $x(t)$ of (44) defined on $[0,T)$, we have the following implication:

$$\|h(x(t),\widehat{p}(t))\| \le \alpha \text{ and } \widehat{p}(t) \in \mathcal{K}\ \forall t \in [0,T) \quad \to \quad x(t) \in \Gamma\ \forall t \in [0,T)\,. \tag{420}$$

Namely, the boundedness of the full state vector x is "observable" from the boundedness of the "output" function h.

Assumption RURS (Refined Uniform Reduced-Order Stab.) (421)
There exists a positive constant c and two functions:

$$f : \mathbb{R}^k \times \Pi \to \mathbb{R}^k \text{ of class } C^1 \qquad \text{and} \qquad U : \mathbb{R}^k \times \Pi \to \mathbb{R}_+ \text{ of class } C^2\,,$$

with U known *and:*

1. $f(O,p) = 0 \quad \forall p \in \Pi$
2. $U(h,p) = 0 \iff h = 0$
3. $\forall \alpha \ge 0$, $\forall \mathcal{K}$ *compact subset of Π, the set*

$$\{h \mid U(h,p) \le \alpha \text{ and } p \in \mathcal{K}\} \text{ is a compact subset of } \mathbb{R}^k\,,$$

such that, for all (x,p,h) in $\mathbb{R}^n \times \Pi \times \mathbb{R}^k$, we have:

$$\begin{aligned} &1.\ f(h(x,p)\,,\,p) = \frac{\partial h}{\partial x}(x,p)\,[a(x,u_{\mathrm{n}}(x,p)) + A(x,u_{\mathrm{n}}(x,p))\,p] \\ &2.\ \frac{\partial U}{\partial h}(h,p)\,f(h,p) \le -c\,U(h,p)\,. \end{aligned} \tag{422}$$

The meaning of assumption RURS is that, for any fixed vector p in Π, the time derivative of h along the solutions x of (S_p) in closed loop with the control

$$u = u_n(x,p), \tag{423}$$

is simply:

$$\dot{h} = f(h,p), \tag{424}$$

i.e., this control makes the reduced-order system, consisting of the components of h, autonomous and decoupled from the other components of the full state vector x. Moreover, U is a Lyapunov function for this reduced-order system implying that it admits $0 \in \mathbb{R}^k$ as a globally asymptotically stable equilibrium point.

Assumption RCO (Refined Convergence Observability) (425)

For any bounded C^1 time functions $\widehat{p} : \mathbb{R}_+ \to \Pi$ and $u : \mathbb{R}_+ \to \mathbb{R}^m$ with $\dot{\widehat{p}}$ also bounded and for any solution $x(t)$ of (44) defined on $[0,+\infty)$, we have the following implication:

$$\left.\begin{array}{l} \lim_{t\to\infty} h(x(t),\widehat{p}(t)) \text{ exists and is zero} \\ \text{and} \quad x(t) \text{ is bounded on } [0,+\infty) \end{array}\right\} \implies \lim_{t\to\infty} x(t) \text{ exists and is equal to } \mathcal{E}.$$

Assumption RMC (Refined Matching Condition) (426)

The functions a and A are affine in u, the function U does not depend on p and there exists a known *C^1 function $v(x,p,q,\partial)$ from $\mathbb{R}^n \times \Pi \times \Pi \times \mathbb{R}^l$ to $\mathbb{R}^m$ satisfying:*

$$\frac{\partial h}{\partial p}(x,p)\,\partial + \frac{\partial h}{\partial x}(x,p)v \odot [b(x) + B(x)q] = 0\,. \tag{427}$$

Note the strong restriction on U in assumption RMC.

Example: System (17) Continued (428)

For the system (17) rewritten as (50) in Example (49), the refined assumptions RBO, RCO, RURS and RMC are satisfied if we choose:

$$h(x_1,x_2,p_1,p_2) = x_1 \quad , \qquad u_n(x_1,x_2,p_1,p_2) = -\frac{x_2^2 + p_1 + x_1}{p_2}\,, \tag{429}$$

and:

$$\begin{aligned} U(h,p_1,p_2) &= \left(h + \frac{2}{3}\right)^2 && \text{if } \ h \le -1 \\ &= \frac{1}{9}\exp\left(3\left(1 - \frac{1}{h^2}\right)\right) && \text{if } \ -1 < h < 1 \\ &= \left(h - \frac{2}{3}\right)^2 && \text{if } \ 1 \le h\,, \end{aligned} \tag{430}$$

with:

$$f(h,p_1,p_2) = -h^3\,, \qquad \Pi = \mathbb{R} \times (\mathbb{R}_+ - \{0\}) \tag{431}$$

and

$$\Gamma = \left\{x \in \mathbb{R}^2 \mid \|x\| \le \|x(0)\| + \alpha(1 + |c_3|)\right\}\,. \tag{432}$$

□

5.1 Equation Error Filtering

Let $\widehat{p}$ be obtained from the equation error filtering technique (230)–(232). If the control is given by:

$$u = u_{\mathrm{n}}(x, \widehat{p}), \tag{433}$$

equation (230) gives with the help of (422) in assumption RBO:

$$\dot{\widehat{h}} = f(\widehat{h}, \widehat{p}) + \left[-r(e, x, \widehat{p})\, e + \left(f(\widehat{h} - e, \widehat{p}) - f(\widehat{h}, p)\right) + \frac{\partial h}{\partial p}(x, \widehat{p})\, \dot{\widehat{p}}\right]. \tag{434}$$

If, instead, assumption RMC holds, let $\widehat{p}$ and $\widehat{q}$ be obtained from (234) and use the control:

$$u = u_{\mathrm{n}}(x, \widehat{p}) + v\left(x, \widehat{p}, \widehat{q}, \dot{\widehat{p}}\right), \tag{435}$$

with v given by (427) in RMC with $\partial = \dot{\widehat{p}}$. Then (434) reduces to:

$$\dot{\widehat{h}} = f(\widehat{h}, \widehat{p}) + \left[-r(e, x, \widehat{p})\, e + \left(f(\widehat{h} - e, \widehat{p}) - f(\widehat{h}, p)\right)\right]. \tag{436}$$

In view of assumption RURS, the term enclosed in brackets in both (434) and (436) should be considered as perturbations. To be able to apply techniques a la Total Stability (see Lakshmikantham and Leela [11]), we need to compare the components of this perturbation with the "Lyapunov function" U. This motivates the following assumption:

Assumption GC1 (Growth Conditions 1) **(437)**
There exist two positive continuous functions d_3, defined on $\Pi \times \mathbb{R}^k$ and d_4, defined on Π and a known *positive real number λ, with $\lambda \leq 1$, such that, for all (h, p) in $\mathbb{R}^k \times \Pi$:*

1. $\left|\frac{\partial U}{\partial h}(h, p)\left(f(h - e, p) - f(h, p)\right)\right| \leq d_3(p, e) \max\left\{1, U(h, p)^{2-\lambda}\right\} \|e\|$
2. $\left\|\frac{\partial U}{\partial h}(h, p)\right\| \leq d_4(p) \max\left\{1, U(h, p)^{\lambda}\right\}$.

(438)

Moreover, if assumption RMC does not hold, then $\lambda < 1$ and there exist three positive continuous functions d_i, $i = 5, 7$ defined on Π and two positive real numbers ω and κ, with $\omega + \kappa \leq 2 - \lambda$ such that, for all (x, p) in $\mathbb{R}^n \times \Pi$:

3. $\left\|\frac{\partial h}{\partial p}(x, p)\right\| \left\|\left[\frac{\partial h}{\partial x}(x, p)\, A(x, u_{\mathrm{n}}(x, p))\right]^{\mathrm{T}}\right\| \leq d_5(p) \max\{1, U(h(x, p), p)^{2(1-\lambda)}\}$
4. $\left\|\frac{\partial U}{\partial p}(h, p)\right\| \leq d_6(p) \max\{1, U(h, p)^{\omega}\} \qquad \forall h \in \mathbb{R}^k$
5. $\left\|\frac{\partial h}{\partial x}(x, p)\, A(x, u_{\mathrm{n}}(x, p))\right\| \leq d_7(p) \max\{1, U(h(x, p), p)^{\kappa}\}$.

(439)

We have:

Proposition (440)
Let assumptions A-LP, RBO, RURS, ICS and GC1 hold in the global case with

$$A(x,t) = \frac{\partial h}{\partial x}(x, \widehat{p}(t)), \tag{441}$$

and choose:

$$r(e,x,\widehat{p}) = \left(1 + U(\widehat{h},\widehat{p})^{(1-\lambda)}\right)^2. \tag{442}$$

All the corresponding solutions $(x(t),\widehat{p}(t),\widehat{h}(t))$ of (S_{p^*})-(230)-(231) are unique, bounded and well-defined on $[0,+\infty)$ and:

$$\lim_{t\to\infty} h(x(t),\widehat{p}(t)) = 0. \tag{443}$$

It follows that the Adaptive Stabilization problem is solved if assumption RCO holds also.

Proof.
Case: Assumption RMC holds: The system we consider is, with notation (13):

$$\begin{aligned}
\dot{x} &= a\left(x\,,\, u_{\mathrm{n}}(x,\widehat{p}) + v(x,\widehat{p},\widehat{q},\dot{\widehat{p}})\right) + A\left(x\,,\, u_{\mathrm{n}}(x,\widehat{p}) + v(x,\widehat{p},\widehat{q},\dot{\widehat{p}})\right) p^* \\
\dot{\widehat{h}} &= -r(e,x,\widehat{p})\, e + \frac{\partial h}{\partial x}(x,\widehat{p})\left[a_0(x) + A_0(x)\widehat{p} + u_{\mathrm{n}}(x,\widehat{p}) \odot (b(x) + B(x)\widehat{p})\right] \\
&\quad + \frac{\partial h}{\partial x}(x,\widehat{p})\, v\left(x,\widehat{p},\widehat{q},\dot{\widehat{p}}\right) \odot (b(x) + B(x)\widehat{q}) + \frac{\partial h}{\partial p}(x,\widehat{p})\,\dot{\widehat{p}} \\
e &= \widehat{h} - h(x,\widehat{p}) \\
\dot{\widehat{p}} &= \mathrm{Proj}\left(I,\widehat{p}, -\left[\frac{\partial h}{\partial x}(x,\widehat{p})\left(A_0(x) + u_{\mathrm{n}}(x,\widehat{p}) \odot B(x)\right)\right]^{\mathrm{T}} e\right), \quad \widehat{p}(0) \in \Pi_1 \\
\dot{\widehat{q}} &= \mathrm{Proj}\left(I,\widehat{q}, -\left[\frac{\partial h}{\partial x}(x,\widehat{p})\, v(x,\widehat{p},\widehat{q},\dot{\widehat{p}}) \odot B(x)\right]^{\mathrm{T}} e\right), \quad \widehat{q}(0) \in \Pi_1\,,
\end{aligned} \tag{444}$$

with $r(e,x,\widehat{p})$ defined by (442) and, with notation (13):

$$\frac{\partial h}{\partial p}(x,\widehat{p})\,\dot{\widehat{p}} + \frac{\partial h}{\partial x}(x,\widehat{p})\, v(x,\widehat{p},\widehat{q},\dot{\widehat{p}}) \odot \left[b(x) + B(x)\,\widehat{q}\right] = 0. \tag{445}$$

From our smoothness assumptions on the various functions and with Point 1 of Lemma (103), this system has a locally Lipschitz-continuous right-hand side in the open set defined by:

$$\left(x,\widehat{p},\widehat{q},\widehat{h}\right) \in \mathbb{R}^n \times \Pi \times \Pi \times \mathbb{R}^k. \tag{446}$$

It follows that, for any initial condition in this open set, there exists a unique solution $(x(t),\widehat{p}(t),\widehat{q}(t),\widehat{h}(t))$, defined on a right maximal interval $[0,T)$, with T

possibly infinite and satisfying (446). Applying Points 3 and 4 of Lemma (235), we also know that, for all t in $[0,T)$:

$$\widehat{p}(t) \in \Pi_1 \quad \text{and} \quad \widehat{q}(t) \in \Pi_1$$

$$\left\| \begin{matrix} \widehat{p}(t) - p^* \\ \widehat{q}(t) - p^* \end{matrix} \right\|^2 + \|e(t)\|^2 + 2\int_0^t r\,\|e\|^2 \leq \left\| \begin{matrix} \widehat{p}(0) - p^* \\ \widehat{q}(0) - p^* \end{matrix} \right\|^2 + \|e(0)\|^2 \stackrel{\text{def}}{=} \beta^2 . \tag{447}$$

Now, from assumption RMC, the function U given by assumption RURS depends on h only. Then, letting:

$$\widehat{U}(t) = U(\widehat{h}(t)), \tag{448}$$

we look at the time derivative of $\widehat{U}$ along the solutions of (444) (see also (436)). From (422) in assumption RURS, (438) in assumption GC1 and (442), we get successively:

$$\begin{aligned}
\dot{\widehat{U}} &\leq -c\widehat{U} + \frac{\partial U}{\partial h}(\widehat{h}) \left[-r(e,x,\widehat{p})\,e + \left(f(\widehat{h} - e\,,\,\widehat{p}) - f(\widehat{h},p)\right)\right] \\
&\leq -c\widehat{U} + d_4(\widehat{p}) \max\left\{1, \widehat{U}^{\lambda}\right\} r\,\|e\| \\
&\qquad + d_3(\widehat{p},e) \max\left\{1, \widehat{U}^{2-\lambda}\right\} \|e\| \\
&\leq -c\widehat{U} + d_3(\widehat{p},e) \left(1 + \widehat{U}^{1-\lambda}\right) \|e\| \left(1 + \widehat{U}\right) \\
&\qquad + d_4(\widehat{p}) \max\left\{1, \widehat{U}^{\lambda}\right\} \left(1 + \widehat{U}^{1-\lambda}\right)^2 \|e\| \\
&\leq -c\widehat{U} + \left[(d_3(\widehat{p},e) + 2\,d_4(\widehat{p})) \left(1 + \widehat{U}^{1-\lambda}\right) \|e\|\right] \left(1 + \widehat{U}\right) \\
&\leq -\left[c - k\left(1 + \widehat{U}^{1-\lambda}\right) \|e\|\right] \widehat{U} + k\left(1 + \widehat{U}^{1-\lambda}\right) \|e\| ,
\end{aligned} \tag{449}$$

where the constant k depends only on the initial conditions and satisfies:

$$d_3(\widehat{p}(t), e(t)) + 2\,d_4(\widehat{p}(t)) \leq k \qquad \forall t \in [0,T) . \tag{450}$$

Such an inequality holds since we have (447) and the functions d_3 and d_4 are continuous. With (442), inequality (449) implies that the assumption of Lemma (583) in Appendix B is satisfied with:

$$X = \widehat{U} \tag{451}$$

and:

$$\begin{aligned}
\vartheta_1 &= k\left(1 + \widehat{U}^{1-\lambda}\right) \|e\| ,\ \sigma_1 = 2,\ S_{11} = k^2\,\frac{\beta^2}{2} \\
\varpi_1 &= k\left(1 + \widehat{U}^{1-\lambda}\right) \|e\| ,\ \zeta_1 = 2,\ S_{21} = k^2\,\frac{\beta^2}{2} .
\end{aligned} \tag{452}$$

With the property 3 of U in assumption RURS, it follows that there exists a constant Υ depending only on $e(0)$, $\widehat{p}(0)$ and $\widehat{q}(0)$ such that, for all t in $[0,T)$, we have:

$$\left\|\widehat{h}(t)\right\| \leq \Upsilon . \tag{453}$$

Then, with (447), this implies the existence of a constant α depending only on $e(0)$, $\widehat{p}(0)$ and $\widehat{q}(0)$ such that, for all t in $[0,T)$:

$$\begin{aligned} \|h(x(t),\widehat{p}(t))\| &\leq \left\|\widehat{h}(t)\right\| + \|e(t)\| && (454)\\ &\leq \alpha\,, && (455)\end{aligned}$$

and we also know from (447) that $\widehat{p}(t) \in \mathcal{K}$ and $\widehat{q}(t) \in \mathcal{K}$, where $\mathcal{K}$ is the following compact subset of Π:

$$\mathcal{K} = \{p \mid \|p - p^*\| \leq \beta\} \bigcap \Pi_1 . \tag{456}$$

Then, from assumption RBO, there exists of a compact subset Γ of $\mathbb{R}^n$ such that:

$$x(t) \in \Gamma \quad \forall t \in [0,T) . \tag{457}$$

With (447) and (453), we have established that the solution remains in a compact subset of the open set defined by (446). This implies by contradiction that $T = +\infty$ and that $x(t)$, $\widehat{p}(t)$, $u(t)$ and $\dot{\widehat{p}}(t)$ are bounded on $[0,+\infty)$.

Then, using the second conclusion of Lemma (583) and the properties of the function U, we have:

$$\lim_{t\to+\infty} \widehat{h}(t) = 0 . \tag{458}$$

Also, from (229) and the fact that the solution is bounded, we deduce that $\dot{e}$ is bounded. Since, from (447), e is in $L^2([0,+\infty))$, we have also:

$$\lim_{t\to+\infty} e(t) = 0 . \tag{459}$$

This yields:

$$\lim_{t\to+\infty} h(x(t),\widehat{p}(t)) = \lim_{t\to+\infty} \widehat{h}(t) - \lim_{t\to+\infty} e(t) = 0 . \tag{460}$$

With assumption RCO this implies finally:

$$\lim_{t\to+\infty} x(t) = \mathcal{E} . \tag{461}$$

Case: Assumption RMC does not hold: The system we consider is:

$$\begin{aligned} \dot{x} &= a(x\,,\, u_{\mathrm{n}}(x,\widehat{p})) + A(x\,,\, u_{\mathrm{n}}(x,\widehat{p}))\, p^* \\ \dot{\widehat{h}} &= -r(e,x,\widehat{p})\, e + \frac{\partial V}{\partial x}(x,\widehat{p})\, A(x,u_{\mathrm{n}}(x,\widehat{p}))\, \widehat{p} + \frac{\partial V}{\partial x}(x,\widehat{p})\, a(x,u_{\mathrm{n}}(x,\widehat{p})) \\ &\quad + \frac{\partial V}{\partial p}(x,\widehat{p})\, \dot{\widehat{p}} \\ e &= \widehat{h} - V(x,\widehat{p}) \\ \dot{\widehat{p}} &= \mathrm{Proj}\left(I\,,\, \widehat{p}, -\left[\frac{\partial h}{\partial x}(x,\widehat{p})\, A(x,u_{\mathrm{n}}(x,\widehat{p}))\right]^{\mathrm{T}} e\right), \qquad \widehat{p}(0) \in \Pi_1\,, \end{aligned} \tag{462}$$

with $r(e,x,\widehat{p})$ defined by (442). From our smoothness assumptions on the various functions and from Point 1 of Lemma (103), this system has a locally Lipschitz-continuous right-hand side in $\mathbb{R}^n \times \Pi \times \mathbb{R}^k$. It follows that, for any initial condition in this open set, there exists a unique solution $(x(t),\widehat{p}(t),\widehat{h}(t))$, defined on a right maximal interval $[0,T)$, with T possibly infinite.

Applying Points 1 and 2 of Lemma (235), we also know that, for all t in $[0,T)$

$$\begin{aligned}
&\widehat{p}(t) \in \Pi_1 \\
&\|\widehat{p}(t)-p^*\|^2 + \|e(t)\|^2 + 2\int_0^t r\,\|e\|^2 \;\le\; \|\widehat{p}(0)-p^*\|^2 + \|e(0)\|^2 \stackrel{\text{def}}{=} \beta^2 .
\end{aligned} \tag{463}$$

Then, as in the previous case, we study the evolution of the time derivative of $\widehat{U}$ along the solutions of (462) (see also (434)), with:

$$\widehat{U}(t) = U(\widehat{h}(t),\widehat{p}(t)) . \tag{464}$$

From (422) in assumption RURS, Point 2 of Lemma (103), (438) and (439) in assumption GC1, (442), and the expression of $\dot{\widehat{p}}$ in (462), we get successively:

$$\begin{aligned}
\dot{\widehat{U}} \le& -c\widehat{U} + \frac{\partial U}{\partial h}(\widehat{h},\widehat{p})\left[-r\,e + \left(f(\widehat{h}-e\,,\widehat{p}) - f(\widehat{h},p)\right) + \frac{\partial h}{\partial p}(x,\widehat{p})\dot{\widehat{p}}\right] \\
&+ \frac{\partial U}{\partial p}(\widehat{h},\widehat{p})\dot{\widehat{p}} \\
\le& -c\widehat{U} \\
&+ \left\|\frac{\partial U}{\partial h}(\widehat{h},\widehat{p})\right\| \left[r\,\|e\| + \left\|\frac{\partial h}{\partial p}(\widehat{h},\widehat{p})\right\| \left\|\left[\frac{\partial h}{\partial x}(x,\widehat{p})\,A(x,u_{\mathrm{n}}(x,\widehat{p}))\right]^{\mathrm{T}}\right\| \|e\|\right] \\
&+ \left|\frac{\partial U}{\partial h}(\widehat{h},\widehat{p})\left(f(\widehat{h}-e\,,\widehat{p}) - f(\widehat{h},p)\right)\right| \\
&+ \left\|\frac{\partial U}{\partial p}(\widehat{h},\widehat{p})\right\| \left\|\left[\frac{\partial h}{\partial x}(x,\widehat{p})\,A(x,u_{\mathrm{n}}(x,\widehat{p}))\right]^{\mathrm{T}}\right\| \|e\| \\
\le& -c\widehat{U} \\
&+ d_4(\widehat{p})\max\left\{1,\widehat{U}^{\lambda}\right\}\left[r\,\|e\| + d_5(\widehat{p})\max\{1\,,\,U(h(x,\widehat{p}),\widehat{p})^{2(1-\lambda)}\}\,\|e\|\right] \\
&+ d_3(\widehat{p},e)\max\left\{1,\widehat{U}^{2-\lambda}\right\}\|e\| \\
&+ d_6(\widehat{p})\max\{1\,,\,\widehat{U}^{\omega}\}\,d_7(\widehat{p})\max\{1\,,\,U(h(x,\widehat{p}),\widehat{p})^{\kappa}\}\,\|e\| .
\end{aligned} \tag{465}$$

A difficulty appearing in this inequality and which we have not encountered yet is the distinction between $\widehat{U} = U(\widehat{h},\widehat{p})$ and $U(h,\widehat{p}) = U(\widehat{h}-e\,,\widehat{p})$. As proved in Appendix C, thanks to point 2 in (438) of assumption GC1, these two quantities are related by:

$$\max\{1\,,\,U(h(x,\widehat{p}),\widehat{p})^{\gamma}\} \le \delta\left[\max\{1\,,\,\widehat{U}^{\gamma}\} + \max\{1\,,\,\widehat{U}^{\lambda\gamma}\}\left(\|e\|^{\gamma} + \|e\|^{\frac{\gamma}{1-\lambda}}\right)\right], \tag{466}$$

where γ is any positive real number and $\delta \geq 1$ depends only on $d_4(\widehat{p})$. Since the functions d_i, $i = 3, 7$ are continuous and e and $\widehat{p}$ are bounded from (463), there exists a constant k depending only on the initial conditions such that inequality (465) yields:

$$\begin{aligned}
\dot{\widehat{U}} \leq & -c\widehat{U} \\
& + k \max\left\{1, \widehat{U}^{\lambda}\right\} \left(1 + \widehat{U}^{1-\lambda}\right)^2 \|e\| \\
& + k \max\left\{1, \widehat{U}^{\lambda}\right\} \\
& \qquad \times \left[\max\left\{1, \widehat{U}^{2(1-\lambda)}\right\} + \max\left\{1, \widehat{U}^{2\lambda(1-\lambda)}\right\} \left(\|e\|^{2(1-\lambda)} + \|e\|^2\right)\right] \|e\| \\
& + k \max\left\{1, \widehat{U}^{2-\lambda}\right\} \|e\| \\
& + k \max\left\{1, \widehat{U}^{\omega}\right\} \\
& \qquad \times \left[\max\left\{1, \widehat{U}^{\kappa}\right\} + \max\left\{1, \widehat{U}^{\lambda\kappa}\right\} \left(\|e\|^{\kappa} + \|e\|^{\frac{\kappa}{1-\lambda}}\right)\right] \|e\| .
\end{aligned} \tag{467}$$

Since assumption GC1 gives

$$\omega + \kappa \leq 2 - \lambda , \tag{468}$$

we have:

$$\max\{1, \widehat{U}^{\omega}\} \max\{1, \widehat{U}^{\kappa}\} \leq \left(1 + \widehat{U}\right) \left(1 + \widehat{U}\right)^{1-\lambda} . \tag{469}$$

Hence, (467) can be simplified in:

$$\begin{aligned}
\dot{\widehat{U}} \leq & -c\widehat{U} \\
& + 5k \left(1 + \widehat{U}\right) \left(1 + \widehat{U}^{1-\lambda}\right) \|e\| \\
& + k \max\{1, \widehat{U}^{\lambda}\} \left[\max\{1, \widehat{U}^{2\lambda(1-\lambda)}\} \left(\|e\|^{2(1-\lambda)+1} + \|e\|^3\right)\right] \stackrel{\text{def}}{=} \text{(a)} \\
& + k \max\{1, \widehat{U}^{\omega}\} \left[\max\{1, \widehat{U}^{\lambda\kappa}\} \left(\|e\|^{\kappa+1} + \|e\|^{1+\frac{\kappa}{1-\lambda}}\right)\right] \stackrel{\text{def}}{=} \text{(b)} .
\end{aligned} \tag{470}$$

Let us now show that the terms (a) and (b) defined in this inequality can be bounded from above by terms of the form:

$$\left(1 + \widehat{U}\right) \left[\left(1 + \widehat{U}^{1-\lambda}\right) \|e\|\right]^{\gamma} ,$$

with $0 < \gamma \leq 2$.

(a) If $2\lambda - 1 > 0$, and since

$$\lambda + 2\lambda(1-\lambda) = 1 + (2\lambda - 1)(1-\lambda) , \tag{471}$$

we get:

$$\text{(a)} \leq k \left(1 + \widehat{U}\right) \left[\left(1 + \widehat{U}^{1-\lambda}\right) \|e\|\right]^{2\lambda-1} \left(\|e\|^{4(1-\lambda)} + \|e\|^{4-2\lambda}\right) , \tag{472}$$

where, knowing from (463) that $\|e\|$ is bounded, the same holds for the last term between parentheses, since we have positive powers.
If $2\lambda - 1 \leq 0$, and since

$$\lambda + 2\lambda(1-\lambda) \leq 2(1-\lambda), \tag{473}$$

we get:

$$(a) \leq k\left[\left(1+\widehat{U}^{1-\lambda}\right)\|e\|\right]^2 \left(\|e\|^{1-2\lambda} + \|e\|\right) \tag{474}$$

$$\leq k\left(1+\widehat{U}\right)\left[\left(1+\widehat{U}^{1-\lambda}\right)\|e\|\right]^2 \left(\|e\|^{1-2\lambda} + \|e\|\right). \tag{475}$$

(b) If $\kappa < 1$, and since

$$\omega + \lambda\kappa \leq 1 + (1-\kappa)(1-\lambda), \tag{476}$$

we get:

$$(b) \leq k\left(1+\widehat{U}\right)\left[\left(1+\widehat{U}^{1-\lambda}\right)\|e\|\right]^{1-\kappa} \left(\|e\|^{2\kappa} + \|e\|^{\frac{\kappa}{1-\lambda}+\kappa}\right). \tag{477}$$

If $\kappa \geq 1$, and since

$$\omega + \lambda\kappa \leq 1 + 2(1-\lambda), \tag{478}$$

we get:

$$(b) \leq k\left(1+\widehat{U}\right)\left[\left(1+\widehat{U}^{1-\lambda}\right)\|e\|\right]^2 \left(\|e\|^{\kappa-1} + \|e\|^{\frac{\kappa}{1-\lambda}-1}\right). \tag{479}$$

With these inequalities, and the expression (442) for r, (470) yields an inequality we write formally as:

$$\begin{aligned}\dot{\widehat{U}} \leq& -c\widehat{U} + 5k\left(1+\widehat{U}\right)\sqrt{r}\|e\| \\ &+ \left(1+\widehat{U}\right)\left[\overline{k}_{\mathrm{a}}\left(\sqrt{r}\|e\|\right)^{\gamma_{\mathrm{a}}} + \overline{k}_{\mathrm{b}}\left(\sqrt{r}\|e\|\right)^{\gamma_{\mathrm{b}}}\right],\end{aligned} \tag{480}$$

where the last term represents the sum (a)+(b), with $\overline{k}_{\mathrm{a}}$ and $\overline{k}_{\mathrm{b}}$ two constants depending only on the initial conditions, and

$$0 < \gamma_{\mathrm{a}} \leq 2, \qquad 0 < \gamma_{\mathrm{b}} \leq 2. \tag{481}$$

This inequality can be rewritten as:

$$\begin{aligned}\dot{\widehat{U}} \leq& -\left(c - 5k\sqrt{r}\|e\| - \left[\overline{k}_{\mathrm{a}}\left(\sqrt{r}\|e\|\right)^{\gamma_{\mathrm{a}}} + \overline{k}_{\mathrm{b}}\left(\sqrt{r}\|e\|\right)^{\gamma_{\mathrm{b}}}\right]\right)\widehat{U} \\ &+ 5k\sqrt{r}\|e\| + \left[\overline{k}_{\mathrm{a}}\left(\sqrt{r}\|e\|\right)^{\gamma_{\mathrm{a}}} + \overline{k}_{\mathrm{b}}\left(\sqrt{r}\|e\|\right)^{\gamma_{\mathrm{b}}}\right].\end{aligned} \tag{482}$$

We may now apply Lemma (583) in Appendix B with

$$X = \widehat{U} \tag{483}$$

and

$$
\begin{aligned}
\vartheta_1 &= 5\,k\,\sqrt{r}\|e\|\,, & \sigma_1 &= 2, & S_{11} &= 25\,k^2\,\frac{\beta^2}{2} \\
\vartheta_2 &= \overline{k}_{\mathrm{a}}\,(\sqrt{r}\|e\|)^{\gamma_{\mathrm{a}}}\,, & \sigma_2 &= \frac{2}{\gamma_{\mathrm{a}}}\,, & S_{12} &= \overline{k}_{\mathrm{a}}{}^2\,\frac{\beta^2}{2} \\
\vartheta_3 &= \overline{k}_{\mathrm{b}}\,(\sqrt{r}\|e\|)^{\gamma_{\mathrm{b}}}\,, & \sigma_3 &= \frac{2}{\gamma_{\mathrm{b}}}\,, & S_{13} &= \overline{k}_{\mathrm{b}}{}^2\,\frac{\beta^2}{2} \\
\varpi_1 &= 5\,k\,\sqrt{r}\|e\|\,, & \zeta_1 &= 2, & S_{21} &= 25\,k^2\,\frac{\beta^2}{2} \\
\varpi_2 &= \overline{k}_{\mathrm{a}}\,(\sqrt{r}\|e\|)^{\gamma_{\mathrm{a}}}\,, & \zeta_2 &= \frac{2}{\gamma_{\mathrm{a}}}\,, & S_{22} &= \overline{k}_{\mathrm{a}}{}^2\,\frac{\beta^2}{2} \\
\varpi_3 &= \overline{k}_{\mathrm{b}}\,(\sqrt{r}\|e\|)^{\gamma_{\mathrm{a}}}\,, & \zeta_3 &= \frac{2}{\gamma_{\mathrm{b}}}\,, & S_{23} &= \overline{k}_{\mathrm{b}}{}^2\,\frac{\beta^2}{2}\,.
\end{aligned}
\tag{484}
$$

From here, we conclude the proof exactly as in the previous case. □

This proposition shows that, in fact, the observation h cannot be chosen independently of V. The refined assumptions RCO and RURS specify the link between these two functions with U playing the role of V. However, not taking $h = V$ creates some difficulties. Even when the Refined Matching Condition (RMC) is met, some extra conditions are required – the increment condition in point 1 and the growth condition in point 2 of assumption GC1. Nevertheless, as with the other estimation designs, we may have a solution to the Adaptive Stabilization problem when assumption RMC does not hold.

Corollary [Campion and Bastin[2]] **(485)**
Let, in equation (S_p), the functions a and A be known *and affine in u and let Π be an open subset of $\mathbb{R}^l$ which satisfies assumption ICS. Assume there exist three* known *functions:*

$$
\begin{array}{lll}
h & : \mathbb{R}^n \times \Pi \;\longrightarrow\; \mathbb{R}^n & \textit{of class } C^2, \textit{ a diffeomorphism for each } p, \\
u_{\mathrm{n}} & : \mathbb{R}^n \times \Pi \;\longrightarrow\; \mathbb{R}^m & \textit{of class } C^1\,, \textit{ and} \\
v & : \mathbb{R}^n \times \Pi \times \Pi \times \mathbb{R}^l \;\longrightarrow\; \mathbb{R}^m & \textit{of class } C^1\,,
\end{array}
$$

such that:

1. *h satisfies assumption RCO,*
2. *for all (x,p,q,∂) in $\mathbb{R}^n \times \Pi \times \Pi \times \mathbb{R}^l$, we have:*

$$
\frac{\partial h}{\partial p}(x,p)\,\partial + \frac{\partial h}{\partial x}(x,p)\,v \odot [b(x) + B(x)\,q] = 0\,, \tag{486}
$$

3. *and by letting*

$$
\varphi = h(x,p)\,, \tag{487}
$$

the time derivative of φ along the solutions of (S_p) with $u = u_{\mathrm{n}}$ satisfies:

$$
\dot{\varphi} = C\,\varphi\,, \tag{488}
$$

where C is an $n \times n$ matrix satisfying:

$$PC + C^{\mathrm{T}} P = -I\,, \tag{489}$$

with P a symmetric positive definite matrix.

Under these conditions, the Adaptive Stabilization problem is solved by the dynamic controller consisting of (234) and:

$$u = u_{\mathrm{n}}(x, \widehat{p}) + v\left(x, \widehat{p}, \widehat{q}, \dot{\widehat{p}}\right). \tag{490}$$

Proof. Let us first notice that, the functions a, A and h being known, assumption Λ-LP holds with:

$$\Lambda(x,t) = \frac{\partial h}{\partial x}(x, \widehat{p}(t))\,. \tag{491}$$

Then, we define two functions U and f as follows:

$$U(h) = h^{\mathrm{T}} P\, h \tag{492}$$

and

$$f(h\,,\, p) = C\, h\,. \tag{493}$$

From the assumptions, conditions Λ-LP, RURS and RMC are satisfied.

Now, define the function F by:

$$\begin{array}{rccc} F : & \mathbb{R}^n \times \Pi & \longrightarrow & \mathbb{R}^n \times \Pi\,. \\ & (x,p) & & (h(x,p)\,,\, p) \end{array} \tag{494}$$

This function is a diffeomorphism. Hence, for all compact subsets $\mathcal{K}$ of Π and all positive real numbers α, the set:

$$F^{-1}\left\{(\varphi, p) \;|\, p \in \mathcal{K} \quad \text{and} \quad \|\varphi\| \leq \alpha\right\}$$

is a compact subset of $\Omega \times \Pi$, and therefore its projection:

$$\Gamma_{\alpha,\mathcal{K}} = \{x \mid \exists p \in \mathcal{K} : \|h(x,p)\| \leq \alpha\} \tag{495}$$

is a compact subset of $\mathbb{R}^n$. It follows that assumption RBO holds.

Finally, let us check that (438) in assumption GC1 is satisfied. We have:

$$\left\|\frac{\partial U}{\partial h}(h,p)\right\| = \|h^{\mathrm{T}} P\| \tag{496}$$

$$\leq \sqrt{\lambda_{\max}\{P\}}\, U(h)^{\frac{1}{2}}\,, \tag{497}$$

and also:

$$\left|\frac{\partial U}{\partial h}(h,p)\,(f(h-e\,,\,p) - f(h\,,\,p))\right| = |h^{\mathrm{T}} P\, C e| \tag{498}$$

$$\leq U(h)^{\frac{1}{2}}\, \sqrt{e^{\mathrm{T}} C^{\mathrm{T}} P\, C\, e} \tag{499}$$

$$\leq \sqrt{\lambda_{\max}\{C^{\mathrm{T}} P\, C\}}\, U(h)^{\frac{1}{2}}\, \|e\|\,. \tag{500}$$

Hence, (438) holds with any λ satisfying:

$$\frac{1}{2} \leq \lambda \leq 1. \tag{501}$$

So we choose $\lambda = 1$ in order to simplify the expression of r in (442). □

Note that in Corollary (485) the Matching Condition (486) can be replaced by Points 3 and 5 of assumption GC1, namely:

There exist two positive continuous functions d_5 and d_7 defined on Π such that, for all (x,p) in $\mathbb{R}^n \times \Pi$:

$$\begin{aligned} &\left\|\frac{\partial h}{\partial p}(x,p)\right\| \cdot \left\|\left[\frac{\partial h}{\partial x}(x,p)\, A(x,u_{\mathrm{n}}(x,p))\right]^{\mathrm{T}}\right\| \leq d_5(p)\, \max\{1\,,\, \|h(x,p)\|^2\} \\ &\left\|\frac{\partial h}{\partial x}(x,p)\, A(x,u_{\mathrm{n}}(x,p))\right\| \leq d_7(p)\, \max\{1\,,\, \|h(x,p)\|^3\}\,. \end{aligned} \tag{502}$$

5.2 Regressor Filtering

When the regressor filtering technique (258)–(261) is applied, by letting:

$$\widehat{h} = h(x,\widehat{p}) + e = Z_{\mathrm{f}}\widehat{p} + \widehat{z}_{\mathrm{f}}\,, \tag{503}$$

and using (259) and (261), we get:

$$\dot{\widehat{h}} = f(\widehat{h}\,,\,\widehat{p}) + \left[-\rho\, e + \left(f(\widehat{h} - e\,,\,\widehat{p}) - f(\widehat{h},\widehat{p})\right) + \left(\frac{\partial h}{\partial p} + Z_{\mathrm{f}}\right)\dot{\widehat{p}}\right], \tag{504}$$

if the control u is:

$$u = u_{\mathrm{n}}(x,\widehat{p})\,. \tag{505}$$

And we get:

$$\dot{\widehat{h}} = f(\widehat{h}\,,\,\widehat{p}) + \left[-\rho\, e + \left(f(\widehat{h} - e\,,\,\widehat{p}) - f(\widehat{h},\widehat{p})\right) + Z_{\mathrm{f}}\dot{\widehat{p}}\right], \tag{506}$$

if assumption RMC holds, A does not depend on u and we use:

$$u = u_{\mathrm{n}}(x,\widehat{p}) + v(x,\widehat{p},\dot{\widehat{p}})\,, \tag{507}$$

where, with notation (13):

$$\frac{\partial h}{\partial p}(x,\widehat{p})\,\dot{\widehat{p}} + \frac{\partial h}{\partial x}(x,\widehat{p}) v(x,\widehat{p},\dot{\widehat{p}}) \odot b(x) = 0\,. \tag{508}$$

Compared with the equation error filtering case, we have the extra term $Z_{\mathrm{f}}\dot{\widehat{p}}$ in both (504) and (506). However, we know that, by choosing ρ appropriately, Z_{f} is bounded. And, by using a vanishing observation gain, $\dot{\widehat{p}}$ is absolutely integrable. This latter property being different from what we have in the error filtering case, different growth conditions are needed:

Assumption GC2 (Growth Conditions 2) (509)
There exist three positive continuous functions d_i, $i = 8, 10$, with d_8 defined on $\Pi \times \mathbb{R}^k$ and d_9 and d_{10} defined on Π, two positive real numbers λ and κ, with $\lambda < 1$ and κ known, *such that, for all (φ, x, p) in $\mathbb{R}^k \times \mathbb{R}^n \times \Pi$:*

$$
\begin{aligned}
&1.\ \left|\frac{\partial U}{\partial h}(\varphi,p)\,(f(\varphi - e\,,\,p) - f(\varphi\,,\,p))\right| \le d_8(p,e)\,\max\left\{1, U(\varphi,p)^{\kappa+\lambda}\right\}\,\|e\| \\
&2.\ \left\|\frac{\partial U}{\partial h}(\varphi,p)\right\| \le d_9(p)\,\max\left\{1, U(\varphi,p)^{\lambda}\right\} \\
&3.\ \left\|\frac{\partial h}{\partial x}(x,p)\,A(x,u_{\mathrm{n}}(x,p))\right\| \le d_{10}(p)\,\max\left\{1\,,\,U(h(x,p),p)^{\kappa}\right\}.
\end{aligned}
\tag{510}
$$

Moreover, if assumption RMC does not hold, there exist two positive continuous functions d_{11} and d_{12}, defined on Π such that, for all (φ, x, p) in $\mathbb{R}^k \times \mathbb{R}^n \times \Pi$:

$$
\begin{aligned}
&4.\ \left\|\frac{\partial h}{\partial p}(x,p)\right\| \le d_{11}(p)\,\max\left\{1\,,\,U(h(x,p),p)^{(1-\lambda)}\right\} \\
&5.\ \left\|\frac{\partial U}{\partial p}(\varphi,p)\right\| \le d_{12}(p)\,\max\left\{1\,,\,U(\varphi,p)\right\}.
\end{aligned}
\tag{511}
$$

Note that there is no constraint on κ besides its existence. We have:

Proposition (512)
Let assumptions Λ-LP, RBO, RURS, ICS and GC2 hold in the global case with:

$$
\Lambda(x,t) = \frac{\partial h}{\partial x}(x,\widehat{p}(t)).
\tag{513}
$$

Choose:

$$
r(x,\widehat{p}) = \left(1 + U(\widehat{h},\widehat{p})^{\kappa}\right)^2
\tag{514}
$$

$$
\rho(x,\widehat{p},u) = 1 + \left\|\frac{\partial h}{\partial x}(x,p)\,A(x,u)\right\|,
\tag{515}
$$

and, in (262):

$$
\dot{M} = G(M\,,\,Z_{\mathrm{f}}\,,\,r), \qquad \varepsilon_3\,M(0) > I\,,
\tag{516}
$$

where G is a negative symmetric matrix, depending locally-Lipschitz-continuously on its arguments and satisfying (see (262) and (263)):

$$
-\,\varepsilon_4\,M\,Z_{\mathrm{f}}^{\mathrm{T}} Z_{\mathrm{f}}\,M\,r \ge G \ge -(2-\varepsilon_1)\,M\,Z_{\mathrm{f}}^{\mathrm{T}} Z_{\mathrm{f}}\,M\,r\,.
\tag{517}
$$

Assume that A does not depend on u when u is chosen to depend on $\dot{\widehat{p}}$. Under these conditions, all the solutions $(x(t), \widehat{p}(t), \widehat{z}_{\mathrm{f}}(t), Z_{\mathrm{f}}(t)), M(t)$ of (S_{p^*}), (259), (261), (262), (516) are well-defined on $[0, +\infty)$, unique, bounded and:

$$
\lim_{t\to\infty} h(x(t),\widehat{p}(t)) = 0\,.
\tag{518}
$$

It follows that the Adaptive Stabilization problem is solved if assumption RCO holds also.

Proposition (512) generalizes a result established by Pomet and Praly [22] who choose the state vector x as the observation function h.

Proof.
Case: Assumption MC holds: In this case, we assume also that $A(x,u)$ does not depend on u, i.e., with notation (13), we have:

$$A(x,u) = A_0(x). \tag{519}$$

The system we consider is:

$$\begin{aligned}
\dot{x} &= a(x\,,\,u) + A_0(x)\,p^* \\
\dot{\widehat{z}}_{\mathrm{f}} &= \rho(x,\widehat{p},u)\,z_{\mathrm{f}} + \frac{\partial h}{\partial x}(x,\widehat{p})\,a(x,u) + \frac{\partial h}{\partial p}(x,\widehat{p})\,\dot{\widehat{p}}\,, \quad \widehat{z}_{\mathrm{f}}(0) = h(x(0),\widehat{p}(0)) \\
z_{\mathrm{f}} &= h(x,\widehat{p}) - \widehat{z}_{\mathrm{f}} \\
\dot{Z}_{\mathrm{f}} &= -\rho(x,\widehat{p})\,Z_{\mathrm{f}} + \frac{\partial h}{\partial x}(x,\widehat{p})\,A_0(x)\,, \quad Z_{\mathrm{f}}(0) = 0 \\
e &= Z_{\mathrm{f}}\,\widehat{p} - z_{\mathrm{f}} \\
\dot{\widehat{p}} &= \mathrm{Proj}\left(M\,,\,\widehat{p}\,,\,-M\,Z_{\mathrm{f}}^{\mathrm{T}} r(x,\widehat{p})\,e\right), \quad \widehat{p}(0) \in \Pi_1 \\
\dot{M} &= G(M\,,\,Z_{\mathrm{f}}\,,\,r(x,\widehat{p}))\,, \quad M(0) > 0\,,
\end{aligned} \tag{520}$$

with r given by (514), ρ given by (515) and:

$$u = u_{\mathrm{n}}(x,\widehat{p}) + v(x,\widehat{p},\dot{\widehat{p}})\,, \tag{521}$$

where v satisfies (515). Since r does not depend on u, $\dot{\widehat{p}}$ is explicitly defined. Then, from our smoothness assumptions on the various functions and with Lemma (103), this system has a locally Lipschitz-continuous right-hand side in the open set defined by:

$$(x,\widehat{z}_{\mathrm{f}},Z_{\mathrm{f}},\widehat{p},M) \in \mathbb{R}^n \times \mathbb{R}^k \times \mathcal{M}_{kl}(\mathbb{R}) \times \Pi \times \mathcal{M}\,. \tag{522}$$

It follows that, for any initial condition in this open set, there exists a unique solution $(x(t),\widehat{z}_{\mathrm{f}}(t),Z_{\mathrm{f}}(t),\widehat{p}(t),M(t))$, defined on a right maximal interval $[0,T)$, with T possibly infinite, and satisfying (522). Then, as in the proof of Proposition

(375), we have, for all t in $[0,T)$:

$$
\begin{aligned}
&\widehat{p}(t) \in \Pi_1 \\
&\varepsilon_2 \left\|\widehat{p}(t) - p^*\right\|^2 + \varepsilon_1 \int_0^t r \left\|e\right\|^2 \leq \varepsilon_3 \left\|\widehat{p}(0) - p^*\right\|^2 \overset{\text{def}}{=} \beta^2 \\
&\|Z_{\mathrm{f}}\| \leq 1 \qquad \text{and} \qquad \|e\| \leq \sqrt{\frac{\varepsilon_3}{\varepsilon_2}} \left\|\widehat{p}(0) - p^*\right\| \\
&\|z_{\mathrm{f}}\| \leq \|p^*\| \\
&\frac{I}{\varepsilon_3 + (2-\varepsilon_1)\int_0^T r} \leq M(t) \leq M(0)\,.
\end{aligned}
\tag{523}
$$

Now, as in the proof of Proposition (440), we look at the time derivative along the solutions of (520) (see also (506)) of $\widehat{U}$ defined in (448). From (510) in assumption GC2, (422) in assumption RURS, (514), (515), and (523), we get successively:

$$
\begin{aligned}
\dot{\widehat{U}} \leq & -c\widehat{U} + \frac{\partial U}{\partial h}(\widehat{h}) \left[-\rho(x,\widehat{p},u)\, e + \left(f(\widehat{h}-e\,,\widehat{p}) - f(\widehat{h},p)\right) + Z_{\mathrm{f}}\,\dot{\widehat{p}}\right] \\
\leq & -c\widehat{U} \\
& + d_9(\widehat{p}) d_{10}(\widehat{p}) \max\left\{1, \widehat{U}^{\lambda}\right\} (1 + \max\{1\,, U(h(x,\widehat{p}))^{\kappa}\}) \|e\| \\
& + d_8(\widehat{p}, e) \max\left\{1, \widehat{U}^{\kappa+\lambda}\right\} \|e\| \\
& + d_9(\widehat{p}) \max\left\{1\,, \widehat{U}^{\lambda}\right\} \left\|\dot{\widehat{p}}\right\| .
\end{aligned}
\tag{524}
$$

And, with Appendix C, we have the following inequality:

$$
\begin{aligned}
\max\{1\,, U(h(x,\widehat{p}),\widehat{p})^{\kappa}\} \leq & \ \delta \max\left\{1\,, \widehat{U}^{\kappa}\right\} \\
& + \delta \left[\max\left\{1\,, \widehat{U}^{\lambda\kappa}\right\} \left(\|e\|^{\kappa} + \|e\|^{\frac{\kappa}{1-\lambda}}\right)\right]
\end{aligned}
\tag{525}
$$

for some constant $\delta \geq 1$ depending only on $d_9(\widehat{p})$. Since the functions d_i, $i = 8, 10$ are continuous and e and $\widehat{p}$ are bounded from (523), there exists a constant k depending only on the initial conditions such that:

$$
\begin{aligned}
\dot{\widehat{U}} \leq & -c\widehat{U} \\
& + k\left(1 + \widehat{U}^{\lambda}\right) \left[\|e\| + 2\left(1 + \widehat{U}^{\kappa}\right) \|e\|\right] \\
& + k\left(1 + \widehat{U}^{\lambda}\right) \left[\left(1 + \widehat{U}^{\kappa}\right) \|e\|\right]^{\lambda} \left(\|e\|^{1+\kappa-\lambda} + \|e\|^{\frac{\kappa}{1-\lambda}+1-\lambda}\right) \\
& + k\left(1 + \widehat{U}^{\lambda}\right) \left\|\dot{\widehat{p}}\right\| .
\end{aligned}
\tag{526}
$$

Finally, since r satisfies (514), e is bounded, and from GC2 we have $\lambda < 1$, we get more simply:

$$\dot{\widehat{U}} \leq -\left[c - \left(k\|e\| + 2k\sqrt{r}\|e\| + \overline{k}\left(\sqrt{r}\|e\|\right)^{\lambda} + k\left\|\dot{\widehat{p}}\right\|\right)\right]\widehat{U}$$
$$+ 2k\sqrt{r}\|e\| + \overline{k}\left(\sqrt{r}\|e\|\right)^{\lambda} + k\left\|\dot{\widehat{p}}\right\| . \tag{527}$$

We may now apply Lemma (583) in Appendix B with:

$$X = \widehat{U} \tag{528}$$

and:

$$\begin{aligned}
\vartheta_1 &= k\|e\| + 2k\sqrt{r}\|e\|, & \sigma_1 &= 2, & S_{11} &= 10\,k^2\frac{\beta^2}{\varepsilon_1}\\
\vartheta_2 &= \overline{k}\left(\sqrt{r}\|e\|\right)^{\lambda}, & \sigma_2 &= \frac{2}{\lambda}, & S_{12} &= \overline{k}^2\,\frac{\beta^2}{\varepsilon_1}\\
\vartheta_3 &= k\left\|\dot{\widehat{p}}\right\|, & \sigma_3 &= 1, & S_{13} &= k^2\,k_1(p^*, \widehat{p}(0))\\
\varpi_1 &= 2k\sqrt{r}\|e\|, & \zeta_1 &= 2, & S_{21} &= 4\,k^2\,\frac{\beta^2}{\varepsilon_1}\\
\varpi_2 &= \overline{k}\left(\sqrt{r}\|e\|\right)^{\lambda}, & \zeta_2 &= \frac{2}{\lambda}, & S_{22} &= \overline{k}^2\,\frac{\beta^2}{\varepsilon_1}\\
\varpi_3 &= k\left\|\dot{\widehat{p}}\right\|, & \zeta_3 &= 1, & S_{23} &= k^2\,k_1(p^*, \widehat{p}(0)) .
\end{aligned} \tag{529}$$

With property 3 of U in assumption RURS, it follows that there exists a constant Υ depending only on the initial conditions such that, for all t in $[0, T)$, we have:

$$\left\|\widehat{h}(t)\right\| \leq \Upsilon . \tag{530}$$

Then, with (523), this implies the existence of a constant α, depending only on the initial conditions, such that, for all t in $[0, T)$:

$$\|h(x(t), \widehat{p}(t))\| \leq \left\|\widehat{h}(t)\right\| + \|e(t)\| \tag{531}$$
$$\leq \alpha . \tag{532}$$

We also know from (447) that $\widehat{p}(t) \in \mathcal{K}$, where $\mathcal{K}$ is the following compact subset of Π:

$$\mathcal{K} = \left\{p \;\middle|\; \|p - p^*\| \leq \frac{\beta}{\sqrt{\varepsilon_2}}\right\} \bigcap \Pi_1 . \tag{533}$$

With assumption RBO, this proves the existence of a compact subset Γ of $\mathbb{R}^n$ such that:

$$x(t) \in \Gamma \quad \forall t \in [0, T) . \tag{534}$$

Finally, with the continuity of the functions r and h and the fact that:

$$\widehat{z}_{\mathrm{f}} = h(x, \widehat{p}) - z_{\mathrm{f}} , \tag{535}$$

we have established that the solution remains in a compact subset of the open set defined in (522). It follows by contradiction that $T = +\infty$ and that the time functions $x(t)$, $\widehat{p}(t)$, $u(t)$ and $\dot{\widehat{p}}(t)$ are bounded on $[0, +\infty)$. From here, we conclude the proof exactly as in the proof of Proposition (440).

Case: Assumption RMC does not hold: The only difference with the previous case, is that we use

$$u = u_{\mathrm{n}}(x, \widehat{p}) \tag{536}$$

instead of (521), and the fact that U may depend on p. Hence, everything remains the same up to, but not including, equation (524). To get the equivalent of (524), we have to evaluate the time derivative of:

$$\widehat{U}(t) = U(h(x(t), \widehat{p}(t))) \tag{537}$$

along the solutions of (504). We get successively:

$$\begin{aligned}\dot{\widehat{U}} \leq & -c\widehat{U} + \frac{\partial U}{\partial p}(\widehat{h}, \widehat{p})\,\dot{\widehat{p}} \\ & + \frac{\partial U}{\partial h}(\widehat{h}, \widehat{p})\left[-\rho(x, \widehat{p}, u)\, e + \left(f(\widehat{h} - e\,, \widehat{p}) - f(\widehat{h}, p)\right) + \left(\frac{\partial h}{\partial p} + Z_{\mathrm{f}}\right)\dot{\widehat{p}}\right] \\ \leq & -c\widehat{U} \\ & + d_9(\widehat{p}) d_{10}(\widehat{p}) \max\left\{1, \widehat{U}^{\lambda}\right\}(1 + \max\{1\,, U(h(x, \widehat{p}))^{\kappa}\})\, \|e\| \\ & + d_8(\widehat{p}, e) \max\left\{1, \widehat{U}^{\kappa+\lambda}\right\} \|e\| \\ & + d_9(\widehat{p}) \max\left\{1\,, \widehat{U}^{\lambda}\right\} \left\|\dot{\widehat{p}}\right\| \\ & + d_9(\widehat{p})\, d_{11}(\widehat{p}) \max\left\{1\,, \widehat{U}^{\lambda}\right\} \max\{1\,, U(h(x, \widehat{p}))^{1-\lambda}\} \left\|\dot{\widehat{p}}\right\| \\ & + d_{12}(\widehat{p}) \max\{1\,, \widehat{U}\} \left\|\dot{\widehat{p}}\right\|.\end{aligned} \tag{538}$$

With Appendix C, we get similarly to (525):

$$\begin{aligned}\max\{1\,, U(h(x, \widehat{p}), \widehat{p})^{1-\lambda}\} \leq & \ \delta \max\{1\,, \widehat{U}^{1-\lambda}\} \\ & + \delta\left[\max\{1\,, \widehat{U}^{\lambda(1-\lambda)}\}\left(\|e\|^{1-\lambda} + \|e\|\right)\right].\end{aligned} \tag{539}$$

And, since $0 \leq \lambda < 1$ implies:

$$\lambda(2 - \lambda) < 1\,, \tag{540}$$

we have also:

$$\max\left\{1, \widehat{U}^{\lambda}\right\} \max\left\{1, \widehat{U}^{\lambda(1-\lambda)}\right\}\left(\|e\|^{1-\lambda} + \|e\|\right) \leq \left(1 + \widehat{U}\right)\left(\|e\|^{1-\lambda} + \|e\|\right). \tag{541}$$

As in the previous case, this implies the existence of a constant k depending only on the initial conditions such that:

$$\dot{\widehat{U}} \leq -\left[c - k\left(\|e\| + \sqrt{r}\|e\| + (\sqrt{r}\|e\|)^{\lambda} + \left\|\dot{\widehat{p}}\right\|\right)\right]\widehat{U} + k\left(\|e\| + \sqrt{r}\|e\| + (\sqrt{r}\|e\|)^{\lambda} + \left\|\dot{\widehat{p}}\right\|\right). \tag{542}$$

We can now apply Lemma (583) in Appendix B with:

$$X = \widehat{U} \tag{543}$$

and:

$$\begin{aligned}
\vartheta_1 &= k\left(\|e\| + \sqrt{r}\|e\|\right), & \sigma_1 &= 2, & S_{11} &= 2k^2\frac{\beta^2}{\varepsilon_1} \\
\vartheta_2 &= k\left(\sqrt{r}\|e\|\right)^{\lambda}, & \sigma_2 &= \frac{2}{\lambda}, & S_{12} &= k^2\frac{\beta^2}{\varepsilon_1} \\
\vartheta_3 &= k\left\|\dot{\widehat{p}}\right\|, & \sigma_3 &= 1, & S_{13} &= k^2 k_1(p^*, \widehat{p}(0)) \\
\varpi_1 &= k\left(\|e\| + \sqrt{r}\|e\|\right), & \zeta_1 &= 2, & S_{21} &= 2k^2\frac{\beta^2}{\varepsilon_1} \\
\varpi_2 &= k\left(\sqrt{r}\|e\|\right)^{\lambda}, & \zeta_2 &= \frac{2}{\lambda}, & S_{22} &= k^2\frac{\beta^2}{\varepsilon_1} \\
\varpi_3 &= k\left\|\dot{\widehat{p}}\right\|, & \zeta_3 &= 1, & S_{23} &= k^2 k_1(p^*, \widehat{p}(0)).
\end{aligned} \tag{544}$$

From here, we conclude as in the previous case. □

Again, in practice, to apply Proposition (512), we must be allowed to choose a particular value for the initial condition $\widehat{z}_f(0)$, and a vanishing observer gain K ((263) is assumed). However, the main feature of this result together with Proposition (440) is that, when the observation function h is not related to V and assumption RMC is not satisfied, it provides solutions to the Adaptive Stabilization problem under different growth conditions.

Corollary **(545)**

Let, in equation (S_p), the functions a and A be known *and affine in u and let Π be an open subset of $\mathbb{R}^l$ which satisfies assumption ICS. Assume there exist two* known *functions:*

$h : \mathbb{R}^n \times \Pi \to \mathbb{R}^n$ *of class C^2 which is a diffeomorphism for each p, and*
$u_n : \mathbb{R}^n \times \Pi \to \mathbb{R}^m$ *of class C^1,*

a positive real number κ and a positive continuous function d_{10}, defined on Π, such that:

1. *h satisfies assumption RCO,*
2. *for all (x,p) in $\mathbb{R}^n \times \Pi$, we have:*

$$\left\|\frac{\partial h}{\partial x}(x,p)\, A(x, u_n(x,p))\right\| \leq d_{10}(p) \max\{1, \|h(x,p)\|^{2\kappa}\}, \tag{546}$$

3. by letting:

$$\varphi = h(x,p)\,, \tag{547}$$

the time derivative of φ along the solutions of (S_p) with $u = u_n$ satisfies:

$$\dot{\varphi} = C\varphi\,, \tag{548}$$

where C is an $n \times n$ matrix satisfying:

$$PC + C^{\mathrm{T}}P = -I\,, \tag{549}$$

with P a symmetric positive definite matrix.

Under these conditions, if

either *the function A does not depend on u and there exists a* known *function:*

$$v : \mathbb{R}^n \times \Pi \times \mathbb{R}^l \to \mathbb{R}^m \quad \textit{of class } C^1\,, \tag{550}$$

such that, for all (x,p,∂) in $\mathbb{R}^n \times \Pi \times \mathbb{R}^l$, we have:

$$\frac{\partial h}{\partial p}(x,p)\,\partial + \frac{\partial h}{\partial x}(x,p)\,v \odot b(x) = 0\,, \tag{551}$$

or *there exists a positive continuous function d_{11}, defined on Π such that, for all (x,p) in $\mathbb{R}^n \times \Pi$:*

$$\left\| \frac{\partial h}{\partial p}(x,p) \right\| \le d_{11}(p)\max\{1\,, \|h(x,p)\|^2\}\,, \tag{552}$$

then the Adaptive Stabilization problem is solved by the dynamic controller consisting of (259), (261), (262, (516), and:

$$u = u_n(x,\widehat{p}) + v\left(x,\widehat{p},\widehat{q},\dot{\widehat{p}}\right) \tag{553}$$

if v exists, or:

$$u = u_n(x,\widehat{p}) \tag{554}$$

if not.

Proof. This proof follows exactly the same lines as the proof of Corollary (485). In particular, the above growth conditions are nothing but GC2 with:

$$\lambda = \frac{1}{2}\,, \quad U(h) = h^{\mathrm{T}}Ph \quad \text{and} \quad f(h\,,p) = Ch\,. \tag{555}$$

□

This Corollary (545) should be compared with the result of Nam and Arapostathis in [16]. In the same context of adaptive feedback linearization, they propose the same dynamic controller except that:

1. the Matching Condition (551) is not assumed,
2. they do not restrict the choice of $\widehat{z}_f(0)$,

3. the observer gain is not allowed to go to zero, namely, (263) does not hold,
4. ρ is kept constant,
5. finally r is given by:

$$r = \left(1 + \left\|\frac{\partial h}{\partial x}(x,p)\, A(x, u_{\mathrm{n}}(x,p))\right\|^2\right)^{-1}, \tag{556}$$

i.e., they impose $\kappa = 1$.

As a consequence, Nam and Arapostathis get a solution to the Adaptive Stabilization problem under more restrictive Growth Conditions. Namely, instead of (546), they assume:

$$\left\|\frac{\partial h}{\partial x}(x,p)\, A(x, u_{\mathrm{n}}(x,p))\right\| \leq d_{10}(p) \max\{1\,, \|h(x,p)\|\}\,, \tag{557}$$

and, instead of (552), they have:

$$\left\|\frac{\partial h}{\partial p}(x,p)\right\| \leq d_{11}(p) \max\{1\,, \|h(x,p)\|\}\,. \tag{558}$$

6 Conclusion

In this paper, we have given a unified and generalizing overview of most (not all!) of the presently proposed approaches to stabilize an equilibrium point of a nonlinear system described by a differential equation containing unknown parameters. Table 3 summarizes our results.

A key assumption is the fact that the right-hand side of the differential equation describing the system depends linearly on these unknown parameters or at least on those actually needed for the control (see Example (49)). To meet this assumption, called Λ-Linear Parameterization Λ-LP, we have mentioned the fact that it may be useful not to work with the a priori given coordinates and parameterization: a parameter-dependent change of coordinates and a reparameterization, i.e., a transformation of the parameters, are allowed (see [14] and [29]). Also, Middleton and Goodwin [15], Pomet and Praly [22] and Slotine and Li [30], for example, have shown that the proposed results extend in some cases to a more general case called Implicit Linear Parameterization in [19]. Finally, Mareels, Penfold and Evans [13] have shown that it is also possible to follow a non parametric approach to solve the problem of stabilizing an equilibrium point of a system whose dynamics are not completely known.

Another important assumption is assumption PRS and its more restrictive versions URS and RURS. It guarantees not only that the system is stabilizable, but also that a parameter-dependent control law for this stabilization is known. It follows that the only problem addressed here is: how can we use this control law when the parameters are unknown, i.e., how can we make this control law adaptive?

Table 3 shows the ten routes we have studied for designing and analyzing adaptive controllers of nonlinear systems. It shows the interplay between structural assumptions, control design techniques and estimation algorithms. In particular, it emphasizes the fact that any stabilizing control law cannot necessarily be made adaptive. It has to give the closed-loop system properties which differ depending on which control design and estimation technique is used. However, a general very desirable property is the fact that the so-called Matching Condition (MC) can be satisfied.

Implicit in all what has been presented was the assumption that the state is completely measured. It follows that the problem we are addressing is a very particular case of the Error Feedback Regulator problem stated by Isidori [6, Sect. 7.2]. But thanks to this particularity, we have solved the problem under less restrictive assumptions. It is also worth mentioning that the results established by Kanellakopoulos, Kokotovic and Middleton [10] lead us to expect that relaxation of the assumptions is also possible in some cases when the state is not completely measured.

Appendices

A: Proof of Lemma (103)

In this proof, we denote by S the following open subset of $\Pi \times \mathbb{R}^l$:

$$S = \left\{ (p, y) \,\middle|\, \mathcal{P}(p) > 0 \text{ and } \frac{\partial \mathcal{P}}{\partial p}(p)\, y > 0 \right\}. \tag{559}$$

$\mathrm{Proj}(M, p, y)$ differs from y if and only if (p, y) belongs to S.

Point 1: We make the following preliminary remark:
Since $0 \leq \mathcal{P}(p)$ and $M \in \mathcal{M}$ imply:

$$\frac{\partial \mathcal{P}}{\partial p}(p)\, M\, \frac{\partial \mathcal{P}}{\partial p}(p)^{\mathrm{T}} > 0, \tag{560}$$

and $\mathcal{P}$ is a twice continuously differentiable function,

1. the function $\mathrm{Proj}(M, p, y)$ is continuously differentiable in the set $\mathcal{M} \times S$,
2. $\mathrm{Proj}(M, p, y)$ tends to y as $\mathcal{P}(p)$ or $\dfrac{\partial \mathcal{P}}{\partial p}(p)\, y$ tends to 0,
3. for any compact subset $\mathcal{C}$ of

$$\mathcal{M} \times \left\{ (p, y) \,\middle|\, \mathcal{P}(p) \geq 0 \text{ and } \frac{\partial \mathcal{P}}{\partial p}(p)\, y \geq 0 \right\},$$

there exists a constant $k_{\mathcal{C}}$ bounding the Jacobian matrix:

$$\|\nabla\, \mathrm{Proj}(M, p, y)\| \leq k_{\mathcal{C}} \qquad \forall (M, p, y) \in \mathcal{C}. \tag{561}$$

Table 3. Summary of results

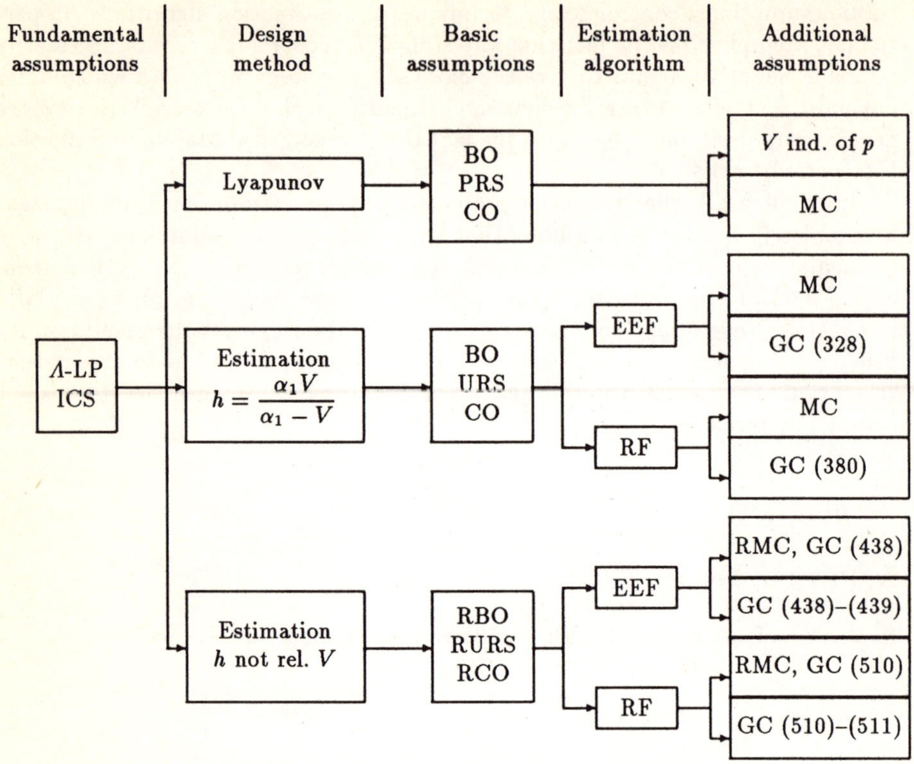

BO	: Boundedness Observability		CO	: Convergence Observability
EEF	: Equation Error Filtering		Λ-LP	: Λ-Linear Parameterization
GC	: Growth Condition		ICS	: Imbedded Convex Sets
MC	: Matching Condition		PRS	: Pointwise Reduced-order Stabilizability
R	: Refined		RF	: Regressor Filtering
			URS	: Uniform Reduced-order Stabilizability

Now, let (M_1, p_1, y_1) and (M_0, p_0, y_0) be two points such that, for any α in $[0, 1]$, the point $(M_\alpha, p_\alpha, y_\alpha)$ is in the set $\mathcal{M} \times \Pi \times \mathbb{R}^l$, with:

$$\begin{aligned} M_\alpha &= \alpha\, M_1 + (1-\alpha)\, M_0 \\ p_\alpha &= \alpha\, p_1 + (1-\alpha)\, p_0 \\ y_\alpha &= \alpha\, y_1 + (1-\alpha)\, y_0 \,. \end{aligned} \tag{562}$$

Four cases must be distinguished:

Case 1: (p_1, y_1) and (p_0, y_0) are not in S. Then, we have trivially:

$$\|\mathrm{Proj}(M_1, p_1, y_1) - \mathrm{Proj}(M_0, p_0, y_0)\| = \|y_1 - y_0\| \,. \tag{563}$$

Case 2: For all α in $[0,1]$, (p_α, y_α) lies in S. Then, from the above preliminary remark and the Mean Value Theorem, we get:

$$\|\mathrm{Proj}(M_1, p_1, y_1) - \mathrm{Proj}(M_0, p_0, y_0)\| \leq k\,[\|M_1 - M_0\| + \|p_1 - p_0\| + \|y_1 - y_0\|] \tag{564}$$

with the constant k given by (561).

Case 3: Say (p_0, y_0) belongs to S but (p_1, y_1) does not. Then, we define α^* by:

$$\alpha^* = \min_{\substack{0 \leq \alpha \leq 1 \\ (p_\alpha, y_\alpha) \notin S}} \alpha\,. \tag{565}$$

Since S is open, all the points of the segment $[(M_0, p_0, y_0), (M_{\alpha^*}, p_{\alpha^*}, y_{\alpha^*}))$ have their (p, y) component lying in S. But $(p_{\alpha^*}, y_{\alpha^*})$ is not in S. With (561) and (562), this implies:

$$\begin{aligned}&\|\mathrm{Proj}(M_{\alpha^*}, p_{\alpha^*}, y_{\alpha^*}) - \mathrm{Proj}(M_0, p_0, y_0)\| \\ &\qquad \leq k\,[\|M_1 - M_0\| + \|p_1 - p_0\| + \|y_1 - y_0\|]\,. \end{aligned}\tag{566}$$

We also have:

$$\|\mathrm{Proj}(M_1, p_1, y_1) - \mathrm{Proj}\,(M_{\alpha^*}, p_{\alpha^*}, y_{\alpha^*})\| = \|y_1 - y_{\alpha^*}\| \leq \|y_1 - y_0\|\,. \tag{567}$$

This yields:

$$\begin{aligned}&\|\mathrm{Proj}(M_1, p_1, y_1) - \mathrm{Proj}(M_0, p_0, y_0)\| \\ &\qquad \leq (1+k)\,[\|M_1 - M_0\| + \|p_1 - p_0\| + \|y_1 - y_0\|]\,. \end{aligned}\tag{568}$$

Case 4: Finally, when both (p_1, y_1) and (p_0, y_0) belong to S, but there are some α in $(0,1)$ for which (p_α, y_α) is not in S, we define α^* as above and let:

$$\beta^* = \max_{\substack{0 \leq \beta \leq 1 \\ (p_\beta, y_\beta) \notin S}} \beta\,. \tag{569}$$

Then all the points of the segments $[(M_0, p_0, y_0), (M_{\alpha^*}, p_{\alpha^*}, y_{\alpha^*}))$ and $((M_{\beta^*}, p_{\beta^*}, y_{\beta^*}), (M_1, p_1, y_1)]$ have their (p, y) component in S. But the points $(p_{\alpha^*}, y_{\alpha^*})$ and $(p_{\beta^*}, y_{\beta^*})$ are not in S. With (561), we get:

$$\begin{aligned}&\|\mathrm{Proj}(M_1, p_1, y_1) - \mathrm{Proj}(M_{\beta^*}, p_{\beta^*}, y_{\beta^*})\| \\ &+ \|\mathrm{Proj}(M_{\alpha^*}, p_{\alpha^*}, y_{\alpha^*}) - \mathrm{Proj}(M_0, p_0, y_0)\| \\ &\qquad \leq 2k\,[\|M_1 - M_2\| + \|p_1 - p_2\| + \|y_1 - y_2\|]\,. \end{aligned}\tag{570}$$

The conclusion follows, since we have trivially:

$$\mathrm{Proj}(M_{\beta^*}, p_{\beta^*}, y_{\beta^*}) - \mathrm{Proj}(M_{\alpha^*}, p_{\alpha^*}, y_{\alpha^*}) = y_{\beta^*} - y_{\alpha^*}\,. \tag{571}$$

Point 2: For (p,y) not in S the inequality of point 2 is trivial. If (p,y) is in S, a direct computation gives:

$$\mathrm{Proj}(M,p,y)^{\mathrm{T}}M^{-1}\mathrm{Proj}(M,p,y) = y^{\mathrm{T}}M^{-1}y - \frac{\mathcal{P}(p)\,(2-\mathcal{P}(p))\,\left(\frac{\partial\mathcal{P}}{\partial p}(p)y\right)^2}{\frac{\partial\mathcal{P}}{\partial p}(p)M\frac{\partial\mathcal{P}}{\partial p}(p)^{\mathrm{T}}}. \tag{572}$$

The conclusion follows since, by definition, for all (p,y) in S with p in Π_1, we have:

$$\mathcal{P}(p)\,(2-\mathcal{P}(p)) > 0. \tag{573}$$

Point 3: For any p satisfying $\mathcal{P}(p) \geq 0$, let q be the orthogonal projection of p^* on the hyperplane through p and orthogonal to $\frac{\partial\mathcal{P}}{\partial p}(p)$, i.e.,

$$q = p^* - \frac{\frac{\partial\mathcal{P}}{\partial p}(p)(p-p^*)}{\left\|\frac{\partial\mathcal{P}}{\partial p}(p)\right\|^2}\frac{\partial\mathcal{P}}{\partial p}(p)^{\mathrm{T}}. \tag{574}$$

Since $\mathcal{P}$ is a convex function we have:

$$\begin{aligned}\mathcal{P}(q) &\geq \mathcal{P}(p) + \frac{\partial\mathcal{P}}{\partial p}(p)\,(q-p) &(575)\\ &\geq \mathcal{P}(p) \geq 0. &(576)\end{aligned}$$

It follows that q is not an interior point of the set Π_0 and, therefore, $\|q-p^*\|$ is larger or equal to D^*, the distance from p^* to the boundary of the closed set $\Pi_0 = \{p \,|\, \mathcal{P}(p) = 0\}$, i.e., we have:

$$\frac{\left(\frac{\partial\mathcal{P}}{\partial p}(p)(p-p^*)\right)^2}{\left\|\frac{\partial\mathcal{P}}{\partial p}(p)\right\|^2} \geq D^{*2}. \tag{577}$$

The conclusion follows from the fact that, p^* being an interior point of the set Π_0 (see (95) and (93)), the Basic Separation Hahn-Banach Theorem [4, Theorem V.1.12] implies:

$$\frac{\partial\mathcal{P}}{\partial p}(p)(p-p^*) > 0 \quad \forall p : \mathcal{P}(p) \geq 0. \tag{578}$$

Point 4: Again, point 4 is clear for all (p,y) not in S. And, for (p,y) in S, we get, using point 3:

$$\begin{aligned}&(p-p^*)^{\mathrm{T}}M^{-1}\mathrm{Proj}(M,p,y)\\ &\quad= (p-p^*)^{\mathrm{T}}M^{-1}y - \frac{\mathcal{P}(p)\left(\frac{\partial\mathcal{P}}{\partial p}(p)y\right)\left(\frac{\partial\mathcal{P}}{\partial p}(p)(p-p^*)\right)}{\frac{\partial\mathcal{P}}{\partial p}(p)M\frac{\partial\mathcal{P}}{\partial p}(p)^{\mathrm{T}}}\\ &\quad\leq (p-p^*)^{\mathrm{T}}M^{-1}y. &(579)\end{aligned}$$

Point 5: Let us compute the time derivative of $\mathcal{P}(p(t))$ along a solution of (104). We get:

$$\begin{aligned}\overline{\mathcal{P}(\widehat{p}(t))}^{\,\cdot} &= \tfrac{\partial \mathcal{P}}{\partial p}(\widehat{p}(t))y(t) && \text{if } \mathcal{P}(\widehat{p}(t)) \leq 0 \text{ or } \tfrac{\partial \mathcal{P}}{\partial p}(\widehat{p}(t))\, y(t) \leq 0 \\ &= \tfrac{\partial \mathcal{P}}{\partial p}(\widehat{p}(t))y(t)\,(1-\mathcal{P}(\widehat{p}(t))) && \text{if } \mathcal{P}(\widehat{p}(t)) > 0 \text{ and } \tfrac{\partial \mathcal{P}}{\partial p}(\widehat{p}(t))\, y(t) > 0\,.\end{aligned} \tag{580}$$

Therefore, we have:

$$\overline{\mathcal{P}(\widehat{p}(t))}^{\,\cdot} \leq 0 \qquad \text{if} \qquad \mathcal{P}(\widehat{p}(t)) \geq 1\,. \tag{581}$$

Since the initial condition satisfies:

$$\mathcal{P}(\widehat{p}(0)) \leq 1\,, \tag{582}$$

a continuity argument proves that the same holds for all t where $\widehat{p}(t)$ is defined. □

B: A Useful Technical Lemma

Lemma: (see also [3, Theorem IV.1.9]) (583)

Let X be a C^1 time function defined on $[0\,,\,T)$ $(0 < T \leq +\infty)$, satisfying:

$$\dot{X} \leq -c\,X + \sum_i \vartheta_i(t)\,X(t) + \sum_j \varpi_j(t)\,, \tag{584}$$

where c is a strictly positive constant, $\sum_i$ and $\sum_j$ are finite sums and ϑ_i, and ϖ_j are positive time functions satisfying:

$$\int_0^T \vartheta_i{}^{\sigma_i} \leq S_{1i} \qquad and \qquad \int_0^T \varpi_j{}^{\zeta_j} \leq S_{2j}\,, \tag{585}$$

where $\sigma_i \geq 1$ and $\zeta_j \geq 1$. Under this assumption, $X(t)$ is bounded from above on $[0,T)$ and, precisely:

$$X(t) \leq K_1\,X(0) + K_2 \qquad \forall t \in [0,T)\,, \tag{586}$$

with K_1 and K_2 depending only on σ_i, ζ_j, S_{1i} and S_{2j}.
Moreover, if T is infinite then:

$$\limsup_{t\to\infty} X(t) \leq 0\,. \tag{587}$$

Proof. This is a straightforward consequence of a known result on differential inequalities. From (584), one derives (see [5, Theorem I.6.1]):

$$X(t) \leq X(0)\, e^{-ct + \int_0^t \sum_i \vartheta_i} + \int_0^t e^{\left[-c(t-\tau) + \int_\tau^t \sum_i \vartheta_i\right]} \sum_j \varpi_j(\tau)\, \mathrm{d}\tau\,. \tag{588}$$

Let $\theta_i > 1$ and $\eta_j > 1$ be defined by:

$$\frac{1}{\sigma_i} + \frac{1}{\theta_i} = 1 \qquad \text{and} \qquad \frac{1}{\zeta_j} + \frac{1}{\eta_j} = 1\,. \tag{589}$$

Inequalities (585) and the Hölder Inequality yield, for any positive t and τ, $t \geq \tau$:

$$\int_\tau^t \vartheta_i \leq (t-\tau)^{\frac{1}{\theta_i}} \left(\int_\tau^t {\vartheta_i}^{\sigma_i}\right)^{\frac{1}{\sigma_i}} \tag{590}$$

$$\leq (t-\tau)^{\frac{1}{\theta_i}}\, {S_{1i}}^{\frac{1}{\sigma_i}}\,. \tag{591}$$

Similarly, we get:

$$\begin{aligned} \int_0^t e^{-\frac{c}{2}(t-\tau)} \varpi_j(\tau) \mathrm{d}\tau &\leq \left(\frac{2}{c}\right)^{\frac{1}{\eta_j}} \left(\int_0^t e^{-\frac{c}{2}(t-\tau)} \varpi_j(\tau)^{\zeta_j} \mathrm{d}\tau\right)^{\frac{1}{\zeta_j}} \\ &\leq \left(\frac{2}{c}\right)^{\frac{1}{\eta_j}} \left(e^{-\frac{c}{2}\frac{t}{2}} \int_0^{\frac{t}{2}} \varpi_j(\tau)^{\zeta_j} \mathrm{d}\tau + \int_{\frac{t}{2}}^t \varpi_j(\tau)^{\zeta_j} \mathrm{d}\tau\right)^{\frac{1}{\zeta_j}} \\ &\leq \left(\frac{2}{c}\right)^{\frac{1}{\eta_j}} \left(e^{-\frac{c}{2}\frac{t}{2}} S_{2j} + \int_{\frac{t}{2}}^t \varpi_j(\tau)^{\zeta_j} \mathrm{d}\tau\right)^{\frac{1}{\zeta_j}}\,. \end{aligned} \tag{592}$$

Then, let us note that the function:

$$f(x) = -\frac{c}{2} x + \sum_i {S_{1i}}^{\frac{1}{\sigma_i}} x^{\frac{1}{\theta_i}} \tag{593}$$

is well-defined and continuous on $[0, +\infty)$, with:

$$f(0) = 0 \qquad \text{and} \qquad f(+\infty) = -\infty\,. \tag{594}$$

This implies the existence of a constant K_1 depending only on S_{1i} and σ_i such that, for all $0 \leq \tau \leq t$:

$$\exp\left(-\frac{c}{2}(t-\tau) + \sum_i (t-\tau)^{\frac{1}{\theta_i}} {S_{1i}}^{\frac{1}{\sigma_i}}\right) \leq K_1\,. \tag{595}$$

With this bound, (588) and (592) we have established:

$$X(t) \leq K_1 \left[e^{-\frac{c}{2}t} X(0) + \sum_j \left(\frac{2}{c}\right)^{\frac{1}{\eta_j}} \left(e^{-\frac{c}{2}\frac{t}{2}} S_{2j} + \int_{\frac{t}{2}}^t \varpi_j(\tau)^{\zeta_j} \mathrm{d}\tau\right)^{\frac{1}{\zeta_j}}\right]. \tag{596}$$

The proof is completed by noticing that (585) implies that, if $T = +\infty$, we have:

$$\lim_{t \to +\infty} \int_{\frac{t}{2}}^{t} \varpi_j(\tau)^{\zeta_j} d\tau = 0 . \tag{597}$$

□

C: An Inequality

Lemma: **(598)**

Let $U : \mathbb{R}^k \to \mathbb{R}_+$ be a C^1 function such that:

$$\left\| \frac{\partial U}{\partial h}(h) \right\| \leq d \max\left\{1, U(h)^\lambda\right\} \qquad \forall h \in \mathbb{R}^k , \tag{599}$$

with $0 \leq \lambda < 1$ and d a positive constant. For any positive real number γ, there exists a constant δ such that, for all (h,e) in $\mathbb{R}^k \times \mathbb{R}^k$, we have:

$$\begin{aligned} \max\{1\,,\, U(h-e)^\gamma\} \leq \delta \max\{1\,,\, U(h)^\gamma\} \\ +\delta\left[\max\{1\,,\, U(h)^{\lambda\gamma}\}\left(\|e\|^\gamma + \|e\|^{\frac{\gamma}{1-\lambda}}\right)\right] . \end{aligned} \tag{600}$$

Proof. Let us define a function W as follows:

$$W(h) = \max\left\{1\,,\, U(h)^{1-\lambda}\right\} . \tag{601}$$

This function is of class C^1 in the open set $\{h \,|\, U(h) > 1\}$ with:

$$\left\| \frac{\partial W}{\partial h}(h) \right\| \leq d(1-\lambda) . \tag{602}$$

Then, for all (h,e) in $\mathbb{R}^k \times \mathbb{R}^k$, we have:

$$W(h-e) - W(h) \leq d(1-\lambda)\|e\| . \tag{603}$$

This is proved by breaking the segment $[h-e\,,\, h]$ into pieces depending on whether U is larger than 1 or not and by noting that $W(h-e) - W(h)$ is equal to the sum of the variations of W on the two extreme pieces only (see the proof of Point 1 in Appendix A). This yields:

$$\max\{1, U(h-e)^\gamma\} - \max\{1, U(h)^\gamma\} \leq (W(h) + d(1-\lambda)\|e\|)^{\frac{\gamma}{1-\lambda}} - W(h)^{\frac{\gamma}{1-\lambda}} . \tag{604}$$

Let us now note that, for all $\tau \geq 0$, the function:

$$f(x) = \frac{(1+x)^\tau - x^\tau}{x^\tau + x^{\lambda\tau} + 1} \tag{605}$$

is positive, well-defined and continuous on $[0, +\infty)$, with:

$$f(0) = 1 \qquad \text{and} \qquad f(+\infty) = 0 . \tag{606}$$

This implies the existence of a constant K depending only on γ and λ such that:

$$\begin{aligned} &\max\{1\,,\, U(h-e)^\gamma\} - \max\{1\,,\, U(h)^\gamma\} \\ &\leq K\left[W(h)^{\frac{\gamma}{1-\lambda}} + W(h)^{\frac{\lambda\gamma}{1-\lambda}}(d(1-\lambda)\|e\|)^\gamma + (d(1-\lambda)\|e\|)^{\frac{\gamma}{1-\lambda}}\right] . \end{aligned} \tag{607}$$

The conclusion follows readily. □

References

1. B. d'Andréa-Novel, J.-B. Pomet, and L. Praly, "Adaptive stabilization for nonlinear systems in the plane," *Proc. 11th IFAC World Congress*, Talinn, 1990.
2. G. Campion and G. Bastin, "Indirect adaptive state feedback control of linearly parametrized nonlinear systems," *Int. J. Adapt. Control Sig. Proc.*, vol. 4, pp. 345–358, Sept. 1990.
3. C. A. Desoer and M. Vidyasagar, *Feedback Systems: Input-Output Properties*, Academic Press, 1975.
4. N. Dunford and J. T. Schwartz, *Linear Operators, Part I: General Theory*, Interscience Publishers, 1957.
5. J. K. Hale, *Ordinary Differential Equations*, Wiley-Interscience, 1969.
6. A. Isidori, *Nonlinear control systems*, 2nd ed., Springer-Verlag, 1989.
7. T. Kailath, *Linear Systems*, Prentice Hall, 1980.
8. I. Kanellakopoulos, P. V. Kokotovic, and R. Marino, "Robustness of adaptive nonlinear control under an extended matching condition," *Prepr. IFAC Symp. Nonlinear Control Syst. Design*, pp. 192–197, Capri, Italy, 1989.
9. I. Kanellakopoulos, P. V. Kokotovic, and R. Marino, "An extended direct scheme for robust adaptive nonlinear control," *Automatica*, to appear, March 1991.
10. I. Kanellakopoulos, P. V. Kokotovic, and R. H. Middleton, "Observer-based adaptive control of nonlinear systems under matching conditions," *Proc. 1990 Amer. Control Conf.*, pp. 549–555, San Diego, CA, May 1990.
11. V. Lakshmikantham and S. Leela, *Differential and Integral Inequalities: Theory and Applications, Volume 1: Ordinary Differential Equations*, Academic Press, 1969.
12. I. D. Landau, *Adaptive Control: The Model Reference Approach*, Control and Systems Theory, vol. 8, Dekker, 1979.
13. I. M. Y Mareels, H. B. Penfold, and R. J. Evans, "Controlling nonlinear time-varying systems via Euler approximations," Technical Report EE 8939, Dept. of Electrical Engineering and Computer Science, University of Newcastle, Australia, Oct. 1989.
14. H. Mayeda, K. Osuka, and A. Kangawa, "A new identification method for serial manipulators arms," *Proc. 9th IFAC World Congress*, Sect. 08.2/B-3, July 1984.
15. R. H. Middleton and G. C. Goodwin, "Adaptive computed torque control for rigid link manipulators," *Syst. Control Lett.*, vol. 10, no. 1, pp. 9–16, 1988.
16. K. Nam and A. Arapostathis, "A model-reference adaptive control scheme for pure-feedback nonlinear systems," *IEEE Trans. Aut. Control*, vol. 33, pp. 803–811, Sept. 1988.
17. K. S. Narendra, L. S. Valavani, "A comparison of Lyapunov and hyperstability approaches to adaptive control of continuous systems," *IEEE Trans. Aut. Control*, April 1980.
18. P. C. Parks, "Lyapunov redesign of model reference adaptive control systems," *IEEE Trans. Aut. Control*, vol. AC-11, pp. 362–367, 1966.
19. J.-B. Pomet, *Sur la commande adaptative des systèmes non linéaires*, Thèse de l'École des Mines de Paris en Mathématiques et Automatique, 1989.
20. J.-B. Pomet and L. Praly, "Indirect adaptive nonlinear control," *Proc. 27th IEEE Conf. Dec. Control*, pp. 2414–2415, Austin, TX, Dec. 1988.
21. J.-B. Pomet and L. Praly, "Adaptive nonlinear regulation: equation error from the Lyapunov equation," *Proc. 28th IEEE Conf. Dec. Control*, pp. 1008–1013, Tampa, FL, Dec. 1989.

22. J.-B. Pomet and L. Praly, "Adaptive nonlinear control: an estimation-based algorithm," in *New Trends in Nonlinear Control Theory*, J. Descusse, M. Fliess, A. Isidori and D. Leborgne Eds., Springer-Verlag, Berlin, 1989.
23. J.-B. Pomet and L. Praly, "Adaptive non-linear stabilization: estimation from the Lyapunov equation," CAI Report 232, submitted for publication, Feb. 1990.
24. J.-B. Pomet and L. Praly, "Adaptive nonlinear control of feedback equivalent systems," *Proc. 9th Int. Conf. Analysis and Optimization of Systems*, Antibes, France, June 1990.
25. J.-B. Pomet and I. Kupka, "Feedback equivalence of a parametrized family of control systems," in preparation, Department of Mathematics, University of Toronto, Canada, 1990.
26. V. M. Popov, *Hyperstability of Control Systems*, Springer-Verlag, 1973.
27. L. Praly, B. d'Andréa-Novel, and J.-M. Coron, "Lyapunov design of stabilizing controllers for cascaded systems," *IEEE Trans. Aut. Control*, to appear, see also *Proc. 28th IEEE Conf. Dec. Control*, Tampa, FL, Dec. 1989.
28. S. S. Sastry and P. V. Kokotovic, "Feedback linearization in the presence of uncertainties," *Int. J. Adapt. Control Sig. Proc.*, vol. 2, pp. 327–346, 1988.
29. S. S. Sastry and A. Isidori, "Adaptive control of linearizable systems," *IEEE Trans. Aut. Control*, vol. 34, pp. 1123–1131, Nov. 1989.
30. J.-J. E. Slotine and W. Li, "Adaptive manipulator control: a case study," *IEEE Trans. Aut. Control*, vol. 33, pp. 995-1003, Nov. 1988.
31. D. G. Taylor, P. V. Kokotovic, R. Marino, and I. Kanellakopoulos, "Adaptive regulation of nonlinear systems with unmodeled dynamics," *IEEE Trans. Aut. Control*, vol. 34, pp. 405–412, April 1989.

Adaptive Nonlinear Control of Induction Motors via Extended Matching *

Riccardo Marino,[1] *Sergei Peresada,*[2] *and Paolo Valigi*[1]

[1] Seconda Università di Roma, Dipartimento di Ingegneria Elettronica Via O. Raimondo, 00173 Roma, ITALY.

[2] Kiev Polytechnical Institute, Department of Electrical Engineering Prospect Pobedy 37, Kiev 252056, USSR.

Abstract. A nonlinear adaptive state-feedback input-output linearizing control is designed for a fifth order model of an induction motor which includes both electrical and mechanical dynamics under the assumptions of linear magnetic circuits. The control algorithm contains a nonlinear identification scheme which asymptotically tracks the true values of the load torque and rotor resistance, which are assumed to be constant but unknown. Once those parameters are identified, the two control goals of regulating rotor speed and rotor flux amplitude are decoupled. Full state measurements are required.

1 Introduction

In the last decade, significant advances have been made in the theory of nonlinear state-feedback control (see [1] and [2] for a comprehensive introduction to nonlinear geometric control): in particular, feedback linearization and input-output decoupling techniques have proved useful in applications. More recently the problems of feedback linearization and input-output linearization have been generalized allowing for some parameters not to be known [3,4,5,6]. In this paper we address the problem of adaptive speed regulation for induction motors with load torque and rotor resistance being unknown but constant parameters. Nonadaptive input-output decoupling controls were presented in [7,8,9] using geometric techniques. We develop an adaptive version of the controller presented in [9], assuming that load torque and rotor resistance are unknown parameters. A preliminary version of this work can be found in [10]. The paper is organized as follows. In Section 2 a fifth-order state-space model of an induction motor, which includes both electrical and mechanical dynamics, is given. In Section 3 previous control schemes are reviewed and it is shown that field oriented control can be viewed as a feedback transformation which achieves asymptotic input-output decoupling and linearization. In Section 4, following the results presented in [3,6], we develop an adaptive version of the exact decoupling and linearizing control given in [9] which covers the more realistic situation in which the load

* We would like to thank Prof. A. Bellini for providing us the data of the motor and for useful discussions. This work was supported in part by Ministero della Università e della Ricerca Scientifica e Tecnologica (fondi 40%).

torque and the rotor resistance are not known. We present a second-order nonlinear identification scheme which asymptotically tracks the correct value of load torque and, when electric torque is different than zero, the correct value of rotor resistance as well. The adaptive state-feedback linearizing control achieves full decoupling in speed and rotor flux magnitude regulation as soon as the identification scheme has converged to the true parameter values. This is shown in Section 5, where the effects of parameter uncertainties are analyzed and some simulation results illustrate the performance of the proposed control algorithm, which has some advantages over the classical scheme of field oriented control: with a comparable complexity, two critical parameters are identified and exact decoupling is achieved.

2 Induction Motor Model

The reader is referred to [11] and [12] for the general theory of electric machines and induction motors and to [13] for related control problems. The symbols used and their meaning are collected in the Appendix. A is made up of three stator windings and three rotor windings. Krause [14] introduced a two-phase equivalent machine representation of an induction motor with two rotor windings and two stator windings. The dynamics of an induction motor under the assumptions of equal mutual inductances and linear magnetic circuit are given by the fifth-order model:

$$\begin{aligned}
\frac{d\omega}{dt} &= \frac{n_p M}{J L_r}(\psi_{ra} i_{sb} - \psi_{rb} i_{sa}) - \frac{T_L}{J} \\
\frac{d\psi_{ra}}{dt} &= -\frac{R_r}{L_r}\psi_{ra} - n_p \omega \psi_{rb} + \frac{R_r}{L_r} M i_{sa} \\
\frac{d\psi_{rb}}{dt} &= n_p \omega \psi_{ra} - \frac{R_r}{L_r}\psi_{rb} + \frac{R_r}{L_r} M i_{sb} \\
\frac{di_{sa}}{dt} &= \frac{M R_r}{\sigma L_r^2}\psi_{ra} + \frac{n_p M}{\sigma L_r}\omega \psi_{rb} - \gamma^* i_{sa} + \frac{1}{\sigma} u_{sa} \\
\frac{di_{sb}}{dt} &= -\frac{n_p M}{\sigma L_r}\omega \psi_{ra} + \frac{M R_r}{\sigma L_r^2}\psi_{rb} - \gamma^* i_{sb} + \frac{1}{\sigma} u_{sb} ,
\end{aligned} \tag{1}$$

where i, ψ, u_s denote current, flux linkage and stator voltage input to the machine; the subscripts s and r stand for stator and rotor; (a, b) denote the components of a vector with respect to a fixed stator reference frame and $\sigma = L_s - \frac{M^2}{L_r}$, $\gamma^* = \left(\frac{M^2 R_r + L_r^2 R_s}{\sigma L_r^2}\right)$.

From now on we will drop the subscripts r and s since we will only use rotor fluxes (ψ_{ra}, ψ_{rb}) and stator currents (i_{sa}, i_{sb}). Let

$$x = (\omega, \psi_a, \psi_b, i_a, i_b)^T \tag{2}$$

be the state vector and let

$$p = (p_1, p_2)^T = (T_L - T_{LN}, R_r - R_{rN})^T \tag{3}$$

be the unknown parameter deviations from the nominal values T_{LN} and R_{rN} of load torque T_L and rotor resistance R_r. T_L is typically unknown whereas R_r may have a range of variations of ±50% around its nominal value (see [13, p. 224]) due to rotor heating. Let $u = (u_a, u_b)^T$ be the control vector. Let $\alpha = \frac{R_{rN}}{L_r}$, $\beta = \frac{M}{\sigma L_r}$, $\gamma = \frac{M^2 R_{rN}}{\sigma L_r^2} + \frac{R_s}{\sigma}$, $\mu = \frac{n_p M}{J L_r}$, be a reparametrization of the induction motor model, where α, β, γ, μ are known parameters depending on the nominal value R_{rN}.

The system (1) can be rewritten in compact form as

$$\dot{x} = f(x) + u_a g_a + u_b g_b + p_1 f_1 + p_2 f_2(x)\,, \tag{4}$$

where the vector fields f, g_a, g_b, f_1, f_2 are

$$f(x) = \begin{pmatrix} \mu(\psi_a i_b - \psi_b i_a) - \frac{T_{LN}}{J} \\ -\alpha\psi_a - n_p\omega\psi_b + \alpha M i_a \\ n_p\omega\psi_a - \alpha\psi_b + \alpha M i_b \\ \alpha\beta\psi_a + n_p\beta\omega\psi_b - \gamma i_a \\ -n_p\beta\omega\psi_a + \alpha\beta\psi_b - \gamma i_b \end{pmatrix}, \tag{5}$$

$$g_a = \begin{pmatrix} 0 \\ 0 \\ 0 \\ \frac{1}{\sigma} \\ 0 \end{pmatrix}, \quad g_b = \begin{pmatrix} 0 \\ 0 \\ 0 \\ 0 \\ \frac{1}{\sigma} \end{pmatrix}, \tag{6}$$

$$f_1(x) = \begin{pmatrix} -\frac{1}{J} \\ 0 \\ 0 \\ 0 \\ 0 \end{pmatrix}, \quad f_2(x) = \begin{pmatrix} 0 \\ -\frac{1}{L_r}\psi_a + \frac{M}{L_r} i_a \\ -\frac{1}{L_r}\psi_b + \frac{M}{L_r} i_b \\ \frac{M}{\sigma L_r^2}\psi_a - \frac{M^2}{\sigma L_r^2} i_a \\ \frac{M}{\sigma L_r^2}\psi_b - \frac{M^2}{\sigma L_r^2} i_b \end{pmatrix}. \tag{7}$$

3 Induction Motor Control

3.1 Field Oriented Control

A classical control technique for induction motors is the field oriented control. First introduced by Blaschke [15,16] in 1971, it involves the transformation of the vectors (i_a, i_b), (ψ_a, ψ_b) in the fixed stator frame (a, b) into vectors in a frame (d, q) which rotates along with the flux vector (ψ_a, ψ_b); if one defines

$$\rho = \arctan\frac{\psi_b}{\psi_a}\,, \tag{8}$$

the transformations are

$$\begin{pmatrix} i_\mathrm{d} \\ i_\mathrm{q} \end{pmatrix} = \begin{bmatrix} \cos\rho & \sin\rho \\ -\sin\rho & \cos\rho \end{bmatrix} \begin{pmatrix} i_\mathrm{a} \\ i_\mathrm{b} \end{pmatrix}, \tag{9}$$

$$\begin{pmatrix} \psi_\mathrm{d} \\ \psi_\mathrm{q} \end{pmatrix} = \begin{bmatrix} \cos\rho & \sin\rho \\ -\sin\rho & \cos\rho \end{bmatrix} \begin{pmatrix} \psi_\mathrm{a} \\ \psi_\mathrm{b} \end{pmatrix}. \tag{10}$$

We now reinterpret field oriented control as a state-feedback transformation (involving a state-space change of coordinates and a nonlinear state feedback) to a control system of simpler structure. If we define the state-space change of coordinates

$$\begin{aligned} \omega &= \omega \\ \psi_\mathrm{d} &= \sqrt{\psi_\mathrm{a}^2 + \psi_\mathrm{b}^2} \\ \rho &= \arctan\frac{\psi_\mathrm{b}}{\psi_\mathrm{a}} \\ i_\mathrm{d} &= \frac{\psi_\mathrm{a} i_\mathrm{a} + \psi_\mathrm{b} i_\mathrm{b}}{|\psi|} \\ i_\mathrm{q} &= \frac{\psi_\mathrm{a} i_\mathrm{b} - \psi_\mathrm{b} i_\mathrm{a}}{|\psi|}, \end{aligned} \tag{11}$$

and the state feedback

$$\begin{pmatrix} u_\mathrm{a} \\ u_\mathrm{b} \end{pmatrix} = \sigma|\psi| \begin{bmatrix} \psi_\mathrm{a} & \psi_\mathrm{b} \\ -\psi_\mathrm{b} & \psi_\mathrm{a} \end{bmatrix}^{-1} \begin{pmatrix} -n_\mathrm{p}\omega i_\mathrm{q} - \alpha M \dfrac{i_\mathrm{q}^2}{\psi_\mathrm{d}} + v_\mathrm{d} \\ n_\mathrm{p}\beta\omega\psi_\mathrm{d} + n_\mathrm{p}\omega i_\mathrm{d} + \alpha M \dfrac{i_\mathrm{d} i_\mathrm{q}}{\psi_\mathrm{d}} + v_\mathrm{q} \end{pmatrix}, \tag{12}$$

the closed-loop system (1), (12) becomes in the new coordinates

$$\begin{aligned} \frac{d\omega}{dt} &= \mu\psi_\mathrm{d} i_\mathrm{q} - \frac{T_\mathrm{LN}}{J} \\ \frac{di_\mathrm{q}}{dt} &= -\gamma i_\mathrm{q} + v_\mathrm{q} \\ \frac{d\psi_\mathrm{d}}{dt} &= -\alpha\psi_\mathrm{d} + \alpha M i_\mathrm{d} \\ \frac{di_\mathrm{d}}{dt} &= -\gamma i_\mathrm{d} + \alpha\beta\psi_\mathrm{d} + v_\mathrm{d} \\ \frac{d\rho}{dt} &= n_\mathrm{p}\omega + \alpha M \frac{i_\mathrm{q}}{\psi_\mathrm{d}}. \end{aligned} \tag{13}$$

In other words, the system (1) is transformed into (13) by the feedback transformation (11)–(12). The system (13) has a simpler structure: its flux amplitude dynamics are linear

$$\begin{aligned} \frac{d\psi_\mathrm{d}}{dt} &= -\alpha\psi_\mathrm{d} + \alpha M i_\mathrm{d} \\ \frac{di_\mathrm{d}}{dt} &= -\gamma i_\mathrm{d} + \alpha\beta\psi_\mathrm{d} + v_\mathrm{d}, \end{aligned} \tag{14}$$

and can be independently controlled by $v_{\rm d}$, for instance via a PI controller, as proposed in [13]

$$v_{\rm d} = -k_{\rm d1}\,(\psi_{\rm d} - \psi_{\rm d\,ref}) - k_{\rm d2}\int_0^t (\psi_{\rm d}(\tau) - \psi_{\rm d\,ref})\,d\tau\,. \tag{15}$$

When the flux amplitude $\psi_{\rm d}$ is regulated to the constant reference value $\psi_{\rm d\,ref}$, rotor speed dynamics are also linear

$$\begin{aligned} \frac{d\omega}{dt} &= \mu\psi_{\rm d\,ref}\,i_{\rm q} - \frac{T_{\rm L}}{J} \\ \frac{di_{\rm q}}{dt} &= -\gamma i_{\rm q} + v_{\rm q}\,, \end{aligned} \tag{16}$$

and can be independently controlled by $v_{\rm q}$, for instance by two nested loops of PI controllers, as proposed in [13]

$$\begin{aligned} v_{\rm q} &= -k_{\rm q1}(T - T_{\rm ref}) - k_{\rm q2}\int_0^t (T(\tau) - T_{\rm ref}(\tau))\,d\tau \\ T_{\rm ref} &= -k_{\rm q3}(\omega - \omega_{\rm ref}) - k_{\rm q4}\int_0^t (\omega(\tau) - \omega_{\rm ref})\,d\tau \\ T &= \mu\psi_{\rm d} i_{\rm q}\,. \end{aligned} \tag{17}$$

If ω and $\psi_{\rm d}$ are defined as outputs, field oriented control achieves asymptotic input-output linearization and decoupling via the nonlinear state feedback (12), (15), (17): PI controllers are used to counteract parameter variations.

During the flux transient the nonlinearity $\psi_{\rm d} i_{\rm q}$ in (13) makes the first four equations in (13) still nonlinear and coupled. Flux transients occur when the motor has to be operated above the nominal speed. In this case flux weakening, (for instance $\psi_{\rm ref} = \frac{k}{\omega_{\rm ref}}$) is required in order to keep applied voltage within inverter ceiling limits [13, p.217] and the speed transients of the closed-loop system (13), (15), (17) are difficult to evaluate and may be unsatisfactory. It should also be mentioned that flux measurements, which are required in (12), are difficult to obtain (see [17,18]), even though flux observers from stator currents and rotor speed measurements have been determined [19].

3.2 Input-Output Decoupling

As shown in [9], one can improve field oriented control by achieving exact input-output decoupling and linearization via a nonlinear state-feedback control which is not more complex than (12).

We will use the following notation for the directional (or Lie) derivative of state function $\phi(x) : \mathbb{R}^n \to \mathbb{R}$ along a vector field $f(x) = (f_1(x), \cdots, f_n(x))$

$$L_f\phi = \sum_{i=1}^{n} \frac{\partial\phi}{\partial x_i} f_i(x)\,. \tag{18}$$

Iteratively we define $L_f^i\phi = L_f(L_f^{(i-1)}\phi)$.

The outputs to be controlled are ω and $\psi_a^2 + \psi_b^2$. Let us define the change of coordinates

$$\begin{aligned} y_1 &= \phi_1(x) = \omega \\ y_2 &= L_f\phi_1(x) = \mu(\psi_a i_b - \psi_b i_a) - \frac{T_{LN}}{J} \\ y_3 &= \phi_2(x) = \psi_a^2 + \psi_b^2 \\ y_4 &= L_f\phi_2(x) = -2\alpha(\psi_a^2 + \psi_b^2) + 2\alpha M(\psi_a i_a + \psi_b i_b) \\ y_5 &= \arctan\left(\frac{\psi_b}{\psi_a}\right) = \phi_3\,, \end{aligned} \tag{19}$$

which is one to one in $\Omega = \{x \in \mathrm{IR}^5 : \psi_a^2 + \psi_b^2 \neq 0\}$ but it is onto only for $y_3 > 0,\ -90 \leq y_5 \leq 90$. The inverse transformation is defined in as

$$\begin{aligned} \omega &= y_1 \\ \psi_a &= \sqrt{y_3}\cos y_5 \\ \psi_b &= \sqrt{y_3}\sin y_5 \\ i_a &= \frac{1}{\sqrt{y_3}}\left(\cos y_5\left(\frac{y_4 + 2\alpha y_3}{2\alpha M}\right) - \frac{1}{\mu}\sin y_5\left(y_2 + \frac{T_{LN}}{J}\right)\right) \\ i_b &= \frac{1}{\sqrt{y_3}}\left(\sin y_5\left(\frac{y_4 + 2\alpha y_3}{2\alpha M}\right) + \frac{1}{\mu}\cos y_5\left(y_2 + \frac{T_{LN}}{J}\right)\right)\,. \end{aligned} \tag{20}$$

The dynamics of the induction motor with nominal parameters are given in the new coordinates by

$$\begin{aligned} \dot y_1 &= y_2 \\ \dot y_2 &= L_f^2\phi_1 + L_{g_a}L_f\phi_1 u_a + L_{g_b}L_f\phi_1 u_b \\ \dot y_3 &= y_4 \\ \dot y_4 &= L_f^2\phi_2 + L_{g_a}L_f\phi_2 u_a + L_{g_b}L_f\phi_2 u_b \\ \dot y_5 &= L_f\phi_3\,. \end{aligned} \tag{21}$$

The first four equations in (21) can be rewritten as

$$\begin{pmatrix} \ddot y_1 \\ \ddot y_3 \end{pmatrix} = \begin{pmatrix} L_f^2\phi_1 \\ L_f^2\phi_2 \end{pmatrix} + D(x)\begin{pmatrix} u_a \\ u_b \end{pmatrix}, \tag{22}$$

where $D(x)$ is the decoupling matrix given by

$$D(x) = \begin{bmatrix} L_{g_a}L_f\phi_1 & L_{g_b}L_f\phi_1 \\ L_{g_a}L_f\phi_2 & L_{g_b}L_f\phi_2 \end{bmatrix} = \begin{bmatrix} -\frac{\mu}{\sigma}\psi_b & \frac{\mu}{\sigma}\psi_a \\ \frac{2\alpha M}{\sigma}\psi_a & \frac{2\alpha M}{\sigma}\psi_b \end{bmatrix}. \tag{23}$$

Since

$$\det[D] = -2\frac{\alpha M\mu}{\sigma^2}(\psi_a^2 + \psi_b^2), \tag{24}$$

$D(x)$ is nonsingular everywhere in Ω.

The terms $L_f^2\phi_1$ and $L_f^2\phi_2$ are given by

$$\begin{aligned} L_f^2\phi_1 &= -\mu\beta n_\mathrm{p}\omega(\psi_\mathrm{a}^2+\psi_\mathrm{b}^2) - \mu(\alpha+\gamma)(\psi_\mathrm{a}i_\mathrm{b}-\psi_\mathrm{b}i_\mathrm{a}) - \mu n_\mathrm{p}\omega(\psi_\mathrm{a}i_\mathrm{a}+\psi_\mathrm{b}i_\mathrm{b}), \\ L_f^2\phi_2 &= (4\alpha^2+2\alpha^2\beta M)(\psi_\mathrm{a}^2+\psi_\mathrm{b}^2) + 2\alpha M n_\mathrm{p}\omega(\psi_\mathrm{a}i_\mathrm{b}-\psi_\mathrm{b}i_\mathrm{a}) \\ &\quad -(6\alpha^2 M + 2\alpha\gamma M)(\psi_\mathrm{a}i_\mathrm{a}+\psi_\mathrm{b}i_\mathrm{b}) + 2\alpha^2M^2(i_\mathrm{a}^2+i_\mathrm{b}^2). \end{aligned} \tag{25}$$

The dynamics of the flux angle $y_5 = \phi_3(x)$ are

$$\frac{d\phi_3}{dt} = \frac{dy_5}{dt} = n_\mathrm{p}\omega + \frac{\alpha M}{\psi_\mathrm{a}^2+\psi_\mathrm{b}^2}(\psi_\mathrm{a}i_\mathrm{b}-\psi_\mathrm{b}i_\mathrm{a}) = n_\mathrm{p}y_1 + \frac{R_\mathrm{rN}}{n_\mathrm{p}\,y_3}(Jy_2+T_\mathrm{LN}). \tag{26}$$

The difference between flux angular speed $\dot\phi_3$ and rotor speed $n_\mathrm{p}\omega$ is usually called slip speed, ω_s, which can be expressed, recalling the expression of α, as

$$\dot\phi_3 - n_\mathrm{p}\omega = \omega_\mathrm{s} = \frac{R_\mathrm{rN}M}{L_\mathrm{r}}\frac{\psi_\mathrm{a}i_\mathrm{b}-\psi_\mathrm{b}i_\mathrm{a}}{\psi_\mathrm{a}^2+\psi_\mathrm{b}^2} = \frac{R_\mathrm{rN}}{n_\mathrm{p}}\frac{T}{|\psi|^2}, \tag{27}$$

where

$$T = \frac{n_\mathrm{p}M}{L_\mathrm{r}}(\psi_\mathrm{ra}i_\mathrm{sb}-\psi_\mathrm{rb}i_\mathrm{sa}) \tag{28}$$

represents the electric torque.

The input-output linearizing feedback for the system (21) is given by

$$\begin{pmatrix} u_\mathrm{a} \\ u_\mathrm{b} \end{pmatrix} = D^{-1}(x)\left[\begin{pmatrix} -L_f^2\phi_1 \\ -L_f^2\phi_2 \end{pmatrix} + \begin{pmatrix} v_\mathrm{a} \\ v_\mathrm{b} \end{pmatrix}\right], \tag{29}$$

where $v = (v_\mathrm{a}, v_\mathrm{b})^\mathrm{T}$ is the new input vector. Substituting the state feedback (29) in (21) the closed-loop dynamics become, in the y−coordinates,

$$\begin{aligned} \dot y_1 &= y_2 \\ \dot y_2 &= v_\mathrm{a} \\ \dot y_3 &= y_4 \\ \dot y_4 &= v_\mathrm{b} \\ \dot y_5 &= n_\mathrm{p}y_1 + \frac{R_\mathrm{rN}}{n_\mathrm{p}}\frac{1}{y_3}(Jy_2+T_\mathrm{LN}). \end{aligned} \tag{30}$$

Equation (26) or (27) represents the dynamics which have been made unobservable from the outputs by the state feedback control (29).

In order to track desired reference signals $\omega_\mathrm{ref}(t)$ and $|\psi|^2_\mathrm{ref}(t)$ for the speed $y_1 = \omega$ and the square of the flux modulus $y_3 = \psi_\mathrm{a}^2+\psi_\mathrm{b}^2$, the input signals v_a and v_b in (29) are designed as

$$\begin{aligned} v_\mathrm{a} &= -k_\mathrm{a1}(y_1-\omega_\mathrm{ref}(t)) - k_\mathrm{a2}(y_2-\dot\omega_\mathrm{ref}(t)) + \ddot\omega_\mathrm{ref}(t) \\ v_\mathrm{b} &= -k_\mathrm{b1}(y_3-|\psi|^2_\mathrm{ref}) - k_\mathrm{b2}(y_4-\dot{|\psi|}^2_\mathrm{ref}) + \ddot{|\psi|}^2_\mathrm{ref}, \end{aligned} \tag{31}$$

where (k_{a1}, k_{a2}) and (k_{b1}, k_{b2}) are constant design parameters to be assigned in order to shape the response of the decoupled, linear second-order systems

$$\frac{d^2}{dt^2}(\omega - \omega_{ref}) = -k_{a1}(\omega - \omega_{ref}) - k_{a2}\frac{d}{dt}(\omega - \omega_{ref})$$

$$\frac{d^2}{dt^2}(|\psi|^2 - |\psi|^2_{ref}) = -k_{b1}(|\psi|^2 - |\psi|^2_{ref}) - k_{b2}\frac{d}{dt}(|\psi|^2 - |\psi|^2_{ref}). \quad (32)$$

Remarks

1) The system (30) is input-output decoupled; the input-output mapping is a pair of second-order linear systems. This allows for an independent regulation (or tracking) of the outputs according to (32). Transient responses are now decoupled also when flux weakening is performed. This is an improvement over the field oriented control (see also [9]).
2) State-space changes of coordinates both in the field oriented control and in the decoupling control (i.e., (11) and (20)) are valid in the open set $\Omega = \{x \in \mathbb{R}^5 : \psi_a^2 + \psi_b^2 \neq 0\}$; notice that $\psi_a^2 + \psi_b^2 = 0$ is a physical singularity of the motor in starting conditions.
3) While measurements of (ω, i_a, i_b) are available, measurements of (ψ_a, ψ_b) pose some problems (see [17]). As far as parameters are concerned, variation in load torque T_L and rotor resistance R_r cause a loss of input-output decoupling and steady-state regulation errors. This calls for an adaptive version of the control (29),(31), which is given in the next Section.
4) Easy computations show that the induction motor model (1) is not feedback linearizable. The necessary and sufficient conditions given in [20] fail; in fact the distribution $\mathcal{G}_1 = \text{span } \{g_a, g_b, ad_f g_a, ad_f g_b\}$ is not involutive since the vector field $[ad_f g_a, ad_f g_b]$ does not belong to $\mathcal{G}_1$ ($ad_X Y$ or $[X, Y]$ denotes the Lie bracket of two vector fields; one defines recursively $ad^i_X Y = ad_X(ad^{i-1}_X Y)$). Following the results in [21], since $\mathcal{G}_0 = \text{span } \{g_a, g_b\}$ is involutive and rank $\mathcal{G}_1 = 4$, it turns out that the largest feedback linearizable subsystem has dimension 4. This shows that the control (29),(31) provides the largest linearizable subsystem in the closed loop.
5) The state-feedback control (29),(31) is essentially the one proposed in [9]. It is made clear that the decoupling control makes the angle ϕ_3 unobservable from the outputs, and that (1) is not feedback linearizable.

Exact input-output decoupling controls for induction motors are proposed also in [7,8] with reference to a simplified model: the mechanical dynamics in (1) are not considered and ω is viewed as a parameter in the last four equations of (1).

4 Adaptive Input-Output Linearization

In this section we develop an adaptive version of the decoupling control (29) under the assumptions that T_L and R_r are unknown constant parameters. Let us rewrite the system (4) in the y-coordinates defined by (20); since the Lie derivatives $L_{f_2}\phi_1$, $L_{f_1}L_f\phi_1$, $L_{f_1}\phi_2$, $L_{f_1}L_f\phi_2$, $L_{f_1}\phi_3$, $L_{f_1}L_{f_2}\phi_2$, $L_{g_a}\phi_3$, $L_{g_b}\phi_3$ all vanish, we have

$$\begin{aligned}
\dot{y}_1 &= y_2 + p_1 L_{f_1}\phi_1 \\
\dot{y}_2 &= L_f^2\phi_1 + p_2 L_{f_2} L_f \phi_1 + L_{g_a} L_f \phi_1 u_a + L_{g_b} L_f \phi_1 u_b \\
\dot{y}_3 &= y_4 + p_2 L_{f_2}\phi_2 \\
\dot{y}_4 &= L_f^2\phi_2 + p_2 L_{f_2} L_f \phi_2 + L_{g_a} L_f \phi_2 u_a + L_{g_b} L_f \phi_2 u_b \\
\dot{y}_5 &= L_f\phi_3 + p_2 L_{f_2}\phi_3 \,.
\end{aligned} \tag{33}$$

Let $\hat{p}(t) = (\hat{p}_1(t), \hat{p}_2(t))^{\mathrm{T}}$ be a time varying estimate of the parameters and let

$$e_p = \begin{pmatrix} e_{p_1} \\ e_{p_2} \end{pmatrix} = \begin{pmatrix} p_1 - \hat{p}_1(t) \\ p_2 - \hat{p}_2(t) \end{pmatrix} \tag{34}$$

be the parameter error. Following [3,6] we now introduce a time-varying state-space change of coordinates depending on the parameters estimate $\hat{p}(t)$

$$\begin{aligned}
z_1 &= y_1 \\
z_2 &= y_2 + \hat{p}_1 L_{f_1}\phi_1 \\
z_3 &= y_3 \\
z_4 &= y_4 + \hat{p}_2 L_{f_2}\phi_2 \\
z_5 &= y_5 \,.
\end{aligned} \tag{35}$$

In the z-coordinates the system (4) becomes

$$\begin{aligned}
\dot{z}_1 &= z_2 + e_{p_1} L_{f_1}\phi_1 \\
\dot{z}_2 &= L_f^2\phi_1 + p_2 L_{f_2} L_f\phi_1 + \frac{d\hat{p}_1}{dt} L_{f_1}\phi_1 + L_{g_a} L_f\phi_1 u_a + L_{g_b} L_f \phi_1 u_b \\
\dot{z}_3 &= z_4 + e_{p_2} L_{f_2}\phi_2 \\
\dot{z}_4 &= L_f^2\phi_2 + p_2 L_{f_2} L_f \phi_2 + \frac{d\hat{p}_2}{dt} L_{f_2}\phi_2 + \hat{p}_2 L_f L_{f_2}\phi_2 + \hat{p}_2 p_2 L_{f_2}^2 \phi_2 \\
&\quad + u_a \left(L_{g_a} L_f \phi_2 + \hat{p}_2 L_{g_a} L_{f_2}\phi_2\right) + u_b \left(L_{g_b} L_f\phi_2 + \hat{p}_2 L_{g_b} L_{f_2}\phi_2\right) \\
\dot{z}_5 &= L_f\phi_3 + p_2 L_{f_2}\phi_3 \,.
\end{aligned} \tag{36}$$

Let

$$\begin{pmatrix} u_a \\ u_b \end{pmatrix} = D(x, \hat{p}_2)^{-1} \left[\begin{pmatrix} \xi_1 \\ \xi_2 \end{pmatrix} + \begin{pmatrix} v_a \\ v_b \end{pmatrix} \right], \tag{37}$$

where

$$\begin{aligned}
D(x, \hat{p}_2) &= \begin{bmatrix} L_{g_a} L_f \phi_1 & L_{g_b} L_f \phi_1 \\ L_{g_a} L_f\phi_2 + \hat{p}_2 L_{g_a} L_{f_2}\phi_2 & L_{g_b} L_f \phi_2 + \hat{p}_2 L_{g_b} L_{f_2}\phi_2 \end{bmatrix} \\
\xi_1 &= -L_f^2\phi_1 - \hat{p}_2 L_{f_2} L_f \phi_1 - \frac{d\hat{p}_1}{dt} L_{f_1}\phi_1 \\
\xi_2 &= -L_f^2\phi_2 - \hat{p}_2 L_{f_2} L_f\phi_2 - \frac{d\hat{p}_2}{dt} L_{f_2}\phi_2 - \hat{p}_2 L_f L_{f_2}\phi_2 - \hat{p}_2^2 L_{f_2}^2 \phi_2 \\
v_a &= -k_{a1}(z_1 - z_{1\,\mathrm{ref}}) - k_{a2} z_2 \\
v_b &= -k_{b1}(z_3 - z_{3\mathrm{ref}}) - k_{b2} z_4 \,,
\end{aligned} \tag{38}$$

(k_{a1}, k_{a2}), (k_{b1}, k_{b2}) are control parameters to be designed and $z_{1\,\mathrm{ref}}$ and $z_{3\,\mathrm{ref}}$ are the desired values for the rotor speed and the square of the rotor flux amplitude respectively. Since

$$\det D(x, \hat{p}_2) = -\frac{2M\mu}{\sigma^2}(R_{\mathrm{rN}} + \hat{p}_2)(\psi_{\mathrm{a}}^2 + \psi_{\mathrm{b}}^2), \tag{39}$$

the decoupling matrix is singular not only when $(\psi_{\mathrm{a}}^2 + \psi_{\mathrm{b}}^2) = 0$ as in the non-adaptive case but also when $\hat{p}_2(t) = -R_{\mathrm{rN}}$; this additional singularity has to be taken into account in the design of the adaptive algorithm.

Defining the regulation error

$$e = (z_1 - z_{1\,\mathrm{ref}}, z_2, z_3 - z_{3\,\mathrm{ref}}, z_4)^{\mathrm{T}}, \tag{40}$$

the closed-loop system becomes

$$\begin{aligned}
\dot{e}_1 &= e_2 + e_{p_1} L_{f_1}\phi_1 \\
\dot{e}_2 &= -k_{a1}e_1 - k_{a2}e_2 + e_{p_2} L_{f_2} L_f \phi_1 \\
\dot{e}_3 &= e_4 + e_{p_2} L_{f_2}\phi_2 \\
\dot{e}_4 &= -k_{b1}e_3 - k_{b2}e_4 + e_{p_2}\left(L_{f_2} L_f \phi_2 + \hat{p}_2 L_{f_2}^2 \phi_2\right) \\
\dot{z}_5 &= L_f\phi_3 + p_2 L_{f_2}\phi_3 .
\end{aligned} \tag{41}$$

While the dynamics of z_5 are

$$\dot{z}_5 = n_{\mathrm{p}} z_1 + \frac{R_{\mathrm{r}}}{n_{\mathrm{p}}} \frac{J z_2 + T_{\mathrm{L}} - e_{p_1}}{z_3}, \tag{42}$$

the dynamics of the vector e can be rearranged as

$$\begin{aligned}
\begin{pmatrix} \dot{e}_1 \\ \dot{e}_2 \\ \dot{e}_3 \\ \dot{e}_4 \end{pmatrix} &= K \begin{pmatrix} e_1 \\ e_2 \\ e_3 \\ e_4 \end{pmatrix} + \begin{bmatrix} L_{f_1}\phi_1 & 0 \\ 0 & L_{f_2} L_f \phi_1 \\ 0 & L_{f_2}\phi_2 \\ 0 & L_{f_2} L_f \phi_2 + \hat{p}_2 L_{f_2}^2 \phi_2 \end{bmatrix} \begin{pmatrix} e_{p_1} \\ e_{p_2} \end{pmatrix} \\
&= K \begin{pmatrix} e_1 \\ e_2 \\ e_3 \\ e_4 \end{pmatrix} + W(z, \hat{p}_2) \begin{pmatrix} e_{p_1} \\ e_{p_2} \end{pmatrix},
\end{aligned} \tag{43}$$

where

$$\begin{aligned}
K &= \text{block diag}\,(K_{\mathrm{a}}, K_{\mathrm{b}}), \\
K_{\mathrm{a}} &= \begin{bmatrix} 0 & 1 \\ -k_{a1} & -k_{a2} \end{bmatrix}, \quad K_{\mathrm{b}} = \begin{bmatrix} 0 & 1 \\ -k_{b1} & -k_{b2} \end{bmatrix}, \\
L_{f_1}\phi_1 &= -\frac{1}{J}, \\
L_{f_2} L_f \phi_1 &= -\frac{\mu}{L_{\mathrm{r}}}\left(1 + \frac{M^2}{\sigma L_{\mathrm{r}}}\right)(\psi_{\mathrm{a}} i_{\mathrm{b}} - \psi_{\mathrm{b}} i_{\mathrm{a}}) = -\left(1 + \frac{M^2}{\sigma L_{\mathrm{r}}}\right)\frac{T}{J L_{\mathrm{r}}},
\end{aligned} \tag{44}$$

$$
\begin{aligned}
L_{f_2}\phi_2 &= \frac{2}{L_\mathrm{r}}\left(M(\psi_\mathrm{a} i_\mathrm{a} + \psi_\mathrm{b} i_\mathrm{b}) - (\psi_\mathrm{a}^2 + \psi_\mathrm{b}^2)\right) , \\
L_{f_2} L_f \phi_2 &= R_\mathrm{rN}\eta , \\
L_{f_2}^2 \phi_2 &= \eta , \\
\eta &= \frac{4\sigma L_\mathrm{r} + 2M^2}{\sigma L_\mathrm{r}^3}(\psi_\mathrm{a}^2 + \psi_\mathrm{b}^2) + \frac{2M^2}{L_\mathrm{r}^2}(i_\mathrm{a}^2 + i_\mathrm{b}^2) \\
&\quad - \frac{6\sigma M L_\mathrm{r} + 2M^3}{\sigma L_\mathrm{r}^3}(\psi_\mathrm{a} i_\mathrm{a} + \psi_\mathrm{b} i_\mathrm{b}) ,
\end{aligned}
$$

$W(z, \hat{p}_2)$ is called ***the regressor matrix*** and is a function of the x-variables (and therefore of the z-variables).

Let $P =$ block diag $(P_\mathrm{a}, P_\mathrm{b})$ be the positive definite symmetric solution to the Lyapunov equation

$$
K^\mathrm{T} P + PK = -Q , \tag{45}
$$

with $Q =$ block diag $(Q_\mathrm{a}, Q_\mathrm{b})$, Q_a and Q_b positive definite symmetric matrices. Consider the quadratic function

$$
V = e^\mathrm{T} Pe + e_p^\mathrm{T} \Gamma^{-1} e_p , \tag{46}
$$

where Γ is a positive definite symmetric matrix. The time derivative of V is

$$
\frac{dV}{dt} = e^\mathrm{T}\left(K^\mathrm{T} P + PK\right) e + 2e_p^\mathrm{T}\left[W^\mathrm{T} Pe + \Gamma^{-1}\frac{de_p}{dt}\right] . \tag{47}
$$

If we now define

$$
\frac{de_p}{dt} = -\Gamma W^\mathrm{T} Pe , \tag{48}
$$

or equivalently

$$
\frac{d\hat{p}}{dt} = \Gamma W^\mathrm{T} Pe , \tag{49}
$$

which defines the dynamics of the parameter estimate $\hat{p}(t)$, and use (45), equation (47) becomes

$$
\frac{dV}{dt} = -e^\mathrm{T} Qe . \tag{50}
$$

This guarantees that $e(t)$ and $\hat{p}(t)$ are bounded and that $e(t)$ is an L^2 signal; it follows from (43) that the first four state variables $(z_1, \ldots, z_4)$ are bounded. We are guaranteed to avoid the singularities $z_3 = 0$ and $\hat{p}_2 = -R_\mathrm{rN}$ for the decoupling matrix, and therefore for the control (37) as well, if the initial conditions $(e(0), e_p(0))$ are in $S = \{(e, e_p) \in \mathrm{IR}^6 : e^\mathrm{T} Pe + e_p^\mathrm{T}\Gamma^{-1} e_p \leq c\}$, the largest set entirely contained in $\{(e, e_p) \in \mathrm{IR}^6 : e_{p_2} < R_\mathrm{r} + \alpha_1,\ e_3 > \alpha_2 - z_{3\,\mathrm{ref}}\}$, where $\alpha_1 > 0$ and $\alpha_2 > 0$ are arbitrary. Since $W(z, \hat{p}_2)$ is continuous, contains only bounded functions of z_5 (sine and cosine), and $(z_1, z_2, z_3, z_4, \hat{p}_2)$ are bounded, it follows that $W(z, \hat{p}_2)$ is bounded and therefore $\dot{e}$ and $\dot{\hat{p}}$ are bounded as well; since, according to (42), $\dot{z}_5$ is bounded for $(e, e_p) \in S$, it follows that

$\ddot{e} = \frac{d}{dt}[Ke + W(x, \hat{p}_2)e_p]$ is bounded as well. Now, since e is a bounded L^2 signal with bounded derivative $\dot{e}$, by Barbalat lemma ([22], p. 211) it follows that

$$\lim_{t\to\infty} \|e(t)\| = 0\,, \tag{51}$$

i.e. zero steady-state regulation error is achieved. Since $\ddot{e}$ is bounded as well, $\dot{e}$ is uniformly continuous and (51) implies by Barbalat lemma again that

$$\lim_{t\to\infty} \|\dot{e}(t)\| = 0\,. \tag{52}$$

Therefore, it must be that

$$\lim_{t\to\infty} \left\| W\left(z(t), \hat{p}_2(t)\right) \begin{pmatrix} e_{p_1}(t) \\ e_{p_2}(t) \end{pmatrix} \right\| = 0\,. \tag{53}$$

Equation (53) implies, from (44), that

$$\begin{aligned} \lim_{t\to\infty} L_{f_1}\phi_1 e_{p_1}(t) &= \lim_{t\to\infty} \left(-\frac{1}{J} e_{p_1}(t)\right) = 0\,, \\ \lim_{t\to\infty} L_{f_2} L_f \phi_1 e_{p_2}(t) &= \lim_{t\to\infty} \left(-\frac{1}{JL_{\mathrm{r}}}\left(1 + \frac{M^2}{\sigma L_{\mathrm{r}}}\right) T(t) e_{p_2}(t)\right) = 0\,, \end{aligned} \tag{54}$$

i.e.,

$$\lim_{t\to\infty} e_{p_1}(t) = 0\,, \tag{55}$$

and, since $\lim_{t\to\infty} T(t) = T_{\mathrm{L}}$, whenever $T_{\mathrm{L}} \neq 0$, i.e. in any physical situation,

$$\lim_{t\to\infty} e_{p_2}(t) = 0\,, \tag{56}$$

that is, parameter convergence is achieved.

The difficulty in identifying rotor resistance under no-load condition is a common problem ([23]) and it is related to physical reasons. If the motor is unloaded, when speed and rotor flux regulation is achieved, the slip frequency in (27) is zero so that the flux vector rotates at speed $n_{\mathrm{p}}\omega$ and we have $R_{\mathrm{r}} i_{\mathrm{rd}} = 0\,, R_{\mathrm{r}} i_{\mathrm{rq}} = 0$; it follows that rotor currents are zero and therefore rotor resistance is not identifiable in steady-state. It is proposed in [23] to track a sinusoidal reference signal for $\psi_{\mathrm{a}}^2 + \psi_{\mathrm{b}}^2$ so that rotor currents are different than zero and rotor resistance can be identified.

In summary, we have shown that the adaptive feedback control (37), (49) gives the closed-loop system

$$\begin{aligned} \dot{e} &= Ke + We_p \\ \dot{e}_p &= -\Gamma W^{\mathrm{T}} P e\,. \end{aligned} \tag{57}$$

If the initial conditions $(e(0), e_p(0)) \in S$ we have

$$\begin{aligned} &\lim_{t\to\infty} \|e(t)\| = 0\,, \\ &\lim_{t\to\infty} |e_{p_1}(t)| = 0\,. \end{aligned} \tag{58}$$

Moreover, if $T_L \neq 0$ we also have

$$\lim_{t\to\infty} |e_{p_2}(t)| = 0 . \tag{59}$$

From (51) and (52) it follows that in any case we have

$$\begin{aligned} &\lim_{t\to\infty} |\omega(t)| = \omega_{\text{ref}} , \\ &\lim_{t\to\infty} |\dot{\omega}| = 0 , \\ &\lim_{t\to\infty} |\psi| = |\psi|_{\text{ref}} , \\ &\lim_{t\to\infty} |\dot{\psi}| = 0 . \end{aligned} \tag{60}$$

5 Simulations

The proposed control algorithm has been extensively simulated for a 15 kW motor, with nominal torque of 70 Nm and nominal speed of 220 rad/sec, whose parameter values are reported in the Appendix.

The simulation test reported involves the following operating sequences: the unloaded motor is required to reach the nominal speed and the nominal value of 1.3 Wb for the rotor flux amplitude $|\psi|$, with the initial estimate of rotor resistance R_r in error of +50 %. At $t = 2$ sec, a 40 Nm load torque, which is unknown to the controller, is applied. At $t = 5$ sec the speed is required to reach 300 rad/sec, well above the nominal value, and rotor flux amplitude reference is weakened accordingly to the rule $|\psi| = \frac{k}{\omega_{\text{ref}}}$. The reference signals for flux amplitude and speed, reported in Fig. 1, consist of step functions. A small time delay at the beginning of the speed reference trajectory is introduced in order to avoid time overlapping of flux and speed transients.

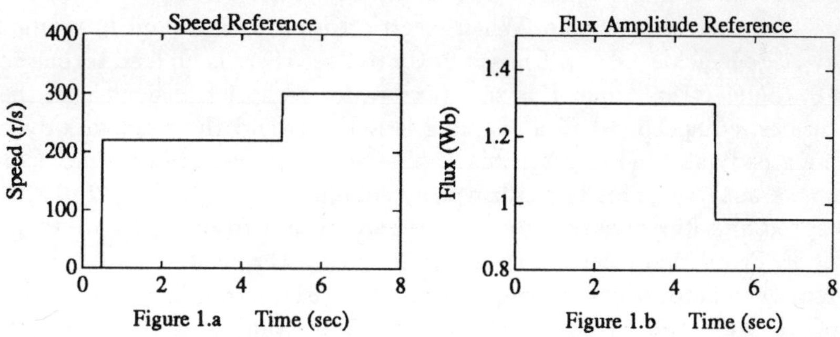

Fig. 1. Reference trajectories.

Both the non-adaptive (29), (31) and adaptive (37), (49) control laws have been simulated.

In the non adaptive case, we observe from simulations that parameter errors cause a steady-state error both in speed and flux regulation and a coupling between speed and rotor flux which is noticeable both during speed transient and at load insertion (see Fig. 2, where dashed lines stand for reference trajectories, solid lines correspond to simulated behavior and the dotted line represents the electric torque). To clarify the effects of unknown parameters, consider the closed-loop system, obtained applying feedback control (29), (31) to the motor (4) (recall (33)):

$$\begin{aligned}\begin{pmatrix}\dot e_1\\ \dot e_2\\ \dot e_3\\ \dot e_4\end{pmatrix} &= K\begin{pmatrix}e_1\\ e_2\\ e_3\\ e_4\end{pmatrix} + \begin{bmatrix}L_{f_1}\phi_1 & 0\\ 0 & L_{f_2}L_f\phi_1\\ 0 & L_{f_2}\phi_2\\ 0 & L_{f_2}L_f\phi_2\end{bmatrix}\begin{pmatrix}p_1\\ p_2\end{pmatrix}\\ &= K\begin{pmatrix}e_1\\ e_2\\ e_3\\ e_4\end{pmatrix} + W^*(y)\,p \qquad (61)\\ \dot y_5 &= L_f\phi_3\,,\end{aligned}$$

where $e = (e_1, e_2, e_3, e_4)^{\mathrm{T}} = (y_1 - y_{1\mathrm{ref}}, y_2, y_3 - y_{3\mathrm{ref}}, y_4)^{\mathrm{T}}$ is the regulation error vector and $y_{1\mathrm{ref}}$ and $y_{3\mathrm{ref}}$ are the desired values for the speed and the square of the flux amplitude respectively, K has the structure given in (44) and represents the linear part corresponding to (32), while W^*p takes into account the effects of parameter uncertainties (see (3)). Matrix W^* entries are given in (44), from which we see that $L_{f_1}\phi_1$ is constant and $L_{f_2}L_f\phi_1$ is proportional to the electric torque T. The entry $L_{f_2}\phi_2$ is proportional, via a non-zero constant, to the derivative of the squared flux amplitude and therefore, once flux steady-state is achieved, $L_{f_2}\phi_2 = 0$. The entry $L_{f_2}L_f\phi_2$ can be rewritten as $L_{f_2}L_f\phi_2 = c_1\frac{d|\psi|^2}{dt} + c_2T^2$, with c_1 and c_2 nonzero constants, that is, as the sum of a term proportional to the derivative of the squared flux amplitude plus a term proportional to the square of the electric torque. When electric torque is zero and flux amplitude steady-state is achieved, $L_{f_2}L_f\phi_2 = 0$. Up to 2 sec there is no load torque so that $p_1 = 0$, the electric torque T is zero (except for a short transient after the first step in desired speed, when a coupling is noticed) and the rotor flux dynamics reach a steady state (Fig. 2.b): this implies zero steady-state error according to the above analysis. This is confirmed by simulation; we see (Fig. 2, $0 \le t \le 2$) that speed and flux steady-state error is zero even if rotor resistance is in error of +50 %. From load insertion (from $t = 2$ sec) on, the electric torque and p_1 are different than zero, which causes, according to (61), a coupling and steady-state errors, as confirmed by Figs. 2.a and 2.b. Notice that even if the load torque were known (and therefore $p_1 = 0$), rotor resistance error ($p_2 \neq 0$) would still cause a speed steady-state error due to the entry $L_{f_2}L_f\phi_1$ which is proportional to the electric torque (see Figs. 3.a and 3.b). The dynamic responses when both parameters are known are reported in Fig. 4 for comparison.

The adaptive case simulations are reported in Figs. 5: speed and flux amplitude behavior are shown in Figs. 5.a and 5.b respectively, where solid lines

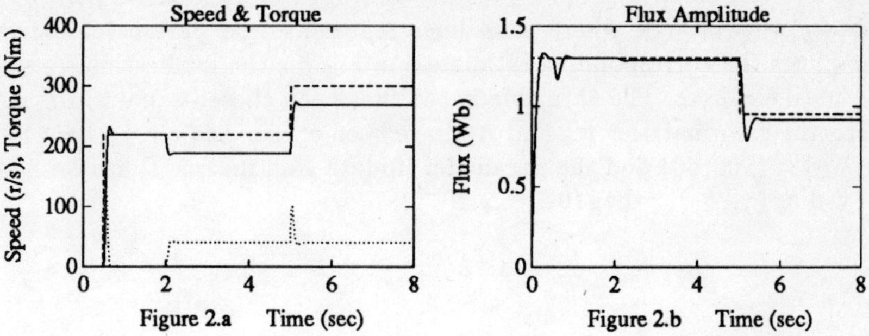

Fig. 2. Nonadaptive case (with both load torque and rotor resistance error).

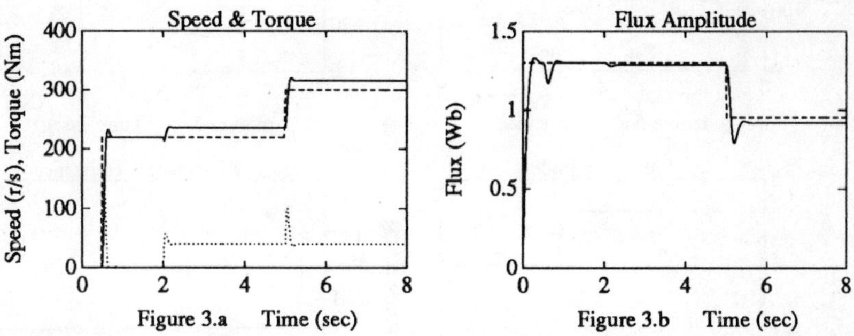

Fig. 3. Nonadaptive case (with rotor resistance error).

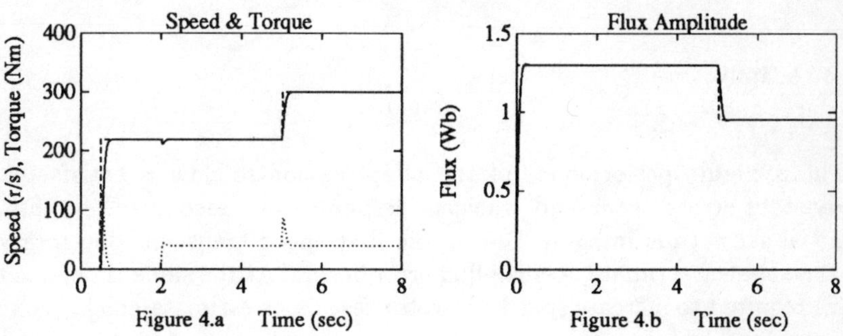

Fig. 4. Nonadaptive case (without parameter error).

represent actual variables, dashed lines the corresponding reference values and dotted line the electric torque; Figs. 5.c and 5.d report load torque and rotor resistance respectively, where solid lines represent true parameter values and dashed lines the corresponding estimates. In Fig. 5.c the load estimator was disabled until $t = 2\,\text{sec}$. The Q matrix in (45) has been chosen equal to the identity matrix, the gain matrices K_a and K_b have been chosen as $(k_{a1}, k_{a2}) = (400, 40)$, $(k_{b1}, k_{b2}) = (900, 60)$ and the parameter update gain matrix Γ has been chosen as $\Gamma = \text{diag}\,(\gamma_1, \gamma_2) = \text{diag}\,(0.1,\ 7 \times 10^{-8})$.

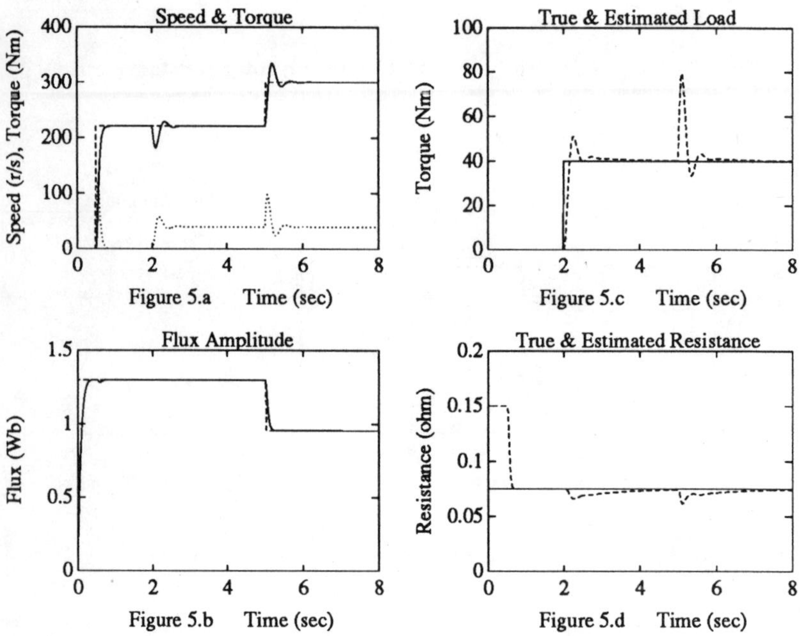

Fig. 5. Adaptive case.

The dynamic performance of the adaptive control law is satisfactory: no steady-state errors occur and transient responses are decoupled, excepting for an initial short time interval. During the first speed transient, due to a wrong initial resistance estimate, a small flux error occurs. At the same time, due to the torque required to increase speed, the rotor resistance estimate quickly converges to the true value and complete decoupling is achieved: no flux error occurs during the subsequent load insertion. Notice that the estimation algorithm (49) is driven by the regulation error e defined by (40); more precisely, in view of the structure of the regressor matrix W and the structure of the P matrix solution of (45), the load estimate dynamics are driven by the regulation error on speed and speed time derivative, i.e. by e_1 and e_2, while the rotor resistance estimate dynamics are driven by the whole error vector e. It is clearly impossible for the adaptation law to distinguish between regulation error due to change in reference value (e.g., step function references) and regulation error due to parameter variations. As a result, the rotor resistance estimate slightly varies both after the load insertion (at $t = 2\,\text{sec}$) and after a step change in speed reference (at $t = 5\,\text{sec}$). Similarly, the step change in speed reference at $t = 5\,\text{sec}$ causes a peak in load torque estimate, even though actual load does not change.

Figures 6 show the control input signal u_a. Control action consists in varying amplitude and frequency of the applied voltage. The nonlinear state-feedback control law results in voltage supply signals which are well within the capabilities of actual inverters and therefore can be easily implemented by current power electronic technology.

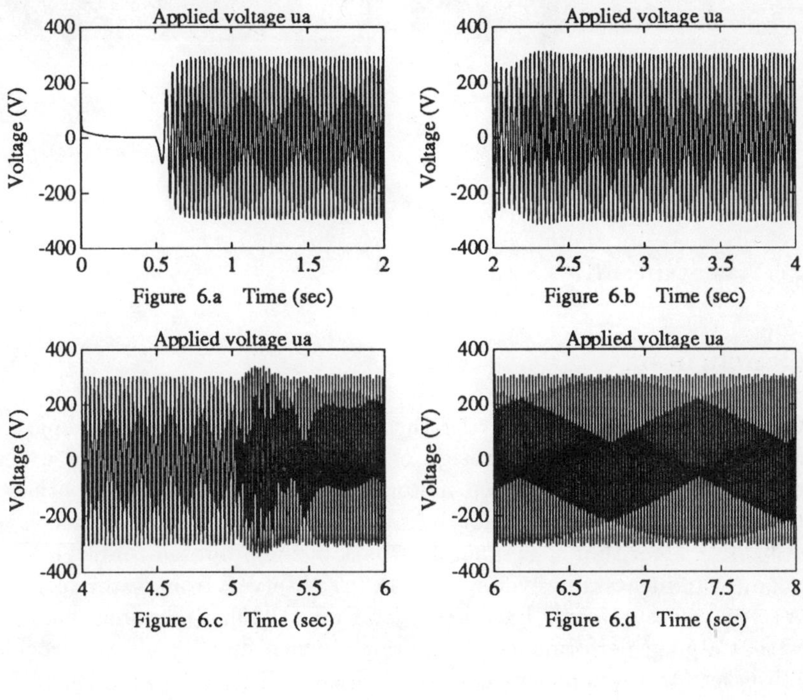

Fig. 6. Applied input voltage u_a.

Figures 7 show the i_a current waveform. The step functions used as reference trajectories in the simulation clearly are the "worst case" signals. In a real implementation smoother reference signals should be used, improving transient responses and reducing stator current and stator voltage peaks.

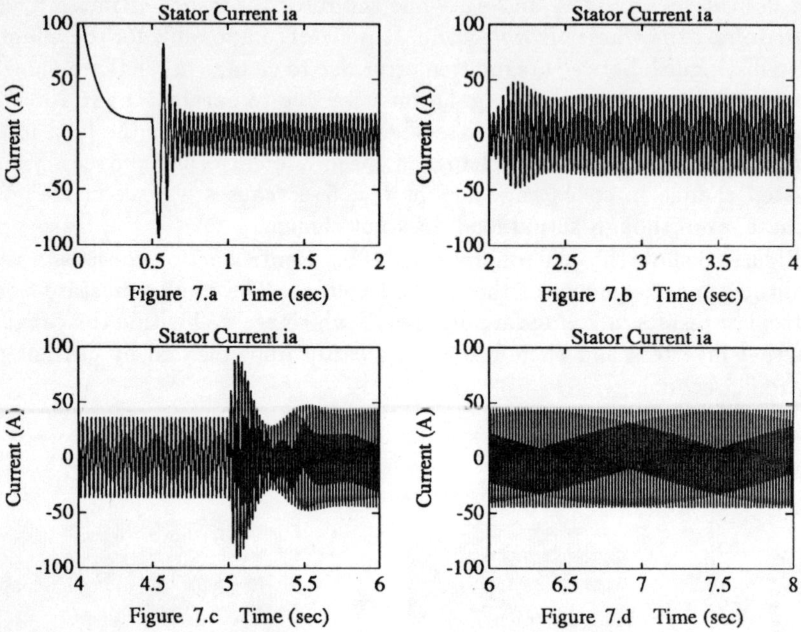

Fig. 7. Stator current i_a.

6 Conclusions

In this paper it is shown how the theory of input-output decoupling and its adaptive versions lead to the design of a satisfactory controller for a detailed nonlinear model of an induction motor. The control is adaptive with respect to two parameters which cannot be measured and is based on a converging identification algorithm. The main drawback of the proposed control is the need of flux measurements. However, nonlinear flux observers from stator currents and rotor speed measurements have been determined [19]. Preliminary simulations show that a good performance is maintained when flux signals are provided by the observers to the adaptive control algorithm. This is a direction of further investigation. Another direction of research is the real implementation of the control in order to verify the influence of sampling rate, truncation errors in digital implementation, measurement noise, simplifying modeling assumptions, unmodeled dynamics and saturations.

Appendix

Induction Motor Parameters (two-phase data)	
R_s	stator resistance
R_r	rotor resistance
i_s	stator current
ψ_s	stator flux linkage
i_r	rotor current
ψ_r	rotor flux linkage
u	voltage input
ω	angular speed
n_p	number of pole pairs
δ	angle of rotation
L_s	stator inductance
L_r	rotor inductance
M	mutual inductance
J	rotor inertia
T_L	load torque
T	electric motor torque

Parameter Values
$R_s = 0.18\,\Omega$
$R_r = 0.15\,\Omega$
$L_s = 0.0699\,\text{H}$
$L_r = 0.0699\,\text{H}$
$M = 0.068\,\text{H}$
$J = 0.0586\,\text{Kgm}^2$

References

1. A. Isidori, *Nonlinear Control Systems*, 2nd ed., Berlin, Springer-Verlag, 1989.
2. H. Nijmeijer and A. van der Schaft, *Nonlinear Dynamical Control Systems*, Berlin, Springer-Verlag, 1990.
3. R. Marino, I. Kanellakopoulos, and P. Kokotovic, "Adaptive tracking for feedback linearizable SISO systems," in *Proc. 28th IEEE Conf. Dec. Control*, Tampa, FL, pp. 1002–1007, Dec. 1989.
4. S. Sastry and A. Isidori, "Adaptive control of linearizable systems," *IEEE Trans. Aut. Control*, vol. 34, pp. 1123–1131, Nov. 1989.
5. D. Taylor, P. Kokotovic, R. Marino, and I. Kanellakopoulos, "Adaptive regulation of nonlinear systems with unmodeled dynamics," *IEEE Trans. Aut. Control*, vol. 34, pp. 405–412, April 1989.
6. I. Kanellakopoulos, P. Kokotovic, and R. Marino, "An extended direct scheme for robust adaptive nonlinear control," *Automatica*, to appear, March 1991.

7. A. De Luca and G. Ulivi, "Dynamic decoupling of voltage frequency controlled induction motors," in *8th Int. Conf. on Analysis and Optimization of Systems*, INRIA, Antibes, pp. 127–137, 1988.
8. A. De Luca and G. Ulivi, "Design of exact nonlinear controller for induction motors," *IEEE Trans. Aut. Control*, vol. AC-34, pp. 1304–1307, Dec. 1989.
9. Z. Krzeminski, "Nonlinear control of induction motors," in *10th IFAC World Congress*, Münich, pp. 349–354, 1987.
10. R. Marino, S. Peresada, and P. Valigi, "Adaptive partial feedback linearization of induction motors," in *Proc. 29th IEEE Conf. Dec. Control*, Honolulu, HI, Dec. 1990.
11. A. Fitzgerald, C. Kingsley Jr, and S. Umans, *Electric Machinery*, McGraw-Hill, 1983.
12. P. Krause, *Analysis of Electric Machinery*, McGraw-Hill, 1986.
13. W. Leonhard, *Control of Electrical Drives*, Berlin, Springer-Verlag, 1985.
14. P. Krause and C. Thomas, "Simulation of symmetrical induction machinery," *IEEE Trans. Power Apparatus and Systems*, vol. PAS-84, no. 11, pp. 1038–1053, 1965.
15. F. Blaschke, "Das Prinzip der Feldorientierung, die Grundlage für die Transvector Regelung von Asynchronmaschienen," *Siemens-Zeitschrift*, vol. 45, pp. 757–760, 1971.
16. F. Blaschke, "The principle of field orientation applied to the new transvector closed-loop control system for rotating field machines," *Siemens-Rev.*, vol. 39, pp. 217–220, 1972.
17. R. Gabriel and W. Leonhard, "Microprocessor control of induction motors," in *Proc. IEEE/IAS Int. Semiconductor Power Converter Conf.*, Orlando, FL, pp. 385–396, 1982.
18. W. Leonhard, "Microcomputer control of high dynamic performance ac-drives: a survey," *Automatica*, vol. 22, no. 1, pp. 1–19, Jan. 1986.
19. G. Verghese and S. Sanders, "Observers for flux estimation in induction machines," *IEEE Trans. Industr. Electronics*, vol. 35, pp. 85–94, Feb. 1988.
20. B. Jakubczyk and W. Respondek, "On linearization of control systems," *Bull. Acad. Pol. Sci., Ser. Sci. Math.*, vol. 28, no. 9-10, pp. 517–522, 1980.
21. R. Marino, "On the largest feedback linearizable subsystem," *Syst. Control Lett.*, vol. 6, pp. 345–351, Jan. 1986.
22. V. Popov, *Hyperstability of Control Systems*, Berlin, Springer-Verlag, 1973.
23. H. Sugimoto and S. Tamai, "Secondary resistance identification of an induction-motor applied model reference adaptive system and its characteristics," *IEEE Trans. Industr. Application*, vol. IA-23, pp. 296–303, March 1987.

Global Adaptive Observers and Output-Feedback Stabilization for a Class of Nonlinear Systems*

Riccardo Marino and Patrizio Tomei

Seconda Università di Roma, "Tor Vergata"
Dipartimento di Ingegneria Elettronica
Via O. Raimondo, 00173 Roma, ITALY.

Abstract. We address the problem of determining global adaptive observers for a class of single-output nonlinear systems which are linear with respect to an unknown constant parameter vector. Sufficient conditions are given for the construction of a global adaptive observer of an equivalent state, without persistency of excitation. Under additional geometric conditions the original (physical) state can be asymptotically observed as well. The results obtained are based on nonlinear changes of coordinates driven by auxiliary filters (filtered transformations). When only a single input is allowed and it is assumed to enter linearly in the state equations, we determine via geometric conditions a more restricted class of nonlinear single-input, single-output systems which can be globally stabilized by a dynamic (adaptive) observer-based output feedback control. Linear minimum-phase systems with unknown poles and zeroes, known sign of the high-frequency gain and known relative degree belong to such a class of systems. Systems which are not feedback linearizable may belong to such a class as well.

1 Introduction

In the last few years several papers have developed adaptive versions of nonlinear control algorithms designed via the so-called geometric approach to nonlinear systems (see [1] and [2] for an introduction to the field). Those nonlinear algorithms very often rely on the cancellation of nonlinear terms by feedback or output injection: when those terms contain unknown parameters the development of adaptive versions is very much needed as simple examples can show.

Under the assumption that unknown parameters enter linearly in the nonlinearities to be cancelled, adaptive versions of state feedback linearizing controls have been developed in [3-7], of input-output linearizing controls were reported in [8], and of nonlinear observers in [9,10]. Nonlinear adaptive stabilization using the "control Lyapunov function" technique was studied in [11,12]. More recently, an adaptive version of a static output-feedback linearizing control given in [26] has been proposed in [13] and [14], under the restriction of sector-type nonlinearities and of so-called output matching conditions.

* This work was supported in part by Ministero della Università e della Ricerca Scientifica e Tecnologica and in part by the U.S. Air Force Office of Scientific Research under Grant AFOSR 90-0011.

In this paper, we address the problem of global adaptive output feedback stabilization of a class of nonlinear systems, that is, the design of a dynamic output feedback control (compensator) such that for any initial condition of the closed-loop system and for any unknown value of the parameter vector, the closed-loop system has the property that the state of the system tends to zero and the state of the compensator is bounded.

The main result presented determines via geometric conditions a class of nonlinear single-input single-output systems, linear with respect to parameters, which are globally stabilizable by an (adaptive) dynamic observer-based output feedback control for every unknown constant value of the parameter vector. Those systems are characterized by a global, asymptotically stable, linear zero dynamics and are such that global observers with linear error dynamics [19] can be designed. The only restriction imposed on the nonlinearities is their dependence, in suitable state coordinates, on the output only. No sector-type restrictions or output matching conditions [13,14] are required. Systems which are not feedback linearizable may be included in such a class. Linear minimum-phase systems with known relative degree, known sign of the high-frequency gain, known upper bound on the number of poles but unknown poles and zeroes are included in the class of nonlinear systems considered.

We make use of nonlinear adaptive observers developed in [10,15] and of filtered transformations into adaptive observer forms. In the case of relative degree greater than one, we use techniques recently developed in [16-18] for the construction of Lyapunov functions of cascade systems and, following [5], we employ several "parameter estimators" in the design of nonlinear adaptive controls.

The paper is organized as follows. In Section 2 some preliminary results on adaptive observers and on the adaptive observer form are recalled. In Section 3 we recall from [15] the concept of filtered transformation and show that a class of nonlinear systems can always be globally transformed into an adaptive observer form by a filtered transformation so that global adaptive observers can be designed. This generalizes a well-known result on adaptive observers for linear systems [21]. Section 4 addresses the problem of observing the original "physical" state. In Section 5 we further restrict the class of nonlinear systems considered. In Section 6 we show how to design an adaptive output-feedback stabilizing control, in the relative-degree-one case, by using filtered transformations. In Section 7 we show how to design adaptive output-feedback stabilizing controls for the class of nonlinear systems determined in Section 5, when the relative degree is greater than one. The results presented in Sections 6 and 7 generalize well-established results for linear minimum-phase systems [21].

2 Adaptive Observers: Basic Results

We consider nonlinear single-output systems

$$
\begin{aligned}
\dot{x} &= f(x) + q_0(x,u) + \sum_{i=1}^{p} \theta_i q_i(x,u) \\
&\triangleq f(x) + q_0(x,u) + Q(x,u)\theta, \quad x \in \mathbb{R}^n, u \in \mathbb{R}^m, \theta \in \mathbb{R}^p \qquad (\Sigma) \\
y &= h(x), \quad y \in \mathbb{R},
\end{aligned}
$$

with $q_i : \mathbb{R}^n \times \mathbb{R}^m \rightarrow \mathbb{R}^n, 0 \leq i \leq p$, $f : \mathbb{R}^n \rightarrow \mathbb{R}^n$, $h : \mathbb{R}^n \rightarrow \mathbb{R}$ smooth functions, $h(x_0) = 0$, $q_0(x,0) = 0$ for every $x \in \mathbb{R}^n$; x is the state, $u(t) : \mathbb{R}^+ \rightarrow \mathbb{R}^m$ is the control, θ is the parameter vector which is assumed to be constant, and y is the scalar output. We assume that the system (Σ) with $u = 0$ and $\theta = 0$:

$$
\begin{aligned}
\dot{x} &= f(x) \\
y &= h(x),
\end{aligned}
$$

satisfies the condition

$$
\operatorname{rank}\{d(L_f^j h(x)) : 0 \leq j \leq n-1\} = n, \quad \forall x \in \mathbb{R}^n,
$$

which implies local observability for every x [2, p. 95].

The assumption of linear dependence on the parameter vector θ is certainly a restrictive one. We further require that each system (Σ) be transformable by a parameter independent, global state-space change of coordinates in $\mathbb{R}^n$,

$$
\zeta = T(x), \quad T(x_0) = 0, \tag{2.1}
$$

into the system

$$
\begin{aligned}
\dot{\zeta} &= A_c\zeta + \psi_0(y,u) + \sum_{i=1}^{p} \theta_i \psi_i(y,u) \\
&\triangleq A_c\zeta + \psi_0(y,u) + \Psi(y,u)\theta \qquad (S) \\
y &= C_c\zeta,
\end{aligned}
$$

with

$$
A_c = \begin{bmatrix} 0\,1\,0 \cdots 0 \\ 0\,0\,1 \cdots 0 \\ \vdots\;\vdots\;\vdots \ddots \vdots \\ 0\,0\,0 \cdots 1 \\ 0\,0\,0 \cdots 0 \end{bmatrix}
$$

$$
C_c = \begin{bmatrix} 1\,0\,0 \cdots 0 \end{bmatrix},
$$

and $\psi_i : R \times \mathbb{R}^m \rightarrow \mathbb{R}^n$ smooth functions for $i = 0, \ldots, p$. There are two reasons to restrict ourselves to such a class of systems. The first one is that for each known parameter vector θ, asymptotic observers are available for system (S) and therefore for (Σ) [19]. The second reason is that necessary and sufficient conditions which characterize those systems (Σ) which are locally transformable into (S) by (2.1) are known (see [10]): they are an immediate consequence of the main result in [19]. Following [20] and [25] the global version of such a result can be stated as follows.

Theorem 2.1 *There exists a parameter-independent global state-space diffeomorphism $\zeta = T(x)$, with $T(x_0) = 0$, transforming the system (Σ) into (S) if and only if*

(i) $[\mathrm{ad}_f^i r, \mathrm{ad}_f^j r] = 0, \qquad 0 \le i, j \le n-1,$
(ii) $[q_i(u), \mathrm{ad}_f^j r] = 0, \qquad 0 \le i \le p,\ 0 \le j \le n-2, \quad \forall u \in \mathbb{R}^m,$
(iii) the vector fields f and r are complete,

where r is the vector field satisfying

$$< \begin{bmatrix} dh \\ \vdots \\ d(L_f^{n-1}h) \end{bmatrix}, r > = \begin{bmatrix} 0 \\ \vdots \\ 1 \end{bmatrix}.$$

Proof. Conditions (i) are shown in [19] to be necessary and sufficient for (Σ) with $u = 0$ and $\theta = 0$ to be transformable via a local diffeomorphism (2.10) in U_{x_0} into the system

$$\begin{aligned} \dot{\zeta} &= A_c\zeta + \psi_0(y) \\ y &= C_c\zeta\,. \end{aligned}$$

The needed local coordinates $\zeta = T(x)$ are defined, by virtue of assumption (i), as those in which

$$ad_f^i r = (-1)^i \frac{\partial}{\partial \zeta_{n-i}}\,, \qquad 0 \le i \le n-1\,,$$

that is, the vector fields $ad_f^i r$, $0 \le i \le n-1$, are simultaneously rectified. Consequently, conditions (ii) guarantee that the vector fields $q_i(x, u)$ depend, in the ζ- coordinates, on the output y only. Condition (iii) is necessary and sufficient, according to [20], for the above change of coordinates to be a global one. When condition (iii) fails we only have a local change of coordinates. □

For every known parameter vector θ, the system

$$\begin{aligned} \dot{\hat{\zeta}} &= A_c\hat{\zeta} + \psi_0(y, u) + \sum_{i=1}^{p} \theta_i\psi_i(y, u) + K(y - C_c\hat{\zeta}) \\ \hat{x} &= T^{-1}(\hat{\zeta}) \end{aligned} \tag{2.2}$$

is, with a suitable choice of K, an asymptotic observer (provided that the state $x(t)$ is bounded) with linear error dynamics ($e = \zeta - \hat{\zeta}$) (see [19]):

$$\dot{e} = (A_c - KC_c)e = \begin{bmatrix} -k_1 & 1\,0 \cdots 0 \\ -k_2 & 0\,1 \cdots 0 \\ \vdots & \vdots\,\vdots\,\ddots\,\vdots \\ -k_{n-1} & 0\,0 \cdots 1 \\ -k_n & 0\,0 \cdots 0 \end{bmatrix} e \,.$$

The eigenvalues of $A_c - KC_c$ can be arbitrarily placed by choosing the vector K, since they coincide with the zeroes of the polynomial $s^n + k_1 s^{n-1} + \cdots + k_n$.

In order to obtain a linear error dynamics the parameter vector θ has to be exactly known. If its estimate $\hat{\theta}$ does not coincide with θ, the following error dynamics results ($e_\theta = \theta - \hat{\theta}$)

$$\dot{e} = (A_c - KC_c)e + \Psi(y,u)e_\theta \,,$$

that is, we are not guaranteed that $\lim_{t\to\infty} \|e(t)\| = 0$. This situation calls for an adaptive version of the observer (2.2).

In the literature there are available sufficient conditions of different nature which guarantee the existence of adaptive asymptotic observers for system (Σ). In [9] an adaptive observer, driven by a $p(n-1)$ dimensional auxiliary filter, is shown to converge under persistency of excitation for any system (Σ) satisfying conditions (i)–(iii). This is a general result: the only drawback is that persistency of excitation cannot be a priori checked. In [10] further structural geometric conditions, which are a priori checkable, are imposed on the class of systems (Σ) and guarantee the construction of an adaptive observer without the need of auxiliary filters and without any persistency of excitation. However, only for a very special subset of systems (Σ), which is identified by necessary and sufficient geometric conditions, can such a simple adaptive observer be used. In [15] it is shown that conditions (i)–(iii) are sufficient for the existence of an adaptive asymptotic observer for an equivalent state of system (Σ). We will now state some basic definitions and results which are needed in order to recall the main results on adaptive observers given in [15].

Definition 2.1 *A global adaptive observer for system (Σ) is a finite dimensional system*

$$\begin{aligned} \dot{w} &= \alpha(w, y(t), u(t)), \quad & w(0) = w_0, w \in \mathbb{R}^r \\ \hat{x} &= \gamma(w, y(t), u(t)), \quad & \hat{x} \in \mathbb{R}^n, \end{aligned}$$

driven by the inputs $u(t), y(t)$, such that

(i) $\|x(t) - \hat{x}(t)\|$ *is bounded,* $\forall t \geq 0, \forall w_0 \in \mathbb{R}^r$
(ii) $\lim_{t\to\infty} \|x(t) - \hat{x}(t)\| = 0, \quad \forall w_0 \in \mathbb{R}^r, \forall x(0) \in \mathbb{R}^n$.

Definition 2.2 *The constant integer* ν *such that*

$$L_{q_i} L^j_{(f+q_0)} h = 0\,, \forall i, 1 \leq i \leq p$$

$$\forall j, 0 \leq j \leq \nu - 2$$

$$\forall x \in \mathbb{R}^n, \quad \forall u \in \mathbb{R}^m$$

$$L_{q_i} L^{\nu-1}_{(f+q_0)} h \neq 0, \text{ for some } i, 1 \leq i \leq p$$

$$\text{for some } x \in \mathbb{R}^n\,,$$

is called the relative degree between the output y *and the parameter* θ *for the system* (Σ).

Definition 2.3 *A vector* $b = [b_1, \ldots, b_n]^{\mathrm{T}}$ *is said to be Hurwitz of degree* ρ *if the associated polynomial*

$$b_1 s^{n-1} + b_2 s^{n-2} + \cdots + b_n$$

is of degree ρ $(b_1 = \cdots = b_{\rho-1} = 0, b_\rho \neq 0)$ *and Hurwitz, i.e., all its zeroes have real part less than zero.*

Lemma (Meyer-Kalman-Yakubovich). *Given an* $n \times n$ *Hurwitz matrix* A, *two* $n \times 1$ *vectors* b *and* c *and an* $n \times n$ *symmetric positive definite matrix* Q, *if*

$$Re[c^{\mathrm{T}}(j\omega I - A)^{-1} b] > 0, \quad \forall \omega \in \mathbb{R}\,, \tag{2.3}$$

then there exist a positive scalar ϵ, *a vector* l *and a symmetric positive definite matrix* P *such that*

$$\begin{aligned} A^{\mathrm{T}} P + PA &= -ll^{\mathrm{T}} - \epsilon Q \\ Pb &= c\,. \end{aligned} \tag{2.4}$$

Proof. See for instance [21, p. 66].

We now recall a basic result from the theory of linear adaptive observers [21, Chap. 4] which shows how to construct adaptive observers for a class of systems said to be “in adaptive observer form”.

Theorem 2.2 *Consider the following system in adaptive observer form*

$$\begin{aligned} \dot{z} &= A_c z + \psi_0(y, u) + b\beta^{\mathrm{T}}(t)\theta \\ y &= C_c z\,, \end{aligned} \tag{2.5}$$

with $z \in \mathbb{R}^n$, $y \in \mathbb{R}$, $\beta : \mathbb{R}^+ \to \mathbb{R}^p$ *a known and piecewise continuous vector and* $b = [b_1, \ldots, b_n]^{\mathrm{T}}$. *If* $\beta(t)$ *has bounded components and if* b *is Hurwitz with degree one, then the following system is an adaptive observer for (2.5),*

$$\dot{\hat{z}} = (A_c - KC_c)\hat{z} + \psi_0(y, u) + b\beta^{\mathrm{T}}(t)\hat{\theta} + Ky \tag{2.6}$$

$$\dot{\hat{\theta}} = \Gamma\beta(t)(y - C_c\hat{z})\,\mathrm{sgn}(b_1)\,, \tag{2.7}$$

where $\hat{z} \in \mathbb{R}^n$, $\hat{\theta} \in \mathbb{R}^p$, Γ is a symmetric positive definite matrix and $K = [k_1, \dots, k_n]^{\mathrm{T}}$ is given by

$$K = \frac{1}{b_1}(A_c b + \lambda b)\,, \tag{2.8}$$

with λ an arbitrary positive scalar.

Proof. The error system is

$$\begin{aligned} \dot{e} &= Ae + b\,\mathrm{sgn}(b_1)(\beta^{\mathrm{T}}(t)\,\mathrm{sgn}(b_1))e_\theta \\ \dot{e}_\theta &= -\Gamma(\beta(t)\,\mathrm{sgn}(b_1))C_c e\,, \end{aligned}$$

where $A = A_c - KC_c$, $e = z - \hat{z}$ and $e_\theta = \theta - \hat{\theta}$. Since by assumption $b_1 \neq 0$, by virtue of (2.8) we can write

$$|b_1|(s^n + k_1 s^{n-1} + \cdots + k_n) = (s + \lambda)(b_1 s^{n-1} + \cdots + b_n)\,\mathrm{sgn}(b_1)\,,$$

which implies that the transfer function

$$C_c(sI - A)^{-1} b\,\mathrm{sgn}(b_1) = \frac{1}{s + \lambda} \tag{2.9}$$

satisfies condition (2.3). There are $n - 1$ pole-zero cancellations in (2.9). The matrix A is Hurwitz, since $\lambda > 0$ and by assumption the vector b is Hurwitz. The thesis is proved once the following claim is proved.

Claim: Consider the system

$$\begin{aligned} \dot{e} &= Ae + b\beta^{\mathrm{T}}(t)e_\theta\,, \quad e \in \mathbb{R}^n, e_\theta \in \mathbb{R}^p \\ \dot{e}_\theta &= -\Gamma\beta(t)c^{\mathrm{T}}e\,, \end{aligned}$$

with A a Hurwitz matrix and Γ a $p \times p$ symmetric positive definite matrix. If the transfer function $c^{\mathrm{T}}(sI - A)^{-1}b$ satisfies condition (2.3) and the components of $\beta(t)$ are bounded, piecewise continuous functions, then $e = 0, e_\theta = 0$ is a uniformly stable equilibrium point, the components of $e(t)$ and $e_\theta(t)$ are bounded, and

$$\lim_{t \to \infty} \|e(t)\| = 0\,. \tag{2.10}$$

Proof of the Claim (it is a standard result, see for instance [21, p. 142]). Since $c^{\mathrm{T}}(sI - A)^{-1}b$ satisfies the conditions of the lemma, there exist two symmetric positive definite matrices P and Q such that

$$\begin{aligned} A^{\mathrm{T}}P + PA &= -Q \\ Pb &= c\,. \end{aligned}$$

Choose as candidate Lyapunov function

$$V(e, e_\theta, t) = e^{\mathrm{T}}Pe + e_\theta^{\mathrm{T}}\Gamma^{-1}e_\theta\,. \tag{2.11}$$

The time derivative of V is

$$\dot{V}(e, e_\theta, t) = -e^{\mathrm{T}} Q e \,. \tag{2.12}$$

It follows that $e = 0, e_\theta = 0$ is uniformly stable [22, p. 199], and $e(t)$, $e_\theta(t)$ are bounded. Since, by hypothesis, $\beta(t)$ is bounded, $\dot{e}$ is also bounded and, as a consequence, $e(t)$ is uniformly continuous. Equations (2.11) and (2.12) imply that $e(t) \in L^2$. By Barbalat's Lemma [23, p. 210], (2.10) follows. □

In the next section we shall discuss the transformation of (Σ) into adaptive observer form.

3 Filtered transformations into adaptive observer form

The system (S) is not in general in adaptive observer form: (S) coincides with (2.5) only when $\psi_i(y,u) = b\beta_i(y,u), 1 \le i \le p$ where $\beta_i(y,u)$: $R \times \mathrm{IR}^m \to R$ and b is a constant vector. The goal of this section is to transform (Σ) into an adaptive observer form. We first recall the main result in [10] which gives necessary and sufficient conditions for system (Σ) to be transformed into an adaptive observer form by a local state-space change of coordinates which is independent of θ.

Theorem 3.1 [10] *There exists a global change of coordinates $z = T(x)$, with $T(x_0) = 0$, transforming system (Σ) into the adaptive observer form*

$$\begin{aligned} \dot{z} &= A_c z + \psi_0(y,u) + b\beta^{\mathrm{T}}(y,u)\theta \\ y &= C_c z \,, \end{aligned} \tag{3.1}$$

if and only if conditions (i), (ii) and (iii) of Theorem 2.1 are satisfied and, in addition,

$$\textit{(iv)}\ q_i(x,u) = \beta_i(h(x),u)\sum_{j=0}^{n-1} b_{n-j}(-1)^j \mathrm{ad}_f^j r, \quad \forall x \in \mathrm{IR}^n, \forall u \in \mathrm{IR}^m, 1 \le i \le p.$$

Remark 3.1. Condition (iv) is a simplification of condition (iv) stated in [10, Theorem 3.1]; a slightly different notation is used here. The above result is the global version of the one presented in [10]. □

Remark 3.2. If Theorem 3.1 applies and the assumptions of Theorem 2.2 are met, namely the vector b is Hurwitz of degree one and the vector $\beta(y(t),u(t))$ has bounded components (for instance, it suffices that the system (Σ) be bounded-input-bounded-state (BIBS) stable for every $\theta \in \mathrm{IR}^p$ and that $u(t)$ be bounded), then the system (2.6)–(2.7) with output $\hat{x} = T^{-1}(\hat{z})$ is a global adaptive observer for (Σ). □

Suppose now that condition (iv) of Theorem 3.1 either fails or it holds for a constant vector b not satisfying the assumptions of Theorem 2.2; since the conditions of Theorem 3.1 are necessary and sufficient, in order to apply Theorem 2.2

we need to determine a class of transformations taking (Σ) into an adaptive observer form (with a Hurwitz vector b of degree one) which is more general than a parameter independent state-space change of coordinates. We now recall from [15] the definition of filtered transformations which will be shown to be able to transform system (Σ) satisfying assumptions (i), (ii) and (iii) into an adaptive observer form with a desired b.

Definition 3.1 *A filtered transformation is defined as a time-varying global state diffeomorphism, which may depend on the unknown θ:*

$$z = S(x, \xi(t), \theta), \quad z \in \mathbb{R}^n, x \in \mathbb{R}^n, \xi \in \mathbb{R}^s, \theta \in \mathbb{R}^p,$$

such that there exists a C^∞ function $S^{-1}(z, \xi(t), \theta)$ satisfying

$$z = S(S^{-1}(z, \xi(t), \theta), \xi(t), \theta), \qquad \forall \xi \in \mathbb{R}^s, \forall \theta \in \mathbb{R}^p,$$

where $\xi(t)$ is a signal generated by the auxiliary linear asymptotically stable filter (Λ is Hurwitz)

$$\dot{\xi} = \Lambda \xi + \delta(y(t), u(t)), \quad \xi(0) = 0,$$

driven by the inputs $y(t), u(t)$.

Lemma 3.2 [15] *If conditions (i)–(iii) of Theorem 2.1 are satisfied for the system (Σ), then (Σ) is globally transformable by a filtered transformation into an adaptive observer form (2.5), with $b = [1, b_2, \ldots, b_n]^{\mathrm{T}}$ the vector of coefficients of the Hurwitz polynomial*

$$s^{n-1} + b_2 s^{n-2} + \cdots + b_n = (s + \lambda_1) \cdots (s + \lambda_{n-1}),$$

where λ_i, $1 \le i \le n-1$, are arbitrary positive scalars.

Proof. If conditions (i)–(iii) of Theorem 2.1 hold, then Theorem 2.1 guarantees that (Σ) is transformable by a global change of coordinates $\zeta = T(x)$ into (S). Consider the $n \times n$ matrix $D = [d_1 \; d_2 \; \cdots \; d_n]$, where

$$\begin{aligned} d_n &= [0 \; 0 \; \cdots \; 1]^{\mathrm{T}} \\ d_j &= A_c d_{j+1} + \lambda_j d_{j+1}, \qquad 1 \le j \le n-1. \end{aligned} \tag{3.2}$$

By construction the vectors $d_1, \ldots, d_n$ are linearly independent: in fact the matrix

$$D = \begin{bmatrix} 1 & 0 & 0 & \cdots & 0 \\ d_{12} & 1 & 0 & \cdots & 0 \\ d_{13} & d_{23} & 1 & \cdots & 0 \\ \vdots & \vdots & \vdots & \ddots & \vdots \\ d_{1n} & d_{2n} & d_{3n} & \cdots & 1 \end{bmatrix} \tag{3.3}$$

is nonsingular, and its entries d_{ij} satisfy the relations

$$\begin{aligned} s^{n-1} + d_{12}s^{n-2} + \cdots + d_{1n} &= (s+\lambda_1)\cdots(s+\lambda_{n-1}) \\ s^{n-2} + d_{23}s^{n-3} + \cdots + d_{2n} &= (s+\lambda_2)\cdots(s+\lambda_{n-1}) \\ &\vdots \\ s + d_{n-1,n} &= s + \lambda_{n-1}\,. \end{aligned}$$

Each vector field ψ_i in (S), $1 \le i \le p$, is uniquely expressed as $\psi_i = \sum_{j=1}^{n} \phi_{ij} d_j$, which, substituted in (S), gives

$$\begin{aligned} \dot{\zeta} &= A_c\zeta + \psi_0(y,u) + \sum_{j=1}^{n} d_j \sum_{i=1}^{p} \phi_{ij}\theta_i \\ &= A_c\zeta + \psi_0(y,u) + \sum_{j=1}^{n} d_j \phi_j^{\mathrm{T}}(y,u)\theta\,. \end{aligned} \tag{3.4}$$

where ϕ_j^{T}, $1 \le j \le n$, are the rows of the $n \times p$ matrix

$$\Phi(y,u) = [\phi_{ij}] = \begin{bmatrix} \phi_1^{\mathrm{T}} \\ \vdots \\ \phi_n^{\mathrm{T}} \end{bmatrix}. \tag{3.5}$$

Now consider the following linear, time-invariant, asymptotically stable system in $\mathbb{R}^{p(n-1)}$

$$\begin{aligned} &\dot{\xi}_i = -\lambda_i\xi_i + \xi_{i+1} + \phi_{i+1}(y,u), \quad \xi_i(0) = 0, \quad \xi_i \in \mathbb{R}^p, \quad 1 \le i \le n-1 \\ &\xi_n = 0, \qquad \xi_n \in \mathbb{R}^p. \end{aligned} \tag{3.6}$$

Define the time-varying affine change of coordinates that depends on the unknown parameter vector θ

$$z = \zeta - \sum_{j=2}^{n} d_j \xi_{j-1}^{\mathrm{T}}\theta\,, \tag{3.7}$$

which admits the global inverse

$$\zeta = z + \sum_{j=2}^{n} d_j \xi_{j-1}^{\mathrm{T}}\theta\,.$$

Differentiating (3.7) with respect to time and using (3.6) and (3.4), we obtain

$$\dot{z} = A_c z + \psi_0(y,u) + \sum_{j=1}^{n} d_j\phi_j^{\mathrm{T}}(y,u)\theta + A_c\sum_{j=2}^{n} d_j\xi_{j-1}^{\mathrm{T}}\theta - \sum_{j=2}^{n} d_j\dot{\xi}_{j-1}^{\mathrm{T}}\theta$$

$$= A_c z + \psi_0(y,u) + \sum_{j=1}^{n} d_j \phi_j^{\mathrm{T}}(y,u)\theta + A_c \sum_{j=2}^{n} d_j \xi_{j-1}^{\mathrm{T}}\theta$$

$$+ \sum_{j=2}^{n} d_j(\lambda_{j-1}\xi_{j-1}^{\mathrm{T}} - \phi_j^{\mathrm{T}}(y,u))\theta - \sum_{j=2}^{n-1} d_j \xi_j^{\mathrm{T}}\theta$$

$$= A_c z + \psi_0(y,u) + \sum_{j=2}^{n}(A_c d_j + \lambda_{j-1}d_j)\xi_{j-1}^{\mathrm{T}}\theta - \sum_{j=3}^{n} d_{j-1}\xi_{j-1}^{\mathrm{T}}\theta + d_1\phi_1^{\mathrm{T}}\theta .$$

Taking (3.2) into account, we have

$$\dot{z} = A_c z + \psi_0(y,u) + d_1(\phi_1^{\mathrm{T}}(y,u) + \xi_1^{\mathrm{T}})\theta . \tag{3.8}$$

In addition, since, according to (3.3), $C_c d_j = 0$, $2 \le j \le n$, from (3.7) we have

$$y = C_c\zeta = C_c z + C_c \sum_{j=2}^{n} d_j \xi_{j-1}^{\mathrm{T}}\theta = C_c z . \tag{3.9}$$

The system (3.8)–(3.9) is in the adaptive observer form (2.5) with $b = d_1$ and $\beta(t) = \phi_1(y(t),u(t)) + \xi_1(t)$. The filtered transformation is therefore given by the composition of $\zeta = T(x)$ with (3.7), where the dynamics of $\xi_{j-1}(t)$, $2 \le j \le n$, are given by (3.6). □

Remark 3.3. Assume there exists a choice for λ_i, $1 \le i \le n-1$, such that in (3.4) $\phi_i(y,u) = 0$, $i^* \le i \le n$ (a necessary condition for such a choice to exist is that $\mathrm{rank}\{\Psi(y,u)\} = i^* - 1$, $\forall y,u$). Then, the system (3.6) becomes

$$\begin{aligned} \dot{\xi}_i &= -\lambda_i\xi_i + \xi_{i+1} + \phi_{i+1}(y,u), \qquad 1 \le i \le i^* - 2 \\ \xi_{i^*-1} &= 0 . \end{aligned}$$

Indeed, from (3.6) we obtain $\xi_i(t) = 0$, $i^* - 1 \le i \le n-1$, and therefore the auxiliary filter is of reduced dimension $(i^* - 2)p$. Consequently, the transformation (3.7) becomes

$$z = \zeta - \sum_{j=2}^{i^*-1} d_j \xi_{j-1}^{\mathrm{T}}\theta . \tag{3.10}$$

□

We are now ready to state the main result of this section.

Theorem 3.3 [15] *If the system (Σ) is bounded-input-bounded-output (BIBO) stable and conditions (i)–(iii) of Theorem 2.1 are satisfied, then for every bounded input $u(t)$ there exists a global adaptive observer for a system whose state is equivalent to the physical state x of (Σ).*

Proof. Lemma 3.2 applies since conditions (i)–(iii) of Theorem 2.1 are satisfied and, consequently, there exists a filtered transformation (3.6)–(3.7) which takes (Σ) into (3.8)–(3.9) which is in adaptive observer form. Since (Σ) is BIBO stable, for every bounded $u(t)$ the functions $\phi_i(y,u)$, $1 \leq i \leq n$, are bounded, ϕ_i being continuous, and ξ_1 is bounded since (3.6) is exponentially stable. It follows that $\phi_1(y,u)+\xi_1$ is bounded and therefore Theorem 2.2 applies to the system (3.8)–(3.9), guaranteeing the construction of an adaptive observer for the state z of (3.8)–(3.9) which is, according to (2.1) and (3.7), an equivalent state for (Σ): in fact, z is related to x by a time-varying nonlinear change of coordinates depending on the unknown parameter vector θ. □

Remark 3.4. Let us examine how Theorem 3.3 specializes for single-output linear systems with m inputs

$$\dot{\zeta} = \begin{bmatrix} 0 & 1 & 0 & \cdots & 0 \\ 0 & 0 & 1 & \cdots & 0 \\ \vdots & \vdots & \vdots & \ddots & \vdots \\ 0 & 0 & 0 & \cdots & 1 \\ 0 & 0 & 0 & \cdots & 0 \end{bmatrix} \zeta - \begin{bmatrix} a_1 \\ a_2 \\ \vdots \\ a_{n-1} \\ a_n \end{bmatrix} y + \sum_{i=1}^{m} \begin{bmatrix} b_1^i \\ b_2^i \\ \vdots \\ b_{n-1}^i \\ b_n^i \end{bmatrix} u_i \qquad (L)$$

$$y = [1\ 0\ 0\ \cdots\ 0]\zeta\,,$$

which are the realizations in observer form of the transfer matrix

$$W(s) = [W_1(s)\ W_2(s)\ \cdots\ W_m(s)]\,,$$

with

$$W_i(s) = \frac{b_1^i s^{n-\rho} + \cdots + b_n^i}{s^n + a_1 s^{n-1} + \cdots + a_n}\,.$$

If we are given the transfer matrix $W(s)$ with unknown poles and zeroes, then

$$\theta = [a_1, \ldots, a_n, b_1^1, \ldots, b_n^1, \ldots, b_1^m, \ldots, b_n^m]^{\mathrm{T}}\,.$$

is the $(m+1)n$-dimensional unknown constant parameter vector and system (L) is of type (S). Under the assumption of BIBO stability Theorem 3.3 applies and provides a global adaptive observer for the system described by the transfer matrix $W(s)$ with unknown poles and zeroes. We then reobtain, as a corollary of Theorem 3.3, a well-known result of adaptive control theory for linear systems (see [21, Chap. 4] for an overview). Filtered transformations can be viewed as a tool to produce a nonminimal realization of the transfer matrix $W(s)$. As an interpretation we can observe that nonminimal realizations of $W(s)$ are needed in order to achieve an adaptive observer form which, as discussed, may be nonminimal. Theorem 3.1 gives sufficient conditions under which filtered transformations are not needed and therefore there is no need for nonminimal realizations. □

Remark 3.5. Theorem 3.3 does not guarantee an estimate for the physical state x, since z is related to x through the unknown parameter θ. If, in addition to the hypotheses of Theorem 3.3, we assume the following persistency of excitation condition

$$\infty > c_1 I \geq \int_t^{t+T} (\phi_1 + \xi_1)(\phi_1 + \xi_1)^{\mathrm{T}}\, d\tau \geq c_2 I > 0\,, \tag{3.11}$$

where c_1, c_2 and T are positive constants, and ϕ_1, ξ_1 have bounded time-derivative for all $t \geq 0$, except possibly for a countable number of points of fixed minimum separation, then the parameter estimates given by the adaptive observer of Theorem 2.2 converge to the true parameter values. This result can be proved by a direct application of Theorem 2.3 of [24, p. 44]. □

Remark 3.6. It remains open the problem of determining sufficient conditions under which the system $\dot{x} = f(x, u, \theta)$, which is more general than the system (Σ), since the parameters may enter in a nonlinear way, can be transformed by filtered transformations (or more general ones) into adaptive observer form. Any result in this direction would enlarge the applicability of the methods discussed in this section. □

4 Adaptive observers for the original coordinates

In this section we provide sufficient conditions for the existence of adaptive observers which asymptotically track the physical coordinates of system (Σ), without requiring persistency of excitation.

Theorem 4.1 [15] *Assume that the system (Σ) is BIBS stable and satisfies the conditions of Theorem 3.1, with $b = [b_1, \ldots, b_n]$ Hurwitz. Then, for every bounded input $u(t)$ there exists a global adaptive observer for (Σ).*

Proof. Since the conditions of Theorem 3.1 (and of Theorem 2.1) hold, the system (Σ) is globally transformable by a parameter-independent change of coordinates $\zeta = T(x)$ into the adaptive observer form (3.1). Let ν be the relative degree of (Σ). The system (3.1) can be rewritten in the form (3.4)

$$\begin{aligned} \dot{\zeta} &= A_c \zeta + \psi_0(y, u) + d_\nu \phi_\nu^{\mathrm{T}} \theta \\ y &= C_c \zeta\,, \end{aligned} \tag{4.1}$$

with

$$d_\nu = \frac{1}{b_\nu} b = [0, \ldots, 1, b_{\nu+1}, \ldots, b_n]^{\mathrm{T}}, \quad \phi_\nu(y, u) = b_\nu \beta(y, u)\,.$$

Define the constant vectors

$$d_{j-1} = A_c d_j + \lambda_{j-1} d_j\,, \qquad 2 \leq j \leq \nu\,, \tag{4.2}$$

where λ_j, $1 \leq j \leq \nu - 1$, are arbitrary positive scalars. As in the proof of Lemma 3.2, we introduce the auxiliary linear filter

$$\begin{aligned} \dot{\xi}_i &= -\lambda_i \xi_i + \xi_{i+1}, \qquad & \xi_i \in \mathbb{R}^p, 1 \leq i \leq \nu - 2, \xi_i(0) = 0 \\ \dot{\xi}_{\nu-1} &= -\lambda_{\nu-1}\xi_{\nu-1} + \phi_\nu(y,u), \qquad & \xi_{\nu-1} \in \mathbb{R}^p, \xi_{\nu-1}(0) = 0 \end{aligned} \tag{4.3}$$

(note that no filter is needed when $\nu = 1$) and define the time-varying change of coordinates

$$z = \zeta - \sum_{j=2}^{\nu} d_j \xi_{j-1}^{\mathrm{T}} \theta \tag{4.4}$$

which transforms the system (4.1) into the adaptive observer form

$$\begin{aligned} \dot{z} &= A_c z + \psi_0(y,u) + d_1 \left(\phi_1^{\mathrm{T}}(y,u) + \xi_1^{\mathrm{T}}\right) \theta \\ y &= C_c z \,, \end{aligned} \tag{4.5}$$

with $\phi_1(y,u) = 0$ if $\nu > 1$ ($\phi_1(y,u) = \beta(y,u)$ if $\nu = 1$) and d_1 satisfying

$$[s^{n-1}\ s^{n-2}\ \cdots\ 1]d_1 = (s+\lambda_1)\cdots(s+\lambda_{\nu-1})(s^{n-\nu}+b_{\nu+1}s^{n-\nu-1}+\cdots+b_n)\,. \tag{4.6}$$

Since λ_i, $1 \leq i \leq \nu - 1$ are positive scalars and by hypothesis the vector b is Hurwitz, the left-hand side of (4.6) is a Hurwitz polynomial of degree $n - 1$. Since, in addition, (Σ) is assumed to be BIBS stable and therefore BIBO stable for every $\theta \in \mathbb{R}^p$, $\phi_1 + \xi_1$ in (4.5) is bounded for every $u(t)$ bounded and Theorem 2.2 applies to (4.5). We consider the adaptive observer

$$\dot{\hat{z}} = (A_c - KC_c)\hat{z} + \psi_0(y,u) + d_1(\phi_1^{\mathrm{T}}(y,u) + \xi_1^{\mathrm{T}})\hat{\theta} + Ky \tag{4.7}$$

$$\dot{\hat{\theta}} = \Gamma \xi_1 (y - C_c \hat{z})\,, \tag{4.8}$$

with $K = (A_c + \lambda I)d_1$, λ being an arbitrary positive scalar. The error system is given by

$$\dot{e} = (A_c - KC_c)e + d_1(\phi_1^{\mathrm{T}}(y,u) + \xi_1^{\mathrm{T}})e_\theta \tag{4.9}$$

$$\dot{e}_\theta = -\Gamma(\phi_1^{\mathrm{T}}(y,u) + \xi_1^{\mathrm{T}})C_c e\,. \tag{4.10}$$

For every bounded input $u(t)$, Theorem 2.2 guarantees that $e(t)$ and $e_\theta(t)$ are bounded, and

$$\lim_{t\to\infty} \|e(t)\| = 0\,. \tag{4.11}$$

Now, using the transformation (4.4), define the estimate of the state x of system (Σ) as

$$\hat{x} = T^{-1}\left(\hat{z} + \sum_{j=2}^{\nu} d_j \xi_{j-1}^{\mathrm{T}} \hat{\theta}\right), \tag{4.12}$$

so that (4.12) is the output of the adaptive observer (4.3), (4.7), (4.8). The estimation error is given by

$$x - \hat{x} = T^{-1}(\zeta) - T^{-1}\left(\zeta - e - \sum_{j=2}^{\nu} d_j \xi_{j-1}^{\mathrm{T}} e_\theta\right). \tag{4.13}$$

If $\nu = 1$, then no filter is needed and the transformation (4.4) coincides with the identity transformation. Consequently, (4.11) and the hypothesis of bounded state $x(t)$ imply

$$\lim_{t\to\infty} \|x(t) - \hat{x}(t)\| = 0\,. \tag{4.14}$$

If $\nu > 1$, differentiating (4.9) with respect to time and recalling that $\phi_1(y,u) = 0$, we obtain

$$\ddot{e} = (A_c - KC_c)\dot{e} + d_1\left(\dot{\xi}_1^{\mathrm{T}} e_\theta + \xi_1^{\mathrm{T}} \dot{e}_\theta\right).$$

By (4.3), (4.9) and (4.10), the signals $\dot{\xi}_1$, $\dot{e}$ and $\dot{e}_\theta$ are bounded. It follows that $\ddot{e}$ is also bounded which, in turn, implies that $\dot{e}$ is uniformly continuous. Besides,

$$\lim_{t\to\infty} \int_0^t \dot{e}_i(\tau)d\tau = \lim_{t\to\infty} e_i(t) - e_i(0) = -e_i(0) < \infty\,, \qquad 1 \le i \le n\,.$$

Applying Barbalat's Lemma [23, p. 210], we have

$$\lim_{t\to\infty} \|\dot{e}(t)\| = 0\,. \tag{4.15}$$

From (4.9), (4.11) and (4.15), we obtain

$$\lim_{t\to\infty} \xi_1^{\mathrm{T}}(t) e_\theta(t) = 0\,. \tag{4.16}$$

The time derivatives of $\xi_1^{\mathrm{T}} e_\theta$ are

$$\begin{aligned} \frac{d}{dt}(\xi_1^{\mathrm{T}} e_\theta) &= \dot{\xi}_1^{\mathrm{T}} e_\theta + \xi_1^{\mathrm{T}} \dot{e}_\theta \\ &= (-\lambda_1\xi_1 + \xi_2)^{\mathrm{T}} e_\theta + \xi_1^{\mathrm{T}}(-G\xi_1 C_c e) \\ \frac{d^2}{dt^2}(\xi_1^{\mathrm{T}} e_\theta) &= \ddot{\xi}_1^{\mathrm{T}} e_\theta + 2\dot{\xi}_1^{\mathrm{T}} \dot{e}_\theta + \xi_1^{\mathrm{T}} \ddot{e}_\theta \\ &= (-\lambda_1\dot{\xi}_1 + \dot{\xi}_2)^{\mathrm{T}} e_\theta + 2\dot{\xi}_1^{\mathrm{T}} \dot{e}_\theta + \xi_1^{\mathrm{T}}(-G\dot{\xi}_1 C_c e - G\xi_1 C_c \dot{e})\,. \end{aligned} \tag{4.17}$$

Since $\dot{\xi}_1$, $\dot{\xi}_2$, $\dot{e}$ and $\dot{e}_\theta$ are bounded, also $\frac{d^2}{dt^2}(\xi_1^{\mathrm{T}} e_\theta)$ is bounded and, as a consequence, $\frac{d}{dt}(\xi_1^{\mathrm{T}} e_\theta)$ is uniformly continuous. By (4.16) and using Barbalat's Lemma it follows that

$$\lim_{t\to\infty} \frac{d}{dt}(\xi_1^{\mathrm{T}} e_\theta) = 0\,, \tag{4.18}$$

which, in view of (4.11), (4.16), (4.17) and (4.18), implies

$$\lim_{t\to\infty} \xi_2^{\mathrm{T}}(t) e_\theta(t) = 0\,.$$

Iterating the above argument it is easy to show that

$$\lim_{t\to\infty} \xi_i^{\mathrm{T}}(t) e_\theta(t) = 0\,, \qquad 1 \le i \le \nu - 1\,. \tag{4.19}$$

From (4.13) and (4.19) we can see that (4.14) holds, for every value of ν. Therefore, the dynamic equations (4.3), (4.7), (4.8) and (4.12) represent a global adaptive observer for system (Σ), for every bounded input $u(t)$. □

Remark 4.1. Consider the system (Σ) and suppose that Theorem 4.1 does not apply. However, if Theorem 3.3 applies we can construct an adaptive observer that asymptotically estimates an equivalent state. Let ν be the relative degree of system (Σ). It is easy to see, by using the same arguments in the proof of Theorem 4.1, that the estimates obtained by the inverse transformation $\hat{x} = T^{-1}(\hat{z} + \sum_{j=2}^{n} d_j \xi_{j-1}^{\mathrm{T}} \hat{\theta})$ are such that

$$\lim_{t \to \infty} |x_i(t) - \hat{x}_i(t)| = 0\,, \qquad 1 \leq i \leq \nu\,, \tag{4.20}$$

where x_i is the i-th component of x. □

5 Adaptive output-feedback stabilization: preliminary results

In this section we restrict the class of systems (Σ) introduced in Section 2 by assuming that only a single input, which appears linearly in the state equations, is available:

$$\begin{aligned} \dot{x} &= f(x) + \left[g_0(x) + \sum_{i=1}^{p} \theta_i g_i(x) \right] u + \sum_{i=1}^{p} \theta_i q_i(x), \quad x \in \mathbb{R}^n, u \in \mathbb{R}, \theta \in \Omega \subset \mathbb{R}^p \\ y &= h(x), \quad y \in \mathbb{R}\,, \end{aligned} \tag{ΣR}$$

where x is the state vector, u is the control, $\theta = [\theta_1, \ldots, \theta_p]^{\mathrm{T}}$ is the (constant) parameter vector belonging to Ω, a closed subset of $\mathbb{R}^p$, $h : C^\infty(\mathbb{R}^n, R)$, $h(0) = 0$, is the output function, f, g_0, g_i, q_i, $1 \leq i \leq p$, are smooth vector fields on $\mathbb{R}^n$, with $f(0) = 0$, $q_i(0) = 0$, $1 \leq i \leq p$, and

$$g_0(x) + \sum_{i=1}^{p} \theta_i g_i(x) \triangleq g(x;\theta) \neq 0\,, \quad \forall x \in \mathbb{R}^n\,, \ \forall \theta \in \Omega\,.$$

From now on we shall assume without loss of generality that the origin is an equilibrium point.

We now further restrict the class of systems (ΣR) determined by Theorem 2.1.

Theorem 5.1 [18] *The system (ΣR) is transformable by a global state-space diffeomorphism (independent of θ)*

$$\zeta = T(x), \quad T(0) = 0\,, \tag{5.1}$$

into (let $\theta_0 = 1$ for convenience of notation)

$$\begin{aligned} \dot{\zeta} &= A_c \zeta + (b_0 + \sum_{i=1}^{p} \theta_i b_i) \sigma(y) u + \psi_0(y) + \sum_{i=1}^{p} \theta_i \psi_i(y) \\ &\triangleq A_c \zeta + b(\theta) \sigma(y) u + \sum_{i=0}^{p} \psi_i(y) \theta_i \\ y &= C_c \zeta\,, \end{aligned} \tag{SR}$$

with (A_c, b, C_c) in observer canonical form

$$A_c = \begin{bmatrix} 0\,1\,0 \cdots 0 \\ 0\,0\,1 \cdots 0 \\ \vdots\;\vdots\;\vdots \;\ddots\; \vdots \\ 0\,0\,0 \cdots 1 \\ 0\,0\,0 \cdots 0 \end{bmatrix}, \quad b(\theta) = \begin{bmatrix} b_1(\theta) \\ b_2(\theta) \\ \vdots \\ b_{n-1}(\theta) \\ b_n(\theta) \end{bmatrix}$$

$$C_c = [1\,0\,0 \cdots 0] \,,$$

if and only if

(i) $[\mathrm{ad}_f^i r, \mathrm{ad}_f^j r] = 0, \qquad 0 \le i, j \le n-1$
(ii) $[q_i, \mathrm{ad}_f^k r] = 0, \qquad 1 \le i \le p,\ 0 \le k \le n-2$
(iii) $[g_j, \mathrm{ad}_f^k r] = 0, \qquad 0 \le k \le n-2, \quad 0 \le j \le p$
(iv) $g_j = \sigma(y) \sum_{i=1}^{n} b_{ji} (-1)^{n-i} \mathrm{ad}_f^{n-i} r, \qquad 0 \le j \le p$
(v) *the vector fields f and r are complete,*

where r is the vector field satisfying

$$< \begin{bmatrix} dh \\ \vdots \\ d(L_f^{n-1} h) \end{bmatrix}, r >= \begin{bmatrix} 0 \\ \vdots \\ 1 \end{bmatrix}.$$

Proof. Conditions (i) are shown in [19] to be necessary and sufficient for the existence of a local state-space change of coordinates

$$\zeta_i = t_i(x), \quad 1 \le i \le n \,, \tag{5.2}$$

in which $\mathrm{ad}_f^i r = (-1)^i \dfrac{\partial}{\partial \zeta_{n-i}}$, $0 \le i \le n-1$, and the system

$$\begin{aligned} \dot{x} &= f(x) \\ y &= h(x) \end{aligned} \tag{5.3}$$

takes the form

$$\begin{aligned} \dot{\zeta} &= A_c \zeta + \psi_0(y) \\ y &= C_c \zeta \,. \end{aligned} \tag{5.4}$$

Conditions (i), (ii) and (iii) are shown in [10] to be necessary and sufficient for the existence of a local, parameter independent, state-space diffeomorphism (5.2) transforming (ΣR) into

$$\begin{aligned} \dot{\zeta} &= A_c \zeta + \left[\beta_0(y) + \sum_{j=1}^{p} \theta_j \beta_j(y) \right] u + \psi_0(y) + \sum_{i=1}^{p} \theta_i \psi_i(y) \\ y &= C_c \zeta \,. \end{aligned} \tag{5.5}$$

Condition (iv) implies that the vector fields $\beta_i(y)$ in (5.5) become

$$\beta_i(y) = \sigma(y) \begin{bmatrix} b_{i1} \\ b_{i2} \\ \vdots \\ b_{i,n-1} \\ b_{i,n} \end{bmatrix}, \qquad 0 \le i \le p. \tag{5.6}$$

The proof of the above statement is an easy consequence of the main result in [19]: in the coordinates $\zeta = T(x)$ in which the system (5.3) takes the form (5.4), we have

$$\mathrm{ad}_f^i r = (-1)^i \frac{\partial}{\partial \zeta_{n-i}},$$

and (iv) becomes in the ζ-coordinates

$$\text{(iv) } g_j = \sigma(y) \sum_{i=1}^{n} b_{ji} \frac{\partial}{\partial \zeta_i}, \qquad 0 \le j \le p,$$

i.e., (5.6). On the other hand, the necessity of conditions (i), (ii), (iii) and (iv), which are coordinate free, is readily verified on (SR). Condition (v) is necessary and sufficient, according to [20], for the above change of coordinates to be global. When (v) is violated the change of coordinates is a local one. □

Let us recall [1,2] the definition of relative degree.

Definition 5.1 *Let ρ, the (strong) relative degree of the output y with respect to the control u for the system (ΣR) with $\theta = 0$ (nominal system), be defined as*

$$L_g L_f^i h(x) = 0, \quad 0 \le i \le \rho - 2, \forall x \in \mathbb{R}^n$$
$$L_g L_f^{\rho-1} h(x) \ne 0, \quad \textit{for some } x \in \mathbb{R}^n.$$

□

The vector field $g(x;\theta)$ becomes in the ζ-coordinates

$$g = \sigma(y) \sum_{i=1}^{n} \sum_{j=0}^{p} \theta_j b_{ji} \frac{\partial}{\partial \zeta_i} = \sigma(y) \sum_{i=1}^{n} b_i(\theta) \frac{\partial}{\partial \zeta_i},$$

At this point we introduce a different parametrization. We consider the $n - \rho + 1$ components of $b(\theta) = [b_1(\theta), \ldots, b_n(\theta)]^{\mathrm{T}}$ as unknown parameters:

$$\begin{aligned} \dot{\zeta} &= A_c \zeta + b\sigma(y)u + \sum_{i=0}^{p} \theta_i \psi_i(y) \\ &\triangleq A_c \zeta + b\sigma(y)u + \Psi(y,u)\bar{\theta} \\ y &= C_c \zeta, \end{aligned}$$

where $\bar{\theta} = [\theta_0, \theta_1, \ldots, \theta_p]^{\mathrm{T}}$ with $\theta_0 = 1$ for convenience of notation. The above system contains $p + n - \rho + 1$ unknown parameters.

In what follows we assume (recall Definition 2.3) that

(vi) the vector $b(\theta) = b_0 + \sum_{i=1}^{p} b_i(\theta)$, which appears in condition (iv), is Hurwitz with constant degree ρ for every $\theta \in \Omega$, and the sign of the scalar $b_{0\rho} + \sum_{i=1}^{p} b_{i\rho}\theta_i$ is known and constant in Ω.

The class of systems (ΣR) satisfying conditions (i)–(v) of Theorem 5.1 and, in addition, assumption (vi), have the remarkable property that the zero dynamics (see [1, Chap. 4], for a definition) can be expressed, in suitable global coordinates, by a linear asymptotically stable system.

Theorem 5.2 *The zero dynamics of the system (ΣR) with relative degree ρ satisfying conditions (i)–(vi), can be expressed in suitable global coordinates as*

$$\dot{z} = \begin{bmatrix} -\frac{b_{\rho+1}}{b_\rho} & 1 & 0 & \cdots & 0 \\ -\frac{b_{\rho+2}}{b_\rho} & 0 & 1 & \cdots & 0 \\ \vdots & \vdots & \vdots & \ddots & \vdots \\ -\frac{b_{n-1}}{b_\rho} & 0 & 0 & \cdots & 1 \\ -\frac{b_n}{b_\rho} & 0 & 0 & \cdots & 0 \end{bmatrix} z . \tag{5.7}$$

Proof. Conditions (i)–(v) guarantee by Theorem 5.1 the existence of a global change of coordinates (5.1) in which the system (ΣR) is expressed as (SR). Since the relative degree is ρ, we have $b_1 = \cdots = b_{\rho-1} = 0, b_\rho \neq 0$ and, by assumption (vi), the polynomial

$$b_\rho s^\rho + \cdots + b_{n-1}s + b_n \tag{5.8}$$

is Hurwitz. By choosing the control law

$$u = \frac{1}{b_\rho \sigma(y)}(-\zeta_{\rho+1}), \tag{5.9}$$

the manifold $\{\zeta \in \mathbb{R}^n : \zeta_1 = \cdots = \zeta_\rho = 0\}$ is made invariant and the dynamics on it (the zero dynamics according to [1, Chap. 4]) is given by (5.7) and is asymptotically stable since (5.8) is Hurwitz. □

Definition 5.2 *A global (adaptive) output-feedback stabilizing controller for system (ΣR) is a finite dimensional system*

$$\begin{aligned} \dot{w} &= \alpha(w, y(t)), \quad w(0) = w_0,\ w \in \mathbb{R}^r, \\ u &= \beta(w, y(t)), \quad u \in \mathbb{R}, \end{aligned}$$

such that for every $x(0) \in \mathbb{R}^n$ and $w_0 \in \mathbb{R}^r$, $\|x(t)\|$ and $\|w(t)\|$ are bounded $\forall t \geq 0$, and

$$\lim_{t\to\infty} \|x(t)\| = 0 ,$$

for the closed-loop system. □

We now first establish a technical result which generalizes Lemma 3.2 and will be one of the basic tools in the next two sections.

Lemma 5.3 *Consider the system*

$$\begin{aligned} \dot{x} &= A_c x + f_0(y,u,\theta) + \sum_{i=1}^{p} f_i(y,u)\theta_i \\ y &= C_c\, x\,. \end{aligned} \tag{5.10}$$

Let f_{ij}, $1 \le j \le n$, be the components of the vector f_i. If

$$f_{ij}(y,u) = 0, \qquad \forall y \in \mathbb{R}, \forall u \in \mathbb{R}, \quad 1 \le i \le p,\ 1 \le j \le \rho_1 - 1 < n, \tag{5.11}$$

then there exists a filtered transformation

$$\begin{aligned} z &= S(x,\xi(t),\theta) \\ \dot{\xi} &= \Lambda\xi + \delta(y,u), \qquad \xi(0) = 0, \end{aligned} \tag{5.12}$$

taking (5.10) into

$$\begin{aligned} \dot{z} &= A_c z + f_0(y,u,\theta) + b\sum_{i=1}^{p} \gamma_i(y,u,\xi)\theta_i \\ y &= C_c\, z\,, \end{aligned} \tag{5.13}$$

in which $b = [0,\ldots,b_{\rho_2},\ldots,b_n]^{\mathrm{T}}$ is a Hurwitz vector of degree $\rho_2 \le \rho_1$, such that

$$b_{\rho_2}s^{n-\rho_2} + \cdots + b_n = b_{\rho_2}(s+\lambda_1)\cdots(s+\lambda_{n-\rho_2})$$

with λ_i, $1 \le i \le n-\rho_2$, arbitrary positive scalars.

Proof. Consider the $n \times (n-\rho_2+1)$ matrix

$$D = \begin{bmatrix} 0 & 0 & \cdots 0 \\ \vdots & \vdots & \vdots\ \vdots \\ 1 & 0 & \cdots 0 \\ d_{\rho_2,\rho_2+1} & 1 & \cdots 0 \\ \vdots & \vdots & \ddots\ \vdots \\ d_{\rho_2,n} & d_{\rho_2+1,n} & \cdots 1 \end{bmatrix} \triangleq \left[\, d_{\rho_2} \cdots d_n \,\right], \tag{5.14}$$

where the columns d_i, $\rho_2 \le i \le n$, which are linearly independent, are related by the recursive formula

$$\begin{aligned} d_n &= \left[0 \cdots 0\ 1\right]^{\mathrm{T}} \\ d_{j-1} &= A_c d_j + \lambda_{j-1} d_j, \qquad n \ge j \ge 2\,. \end{aligned} \tag{5.15}$$

From (5.14) and (5.15) it follows that

$$s^{n-\rho_2} + d_{\rho_2,\rho_2+1}s^{n-\rho_2-1} + \cdots + d_{\rho_2,n} = (s+\lambda_{n-1})\cdots(s+\lambda_1)$$

$$s^{n-\rho_2-1} + d_{\rho_2+1,\rho_2+2}s^{n-\rho_2-2} + \cdots + d_{\rho_2+1,n} = (s+\lambda_{n-1})\cdots(s+\lambda_2)$$

$$\vdots$$

$$s + d_{n-1,n} = s + \lambda_{n-1}\,.$$

Note that $d_{\rho_2} = \frac{1}{b_{\rho_2}}b$. Condition (5.11) guarantees that the first $\rho_1 - 1$ components of the vectors f_i, $1 \le i \le p$, are zero: each vector field $f_i(y,u)$ can be uniquely expressed as

$$f_i(y,u) = \sum_{k=\rho_2}^{n} d_k\phi_{ik}(y,u), \quad 1 \le i \le p\,. \tag{5.16}$$

Substituting the expression (5.16) in (5.10), we obtain

$$\begin{aligned} \dot{x} &= A_c x + \sum_{i=1}^{p}\theta_i \sum_{k=\rho_2}^{n} d_k\phi_{ik}(y,u) + f_0(y,u,\theta) \\ &= A_c x + \sum_{k=\rho_2}^{n} d_k\phi_k^{\mathrm{T}}(y,u)\theta + f_0(y,u,\theta)\,, \end{aligned} \tag{5.17}$$

where the $p \times 1$ vectors ϕ_k are given by $\phi_k = [\phi_{1k}, \ldots, \phi_{pk}]^{\mathrm{T}}$, $\rho_2 \le k \le n$. Now consider the filtered transformation consisting of the linear time-invariant asymptotically stable filters

$$\begin{aligned} \dot{\xi}_j &= -\lambda_j\xi_j + \xi_{j+1} + \phi_{j+1}(y,u)\,, \qquad \rho_2 \le j \le n-1 \\ \xi_n &= 0\,, \end{aligned} \tag{5.18}$$

where $\xi_i \in \mathbb{R}^p$, $1 \le i \le n$, and the time-varying affine change of coordinates

$$z = x - \sum_{j=\rho_2+1}^{n} d_j\xi_{j-1}^{\mathrm{T}}\theta\,, \tag{5.19}$$

which has the global inverse

$$x = z + \sum_{j=\rho_2+1}^{n} d_j\xi_{j-1}^{\mathrm{T}}\theta\,.$$

Differentiating (5.19) with respect to time, and using (5.17) and (5.18), we obtain

$$\begin{aligned}
\dot{z} &= \dot{x} - \sum_{j=\rho_2+1}^{n} d_j \dot{\xi}_{j-1}^{\mathrm{T}} \theta \\
&= A_c z + \sum_{j=\rho_2+1}^{n} A_c d_j \xi_{j-1}^{\mathrm{T}} \theta + \sum_{j=\rho_2}^{n} d_j \phi_j^{\mathrm{T}}(y,u)\theta - \sum_{j=\rho_2+1}^{n} d_j \dot{\xi}_{j-1}^{\mathrm{T}} \theta + f_0(y,u,\theta) \\
&= A_c z + f_0(y,u,\theta) + \sum_{j=\rho_2+1}^{n} A_c d_j \xi_{j-1}^{\mathrm{T}} \theta + \sum_{j=\rho_2}^{n} d_j \phi_j^{\mathrm{T}}(y,u)\theta \\
&\quad + \sum_{j=\rho_2+1}^{n} d_j \left[\lambda_{j-1}\xi_{j-1}^{\mathrm{T}} - d_j \phi_j^{\mathrm{T}}(y,u)\right]\theta - \sum_{j=\rho_2+1}^{n-1} d_j \xi_j^{\mathrm{T}} \theta \\
&= A_c z + f_0(y,u,\theta) + \sum_{j=\rho_2+1}^{n} (A_c d_j + \lambda_{j-1} d_j)\xi_{j-1}^{\mathrm{T}} \theta - \sum_{j=\rho_2+2}^{n} d_{j-} \xi_{j-1}^{\mathrm{T}} \theta \\
&\quad + d_{\rho_2} \phi_{\rho_2}^{\mathrm{T}}(y,u)\theta \,,
\end{aligned}$$

which, taking (5.15) into account, becomes

$$\dot{z} = A_c z + f_0(y,u,\theta) + d_{\rho_2}\left[\phi_{\rho_2}^{\mathrm{T}}(y,u) + \xi_{\rho_2}^{\mathrm{T}}\right]\theta \,. \tag{5.20}$$

In addition, since according to (5.14), $C_c d_j = 0$, $\rho_2 + 1 \le j \le n$, from (5.19) we obtain

$$y = C_c x = C_c z + \sum_{j=\rho_2+1}^{n} C_c d_j \xi_{j-1}^{\mathrm{T}} \theta = C_c z \,. \tag{5.21}$$

Recalling that $d_{\rho_2} = \frac{1}{b_{\rho_2}} b$, we can finally express (5.20)–(5.21) in the form (5.13), defining $\gamma_i(y,u,\xi)$ as

$$\gamma_i(y,u,\xi) = \frac{1}{b_{\rho_2}}\left[\phi_{i\rho_2}(y,u) + \xi_{\rho_2 i}\right], \qquad 1 \le i \le p \,.$$

with $\xi_{\rho_2 i}$ being the i-th component of the vector ξ_{ρ_2}. □

6 Adaptive output-feedback stabilization: relative-degree-one case

Theorem 6.1 *Consider the system (ΣR) with relative degree $\rho = 1$. If assumptions (i)–(vi) are satisfied, then there exists a global adaptive output-feedback stabilizing controller for every unknown value of the parameter vector $\theta \in \Omega$.*

Proof. Since assumptions (i)–(v) are satisfied, Theorem 5.1 applies and guarantees the existence of a global change of coordinates $\zeta = T(x)$ transforming (ΣR) into

$$\begin{aligned}\dot{\zeta} &= A_c\zeta + b\sigma(y)u + \sum_{i=0}^{p}\theta_i\psi_i(y) \\ &\triangleq A_c\zeta + b\sigma(y)u + \Psi(y)\bar{\theta} \\ y &= C_c\zeta\,,\end{aligned} \tag{6.1}$$

which can be rewritten as

$$\begin{aligned}\dot{\zeta} &= A_c\zeta + \frac{1}{k}\bar{b}\,\mathrm{sgn}(b_1)\sigma(y)u + \sum_{i=0}^{p}\theta_i\psi_i(y) \\ y &= C_c\zeta\,,\end{aligned} \tag{6.2}$$

where $k = \dfrac{1}{|b_1|}$ and $\bar{b} = \left[1, \dfrac{b_2}{b_1}, \ldots, \dfrac{b_n}{b_1}\right]^{\mathrm{T}}$. Consider the matrix

$$D = \begin{bmatrix} 1 & 0 & \cdots & 0 \\ d_{12} & 1 & \cdots & 0 \\ \vdots & \vdots & \ddots & \vdots \\ d_{1n} & d_{2n} & \cdots & 1 \end{bmatrix} \triangleq [\,d_1 \cdots d_n\,]\,, \tag{6.3}$$

where the vectors d_i, $1 \le i \le n$, which are linearly independent, are related by the recursive formula

$$\begin{aligned} d_n &= [\,0 \cdots 0\;1\,]^{\mathrm{T}} \\ d_{j-1} &= A_c d_j + \lambda_{j-1} d_j\,, \qquad n \ge j \ge 2\,, \end{aligned} \tag{6.4}$$

with λ_j, $1 \le j \le n-1$, being positive scalars. The system (6.2) can be rewritten as

$$\begin{aligned}\dot{\zeta} &= A_c\zeta + \left(\frac{1}{k}d_1 + \sum_{j=2}^{n}\frac{1}{k}d_j\bar{\beta}_{j-1}\right)\mathrm{sgn}(b_1)\sigma(y)u + \sum_{i=0}^{p}\theta_i\psi_i(y) \\ y &= C_c\zeta\,,\end{aligned} \tag{6.5}$$

where $\bar{\beta} = [1, \bar{\beta}_1, \ldots, \bar{\beta}_{n-1}]^{\mathrm{T}}$ is such that

$$\bar{b} = D\bar{\beta}\,.$$

Defining $\beta_j = \dfrac{\bar{\beta}_j}{k}$, $1 \le j \le n-1$, we can rewrite (6.5) as

$$\begin{aligned}\dot{\zeta} &= A_c\zeta + \frac{1}{k}d_1\,\mathrm{sgn}(b_1)\sigma(y)u + \sum_{j=1}^{n-1} d_{j+1}\beta_j\,\mathrm{sgn}(b_1)\sigma(y)u + \sum_{i=0}^{p}\theta_i\psi_i(y) \\ y &= C_c\zeta\,.\end{aligned} \tag{6.6}$$

Now, consider the following filtered transformation consisting of the filters

$$\begin{aligned}
\dot{\mu}_{n-1,1} &= -\lambda_1 \mu_{n-1,1} + \mu_{n-1,2} \\
&\vdots \\
\dot{\mu}_{n-1,n-1} &= -\lambda_{n-1} \mu_{n-1,n-1} + \operatorname{sgn}(b_1)\sigma(y)u \\
\dot{\mu}_{n-2,1} &= -\lambda_1 \mu_{n-2,1} + \mu_{n-2,2} \\
&\vdots \\
\dot{\mu}_{n-2,n-2} &= -\lambda_{n-2} \mu_{n-2,n-2} + \operatorname{sgn}(b_1)\sigma(y)u \\
&\vdots \\
\dot{\mu}_{1,1} &= -\lambda_1 \mu_{1,1} + \operatorname{sgn}(b_1)\sigma(y)u
\end{aligned} \tag{6.7}$$

and of the transformation

$$\bar{z} = \zeta - \sum_{i=1}^{n-1} \beta_i \sum_{j=2}^{i+1} d_j \mu_{i,j-1} \,. \tag{6.8}$$

In these new coordinates, we have

$$\begin{aligned}
\dot{\bar{z}} &= A_c \bar{z} + \frac{1}{k} d_1 \operatorname{sgn}(b_1)\sigma(y)u + \sum_{i=0}^{p} \theta_i \psi_i(y) + d_1 \sum_{i=1}^{n-1} \beta_i \mu_{i,1} \\
y &= C_c \bar{z} \,.
\end{aligned} \tag{6.9}$$

We now apply Lemma 5.3 which guarantees that there exists a filtered transformation

$$\begin{aligned}
z &= S(\bar{z}, \xi(t), \bar{\theta}) \\
\dot{\xi} &= \Lambda \xi + \delta(y), \qquad \xi(0) = 0 \,,
\end{aligned} \tag{6.10}$$

such that (6.9) is transformed into

$$\begin{aligned}
\dot{z} &= A_c z + \frac{1}{k} d_1 \operatorname{sgn}(b_1)\sigma(y)u + d_1 \sum_{i=1}^{n-1} \beta_i \mu_{i,1} + d_1 \sum_{i=0}^{p} \gamma_i(y,\xi)\theta_i \\
&\triangleq A_c z + \frac{1}{k} d_1 \operatorname{sgn}(b_1)\sigma(y)u + d_1 \mu_1^{\mathrm{T}} \beta + d_1[\gamma_0(y,\xi) + \gamma^{\mathrm{T}}(y,\xi)\theta)] \\
y &= C_c z \,,
\end{aligned} \tag{6.11}$$

where $\beta = [\beta_1, \ldots, \beta_{n-1}]^{\mathrm{T}}$, $\mu_1 = [\mu_{1,1}, \ldots, \mu_{n-1,1}]^{\mathrm{T}}$, $\gamma = [\gamma_1, \ldots, \gamma_p]^{\mathrm{T}}$ and $\gamma_0(0,0) = 0$, $\gamma(0,0) = 0$. Let the control u be defined as

$$u = -\hat{k} \frac{1}{\sigma(y)} \operatorname{sgn}(b_1) \left[K_c \hat{z} + \mu_1^{\mathrm{T}} \hat{\beta} + \gamma_0(y,\xi) + \gamma^{\mathrm{T}}(y,\xi)\hat{\theta} \right] , \tag{6.12}$$

where $\sigma(y) \neq 0$, $\forall y \in \mathbb{R}$, by virtue of the hypothesis that $g(x;\theta) \neq 0$, $\forall x \in \mathbb{R}^n$, $\forall \theta \in \Omega$, and $\mathrm{sgn}(b_1)$ is known by assumption (vi). The matrix K_c is chosen so that (λ_c is a positive scalar)

$$\det[sI - (A_c - d_1 K_c)] = (s + \lambda_c)(s^{n-1} + d_{12}s^{n-2} + \cdots + d_{1n}), \tag{6.13}$$

i.e., the eigenvalues of the matrix $(A_c - d_1 K_c)$ are the zeroes of the polynomial $s^{n-1}+d_{12}s^{n-2}+\cdots+d_{1n}$, which is Hurwitz by (6.4), in addition to $-\lambda_c$. Such a K_c exists since the pair (A_c, d_1) is controllable (since d_1 is Hurwitz by construction and, therefore, $d_{1n} \neq 0$). The state estimate $\hat{z}$ in (6.12) is provided by the adaptive observer

$$\dot{\hat{z}} = A_c\hat{z} + \hat{b}_1 d_1 \sigma(y)u + d_1 \left[\mu_1^{\mathrm{T}}\hat{\beta} + \gamma_0(y,\xi) + \gamma^{\mathrm{T}}(y,\xi)\hat{\theta}\right] + K_o(y - C_c\hat{z}), \tag{6.14}$$

in which K_o is chosen so that (λ_o is a positive scalar)

$$\det[sI - (A_c - K_o C_c)] = (s + \lambda_o)(s^{n-1} + d_{12}s^{n-2} + \cdots + d_{1n}). \tag{6.15}$$

Such a K_o exists since the pair (C_c, A_c) is observable. The parameter update laws are chosen to be

$$\dot{\hat{\theta}} = G_\theta[y + \eta(y - C_c\hat{z})]\gamma(y,\xi) \tag{6.16}$$

$$\dot{\hat{\beta}} = G_\beta[y + \eta(y - C_c\hat{z})]\mu_1 \tag{6.17}$$

$$\dot{\hat{k}} = g_k \left[K_c\hat{z} + \mu_1^{\mathrm{T}}\hat{\beta} + \gamma_0(y,\xi) + \gamma(y,\xi)^{\mathrm{T}}\hat{\theta}\right] y \tag{6.18}$$

$$\dot{\hat{b}}_1 = g_b(y - C_c\hat{z})\sigma(y)u, \tag{6.19}$$

with η, g_k and g_b being positive scalars, G_θ and G_β being symmetric positive definite matrices. From (6.11), (6.12), (6.14) and (6.16)–(6.19), we obtain the extended dynamics (we use the notations: $\tilde{z} = z - \hat{z}$, $\tilde{\theta} = \theta - \hat{\theta}$, $\tilde{\beta} = \beta - \hat{\beta}$, $\tilde{k} = k - \hat{k}$, $\tilde{b}_1 = b_1 - \hat{b}_1$)

$$\begin{aligned} \dot{z} &= (A_c - d_1K_c)z + d_1K_c\tilde{z} + d_1\left[\mu_1^{\mathrm{T}}\tilde{\beta} + \gamma^{\mathrm{T}}(y,\xi)\tilde{\theta}\right] \\ &\quad + d_1\frac{\tilde{k}}{k}\left[K_c\hat{z} + \mu_1^{\mathrm{T}}\hat{\beta} + \gamma_0(y,\xi) + \gamma^{\mathrm{T}}(y,\xi)\hat{\theta}\right] \end{aligned} \tag{6.20}$$

$$\dot{\tilde{z}} = (A_c - K_oC_c)\tilde{z} + d_1\tilde{b}_1\sigma(y)u + d_1[\mu_1^{\mathrm{T}}\tilde{\beta} + \gamma^{\mathrm{T}}(y,\xi)\tilde{\theta}] \tag{6.21}$$

$$\begin{aligned} \dot{\tilde{\theta}} &= -G_\theta C_c(z + \eta\tilde{z})\gamma(y,\xi) \\ \dot{\tilde{\beta}} &= -G_\beta C_c(z + \eta\tilde{z})\mu_1 \\ \dot{\tilde{k}} &= -g_k C_c z[K_c\hat{z} + \gamma_0(y,\xi) + \mu_1^{\mathrm{T}}\hat{\beta} + \gamma(y,\xi)^{\mathrm{T}}\hat{\theta}] \\ \dot{\tilde{b}}_1 &= -g_b C_c\tilde{z}\sigma(y)u. \end{aligned} \tag{6.22}$$

Consider the function

$$V(z,\tilde{z},\tilde{\theta},\tilde{\beta},\tilde{k},\tilde{b}_1) = z^{\mathrm{T}}P_z z + \eta\tilde{z}^{\mathrm{T}}P_{\mathrm{e}}\tilde{z} + \tilde{\theta}^{\mathrm{T}}G_{\theta}^{-1}\tilde{\theta} + \tilde{\beta}^{\mathrm{T}}G_{\beta}^{-1}\tilde{\beta} + \frac{1}{k}g_k^{-1}\tilde{k}^2 + \eta g_b^{-1}\tilde{b}_1^2\,, \tag{6.23}$$

where $k = \dfrac{1}{|b_1|}$ is a positive unknown constant and P_z and P_{e} are, respectively, the symmetric positive definite solutions of

$$\begin{aligned}(A_{\mathrm{c}} - d_1K_{\mathrm{c}})^{\mathrm{T}}P_z + P_z(A_{\mathrm{c}} - d_1K_{\mathrm{c}}) &= -l_zl_z^{\mathrm{T}} - \epsilon_z I\\ P_zd_1 &= C_{\mathrm{c}}^{\mathrm{T}}\end{aligned} \tag{6.24}$$

and

$$\begin{aligned}(A_{\mathrm{c}} - K_{\mathrm{o}}C_{\mathrm{c}})^{\mathrm{T}}P_{\mathrm{e}} + P_{\mathrm{e}}(A_{\mathrm{c}} - K_{\mathrm{o}}C_{\mathrm{c}}) &= -l_{\mathrm{e}}l_{\mathrm{e}}^{\mathrm{T}} - \epsilon_{\mathrm{e}} I\\ P_{\mathrm{e}}d_1 &= C_{\mathrm{c}}^{\mathrm{T}}\,,\end{aligned} \tag{6.25}$$

with ϵ_z and ϵ_{e} positive scalars. Such solutions exist since, according to (6.13) and (6.15), the triples $(A_{\mathrm{c}} - d_1K_{\mathrm{c}}, d_1, C_{\mathrm{c}})$ and $(A_{\mathrm{c}} - K_{\mathrm{o}}C_{\mathrm{c}}, d_1, C_{\mathrm{c}})$ satisfy the Meyer-Kalman-Yakubovich Lemma [21, p. 66]. The time derivative of (6.23) with respect to the dynamics (6.20)–(6.22) satisfies the inequality

$$\dot{V}(z,\tilde{z},\tilde{\theta},\tilde{\beta},\tilde{k},\tilde{b}_1) \le -\begin{bmatrix}\|z\|\\ \|\tilde{z}\|\end{bmatrix}^{\mathrm{T}}\begin{bmatrix}\epsilon_z & \|C_{\mathrm{c}}^{\mathrm{T}}K_{\mathrm{c}}\|\\ \|C_{\mathrm{c}}^{\mathrm{T}}K_{\mathrm{c}}\| & \eta\epsilon_{\mathrm{e}}\end{bmatrix}\begin{bmatrix}\|z\|\\ \|\tilde{z}\|\end{bmatrix}, \tag{6.26}$$

and, therefore, if η is chosen such that

$$\eta > \frac{\|C_{\mathrm{c}}^{\mathrm{T}}K_{\mathrm{c}}\|^2}{\epsilon_z\epsilon_{\mathrm{e}}}\,, \tag{6.27}$$

then there exists a positive constant k_{L} such that

$$\dot{V} \le -k_{\mathrm{L}}\left\|\begin{bmatrix}\|z\|\\ \|\tilde{z}\|\end{bmatrix}\right\|^2. \tag{6.28}$$

Then, $\dot{V}$ is negative semidefinite. From (6.23) and (6.28), we obtain that $\|z(t)\|$, $\|\tilde{z}(t)\|$, $\|\tilde{\theta}(t)\|$, $\|\tilde{\beta}(t)\|$, $\|\tilde{k}(t)\|$ and $\|\tilde{b}_1(t)\|$ are bounded, which imply that also $y(t) = C_{\mathrm{c}}z(t)$ is bounded, $\forall t \ge 0$. From (6.10) it then follows that $\|\xi(t)\|$ is bounded. From (6.1), the input-output map is given by the differential equation

$$\sum_{i=1}^{n} b_i\frac{d^{n-i}}{dt^{n-i}}[\sigma(y)u] = \frac{d^ny}{dt^n} - \sum_{i=1}^{n}\frac{d^{n-i}}{dt^{n-i}}\left[\sum_{j=0}^{p}\psi_{ji}(y)\theta_i\right], \tag{6.29}$$

in which ψ_{ji} denotes the i-th element of the vector ψ_j. From the first set of filters in (6.7), we obtain the differential equation relating $\sigma(y)u$ to $\mu_{n-1,1}$

$$\frac{d^{n-1}}{dt^{n-1}}\mu_{n-1,1} + d_{12}\frac{d^{n-2}}{dt^{n-2}}\mu_{n-1,1} + \cdots + d_{1n}\mu_{n-1,1} = \mathrm{sgn}(b_1)\sigma(y)u\,. \tag{6.30}$$

Substituting (6.30) in (6.29), we have

$$\sum_{i=1}^{2n-1} a_i \frac{d^{2n-1-i}}{dt^{2n-1-i}} \mu_{n-1,1} = \frac{d^n y}{dt^n} - \sum_{i=1}^{n} \frac{d^{n-i}}{dt^{n-i}} \left[\sum_{j=0}^{p} \psi_{ji}(y)\theta_i \right], \tag{6.31}$$

where a_i, $1 \le i \le 2n-1$, are the coefficients of the Hurwitz polynomial defined by

$$\mathrm{sgn}(b_1)(b_1 s^{n-1} + \cdots + b_n)(s^{n-1} + d_{12}s^{n-2} + \cdots + d_{1n}) = a_1 s^{2n-2} + \cdots + a_{2n-1}\,.$$

The system (6.31) can be realized as

$$\dot{\omega} = \begin{bmatrix} -\frac{a_2}{a_1} & 1\,0 \cdots 0 \\ \vdots & \vdots\ \vdots\ \vdots\ \vdots \\ -\frac{a_{2n-2}}{a_1} & 0\,0 \cdots 1 \\ -\frac{a_{2n-1}}{a_1} & 0\,0 \cdots 0 \end{bmatrix} \omega + yE_{n-2} - \sum_{i=1}^{n} E_{n-2+i} \sum_{j=0}^{p} \psi_{ji}(y)\theta_i \tag{6.32}$$

$$\mu_{n-1,1} = \begin{bmatrix} 1\,0\,0 \cdots 0 \end{bmatrix} \omega\,,$$

where ω is the $(2n-2) \times 1$ state vector and E_i represents the i-th column of the $(2n-2) \times (2n-2)$ identity matrix. Since (6.32) is an asymptotically stable linear system whose inputs are bounded (y has been already been shown to be bounded), it follows that the output $\mu_{n-1,1}(t)$ is also bounded. A similar reasoning may be used to prove that also the other filter variables $\mu_{j,1}$, $1 \le j \le n-2$, are bounded. According to (6.12) the control $u(t)$ is bounded as well. Consequently, in (6.7) the state vectors $\mu_i(t)$, $1 \le i \le n-1$, are bounded. From (6.20) and (6.21) we can see that $\|\dot{z}(t)\|$ and $\|\dot{\tilde{z}}(t)\|$ are bounded, which imply that the quadratic form on the right-hand side of (6.32) is uniformly continuous. Moreover,

$$k_L \lim_{t\to\infty} \int_0^t \left\| \begin{bmatrix} \|z(\tau)\| \\ \|\tilde{z}(\tau)\| \end{bmatrix} \right\|^2 d\tau \le \lim_{t\to\infty} \int_0^t -\dot{V}\, d\tau = V(0) - \lim_{t\to\infty} V(t) < \infty\,.$$

Applying Barbalat's Lemma [23, p. 210] it follows that

$$\begin{aligned} \lim_{t\to\infty} \|z(t)\| &= 0 \\ \lim_{t\to\infty} \|\tilde{z}(t)\| &= 0\,, \end{aligned} \tag{6.33}$$

and, since $y = C_c z$,

$$\lim_{t\to\infty} y(t) = 0\,. \tag{6.34}$$

Now consider the systems (6.10) and (6.32). Since $y(t)$ is bounded and (6.34) holds, we can apply a result in [22, p. 59] to establish that

$$\begin{aligned} \lim_{t\to\infty} \|\xi(t)\| &= 0 \\ \lim_{t\to\infty} \mu_{n-1,1}(t) &= 0\,, \end{aligned} \tag{6.35}$$

and, similarly,

$$\lim_{t\to\infty} \mu_{i,1}(t) = 0, \qquad 1 \le i \le n-2. \tag{6.36}$$

Taking (6.33)–(6.36) into account, we obtain from (6.12) that

$$\lim_{t\to\infty} u(t) = 0\,, \tag{6.37}$$

which allows us to apply again the result in [22, p. 59] to the filters (6.7), yielding

$$\lim_{t\to\infty} \|\mu_i(t)\| = 0\,, \qquad 1 \le i \le n-1. \tag{6.38}$$

Since $\zeta(t)$ is related to $z(t)$ by the transformations (6.8) and (6.10), and (6.33)–(6.38) hold, we finally have that

$$\lim_{t\to\infty} \|\zeta(t)\| = 0\,,$$

which implies

$$\lim_{t\to\infty} \|x(t)\| = 0\,.$$

□

Example 6.1. Consider the system

$$\begin{aligned}
\dot{x}_1 &= x_2 + \theta(e^y - 1)\\
\dot{x}_2 &= x_3\\
\dot{x}_3 &= u\\
y &= c_1x_1 + c_2x_2 + c_3x_3\,,
\end{aligned}$$

with $c_3 > 0$ and $[c_1, c_2, c_3]$ an unknown Hurwitz vector. For every value of the constant parameter $\theta \neq 0$, this system is not state-feedback linearizable; hence, the adaptive algorithms proposed in [3,4,6,8] do not apply. On the other hand, the global Lipschitz conditions required in [7] are not satisfied either. However, this example is a relative-degree-one system which satisfies the assumptions of Theorem 6.1 and can therefore be globally adaptively stabilized by output feedback. □

7 Adaptive output-feedback stabilization: general case ($\rho \ge 2$)

In this section we generalize Theorem 6.1 to the case of any relative degree ρ, by using an adaptive scheme introduced in [5] and a recursive procedure for the construction of Lyapunov functions for cascade nonlinear systems given in [16–18].

Theorem 7.1 *Consider the system (ΣR) with relative degree ρ, $2 \le \rho \le n$. If assumptions (i)–(vi) are satisfied, then there exists a global adaptive output-feedback stabilizing controller for every unknown value of the parameter vector $\theta \in \Omega$.*

Proof. Since assumptions (i)–(v) are satisfied, Theorem 5.1 applies and guarantees the existence of a global change of coordinates $\zeta = T(x)$ transforming (ΣR) into (SR) with $b = [0, \dots, 0, b_\rho, \dots, b_n]^{\mathrm{T}}$. Now, consider the filtered transformation consisting of the filters

$$\begin{aligned} \dot{\varphi}_1 &= -\bar{\lambda}_1 \varphi_1 + \varphi_2 \\ &\vdots \\ \dot{\varphi}_{\rho-1} &= -\bar{\lambda}_{\rho-1} \varphi_{\rho-1} + \sigma(y) u , \end{aligned} \tag{7.1}$$

with $\varphi_i \in \mathbb{R}$, $1 \le i \le \rho - 1$, and of the transformation

$$\bar{z} = \zeta - \sum_{i=2}^{\rho} (\bar{d}_i \varphi_{i-1}) , \tag{7.2}$$

where $\bar{\lambda}_i$, $1 \le i \le \rho - 1$, are arbitrary positive scalars and $\bar{d}_i$, $1 \le i \le \rho$, are unknown vectors given by the recursive formula

$$\begin{aligned} \bar{d}_\rho &= b \\ \bar{d}_{j-1} &= A_c \bar{d}_j + \bar{\lambda}_{j-1} \bar{d}_j , \qquad \rho \ge j \ge 2 . \end{aligned} \tag{7.3}$$

Note that, since the vector b is Hurwitz by assumption (vi), $\bar{d}_1$ is also Hurwitz and its first element is $\bar{d}_{11} = b_\rho$. The dynamics in the $\bar{z}$-coordinates is given by

$$\begin{aligned} \dot{\bar{z}} &= A_c \bar{z} + \bar{d}_1 \varphi_1 + \sum_{i=0}^{p} \psi_i(y) \theta_i \\ &\triangleq A_c \bar{z} + \bar{d}_1 \varphi_1 + \bar{\Psi}(y) \bar{\theta} \\ y &= C_c \bar{z} , \end{aligned} \tag{7.4}$$

where $\bar{d}_1$ is an unknown Hurwitz vector, of degree one by construction, whose first element b_ρ is of known sign. We now apply Theorem 6.1 to (7.4) and, following [16–18], consider φ_1 as the control variable. According to the proof of Theorem 6.1, we rewrite (7.4) as

$$\begin{aligned} \dot{\bar{z}} &= A_c \bar{z} + \frac{1}{k} \bar{\bar{d}}_1 \varphi_1 \operatorname{sgn}(b_\rho) + \bar{\Psi}(y) \bar{\theta} \\ y &= C_c \bar{z} , \end{aligned} \tag{7.5}$$

where $k = \frac{1}{|b_\rho|}$ and $\bar{\bar{d}}_1 = \left[1, \frac{\bar{d}_{12}}{b_\rho}, \ldots, \frac{\bar{d}_{1n}}{b_\rho}\right]^{\mathrm{T}}$. We introduce the filtered transformation

$$\begin{aligned}
\dot{\mu}_{n-1,1} &= -\lambda_1 \mu_{n-1,1} + \mu_{n-1,2} \\
&\vdots \\
\dot{\mu}_{n-1,n-1} &= -\lambda_{n-1} \mu_{n-1,n-1} + \operatorname{sgn}(b_\rho)\varphi_1 \\
\dot{\mu}_{n-2,1} &= -\lambda_1 \mu_{n-2,1} + \mu_{n-2,2} \\
&\vdots \\
\dot{\mu}_{n-2,n-2} &= -\lambda_{n-2} \mu_{n-2,n-2} + \operatorname{sgn}(b_\rho)\varphi_1 \\
&\vdots \\
\dot{\mu}_{1,1} &= -\lambda_1 \mu_{1,1} + \operatorname{sgn}(b_\rho)\varphi_1
\end{aligned} \tag{7.6}$$

$$\bar{z}' = \bar{z} - \sum_{i=1}^{n-1} \beta_i \sum_{j=2}^{i+1} d_j \mu_{i,j-1}\,, \tag{7.7}$$

where β_i, $1 \le i \le n-1$, and d_i, $1 \le i \le n$, are defined as in the proof of Theorem 6.1. In the $\bar{z}'$-coordinates, we have

$$\begin{aligned}
\dot{\bar{z}}' &= A_c \bar{z}' + \frac{1}{k} d_1 \operatorname{sgn}(b_\rho)\varphi_1 + d_1 \mu_1^{\mathrm{T}} \beta + \bar{\Psi}(y)\bar{\theta} \\
y &= C_c \bar{z}'\,,
\end{aligned} \tag{7.8}$$

where $\beta = [\beta_1, \ldots, \beta_{n-1}]^{\mathrm{T}}$, $\mu_1 = [\mu_{1,1}, \ldots, \mu_{n-1,1}]^{\mathrm{T}}$. We now apply Lemma 5.3, which guarantees that there exists a filtered transformation

$$\begin{aligned}
z &= S(\bar{z}', \xi(t), \bar{\theta}) \\
\dot{\xi} &= \Lambda\xi + \delta(y), \qquad \xi(0) = 0\,,
\end{aligned} \tag{7.9}$$

such that (7.8) is transformed into

$$\begin{aligned}
\dot{z} &= A_c z + \frac{1}{k} d_1 \operatorname{sgn}(b_\rho)\varphi_1 + d_1 \mu_1^{\mathrm{T}} \beta + d_1 \left[\gamma_0(y,\xi) + \gamma^{\mathrm{T}}(y,\xi)\theta\right] \\
y &= C_c z\,,
\end{aligned} \tag{7.10}$$

where $\gamma = [\gamma_1, \ldots, \gamma_p]^{\mathrm{T}}$ with $\gamma_0(0,0) = 0$ and $\gamma(0,0) = 0$. Following the proof of Theorem 6.1, φ_1 should be designed as follows

$$\varphi_1^* = -\operatorname{sgn}(b_\rho)\hat{k}\left[K_c \hat{z} + \mu_1^{\mathrm{T}} \hat{\beta}_1 + \gamma_0(y,\xi) + \gamma^{\mathrm{T}}(y,\xi)\hat{\theta}_1\right], \tag{7.11}$$

where the estimates of the states and of the parameters are given by

$$\dot{\hat{z}} = A_c\hat{z} + \hat{b}_{\rho_1} d_1 \varphi_1^* + d_1 \left[\mu_1^T \hat{\beta}_1 + \gamma_0(y,\xi) + \gamma^T(y,\xi)\hat{\theta}_1\right] + K_o(y - C_c\hat{z}) \quad (7.12)$$

$$\dot{\hat{\theta}}_1 = G_{\theta_1}\left[y + \eta(y - C_c\hat{z})\right]\gamma(y,\xi) \quad (7.13)$$

$$\dot{\hat{\beta}}_1 = G_{\beta_1}\left[y + \eta(y - C_c\hat{z})\right]\mu_1 \quad (7.14)$$

$$\dot{\hat{k}} = g_k y \left[K_c\hat{z} + \mu_1^T\hat{\beta}_1 + \gamma_0(y,\xi) + \gamma(y,\xi)^T\hat{\theta}_1\right] \quad (7.15)$$

$$\dot{\hat{b}}_{\rho_1} = g_{b_1}(y - C_c\hat{z})\varphi_1^* . \quad (7.16)$$

For the closed-loop system (7.10)–(7.16), with $\varphi_1 = \varphi_1^*$, the function

$$V_0 = z^T P_z z + \eta\tilde{z}^T P_e \tilde{z} + \tilde{\theta}_1^T G_{\theta_1}^{-1}\tilde{\theta}_1 + \tilde{\beta}_1^T G_{\beta_1}^{-1}\tilde{\beta}_1 + \frac{1}{k} g_k^{-1}\tilde{k}^2 + \eta g_{b_1}^{-1}\tilde{b}_{\rho_1}^2 , \quad (7.17)$$

with $k = \dfrac{1}{|b_\rho|}$ a positive unknown constant and P_z and P_e solutions of (6.24) and (6.25), is such that its time derivative, denoted by $\dot{V}_0^*$, satisfies the inequality (recall (6.28))

$$\dot{V}_0^* \le -k_L \left\| \begin{bmatrix} \|z\| \\ \|\tilde{z}\| \end{bmatrix} \right\|^2 , \quad (7.18)$$

where k_L is a positive scalar. Consider the extended system consisting of (7.10), (7.13)–(7.15) and

$$\dot{\hat{z}} = A_c\hat{z} + \hat{b}_{\rho_1} d_1 \varphi_1 + d_1 \left[\mu_1^T \hat{\beta}_1 + \gamma_0(y,\xi) + \gamma^T(y,\xi)\hat{\theta}_1\right] + K_o(y - C_c\hat{z}) \quad (7.19)$$

$$\dot{\hat{b}}_{\rho_1} = g_{b_1}(y - C_c\hat{z})\varphi_1 \quad (7.20)$$

$$\dot{\varphi}_1 = -\bar{\lambda}_1\varphi_1 + \varphi_2 . \quad (7.21)$$

Perform the change of coordinates

$$\tilde{\varphi}_1 = \varphi_1 - \varphi_1^* . \quad (7.22)$$

In the new coordinates, we have

$$\begin{aligned}
\dot{z} &= A_c z + \operatorname{sgn}(b_\rho)\frac{1}{k} d_1(\varphi_1^* + \tilde{\varphi}_1) + d_1\left[\mu_1^T\beta + \gamma_0(y,\xi) + \gamma^T(y,\xi)\theta)\right] \\
\dot{\hat{z}} &= A_c\hat{z} + \hat{b}_{\rho_1} d_1(\varphi_1^* + \tilde{\varphi}_1) + d_1\left[\mu_1^T\hat{\beta}_1 + \gamma_0(y,\xi) + \gamma^T(y,\xi)\hat{\theta}_1\right] + K_o(y - C_c\hat{z}) \\
\dot{\hat{\theta}}_1 &= G_{\theta_1}[y + \eta(y - C_c\hat{z})]\gamma(y,\xi) \\
\dot{\hat{\beta}}_1 &= G_{\beta_1}[y + \eta(y - C_c\hat{z})]\mu_1 \\
\dot{\hat{k}} &= g_k y\left[K_c\hat{z} + \mu_1^T\hat{\beta}_1 + \gamma_0(y,\xi) + \gamma(y,\xi)^T\hat{\theta}_1\right] \\
\dot{\hat{b}}_{\rho_1} &= g_{b_1}(y - C_c\hat{z})(\varphi_1^* + \tilde{\varphi}_1) \\
\dot{\tilde{\varphi}}_1 &= -\bar{\lambda}_1\tilde{\varphi}_1 + \varphi_2 - \bar{\lambda}_1\varphi_1^* - \dot{\varphi}_1^* .
\end{aligned} \quad (7.23)$$

We now consider φ_2 as a control input which is to be designed using the function

$$\bar{V}_1 = V_0 + \tilde{\varphi}_1^2 + g_{b_x}^{-1}\tilde{b}_{\rho_x}^2 , \tag{7.24}$$

where $\hat{b}_{\rho_x}$ is an auxiliary estimate of b_ρ ($\tilde{b}_{\rho_x} = b_\rho - \hat{b}_{\rho_x}$) and g_{b_x} is a positive constant. The time derivative of $\bar{V}_1$ with respect to (7.23) is

$$\begin{aligned}\dot{\bar{V}}_1 &= \dot{V}_0^* + 2z^{\mathrm{T}}P_z d_1\tilde{\varphi}_1 \operatorname{sgn}(b_\rho)\frac{1}{k} + 2\tilde{\varphi}_1\dot{\tilde{\varphi}}_1 + 2g_{b_x}^{-1}\tilde{b}_{\rho_x}\dot{\tilde{b}}_{\rho_x} \\ &= \dot{V}_0^* + 2y\tilde{\varphi}_1 \operatorname{sgn}(b_\rho)\frac{1}{k} + 2\tilde{\varphi}_1\dot{\tilde{\varphi}}_1 + 2g_{b_x}^{-1}\tilde{b}_{\rho_x}\dot{\tilde{b}}_{\rho_x} ,\end{aligned} \tag{7.25}$$

which can be rewritten as

$$\dot{\bar{V}}_1 = \dot{V}_0^* + 2\tilde{b}_{\rho_x}y\tilde{\varphi}_1 + 2\hat{b}_{\rho_x}y\tilde{\varphi}_1 + 2\tilde{\varphi}_1\dot{\tilde{\varphi}}_1 + 2g_{b_x}^{-1}\tilde{b}_{\rho_x}\dot{\tilde{b}}_{\rho_x} . \tag{7.26}$$

Factoring out $\tilde{\varphi}_1$ in (7.26), using the last of (7.23), and choosing $\dot{\hat{b}}_{\rho_x}$ as

$$\dot{\hat{b}}_{\rho_x} = g_{b_x}y\tilde{\varphi}_1 , \tag{7.27}$$

the function φ_2 should be designed as

$$\varphi_2^{id} = \dot{\varphi}_1^* - y\hat{b}_{\rho_x} + \bar{\lambda}_1\varphi_1^* , \tag{7.28}$$

so that substituting $\varphi_2 = \varphi_2^{id}$ in (7.23), the time derivative of $\bar{V}_1$, denoted by $\dot{\bar{V}}_1^{id}$, would become

$$\dot{\bar{V}}_1^{id} = \dot{V}_0^* - 2\bar{\lambda}_1\tilde{\varphi}_1^2 \le -k_{\mathrm{L}}\left\| \begin{bmatrix} \|z\| \\ \|\tilde{z}\| \\ \|\tilde{\varphi}_1\| \end{bmatrix} \right\|^2 . \tag{7.29}$$

Let us examine $\dot{\varphi}_1^*$, which is given by

$$\begin{aligned}\dot{\varphi}_1^* &= \frac{\partial\varphi_1^*}{\partial y}\dot{y} + \frac{\partial\varphi_1^*}{\partial\xi}\dot{\xi} + \frac{\partial\varphi_1^*}{\partial\hat{\theta}_1}\dot{\hat{\theta}}_1 + \frac{\partial\varphi_1^*}{\partial\hat{\beta}_1}\dot{\hat{\beta}}_1 + \frac{\partial\varphi_1^*}{\partial\mu_1}\dot{\mu}_1 + \frac{\partial\varphi_1^*}{\partial\hat{k}}\dot{\hat{k}} + \frac{\partial\varphi_1^*}{\partial\hat{z}}\dot{\hat{z}} \\ &\triangleq f_1(y,\xi,\hat{z},\hat{k},\hat{\beta}_1,\hat{\theta}_1,\hat{b}_{\rho_1},\mu_1,\mu_2,\tilde{\varphi}_1) + \frac{\partial\varphi_1^*}{\partial y}\dot{y} ,\end{aligned} \tag{7.30}$$

where $\mu_2 = [\mu_{2,2}, \dots, \mu_{n-1,2}]^{\mathrm{T}}$. Since $\dot{y}$ is not available, it follows that $\dot{\varphi}_1^*$ and, consequently, φ_2^{id} *are not known*. Noting that

$$\begin{aligned}\dot{y} = C_c\dot{z} &= C_cA_cz + C_cd_1\left[\frac{1}{k}(\tilde{\varphi}_1 + \varphi_1^*)\operatorname{sgn}(b_\rho) + \gamma_0(y,\xi) + \gamma^{\mathrm{T}}(y,\xi)\theta + \mu_1^{\mathrm{T}}\beta\right] \\ &= C_cA_cz + \frac{1}{k}(\tilde{\varphi}_1 + \varphi_1^*)\operatorname{sgn}(b_\rho) + \gamma_0(y,\xi) + \gamma^{\mathrm{T}}(y,\xi)\theta + \mu_1^{\mathrm{T}}\beta ,\end{aligned} \tag{7.31}$$

we will replace $\dot{y}$ by (the use of several "parameter estimators" was introduced in [5])

$$\hat{\dot{y}}_1 = C_cA_c\hat{z} + \hat{b}_{\rho_2}(\tilde{\varphi}_1 + \varphi_1^*) + \gamma_0(y,\xi) + \gamma^{\mathrm{T}}(y,\xi)\hat{\theta}_2 + \mu_1^{\mathrm{T}}\hat{\beta}_2 - \alpha_1(t) , \tag{7.32}$$

in which the dynamics of $\hat{b}_{\rho_2}$, $\hat{\beta}_2$ and $\hat{\theta}_2$ and the function $\alpha_1(t)$ are yet to be defined; replacing $\dot{y}$ by $\hat{\dot{y}}_1$ in (7.28), we obtain

$$\begin{aligned}\varphi_2^* &= -y\hat{b}_{\rho_x} + \bar{\lambda}_1\varphi_1^* + f_1 + \frac{\partial\varphi_1^*}{\partial y}\hat{\dot{y}}_1 \\ &= \varphi_2^{id} - \frac{\partial\varphi_1^*}{\partial y}\left[C_cA_c\tilde{z} + \tilde{b}_{\rho_2}(\varphi_1^* + \tilde{\varphi}_1) + \mu_1^T\tilde{\beta}_2 + \gamma^T(y,\xi)\tilde{\theta}_2 + \alpha_1\right].\end{aligned} \tag{7.33}$$

Consider the function

$$V_1 = \bar{V}_1 + g_{b_2}^{-1}\tilde{b}_{\rho_2}^2 + \tilde{\beta}_2^T G_{\beta_2}^{-1}\tilde{\beta}_2 + \tilde{\theta}_2^T G_2^{-1}\tilde{\theta}_2\,, \tag{7.34}$$

where g_{b_2} is a positive constant and G_{β_2}, G_{θ_2} are symmetric positive definite matrices. The time derivative of (7.34) with respect to the dynamics (7.23),(7.27), with $\varphi_2 = \varphi_2^*$, is

$$\begin{aligned}\dot{V}_1^* = \dot{V}_0^* &- 2\tilde{\varphi}_1\frac{\partial\varphi_1^*}{\partial y}\left[C_cA_c\tilde{z} + \tilde{b}_{\rho_2}(\tilde{\varphi}_1 + \varphi_1^*) + \mu_1^T\tilde{\beta}_2 + \gamma^T(y,\xi)\tilde{\theta}_2 + \alpha_1\right] \\ &-2\bar{\lambda}_1\tilde{\varphi}_1^2 + 2g_{b_2}^{-1}\dot{\tilde{b}}_{\rho_2}\tilde{b}_{\rho_2} + 2\tilde{\beta}_2^T G_{\beta_2}^{-1}\dot{\tilde{\beta}}_2 + 2\tilde{\theta}_2^T G_{\theta_2}^{-1}\dot{\tilde{\theta}}_2\,.\end{aligned} \tag{7.35}$$

We choose

$$\begin{aligned}\dot{\hat{b}}_{\rho_2} &= -g_{b_2}(\varphi_1^* + \tilde{\varphi}_1)\frac{\partial\varphi_1^*}{\partial y}\tilde{\varphi}_1 \\ \dot{\hat{\beta}}_2 &= -G_{\beta_2}\mu_1\frac{\partial\varphi_1^*}{\partial y}\tilde{\varphi}_1 \\ \dot{\hat{\theta}}_2 &= -G_{\theta_2}\gamma(y,\xi)\frac{\partial\varphi_1^*}{\partial y}\tilde{\varphi}_1 \\ \alpha_1 &= \frac{1}{2}k_{\alpha_1}\tilde{\varphi}_1\frac{\partial\varphi_1^*}{\partial y}\,,\end{aligned} \tag{7.36}$$

with k_{α_1} a positive scalar, so that (7.35) becomes

$$\dot{V}_1^* = \dot{V}_0^* - \tilde{\varphi}_1^2\left[2\bar{\lambda}_1 + k_{\alpha_1}\left(\frac{\partial\varphi_1^*}{\partial y}\right)^2\right] - 2\tilde{\varphi}_1\frac{\partial\varphi_1^*}{\partial y}C_cA_c\tilde{z}\,, \tag{7.37}$$

and, in view of (7.18), a positive finite value of k_{α_1} exists such that

$$\dot{V}_1^* \leq -k_L\left\|\begin{bmatrix}\|z\| \\ \|\tilde{z}\| \\ \|\tilde{\varphi}_1\|\end{bmatrix}\right\|^2. \tag{7.38}$$

From (7.11), (7.30), (7.32), (7.33) and (7.36), we have

$$\begin{aligned}\varphi_2^* &= -y\hat{b}_{\rho_x} + \bar{\lambda}_1\varphi_1^* + f_1 + \frac{\partial\varphi_1^*}{\partial y} \\ &\quad\times\left[C_cA_c\hat{z} + \hat{b}_{\rho_2}(\tilde{\varphi}_1 + \varphi_1^*) + \gamma_0(y,\xi) + \mu_1^T\hat{\beta}_2 + \gamma^T(y,\xi)\hat{\theta}_2 - \frac{1}{2}\tilde{\varphi}_1\frac{\partial\varphi_1^*}{\partial y}k_{\alpha_1}\right] \\ &\triangleq \varphi_2^*(y,\xi,\hat{z},k,\hat{b}_{\rho_x},\hat{\beta}_1,\hat{\theta}_1,\hat{b}_{\rho_1},\mu_1,\hat{\beta}_2,\hat{\theta}_2,\hat{b}_{\rho_2},\mu_2,\tilde{\varphi}_1)\,.\end{aligned} \tag{7.39}$$

If $\rho = 2$, φ_2^* represents the final control $\tilde{\varphi}_2^* = \sigma(y)u$, so that the input u is given by

$$u = \sigma^{-1}(y)\varphi_2^* \triangleq u(y, \xi, \hat{z}, \hat{k}, \hat{b}_{\rho_x}, \hat{\beta}_1, \hat{\theta}_1, \hat{b}_{\rho_1}, \mu_1, \hat{\beta}_2, \hat{\theta}_2, \hat{b}_{\rho_2}, \mu_2, \tilde{\varphi}_1) . \tag{7.40}$$

If $\rho > 2$, we make a further change of coordinates $\tilde{\varphi}_2 = \varphi_2 - \varphi_2^*$, consider also the dynamics of $\tilde{\varphi}_2$

$$\dot{\tilde{\varphi}}_2 = -\bar{\lambda}_2\tilde{\varphi}_2 + \varphi_3 - \bar{\lambda}_2\varphi_2^* - \dot{\varphi}_2^* , \tag{7.41}$$

and consider φ_3 as a control variable which is designed according to the function

$$\bar{V}_2 = V_1 + \tilde{\varphi}_2^2 , \tag{7.42}$$

whose time derivative satisfies the inequality

$$\dot{\bar{V}}_2 \leq -k_{\mathrm{L}} \left\| \begin{bmatrix} \|z\| \\ \|\tilde{z}\| \\ \|\tilde{\varphi}_1\| \end{bmatrix} \right\|^2 + 2\tilde{\varphi}_1\tilde{\varphi}_2 + \tilde{\varphi}_2\dot{\tilde{\varphi}}_2 . \tag{7.43}$$

From (7.41) and (7.43) we can obtain the ideal expression of φ_3

$$\varphi_3^{id} = \bar{\lambda}_2\varphi_2^* + \dot{\varphi}_2^* - \tilde{\varphi}_1 , \tag{7.44}$$

and proceed as we did in the design of φ_2. At the end of this step, we obtain an expression of the desired φ_3 in the following form

$$\varphi_3^* = \varphi_3^*(y, \xi, \hat{z}, \hat{k}, \hat{b}_{\rho_x}, \hat{\beta}_1, \hat{\theta}_1, \hat{b}_{\rho_1}, \mu_1, \hat{\beta}_2, \hat{\theta}_2, \hat{b}_{\rho_2}, \mu_2, \tilde{\varphi}_1, \hat{b}_{\rho_3}, \hat{\beta}_3, \hat{\theta}_3, \mu_3, \tilde{\varphi}_2) , \tag{7.45}$$

having introduced a new estimate $\hat{y}_2$ of the time derivative of the output along with new parameter estimators $\hat{b}_{\rho_3}$, $\hat{\beta}_3$, $\hat{\theta}_3$ and a new function $\alpha_2(t)$. Proceeding the same way, we can obtain the final control u, which will be of the form

$$u = u(y, \xi, \hat{z}, \hat{k}, \hat{b}_{\rho_x}, \hat{\beta}_1, \ldots, \hat{\beta}_\rho, \hat{\theta}_1, \ldots, \hat{\theta}_\rho, \hat{b}_{\rho_1}, \ldots, \hat{b}_{\rho_\rho}, \mu_1, \ldots, \mu_\rho, \tilde{\varphi}_1, \ldots, \tilde{\varphi}_{\rho-1}) , \tag{7.46}$$

where $\mu_\rho = [\mu_{\rho,\rho}, \ldots, \mu_{n-1,\rho}]^{\mathrm{T}}$, with the associated function

$$V_{\rho-1} = V_0 + g_{b_x}^{-1}\tilde{b}_{\rho_x}^2 + \sum_{i=1}^{\rho-1} \tilde{\varphi}_i^2 + \sum_{i=2}^{\rho} g_{b_i}^{-1}\tilde{b}_{\rho_i} + \sum_{i=2}^{\rho} \tilde{\beta}_i^{\mathrm{T}} G_{\beta_i}^{-1}\tilde{\beta}_i + \sum_{i=2}^{\rho} \tilde{\theta}_i^{\mathrm{T}} G_{\theta_i}^{-1}\tilde{\theta}_i , \tag{7.47}$$

whose time derivative satisfies the inequality

$$\dot{V}_{\rho-1} \leq -k_{\mathrm{L}} \left\| \begin{bmatrix} \|z\| \\ \|\tilde{z}\| \\ \|\tilde{\varphi}\| \end{bmatrix} \right\|^2 , \tag{7.48}$$

where $\tilde{\varphi} = [\tilde{\varphi}_1, \ldots, \tilde{\varphi}_{\rho-1}]^{\mathrm{T}}$. From (7.47)–(7.48) it follows that $\|z\|$,$\|\tilde{z}\|$, $\|\tilde{\varphi}\|$, $\hat{k}$, $\hat{b}_{\rho_x}$, $\hat{b}_{\rho_i}$, $\hat{\beta}_i$, $\hat{\theta}_i$, $1 \leq i \leq \rho$, and, consequently, $y = C_{\mathrm{c}}z$ are bounded. The boundedness of $\|\xi(t)\|$ follows from (7.9). From (7.4), the differential equation relating φ_1 to y is

$$\sum_{i=1}^{n} \bar{d}_{1i} \frac{d^{n-i}}{dt^{n-i}} \varphi_1 = \frac{d^n y}{dt^n} - \sum_{i=1}^{n} \frac{d^{n-i}}{dt^{n-i}} \left[\sum_{j=0}^{p} \psi_{ji}(y) \theta_i \right]. \tag{7.49}$$

The first set of filters in (7.6) can be written as

$$\frac{d^{n-1}}{dt^{n-1}} \mu_{n-1,1} + d_{12} \frac{d^{n-2}}{dt^{n-2}} \mu_{n-1,1} + \cdots + d_{1n} \mu_{n-1,1} = \varphi_1 \operatorname{sgn}(b_\rho), \tag{7.50}$$

which substituted in (7.49), gives

$$\operatorname{sgn}(b_\rho) \sum_{i=1}^{2n-1} a_i \frac{d^{2n-1-i}}{dt^{2n-1-i}} \mu_{n-1,1} = \frac{d^n y}{dt^n} - \sum_{i=1}^{n} \frac{d^{n-i}}{dt^{n-i}} \left[\sum_{j=0}^{p} \psi_{ji}(y) \theta_i \right], \tag{7.51}$$

where a_i, $1 \leq i \leq 2n-1$, are the coefficients of the Hurwitz polynomial defined by

$$(\bar{d}_{11} s^{n-1} + \cdots + \bar{d}_{1n})(s^{n-1} + d_{12} s^{n-2} + \cdots + d_{1n}) = a_1 s^{2n-2} + \cdots + a_{2n-1} \,. \tag{7.52}$$

As shown in the proof of Theorem 6.1, (7.51) implies that $\mu_{n-1,1}$ is bounded. Similarly, we can prove that $\mu_{j,1}$, $1 \leq j \leq n-2$, are bounded. From (7.11), we can see that φ_1^* is bounded, and therefore φ_1 is also bounded, since $\tilde{\varphi}_1$ has been already proved to be bounded. Hence, the input to the filters (7.6) is bounded, which implies that all the state vectors μ_j, $1 \leq j \leq n-1$ and, from (7.46), the input u, are bounded. Thus, from (7.1), (7.9), (7.10), (7.13)–(7.15), (7.19) and (7.21), $\|\dot{z}\|$, $\|\dot{\tilde{z}}\|$ and $\|\dot{\tilde{\varphi}}\|$ are bounded. Recalling (7.48) and applying Barbalat's Lemma, we obtain

$$\begin{aligned} \lim_{t\to\infty} \|z(t)\| &= 0 \\ \lim_{t\to\infty} \|\tilde{z}(t)\| &= 0 \\ \lim_{t\to\infty} \|\tilde{\varphi}(t)\| &= 0 \,, \end{aligned} \tag{7.53}$$

and, consequently,

$$\lim_{t\to\infty} y(t) = 0 \,. \tag{7.54}$$

Now consider the systems (7.9) and (7.51). By virtue of (7.54), we can apply [22, p. 59], which yields

$$\begin{aligned} \lim_{t\to\infty} \|\xi(t)\| &= 0 \\ \lim_{t\to\infty} \mu_{n-1,1}(t) &= 0 \,, \end{aligned} \tag{7.55}$$

and, similarly,

$$\lim_{t\to\infty} \mu_{i,1}(t) = 0, \qquad 1 \leq i \leq n-2 \,, \tag{7.56}$$

which, along with (7.46) and (7.53)–(7.55), imply

$$\lim_{t\to\infty} u(t) = 0 \,.$$

Hence, we can apply again [22, p. 59] to (7.1) and (7.6), to obtain

$$\lim_{t\to\infty} \|\mu_i(t)\| = 0, \qquad 1 \le i \le n-1$$
$$\lim_{t\to\infty} \varphi_i(t) = 0, \qquad 1 \le i \le \rho-1\,.$$

Since $\zeta(t)$ is related to $z(t)$ by the transformations (7.2), (7.7) and (7.9), in view of (7.53)–(7.55) and (7.56), we finally have that

$$\lim_{t\to\infty} \|\zeta(t)\| = 0\,,$$

which, since $\zeta = T(x)$ with $T(0) = 0$, implies

$$\lim_{t\to\infty} \|x(t)\| = 0\,.$$

□

Example 7.1. Consider the system

$$\begin{aligned} \dot{x}_1 &= x_2 + \theta x_1^2 \\ \dot{x}_2 &= x_3 \\ \dot{x}_3 &= bu, \qquad b > 0 \\ y &= x_1\,. \end{aligned}$$

This system is both globally state-feedback linearizable and input-output linearizable by state feedback for any known value of the parameters θ and b. Nevertheless, the adaptive algorithms proposed in [3,4,8] do not apply. A global adaptive stabilizing state-feedback control for a simpler example (b=1) was first obtained in [11]; this system belongs to the class of systems in the so-called "pure feedback form" which can be globally adaptively stabilized by state feedback via the systematic algorithm proposed in [5]. According to Theorem 7.1 this example can be adaptively globally stabilized by output feedback: in fact, the example is of relative degree $\rho = 3$ and belongs to the class of systems (SR), with $b = [0, 0, 1]^{\mathrm{T}}$ a Hurwitz vector of degree $\rho = 3$. □

Remark 7.1. Let us examine how Theorem 7.1 applies to a single-input, single-output linear system characterized by the transfer function of known relative degree ρ

$$W(s) = \frac{b_\rho s^{n-\rho} + \cdots + b_n}{s^n + a_1 s^{n-1} + \cdots + a_n} = \frac{n(s)}{d(s)}\,, \tag{7.57}$$

where the numerator is a Hurwitz polynomial with unknown coefficients but with b_ρ of known sign, while the denominator is a polynomial with unknown coefficients a_i, $1 \le i \le n$. The above assumptions are standard ones in adaptive control of linear systems [21, p. 183]. A realization in observer form of (7.57) can be written as

$$\dot{\zeta} = \begin{bmatrix} 0\,1\,0\cdots 0 \\ 0\,0\,1\cdots 0 \\ \vdots\,\vdots\,\vdots\,\ddots\,\vdots \\ 0\,0\,0\cdots 1 \\ 0\,0\,0\cdots 0 \end{bmatrix} \zeta + \begin{bmatrix} 0 \\ \vdots \\ b_\rho \\ \vdots \\ b_n \end{bmatrix} u - \begin{bmatrix} a_1 \\ a_2 \\ a_3 \\ \vdots \\ a_n \end{bmatrix} y \tag{7.58}$$

$$y = [1\,0\,0\cdots 0]\,\zeta .$$

If we set $\theta = [a_1, \ldots, a_n, b_\rho, \ldots, b_n]^{\mathrm{T}}$ as the $(2n-\rho+1)$-dimensional vector of unknown parameters, the system (7.58) is of type (SR) and Theorem 7.1 (or Theorem 6.1 when $\rho = 1$) applies, i.e., there exists an adaptive output-feedback control which drives the state $\zeta(t)$ to zero. □

8 Conclusions

We have determined in Theorem 2.1 a class of single-output nonlinear systems, linear with respect to an unknown parameter vector, for which adaptive observers for equivalent states can be designed without the assumption of persistency of excitation (Theorem 3.3). Theorem 3.3 generalizes to a class of nonlinear systems the available theory for linear systems. Theorem 4.1 gives sufficient conditions under which the adaptive observer tracks the original physical coordinates.

The observers obtained in Theorem 3.3 and filtered transformations are then used to design adaptive output-feedback stabilizing controls for a more restricted class of nonlinear systems characterized by the geometric conditions given in Theorem 5.1. Theorem 5.2 shows that such a class of systems is characterized by a global linear asymptotically stable zero dynamics. Theorem 6.1 (for systems with relative degree equal to one) and Theorem 7.1 (for systems with relative degree greater than one) provide the construction of globally stabilizing adaptive output-feedback controls. Linear minimum-phase systems with unknown poles and zeroes, known sign of the high-frequency gain and known relative degree belong to the class of systems determined. As shown by Example 6.1, systems which are not feedback linearizable for every value of the unknown parameter vector may belong to such a class and therefore be adaptively stabilized by output feedback.

The theory presented in this paper for adaptive stabilization also leads to the design of output-tracking adaptive controls by output feedback by introducing a suitable linear reference model.

Acknowledgements

The authors would like to thank P. V. Kokotovic and I. Kanellakopoulos for helpful discussions on the proof of Theorem 7.1 and for making them available a preprint of [5].

References

1. A. Isidori, *Nonlinear Control Systems*, 2nd ed., Berlin, Springer-Verlag, 1989.
2. H. Nijmeijer and A. van der Schaft, *Nonlinear Dynamical Control Systems*, Berlin, Springer-Verlag, 1990.
3. D. Taylor, P. Kokotovic, R. Marino, and I. Kanellakopoulos, "Adaptive regulation of nonlinear systems with unmodeled dynamics," *IEEE Trans. Aut. Control*, vol. 34, pp. 405–412, April 1989.
4. I. Kanellakopoulos, P. Kokotovic, and R. Marino, "An extended direct scheme for robust adaptive nonlinear control," *Automatica*, to appear, March 1991.
5. I. Kanellakopoulos, P. V. Kokotovic, and A. S. Morse, "Systematic design of adaptive controllers for feedback linearizable systems," Report UILU-ENG-90-2249, DC-124, University of Illinois at Urbana Champaign, October 1990.
6. K. Nam and A. Arapostathis, "A model-reference adaptive control scheme for pure-feedback nonlinear systems," *IEEE Trans. Aut. Control*, vol. 33, pp. 803–811, Sept. 1988.
7. G. Campion and G. Bastin, "Indirect adaptive state feedback control of linearly parametrized nonlinear systems," *Int. J. Adapt. Control Sig. Proc.*, vol. 4, pp. 345–358, 1990.
8. S. S. Sastry and A. Isidori, "Adaptive control of linearizable systems," *IEEE Trans. Aut. Control*, vol. 34, pp. 1123–1131, Nov. 1989.
9. G. Bastin and M. R. Gevers, "Stable adaptive observers for nonlinear time-varying systems," *IEEE Trans. Aut. Control*, vol. 33, pp. 650–658, July 1988.
10. R. Marino, "Adaptive observers for single output nonlinear systems," *IEEE Trans. Aut. Control*, vol. 35, pp. 1054–1058, Sept. 1990.
11. J. B. Pomet and L. Praly, "Adaptive nonlinear regulation: equation *Proc. 8th IEEE Conf. Dec. Control*, pp. 1008–1013, Tampa, FL, 1989.
12. L. Praly, G. Bastin and J. B. Pomet, "Adaptive stabilization of nonlinear systems," *Proc. Conf. Analysis of controlled dynamical systems*, Lyon, France, 1990.
13. I. Kanellakopoulos, P. V. Kokotovic and R. H. Middleton, "Observer-based adaptive control of nonlinear systems under matching conditions," *Proc. 1990 Amer. Control Conf.*, San Diego, CA, pp. 549–555.
14. I. Kanellakopoulos, P. V. Kokotovic, and R. H. Middleton, "Indirect adaptive output-feedback control of a class of nonlinear systems," *Proc. 29th IEEE Conf. Dec. Control*, pp. 2714–2719, Honolulu, HI, 1990.
15. R. Marino and P. Tomei, "Global adaptive observers for nonlinear systems via filtered transformations," Report R-90.06, University of Rome "Tor Vergata", September 1990.
16. C. I. Byrnes and A. Isidori, "New results and examples in nonlinear feedback stabilization," *Syst. Control Lett.*, vol. 12, pp. 437–442, 1989.
17. P. V. Kokotovic and H. J. Sussmann, "A positive real condition for global stabilization of nonlinear systems," *Syst. Control Lett.*, vol. 12, pp. 125–134, 1989.
18. J. Tsinias, "Sufficient Liapunov-like conditions for stabilization," *Math. Control Signals Syst.*, vol. 2, pp. 343–357, 1989.
19. A. J. Krener and A. Isidori, "Linearization by output injection and nonlinear observers," *Syst. Control Lett.*, vol. 3, pp. 47–52, 1983.
20. W. Respondek, "Global aspects of linearization, equivalence to polynomial forms and decomposition of nonlinear control systems," in *Algebraic and Geometric Methods in Nonlinear Control Theory*, M. Fliess and M. Hazewinkel, eds., pp. 257–284, D. Reidel Publishing Co., Dordrecht, 1986.

21. K. S. Narendra and A. M. Annaswamy, *Stable Adaptive Systems*, Prentice-Hall, Inc., Englewood Cliffs, NJ, 1989.
22. C. A. Desoer and M. Vidyasagar, *Feedback Systems: Input-Output Properties*, Academic Press, Orlando, 1975.
23. V. Popov, *Hyperstability of Control Systems*, Berlin, Springer-Verlag, 1973.
24. B. D. O. Anderson, R. R. Bitmead, C. R. Johnson, Jr., P. V. Kokotovic, R. L. Kosut, I. M. Y. Mareels, L. Praly, and B. D. Riedle, *Stability of Adaptive Systems: Passivity and Averaging Analysis*, MIT Press, Cambridge, MA, 1986.
25. W. P. Dayawansa, W. M. Boothby, and D. L. Elliott, "Global state and feedback equivalence of nonlinear systems," *Syst. Control Lett.*, vol. 6, pp. 229–234, 1985.
26. W. Respondek, "Linearization, feedback and Lie brackets," *Proc. Int. Conf. Geometric Theory of Nonlinear Control Systems*, Wroclaw Technical University Press, Wroclaw, 1985, pp. 131–166.

Adaptive Output-Feedback Control of Systems with Output Nonlinearities*

I. Kanellakopoulos,[1] *P. V. Kokotović,*[1] *and A. S. Morse*[2]

[1] Coordinated Science Laboratory
University of Illinois, Urbana, IL 61801, USA.
[2] Department of Electrical Engineering
Yale University, New Haven, CT 06520-1968, USA.

Abstract. For a class of single-input single-output nonlinear systems with unknown constant parameters, we present a direct model-reference adaptive control scheme, which requires only output, rather than full-state, measurement. The nonlinearities are not required to satisfy any growth conditions. The assumptions on the linear part of the nonlinear system are the same as in the standard adaptive control problem for linear systems, which now appears as a special case of the nonlinear problem solved in this paper.

1 Introduction

The two most common assumptions employed in adaptive nonlinear control are those of linear parametrization [1–13] and full-state feedback [1–11]. The purpose of this paper is to avoid the full-state feedback assumption and to remove the specific restrictions of previous output-feedback results [12,13].

In the linear case, the adaptive output-feedback designs follow either a direct model-reference path or an indirect path via adaptive observers. Current research on adaptive observers for nonlinear systems [14–16] indicates that the indirect path may become promising for adaptive nonlinear control. However, the major stumbling block along this path continues to be its linear-like proof of stability which imposes restrictive conic conditions on the nonlinearities [12,13]. Under such linear growth constraints the actual nonlinear problem is, in fact, not addressed.

In this paper we formulate and solve a truly nonlinear output-feedback problem by following the direct model-reference path of Feuer and Morse [17]. In contrast to other more popular adaptive linear control methods [18,19], the method of Feuer and Morse offers a possibility to prove stability without any growth constraints. In a companion paper [20] we have exploited this possibility to solve a full-state-feedback adaptive nonlinear control problem. In this paper we present an adaptive output-feedback result without nonlinearity growth constraints.

* The work of the first two authors was supported in part by the National Science Foundation under Grant ECS-87-15811 and in part by the Air Force Office of Scientific Research under Grant AFOSR 90-0011. The work of the third author was supported by the National Science Foundation under Grants ECS-88-05611 and ECS-90-12551.

The results of this paper apply to nonlinear input-output models consisting of a linear transfer function and output-dependent nonlinearities. The coefficients of the transfer function and the parameters multiplying the nonlinearities are unknown. For the linear part, the assumptions of minimum phase and known sign of the high-frequency gain are the same as in the adaptive linear control theory, which now appears as a special case of the nonlinear theory presented in this paper.

For easier understanding, the new adaptive scheme is first designed for a particular system of sufficient complexity to be illustrative of both the design procedure and the stability properties of the resulting closed-loop adaptive system. In Section 2 we design the adaptive scheme for this system and then prove the stability and tracking properties of the resulting adaptive system in full detail. The design procedure for the general case is presented in Section 3, and the proof of stability and tracking is given in Section 4.

Nonlinear input-output models are intimately tied to state-space equations which originate from nonlinear physical laws expressed in specific state coordinates. In Section 5 we give a state-space form of the class of nonlinear plants which have the desired input-output representation, and characterize this class of plants via a set of geometric conditions.

2 Adaptive Scheme Design: An Example

The purpose of this section is to make both the proposed adaptive scheme and the main features of the Feuer-Morse method more easily accessible to the reader with the usual background in control theory and limited familiarity with adaptive linear control.

2.1. Nonlinear system properties. The nonlinear system is assumed to be minimum-phase [21, Chap. 4] and its nonlinearities depend only on the output variable. This implies that the nonlinear system is linearizable by output injection [22]. The input-output description of a typical nonlinear system of this kind is given by

$$D^5 y = (D^2 + 2D + 1)u + \theta\,\left[D^2 p_2(y) + Dp_1(y) + p_0(y)\right], \tag{2.1}$$

where u and y are the scalar control and output, respectively, $D = \frac{d}{dt}$, and θ is an unknown constant parameter. To address a truly nonlinear problem, we choose the nonlinearities which do not satisfy linear growth constraints:

$$p_0(y) = y^3\,, \quad p_1(y) = y^2 + 2y^3\,, \quad p_2(y) = ye^y + 2y^2 + y^3\,. \tag{2.2}$$

It is important to notice that these nonlinearities are not in the span of u, and, hence, the system (2.1) is not full-state linearizable by static output feedback, or even by static full-state feedback, as shown in Sect. 5. However, it is input-output linearizable by full-state feedback [21, Chap. 4].

The above structural and growth properties of (2.1) and its relative degree [21, Chap. 4] show that (2.1) is a nonlinear system of considerable complexity. However, this system also satisfies a structural constraint under which the results of this paper are applicable: *the nonlinearities do not enter the system before the control input u.*

2.2. Augmenting the CE control. As in most adaptive designs, our first step is to find a dynamic output-feedback control that guarantees the specified stability and tracking properties *when the parameter θ is known.* Most adaptive schemes then replace the unknown θ with its estimate $\hat{\theta}$ and implement the so formed "certainty-equivalence" control. Such certainty-equivalence designs have been satisfactory in adaptive linear control, but have failed to produce truly nonlinear results because of their inherent linear growth constraints [12,13]. To avoid this difficulty we must go beyond the certainty-equivalence approach. Following Feuer and Morse [17], we will *augment the certainty-equivalence control* by an additive term $\bar{u}$ which will counteract the effects of rapidly growing nonlinearities. It will also provide us with additional flexibility in the proof of stability.

The certainty-equivalence part of our control will be designed to match a reference model of the same relative degree as that of the nonlinear plant (2.1). As this plant is input-output linearizable by full-state feedback, we will choose the simplest linear reference model of relative degree three:

$$(D+1)^3 y_r = r\,. \tag{2.3}$$

The first step in matching this reference model is to filter the plant equation (2.1) by the strictly proper stable filter F/E_2, where F is a monic polynomial of degree 2, and E_2 is a monic Hurwitz polynomial of degree 4. This results in

$$\frac{FA}{E_2}y = \frac{FB}{E_2}u + \theta\left[\frac{FD^2}{E_2}p_2(y) + \frac{FD}{E_2}p_1(y) + \frac{F}{E_2}p_0(y)\right], \tag{2.4}$$

where $A = D^5$, $B = D^2 + 2D + 1$ as in (2.1). It is now straightforward to verify that the desired matching is achieved by the control

$$u^* = -\frac{G}{E_2}y + r - \theta\nu(y) - \frac{FB - E_2}{E_2}u^*\,, \tag{2.5}$$

provided that

$$\nu(y) = \frac{FD^2}{E_2}p_2(y) + \frac{FD}{E_2}p_1(y) + \frac{F}{E_2}p_0(y)\,, \tag{2.6}$$

and that G, a polynomial of degree 4, and F satisfy the polynomial equation

$$FD^5 + G = (D+1)^3 E_2\,. \tag{2.7}$$

Note that the polynomial $FB - E_2$ in (2.5) is of degree 3, since FB and E_2 are both monic polynomials of degree 4. As an illustration, the choice $E_2 = (D+2)^4$ yields the following solution of (2.7):

$$F = D^2 + 11D + 51 \tag{2.8}$$

$$G = 129D^4 + 192D^3 + 168D^2 + 80D + 16\,. \tag{2.9}$$

When the control (2.5) is applied to the system (2.1) and the initial conditions of the filters used in (2.4), (2.5) and (2.6) are exactly matched with those of the system (2.1), then (2.5) achieves the exact tracking $y(t) = y_r(t)$ for all $t \geq 0$. However, the initial conditions of (2.1) are unknown and the tracking can be achieved only asymptotically, that is,

$$y(t) = y_r(t) + \epsilon(t) \rightarrow y_r(t) \quad \text{as } t \rightarrow \infty, \tag{2.10}$$

where $\epsilon(t)$ is the exponentially decaying tracking error caused by the mismatch of the initial conditions.

When the parameter θ is unknown, we replace it in (2.5) by its estimate $\hat{\theta}$, to be obtained from a parameter update law. To this "certainty-equivalence" part of our control we add a term $\bar{u}$ which will be a handy tool later. So, our adaptive control will be of the following form:

$$u = -\frac{G}{E_2}y + r - \hat{\theta}\nu(y) - \frac{FB - E_2}{E_2}u + \bar{u}. \tag{2.11}$$

When applied to the nonlinear plant (2.1), this control yields the following input-output description of the resulting feedback system:

$$y = \frac{1}{(D+1)^3}\left[r + \left(\theta - \hat{\theta}\right)\nu(y) + \bar{u}\right] + \epsilon(t), \tag{2.12}$$

where, as in the case when θ was known, $\epsilon(t)$ contains all the exponentially decaying terms caused by the mismatch of the initial conditions. It should be observed that with an exact estimate $\hat{\theta} = \theta$ the linearization of (2.12) is achieved.

Introducing the error variables

$$e = y - y_r, \quad \tilde{\theta} = \theta - \hat{\theta}, \tag{2.13}$$

and taking the difference between (2.12) and the reference model (2.3), we obtain the tracking error equation:

$$e = \frac{1}{(D+1)^3}\left[\tilde{\theta}\nu(y) + \bar{u}\right] + \epsilon(t). \tag{2.14}$$

2.3. Error augmentation and swapping. Following the standard practice in adaptive control, we now set out to construct an error equation in which the parameter error is filtered only by a strictly positive real (SPR) transfer function. As a first step, we rewrite (2.14) in the form

$$\begin{aligned} e = {} & \frac{1}{(D+1)}\left[\tilde{\theta}\frac{1}{(D+1)^2}\nu(y)\right] + \frac{1}{(D+1)^3}\bar{u} \\ & + \frac{1}{(D+1)^3}\left[\tilde{\theta}\nu(y)\right] - \frac{1}{(D+1)}\left[\tilde{\theta}\frac{1}{(D+1)^2}\nu(y)\right] + \epsilon(t). \end{aligned} \tag{2.15}$$

The first term in (2.15) is in the desired SPR form, while the second term is due to the additional control term $\bar{u}$. As for the third and fourth terms, these are the familiar *swapping terms*, whose presence is caused by the time-varying nature of $\tilde{\theta}$: if $\tilde{\theta}$ were constant, these two terms would cancel out. Let us therefore define the *augmented error* $\bar{e}$ as

$$\bar{e} = e + \eta_0 , \tag{2.16}$$

where the term η_0 represents all the undesirable terms in (2.14):

$$\eta_0 = -\left[\frac{1}{(D+1)^3}\bar{u} + \frac{1}{(D+1)^3}\left[\tilde{\theta}\nu(y)\right] - \frac{1}{(D+1)}\left[\tilde{\theta}\frac{1}{(D+1)^2}\nu(y)\right]\right] . \tag{2.17}$$

The signal multiplying $\tilde{\theta}$ in the first brackets is of particular importance and is denoted by

$$h_1 = \frac{1}{(D+1)^2}\nu(y) . \tag{2.18}$$

Considering $\nu(y)$ as the input and h_1 as the output, we represent (2.18) in the state-space form

$$\begin{bmatrix} \dot{h}_1 \\ \dot{h}_2 \end{bmatrix} = \dot{h} = A_0 h + b\nu(y) , \tag{2.19}$$

where

$$A_0 = \begin{bmatrix} 0 & 1 \\ -1 & -2 \end{bmatrix} , \quad b = \begin{bmatrix} 0 \\ 1 \end{bmatrix} . \tag{2.20}$$

It can now be verified that η_0 is the output of the third order system

$$\dot{\eta}_0 = -\eta_0 + \eta_1 \tag{2.21}$$

$$\begin{bmatrix} \dot{\eta}_1 \\ \dot{\eta}_2 \end{bmatrix} = \dot{\eta} = A_0\eta - b\bar{u} - h\dot{\hat{\theta}} . \tag{2.22}$$

The variables h_1 and η_1 from (2.19) and (2.22) allow us to express the tracking error as

$$e = \frac{1}{D+1}\left[\tilde{\theta}h_1 - \eta_1\right] + \epsilon(t) . \tag{2.23}$$

The analogous expression for the augmented error is

$$\bar{e} = \frac{1}{D+1}\left[\tilde{\theta}h_1\right] + \epsilon(t) , \tag{2.24}$$

and it has the desired SPR form: the parameter error $\tilde{\theta}$ multiplied by the "regressor" h_1 is the input into the SPR filter $1/(D+1)$.

2.4. Update law. From this point on, the route prescribed by most of the adaptive linear control literature is to choose a normalized gradient update law and to set $\bar{u} \equiv 0$ (thus returning to a pure certainty equivalence control). In the case of adaptive linear systems, boundedness of the closed-loop signals can then be established using the Gronwall lemma or some type of small-gain argument. Attempts to apply this type of stability proof to nonlinear systems have so far been successful only when conic constraints are imposed on the nonlinearities. Without such linear growth constraints, the term $\left(\theta - \hat{\theta}\right) \nu(y)$ can cause some signals to escape to infinity in finite time if the parameter error $\theta - \hat{\theta}$ is not rapidly decreased. The difficulty with normalizations of update laws is that they don't allow a rapid enough decrease of the parameter error when this error is most harmful. A simulation example of instability of a full-state-feedback scheme with normalization [10] is given in our companion paper [20], where it is also shown that an unnormalized update law preserves global stability. We are, therefore, motivated to look for an unnormalized update law.

The SPR form of (2.24) suggests the unnormalized gradient update law

$$\dot{\hat{\theta}} = h_1 \bar{e} \,. \tag{2.25}$$

Given the complexity of the nonlinear system (2.1), it is likely that such a simple update law will shift the difficulties in the adaptive design to the proof of stability. Indeed, this is the case. A simple Lyapunov-like function involving $\bar{e}^2$ and $\tilde{\theta}^2$ has a nonpositive derivative, but fails to prove boundedness of y. It clearly shows, however, that η_0 must be taken into account. Our next attempt is with the function

$$V_\eta = \frac{1}{2}\left(\bar{e}^2 + \tilde{\theta}^2 + \int_t^\infty \epsilon^2(\tau) d\tau\right) + \frac{1}{2}\eta^{\mathrm{T}} P \eta \,, \tag{2.26}$$

where P is the positive definite solution of the Lyapunov equation

$$PA_0 + A_0^{\mathrm{T}} P = -2I \Rightarrow P = \begin{bmatrix} 3 & 1 \\ 1 & 1 \end{bmatrix} . \tag{2.27}$$

Using (2.22), (2.24) and (2.25), we compute

$$\dot{V}_\eta = -\frac{1}{2}\left(\bar{e} - \epsilon(t)\right)^2 - \frac{1}{2}\left[\bar{e}^2 + 2\eta^{\mathrm{T}} P h h_1 \bar{e} + 2\eta^{\mathrm{T}} P b \bar{u}\right] , \tag{2.28}$$

and try to render it nonpositive. The tool we have prepared for this task is $\bar{u}$. However, it turns out to be impossible to counteract the effects of the two-dimensional vector $hh_1\bar{e}$ by $\bar{u}$ alone. Hence, we need an additional degree of freedom involving h_1. This prompts us to replace η in (2.26) by the new vector

$$\begin{aligned} \zeta_1 &= \eta_1 \\ \zeta_2 &= \xi_1(h_1)\eta_1 + \eta_2 \,, \end{aligned} \tag{2.29}$$

where the nonlinear function $\xi_1(h_1)$ is at our disposal. With the new variables ζ, the nonnegative function to be used in our proof is

$$V = \frac{1}{2}\left(\bar{e}^2 + \tilde{\theta}^2 + \int_t^\infty \epsilon^2(\tau) d\tau\right) + \frac{1}{2}\zeta^{\mathrm{T}} P \zeta \,. \tag{2.30}$$

To evaluate $\dot{V}$ we need $\dot{\zeta}$, which is obtained by differentiating (2.29) and using (2.19)–(2.22):

$$\begin{aligned}\dot{\zeta}_1 &= \zeta_2 - h_1\dot{\hat{\theta}} - \xi(h_1)\zeta_1 \\ \dot{\zeta}_2 &= -\zeta_1 - 2\zeta_2 + 2\xi_1(h_1)\zeta_1 - \bar{u} - h_2\dot{\hat{\theta}} \\ &\quad + \xi_1(h_1)\left[\zeta_2 - h_1\dot{\hat{\theta}} - \xi_1(h_1)\zeta_1\right] + \frac{\partial \xi_1}{\partial h_1}h_2\zeta_1 \,.\end{aligned} \tag{2.31}$$

Introducing the notation

$$w = \begin{bmatrix} -h_1^2 \\ -h_1h_2 - h_1^2\xi_1 \end{bmatrix}, \tag{2.32}$$

we compute the time derivative of V as:

$$\begin{aligned}\dot{V} = -\frac{1}{2}(\bar{e} - \epsilon(t))^2 - \zeta^{\mathrm{T}}\zeta - \Bigg\{ &\frac{1}{2}\bar{e}^2 - \zeta^{\mathrm{T}}Pw\bar{e} \\ &+ \zeta^{\mathrm{T}}P \begin{bmatrix} \xi_1\zeta_1 \\ -2\xi_1\zeta_1 + \bar{u} - \xi_1\zeta_2 + \xi_1^2\zeta_1 - \dfrac{\partial \xi_1}{\partial h_1}h_2\zeta_1 \end{bmatrix} \Bigg\} \,.\end{aligned} \tag{2.33}$$

2.5. Design equation. We now have two tools to make $\dot{V}$ nonpositive: the function $\xi_1(h_1)$ and the control term $\bar{u}$. With these tools we will attempt to represent the quantity enclosed in braces in (2.33) as the sum of two squares. It turns out that this is possible to achieve by decomposing P as $P = P_1 + P_2$ such that the following *design equation* holds:

$$P_1ww^{\mathrm{T}}P_1\zeta + P_2ww^{\mathrm{T}}P_2\zeta = P \begin{bmatrix} \xi_1\zeta_1 \\ -2\xi_1\zeta_1 + \bar{u} - \xi_1\zeta_2 + \xi_1^2\zeta_1 - \dfrac{\partial \xi_1}{\partial h_1}h_2\zeta_1 \end{bmatrix} . \tag{2.34}$$

The substitution of (2.34) into (2.33) yields the desired form for $\dot{V}$:

$$\dot{V} = -\frac{1}{2}(\bar{e} - \epsilon(t))^2 - \zeta^{\mathrm{T}}\zeta - \left(\frac{\bar{e}}{2} - \zeta^{\mathrm{T}}P_1w\right)^2 - \left(\frac{\bar{e}}{2} - \zeta^{\mathrm{T}}P_2w\right)^2 \leq 0 \,. \tag{2.35}$$

Our task now is to find P_1, P_2, $\xi_1(h_1)$ and $\bar{u}$ which satisfy the design equation (2.34). For the example considered here, the systematic procedure of Sect. 3 gives the following solution for P_1 and P_2:

$$P_1 = \begin{bmatrix} 2 & 0 \\ 0 & 0 \end{bmatrix}, \quad P_2 = \begin{bmatrix} 1 & 1 \\ 1 & 1 \end{bmatrix}. \tag{2.36}$$

Substitution of (2.36) into (2.34) results in

$$\begin{aligned}&\begin{bmatrix} 2h_1^4\zeta_1 \\ \left[(h_1h_2 + h_1^2 + \xi_1h_1^2)^2 - 2h_1^4\right]\zeta_1 + (h_1h_2 + h_1^2 + \xi_1h_1^2)^2\zeta_2 \end{bmatrix} = \\ &\qquad = \begin{bmatrix} \xi_1\zeta_1 \\ \left(-2\xi_1 + \xi_1^2 - \dfrac{\partial \xi_1}{\partial h_1}h_2\right)\zeta_1 - \xi_1\zeta_2 + \bar{u} \end{bmatrix},\end{aligned} \tag{2.37}$$

which directly yields the following solutions for ξ_1 and $\bar{u}$:

$$\xi_1 = 2h_1^4 \tag{2.38}$$

$$\begin{aligned}\bar{u} = &\left[8h_1^3h_2 + 2h_1^4 - 4h_1^8 + (h_1h_2 + h_1^2 + 2h_1^6)^2\right]\zeta_1 \\ &+ \left[2h_1^4 + (h_1h_2 + h_1^2 + 2h_1^6)^2\right]\zeta_2 .\end{aligned} \tag{2.39}$$

Using (2.29) and the notation

$$\varphi_1 = 8h_1^3h_2 + 2h_1^4 + (1 + 2h_1^4)(h_1h_2 + h_1^2 + 2h_1^6)^2 \tag{2.40}$$

$$\varphi_2 = 2h_1^4 + (h_1h_2 + h_1^2 + 2h_1^6)^2 , \tag{2.41}$$

$\bar{u}$ can be defined in terms of available signals as

$$\bar{u} = \varphi_1\eta_1 + \varphi_2\eta_2 . \tag{2.42}$$

To summarize, the complete closed-loop adaptive system is

$$\begin{aligned}
&\textit{Plant:}\\
&\qquad D^5y = (D^2 + 2D + 1)u + \theta\left[D^2p_2(y) + Dp_1(y) + p_0(y)\right]\\
&\textit{Control:}\\
&\qquad u = -\frac{G}{E_2}y + r - \hat{\theta}\nu(y) - \frac{FB - E_2}{E_2}u + \varphi_1\eta_1 + \varphi_2\eta_2\\
&\textit{Update law:}\\
&\qquad \dot{\hat{\theta}} = h_1\bar{e}, \quad \bar{e} = e + \eta_0 = y - y_r + \eta_0\\
&\textit{Filters:}\\
&\qquad \dot{h} = A_0h + b\nu(y)\\
&\qquad \dot{\eta} = A_0\eta - b(\varphi_1\eta_1 + \varphi_2\eta_2) - hh_1\bar{e}\\
&\qquad \dot{\eta}_0 = -\eta_0 + \eta_1 ,
\end{aligned} \tag{2.43}$$

where y_r is the output of the reference model (2.3) and $\nu(y)$ is defined in (2.6).

2.6. Stability and tracking. The stability and tracking properties of (2.43) are now established using the nonnegative function V from (2.30), whose derivative, given in (2.35), is nonpositive. Because of the piecewise continuity of $r(t)$ and the smoothness of the nonlinear functions p_0, p_1, p_2, the solution of (2.43) has a maximum interval of definition, which we denote by $[0, t_f)$. We will now show that $t_f = \infty$.

From (2.30) and (2.35) we conclude that $\bar{e}$, ζ, and $\tilde{\theta}$ are bounded on $[0, t_f)$. Solving (2.29) for η and using (2.38) we obtain

$$\begin{aligned}\eta_1 &= \zeta_1\\ \eta_2 &= \zeta_2 - 2h_1^4\zeta_1 .\end{aligned} \tag{2.44}$$

Thus, the boundedness of ζ implies that η_1 is bounded, which, in turn, implies that η_0 is bounded: $\dot{\eta}_0 = -\eta_0 + \eta_1$. Since $\bar{e}$ and η_0 are bounded, e is bounded: $e = \bar{e} - \eta_0$. By the boundedness of y_r, this implies that y is bounded: $y = e + y_r$.

Hence, $\nu(y)$ is bounded, which means that h is bounded: $\dot{h} = A_0 h + b\nu(y)$. The boundedness of h and ζ implies that η and $\bar{u}$ are bounded (cf. (2.44) and (2.39)).

This does not yet prove that u is bounded. From (2.11), to prove the boundedness of u we only need to show that $\frac{FB-E_2}{E_2}u$ is bounded. Since $\frac{FB-E_2}{E_2}$ is of relative degree 1, we can express $\frac{FB-E_2}{E_2}u$ in the form

$$\frac{FB-E_2}{E_2}u = \frac{(D+1)^3(D+\lambda_1)}{E_2}\left[\frac{E_3}{(D+1)^3}u + \frac{1}{D+\lambda_1}\frac{E_4}{(D+1)^3}u\right] + \epsilon(t)\,, \tag{2.45}$$

where λ_1 is a positive constant and E_3, E_4 are polynomials of degree 2. Now (2.45) clearly shows that u is bounded if $\frac{D^i}{(D+1)^3}u$ is bounded for $i = 0, 1, 2$. Since y is bounded and the plant (2.1) is minimum phase and of relative degree 3, it follows that $\frac{1}{(D+1)^3}u$ is bounded. Differentiating (2.14) and substituting $-\hat{\theta}\nu(y) + \bar{u}$ from (2.11), we obtain for $i = 1, 2$:

$$\begin{aligned} e^{(i)} &= \frac{D^i}{(D+1)^3}\left[\tilde{\theta}\nu(y) + \bar{u}\right] + \epsilon(t) \\ &= \frac{D^i}{(D+1)^3}\left[\theta\nu(y) + \frac{FB}{E_2}u + \frac{G}{E_2}y - r\right] + \epsilon(t)\,. \end{aligned} \tag{2.46}$$

Using $h_i = \frac{D^{i-1}}{(D+1)^2}\nu(y)$, $i = 1, 2$, from (2.19) and rearranging terms in (2.46) we get

$$\frac{D^i}{(D+1)^3}\left[\frac{FB}{E_2}u\right] = e^{(i)} - \frac{D^i}{(D+1)^3}\left[\frac{G}{E_2}y - r\right] - \frac{D}{D+1}h_i\theta + \epsilon(t)\,. \tag{2.47}$$

From (2.16), (2.21), (2.22), (2.24) and (2.25) we have

$$\dot{e} = \dot{\bar{e}} - \dot{\eta}_0 = -\bar{e} + \tilde{\theta}h_1 + \eta_0 - \eta_1 + \epsilon(t) \tag{2.48}$$

$$\begin{aligned} \ddot{e} &= -\dot{\bar{e}} + \dot{\tilde{\theta}}h_1 + \tilde{\theta}\dot{h}_1 + \dot{\eta}_0 - \dot{\eta}_1 + \epsilon(t) \\ &= \bar{e} - \tilde{\theta}h_1 - h_1^2\bar{e} + \tilde{\theta}h_2 - \eta_0 + \eta_1 - \eta_2 + h_1^2\bar{e} + \epsilon(t)\,. \end{aligned} \tag{2.49}$$

Since $\bar{e}$, h, η, $\tilde{\theta}$ are bounded, (2.48) and (2.49) imply that $\dot{e}$, $\ddot{e}$ are bounded. Hence, by (2.47), $\frac{D^i}{(D+1)^3}\left[\frac{FB}{E_2}u\right]$ is bounded for $i = 1, 2$. The boundedness of $\frac{D^i}{(D+1)^3}u$ for $i = 1, 2$ then follows from the boundedness of $\frac{1}{(D+1)^3}u$ and the recursive expression

$$\frac{D^i}{(D+1)^3}u = \frac{D^i}{(D+1)^3}\frac{FB}{E_2}u - \frac{D(FB-E_2)}{E_2}\frac{D^{i-1}}{(D+1)^3}u + \epsilon(t)\,. \tag{2.50}$$

Next, we prove that the state of the plant is bounded. From (2.12) it follows that $D^i y$, $0 \leq i \leq 3$ are bounded. Combining this with the fact that the plant is minimum phase, we conclude that the state of any minimal realization of (2.1) is bounded on $[0, t_f)$.

Thus, we have shown that the state of the closed-loop adaptive system (2.43) is bounded on its maximum interval of existence $[0, t_f)$. Therefore, $t_f = \infty$.

Finally, we prove convergence of the tracking error to zero. From $t_f = \infty$ and (2.35) we conclude that $\dot{V}$ is bounded and integrable on $[0, \infty)$. Furthermore, the boundedness of $\dot{\bar{e}}$ (cf. (2.24)), $\dot{\zeta}$ (cf. (2.31)), and $\dot{h}$ (cf. (2.19)), implies that $\ddot{V}$ is bounded. Hence, $\dot{V} \to 0$ as $t \to \infty$, which implies (cf. (2.35)) that $\bar{e} \to 0$, $\zeta \to 0$ and, from (2.44), that $\eta \to 0$ (since h is bounded). This, in turn, implies that $\eta_0 \to 0$ as $t \to \infty$ by (2.21). We conclude that

$$\lim_{t \to \infty} [y(t) - y_r(t)] = \lim_{t \to \infty} [\bar{e}(t) - \eta_0(t)] = 0\,. \tag{2.51}$$

□

3 The Systematic Design Procedure

Even though the expressions in the general case become more complicated than in the preceding section, the main steps of the design procedure remain essentially the same.

3.1. Nonlinear system properties. We consider the class of n-dimensional nonlinear systems which have an input-output description expressed globally by the n-th order scalar differential equation

$$A(D)y = B(D)\,[q(y)u] + \sum_{i=0}^{m} D^i \left[p_{i0}(y) + p_i^{\mathrm{T}}(y)\theta_1\right], \tag{3.1}$$

where

- the coefficients $a_0, \ldots, a_{n-1}$ of the denominator polynomial $A(D) = D^n + a_{n-1}D^{n-1} + \cdots + a_0$ are unknown,
- the coefficients $b_0, \ldots, b_m (m \leq n-1)$ of the numerator polynomial $B(D) = b_m D^m + \cdots + b_0$ are unknown, but $B(D)$ is known to be Hurwitz, and the sign of b_m is known,
- θ_1 is an ℓ-dimensional vector of unknown parameters,
- $q(y)$, $p_{ij}(y)$, $0 \leq i \leq m$, $0 \leq j \leq \ell$, are smooth nonlinearities with $q(y) \neq 0\ \forall y \in \mathbb{R}$, $p_{ij}(0) = 0$, $0 \leq i \leq m$, $0 \leq j \leq \ell$.

Systems in this class are linearizable by output injection, and input-output linearizable by full-state feedback, but not necessarily full-state linearizable, even by full-state feedback, as will be shown in Sect. 5.

3.2. Augmenting the CE control. The design objective of the certainty-equivalence part of our control is to match a reference model of the same or higher relative degree than that of the nonlinear plant (3.1). As this plant is input-output linearizable by output feedback, we choose the linear reference model:

$$E_1(D)E_2(D)y_r = R\,r\,,\; E_2(D) = E_{21}(D)E_{22}(D)\,, \tag{3.2}$$

where $E_1(D)$, $E_2(D)$, $E_{21}(D)$, $E_{22}(D)$ are monic Hurwitz polynomials of degree $n-m$, $n-1$, $n-m-1$, and m, respectively, and $R(D)$ is a polynomial of degree $h \le n-1$. Filtering (3.1) by the strictly proper stable filter F/E_2, where F is a monic polynomial of degree $n-m-1$, we obtain

$$\frac{FA}{E_2}y = \frac{FB}{E_2}\left[q(y)u\right] + \sum_{i=0}^{m}\frac{FD^i}{E_2}\left[p_{i0}(y) + p_i^{\mathrm{T}}(y)\theta_1\right]. \tag{3.3}$$

It is now straightforward to verify that in the case when the coefficients of $A(D)$ and $B(D)$ and the parameters θ_1 are known, the desired matching is achieved by the control

$$u^* = \frac{1}{q(y)}\left\{-\frac{G}{E_2}y + \frac{M}{E_2}r - \sum_{j=0}^{\ell}\left[\frac{T_j}{E_{21}}\sum_{i=0}^{m}\frac{D^i}{E_{22}}p_{ij}(y)\right] - \frac{L_0}{E_2}\left[q(y)u^*\right]\right\}, \tag{3.4}$$

provided that G, a polynomial of degree $n-1$, F, M, T_j and L_0 satisfy the polynomial equations

$$FA + b_m G = E_1E_2 \tag{3.5}$$

$$M = \frac{1}{b_m}R \tag{3.6}$$

$$T_j = \frac{1}{b_m}\theta_{1j}F\,,\quad 0 \le j \le \ell\,,\; \theta_{10} \equiv 1 \tag{3.7}$$

$$L_0 = \frac{1}{b_m}FB - E_2\,. \tag{3.8}$$

Note that L_0 is a polynomial of degree $n-2$, since both $\frac{1}{b_m}FB$ and E_2 are monic polynomials of degree $n-1$. When the control (3.4) is applied to the system (3.1), asymptotic tracking is achieved:

$$y(t) = y_r(t) + \epsilon(t) \rightarrow y_r(t) \quad \text{as } t \rightarrow \infty\,, \tag{3.9}$$

where $\epsilon(t)$ is the exponentially decaying tracking error caused by the mismatch of the initial conditions.

We now rewrite the control (3.4) as

$$u^* = -\frac{1}{q(y)}\left[\phi^{\mathrm{T}}(y, u^*, r)\theta\right], \tag{3.10}$$

where the n_θ-dimensional vectors θ and ϕ (with $n_\theta = 2n + (\ell+1)(n-m)$) are defined as

$$\theta^{\mathrm{T}} = [g_0, \ldots, g_{n-1}, -m_0, t_{00}, \ldots, t_{0n-m-1}, t_{10}, \ldots, t_{1n-m-1}, \ldots, t_{\ell 0}, \ldots, t_{\ell n-m-1}, \ell_{00}, \ldots, \ell_{0n-2}] \tag{3.11}$$

$$\phi^{\mathrm{T}}(y, u^*, r) = \left[\frac{\overline{D}_{n-1}}{E_2} y, \frac{R}{E_2} r, \frac{\overline{D}_{n-m-1}}{E_{21}} \sum_{i=0}^{m} \frac{D^i}{E_{22}} p_{i0}(y), \frac{\overline{D}_{n-m-1}}{E_{21}} \sum_{i=0}^{m} \frac{D^i}{E_{22}} p_{i1}(y), \ldots, \frac{\overline{D}_{n-m-1}}{E_{21}} \sum_{i=0}^{m} \frac{D^i}{E_{22}} p_{i\ell}(y), \frac{\overline{D}_{n-2}}{E_2} [q(y)u^*] \right], \tag{3.12}$$

with $\overline{D}_k$ defined as the $(k+1)$-dimensional row operator $\overline{D}_k = [1, D, \ldots, D^k]$. The form (3.10) is particularly useful in the case where the coefficients of $A(D)$ and $B(D)$ and the components of the vector θ_1 are unknown. Since in that case the parameter vector θ defined in (3.11) and used in (3.10) cannot be computed, it is replaced by an estimate $\hat{\theta}$. The so formed "certainty-equivalence" control is then *augmented by an additive term* $\bar{u}$ which is yet to be designed. Hence, the adaptive control will be of the form:

$$u = \frac{1}{q(y)} \left[-\phi^{\mathrm{T}}(y, u, r)\hat{\theta} + \bar{u} \right]. \tag{3.13}$$

Filtering the system equation (3.1) by the strictly proper stable filter $\frac{F}{E_1 E_2}$, and using (3.5)–(3.8) and (3.11)–(3.12), we obtain:

$$\begin{aligned} \frac{FA}{E_1E_2} y &= \frac{FB}{E_1E_2} [q(y)u] + \sum_{i=0}^{m} \frac{FD^i}{E_1E_2} [p_{i0}(y) + p_i^{\mathrm{T}}(y)\theta_i] + \epsilon(t) \\ y &= \frac{b_m}{E_1} \left\{ \frac{L_0 + E_2}{E_2} [q(y)u] + \frac{G}{E_2} y + \sum_{j=0}^{\ell} \left[\frac{T_j}{E_{21}} \sum_{i=0}^{m} \frac{D^i}{E_{22}} p_{ij}(y) \right] \right\} + \epsilon(t) \\ y &= \frac{b_m}{E_1} \left\{ q(y)u + \phi^{\mathrm{T}}(y, u, r)\theta + \frac{M}{E_2} r \right\} + \epsilon(t), \end{aligned} \tag{3.14}$$

where, as in the case when θ is known, $\epsilon(t)$ denotes a linear combination of exponentially decaying terms caused by the mismatch of the initial conditions. Substitution of (3.13) into (3.14) yields the following description of the resulting feedback system:

$$y = \frac{b_m}{E_1} \left\{ \frac{R}{b_m E_2} r + \phi^{\mathrm{T}}(y, u, r) \left(\theta - \hat{\theta} \right) + \bar{u} \right\} + \epsilon(t). \tag{3.15}$$

Introducing the error variables

$$e = y - y_{\mathrm{r}}, \quad \tilde{\theta} = \theta - \hat{\theta}, \tag{3.16}$$

and taking the difference between (3.15) and the reference model (3.2), we obtain the *tracking error equation*:

$$e = \frac{b_m}{E_1}\left[\phi^{\mathrm{T}}(y,u,r)\tilde{\theta} + \bar{u}\right] + \epsilon(t)\,. \tag{3.17}$$

In the special case of relative degree one ($n-m=1$), the design is extremely simple. Since the transfer function $|b_m|/E_1$ is SPR, the parameter update law

$$\dot{\hat{\theta}} = \mathrm{sgn}(b_m)\Gamma\phi(y,u,r)e\,, \tag{3.18}$$

where $\Gamma = \Gamma^{\mathrm{T}} > 0$ is the ***adaptive gain***, guarantees boundedness of all the closed-loop signals and convergence of the tracking error e to zero [18, Chap. 5], and the control augmentation is not needed: $\bar{u} \equiv 0$.

3.3. Error augmentation and swapping. For relative degree higher than one ($n-m>1$), the design becomes considerably more complicated, since $|b_m|/E_1$ is no longer SPR. We first rewrite (3.17) in the form

$$\begin{aligned} e = {} & \frac{1}{D+\lambda_0}\left[b_m\tilde{\theta}^{\mathrm{T}}\frac{1}{E_0}\phi(y,u,r)\right] + \frac{b_m}{E_1}\bar{u} \\ & + \frac{b_m}{D+\lambda_0}\left[\frac{1}{E_0}\left[\tilde{\theta}^{\mathrm{T}}\phi\right] - \tilde{\theta}^{\mathrm{T}}\frac{1}{E_0}\phi\right] + \epsilon(t)\,, \end{aligned} \tag{3.19}$$

where

$$E_0(D)(D+\lambda_0) = E_1(D)\,. \tag{3.20}$$

In contrast to the example of Sect. 2, where the high-frequency gain b_m was known, here it is unknown. Therefore, using an estimate $\hat{b}_m$ and denoting $\tilde{b}_m = b_m - \hat{b}_m$, we rewrite (3.19) in the form

$$\begin{aligned} e = {} & \frac{1}{D+\lambda_0}\left[b_m\tilde{\theta}^{\mathrm{T}}\frac{1}{E_0}\phi + \tilde{b}_m\left(\frac{1}{E_0}\bar{u} - \tilde{\theta}^{\mathrm{T}}\frac{1}{E_0}\phi + \frac{1}{E_0}\left[\tilde{\theta}^{\mathrm{T}}\phi\right]\right)\right] \\ & + \frac{1}{D+\lambda_0}\left[\hat{b}_m\left(\frac{1}{E_0}\bar{u} - \tilde{\theta}^{\mathrm{T}}\frac{1}{E_0}\phi + \frac{1}{E_0}\left[\tilde{\theta}^{\mathrm{T}}\phi\right]\right)\right] + \epsilon(t)\,. \end{aligned} \tag{3.21}$$

Since the first summand in (3.21) is in the desired SPR form, we define the augmented error

$$\bar{e} = e + \eta_0\,, \tag{3.22}$$

where the term η_0 represents all the undesirable terms in (3.21):

$$\eta_0 = -\frac{1}{D+\lambda_0}\left[\hat{b}_m\left(\frac{1}{E_0}\bar{u} - \tilde{\theta}^{\mathrm{T}}\frac{1}{E_0}\phi + \frac{1}{E_0}\left[\tilde{\theta}^{\mathrm{T}}\phi\right]\right)\right]. \tag{3.23}$$

The vector multiplying $\tilde{\theta}$ in the second term is denoted by

$$\psi = \frac{1}{E_0}\phi(y,u,r)\,. \tag{3.24}$$

Considering $\phi(y,u,r)$ as the input and ψ as the output, we represent (3.24) in the matrix state-space form:

$$\begin{aligned} \dot{H} &= A_0 H + b\phi^{\mathrm{T}}(y,u,r) \\ \psi^{\mathrm{T}} &= c^{\mathrm{T}} H\,, \end{aligned} \tag{3.25}$$

where (c, A_0, b) is a minimal realization of $1/E_0$:

$$c^{\mathrm{T}}(sI - A_0)^{-1}b = \frac{1}{E_0(s)}\,. \tag{3.26}$$

Now η_0 is the output of the $(n-m)$-dimensional system

$$\dot{\eta}_0 = -\lambda_0\eta_0 + \hat{b}_m c^{\mathrm{T}}\eta \tag{3.27}$$

$$\dot{\eta} = A_0\eta - b\bar{u} - H\dot{\hat{\theta}}\,. \tag{3.28}$$

The variables ψ and $c^{\mathrm{T}}\eta$ from (3.25) and (3.28) allow us to express the tracking error as (cf. (3.19))

$$e = \frac{1}{D+\lambda_0}\left[b_m\tilde{\theta}^{\mathrm{T}}\psi - b_m c^{\mathrm{T}}\eta\right] + \epsilon(t)\,. \tag{3.29}$$

The analogous expression for the augmented error is

$$\bar{e} = \frac{1}{D+\lambda_0}\left[b_m\tilde{\theta}^{\mathrm{T}}\psi - \tilde{b}_m c^{\mathrm{T}}\eta\right] + \epsilon(t)\,, \tag{3.30}$$

and it has the desired SPR form: the parameter errors $\tilde{\theta}$ and $\tilde{b}_m$ are filtered only by the SPR filter $1/(D+\lambda_0)$.

3.4. Update law. As in Sect. 2, we choose the unnormalized gradient update laws suggested by the SPR form of (3.30):

$$\dot{\hat{\theta}} = \operatorname{sgn}(b_m)\Gamma\psi\bar{e} \tag{3.31}$$

$$\dot{\hat{b}}_m = -\gamma c^{\mathrm{T}}\eta\bar{e}\,, \tag{3.32}$$

where $\Gamma = \Gamma^{\mathrm{T}} > 0$ and $\gamma > 0$ are the adaptive gains.

From Sect. 2 we know that in the proof of stability there will be a need to balance the interaction between η and H. Therefore, we introduce the new variables ζ:

$$\zeta = S^{-1}\eta\,,\quad S \equiv C_{\bar{n}}^{-1}\Xi C_{\bar{n}}\,, \tag{3.33}$$

where $\bar{n} \equiv n - m - 1$,

$$C_i \equiv \left[c, A_0^{\mathrm{T}}c, \ldots, \left(A_0^{\mathrm{T}}\right)^{i-1}c\right]^{\mathrm{T}}, \quad i = 1, \ldots, \bar{n} \tag{3.34}$$

$$\Xi = I_{\bar{n}\times\bar{n}} + \begin{bmatrix} 0 \\ \xi_1 E_{1,\bar{n}} \\ \xi_2 E_{2,\bar{n}} \\ \vdots \\ \xi_{\bar{n}-1} E_{\bar{n}-1,\bar{n}} \end{bmatrix} \tag{3.35}$$

$$E_{j,i} = [I_{j\times j},\ 0]_{j\times i}\,, \quad 0 < j \le i\,, \tag{3.36}$$

and $I_{i\times i}$ is the $i \times i$ identity matrix. The components of the i-dimensional row vectors ξ_i are nonlinear functions of the elements of H which represent the aforementioned additional degrees of design freedom. In order to show that the matrix S defined in (3.33) is invertible, we note that, because of the structure of the matrices $E_{i,j}$ defined by (3.36), the matrix Ξ in (3.35) is lower triangular with ones on its diagonal. From this it follows that Ξ^{-1} always exists. Furthermore, $C_{\bar{n}}^{-1}$ exists because (c^{T}, A_0) is assumed to be an observable pair.

The nonnegative function to be used in our stability proof is

$$V = \frac{1}{2}\left(\bar{e}^2 + |\, b_m \,|\, \tilde{\theta}^{\mathrm{T}}\Gamma^{-1}\tilde{\theta} + \frac{1}{\gamma}\tilde{b}_m^2 + \frac{1}{\lambda_0}\int_t^\infty \epsilon^2(\tau)d\tau\right) + \frac{\lambda_0}{\bar{n}}\zeta^{\mathrm{T}}P\zeta\,. \tag{3.37}$$

The form of (3.37) is the same as that of (2.30), where P is the positive definite solution of the Lyapunov equation

$$PA_0 + A_0^{\mathrm{T}}P = -Q_0\,. \tag{3.38}$$

To evaluate $\dot{V}$ we need $\dot{\zeta}$, which is obtained by differentiating (3.33) and using (3.28), (3.31) and (3.34)–(3.36):

$$\begin{aligned} \dot{\zeta} &= S^{-1}\left(A_0\eta - b\bar{u} - \mathrm{sgn}(b_m)H\Gamma\psi\bar{e}\right) + \frac{d}{dt}\left(S^{-1}\right)\eta \\ &= S^{-1}A_0S\zeta - S^{-1}(b\bar{u} + \mathrm{sgn}(b_m)H\Gamma\psi\bar{e}) - S^{-1}\dot{S}\zeta \\ &= A_0\zeta - \mathrm{sgn}(b_m)S^{-1}H\Gamma\psi\bar{e} + S^{-1}(A_0S - SA_0 - \dot{S})\zeta - S^{-1}b\bar{u}\,. \end{aligned} \tag{3.39}$$

Introducing the notation

$$w = -\mathrm{sgn}(b_m)S^{-1}H\Gamma\psi \tag{3.40}$$

$$W = S^{-1}\left(SA_0 - A_0S + \dot{S}\right), \tag{3.41}$$

we rewrite (3.39) as

$$\dot{\zeta} = A_0\zeta + w\bar{e} - W\zeta - S^{-1}b\bar{u}\,, \tag{3.42}$$

and compute the time derivative of V as

$$\begin{aligned} \dot{V} = &-\frac{1}{2}\left(\sqrt{\lambda_0}\bar{e} - \frac{1}{\sqrt{\lambda_0}}\epsilon\right)^2 - \frac{\lambda_0}{\bar{n}}\zeta^{\mathrm{T}}Q_0\zeta \\ &-\frac{2\lambda_0}{\bar{n}}\left\{\frac{\bar{n}}{4}\bar{e}^2 - \zeta^{\mathrm{T}}P\left(w\bar{e} - W\zeta - S^{-1}b\bar{u}\right)\right\}. \end{aligned} \tag{3.43}$$

3.5. Design equation. The tools we have at our disposal to make $\dot{V}$ nonpositive are the functions $\xi_i(H)$ and the control term $\bar{u}$. With these tools we will attempt to represent the quantity enclosed in braces in (3.43) as the sum of $\bar{n}$ squares. It turns out that this is possible to achieve by decomposing P as $P = \sum_{i=1}^{\bar{n}} P_i$ such that the following *design equation* holds:

$$\sum_{i=1}^{\bar{n}} P_i w w^{\mathrm{T}} P_i \zeta = P(W\zeta + S^{-1} b \bar{u}) . \tag{3.44}$$

The substitution of (3.44) into (3.43) yields the desired form for $\dot{V}$:

$$\dot{V} = -\frac{1}{2}\left(\sqrt{\lambda_0}\bar{e} - \frac{1}{\sqrt{\lambda_0}}\epsilon\right)^2 - \frac{\lambda_0}{\bar{n}}\zeta^{\mathrm{T}} Q_0 \zeta - \frac{2\lambda_0}{\bar{n}} \sum_{i=1}^{\bar{n}} \left(\frac{\bar{e}}{2} - \zeta^{\mathrm{T}} P_i w\right)^2 \leq 0 . \tag{3.45}$$

Our task now is to find P_i, $\xi_i(H)$ and $\bar{u}$ which satisfy the design equation (3.44). Following the development in [17], we define

$$P_i = C_i^{\mathrm{T}} M_i C_i , \tag{3.46}$$

where

$$\begin{aligned} M_1 &= \left(C_1 P^{-1} C_1\right)^{-1} \\ M_i &= \left(C_i P^{-1} C_i^{\mathrm{T}}\right)^{-1} \left(I_{i\times i} - \sum_{j=1}^{i-1} C_i P^{-1} C_j^{\mathrm{T}} M_j E_{j,i} \right) , \qquad i = 2, \ldots, \bar{n} . \end{aligned} \tag{3.47}$$

In [17, Lemma 1] it is proved that (3.46)–(3.47) result in

$$\sum_{i=1}^{\bar{n}} P_i = P \tag{3.48}$$

$$C_j P^{-1} C_i^{\mathrm{T}} M_i C_i = 0 , \quad 1 \leq j < i \leq \bar{n} . \tag{3.49}$$

This proof is now given for completeness. From (3.47) we have

$$\begin{aligned} I_{i\times i} &= C_i P^{-1} C_i^{\mathrm{T}} M_i + \sum_{j=1}^{i-1} C_i P^{-1} C_j^{\mathrm{T}} M_j E_{j,i} \\ &= C_i P^{-1} \sum_{k=1}^{i} C_k^{\mathrm{T}} M_k E_{k,i} . \end{aligned} \tag{3.50}$$

Postmultiplying both sides of (3.50) by C_i and using the identity

$$C_k = E_{k,i} C_i , \quad k \leq i , \tag{3.51}$$

we obtain

$$C_i = C_i P^{-1} \sum_{k=1}^{i} C_k^{\mathrm{T}} M_k C_k . \tag{3.52}$$

Evaluating (3.52) at $i = \bar{n}$ and using the nonsingularity of $C_{\bar{n}}$ we obtain (3.48). Furthermore, premultiplying (3.52) by $E_{j,i}$, where $j < i$, and using (3.51) again, we obtain

$$C_j = C_j P^{-1} \sum_{k=1}^{i} C_k^{\mathrm{T}} M_k C_k\,, \quad i > j\,. \tag{3.53}$$

But from (3.52) we have

$$C_j = C_j P^{-1} \sum_{k=1}^{j} C_k^{\mathrm{T}} M_k C_k\,,$$

which, combined with (3.53), results in (3.49).

Having established (3.48) and (3.49), we now set out to find $\xi_i(H)$ and $\bar{u}$ which, along with P_i defined by (3.46)–(3.47), satisfy the design equation (3.44). Substituting (3.46) into (3.44) we obtain

$$W\zeta + S^{-1} b\bar{u} = P^{-1} \sum_{i=1}^{\bar{n}} C_i^{\mathrm{T}} M_i C_i w w^{\mathrm{T}} C_i^{\mathrm{T}} M_i^{\mathrm{T}} C_i\,. \tag{3.54}$$

Using (3.41) we rewrite the design equation (3.54) as

$$\left(SA_0 - A_0 S + \dot{S}\right)\zeta + b\bar{u} = SP^{-1} \sum_{i=1}^{\bar{n}} C_i^{\mathrm{T}} M_i C_i w w^{\mathrm{T}} C_i^{\mathrm{T}} M_i^{\mathrm{T}} C_i \zeta\,. \tag{3.55}$$

Premultiplying both sides of (3.55) by $c^{\mathrm{T}} A_0^{i-1}$ for $i = 1, \ldots, \bar{n}$, we obtain

$$c^{\mathrm{T}} A_0^{i-1}(A_0 S\zeta - b\bar{u}) = c^{\mathrm{T}} A_0^{i-1} \left[S \left(A_0 - P^{-1} \sum_{j=1}^{\bar{n}} C_j^{\mathrm{T}} M_j C_j w w^{\mathrm{T}} C_j^{\mathrm{T}} M_j C_j \right) + \dot{S} \right] \zeta + c^{\mathrm{T}} A_0^{i-1} b\bar{u}\,, \qquad 1 \le i \le \bar{n}\,. \tag{3.56}$$

From (3.33), (3.35) and (3.51) we have

$$C_{\bar{n}} S = \Xi C_{\bar{n}} = C_{\bar{n}} + \begin{bmatrix} 0 \\ \xi_1 C_1 \\ \xi_2 C_2 \\ \vdots \\ \xi_{\bar{n}-1} C_{\bar{n}-1} \end{bmatrix}, \tag{3.57}$$

which gives

$$c^{\mathrm{T}} A_0^{i-1} S = c^{\mathrm{T}} A_0^{i-1} + \xi_{i-1} C_{i-1}, \quad 1 \le i \le \bar{n}\,, \tag{3.58}$$

where we have defined for notational convenience

$$\xi_0 = 0\,, \quad C_0 = 0\,. \tag{3.59}$$

Furthermore,

$$c^{\mathrm{T}}A_0^{i-1}\dot{S} = c^{\mathrm{T}}A_0^{i-1}C_{\bar{n}}^{-1}\dot{\Xi}C_{\bar{n}} = c^{\mathrm{T}}A_0^{i-1}C_{\bar{n}}^{-1}\begin{bmatrix} 0 \\ \dot{\xi}_1 C_1 \\ \vdots \\ \dot{\xi}_{\bar{n}-1}C_{\bar{n}-1}\end{bmatrix}$$

$$= \dot{\xi}_{i-1}C_{i-1}, \quad 1 \le i \le \bar{n}, \tag{3.60}$$

where we have used the definition of $C_{\bar{n}}$ to obtain the last equality. Finally, from the definition (3.26) of the triple (c, A_0, b) we have:

$$c^{\mathrm{T}}A_0^{i-1}b = 0, \quad 1 \le i \le \bar{n}-1 \tag{3.61}$$

$$c^{\mathrm{T}}A_0^{\bar{n}-1}b = 1. \tag{3.62}$$

Substituting (3.58)–(3.62) into (3.56) results in

$$\xi_i C_i = \dot{\xi}_{i-1}C_{i-1} + \xi_{i-1}C_{i-1}A_0 - (c^{\mathrm{T}}A_0^{i-1} + \xi_{i-1}C_{i-1}) \times P^{-1}\sum_{j=1}^{\bar{n}} C_j^{\mathrm{T}}M_jC_jww^{\mathrm{T}}C_j^{\mathrm{T}}M_j^{\mathrm{T}}C_j, \quad 1 \le i \le \bar{n}-1 \tag{3.63}$$

$$\bar{u} = \Big[c^{\mathrm{T}}A_0^{\bar{n}}S - \dot{\xi}_{\bar{n}-1}C_{\bar{n}-1} - c^{\mathrm{T}}A_0^{\bar{n}} - \xi_{\bar{n}-1}C_{\bar{n}-1}A_0 + (c^{\mathrm{T}}A_0^{\bar{n}-1} + \xi_{\bar{n}-1}C_{\bar{n}-1}) \times P^{-1}\sum_{j=1}^{\bar{n}} C_j^{\mathrm{T}}M_jC_jww^{\mathrm{T}}C_j^{\mathrm{T}}M_j^{\mathrm{T}}C_j\Big]\zeta. \tag{3.64}$$

At this point, we have almost achieved our goal of finding ξ_i and $\bar{u}$ which satisfy the design equation and thus render $\dot{V}$ nonpositive. Still, (3.63)–(3.64) are in a rather complicated form and, moreover, they involve the time derivatives of the functions ξ_i. Therefore, we now set out to simplify (3.63)–(3.64) and to express $\dot{\xi}_{i-1}$ as explicit functions of available signals.

Motivated by the appearance of the terms C_jw in (3.63)–(3.64), we introduce the i-dimensional column vectors $w_1, \ldots, w_{\bar{n}}$ which are defined as

$$w_i = C_iw, \quad 1 \le i \le \bar{n}. \tag{3.65}$$

Combining (3.65) with (3.51) we see that these vectors satisfy the recursive expressions

$$w_{\bar{n}} = C_{\bar{n}}w \tag{3.66}$$

$$w_i = E_{i,i+1}w_{i+1}, \quad 1 \le i \le \bar{n}-1. \tag{3.67}$$

Using (3.36) we can rewrite (3.67) as

$$w_i^{\mathrm{T}} = [w_{i-1}^{\mathrm{T}},\ w_{i,i}], \quad 1 \le i \le \bar{n}-1. \tag{3.68}$$

We now set out to obtain explicit expressions for $w_1, \ldots, w_{\bar{n}}$ in terms of ξ_i. Substituting (3.40) into (3.66) results in

$$\Xi w_{\bar{n}} = -\mathrm{sgn}(b_m)C_{\bar{n}}H\Gamma\psi \tag{3.69}$$

We then use (3.25), (3.35) and (3.68) to rewrite (3.69) as

$$\left(I_{\bar{n}\times\bar{n}}+\begin{bmatrix}0\\ \xi_1E_{1,\bar{n}}\\ \vdots\\ \xi_{\bar{n}-1}E_{\bar{n}-1,\bar{n}}\end{bmatrix}\right)\begin{bmatrix}w_{\bar{n}-1}\\ w_{\bar{n},\bar{n}}\end{bmatrix}=-\mathrm{sgn}(b_m)\begin{bmatrix}c^{\mathrm{T}}\\ c^{\mathrm{T}}A_0\\ \vdots\\ c^{\mathrm{T}}A_0^{\bar{n}-1}\end{bmatrix}H\Gamma H^{\mathrm{T}}c\,. \tag{3.70}$$

By (3.36), (3.70) is equivalent to

$$\left(I_{(\bar{n}-1)\times(\bar{n}-1)}+\begin{bmatrix}0\\ \xi_1E_{1,\bar{n}-1}\\ \vdots\\ \xi_{\bar{n}-2}E_{\bar{n}-2,\bar{n}-1}\end{bmatrix}\right)w_{\bar{n}-1}=-\mathrm{sgn}(b_m)C_{\bar{n}-1}H\Gamma H^{\mathrm{T}}c \tag{3.71}$$

$$w_{\bar{n},\bar{n}}+\xi_{\bar{n}-1}w_{\bar{n}-1}=-\mathrm{sgn}(b_m)c^{\mathrm{T}}A_0^{\bar{n}-1}H\Gamma H^{\mathrm{T}}c\,. \tag{3.72}$$

Starting from (3.71)–(3.72), one can repeat the above procedure to show that (3.70) is equivalent to

$$\begin{bmatrix}w_1\\ w_{2,2}+\xi_1w_1\\ \vdots\\ w_{\bar{n},\bar{n}}+\xi_{\bar{n}-1}w_{\bar{n}-1}\end{bmatrix}=-\mathrm{sgn}(b_m)\begin{bmatrix}c^{\mathrm{T}}\\ c^{\mathrm{T}}A_0\\ \vdots\\ c^{\mathrm{T}}A_0^{\bar{n}-1}\end{bmatrix}H\Gamma H^{\mathrm{T}}c\,. \tag{3.73}$$

Hence, the explicit expressions for the vectors $w_1,\ldots,w_{\bar{n}}$ are

$$w_1^{\mathrm{T}}=-\mathrm{sgn}(b_m)\left(c^{\mathrm{T}}H\Gamma H^{\mathrm{T}}c\right) \tag{3.74}$$

$$w_i^{\mathrm{T}}=\left[w_{i-1}^{\mathrm{T}},-\mathrm{sgn}(b_m)\left(c^{\mathrm{T}}A_0^{i-1}H\Gamma H^{\mathrm{T}}c\right)-\xi_{i-1}w_{i-1}\right],\quad i=2,\ldots,\bar{n}\,. \tag{3.75}$$

Combining (3.49) and (3.65) we obtain

$$\begin{aligned}C_iP^{-1}\sum_{j=1}^{\bar{n}}C_j^{\mathrm{T}}M_jC_jww^{\mathrm{T}}C_j^{\mathrm{T}}M_j^{\mathrm{T}}C_j&=C_iP^{-1}\sum_{j=1}^{i}C_j^{\mathrm{T}}M_jC_jww^{\mathrm{T}}C_j^{\mathrm{T}}M_j^{\mathrm{T}}C_j\\&=C_i\left[P^{-1}\sum_{j=1}^{i}C_j^{\mathrm{T}}M_jw_jw_j^{\mathrm{T}}M_j^{\mathrm{T}}E_{j,i}\right]C_i\\&=C_iN_iC_i\,,\end{aligned} \tag{3.76}$$

where

$$N_i=P^{-1}\sum_{j=1}^{i}C_j^{\mathrm{T}}M_jw_jw_j^{\mathrm{T}}M_j^{\mathrm{T}}E_{j,i}\,,\quad i=1,\ldots,\bar{n}\,. \tag{3.77}$$

Thus, the last term in (3.63) and (3.64) can be rewritten as

$$
\begin{aligned}
&(c^{\mathrm{T}}A_0^{i-1}+\xi_{i-1}C_{i-1})P^{-1}\sum_{j=1}^{\bar{n}}C_j^{\mathrm{T}}M_jC_jww^{\mathrm{T}}C_j^{\mathrm{T}}M_j^{\mathrm{T}}C_j = \\
&\quad = ([0\ldots 0\,1]+\xi_{i-1}E_{i-1,i})C_iP^{-1}\sum_{j=1}^{\bar{n}}C_jM_jC_jww^{\mathrm{T}}C_j^{\mathrm{T}}M_j^{\mathrm{T}}C_j \\
&\quad = ([0\ldots 0\,1]+\xi_{i-1}E_{i-1,i})C_iN_iC_i \\
&\quad = (c^{\mathrm{T}}A_0^{i-1}+\xi_{i-1}C_{i-1})N_iC_i\,.
\end{aligned}
\tag{3.78}
$$

Introducing the $i\times(i+1)$ matrix $R_{i+1}=[0,\ \ I_{i\times i}]$ and substituting (3.78) into (3.63)–(3.64) we obtain for $i=1,\ldots,\bar{n}-1$:

$$
\begin{aligned}
\xi_iC_i &= \dot{\xi}_{i-1}C_{i-1}+\xi_{i-1}C_{i-1}A_0-(c^{\mathrm{T}}A_0^{i-1}+\xi_{i-1}C_{i-1})N_iC_i \\
&= \left[\dot{\xi}_{i-1}E_{i-1,i}+\xi_{i-1}R_i-(c^{\mathrm{T}}A_0^{i-1}+\xi_{i-1}C_{i-1})N_i\right]C_i
\end{aligned}
\tag{3.79}
$$

$$
\begin{aligned}
\bar{u} = \Big\{c^{\mathrm{T}}A_0^{\bar{n}}S-c^{\mathrm{T}}A_0^{\bar{n}}-\Big[&\dot{\xi}_{\bar{n}-1}E_{\bar{n}-1,\bar{n}}+\xi_{\bar{n}-1}R_{\bar{n}} \\
&-\left(c^{\mathrm{T}}A_0^{\bar{n}-1}+\xi_{\bar{n}-1}C_{\bar{n}-1}\right)N_{\bar{n}}\Big]C_{\bar{n}}\Big\}\,\zeta\,.
\end{aligned}
\tag{3.80}
$$

This form makes it apparent that the design equation (3.54) is satisfied by the recursive expressions

$$\xi_1=-c^{\mathrm{T}}N_1 \tag{3.81}$$

$$\xi_i=\dot{\xi}_{i-1}E_{i-1,i}+\xi_{i-1}R_i-\left(c^{\mathrm{T}}A_0^{i-1}+\xi_{i-1}C_{i-1}\right)N_i\,,\quad i=2,\ldots,\bar{n}\,, \tag{3.82}$$

$$\bar{u}=\left[c^{\mathrm{T}}A_0^{\bar{n}}-\left(c^{\mathrm{T}}A_0^{\bar{n}}+\xi_{\bar{n}}C_{\bar{n}}\right)S^{-1}\right]\eta\,. \tag{3.83}$$

To finally solve the design equation, we need to express $\dot{\xi}_{i-1}$ in (3.82) as an explicit function of available signals. We first show via an induction argument that (3.74), (3.75), (3.77), (3.81) and (3.82) imply that the elements of w_i, N_i and ξ_i are polynomial functions of the elements of C_iH:

- For $i=1$, this fact is obvious from the definitions:

$$
\begin{aligned}
w_1 &= -\mathrm{sgn}(b_m)C_1H\Gamma(C_1H)^{\mathrm{T}} \\
N_1 &= P^{-1}C_1^{\mathrm{T}}M_1w_1w_1^{\mathrm{T}}M_1^{\mathrm{T}} \\
\xi_1 &= -c^{\mathrm{T}}N_1\,.
\end{aligned}
$$

- For $i=k\le\bar{n}-1$, suppose the elements of w_k, N_k and ξ_k are polynomial functions of the elements of C_kH. Then, the derivative of ξ_k can be expressed as

$$\dot{\xi}_k=\sum_{j=1}^{n_\theta}\dot{h}_{k,j}^{\mathrm{T}}\frac{\partial\xi_k}{\partial h_{k,j}}\,, \tag{3.84}$$

where $h_{k,j}$ is the j-th column of C_kH and $\partial\xi_k/\partial h_{k,j}$ is the $k\times j$ matrix of partial derivatives of $\xi_k(C_kH)$ with respect to the elements of $h_{k,j}$. But from (3.25), (3.61), (3.62), and the definition of R_{k+1}, we obtain

$$C_k\dot{H}=C_kA_0H=R_{k+1}C_{k+1}H\,,\quad 1\le k\le\bar{n}-1\,. \tag{3.85}$$

Combining (3.84) and (3.85) we can express $\dot{\xi}_k$ as

$$\dot{\xi}_k = \sum_{j=1}^{n_\theta} h_{k+1,j}^{\mathrm{T}} R_{k+1}^{\mathrm{T}} \frac{\partial \xi_k}{\partial h_{k,j}} . \tag{3.86}$$

– For $i = k+1$, we have

$$\begin{aligned}
w_{k+1}^{\mathrm{T}} &= \left[w_k^{\mathrm{T}}, -\mathrm{sgn}(b_m)[0\dots 0\,1]C_{k+1}H\Gamma(C_1H)^{\mathrm{T}} - \xi_k w_k\right] \\
N_{k+1} &= N_k E_{k,k+1} + P^{-1}C_{k+1}^{\mathrm{T}} M_{k+1} w_{k+1} w_{k+1}^{\mathrm{T}} M_{k+1}^{\mathrm{T}} \\
\xi_{k+1} &= \sum_{j=1}^{n_\theta} h_{k+1,j}^{\mathrm{T}} R_{k+1}^{\mathrm{T}} \frac{\partial \xi_k}{\partial h_{k,j}} E_{k,k+1} + \xi_k R_{k+1} - (c^{\mathrm{T}} A_0^k + \xi_k C_k) N_{k+1} .
\end{aligned}$$

Hence, the elements of w_{k+1}, N_{k+1} and ξ_{k+1} are polynomial functions of the elements of $C_{k+1}H$.

Thus, the term $\dot{\xi}_{i-1}$ in (3.82) can be calculated explicitly from (3.86). Furthermore, the lower triangular form of Ξ (cf. (3.35)) and the polynomial dependence of ξ_i on the elements of C_iH imply that the elements of both Ξ and Ξ^{-1} are polynomial functions of the elements of H.

The design procedure is now complete. The expressions for ξ_i and $\bar{u}$, which guarantee that the nonnegative function V in (3.37) has the nonpositive derivative (3.45), are

$$\begin{aligned}
\xi_1 &= -c^{\mathrm{T}} N_1 \\
\xi_i &= \sum_{j=1}^{n_\theta} h_{i,j}^{\mathrm{T}} R_i^{\mathrm{T}} \frac{\partial \xi_{i-1}}{\partial h_{i-1,j}} E_{i-1,i} + \xi_{i-1} R_i \\
&\quad - \left(c^{\mathrm{T}} A_0^{i-1} + \xi_{i-1} C_{i-1}\right) N_i , \quad i = 2, \dots, \bar{n} \\
\bar{u} &= \varphi^{\mathrm{T}} \eta \\
\varphi^{\mathrm{T}} &= c^{\mathrm{T}} A_0^{\bar{n}} - \left(c^{\mathrm{T}} A_0^{\bar{n}} + \xi_{\bar{n}} C_{\bar{n}}\right) \left(C_{\bar{n}}^{-1} \Xi C_{\bar{n}}\right)^{-1} .
\end{aligned} \tag{3.87}$$

The designed closed-loop adaptive system is:

$$\begin{aligned}
&\textit{Plant:} \\
&\qquad \dot{x} = A_\Sigma x + b_\Sigma q(y) u + \sum_{i=0}^{m} \varepsilon_{n-i} \left[p_{i0}(y) + p_i^{\mathrm{T}}(y)\theta_1\right] \\
&\qquad y = c_\Sigma^{\mathrm{T}} x \\
&\textit{Control:} \\
&\qquad u = \frac{1}{q(y)} \left[-\phi^{\mathrm{T}}(y,u,r)\hat{\theta} + \varphi^{\mathrm{T}}\eta\right] \\
&\textit{Update law:} \\
&\qquad \dot{\hat{\theta}} = \mathrm{sgn}(b_m)\Gamma\psi\bar{e} \\
&\qquad \dot{\hat{b}}_m = -\gamma c^{\mathrm{T}} \eta \bar{e}
\end{aligned} \tag{3.88}$$

$$\bar{e} = e + \eta_0 = y - y_r + \eta_0$$

Filters:

$$\begin{aligned}
\dot{H} &= A_0 H + b\phi^{\mathrm{T}}(y,u,r) \\
\psi^{\mathrm{T}} &= c^{\mathrm{T}} H \\
\dot{\eta}_0 &= -\lambda_0 \eta_0 + \hat{b}_m c^{\mathrm{T}} \eta \\
\dot{\eta} &= A_0 \eta - b\varphi^{\mathrm{T}} \eta - H \mathrm{sgn}(b_m) \Gamma \psi \bar{e} \,,
\end{aligned}$$

where y_r is the output of the reference model (3.2), $\phi(y,u,r)$ is defined in (3.12), and $(c_\Sigma, A_\Sigma, b_\Sigma)$ is a minimal state representation of the plant equation (3.1):

$$A_\Sigma = \begin{bmatrix} -a_{n-1} & \\ \vdots & I \\ -a_0 & 0 \ldots 0 \end{bmatrix}, \quad b_\Sigma = \begin{bmatrix} 0 \\ \vdots \\ 0 \\ b_m \\ \vdots \\ b_0 \end{bmatrix}, \quad c_\Sigma = \begin{bmatrix} 1 \\ 0 \\ \vdots \\ 0 \end{bmatrix}, \tag{3.89}$$

with ε_{n-i} the $(n-i)$-th coordinate vector in $\mathrm{I\!R}^n$.

The stability and tracking properties of (3.88) are established in the next section.

4 Stability and Tracking

We are now ready to state and prove our main result:

Theorem 4.1 *For any uniformly bounded and piecewise continuous reference input r, all the signals in the closed-loop adaptive system (3.88) are well-defined and uniformly bounded on* $[0,\infty)$*, and, in addition,*

$$\lim_{t\to\infty} e(t) = 0, \quad \lim_{t\to\infty} \eta(t) \to 0, \quad \lim_{t\to\infty} \eta_0(t) = 0\,. \tag{4.1}$$

Proof. Due to the piecewise continuity of $r(t)$ and the smoothness of the nonlinear functions appearing in the definitions of various terms in the closed-loop system (3.88), the solution of (3.88) has a maximum interval of existence $[0, t_f)$. On this interval, the time derivative of the nonnegative function V defined in (3.37)

$$V = \frac{1}{2}\left(\bar{e}^2 + |\, b_m \,| \, \tilde{\theta}^{\mathrm{T}} \Gamma^{-1} \tilde{\theta} + \frac{1}{\gamma}\tilde{b}_m^2 + \frac{1}{\lambda_0}\int_t^\infty \epsilon^2(\tau) d\tau\right) + \frac{\lambda_0}{\bar{n}} \zeta^{\mathrm{T}} P \zeta\,,$$

computed along the solutions of (3.88), is given by (3.45):

$$\dot{V} = -\frac{1}{2}\left(\sqrt{\lambda_0}\bar{e} - \frac{1}{\sqrt{\lambda_0}}\epsilon\right)^2 - \frac{\lambda_0}{\bar{n}} \zeta^{\mathrm{T}} Q_0 \zeta - \frac{2\lambda_0}{\bar{n}} \sum_{i=1}^{\bar{n}} \left(\frac{\bar{e}}{2} - \zeta^{\mathrm{T}} P_i w\right)^2 \leq 0\,.$$

We conclude that V, $\bar{e}$, $\tilde{\theta}$, $\tilde{b}_m$ and ζ are bounded on $[0,t_f)$ by constants depending only on the initial conditions of (3.88). This implies that $\hat{\theta}$, $\hat{b}_m$ are bounded on $[0,t_f)$. The boundedness of ζ together with (3.33) and (3.58) evaluated at $i=1$ imply that $c^T\eta$ is bounded; from the definition of η_0 in (3.88) and the boundedness of $\hat{b}_m$ and $c^T\eta$ it follows that η_0 is bounded. But since $e=\bar{e}-\eta_0$, and $\bar{e}$, η_0 are bounded, we have that e is bounded. Now from the boundedness of r we have that y_r is bounded, and, hence, y is bounded, since $y=e+y_r$. The boundedness of y implies that all the nonlinearities appearing are bounded, and, furthermore, that $q(y)$ is bounded away from zero. Filtering the system equation (3.1) with the strictly proper stable filter $1/BE_1$ and rearranging terms, we obtain

$$\frac{1}{E_1}\left[q(y)u\right]=\frac{A}{BE_1}y+\sum_{i=0}^{m}\frac{D^i}{BE_1}\left[p_{i0}(y)+p_i^T(y)\theta_1\right], \tag{4.2}$$

which, by the boundedness of y, implies that $\dfrac{1}{E_1}q(y)u$ is bounded. The boundedness of H is now established by proving that the row vectors $c^TA_0^iH\,,0\le i\le\bar{n}-1$ are bounded. This, in turn, is proved as follows: First, from [17, Lemma 5] we have that the first $\bar{n}$ derivatives of e can be expressed as

$$e^{(i)}=\mu_i\left(\bar{e},\tilde{\theta},\tilde{b}_m,\eta_0,C_iH,C_i\zeta,\epsilon_i\right),\quad i=1,\ldots,\bar{n}, \tag{4.3}$$

where the μ_i's are continuous functions of their arguments and the ϵ_i's are exponentially decaying terms. This is straightforward to show (cf. (2.48)–(2.49) in the example of Sect. 2), starting from $e=\bar{e}-\eta_0$ and using (3.88) and the facts that the derivative of ζ is given by

$$\dot{\zeta}=A_0\zeta+w\bar{e}-P^{-1}\sum_{i=1}^{\bar{n}}C_i^TM_iC_iww^TC_i^TM_i^TC_i\zeta, \tag{4.4}$$

and that the elements of w_i, N_i and ξ_i are polynomial functions of the elements of C_iH. Second, from (3.25) we have

$$c^TA_0^iH=\frac{D^i}{E_0}\phi^T(y,u,r)+\bar{\epsilon}_i(t),\quad 0\le i\le\bar{n}-1, \tag{4.5}$$

where $\bar{\epsilon}_i(t)$ are n_θ-dimensional row vectors of exponentially decaying terms. Then, we use (3.12) to express ϕ as

$$\phi^T(y,u,r)=\left[\overline{v}(y,r)\,,\,\frac{\overline{D}_{n-2}}{E_2}\left[q(y)u\right]\right], \tag{4.6}$$

with

$$\overline{v}(y,r)=\left[\frac{\overline{D}_{n-1}}{E_2}y\,,\,\frac{R}{E_2}r\,,\,\frac{\overline{D}_{n-m-1}}{E_{21}}\sum_{i=0}^{m}\frac{D^i}{E_{22}}p_{i0}(y)\,,\ldots,\,\frac{\overline{D}_{n-m-1}}{E_{21}}\sum_{i=0}^{m}\frac{D^i}{E_{22}}p_{i\ell}(y)\right] \tag{4.7}$$

being bounded, since y and r are bounded. Combining (3.13), (3.17) and (4.5) we obtain the following expressions for the first $\bar{n}$ derivatives of the tracking error e:

$$\begin{aligned} e^{(i)} &= \frac{b_m D^i}{E_1}\left[\phi^{\mathrm{T}}(y,u,r)\tilde{\theta}+\bar{u}\right]+\epsilon(t) \\ &= \frac{b_m D^i}{E_1}\left[q(y)u+\phi^{\mathrm{T}}(y,u,r)\theta\right]+\epsilon(t) \\ &= \frac{b_m D^i}{E_1}\left[q(y)u\right]+\frac{b_m E_0 D}{E_1}\frac{D^{i-1}}{E_0}\phi^{\mathrm{T}}(y,u,r)\theta+\epsilon(t) \\ &= \frac{b_m D^i}{E_1}\left[q(y)u\right]+\frac{b_m E_0 D}{E_1}c^{\mathrm{T}}A_0^{i-1}H\,\theta+\epsilon(t)\,,\quad 1\le i\le \bar{n}-1\,. \end{aligned} \tag{4.8}$$

It is important to note that $b_m E_0 D/E_1$ is stable and proper. The boundedness of $c^{\mathrm{T}}A_0^i H$ is now established for $i=0,\ldots,\bar{n}-1$ by the following induction argument:

- For $i=0$, (4.5)–(4.6) give

$$\begin{aligned} c^{\mathrm{T}}H &= \left[\frac{1}{E_0}\overline{v}(y,r)\,,\ \frac{\overline{D}_{n-2}}{E_0E_2}\left[q(y)u\right]\right] \\ &= \left[\frac{1}{E_0}\overline{v}(y,r)\,,\ \frac{E_1\overline{D}_{n-2}}{E_0E_2}\frac{1}{E_1}\left[q(y)u\right]\right]. \end{aligned} \tag{4.9}$$

  Since $\overline{v}(y,r)$, $\frac{1}{E_1}\left[q(y)u\right]$ are bounded and $\frac{E_1\overline{D}_{n-2}}{E_0E_2}$ is a row of stable proper filters, (4.9) implies that $c^{\mathrm{T}}H$ is bounded. Furthermore, we have already shown that e is bounded.
- For $1\le i\le \bar{n}-1$, assume that $c^{\mathrm{T}}A_0^kH$ and $e^{(k)}$ are bounded for $0\le k\le i-1$. Hence, C_iH is bounded, and, by (4.3), $e^{(i)}$ is bounded. Then, rewriting (4.8) as

$$\frac{D^i}{E_1}\left[q(y)u\right]=\frac{1}{b_m}e^{(i)}-\frac{E_0D}{E_1}c^{\mathrm{T}}A_0^{i-1}H\,\theta+\epsilon(t)\,, \tag{4.10}$$

  we conclude that $\frac{D^i}{E_1}\left[q(y)u\right]$ is bounded. Finally, using (4.5)–(4.7), we obtain

$$\begin{aligned} c^{\mathrm{T}}A_0^iH &= \left[\frac{D^i}{E_0}\overline{v}(y,r)\,,\ \frac{D^i\overline{D}_{n-2}}{E_0E_2}\left[q(y)u\right]\right] \\ &= \left[\frac{D^i}{E_0}\overline{v}(y,r)\,,\ \frac{E_1\overline{D}_{n-2}}{E_0E_2}\frac{D^i}{E_1}\left[q(y)u\right]+\bar{\epsilon}(t)\right]. \end{aligned} \tag{4.11}$$

  Hence, $c^{\mathrm{T}}A_0^iH$ is bounded.

This proves that H is bounded, which, by (4.3), means that $e^{(i)}$, $1\le i\le\bar{n}$, are bounded.

Next, we prove the boundedness of u. From (3.13), the boundedness of $\hat{\theta}$ and the fact that $q(y)$ is bounded away from zero, it follows that u is bounded if $\bar{u}$

and $\phi(y,u,r)$ are bounded. The boundedness of H implies the boundedness of Ξ, Ξ^{-1} and φ. Since $\eta = S\zeta = (C_{\bar{n}})^{-1}\Xi C_{\bar{n}}$, and Ξ and ζ are bounded, η is bounded as well. Hence, $\bar{u}$ is bounded. From (4.6)–(4.7), to prove boundedness of $\phi(y,u,r)$ we only need to show that $\frac{\overline{D}_{n-2}}{E_2}[q(y)u]$. We first rewite (4.8) as

$$\frac{D^i}{E_1}[q(y)u] = \frac{1}{b_m}e^{(i)} - \frac{E_0 D}{E_1}c^{\mathrm{T}}A_0^{i-1}H\,\theta + \epsilon(t)\,,\ 1 \le i \le \bar{n}-1\,, \tag{4.12}$$

which implies that $\frac{D^i}{E_1}[q(y)u]$ is bounded for $i = 1,\ldots,\bar{n}-1$. Combining this with the fact that $\frac{1}{E_1}[q(y)u]$ is bounded, it follows that $\frac{D^i}{E_2}[q(y)u]$ is bounded for $i = 1,\ldots,n-3$. Differentiating (4.12) with $i = \bar{n}-1$ and using (3.25) and (3.62) we obtain

$$\frac{D^{\bar{n}}}{E_1}[q(y)u] = \frac{1}{b_m}e^{(\bar{n})} - \frac{E_0 D}{E_1}\left[c^{\mathrm{T}}A_0^{\bar{n}}H + \phi^{\mathrm{T}}(y,u,r)\right]\theta + \epsilon(t)\,. \tag{4.13}$$

Substituting $\phi^{\mathrm{T}}(y,u,r)$ from (4.6) and rewriting (3.11) as $\theta^{\mathrm{T}} = [\overline{\theta}^{\mathrm{T}}, \ell_{0n-2}]$, we express (4.13) as

$$\begin{aligned}\frac{L}{E_2}[q(y)u] &= \frac{D^{\bar{n}}E_2 + \ell_{0n-2}D^{n-1}E_0}{E_1E_2}[q(y)u]\\ &= \frac{1}{b_m}e^{(\bar{n})} - \frac{E_0D}{E_1}\left[c^{\mathrm{T}}A_0^{\bar{n}}H\,\theta + \left[\overline{\nu}(y,r), \frac{\overline{D}^{n-3}}{E_2}[q(y)u]\right]\overline{\theta}\right] + \epsilon(t),\end{aligned} \tag{4.14}$$

which implies that $\frac{L}{E_2}[q(y)u]$ is bounded. Since L is of degree $n-2$ and $\frac{D^i}{E_2}[q(y)u]$ is bounded for $i = 1,\ldots,n-3$, it follows that $\frac{D^{n-2}}{E_2}[q(y)u]$ is bounded. Hence, $\frac{\overline{D}_{n-2}}{E_2}[q(y)u]$ is bounded, which proves that u is bounded.

In order to show the boundedness of the state of the plant, we note that the boundedness of u, y and r together with (3.14) imply that $D^i y$, $0 \le i \le n-m$, are bounded. From this and the fact that $B(D)$ is Hurwitz, we conclude that the state x in (3.88) is bounded.

We have thus proved that the state of the closed-loop adaptive system (3.88) is bounded on $[0,t_{\mathrm{f}})$. Hence, $t_{\mathrm{f}} = \infty$.

To prove the convergence of the tracking error e to zero, we first note that (3.37) and (3.45) imply that $\dot{V}$ is bounded and integrable on $[0,\infty)$. Furthermore, the boundedness of $\dot{\bar{e}}$ (cf. (3.30)), $\dot{\zeta}$ (cf. (4.4)) and $\dot{H}$ (cf. (3.88)) implies that $\ddot{V}$ is bounded. Hence, $\dot{V} \to 0$ as $t \to \infty$, which, in view of (3.45), proves that $\bar{e} \to 0$, $\zeta \to 0$ as $t \to \infty$. Since $\eta = S\zeta$ and S is bounded, $\eta \to 0$ as $t \to \infty$. Combined with (3.88) and the boundedness of $\hat{b}_m$, this also proves that $\eta_0 \to 0$ as $t \to \infty$. Thus,

$$\lim_{t\to\infty}[y(t)-y_{\mathrm{r}}(t)]=\lim_{t\to\infty}[\bar{e}(t)-\eta_0(t)]=0\,. \tag{4.15}$$

□

5 The Class of Nonlinear Systems

Most models of nonlinear systems are expressed in specific state coordinates. From that state-space form it may not always be obvious whether or not the nonlinear system at hand has the input-output description assumed in Sect. 3. Therefore, we now give coordinate-free geometric conditions which are necessary and sufficient for a single-input single-output nonlinear system of the form

$$\begin{aligned}\dot{z}&=f(z;\alpha)+g(z;\alpha)u\\ y&=h(z;\alpha)\end{aligned} \tag{5.1}$$

to have an input-output description of the form (3.1), which is repeated here for convenience:

$$A(D)y=B(D)\left[q(y)u\right]+\sum_{i=0}^{m}D^i\left[p_{i0}(y)+p_i^{\mathrm{T}}(y)\theta_1\right]. \tag{5.2}$$

In (5.1) $z\in\mathbb{R}^n$ is the state, $u\in\mathbb{R}$ is the input, $y\in\mathbb{R}$ is the output, $\alpha=[\alpha_1\ldots\alpha_l]^{\mathrm{T}}\in\mathbb{R}^l$ is a vector of unknown constant parameters, and f, g, h, are smooth vector fields with $f(0;\alpha)=0$, $h(0;\alpha)=0$, for all $\alpha\in\mathbb{R}^l$, $g(z)\neq 0$ for all $z\in\mathbb{R}^n$. In (5.2)

- the coefficients $a_0,\ldots,a_{n-1}$ of the denominator polynomial $A(D)=D^n+a_{n-1}D^{n-1}+\cdots+a_0$ are unknown,
- the coefficients $b_0,\ldots,b_m (m\leq n-1)$ of the numerator polynomial $B(D)=b_mD^m+\cdots+b_0$ are unknown,
- θ_1 is an ℓ-dimensional vector of unknown parameters, resulting from a possible overparameterization in which products and powers of the original unknown parameters α_i are treated as new parameters (so that $\ell\geq l$),
- $q(y)$, $p_{ij}(y)$, $0\leq i\leq m$, $0\leq j\leq\ell$ are smooth nonlinearities with $q(y)\neq 0$, $\forall y\in\mathbb{R}$, $p_{ij}(0)=0$, $0\leq i\leq m$, $0\leq j\leq\ell$.

We first note that a minimal state representation of (5.2) is given by

$$\begin{aligned}\dot{x}&=A_\Sigma x+b_\Sigma q(y)u+\sum_{i=0}^{m}\varepsilon_{n-i}\left[p_{i0}(y)+p_i^{\mathrm{T}}(y)\theta_1\right]\\ y&=c_\Sigma^{\mathrm{T}}x\,,\end{aligned} \tag{5.3}$$

with A_Σ, b_Σ, c_Σ, ε_{n-i} as defined in (3.89). Hence, the following statement becomes obvious:

Fact 5.1. The nonlinear system (5.1) has an input-output description of the form (5.2) if and only if there exists a global in z, possibly parameter-dependent, diffeomorphism transforming (5.1) into (5.3). □

Using this fact, we now state the following result:

Proposition 5.2 *The system (5.1) has an input-output description of the form (5.2) if and only if the following conditions are satisfied for all $z \in \mathbb{R}^n$ and for the true value of the parameter vector α:*

(C1) the one-forms $dh, dL_f h, \ldots, dL_f^{n-1}h$ are linearly independent,

(C2) $\left[\mathrm{ad}_f^i \bar{g}, \mathrm{ad}_f^j \bar{g}\right] = 0\,,\ \ i,j = 0,\ldots,n-1$, where $\bar{g}$ is uniquely defined by

$$L_{\bar{g}} L_f^i h = \begin{cases} 0, & i = 0,\ldots,n-2 \\ 1, & i = n-1\,, \end{cases}$$

(C3)

$$\mathrm{ad}_f^n \bar{g} = \sum_{j=0}^{n-1} d_j(\alpha)\mathrm{ad}_f^j \bar{g} + \sum_{j=0}^{m} \left[p_{j0}'(y) + {p_j'}^{\mathrm{T}}(y)\theta_1\right] \mathrm{ad}_f^j \bar{g}$$

$$\left[g, \mathrm{ad}_f^j \bar{g}\right] = 0\,,\ \ j = 0,\ldots,n-2$$

$$g = q(y)\sum_{j=0}^{m} c_j(\alpha)\mathrm{ad}_{f_0}^j \bar{g}\,,$$

with $d_j(\alpha), c_j(\alpha)$ polynomial functions of α, θ_1 the new unknown parameter vector, and $p_i(y) = \int_0^y p_i'(v)dv,\ i = 0,\ldots,\ell$, and

(C4) the vector fields f and $\bar{g}$ are complete.

Proof. Using Proposition 3 of [22], it is straightforward to show that conditions (C1)-(C3) are necessary and sufficient for the existence of a local diffeomorphism such that in the new coordinates the system (5.1) is expressed as

$$\dot{x} = \begin{bmatrix} (-1)d_{n-1}(\alpha) & \\ \vdots & I \\ (-1)^n d_0(\alpha) & 0\ldots 0 \end{bmatrix} x + \begin{bmatrix} 0 \\ \vdots \\ 0 \\ (-1)^m c_m \\ \vdots \\ c_0 \end{bmatrix} q(y)u$$

$$+ \begin{bmatrix} 0 \\ \vdots \\ 0 \\ (-1)^{n-m}\left[p_{m0}(y) + p_m^{\mathrm{T}}(y)\theta_1\right] \\ \vdots \\ (-1)^n\left[p_{00}(y) + p_0^{\mathrm{T}}(y)\theta_1\right] \end{bmatrix} \tag{5.4}$$

$$y = x_1\,,$$

which is exactly in the form (5.3), where the coefficients $a_0, \ldots, a_{n-1}, b_0, \ldots, b_m$ depend on the physical parameters α. From [23], condition (C4) is necessary and sufficient for the above diffeomorphism to be global. □

Remark 5.3. The above proposition gives a set of geometric conditions characterizing the class of nonlinear systems to which our adaptive scheme can be applied. Whenever the conditions (C1)–(C4) can be verified a priori, the input-output description (5.2) of the nonlinear system at hand is determined directly from (C3), without the need to compute the diffeomorphism of Proposition 5.2. Unfortunately, the verification of (C1)–(C4) may require some a priori information about the unknown parameter vector α. □

Remark 5.4. The conditions of Proposition 5.2 are satisfied by nonlinear systems that are linearizable by output injection and input-output linearizable by full-state feedback. However, they need not be full-state feedback linearizable. □

We illustrate these two remarks with an example.

Example 5.6. It may not be obvious that (2.1) is the input-output description of the nonlinear system

$$\begin{aligned} \dot{z}_1 &= z_2 + \alpha\left(3ye^y + 2y^2\right) \\ \dot{z}_2 &= z_3 - \alpha\left(2ye^y + y^2\right) \\ \dot{z}_3 &= z_4 + \alpha ye^y \\ \dot{z}_4 &= z_5 + \alpha y^2 \\ \dot{z}_5 &= u + \alpha y^3 \\ y &= z_1 + 2z_2 + z_3\,. \end{aligned} \tag{5.5}$$

However, this can be established by checking the conditions (C1)–(C4). Straightforward calculations show that for (5.5) we have

$$\bar{g} = 5\frac{\partial}{\partial z_1} - 4\frac{\partial}{\partial z_2} + 3\frac{\partial}{\partial z_3} - 2\frac{\partial}{\partial z_4} + \frac{\partial}{\partial z_5} \tag{5.6}$$

$$\mathrm{ad}_f^5\bar{g} = \alpha\left[-3y^2\bar{g} + (2y+6y^2)\mathrm{ad}_f\bar{g} - (ye^y + e^y + 4y + 3y^2)\mathrm{ad}_f^2\bar{g}\right] \tag{5.7}$$

$$g = \bar{g} - 2\mathrm{ad}_f\bar{g} + \mathrm{ad}_f^2\bar{g}\,. \tag{5.8}$$

Hence, the conditions (C1)–(C4) are satisfied for all α and the input-output description of (5.5) is

$$D^5y = (D^2+2D+1)u + \theta\left[D^2(ye^y + 2y^2 + y^3) + D(y^2 + 2y^3) + y^3\right], \tag{5.9}$$

where $\theta = \alpha$. It is important to note that to determine this input-output description no explicit change of coordinates was required. In this simple example, however, one can find the corresponding change of coordinates by inspection:

$$\begin{aligned} x_1 &= z_1 + 2z_2 + z_3 \\ x_2 &= z_2 + 2z_3 + z_4 \\ x_3 &= z_3 + 2z_4 + z_5 \\ x_4 &= z_4 + 2z_5 \\ x_5 &= z_5 \,. \end{aligned} \tag{5.10}$$

In these coordinates, (5.5) becomes

$$\begin{aligned} \dot{x}_1 &= x_2 \\ \dot{x}_2 &= x_3 \\ \dot{x}_3 &= x_4 + u + \alpha\left(ye^y + 2y^2 + y^3\right) \\ \dot{x}_4 &= x_5 + 2u + \alpha\left(y^2 + 2y^3\right) \\ \dot{x}_5 &= u + \alpha y^3 \\ y &= x_1 \,. \end{aligned} \tag{5.11}$$

We immediately see that (5.11) has the input-output description (5.9).

However, it should also be pointed out that (5.5) is *not* full-state feedback linearizable, since the distribution

$$\mathcal{G}^3 = \operatorname{span}\left\{g, \mathrm{ad}_f g, \mathrm{ad}_f^2 g, \mathrm{ad}_f^3 g\right\} \tag{5.12}$$

is not involutive[21]. □

6 Conclusions

This paper has extended the theory of adaptive control for linear systems to a class of systems which are essentially nonlinear in the sense that their nonlinearities are not restricted by any growth constraints. In spite of this absence of growth constraints, all the stability and tracking results are global.

The assumptions on the linear part of the system are the same as in the standard adaptive theory for linear systems. However, to guarantee the aforementioned global properties, the systematic design procedure has departed from the two main ingredients of most adaptive schemes for linear systems: the certainty-equivalence control and the normalization of the update law. In addition to the certainty-equivalence part, the control contains a term which counteracts the effects of rapidly growing nonlinearities. Thanks to the presence of this term, the normalization of the update law is avoided, which allows the rapid decrease of the parameter error. This proved to be crucial in preventing finite escape times common in systems with rapidly growing nonlinearities.

The class of nonlinear systems has been restricted by coordinate-free geometric conditions which are equivalent to the structural requirements that the nonlinearities depend only on the output and do not enter the system before the control does, and that the zero dynamics are linear and exponentially stable. Relaxing these restrictions, and thus enlarging the class of systems that can be adaptively controlled using only output measurement, is a topic of further research.

Acknowledgement

The authors would like to thank Professor Riccardo Marino for many helpful discussions which led to the development of Section 5.

References

1. G. Campion and G. Bastin, "Indirect adaptive state feedback control of linearly parametrized nonlinear systems," *Int. J. Adapt. Control Sig. Proc.*, vol. 4, pp. 345–358, Sept. 1990.
2. I. Kanellakopoulos, P. V. Kokotović, and R. Marino, "Robustness of adaptive nonlinear control under an extended matching condition," *Prepr. IFAC Symp. Nonlinear Control Syst. Design*, pp. 192–197, Capri, Italy, June 1989.
3. I. Kanellakopoulos, P. V. Kokotović, and R. Marino, "An extended direct scheme for robust adaptive nonlinear control," *Automatica*, to appear, March 1991.
4. R. Marino, I. Kanellakopoulos, and P. V. Kokotović, "Adaptive tracking for feedback linearizable SISO systems," *Proc. 28th IEEE Conf. Dec. Control*, pp. 1002–1007, Tampa, FL, Dec. 1989.
5. J.-J. E. Slotine and J. A. Coetsee, "Adaptive sliding controller synthesis for nonlinear systems," *Int. J. Control*, vol. 43, pp. 1631–1651, June 1986.
6. D. G. Taylor, P. V. Kokotović, R. Marino, and I. Kanellakopoulos, "Adaptive regulation of nonlinear systems with unmodeled dynamics," *IEEE Trans. Aut. Control*, vol. 34, pp. 405–412, April 1989.
7. K. Nam and A. Arapostathis, "A model-reference adaptive control scheme for pure-feedback nonlinear systems," *IEEE Trans. Aut. Control*, vol. 33, pp. 803–811, Sept. 1988.
8. J.-B. Pomet and L. Praly, *Adaptive nonlinear control: an estimation-based algorithm*, in *New Trends in Nonlinear Control Theory*, J. Descusse, M. Fliess, A. Isidori and D. Leborgne Eds., Springer-Verlag, Berlin, 1989.
9. J.-B. Pomet and L. Praly, "Adaptive nonlinear regulation: equation error from the Lyapunov equation," *Proc. 28th IEEE Conf. Dec. Control*, pp. 1008–1013, Tampa, FL, Dec. 1989.
10. S. S. Sastry and A. Isidori, "Adaptive control of linearizable systems," *IEEE Trans. Aut. Control*, vol. 34, pp. 1123–1131, Nov. 1989.
11. A. Teel, R. Kadiyala, P. V. Kokotović, and S. S. Sastry, "Indirect techniques for adaptive input output linearization of nonlinear systems," *Int. J. Control*, vol. 53, pp. 193–222, Jan. 1991.
12. I. Kanellakopoulos, P. V. Kokotović, and R. H. Middleton, "Observer-based adaptive control of nonlinear systems under matching conditions," *Proc. 1990 Amer. Control Conf.*, pp. 549–555, San Diego, CA, May 1990.
13. I. Kanellakopoulos, P. V. Kokotović, and R. H. Middleton, "Indirect adaptive output-feedback control of a class of nonlinear systems," *Proc. 29th IEEE Conf. Dec. Control*, pp. 2714–2719, Honolulu, HI, Dec. 1990.

14. G. Bastin and M. R. Gevers, "Stable adaptive observers for nonlinear time-varying systems," *IEEE Trans. Aut. Control*, vol. 33, pp. 650–658, July 1988.
15. R. Marino, "Adaptive observers for single output nonlinear systems," *IEEE Trans. Aut. Control*, vol. 35, pp. 1054–1058, Sept. 1990.
16. R. Marino and P. Tomei, "Global adaptive observers for nonlinear systems via filtered transformations," Submitted to *IEEE Trans. Aut. Control*, 1990.
17. A. Feuer and A. S. Morse, "Adaptive control of single-input single-output linear systems," *IEEE Trans. Aut. Control*, vol. AC-23, pp. 557–569, Aug. 1978.
18. K. S. Narendra and A. M. Annaswamy, *Stable Adaptive Systems*, Prentice-Hall, Inc., Englewood Cliffs, NJ, 1989.
19. S. S. Sastry and M. Bodson, *Adaptive Control: Stability, Convergence and Robustness*, Prentice-Hall, Inc., Englewood Cliffs, NJ, 1989.
20. P. V. Kokotović, I. Kanellakopoulos, and A. S. Morse, "Adaptive feedback linearization of nonlinear systems," this volume, pp. 309–344.
21. A. Isidori, *Nonlinear Control Systems*, 2nd ed., Springer-Verlag, Berlin, 1989.
22. A. J. Krener and A. Isidori, "Linearization by output injection and nonlinear observers," *Syst. Control Lett.*, vol. 3, pp. 47–52, June 1983.
23. W. Respondek, *Global aspects of linearization, equivalence to polynomial forms and decomposition of nonlinear control systems*, in *Algebraic and Geometric Methods in Nonlinear Control Theory*, M. Fliess and M. Hazewinkel, eds., pp. 257–284, D. Reidel Publishing Co., Dordrecht, 1986.

Lecture Notes in Control and Information Sciences

Edited by M. Thoma and A. Wyner

Vol. 137: S. L. Shah, G. Dumont (Eds.)
Adaptive Control Strategies for Industrial Use
Proceedings of a Workshop
Kananaskis, Canada, 1988
VI, 360 pages. 1989

Vol. 138: D. C. McFarlane, K. Glover
Robust Controller Design Using Normalized Coprime Factor Plant Descriptions
X, 206 pages. 1990

Vol. 139: V. Hayward, O. Khatib (Eds.)
Experimental Robotics I
The First International Symposium
Montreal, June 19-21, 1989
XVII, 613 pages. 1990

Vol. 140: Z. Gajic, D. Petkovski, X. Shen
Singularly Perturbed and Weakly Coupled Linear Control Systems
A Recursive Approach
VII, 202 pages. 1990

Vol. 141: S. Gutman
Root Clustering in Parameter Space
VIII, 153 pages. 1990

Vol. 142: A. N. Gündeş, C. A. Desoer
Algebraic Theory of Linear Feedback Systems with Full and Decentralized Compensators
V, 176 pages. 1990

Vol. 143: H.-J. Sebastian, K. Tammer (Eds.)
System Modelling and Optimization
Proceedings of the 14th IFIP Conference
Leipzig, GDR, July 3-7, 1989
X, 960 pages. 1990

Vol. 144: A. Bensoussan, J. L. Lions (Eds.)
Analysis and Optimization of Systems
Proceedings of the 9th International Conference, Antibes, June 12–15, 1990
XII, 992 pages. 1990

Vol. 145: M. B. Subrahmanyam
Optimal Control with a Worst-Case Performance Criterion and Applications
X, 133 pages. 1990

Vol. 146: D. Mustafa, K. Glover
Minimum Entropy H_∞ Control
IX, 144 pages. 1990

Vol. 147: J. P. Zolésio (Ed.)
Stabilization of Flexible Structures
Third Working Conference,
Montpellier, France, January 1989
V, 327 pages, 1990

Vol. 148: In preparation

Vol. 149: K.H. Hoffmann, W. Krabs (Eds.)
Optimal Control of Partial Differential Equations
Proceedings of the IFIP WG 7.2 International Conference, Irsee, April 9-12, 1990
VI, 245 pages. 1991

Vol. 150: L. C. G. J. M. Habets
Robust Stabilization in the Gap-topology
IX, 126 pages. 1991

Vol. 151: J. M. Skowronski, H. Flashner, R. S. Guttalu (Eds.)
Mechanics and Control
Proceedings of the 3rd Workshop on Control Mechanics in Honor of the 65th Birthday of George Leitmann
January 22-24, 1990, University of Southern California
IV, 497 pages. 1991

Vol. 152: J. D. Aplevich
Implicit Linear Systems
XI, 176 pages. 1991

Vol. 153: O. Hájek
Control Theory in the Plane
X, 272 pages. 1991

Vol. 154: A. Kurzhanski, I. Lasiecka (Eds.)
Modelling and Inverse Problems of Control for Distributed Parameter Systems
Proceedings of IFIP (W.G. 7.2)-IIASA Conference
Laxenburg, Austria, July 24-28, 1989
VII, 179 pages. 1991

Vol. 155: M. Bouvet, G. Bienvenu (Eds.)
High-Resolution Methods in Underwater Acoustics
V, 249 pages. 1991

Lecture Notes in Control and Information Sciences

Edited by M. Thoma and A. Wyner

Vol. 117: K.J. Hunt
Stochastic Optimal Control Theory with Application in Self-Tuning Control
X, 308 pages, 1989.

Vol. 118: L. Dai
Singular Control Systems
IX, 332 pages, 1989

Vol. 119: T. Başar, P. Bernhard
Differential Games and Applications
VII, 201 pages, 1989

Vol. 120: L. Trave, A. Titli, A. M. Tarras
Large Scale Systems: Decentralization, Structure Constraints and Fixed Modes
XIV, 384 pages, 1989

Vol. 121: A. Blaquière (Editor)
Modeling and Control of Systems in Engineering, Quantum Mechanics, Economics and Biosciences
Proceedings of the Bellman Continuum Workshop 1988, June 13–14, Sophia Antipolis, France
XXVI, 519 pages, 1989

Vol. 122: J. Descusse, M. Fliess, A. Isidori, D. Leborgne (Eds.)
New Trends in Nonlinear Control Theory
Proceedings of an International Conference on Nonlinear Systems, Nantes, France, June 13–17, 1988
VIII, 528 pages, 1989

Vol. 123: C. W. de Silva, A. G. J. MacFarlane
Knowledge-Based Control with Application to Robots
X, 196 pages, 1989

Vol. 124: A. A. Bahnasawi, M. S. Mahmoud
Control of Partially-Known Dynamical Systems
XI, 228 pages, 1989

Vol. 125: J. Simon (Ed.)
Control of Boundaries and Stabilization
Proceedings of the IFIP WG 7.2 Conference Clermont Ferrand, France, June 20–23, 1988
IX, 266 pages, 1989

Vol. 126: N. Christopeit, K. Helmes M. Kohlmann (Eds.)
Stochastic Differential Systems
Proceedings of the 4th Bad Honnef Conference June 20–24, 1988
IX, 342 pages, 1989

Vol.127: C. Heij
Deterministic Identification of Dynamical Systems
VI, 292 pages, 1989

Vol. 128: G. Einarsson, T. Ericson, I. Ingemarsson, R. Johannesson, K. Zigangirov, C.-E. Sundberg
Topics in Coding Theory
VII, 176 pages, 1989

Vol. 129: W. A.Porter, S. C. Kak (Eds.)
Advances in Communications and Signal Processing
VI, 376 pages, 1989.

Vol. 130: W. A. Porter, S. C. Kak, J. L. Aravena (Eds.)
Advances in Computing and Control
VI, 367 pages, 1989

Vol. 131: S. M. Joshi
Control of Large Flexible Space Structures
IX, 196 pages, 1989.

Vol. 132: W.-Y. Ng
Interactive Multi-Objective Programming as a Framework for Computer-Aided Control System Design
XV, 182 pages, 1989.

Vol. 133: R. P. Leland
Stochastic Models for Laser Propagation in Atmospheric Turbulence
VII, 145 pages, 1989.

Vol. 134: X. J. Zhang
Auxiliary Signal Design in Fault Detection and Diagnosis
XII, 213 pages, 1989

Vol. 135: H. Nijmeijer, J. M. Schumacher (Eds.)
Three Decades of Mathematical System Theory
A Collection of Surveys at the Occasion of the 50th Birthday of Jan C. Willems
VI, 562 pages, 1989

Vol. 136: J. Zabczyk (Ed.)
Stochastic Systems and Optimization
Proceedings of the 6th IFIP WG 7.1 Working Conference, Warsaw, Poland, September 12–16, 1988
VI, 374 pages. 1989